PHOTOCHEMICAL CONVERSION AND STORAGE OF SOLAR ENERGY

Photochemical Conversion and Storage of Solar Energy

Proceedings of the Eighth International Conference on Photochemical Conversion and Storage of Solar Energy, IPS-8, held July 15–20, 1990, in Palermo, Italy

Edited by

E. Pelizzetti
University of Torino, Italy

and

M. Schiavello
University of Palermo, Italy

KLUWER ACADEMIC PUBLISHERS
DORDRECHT / BOSTON / LONDON

Library of Congress Cataloging-in-Publication Data

International Conference on the Photochemical Conversion and Storage of Solar Energy (8th : 1990 : Palermo, Italy)
Photochemical conversion and storage of solar energy : proceedings of the Eighth International Conference on the Photochemical Conversion and Storage of Solar Energy, IPS-8, held July 15-20, 1990, in Palermo, Italy / editors, Ezio Pelizzetti and Mario Schiavello.
p. cm.
Includes indexes.
ISBN 0-7923-1194-9 (U.S. : alk. paper)
1. Solar energy--Congresses. 2. Photochemistry--Congresses. 3. Energy storage--Congresses. I. Pelizzetti, Ezio, 1944- II. Schiavello, Mario, 1935- . III. Title.
TJ810.I465 1990
621.47--dc20 91-2523

ISBN 0-7923-1194-9

Published by Kluwer Academic Publishers,
P.O. Box 17, 3300 AA Dordrecht, The Netherlands.

Kluwer Academic Publishers incorporates
the publishing programmes of
D. Reidel, Martinus Nijhoff, Dr W. Junk and MTP Press.

Sold and distributed in the U.S.A. and Canada
by Kluwer Academic Publishers,
101 Philip Drive, Norwell, MA 02061, U.S.A.

In all other countries, sold and distributed
by Kluwer Academic Publishers Group,
P.O. Box 322, 3300 AH Dordrecht, The Netherlands.

Printed on acid-free paper

Printed in the Netherlands

CONTENTS

FOREWORD

The book collects the lectures and the status reports delivered during the "Eighth International Conference on Photochemical Conversion and Storage of Solar Energy", IPS-8, held in Palermo (Italy) from 15th to 20th of July 1990.

As usual, the main theme of the Conference was that of making the point about the trends and the developments of the studies related to the photochemical exploitation of solar energy and also to report the main lines of potential applications. Therefore the contributions reflect this point; they vary from those reporting basic and fundamental theories to those reporting cases of possible applications.

For the sake of following the logical line which links each other the various contributions, we report the six areas in which the main theme of the conference was devided: (a) Electron and energy transfer in homogeneous and heterogeneous systems; (b) Photosynthesis: organized assemblies and biomimetic systems; (c) Photoelectrochemistry; (d) Photocatalysis: homogeneous and heterogeneous regime; (e) Environment: photochemical and photocatalytic processes; (f) Solar energy materials and photochemical engineering.

It remains now to thank persons and institutions which made possible the organization of the Conference. The persons to thank are all the members of the International and National Organizing Committees and in particular Prof. A.Sclafani and Dr. L.Palmisano whose efforts were essential for the success of the Conference.

Ezio Pelizzetti
University of Parma

Mario Schiavello
University of Palermo

Contributors: *Università di Palermo, Commission of the European Communities (Brussels), Eniricerche (Milano), Italsolar (Roma), Consiglio Nazionale delle Ricerche (Roma), Sicilcassa (Palermo), Assessorato Regionale Siciliano Turismo, Comunicazioni e Trasporti, Presidenza della Regione Siciliana, Assemblea Regionale Siciliana, Assessorato Regionale Siciliano Beni Culturali, Ambientali e Pubblica Istruzione, Azienda Autonoma di Turismo per Palermo e Monreale, IRFIS (Palermo), Cassa Rurale di Risparmio di Palermo e Monreale, Azienda Provinciale del Turismo (Palermo).*

PHOTOINDUCED ELECTRON AND ENERGY TRANSFER IN COVALENTLY LINKED SUPRAMOLECULAR SYSTEMS

F. SCANDOLA and M. T. INDELLI
Department of Chemistry, University of Ferrara
44100 Ferrara, Italy

ABSTRACT. Within the general subject of photoinduced electron and energy transfer, the study of unimolecular processes in covalently linked supramolecular systems constitutes a recent but very active research field. In the present report, the status of this field is briefly outlined. Particular attention is given to the evolution taking place from studies of fundamental interest on simple two-component systems (dyads) to studies of more oriented character involving increasingly complex systems (triads, tetrads, pentads, etc.). The development of this tendency towards the design of photochemical molecular devices is finally pointed out.

1. INTRODUCTION

Photoinduced electron and energy transfer processes play a fundamental role in natural photosynthetic systems [1-3]. A detailed understanding of such processes is also likely to be the key for the development of artificial system for the photochemical conversion of solar energy [4,5]. A remarkable amount of experimental and theoretical work [6] is currently being devoted to the study of photoinduced electron and energy transfer in both homogeneous and heterogeneous systems.

In homogeneous systems, there is a tendency to change from the study of bimolecular processes between reactants in solution (eq 1) to

$$A^* + B \longrightarrow \text{products} \qquad (1)$$

that of <u>unimolecular</u> processes in supramolecular structures [7] in which two (or more) molecular components are held together by various types of chemical forces (eq 2). Aside from reasons of

$$A^*\text{-}B \longrightarrow \text{products} \qquad (2)$$

mechanistic nature, this tendency reflects the recognition of the importance that spatial organization of molecular components has in the function of natural photosynthetic systems [7-12]. In this report, we shall focus our attention on supramolecular systems in which the molecular components are held together by covalent bonds, i.e., on <u>photoinduced electron and energy transfer in covalently linked supramolecular species</u>. The number of studies on such systems is

E. Pelizzetti and M. Schiavello (eds.), Photochemical Conversion and Storage of Solar Energy, 1–14.

growing exponentially, and several of aspects of the field have been recently reviewed [13-19].

In the limited space available, we attempt to provide a perspective overview of the field. First, a number of processes that have been observed to occur in simple two-component (dyad) systems are briefly described. Then, the tendency towards the use of more complex systems (triads, tetrads, pentads, etc.), capable of achieving high efficiencies of photoinduced charge separation or light energy harvesting, is pointed out. Finally, a likely evolution of this research field, i.e., the design and synthesis of "photochemical molecular devices", is indicated. For space reasons, the report is not meant to be exhaustive but rather to provide a guide to representative studies and general references in the field. By the same token, the studies quoted are not discussed in any detail. For chemical formulas, structures, photophysical results, and mechanisms, the reader is referred to the original papers.

2. TWO-COMPONENT SYSTEMS

The relevant processes occurring in a covalently linked two-component system (dyad) can be schematically represented in terms of a ground state and three excited states of different nature (Scheme I) , i.e., two locally excited states and an intercomponent charge transfer state

Scheme I

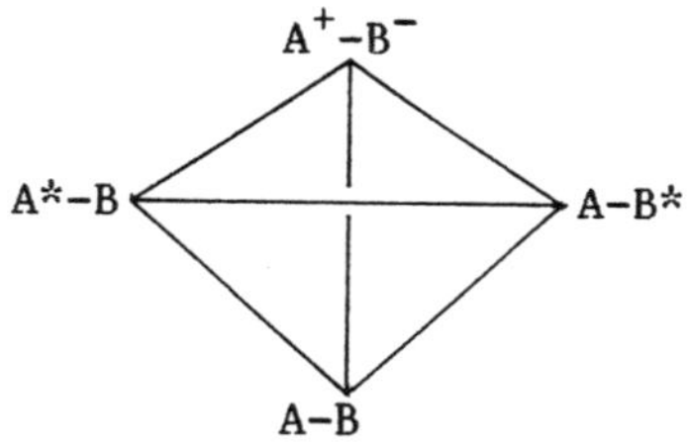

(in this simplified picture, the multiplicity of the locally excited states is not considered, and the charge transfer state of opposite polarity is omitted). The relative ordering of the three types of excited state depends on the specific system investigated, and is crucial to the behavior of the system.

2.1 Photoinduced electron transfer

When the energetics is such that the charge transfer state is the lowest one, excitation of the dyad can give rise to photoinduced electron transfer, i.e., absorption to populate a localized excited state (e.g., process 1 in Scheme II) followed by radiationless conversion to the charge transfer state (e.g., process 2 in Scheme II). The process can be monitored, through not identified, from the quenching of the emission of the excited component. Usually, the process is followed by fast charge recombination (process 3 in Scheme

II) so that very fast spectroscopic techniques are required for direct detection of the charge transfer state. This type of phenomenon has

Scheme II

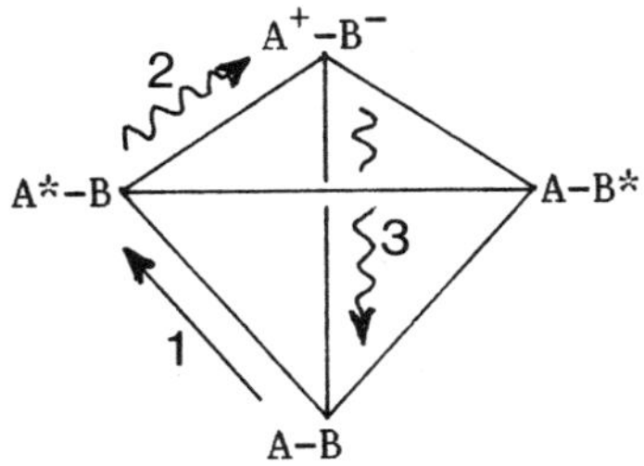

been extensively studied, and a number of reviews are available in the field [13,14,19].

Studies of photoinduced electron transfer in organic systems in which the donor, the acceptor and the covalent bridge are changed in a systematic way [20-24] are of particular interest. Together with some related radiation-induced electron transfer work [25], such studies have provided basic insight into the kinetics and mechanism of electron transfer processes. In particular, the long sought Marcus inverted behavior (highly exergonic electron transfer processes slowing down with increasing driving force) has been definitely established. Also, the nature of the electronic interaction responsible for the electron transfer process has been greatly clarified: it is in most cases of the through-bond (rather than through-space) type, it is sensitive to the nature of the bridge (degree of unsaturation, geometric configuration, stereochemistry of attachment), it decreases exponentially with the number of bonds in the bridge (or, in homogeneous cases, with donor-acceptor distance). It has been demonstrated that, thanks to the through-bond interaction, electrons can be transferred over large distances of the order of 15-20 Å at remarkably high rates (15 Å in ca. 10^{-9} s with fully saturated norbornylogous bridges [26], 25 Å in ca. 10^{-12} s with polyene "molecular wires" as bridges [23]).

2.2 Charge Transfer Absorption

The product of photoinduced electron transfer process within a dyad, e.g., A^+–B^-, is an electronic excited state (of charge transfer character) of the A-B supermolecule. As such, it can in principle be reached, besides by the indirect route of Scheme II, also by direct optical transition from the ground state (process 1 in Scheme III). The corresponding spectroscopic features are usually called <u>charge transfer</u> or (in the inorganic literature) <u>intervalence transfer</u> bands. The problem with the observation of charge transfer absorption bands is their intensity, which, as shown, e.g., by the Hush model of intervalence transfer [27], is a function of the degree of electronic coupling between the two components. In practice, couplings of hundreds

Scheme III

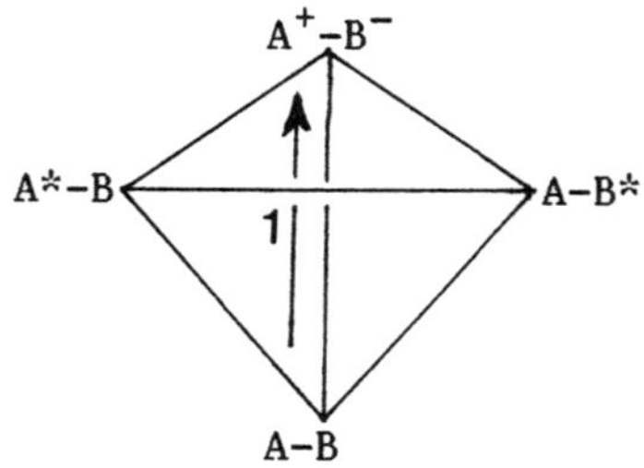

of wavenumbers are required to give sizable absorption bands, while a few wavenumbers are already sufficient to give fast radiationless electron transfer. Thus, in many of the dyads in which photoinduced electron transfer has been studied, optical charge transfer has not be detected. It is only for the more heavily coupled systems, such as organic dyads with short bridges [24] or bimetallic systems with highly delocalizing bridging ligands [28,29], that charge transfer absorption can be measured and studied. The interesting feature of charge transfer absorption is that it can give first-hand information on the magnitude of the electronic couplings.

2.3 Charge Transfer Emission

The reverse radiative process is <u>charge transfer emission</u> (e.g., process 3 in Scheme IV). The problems with the observation of charge transfer emission are related to the lifetime of the excited state,

Scheme IV

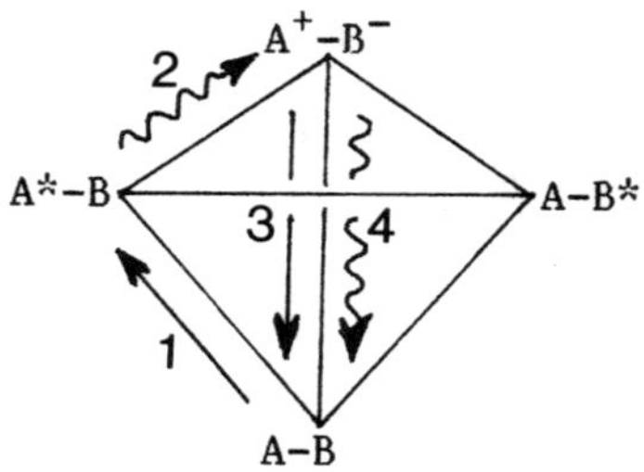

determined by the competition with the radiationless decay (charge recombination, process 4 in Scheme IV). In practice, a very unfavorable nuclear factor for the radiationless charge recombination (i.e., a highly exergonic, strongly "inverted" situation) is required in order for the radiative process to be efficient. A number of reports on charge transfer emission following photoinduced electron transfer (processes 1 and 2 in Scheme IV) in dyad systems have appeared recently [24,30-32]. The characteristics of such emissions (broad band profile, solvatochromic behavior) are very similar to those considered to be typical of "exciplex" emission [33], and the differentiation between the two phenomena in flexible dyads may not be a trivial problem [32].

2.4 Electronic Energy Transfer

When the energetics is such that the charge transfer state is higher than both localized states, excitation of the dyad (process 1 in Scheme V) can give rise to intercomponent electronic energy transfer (process 2 in Scheme V). The common way of identifying such process is

Scheme V

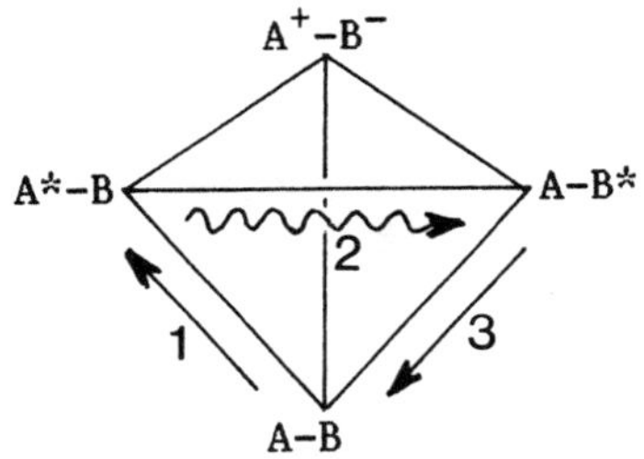

by monitoring the quenching of the emission of the excited component and sensitization of that of the other component (e.g., process 3 in Scheme V). Electronic energy transfer in dyad systems has been extensively studied, and some reviews are available [17,34,35].

Triplet energy transfer has been detected in a number of organic dyads [36,37]. In systems containing steroid-type bridges of various length [37], the triplet energy transfer rate constants have been compared with those of intercomponent electron transfer within the radical cations ("hole transfer") and anions ("electron transfer") of similar dyads [38,39]. This comparison has shown that the exchange interaction responsible for the triplet energy transfer is proportional to the product of the electronic factors for the electron and hole transfer processes [40], leading to a greater distance sensitivity of energy relative to electron transfer.

Singlet energy transfer has been studied in organic dyads with rigid bridges as well, with examples of both coulombic [41,42] and exchange [43,44] mechanisms being found. The detection of exchange energy transfer over intercomponent distances where analogous bimolecular processes would be absolutely inefficient emphasizes the role of the bridge and trough-bond interactions.

A number of studies on energy transfer in inorganic dyads have recently been performed [17]. Particularly interesting cases are those in which a component with good light emitting properties (luminophore) but poor absorption characteristics is coupled with an appropriate chromophoric component. In such systems, energy transfer from the chromophore to the luminophore leads to efficient spectral sensitization of the luminophore emission [45].

2.5. Chemiluminescent charge recombination.

When the lowest excited state is a localized one, and the charge transfer absorption has sufficient intensity to allow direct population of the charge transfer state (process 1 in Scheme VI), an interesting possibility arises. The charge transfer state has two available channels for deactivation: direct charge recombination to the ground state (process 2 in Scheme VI), and charge recombination to give the lowest localized excited state (process 3 in Scheme VI). In

Scheme VI

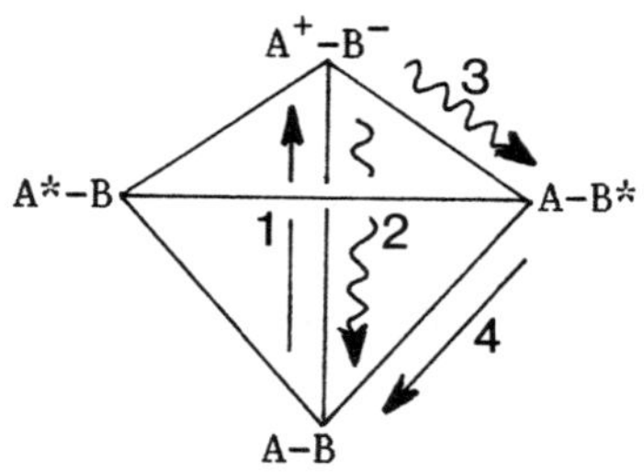

the second case, the localized excited state formed can be detected by its emission (process 4 in Scheme VI). The situation is similar to that encountered in chemiluminescent bimolecular electron transfer reactions, whereby the name of chemiluminescent charge recombination for the pathway shown in Scheme VI. The choice between the two pathways is determined by kinetics, which can favor chemiluminescent charge recombination if the direct one is in a highly "inverted" situation. An example of such a process has been recently reported [46].

3. MULTICOMPONENT SYSTEMS

As shown in the previous section, a simple two-component system gives rise to a variety of interesting photoprocesses. Some of these processes are potentially useful. In particular, photoinduced electron transfer in a dyad brings about charge separation, which is a way to convert light energy into chemical (redox) energy. The well-known problem is that, due to charge recombination, the charge separated state is very short-lived whereas utilization of the chemical energy likely involves relatively slow processes. On the other hand, electronic energy transfer in certain dyads may be a way to circumvent the poor cross section for light excitation of an active (photoreactive or emitting) component, i.e., to perform energy harvesting, or the so-called "antenna effect".

Natural photosynthetic systems feature both long lifetimes of charge separation and high efficiencies of energy harvesting [1-3,7-12]. The key to the high performance of the natural systems is their multicomponent structure, and the consequent multi-step nature of the processes. The idea that some of the architectural features of natural systems could be profitably used to develop artificial (biomimetic) systems for light harvesting and photoinduced charge separation is at the root of the extension from the dyad work towards more complex systems (triads, tetrads, pentads, etc.).

The first triads for photoinduced charge separation have been designed several years ago [47,48], and since then the field has developed into a very active one. The best characterized triads contain a central porphyrin chromophore (P) to which an acceptor (a quinone, Q) and a donor (a carotenoid or an aromatic amine, D) are covalently linked by appropriate bridges (Scheme VII)[49,50]. In such systems, following light absorption by the chromophore (process 1 in Scheme VII) a sequence of two electron transfer steps (processes 2 and

Scheme VII

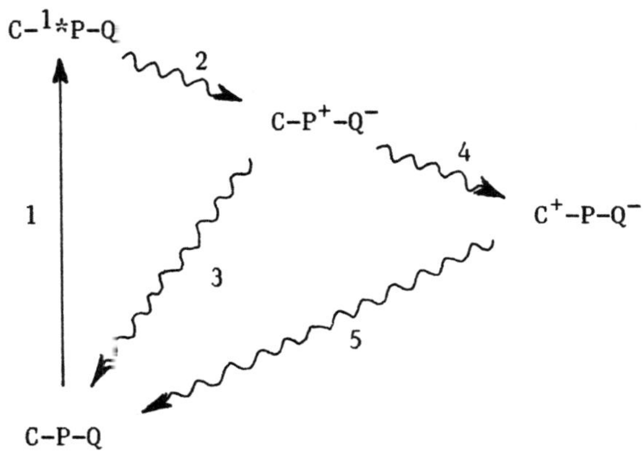

4) leads to charge separation on remote molecular components. The final charge-separated state is long-lived, because the remote charge recombination step (process 5 in Scheme VII) is slowed down by poor electronic factors. The key point in the mechanism is the competition between the second forward electron transfer step (process 4 in Scheme VII) and the primary charge recombination (process 3 in Scheme VII). The competition is favorable to the former process provided that the system has the right energetics, i.e., the forward process is slightly exergonic (in the almost activationless regime) while the back electron transfer step is highly exergonic (in the inverted regime) [50]. Along these basic principles, several triad systems involving a variety of molecular components, bridges, and geometries have been developed [51-60]. Recently, steps towards more complex multicomponent systems for photoinduced charge separation have been performed: two tetrad systems have been developed and studied [61,62], and the first pentad has just been presented [63].

A number of multicomponent systems that mimic the light harvesting function of photosynthetic antennas have been recently designed. An interesting organic example is a pentad made up of four zinc-porphyrin rings attached to a central free-base one, in which the light energy absorbed by the peripheral units is efficiently conveyed to the central one (Scheme VIII) [64]. Inorganic systems featuring the same type of effect have also been synthesized and investigated [17,35]. In particular, triads [65], tetrads [66-68], and a heptad [69] based on metal-polypyridine components have been synthesized and studied.

The tendency towards the study of more and more complex multicomponent systems is self-evident. The progress in this field is tightly bound to the progress in the development of organic and inorganic methods for stepwise synthesis. The driving force is

Scheme VIII

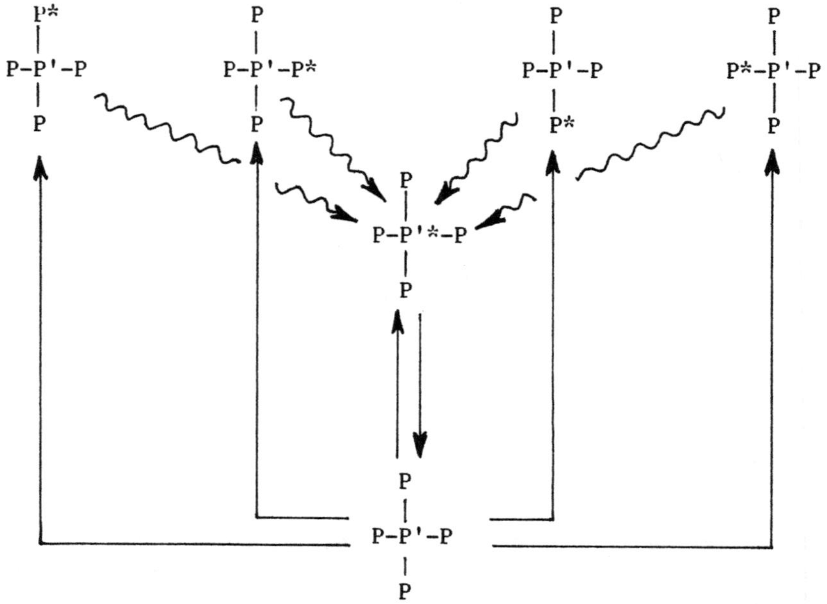

provided by the prospects for practical applications of complex multicomponent systems as "photochemical molecular devices".

4. TOWARDS PHOTOCHEMICAL MOLECULAR DEVICES

The research work outlined in sections 2 and 3 reflects the evolution of the field of photoinduced electron and energy transfer from the study of fundamental aspects in relatively simple donor-acceptor systems towards the design and synthesis of complex supramolecular species capable of performing light-induced functions, such as charge separation or light harvesting.

By extending the macroscopic concept of "device" to the molecular level (a process that is currently the object of lively discussions in the field of microelectronics) [70-87], an assembly of molecular components (i.e., a supramolecular structure) capable of performing a function can be defined as a "molecular device". From this point of view, the multicomponent systems mentioned in Section 3 can be considered as examples of photochemical molecular devices. The concept of photochemical molecular device [76,88] is, however, more general, as a variety of light-induced functions can be conceived besides photoinduced charge separation and light energy harvesting. Design principles for photochemical molecular devices (molecular components, supramolecular architecture, possible functions) have been discussed on general grounds [76,88].

A general problem concerning the utilization of molecular devices (including the photochemical ones) is that of their interfacing with structures of the macroscopic world. In principle, this implies the ability to assembly a great number of individual molecular devices, in an ordered manner, on "terminals" (e.g., electrodes, semiconductor particles) that can be handled with available technologies. Solutions to this rather formidable problem are likely to be found in the rapidly evolving areas of molecular self-organization [89-96] and electrode derivatization [97-109].

A simple example of practical use of a photochemical molecular device is the spectral sensitization of semiconductor electrodes by adsorbed sensitizer-antenna triad systems [110]. An application of self-assembling techniques is the observation of photocurrents from Langmuir-Blodgett films containing amphiphylic derivatives of charge separating triads and antenna pigments [111]. Fascinating, though futuristic, applications of photochemical molecular devices can be imagined in the area of "molecular computing". A "molecular shift register" based on an array of charge separating triads connected in series has been recently proposed [112,113]. In this device, bits of information (represented by the oxidation state of each triad) are shifted stepwise along the device under the action of periodic light pulses used as clock cycle.

It should be recognized that most of the work on photochemical molecular devices is still at a speculative level and that the development of practical devices should be considered, at best, as a long term goal. Nevertheless, the concept is intellectually stimulating and no doubt will provide in the years to come substantial driving force for studies on complex covalently linked supramolecular systems.

5. REFERENCES

[1] Michel-Beyerle, M. E. (ed.) (1985) Antennas and reaction centers in photosynthetic bacteria. Springer

[2] Hader, D. P. and Tevini, M. (1987) General Photobiology. Pergamon

[3] Breton, J. and Vermeglio, H. (eds) (1988) The photosynthetic bacterial reaction center - structure and dynamics. Plenum

[4] Porter, G. (1982). In: Coyle, J.D., Hill, R.R, and Roberts, D. R. (eds) Light, chemical change, and life. The Open University Press (Milton Keynes), p.362

[5] Norris, J. R. and Meisel, D. (eds) (1989) Photochemical energy

conversion. Elsevier
[6] The enormous amount of work on photoinduced elctron transfer is witnessed by two recent multi-volume compendia: Fox, M. A. and Chanon, M. (eds) (1988) Photoinduced electron transfer. Elsevier, Volumes A-D; Mattay, J. (Ed.) (1990) Top. Curr. Chem., Springer-Verlag, Volumes 156-158
[7] Balzani, V. and Scandola, F. (1990) Supramolecular Photochemistry. Horwood, Chap. 3
[8] Deisenhofer, J., Epp, O., Miki, K., Huber, R., and Michel, H. (1984) J. Mol. Biol. **180** 385
[9] Chang, C.-H., Tiede, D. M., Tang, J., Smith, U, Norris, J., and Schiffer, M. (1986) FEBS Lett. **205** 82
[10] Allen, J. P., Feher, G., Yeates, T. O., Komiya, H., and Rees, D. C. (1987) Proc. Natl. Acad. Sci. USA **84** 5730
[11] Deisenhofer, J. and Michel, H. (1989) Angew. Chem. Int. Ed. Engl. **28** 829
[12] Huber, R (1989) Angew. Chem. Int. Ed. Engl. **28** 848
[13] Connolly, J. S. and Bolton, J. R. (1988). In: Fox, M. A. and Chanon, M. (eds) Photoinduced electron transfer. Part D. Elsevier, p. 303
[14] Wasielewski, M. R. (1988) In: Fox, M. A. and Chanon, M. (eds) Photoinduced electron transfer. Part A. Elsevier, p. 161
[15] Gust, D. and Moore, T. A. (eds) (1989) Tetrahedron **45** 4669-4903
[16] Meyer, T. J. (1989) Accounts Chem. Res. **22** 163
[17] Scandola, F., Indelli, M. T., Chiorboli, C., and Bignozzi, C. A. (1990) Top. Curr. Chem. **158** 000
[18] Gust, D. and Moore, T. A. (1989) Science **244** 35
[19] Balzani, V. and Scandola, F. (1990) Supramolecular Photochemistry. Horwood, Chap. 5
[20] Wasielewski, M. R., Niemczyk, M. P., Svec, W. A., and Pewitt, E. B. (1985) J. Am. Chem. Soc. **107** 1080
[21] Joran, A. D., Leland, B. A., Felker, P. M., Zewail, A. H., Hopfield, J. J., and Dervan, P. B. (1987) Nature **327** 508
[22] Finckh, P., Heitele, H., Volk, M., and Michel-Beyerle, M. E. (1988) J. Phys. Chem. **92** 6584
[23] Wasielewski, M. R., Johnson, D. G., Svec, W. A., Kersey, K. M., Cragg, D. E., and Minsek, D. W. (1989). In: Norris, J. R. Jr. and Meisel, D. (eds) Photochemical energy conversion. Elsevier, p. 135
[24] Oevering, H., Verhoeven, J. W., Paddon-Row, M. N., and Warman, J. M. (1989) Tetrahedron **45** 4751
[25] Closs, G. L. and Miller, J. R. (1988) Science **240** 440
[26] Oevering, H., Paddon-Row, M. N., Heppener, M., Oliver, A. M., Cotsaris, E., Verhoeven, J. W., and Hush, N. S. (1987) J. Am. Chem. Soc. **109** 3258
[27] Hush, N. S. (1967) Prog. Inorg. Chem. **8** 391
[28] Bignozzi, C. A., Paradisi, C., Roffia, S., and Scandola, F. (1988) Inorg. Chem. **27** 408
[29] Creutz, C. (1983) Prog. Inorg. Chem. **30** 1
[30] Heitele, H., Finckh, P., and Michel-Beyerle, M. E. (1990) Angew. Che. Int. Ed. Engl. **28** 619

[31] Wasielewski, M. R., Minsek, D. W., Niemczyk, M. P., Svec, W. A., and Yang, N. C. (1990) J. Am. Chem. Soc. **112** 2823
[32] Wegewijs, B., Hermant, R. M., Verhoeven, J. W., De Haas, M. P., and Warman, J. M. (1990) Chem. Phys. Lett. **168** 185
[33] Turro, N. J. (1978) Molecular photochemistry. Benjamin
[34] De Schryver, F.C., Boens, N., and Put, J. (1977) Adv. Photochem. **10** 359
[35] Balzani, V. and Scandola, F. (1990) Supramolecular photochemistry. Horwood, Chap. 6
[36] Speiser, S., Hassoon, S., and Rubin, M.B. (1986) J. Phys. Chem. **90** 5085
[37] Closs, G.L., Piotrowiak, P., McInnis, J.M., and Fleming, G.R. (1988) J. Am. Chem. Soc. **110** 2657
[38] Closs, G.L. and Miller, J.R. (1988) Science **240** 440
[39] Johnson, M.D., Miller, J.R., Green, N.D., and Closs, G.L. (1989) J. Phys. Chem. **93** 1173
[40] Closs, G.L., Piotrowiak, P., and Miller, J.R. (1989). In: Norris, J.R., Jr. and Meisel, D. (eds) Photochemical energy conversion. Elsevier, p.23
[41] Osuka, A., Maruyama, K., Yamazaki, I., and Tamai, N. (1990) Chem. Phys. Lett. **165** 392
[42] Chardon-Noblat, S., Sauvage, J.P., and Mathis, P. (1989) Angew. Chem. Int. Ed. Engl. **28** 593
[43] Hassoon, S., Lustig, H., Rubin, M.B., and Speiser, S. (1984) J. Phys. Chem. **88** 6367
[44] Oevering. H., Verhoeven, J.W., Paddon-Row, M.N., Cotsaris, E., and Hush, N.S. (1988) Chem. Phys. Lett **143** 488
[45] Bignozzi, C.A., Indelli, M.T., and Scandola, F. (1989) J. Am Chem. Soc. **111** 5192
[46] Scandola, F., Bignozzi, C. A., Chiorboli, C., Indelli, M. T., and Rampi, M. A. (1990) Coord. Chem. Rev. **97** 299
[47] Nishitani, S., Kurata, N., Sakata, Y., Misumi, S., Karen, A, Okada, T, and Mataga, N. (1983) J. Am. Chem. Soc. **105** 7771
[48] Gust, D., Mathis, P., Moore, A. L., Liddell, P. A., Nemeth, G. A., Lehman, W. R., Moore, T. A., Bensasson, R. V., Land, E. J., and Chachaty, C. (1983) Photochem. Photobiol. **37S** S46
[49] Gust, D. and Moore, T. A. (1989) Science **244** 35
[50] Wasielewski, M. R., Niemczyk, M. P., Svec, W. A., and Pewitt, E. B. (1985) J. Am. Chem. Soc. **107** 5562
[51] Sanders, G. M., van Dijk, M., van Veldhuizen, A., and van der Plas, H. C. (1986) J. Chem. Soc., Chem. Commun. 1311
[52] Sessler, J. L., Hugdahl, J.,and Johnson, M. R. (1986) J. Org. Chem. **51** 2828
[53] Liddell, P. A., Barrett, D., Makings, L. R., Pessiki, P. J., Gust, D., and Moore, T. A. (1986) J. Am. Chem. Soc. **108** 5350
[54] Sessler, J. L. and Johnson, M. R. (1987) Rec. Trav. Chim. Pays-Bas **106** 222
[55] Cowan, J. A., Sanders J. K. M., Beddard, G. S., and Harrison, R. J. (1987) J. Chem. Soc. Chem. Commun. 55
[56] Danielson, E., Elliott, C. M., Merkert, J. W., and Meyer, T. J. (1987) J. Am. Chem. Soc. **109** 2519

[57] Sessler, J. L., Johnson, M. R., Lin, T. Y., and Creager, E. E. (1988) J. Am. Chem. Soc. **110** 3659
[58] Hofstra, U., Schaafsma, T.j., Sanders, G. M., van Dijk, M., van der Plas, H. C., Johnson, D. G., and Wasielewski, M. R. (1988) Chem. Phys. Lett. **151** 169
[59] Sessler, J. L., Johnson, M. R., and Lin, T. Y. (1989) Tetrahedron **45** 4767
[60] Momenteau, M., Loock, B., Seta, P., Bienvenue, E., and d'Epenoux, B. (1989) Tetrahedron **45** 4893
[61] Gust, D., Moore, T. A., Moore, A. L., Makings, L. R., Seely, G. R, Ma, X., Trier, T. T., and Gao, F. (1988) J. Am. Chem. Soc. **110** 7567
[62] Gust, D., Moore, T. A., Moore, A. L., Barrett, D., Harding, L. O., Makings, L. R., Liddell, P. A., De Schryver, F. C., Van der Auweraer, M., Bensasson, R. V., and Rougée, M. (1988) J. Am. Chem. Soc. **110** 321
[63] Gust, D, Moore, T. A., Moore, A. L., Lee., S. J., Bittersman, E., Luttrull, D. K., DeGraziano, J., Ma, Z. C., and Gao, F. (1990) Abstracts of the VIIIth Conference on Photochemcal Conversion and Storage of Solar Energy, Palermo, p. 105
[64] Davila, J., Harriman, A., and Milgron, L.R. (1987) Chem. Phys. Lett. **136** 427
[65] Bignozzi, C.A., Roffia, S., Chiorboli, C., Davila, J., Indelli, M.T., and Scandola. F. (1989) Inorg. Chem. **28** 4350
[66] Denti, G., Campagna, S., Sabatino, L., Serroni, S., Ciano, M., and Balzani, V. (1990). This volume
[67] Murphy, W.R., Jr., Brewer, K.J., Gettliffe, G., and Petersen, J.D. (1989) Inorg. Chem. **28** 81
[68] Campagna, S., Denti, G.F., Sabatino, L., Serroni, S., Ciano, M., and Balzani, V. (1989) J. Chem. Soc. Chem. Commun. 1500
[69] Denti, G., Campagna, S., Sabatino, L., Serroni, S., Ciano, M., and Balzani, V. (1990) J. Am. Chem. Soc., submitted
[70] Aviram, A. and Ratner, M.A. (1974) Chem. Phys. Lett. **29** 277
[71] Carter, F.L. (ed.) (1982) Molecular electronic devices. Dekker
[72] Carter, F.L. (1984) Physica **10D** 175
[73] Wrighton, M.S. (1985) Comments Inorg. Chem. **4** 269
[74] Joachim, C. and Launay, J.P. (1984) Nouv. J. Chim. **8** 723
[75] Carter, F.L. (ed.) (1987) Molecular electronic devices II. Dekker
[76] Balzani, V., Moggi, L., and Scandola, F. (1987) In: Balzani, V. (ed.) Supramolecular photochemistry. Reidel, p.1
[77] Lehn, J.M. (1988) Angew. Chem. Int. Ed. Engl. **27** 89
[78] Carter, F.L., Siatkowski, R.E., and Wohltjen, H. (eds) (1988) Molecular electronic devices. North-Holland
[79] Stolarczyk, L.Z. and Piela, L. (1984) Chem. Phys. Lett. **85** 451
[80] Haddon, R.C. and Lamola, A. (1985) Proc. Natl. Acad. Sci. USA **82** 1874
[81] Aviram, A. (1988) J. Am. Chem. Soc. **110** 5687
[82] Gilmanshin, R.I. and Lazarev, P.I. (1988) J. Molec. Electronics **4** S83

[83] Metzger, R.M. and Panetta, C.A. (1989) J. Molec. Electronics 5 1
[84] Hong, F.T. (1989) J. Molec. Electronics 5 163
[85] Aviram, A. (1989) Angew. Chem. Int. Ed. Engl. 28 520
[86] Haarer, D. (1989) Angew. Chem. Int. Ed. Engl. 28 1544
[87] Clarkson, M.A. (1989) Byte **May 1989** 268
[88] Balzani, V. and Scandola, F. (1990) Supramolecular photochemistry. Horwood, Chap. 12
[89] Kuhn, H. (1987). In: Carter, F.L. (ed.) Molecular electronic devices II. Dekker, p. 411
[90] Fuhrhop, J.H. and Boettcher, C. (1990) J. Am. Chem. Soc. **112** 1768
[91] Ringsdorf, H., Schlarb, B. and Venzmer, J. (1988) Angew. Chem. Int. Ed. Engl. **27** 113
[92] Eidenschink, R. (1989) Angew. Chem. Int. Ed. Engl. Adv. Mater. **28** 1424
[93] Groh, W., Lupo, D., and Sixl, H. (1989) Angew. Chem. Int. Ed. Engl. Adv. Mater. **28** 1548
[94] Laschewsky, A. (1989) Angew. Chem. Int. Ed. Engl. Adv. Mater. **28** 1574
[95] Simon, J., Bassoul, P., and Norvez, S. (1989) New. J. Chem. **13** 13
[96] Hickman, J.J., Zou, C., Ofer, D., Harvey, P.D., and Wrighton, M.S. (1989) J. Am. Chem. Soc. **111** 7271
[97] Meyer T.J. (1989). In: Norris, J.R., Jr. and Meisel, D. (eds) Photochemical energy conversion. Elsevier, p. 75
[98] Younathan, J.N., McClanahan, S.F., and Meyer, T.J. (1989) Macromolecules **22** 1048
[99] Ennis, P.M. and Kelly, J.M. (1989) J. Phys. Chem. **93** 5735
[100] Shimidzu, T., Iyoda, T., and Honda, K. (1988) Pure Appl. Chem. **60** 1025
[101] Dabestani, R., Bard, A.J., Campion, A., Fox, M.A., Mallouk, T.E., Webber, S.E., and White, J.M. (1988) J. Phys. Chem. **92** 1872
[102] Vlachopoulos, N., Liska, P., Augustynski, J., and Graetzel, M. (1988) J. Am. Chem. Soc. **110** 1216
[103] Vrachnou, E., Graetzel, M., and McEvoy, A.J. (1989) J. Electroanal. Chem. **258** 193
[104] Enea, O., Moser, J., and Graetzel, M., (1989) J. Electroanal. Chem. **259** 59
[105] Andrieux, C.P., Haas, O., and Savéant, J.M. (1986) J. Am. Chem. Soc. **108** 8175
[106] Degani, Y. and Heller, A. (1989) J. Am. Chem. Soc. **111** 2357
[107] Licht, S., Cammarata, V., and Wrighton, M.S. (1989) Science **243** 1176
[108] Jernigan, J.C., Surridge, N.A., Zvanut, M.E., Silver, M., and Murray, R.W. (1989) J. Phys. Chem. **93** 4620
[109] Jernigan, J.C. and Murray, R.W. (1990) J. Am. Chem. Soc. **112** 1034
[110] Amadelli, R., Argazzi, R., Bignozzi, C.A., and Scandola, F. (1990) J. Am. Chem. Soc. **112** 0000
[111] Fujihira, M., Sakomura, M., and Kamei, T. (1989) Thin Solid

Films **180** 43
[112] Hopfield, J.J., Onuchic, J.N., and Beratan, D.N. (1988) Science **241** 817
[113] Hopfield, J.J., Onuchic, J.N., and Beratan, D.N. (1989) J. Phys. Chem. **93** 6350

ELECTRON TRANSFER IN HETEROGENEOUS SYSTEMS
(A Progress Report)

DAN MEISEL
Chemistry Division
Argonne National Laboratory
Argonne, IL 60439, USA

Abstract

Advances in the understanding of heterogeneous energy and electron transfer during the last few years are outlined. Analysis of the kinetics of electron transfer across interfaces at the molecular level lags behind the impressive advances for similar process in homogeneous solutions. The complexity of the interface imposes unrealistic simplifications in the theoretical description and its experimental characterization is still relatively poor. At the phenomenological level electron transfer across interfaces can now be given reasonably accurate predictive descriptions. Effects of free energy in the normal regime (while little has been observed in the inverted region), electrostatic interactions, reduction in dimensionality and effects of size distribution can all be incorporated into the a master equation and can be experimentally verified. Outstanding problems of considerable importance to energy conversion and photocatalysis are the nature of interparticle junctions and the mechanistic delineation of hole transfer from particulate metal oxides to organic substrates.

Introduction

It is impossible and perhaps even inappropriate to attempt to survey the vast field of heterogeneous electron transfer given the time and space constraints of the present format. More fitting for such a task is the voluminous collection of Fox and Chanon [1] or any of the monographs on this subject which appeared in recent years [2-4]. Many other reviews on this multidisciplinary area appeared in the last few years and the interested reader is better advised to consult these for comprehensiveness[5]. Some of these reviews will focus on specific activities in the broad field of heterogeneous electron transfer while others were restricted to particular materials. These include articles on photoinduced electron transfer in organized assemblies and semiconductor surfaces[6], charge separation at interfaces[7], dynamics of charge transfer across semiconductor-electrolyte

E. Pelizzetti and M. Schiavello (eds.), Photochemical Conversion and Storage of Solar Energy, 15–26.

interfaces[8], mechanisms of redox reactions on metallic and semiconductor particles[9] and on quantum size semiconductor particles[10-12]. These reviews will cover the whole gamut of subjects associated with heterogeneous electron transfer from the initial act of electron transfer to the mechanistic details of the follow-up catalytic thermal redox processes[13-14]; from the basic concepts of electron transfer in molecular systems through semiconductors and to complex biological systems[15]; Some will be entirely general to any system others will be confined to a family of microenvironments (e.g. micelles [16] or lipid membranes[17]) or to a particular single material (e.g. TiO_2[18]). In view of this wealth of information we can only hope to indicate in the following few of the advances which occurred in the field during the last couple of years and to point to few of the outstanding problems.

Basic Principles:

The insightful basic understanding at the level which has been accomplished for electron transfer in homogeneous systems, and which was reflected in the progress report by Scandola earlier in this conference, has not yet been achieved for electron transfer across interfaces. It is the complexity of the interface which introduces most of the difficulties, both for the theorist and the experimentalist. Inverted region effects for the rate dependence on free energy have been repeatedly verified in homogeneous solutions, yet only few examples show some decrease in rate for highly exergonic interfacial transfer[19]. Rates of homogeneous electron self-exchange have been measured for many transition metal ions and for radical-ion/molecule couples. The analogous exchange rate for a solid/electrolyte systems has been measured only in limited cases and its dependence on various physical parameters is not yet understood. Conventional wisdom, dating back to the original Marcus theory, predicts that the reorganization energy for the heterogeneous exchange will be equal or smaller than half the reorganization energy for the same homogeneous self exchange. This is because the number of reactants involved in the heterogeneous exchange is only half of those involved in the homogeneous process and, furthermore, some of the ionic charge will be screened by the image charge in the metal. Surprisingly, recent results on homogeneous and heterogeneous exchange rates for tetrathiafulvalene (TTF/TTF^+) reveal that the reorganization energy for the heterogeneous reaction in a large number of solvents is larger than that of the homogeneous exchange[20]. This was attributed to a larger reaction distance in the heterogeneous reaction probably due to counter ion adsorption on the metal. Indeed, the reaction distance in heterogeneous electron transfer is very difficult to quantify. It is safe to assume that before synthetic method are developed which will allow covalent binding of the reactant at particular surface sites the dependence of the rate on distance could not be ascertained.

Other problems associated with understanding the principles that govern interfacial electron transfer are the magnitude of the dielectric constant and the effect of image potentials at the loci of the transfer or in the double layer in general. Commonly this location is identified as the outer Helmholtz plane, approximately a monolayer away from the surface, and sometimes equated with the shear plane where the zeta potential is measured. Conceivably, the effective dielectric constant at this plane, where significant ordering of water dipoles and counter ions occurs will deviate from that of bulk water. Image potential effects, in particular for semiconductor/electrolyte interfaces, have only recently been tackled theoretically utilizing modern solid state - surface physics theories[21]. These studies already show that adsorption, commonly considered a chemical short range interaction, may result from non-local image potential effects. Contrary to expectation from classical theories, these calculations do predict under certain conditions a decrease in solvent reorganization energy as the distance to the solid, semiconductor or metal, decreases[21,22].

While the principles to the understanding of electron transfer at the molecular level are still under active investigations, phenomenological descriptions have been developed during the last few years which adequately delineate the dependence of the rate on macroscopic parameters. The effect of the synthetically inevitable size distribution of particles in the ensemble on the kinetics of charge injection has been incorporated into the analysis by Albery and coworkers[23,24] and this treatment has been adopted in the study of various photo-oxidative processes[25,26]. For charge injection into metal oxide colloids complete phenomenological description has been developed for reaction which are diffusion controlled, activation controlled or intermediate cases[27]. Effects of pH and ionic strength on the kinetics were incorporated and were shown to result from the effects of these parameters on free energy, surface potentials or double layer screening[27,28].

Electron transfer in reduced dimensionality:

A comprehensive unified treatment of energy or electron transfer in one, two and three dimensional disordered systems which includes donor-acceptor and donor-donor (hoping) transfer has been developed by Baumann and Fayer [29]. Anisotropic interactions in the limits of fast (relative to the transfer step) and slow rotational reorientation were incorporated in the averaging procedures and applications to infinite systems, intra layer and inter layer transfer in stacked multilayers were developed. The effects of surface concentration and layer separation distances were analyzed and fractal-like gradual change from two dimensional to three dimensional behavior upon increasing the number of layers has been demonstrated. Polarization grating experiments have been utilized to test the theoretical derivations for both energy transfer and electron transfer[30,31]. The same technique was also utilized to probe the electron transfer dynamics at the TiO_2/water interface by Miller and coworkers[32]. Yamazaki and coworkers[33] studied energy transfer in LB monolayers from long-chain carbazole donor chromophores to anthracene acceptors. Fluorescence decay functions in 2-dimensions could be described as a modified Förster type quenching process[34]. It was noticed that some of the fluorescers will be isolated from the quencher molecules even at high concentrations of quencher and will thus decay at its natural rate. Electron transfer from the carbazole moiety to the stearic acid monolayer support was also discerned in these fluorescence quenching experiments[33]. This process was best described by an Inokuti Hirayama formulation in a three-dimensional space[35]. Three dimensional sequential energy transfer, as well as two dimensional transfer, between carbazole and cyanine dyes and between photosynthetic antenna pigments in stacked multilayers was also demonstrated by Yamazaki et al[36]. Effects of inter layer distance relative to donor-donor and donor-acceptor intra layer distances on the rate of excitation energy transfer were analyzed using the Baumann and Fayer analysis.

Theoretical analysis of transport processes, energy transfer or electron transfer and for direct transfer as well as for a multi step hopping mechanism, in amorphous self-similar fractal structures was given by Klafter and Blumen[37]. Interestingly these authors find that the direct transfer will depend on the fractal dimension[38] while a hopping mechanism will depend on the spectral dimension. Transport of charge or excitation in restricted geometries via direct transfer or via diffusion within pore networks has also been analyzed[39]. Experimentally, energy transfer between rhodamine dyes adsorbed on vesicle surfaces was studied by Tamai et al.[40,41] and the fractal analysis of Klafter and Blumen[37] has been applied. These investigations are important not only for the understanding of the dynamics of energy or charge transport in the complex systems of restricted dimensionality but also in the structural characterization of the system itself. An example of the structural information that may be obtained from fractal analysis of energy transfer studies is given by Pines and Huppert[42].

Excess Charge and its Consequences:

Many examples have by now been reported which clearly show the shift to higher energies in the exciton absorption due to the presence of excess charge in small particles[43-48]. The physical rationalization for this effect is still under investigation. While lifetime considerations indicate that the effect may result from the electric field induced by the excess charge[47], a band filling Burstein-Moss effect seems to fit much of the observed experimental data[44-46].

The effect of excess charge however is not limited to the short time scales. At longer times two major consequences are known to result from the presence excess charge. One is the well known corrosion of the particle and the other is localization of the charge in deep traps which may be utilized as storage entities for reducing or oxidizing equivalents. These stored charges may also lead to significant changes in the properties of the material[49]. Many examples of oxidative dissolution of chalcogenides by excess holes in the semiconductor have been reported and the mechanism for this destructive process is believed to proceed via a semioxidized chalcogenide radical[50-52]. For metal oxide particles the destructive process is reductive dissolution caused by the presence of excess electrons[53]. The mechanism for this process is less understood and will depend on the particular transition metal. Nevertheless, a variety of unstable oxidation states of several transition metals have been observed following electron injection into the oxide using the pulse radiolysis technique. Unstable oxidation state such as Mn(III)[54,55] in MnO_2, Pb(III)[56-58] in PbO_2, were observed as intermediates in the dissolution process.

Interestingly, in all the metal oxide studied so far proton transfer to the particle follows immediately, or nearly immediately, the electron injection process leaving the net charge on the particle unchanged[59-61].

Interparticle Charge Transfer:

Charge transfer between two different particles is a promising route to enhance charge separation. Several example for such charge transfer have been reported since the original observation of electron transfer from CdS to RuO_2/TiO_2 particles[62], while the applications of such processes in photocatalysis have been demonstrated[63-64]. Electron transfer from CdS to TiO_2 and ZnO[65], CdS to Ag_2S[66], Cd_3P_2 to TiO_2 and ZnO[67], ZnS to ZnO[68], AgI and Ag_2S[69], alloys of $Zn_xCd_{(1-x)}S$ and CuS[70] have all been reported. In view of the very short life-times of electron-hole pairs in semiconductor particles, it seems clear that, when no scavenging of one of the carriers takes place, the charge transfer has to occur among preadsorbed particles. Yet, the mechanism of such a process is still unclear. The nature of the junction between the two particles has not been characterized and the pathway for the electron transport is still debated in the literature. Serpone et al.[62] invoke electron transfer from the conduction band of CdS to the CB of TiO_2 and eventually to RuO_2. This route apparently can be shortened. Based on the observation that the yield of hydrogen evolution is independent of the intermediate platinized metal-oxide Sobczynski et al. concluded that the electron transfer can proceeds directly from the CdS particle to the platinized islands[71]. Nevertheless, many of the systems mentioned above were studied by time resolved spectroscopy and there is little doubt that electron transfer between the chalcogenide and the metal oxide can occur. We may expect much more activity on this important question in the future.

Hole Transfer or Radical Mechanisms in Photocatalysis:

A major point of contention, which is of particular relevance to the present conference, is the question of the mechanism of oxidative degradation of organic pollutants on particulate metal oxide colloidal particles. The spectroscopic identity of the positive hole in the most frequently used semiconductor particles, TiO_2, is still controversial. Conflicting reports assign different absorption bands to the absorption by holes created upon bandgap excitation of TiO_2, ranging from ca. 400 nm to 600 nm[72,73]. It is therefore not surprising that the mechanism of the degradation is far from understood. As many of the reports to this conference indicate, much of the literature on this question assumes hole transfer to water molecules as the primary process. This reaction will generate hydroxyl radicals which in turn lead to the degradation of the organic substrate. On the other hand recent results using diffuse reflectance spectroscopy show direct hole transfer to many substrates[26]. Our own results[74], from pulse radiolysis experiments, indicate that hydroxyl radicals will react with TiO_2 particles at a diffusion controlled rate and therefore the release of hydroxyl radical, when photochemically generated, from the surface is rather unfavorable. This question is of considerable importance to the development of efficient photocatalytic systems and is certain to continue to be a focal point of much research activity.

Conclusions:

While studies to unravel the principles that control the dynamics of heterogeneous electron transfer on the molecular level proceed, two other directions seem to evolve in parallel: One is the design of new heterogeneous media for electron transfer process. We have seen examples of such studies in Ottolenghi's presentation on interfacial electron transfer in sol-gel matrices[75]. The study of electron transfer in chalcogenide particles deposited in bilayer membranes[76,77] or vesicles[78] are further examples for the search for new media.

The other direction is the application of new experimental techniques which will allow monitoring of the dynamics of electron transfer in either previously unattainable systems or at faster time resolution than previously attempted. Resonance Raman spectroscopy may provide precise identification of the intermediates and details of their interaction with the solid surface[79] while high pressure studies may provide information on the photophysics of the solid particles[80]. Microwave conductivity is increasingly utilized to follow the dynamics of charge recombination either in the solid phase[81] or in particles suspended in dielectric solvents[82,83]. The use of diffuse reflectance flash photolysis to study oxidation reaction in opaque TiO_2 dispersions has been already been mentioned above[25,26]. Picosecond and sub-picosecond resolution has been applied in the study of TiO2 - adsorbate hole transfer[73] and the competing processes of electron trapping and electron ejection have been recently studied with sub-picosecond time resolution[84]. Combined with theoretical formulations such studies will provide basic foundations for our understanding of electron transfer across interfaces in the future.

Acknowledgment: Work performed under the auspices of the Office of Basic Energy Sciences, US Department of Energy.

References:

1) Fox, M. A.; Chanon, M., *"Photoinduced Electron Transfer"*, Elsevier, Amsterdam, Oxford, New York, Tokyo, **(1988)**.
2) Gratzel, M., *"Heterogeneous Photochemical Electron Transfer"*, CRC Press, Boca Raton, Fl., **(1989)**.
3) Kalyanasundaram, K., *"Photochemistry in Microheterogeneous Systems"*, Academic Press, New York, **(1987)**.
4) Thomas, J. K., *"The Chemistry of Excitation at Interfaces"*, ACS monograph 181, **(1984)**.
5) Serpone, N.; Pelizzetti, E., *"Photocatalysis"*, J. Wiley & Sons, **(1989)**.
6) Fox, M. A., Pure Appl. Chem. **(1988)** *60,* 1013.
7) Chu, D. Y.; Stramel, R.; Hashimoto, S.; Nakamura, T.; Murtagh, J.; Kuczynski, J.; Milosavljevic, B.; Wheeler, J.; Thomas, J. K. in *"NATO ASI Ser., Ser. C 165 (Chem. React. Org. Inorg. Constrained Syst.)"*, Eds. R. Setton; Reidel, Dordrecht, Boston, Lancaster, Tokyo, **(1986)** pp 237.
8) Gratzel, M. in *"Homogeneous and Heterogeneous Photocatalysis, NATO ASI Ser., Ser. C 174,"*, Eds. E. Pelizzetti and N. Serpone; Reidel, Dordrecht, Boston, Lancaster, Tokyo, **(1986)** pp 91.
9) Henglein, A., Top. Curr. Chem. **(1988)** *143,* 113.
10) Steigerwald, M.; Brus, L. E., Ann. Rev. Mater. Sci. **(1989)** *19,* 471.
11) Henglein, A., Chem. Rev. **(1989)** *89,* 1861.
12) Henglein, A.; Weller, H. in *""Photochemical Energy Conversion.""*, Eds. J. Norris and D. Meisel; Elsevier, N.Y., **(1989)** pp 161.
13) Kamat, V. K.; Dimitrijevic, N. M., Solar Energy **(1990)** *44,* 83.

14) Schiavello, M.; Sclafani, A. in "*Photocatalysis*", Eds. N. Serpone and E. Pelizzetti; J. Wiley & Sons, **(1989)** pp 159.
15) McLendon, G.; Conklin, K. J.; Corvan, R.; Johansson, K.; Magner, E.; O'Neil, M.; Pardue, K.; Rogalski, J. S.; Whitten, D. in "*Photochemical Energy Conversion*", Eds. J. Norris and D. Meisel; Elsevier, N.Y., **(1989)** pp 47.
16) von Buenau, G.; Wolff, T., Adv. Photochem. **(1988)** *14*, 273.
17) Zamaraev, K. I.; Lymar, S. V.; Khramov, M. I.; Parmon, V. N., Pure Appl. Chem. **(1988)** *60*, 1039.
18) Anpo, M., Res. Chem. Intermed. **(1989)** *11*, 67.
19) Itoh, K.; Nakajima, Y.; Fujishima, A., Chem. Lett. **(1987)** 2125.
20) Grampp, G.; Kapturkiewicz, A.; Jaenicke, W., Ber. Bunsenges. Phys. Chem. **(1990)** *94*, 439.
21) Smith, B. B.; Koval, C. A., J. Electroanal. Chem. **(1990)** *277*, 43.
22) Dzhavakhidze, P. G.; Kornyshev, A. A.; Krishtalik, L., J. Electroanl. Chem. **(1987)** *228*, 329.
23) Albery, W. J.; Brown, G. T.; Darwent, J. R.; Saievar-Iranizad, E., J. Chem. Soc. Faraday Trans. 1 **(1985)** *81*, 1999.
24) Albery, W. J.; Bartlett, P. N.; Wilde, C. P.; Darwent, J. R., J. Am. Chem. Soc. **(1985)** *107*, 1854.
25) Draper, R. B.; Fox, M. A., J. Phys. Chem. **(1990)** *94*, 4628.
26) Draper, R. B.; Fox, M. A., Langmuir **(1990)** *6*, 1396.
27) Mulvaney, P.; Swayambunathan, V.; Grieser, F.; Meisel, D., Langmuir **(1990)** *6*, 555.
28) Mulvaney, P.; Swayambunathan, V.; Grieser, F.; Meisel, D., J. Phys. Chem. **(1988)** *92*, 6732.
29) Baumann, J.; Fayer, M. D., J. Phys. Chem. **(1986)** *85*, 4087.
30) Eyring, G.; Fayer, M. D., J. Chem. Phys. **(1984)** *81*, 4314.
31) Dorfman, R. C.; Lin, Y.; Zimmt, M. B.; Baumann, J.; Domingue, R. P.; Fayer, M. D., J. Phys. Chem. **(1988)** *92*, 4258.
32) Kasinski, J. J.; Gomez-Jahn, L. A.; Faran, K. J.; Gracewski, S. M.; Miller, R. J. D., J. Chem. Phys. **(1989)** *90*, 1253.
33) Tamai, N.; Yamazaki, T.; Yamazaki, I., J. Phys. Chem. **(1987)** *91*, 841.
34) Hauser, M.; Klein, U. K. A.; Gosele, U., Z. Phys. Chem. (Munich) **(1976)** *101*, S255.
35) Inokuti, M.; Hirayama, F., J. Chem. Phys. **(1965)** *43*, 1978.

36) Yamazaki, I.; Tamai, N.; Yamazaki, T.; Murakami, A.; Mimuro, M.; Fujita, Y., J. Chem. Phys. **(1988)** *92,* 5035.
37) Klafter, J.; Blumen, A., J. Chem. Phys. **(1984)** *80,* 875.
38) Avnir, D., *"The Fractal Approach to Heterogeneous Chemistry: Surfaces, Colloids, Polymers.",* Wiley, Chichester, UK, **(1989)**.
39) Klafter, J.; Drake, J. M.; Levitz, P.; Blumen, A.; Zumofen, G., J. Luminescence **(1990)** *45,* 34.
40) Tamai, N.; Yamazaki, T.; Yamazaki, I.; Mataga, N. in *"Ultrafast Phenomena V.",* Eds. G.R. Fleming and A.E. Siegman; Springer-Verlag, Berlin, **(1986)** pp 449.
41) Yamazaki, I.; Tamai, N.; Yamazaki, T., J. Phys. Chem. **(1990)** *94,* 516.
42) Pines, D.; Huppert, D., Chem. Phys. Lett. **(1989)** *156,* 223.
43) Henglein, A.; Kumar, A.; Janata, E.; Weller, H., Chem. Phys. Lett. **(1986)** *132,* 133.
44) Liu, C.; Bard, A. J., J. Chem. Phys. **(1989)** *93,* 7047.
45) Liu, C.; Bard, A. J., J. Chem. Phys. **(1989)** *93,* 3232.
46) Kamat, P. V.; Dimitrijevic, N. M.; Nozic, A. J., J. Phys. Chem. **(1989)** *93,* 2873.
47) Hilinski, E. F.; Luccas, P. A.; Wang, Y., J. Chem. Phys. **(1988)** *89,* 3435.
48) Hayes, D.; Micic, O. I.; Nenadovic, M. T.; Swayambunathan, V.; Meisel, D., J. Phys. Chem. **(1989)** *93,* 4603.
49) Mulvaney, P.; Cooper, R.; Grieser, F.; Meisel, D., Langmuir **(1988)** *4,* 1206.
50) Henglein, A., Ber. Bunsenges. Phys. Chem. **(1982)** *86,* 301.
51) Dimitrijevic, N. M.; Kamat, P. V., Langmuir **(1987)** *3,* 1004.
52) Dimitrijevic, N. M.; Kamat, P. V., Radiat. Phys. Chem. **(1988)** *32,* 53.
53) Blesa, M. A.; Maroto, A. J. G., J. Chim. Phys. **(1986)** *83,* 757.
54) Lume-Pereira, C.; Baral, S.; Henglein, A.; Janata, E., J. Phys. Chem. **(1985)** *89,* 5772.
55) Baral, S.; Lume-Pereira, C.; Janata, E.; Henglein, A., J. Phys. Chem. **(1985)** *89,* 5779.
56) Kumar, A.; Henglein, A.; Weller, H., J. Phys. Chem. **(1989)** *93,* 2262.
57) Mosseri, S.; Henglein, A.; Janata, E., J. Phys. Chem. **(1990)** *94,* 2722.
58) Mulvaney, P.; Grieser, F.; Meisel, D., J. Phys. Chem. in Press, .
59) Borgarello, E.; Pelizzetti, E.; Mulac, W. A.; Meisel, D., J. Chem. Soc., Faraday Trans. 1 **(1985)** *81,* 143.

60) Mulvaney, P.; Cooper, R.; Grieser, F.; Meisel, D., J. Phys. Chem. **(1990)** in press., .
61) Mulvaney, P.; Grieser, F.; Meisel, D., Langmuir **(1990)** *6,* 567.
62) Serpone, N.; Borgarello, E.; Gratzel, M., J. Chem. Soc. Chem. Comm. **(1984)** 342.
63) Pichat, P.; Disdier, J.; Herrmann, J.-M.; Mozzanega, M.-N.; Hoang-Van, C. in "Photoelectrochemistry and Electrosynthesis on Semiconductor Materials", Eds. D. Ginley,K. Honda,A. Nozik,A. Fujishima,A. Armstrong, et al.; The Electrochemical Society, **(1988)** pp 50.
64) Serpone, N.; Borgarello, E.; Pelizzetti, E., J. Electrochem. Soc. **(1988)** *135,* 2760.
65) Spanhel, L.; Weller, H.; Henglein, A., J. Am. Chem. Soc. **(1987)** *109,* 6632.
66) Spanhel, L.; Weller, H.; Fojtik, A.; Henglein, A., Ber. Bunsenges. Phys. Chem. **(1987)** *91,* 88.
67) Spanhel, L.; Henglein, A.; Weller, H., Ber. Bunsenges. Phys. Chem. **(1987)** *91,* 1359.
68) Rabani, J., J. Phys. Chem. **(1989)** *93,* 7707.
69) Henglein, A.; Gutierrez, M.; Weller, H.; Fojtik, A.; Jirkovsky, J., Ber. Bunsenges. Phys. Chem. **(1989)** *93,* 593.
70) Gruzdkov, Y. A.; Savinov, E. N.; Korolkov, V. V.; Parmon, V. N., React. Kinet. Catal. Lett. **(1988)** *36,* 395.
71) Sobczynski, A.; Bard, A. J.; Campion, A.; Fox, M. A.; Mallouk, T.; Webber, S. E.; White, J. M., J. Phys. Chem. **(1987)** *91,* 3316.
72) Bahnemann, D.; Henglein, A.; Lilie, J.; Spanhel, L., J. Phys. Chem. **(1984)** *88,* 709.
73) Arbour, C.; Sharma, D.; Devendra, K.; Langford, C., J. Phys. Chem. **(1990)** *94,* 331.
74) Lawless, D.; Serpone, N.; Meisel, D., To be published .
75) Slama-Schwok, A.; Avnir, D.; Ottolenghi, M., 8th Internatl. Conf. Photochem. Conversion of Solar Energy (IPS8), Book of Abstracts **(1990)** 78.
76) Fendler, J. H.; Baral, S. in "*Photochemical Energy Conversion*", Eds. J. Norris and D. Meisel; Elsevier, New York, **(1989)** pp 148.
77) Baral, S.; Fendler, J. H., J. Am. Chem. Soc. **(1989)** *111,* 1604.
78) Chang, A.-C.; Fendler, J. H., J. Phys. Chem. **(1989)** *93,* 2538.
79) Brus L. in "*Photoinduced Electron Transfer*", Eds. M.A.F.a.M. Chanon; Elsevier, Amsterdam, Oxford, New York, Tokyo, **(1988)** pp 314.

80) Alivisatos, A. P.; Harris, T. D.; Brus, L. E.; Jayaraman, A., J. Chem. Phys. **(1988)** *89,* 5979.
81) Warman, J. M.; de Haas, M. P.; van Hovell tot Westerflier, S. W. F. M.; Binsma, J. J. M.; Kolar, Z. I., J. Phys. Chem. **(1989)** *93,* 5895.
82) Eden, J.; van Lith, D.; Warman, J. M.; Hummel, A., J. Chem. Soc. Faraday Trans. 1 **(1989)** *85,* 991.
83) Warman, J. M.; de Haas, M. P.; Wentinck, H. M., Radiat. Phys. Chem. **(1989)** *34(4),* 581.
84) Kaschke, M.; Ernsting, N. P.; Mueller, U.; Weller, H., Chem. Phys. Lett. **(1990)** *168,* 543.

TOWARDS AN ARTIFICIAL PHOTOSYNTHESIS. DI-, TRI-, TETRA-, AND HEPTA-NUCLEAR LUMINESCENT AND REDOX-REACTIVE METAL COMPLEXES

G. DENTI,[a] S. CAMPAGNA,[b] L. SABATINO,[c] S. SERRONI,[c] M. CIANO,[d] AND V. BALZANI[de]

[a]Istituto di Chimica Agraria dell'Università, Pisa, Italy;
[b]Dipartimento di Chimica Inorganica dell'Università, Messina, Italy;
[c]Dipartimento di Chimica, Università della Calabria, Rende, Italy;
[d]Istituto FRAE-CNR, Bologna, Italy;
[e]Dipartimento di Chimica "G. Ciamician" dell'Università, Bologna, Italy.

ABSTRACT. We have synthesized more than 30 mono-, di-, tri-, tetra-, and heptanuclear Ru(II) and/or Os(II) complexes containing 2,3-dpp and 2,5-dpp as bridging ligands and bpy and biq as peripheral ligands. These oligonuclear complexes are quite interesting because they accumulate many chromophoric and redox centers in the same "supermolecule" and exhibit luminescence from relatively long-lived excited states.

1. Introduction

Assembly of molecular components that carry appropriate "pieces of information" can be used to build up supramolecular species capable of performing useful functions. In particular, assembly of components that exhibit suitable excited state properties can lead to photochemical molecular devices (PMDs) [1,2] capable of performing light induced energy migration or charge separation processes which are at the basis of important applications, including natural [3] and artificial [4] solar energy conversion. Fig. 1 illustrates the concept of an antenna device for light energy collection [1,2]: an array of chromophores (Pen) absorb light and the excited states so obtained transfer electronic energy to another component where energy may be used for physical purposes (for example, luminescence, Fig. 1) or chemical reactions. Fig. 2 shows schematically a photochemical molecular device for light-induced charge separation, where light energy is converted into chemical redox energy [1,2].

To design and construct photochemical molecular devices, one needs

E. Pelizzetti and M. Schiavello (eds.), Photochemical Conversion and Storage of Solar Energy, 27–45.

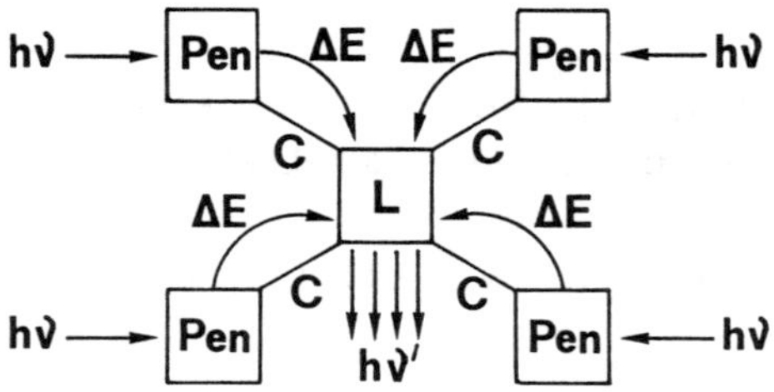

Figure 1. PMD for light energy collection.

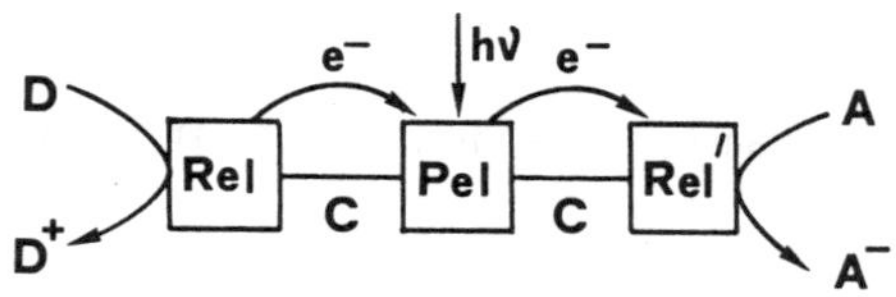

Figure 2. PMD for light-induced charge separation.

suitable active components and bridges. In the field of transition metal chemistry, a great number of Ru(II) [5] and Os(II) [6] polypyridine complexes have been characterized in the past 15 years. Most of them exhibit electrochemical, photochemical, and photophysical properties suitable to play the role of building blocks for photochemical molecular devices capable to perform energy and/or electron transfer. A variety of multidentate (bridging) polypyridine ligands are also available to synthesize mononuclear Ru and Os complexes which exhibit a residual coordination ability and are therefore capable of performing the role of ligands towards other metal units [7]. Using such "complexes as ligands" strategy, illustrated in more detail later on, we have synthesized a number of bi-, tri-, tetra-, and heptanuclear Ru and Os complexes which show quite interesting luminescence and redox properties. The bridging ligands (BL) used are the 2,3 and 2,5 isomers of bis(2-pyridyl)pirazine (2,3-dpp and 2,5-dpp, Fig. 3). As peripheral ligands (L), we have chosen the well known 2,2'-bipyridine (bpy) and 2,2'-biquinoline (biq) (Fig. 3) because they exhibit quite different properties (vide infra). The complexes studied correspond to the following general formulas (M = Ru and/or Os, L = bpy and/or biq, BL = 2,3-dpp or 2,5-dpp, n = 0-3; not all the possible combinations are available):

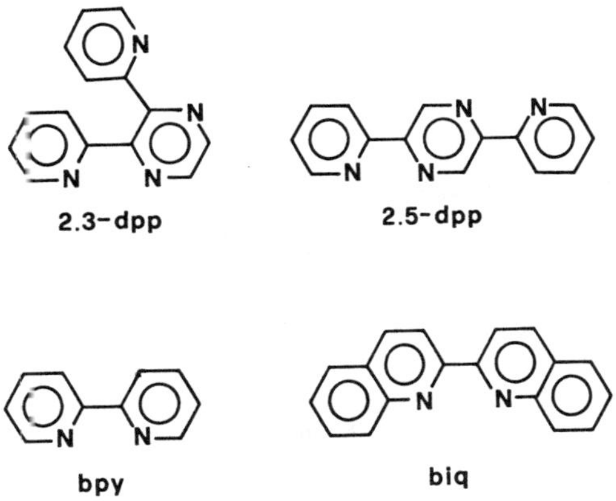

Figure 3. Structural formulas of the ligands.

mononuclear: $M(L)_n(BL)_{3-n}{}^{2+}$

binuclear: $(L)_2M(\mu\text{-}BL)M(L)_2{}^{4+}$

trinuclear: $(L)M[(\mu\text{-}BL)M(L)_2]_2{}^{6+}$

tetranuclear: $M[(\mu\text{-}BL)M(L)_2]_3{}^{8+}$

heptanuclear: $M[(\mu\text{-}BL)M(L)(\mu\text{-}BL)M(L)_2]_3{}^{14+}$

The list of the complexes that have actually been investigated is given in Tables 1 and 2 (some of the complexes have also been studied in other laboratories[8-11]). The structural formulas of the tri-, tetra-, and heptanuclear complexes of 2,3-dpp are schematized in Fig. 4.

2. Results

2.1 SYNTHESES

The synthetic procedures have been described in detail elsewhere [12-17] and are summarized in Scheme 1. As one can see, the complexes of higher nuclearity are built up using smaller complexes as entering ligands on other complexes. This "complexes as ligands" strategy is very effective and can be considered a particularly interesting case of the structure-directed synthetic methods currently used to obtain large artificial molecules [18,19].

For all the novel complexes satisfactory elemental analyses and consistent IR spectra and conductivity values were obtained. The

SCHEME 1

Mononuclear complexes

$$ML_2Cl_2 \; + \; BL \xrightarrow{NH_4PF_6} [ML_2(BL)](PF_6)_2$$

$$[MLCl_3]_n \; + \; 2\;BL \xrightarrow{NH_4PF_6} [ML(BL)_2](PF_6)_2$$

$$MCl_3 \; + \; 3\;BL \xrightarrow{NH_4PF_6} [M(BL)_3](PF_6)_2$$

Binuclear complexes

$$ML_2Cl_2 \; + \; [ML_2(BL)](PF_6)_2 \xrightarrow{NH_4PF_6} [L_2M(\mu\text{-}BL)ML_2](PF_6)_4$$

Trinuclear complexes

$$[ML(BL)_2](PF_6)_2 \; + \; 2\;ML_2Cl_2 \xrightarrow{NH_4PF_6} [L_2M(\mu\text{-}BL)ML(\mu\text{-}BL)ML_2](PF_6)_6$$

Tetranuclear complexes

$$[M(BL)_3](PF_6)_2 \; + \; 3\;ML_2Cl_2 \xrightarrow{NH_4PF_6} \{M[(\mu\text{-}BL)ML_2]_3\}(PF_6)_8$$

$$3\;[ML_2(BL)](PF_6)_2 \; + \; MCl_3 \xrightarrow{NH_4PF_6} \{M[(\mu\text{-}BL)ML_2]_3\}(PF_6)_8$$

$$3\;[ML(BL)_2](PF_6)_2 \; + \; MCl_3 \xrightarrow{NH_4PF_6} \{M[(\mu\text{-}BL)ML(BL)]_3\}(PF_6)_8$$

Heptanuclear complexes

$$\{M[(\mu\text{-}BL)ML(BL)]_3\}(PF_6)_8 \; + \; 3\;ML_2Cl_2 \xrightarrow{NH_4PF_6} \{M[(\mu\text{-}BL)ML(\mu\text{-}BL)ML_2]_3\}(PF_6)_{14}$$

structural characterization of the complexes is an open problem because it was impossible to grow crystals suitable for diffractometric analysis, and the ^{1}H NMR spectra were complicated by the high number of

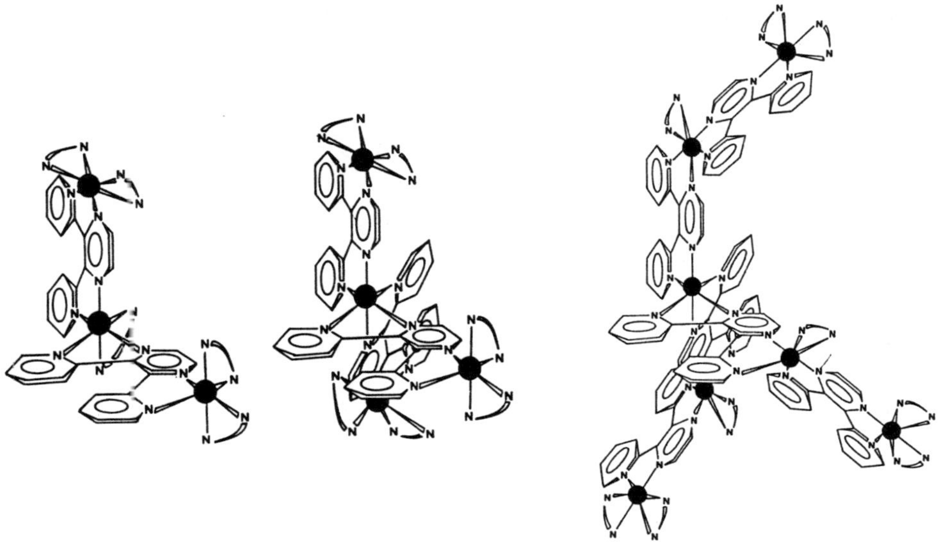

Figure 4. Structural formulas of tri-, tetra-, and heptanuclear complexes of 2,3-dpp.

hydrogens resonant in a narrow spectral region, even in mononuclear species. Furthermore, taking into account that each metal is a chiral center, the products of nuclearity greater than one are mixtures of diastereoisomeric species. Evidences about the structure around the central metal would be necessary in particular for tri-, tetra-, and heptanuclear compounds, where different geometrical arrangements of the chelating ligands can exist. The problem might be dealt with (i) by simplifying NMR spectra using perdeuterio L ligands [20], and (ii) by preparing the complexes starting from enantiomerically resolved precursors and by separating the diastereoisomeric products. If the configurations (optical and geometrical) of precursors are stable enough under the reaction and purification conditions, it would be possible, through a combination of these two ways, to obtain products suitable for NMR analysis. Of course the premises for crystal structure determinations would also be reached. We plan to devote to this problem some experimental effort in the near future.

2.2 SPECTROSCOPIC, PHOTOPHYSICAL, AND ELECTROCHEMICAL PROPERTIES

The absorption spectra in acetonitrile solution show extremely intense (ε_{max} 30000–300000 $M^{-1}cm^{-1}$) bands in the UV region and broad, intense (ε_{max} 7000–75000 $M^{-1}cm^{-1}$) bands in the visible, which account for the beautiful colors (from red to green to blue) exhibited by the complexes. Fig. 5 shows the spectra of the di-, tri-, tetra- and heptametallic complexes which contain the 2,3-dpp bridging ligand (BL) and bpy as peripheral ligand (L).

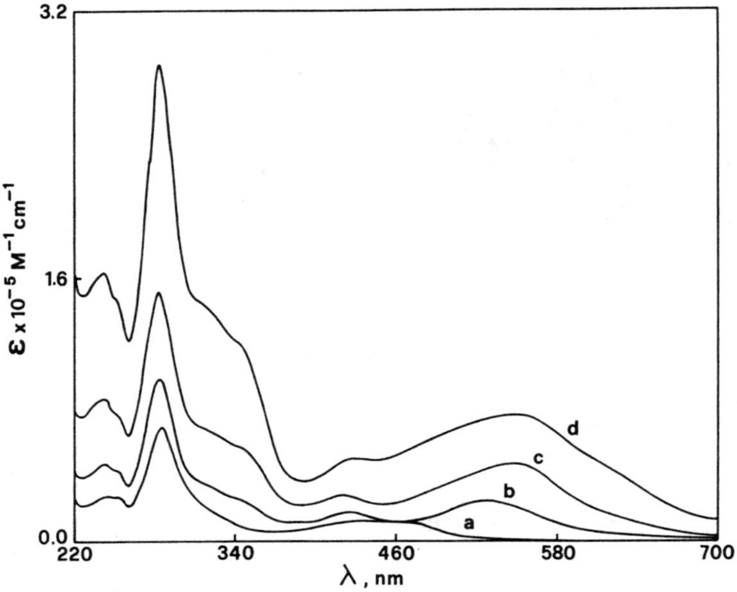

Figure 5. Electronic absorption spectra of complexes **1** (a), **13** (b), **26** (c), and **32** (d) in acetonitrile solution.

Most of the complexes examined are luminescent in rigid matrix at 77 K and several are also luminescent in fluid solution at room temperature. For some complexes, namely **13**, **22**, **25**, **26**, **30**, and **32**, corrected excitation spectra in the visible were also recorded and were found to match closely the corresponding absorption spectra. The luminescence decay was monoexponential in all cases. The corrected maxima of the luminescence bands, the luminescence lifetimes, and luminescence quantum yields are gathered in Table 1.

The electrochemical experiments showed that the complexes undergo several redox processes in the potential window examined. Most of the observed waves are reversible. In the oligometallic complexes wave overlapping is frequent, especially when L=biq. Several waves are di- or tri-electronic in nature. The potential values are gathered in Table 2.

3. Discussion

To a first approximation, the spectroscopic and electrochemical properties of the complexes examined can be discussed within the assumption that the ground as well as the excited and redox states involved can be described by localized molecular orbital configurations [21,22]. With such an assumption, the various spectroscopic transitions are classified as metal-centered (MC), ligand-centered (LC), or charge-transfer (either metal-to-ligand, MLCT, or ligand-to-metal, LMCT) [21], and the oxidation and reduction processes are classified as metal- or ligand-centered [22]. Excited states and redox forms concerning specific metals and ligands can be distinguished in most cases . The 2,3- and 2,5-dpp bridges, however, allow some coupling between the bridged units as will be discussed in more detail later.

3.1 ELECTROCHEMISTRY

The interpretation of the electrochemical behavior of most of the complexes is reported elsewhere [13-17]. The present discussion is limited to a comparison between analogous Ru and Os complexes.

The three binuclear $(bpy)_2M(\mu\text{-}2,3\text{-dpp})M'(bpy)_2^{4+}$ complexes (M=M'=Ru, **13**; M=Ru and M'=Os, **19**; M=M'=Os, **20**) show two one-electron oxidation waves. Comparison between the two homonuclear complexes shows that, as expected, in the heteronuclear complex **19** the first oxidation wave can be assigned to Os oxidation and the second one to Ru oxidation. On reduction, the two homonuclear complexes show the first two waves, which can be assigned to successive one-electron reduction of the bridge, at quite comparable potentials, which indicates that the $Os(bpy)_2^{2+}$ and $Ru(bpy)_2^{2+}$ moieties exhibit comparable electron withdrawing properties. For the two homonuclear complexes, the dielectronic reduction waves which follow correspond to the simultaneous reduction of one bpy on each of the two equivalent metal-containing moieties. As expected these waves occur at less negative potentials for the Os complexes. In the heteronuclear complex each bielectronic wave splits into two components. The four waves so

TABLE 1. ABSORPTION AND LUMINESCENCE DATA

		Abs 298 K[a]	Luminescence				
		λ_{max}, nm[b]	298 K[a]			77 K[c]	
		(ε, 10^3 $M^{-1}cm^{-1}$)	λ_{max}, nm	τ, ns[d]	Φx10^2	λ_{max}, nm	μs
1	$Ru(bpy)_2(2,3\text{-dpp})^{2+}$ [f]	475 (11.5)	691	240(380)	2.3	628	4.18
2	$Ru(bpy)_2(2,5\text{-dpp})^{2+}$	486 (5.2)	695	270(620)	4.1	647	2.40
3	$Ru(biq)_2(2,3\text{-dpp})^{2+}$	528 (7.6)	738	20		715	2.02
4	$Ru(biq)_2(2,5\text{-dpp})^{2+}$	527 (5.6)	737	15		715	2.15
5	$Ru(bpy)(2,3\text{-dpp})_2{}^{2+}$	465 (13.6)	652[g]	253		618[g]	
6	$Ru(bpy)(2,5\text{-dpp})_2{}^{2+}$	473 (11.2)	655[g]	400			
7	$Os(2,3\text{-dpp})_3{}^{2+}$ [f]	475 (18.0)	736[g]	(82)		716[g]	0.85
8	$Os(2,5\text{-dpp})_3{}^{2+}$	478 (13.0)	742[g]	33		733[g]	0.61
9	$Os(bpy)_2(2,3\text{-dpp})^{2+}$ [f]	480 (10.5)	800[g]	20		761[g]	0.51
10	$Os(bpy)_2(2,5\text{-dpp})^{2+}$	498 (12.3)	824[g]	11		787[g]	0.39
11	$Os(biq)_2(2,3\text{-dpp})^{2+}$	544 (14.0)	829[g]	h		828[g]	h
12	$Os(biq)_2(2,5\text{-dpp})^{2+}$	545 (10.6)	830[g]	h		829[g]	h
13	$(bpy)_2Ru(\mu\text{-}2,3\text{-dpp})Ru(bpy)_2{}^{4+}$ [f]	527 (24.2)	802	102(125)	0.3	709	2.00
14	$(biq)_2Ru(\mu\text{-}2,3\text{-dpp})Ru(biq)_2{}^{4+}$	537 (15.6)	789	65		720	1.73
15	$(bpy)_2Ru(\mu\text{-}2,3\text{-dpp})Ru(biq)_2{}^{4+}$	543 (22.4)	799	75		723	1.55

16	$(bpy)_2Ru(\mu\text{-}2,5\text{-}dpp)Ru(bpy)_2^{4+}$	585 (15.9)	824	155		771	0.94
17	$(biq)_2Ru(\mu\text{-}2,5\text{-}dpp)Ru(biq)_2^{4+}$	609 (11.5)	820	170		722	1.82
18	$(bpy)_2Ru(\mu\text{-}2,5\text{-}dpp)Ru(biq)_2^{4+}$	595 (13.7)	830	190		792	1.00
19	$(bpy)_2Ru(\mu\text{-}2,3\text{-}dpp)Os(bpy)_2^{4+}$	543 (23.5)					
20	$(bpy)_2Os(\mu\text{-}2,3\text{-}dpp)Os(bpy)_2^{4+}$	555 (21.8)					
21	$(bpy)_2Os(\mu\text{-}2,5\text{-}dpp)Os(bpy)_2^{4+}$	616 (27.0)					
22	$(bpy)Ru[(\mu\text{-}2,3\text{-}dpp)Ru(bpy)_2]_2^{6+}$	545 (23.5)	804	75(80)	0.1	721	1.78
23	$(bpy)Ru[(\mu\text{-}2,5\text{-}dpp)Ru(bpy)_2]_2^{6+}$	595 (28.1)	831	65		767	0.98
24	$(bpy)Ru[(\mu\text{-}2,3\text{-}dpp)Ru(biq)_2]_2^{6+}$	546 (28.7)	773	140		713	2.18
25	$(bpy)Ru[(\mu\text{-}2,5\text{-}dpp)Ru(biq)_2]_2^{6+}$	591 (22.9)	805	120(190)	0.6	739	1.28
26	$Ru[(\mu\text{-}2,3\text{-}dpp)Ru(bpy)_2]_3^{8+}$ l	545 (46.0)	811	50(60)	0.1	727	1.38
27	$Ru[(\mu\text{-}2,3\text{-}dpp)Ru(biq)_2]_3^{8+}$	610 (41.5)	795	130(190)	0.1	725	1.86
28	$Ru[(\mu\text{-}2,5\text{-}dpp)Ru(bpy)_2]_3^{8+}$	649 (25.4)	810 g	158			
29	$Ru[(\mu\text{-}2,5\text{-}dpp)Ru(biq)_2]_3^{8+}$	699 (30.5)	799 g	100			
30	$Os[(\mu\text{-}2,3\text{-}dpp)Ru(bpy)_2]_3^{8+}$	549 (40.0)	875	(18)		802	0.41
31	$Ru[(\mu\text{-}2,5\text{-}dpp)Os(bpy)_2]_3^{8+}$	618 (24.6)					
32	$Ru[(\mu\text{-}2,3\text{-}dpp)Ru(bpy)(\mu\text{-}2,3\text{-}dpp)Ru(bpy)_2]_3^{14+}$	547 (76.2)	808	(80)	0.09	725	2.00

a) AN solution; b) lowest energy maximum; c) MeOH/EtOH 4:1 rigid matrix; d) aerated solution; deaerated values in parentheses; e) deaerated solution; f) see also refs 8-10; g) uncorrected spectrum; h) not measured, because of the low intensity of the emission; l) see also refs 9 and 13.

TABLE 2. ELECTROCHEMICAL DATA[a]

		oxidation		reduction			
1	$Ru(bpy)_2(2,3\text{-}dpp)^{2+}$ [b]		+1.31	-1.06	-1.55	-1.74	
2	$Ru(bpy)_2(2,5\text{-}dpp)^{2+}$		+1.33	-1.03	-1.52	-1.71	-1.89
3	$Ru(biq)_2(2,3\text{-}dpp)^{2+}$		+1.47	-0.79	-0.98	-1.40	-1.81
4	$Ru(biq)_2(2,5\text{-}dpp)^{2+}$		+1.45	-0.80	-1.31	-1.81	
5	$Ru(bpy)(2,3\text{-}dpp)_2^{2+}$		+1.46	-0.92	-1.18	-1.58	
6	$Ru(bpy)(2,5\text{-}dpp)_2^{2+}$		+1.39	-0.88	-1.05	-1.55	
7	$Os(2,3\text{-}dpp)_3^{2+}$ [b]		+0.92	-0.84	-1.00	-1.36	
8	$Os(2,5\text{-}dpp)_3^{2+}$		+0.96	-0.62	-0.98	-1.32	
9	$Os(bpy)_2(2,3\text{-}dpp)^{2+}$ [b]		+0.79	-0.90	-1.39	-1.57	
10	$Os(bpy)_2(2,5\text{-}dpp)^{2+}$		+0.87	-0.90	-1.32	-1.58	
11	$Os(biq)_2(2,3\text{-}dpp)^{2+}$		+1.02	-0.58	-0.79	-1.34	
12	$Os(biq)_2(2,5\text{-}dpp)^{2+}$		+1.04	-0.62	-0.79	-1.23	
13	$(bpy)_2Ru(\mu\text{-}2,3\text{-}dpp)Ru(bpy)_2^{4+}$ [b]	+1.55	+1.38	-0.67	-1.17	-1.57[c]	-1.89[c]
14	$(biq)_2Ru(\mu\text{-}2,3\text{-}dpp)Ru(biq)_2^{4+}$		(+1.57)[d]	-0.45	-0.81[c]	-0.95	-1.19[c]
15	$(bpy)_2Ru(\mu\text{-}2,3\text{-}dpp)Ru(biq)_2^{4+}$	+1.48	+1.36	-0.68	-1.18	-1.57[c]	-1.81[c]
16	$(bpy)_2Ru(\mu\text{-}2,5\text{-}dpp)Ru(bpy)_2^{4+}$ [e]	+1.54	+1.37	-0.53	-1.08	-1.50[c]	-1.81[c]
17	$(biq)_2Ru(\mu\text{-}2,5\text{-}dpp)Ru(biq)_2^{4+}$		(+1.48)[d]	-0.45	-0.82[c]	-0.99	-1.26[c]
18	$(bpy)_2Ru(\mu\text{-}2,5\text{-}dpp)Ru(biq)_2^{4+}$		(+1.43)[d]	-0.47	overlapping waves		

19	$(bpy)_2Ru(\mu\text{-}2,3\text{-dpp})Os(bpy)_2^{4+}$		+1.52	+0.99	-0.65	-1.11	overlapping waves		
20	$(bpy)_2Os(\mu\text{-}2,3\text{-dpp})Os(bpy)_2^{4+}$ b		+1.20	+0.90	-0.68	-1.10	-1.38[c]	-1.62[c]	
21	$(bpy)_2Os(\mu\text{-}2,5\text{-dpp})Os(bpy)_2^{4+}$		+1.22	+0.92	-0.56	-1.00	-1.46[c]	-1.71[c]	
22	$(bpy)Ru[(\mu\text{-}2,3\text{-dpp})Ru(bpy)_2]_2^{6+}$			+1.48[c]	-0.55	-0.75	-1.17	-1.47[c]	-1.75[c]
23	$(bpy)Ru[(\mu\text{-}2,5\text{-dpp})Ru(bpy)_2]_2^{6+}$			+1.45[c]	-0.48	-0.60	-1.10	-1.30	-1.52
24	$(bpy)Ru[(\mu\text{-}2,3\text{-dpp})Ru(biq)_2]_2^{6+}$			+1.60[c]	-0.47[c]	-0.87[c]	-1.17[c]	-1.50	-1.79[c]
25	$(bpy)Ru[(\mu\text{-}2,5\text{-dpp})Ru(biq)_2]_2^{6+}$			+1.57[c]	-0.47[c]	-0.89[c]	-1.21[c]	-1.53	-1.78[c]
26	$Ru[(\mu\text{-}2,3\text{-dpp})Ru(bpy)_2]_3^{8+}$			+1.50[f]	-0.56	-0.63	-0.70	-1.20	-1.33
27	$Ru[(\mu\text{-}2,3\text{-dpp})Ru(biq)_2]_3^{8+}$			+1.58[f]	(-0.6)[g]	-0.87[f]	-1.15[f]		
28	$Ru[(\mu\text{-}2,5\text{-dpp})Ru(bpy)_2]_3^{8+}$			+1.40[f]	-0.58	-0.65	-0.80	-1.10	-1.30
29	$Ru[(\mu\text{-}2,5\text{-dpp})Ru(biq)_2]_3^{8+}$			+1.67[f]	-0.52	-0.87	-1.10	-1.60	-2.00
30	$Os[(\mu\text{-}2,3\text{-dpp})Ru(bpy)_2]_3^{8+}$		+1.55[f]	+1.25	-0.55	-0.65	-0.89		
31	$Ru[(\mu\text{-}2,5\text{-dpp})Os(bpy)_2]_3^{8+}$	+1.37	+1.10	+0.92	-0.64	-0.80	-0.92	-1.22	-1.38
32	$Ru[(\mu\text{-}2,3\text{-dpp})Ru(bpy)$ $(\mu\text{-}2,3\text{-dpp})Ru(bpy)_2]_3^{14+}$			+1.38[f]	-0.58[f]	overlapping waves			

a) An solution, room temperature; $E_{1/2}$ values vs SCE; monoelectronic waves unless otherwise noted; at more negative potentials other waves, often overlapping and/or irreversible, are present; b) see also refs 8-10; c) dielectronic; d) large wave, presumably dielectronic; e) see also ref. 11; f) trielectronic; g) large wave, presumably trielectronic.

obtained strongly overlap and determination of the peak potentials is therefore prevented.

The tetranuclear Ru complexes **26**, **27**, **28**, and **29** show a trielectronic oxidation wave that can be attributed to the simultaneous oxidation of the three peripheral Ru-ions. Oxidation of the central Ru ion cannot be achieved within the potential window examined. Comparison between the data obtained for $Ru[(\mu\text{-}2,3\text{-dpp})Ru(bpy)_2]_3^{8+}$ (**26**) and $Os[(\mu\text{-}2,3\text{-dpp})Ru(bpy)_2]_3^{8+}$ (**30**) shows clearly that in the latter complex the oxidation of the three peripheral Ru ions occurs after the oxidation of the central Os ion. The data obtained for oxidation of $Ru[(\mu\text{-}2,5\text{-dpp})Os(bpy)_2]_3^{8+}$ (**31**) show that there is a noticeable interaction among the three peripheral Os moieties, in contrast with the behavior of the peripheral Ru moieties of complexes **26-29**.

3.2 ABSORPTION SPECTRA

In the electronic absorption spectra of the complexes examined (Table 1, Fig. 5) one can distinguish two main regions. Below 400 nm there are very intense absorption bands which correspond to ligand centered (LC) transitions: narrow bands at 380 and 360 nm and very intense band at 262 nm can be assigned to the biq ligands, a broad absorption in the 300-350 nm region to the dpp ligands, and a band at 282 nm, whose molar absorptivity reaches a value of about 300000 $M^{-1}cm^{-1}$ in the heptanuclear complex **32**, to the bpy ligands. In this region, dominated by the LC transitions, the absorption spectra are roughly additive.

For $\lambda > 400$ nm the spectra are dominated by moderately intense bands that can be assigned to MLCT transitions. From the spectra of the parent mononuclear homoleptic and heteroleptic complexes and from the above discussed electrochemical data, the energy order of the metal-to-ligand charge transfer transitions is expected to be the following: M--->(bpy) > M--->(2,3-dpp) > M--->(2,5-dpp) > M--->(biq) > M--->(μ-2,3-dpp) > M--->(μ-2,5-dpp) and, of course, Ru--->L > Os--->L and Ru--->BL > Os--->BL. It should be noted, however, that in the oligometallic complexes the metals may be not equivalent because of a different coordination environment, and that the two coordinating moieties of a bridging ligand are not equivalent when they bridge different metal units. For example, in the trimetallic complexes as many as four different types of metal-to-ligand CT transitions between proximate centers are expected (Fig. 6). Under such conditions, a great number of overlapping waves are present and detailed assignments become difficult. Further complications may arise from interchromophoric coupling, as pointed out by De Armond and coworkers [23], and from CT transitions between remote centers [7]. In spite of these difficulties, the general features of each spectrum and the band shifts on changing the type of ligand and/or the number of metals can be well rationalized [13-17].

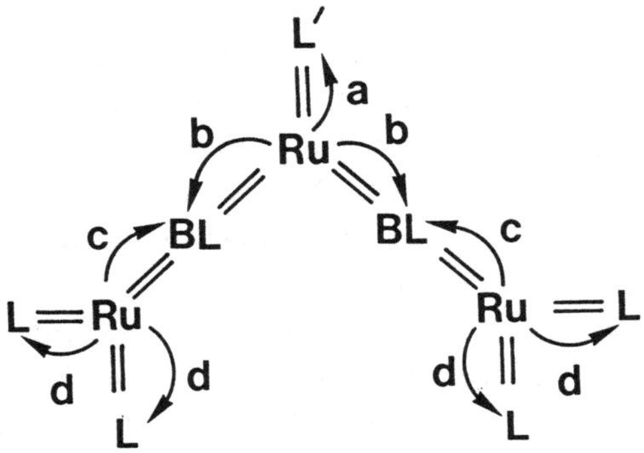

Figure 6. Different types of metal-to-ligand CT transitions in a homonuclear trimetallic complex.

3.3 LUMINESCENCE SPECTRA AND DECAYS

All the Ru complexes examined exhibit an emission band in the red region of the spectrum (Table I). As it usually happens for Ru(II)-polypyridine complexes, such an emission originates from the lowest triplet metal-to-ligand charge transfer excited state.

For the dimetallic Ru complexes, electrochemistry suggests that the lowest excited state is a Ru--->(μ-BL) level in all cases. This is supported by the observation that, at room temperature, the bpy and biq complexes emit at approximately the same wavelengths ($\sim$800 nm for BL = 2,3-dpp and $\sim$825 nm for BL = 2,5-dpp). At 77 K, however, the situation is complicated because the energies of the Ru--->(biq) levels are only slightly dependent on temperature [5], whereas the energies of the Ru--->(BL) levels (expecially for 2,3-dpp) are more sensitive to temperature not only in the mononuclear, but also in the oligonuclear species (Table 1). It should also be noted that the lowest Ru--->(μ-dpp) and Ru--->biq levels involve different Ru ions in the bpy-biq dinuclear complex.

In the trinuclear Ru complexes luminescence can again be assigned to Ru--->(μ-dpp) levels, in agreement with the assignment of the reduction waves. The dielectronic nature of the first oxidation wave suggests that the lowest excited levels involve the external Ru ions. The blue shift of the luminescent Ru--->(μ-dpp) level in going from the bpy to the biq complexes is due to the lower electron donor capacity of biq compared to bpy.

In the tetranuclear Ru complexes **26** and **28** luminescence originates from the peripheral Ru--->(μ-dpp) levels because (i) the bridged dpp ligands are easier to reduce than bpy, and (ii) the peripheral ruthenium ions carry a lower positive charge than the central one owing to the better donor properties of bpy compared with bridged dpp. In the tetranuclear biq complexes **27** and **29**, emission from a Ru--->biq level can be ruled out because it would be expected to occur at λ<738 nm from the behavior of $Ru(biq)_2(2,3\text{-dpp})^{2+}$ complex.

In the heptanuclear Ru complex **32** the luminescence band is again due to a peripheral Ru--->(μ-2,3-dpp) level. Its maximum lies at slighthy higher energy than that of the tetranuclear complex **26** indicating that the $Ru[(\mu\text{-}2,3\text{-dpp})Ru(bpy)(2,3\text{-dpp})]_3^{8+}$ core is slightly less electron withdrawing than the $Ru(2,3\text{-dpp})_3^{2+}$ one.

The Os complexes are known to emit at considerably lower energies than the analogous Ru complexes bacause Os is easier to oxidize than Ru and, as a consequence, the MLCT transitions require less energy. Luminescence was in fact observed only for the mononuclear Os complexes and for the heteronuclear complex **30**. For the other Os complexes the electrochemical data indicate that the lowest Os--->(μ-BL) excited state lies too low in energy to emit luminescence in the wavelength range that can be studied with our equipment.

It should be pointed out that the corrected emission spectra in the visible region of **13**, **22**, **25**, **26**, **30**, and **32** have shown that the luminescence quantum yield is independent of excitation wavelength. This indicates that the luminescent level is populated with the same (presumably unitary) efficiency on excitation of the various CT bands. Thus, in the tri-, tetra-, and hepta-metallic Ru complexes there is energy transfer from the central to the peripheral chromophores. When the central Ru ion is replaced by Os, energy transfer occurs in the reverse direction (see above) and all the energy is collected in the central chromophoric unit (antenna effect, Fig. 7).

It should be recalled [1,7,15,24,25] that in polymetallic complexes the deactivation of upper CT excited states to the lowest one can be viewed either as radiationless transitions between levels of a "large molecule", or as energy or electron transfer processes between various components of a "supramolecular" species. For example, for the

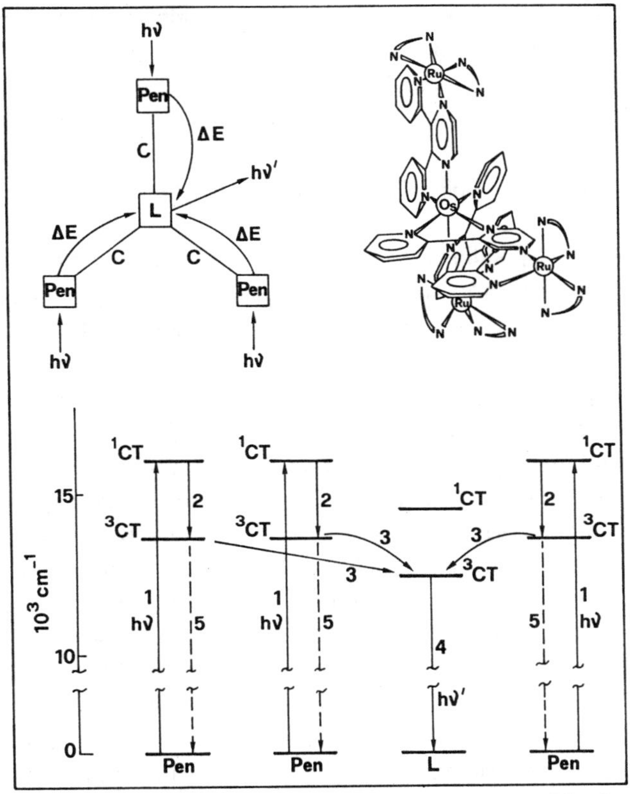

Figure 7. Schematic representation of the antenna effect occurring in **30**.

complexes schematized by Fig. 6, deactivation of the Ru--->L' CT level that can be obtained with excitation a to the lower lying Ru--->BL CT level that can be obtained with excitation c can be described as (i) interconversion between excited states of the large molecule $(L)_2Ru(\mu\text{-}BL)Ru(L')(\mu\text{-}BL)Ru(L)_2^{6+}$ or (ii) energy transfer between the central Ru(L') and the peripheral Ru(BL) chromophoric groups.

All the luminescent complexes show emission lifetimes of the order of μs at 77 K, as expected for MLCT levels of polypyridine complexes. From the measured luminescence quantum yields and lifetimes at room temperature one can evaluate radiative rate constants, k_r, in the range 10^4-10^5 s^{-1}. On the assumption that k_r does not depend on T, it follows that τ(77 K) is largely determined by the rate of radiationless decay, k_{nr}. A plot of $\ln(1/\tau)$ vs the energy of the luminescent level is (roughly) linear, as expected from the energy gap law [25]. At room temperature, the luminescence lifetime is much shorter in all cases, and the lifetimes values do not obey the energy gap law, as it happens when activated radiationless decays, possibly related to photodecomposition reactions, take place [5].

3.4 INTERACTION BETWEEN CHROMOPHORIC OR REDOX CENTERS.

When metal-containing subunits are linked by a bridging ligand, a more or less strong interaction may arise between the redox and chromophoric centers. In going from a mononuclear $L\text{-}M\text{-}(BL)^{2+}$ to a symmetrical dinuclear $L\text{-}M\text{-}(\mu\text{-}BL)\text{-}M_1\text{-}L_1{}^{4+}$ complex (where only one of the two peripheral L ligands is shown and the metal and ligand of the second unit are labelled with 1 for clarity purposes) one can expect changes in the following quantities: (i) reduction potentials of BL; (ii) energy of the M--->(BL) CT transition; (iii) oxidation potential of M; (iv) reduction potential of L; (v) energy of the M--->L CT transition. Furthermore, interaction between the two units in the dinuclear complex may cause: (vi) a splitting between the oxidation potentials of the identical M and M_1 metals; (vii) a splitting between the reduction potentials of the identical L and L_1 ligands. The component most directly interested in the formation of the binuclear complex is, of course, the bridging ligand; therefore, the most important changes concern the quantities related to its electronic situation. The more a component is far from BL, the less its properties will change. Since the added $M_1\text{-}L_1{}^{2+}$ unit is electron withdrawing, one can expect a noticeable increase in the reduction potential of BL (point (i), a smaller effect on the reduction potential of M (point (ii)), and an even smaller effect on the reduction potential of L (point (iv)). The effect on the M--->(BL) CT transition (point(iii)) is expected to be smaller than that on the BL reduction potential because of a compensation effect coming from the increased oxidation potential of M. This is also true for point (v) compared to point (iv). The results obtained substantially agree with these expectations. For example, the ΔE values for the various effects for **1** and **13** are: (i) 3200 cm^{-1}; (ii) 2000 cm^{-1}; (iii) 500 cm^{-1}; (iv) 150 cm^{-1}; (v) undetected. Concerning the splitting of the metal oxidation and ligand reduction in **13**, the values obtained are: (vi) 1370 cm^{-1}; (vii) <250 cm^{-1}.

More generally, comparison between analogous complexes allows us to conclude that: (1) communication between metal-containing units is almost the same through the 2,3-dpp and 2,5-dpp bridges; (2) when the terminal ligands bpy are replaced by biq, communication between the metal units becomes worse; (3) communication between Os metal-units is much better than between Ru metal-units.

4. Conclusions

The mono-, di-, tri-, tetra-, and heptametallic complexes of general formulas $M(L)_2(BL)^{2+}$, $(L)_2M(\mu\text{-}BL)M(L)_2{}^{4+}$, $(L)M[(\mu\text{-}BL)M(L)_2]_2{}^{6+}$, $M[(\mu\text{-}BL)M(L)_2]_3{}^{8+}$ and $M[(\mu\text{-}BL)M(L)(\mu\text{-}BL)M(L)_2]_3{}^{14+}$ (M=Ru or Os, L=bpy or biq; BL=2,3- or 2,5-dpp) exhibit very intense absorption bands in the UV and visible regions, and a very rich electrochemical behavior. All the Ru complexes and the tetranuclear Ru-Os complex **30** exhibit relatively long-lived luminescence both in rigid matrix at 77 K and in fluid solution at room temperature. The electrochemical, spectroscopic, and photophysical properties of this family of complexes can be rationalized on the basis of metal and ligand centered redox orbitals, and ligand centered and metal-to-ligand charge transfer excited states. Such properties can be tuned by a suitable choice of M, L, BL, and number of metal ions. Because of these properties, the complexes examined are good candidates to play the role of light absorption and/or light emission sensitizers [5,26].

Corrected excitation spectra show that all the states populated by excitation in the visible region undergo conversion to the lowest (luminescent) excited state with the same (presumably unitary) efficiency. This property and the very high extinction coefficients all over the visible region exhibited by the tri-, tetra-, and hepta-metallic complexes suggest that these species can be very efficient antenna devices [1,2], for example for photoinduced electron or hole injection on semiconductors [27,28].

It should be pointed out that for each complex replacement of a terminal ligand with a bridging BL ligand gives rise to a new complex that can be used as a "ligand" for the design of larger supramolecular species. This "complexes as ligands" strategy based on components which exhibit the desired properties is a simple way to synthesize photochemical molecular devices [1,2] and more generally can be viewed as a structure-directed synthesis [17,18] to obtain new molecule-controlled materials. By such "complexes as ligands" strategy we plan to design other homo- and hetero-oligometallic species of great theoretical and applicative interest.

Acknowledgments. We thank Dr. F. Stillitani, Mr. L. Minghetti and Mr. G. Gubellini for technical assistance. Financial support from Ministero della Pubblica Istruzione and Progetto Finalizzato CNR Chimica Fine II is gratefully acknowledged.

4. References

[1] Balzani, V.; Moggi, L.; Scandola, F. in Supramolecular Photochemistry, Balzani, V. Ed., Reidel, **1987**.
[2] Balzani, V.; Scandola, F. Supramolecular Photochemistry, Horwood, Chichester, 1990.
[3] Deisenhofer, J.; Michel, H. Angew. Chem. Int. Ed. Engl., **1989**, 28, 829.
[4] Connolly, J. S.; Bolton, J. R. in Photoinduced Electron Transfer, Fox, M. A.; Chanon, M. Eds, Part D, Elsevier, **1988**, p.303.
[5] Juris, A.; Balzani, V.; Barigelletti, F.; Campagna, S.; Belser, P.; von Zelewsky, A. Coord. Chem. Rev., **1988**, 84, 85.
[6] Kober, E. M.; Caspar, J. V.; Sullivan, B. P.; Meyer, T. J. Inorg. Chem., **1988**, 27, 4587.
[7] Scandola, F.; Indelli, M. T.; Chiorboli, C.; Bignozzi, C. A. Topics Curr. Chem., **1990**, 158, in press.
[8] Fuchs, Y.; Lofters, S.; Dieter, T.; Shi, W.; Morgan, R.; Strekas, T. C.; Gafney, H. D.; Baker, A. D. J. Am. Chem. Soc., **1987**, 109, 2691.
[9] (a) Brewer, K. J.; Murphy, W. R. Jr.; Spurlin, S. R.; Petersen, J. D. Inorg. Chem., **1986**, 25, 882; (b) Murphy, W. R. Jr.; Brewer, K. J.; Gettliffe, G.; Petersen, J. D. Inorg. Chem., **1989**, 28, 81.
[10] (a) Kalyanasundaram, K.; Nazeeruddin, Md. K. Chem. Phys. Letts, **1989**, 158, 45; (b) Kalyanasundaram, K.; Nazeeruddin, Md. K. Inorg. Chem., **1990**, 29, 1888.
[11] (a) Ernst, S. D.; Kasack, V.; Kaim, W. Inorg. Chem., **1988**, 27, 1146; (b) Ernst, S. D.; Kaim, W. Inorg. Chem., **1989**, 28, 1520.
[12] Campagna, S.; Denti, G.; De Rosa, G.; Sabatino, L.; Ciano, M.; Balzani, V. Inorg. Chem., **1989**, 28, 2565.
[13] Campagna, S.; Denti, G.; Sabatino, L.; Serroni, S.; Ciano, M.; Balzani, V. J. Chem. Soc. Chem. Comm., **1989**, 1500.
[14] Campagna, S.; Denti, G.; Sabatino, L.; Serroni, S.; Ciano, M.; Balzani, V. Gazz. Chim. Ital., **1989**, 119, 415.
[15] Denti, G.; Campagna, S.; Sabatino, L.; Serroni, S.; Ciano, M.;

Balzani, V., Inorg. Chem., in press.
[16] Denti G.; Campagna, S.; Sabatino, L.; Serroni, S.; Ciano, M.; Balzani, V., Inorg. Chim. Acta, in press.
[17] Denti, G.; Sabatino, L.; Serroni, S.; Ciano, M.; Ricevuto, V.; Campagna, S., manuscript in preparation.
[18] Kohnke, F. H.; Mathias, J. P.; Stoddart, J. F. Angew. Chem. Int. Ed. Engl., **1989**, 28, 1103.
[19] Tomalia, D. A.; Naylor, A. M.; Goddard III, W. A. Angew. Chem. Int. Ed. Engl., **1990**, 29, 138.
[20] Chirayil, S.; Thummel, R. P. Inorg. Chem., **1989**, 28, 85.
[21] Balzani, V.; Carassiti, V. Photochemistry of Coordination Compounds, Academic Press, **1970**.
[22] De Armond, M. K.; Carlin, C. M. Coord. Chem. Rev., **1981**, 36, 325.
[23] (a) Myrick, M. L.; Blakley, R. L.; De Armond, M. K. J. Phys. Chem., **1989**, 93, 3936; (b) Myrick, M. L.; De Armond, M. K.; Blakley, R. L. Inorg. Chem., **1989**, 28, 4077.
[24] Meyer, T. J. Accounts Chem. Res., **1989**, 22, 163.
[25] Chen, P.; Duesing, R.; Tapolsky, G.; Meyer, T. J. J. Am. Chem. Soc., **1989**, 111, 8305.
[26] Balzani, V.; Barigelletti, F.; De Cola, L. Topics Curr. Chem., **1990**, 158, in press.
[27] Vlachopulos, N.; Liska, P.; Augustynski, J.; Graetzel, M. J. Am. Chem. Soc., **1988**, 110, 1216.
[28] Amadelli, R.; Argazzi, R.; Bignozzi, C. A.; Scandola, F. J. Am. Chem. Soc., in press.

UNIDIRECTIONAL ELECTRON TRANSFER IN THE SUPRAMOLECULES: THE CONTROL BY AN EXTERNAL MAGNETIC FIELD

V.Ya.Shafirovich, E.E.Batova
USSR Academy of Sciences Institute of Chemical Physics
Chernogolovka, 142432
U.S.S.R.
P.P.Levin
USSR Academy of Sciences Institute of Chemical Physics
Moscow, 117334
U.S.S.R.

ABSTRACT. The design of the supramolecules: Zn-porphyrin-viologen dyads and Zn-porphyrin-viologen-quinone triads, which give rise to charge separation and stabilization after absorption of visible light in polar media has been developed. The photoinduced charge separation is achieved by the competition of the fast spin-allowed forward and backward spin-forbidden electron transfers (ETs). An external magnetic field is used to control the recombination of the triplet radical ion pair (RIP) state formed in a primary photochemical reaction. The triplet RIP state is stabilized by the subsequent ET to the second electron acceptor (quinone) which is linked to the first one. The mechanism of the triplet-singlet (T-S) evolution of RIPs and the magnetic field effect on their recombination is discussed in respect of the different behaviour of RIPs in several conformers. In the opened conformer of the RIP (with a small value of the exchange interaction (J) between radical centers) the hyperfine coupling (hfc) and the relaxation mechanisms of T-S evolution of the RIP are active; in the closed conformer of the RIP, where J is large enough ($2J \gg A_{hfc}$), the spin-orbit coupling (SOC) dominates.

1. INTRODUCTION

In photosynthesis the storage of solar energy occurs *via* multistep electron transfer (ET) from an excited pigment entity (chlorophyll) to the terminal electron acceptor (quinone). Components of a photosynthetic ET chain are held by a protein matrix of the photosynthetic reaction center (RC) at fixed distances and fixed orientations.

One of the possible ways to mimic the photosynthetic reaction center is to design a supramolecule which consists of chemically linked photosensitizer (S) and two electron acceptors (A_1 and A_2). The

E. Pelizzetti and M. Schiavello (eds.), Photochemical Conversion and Storage of Solar Energy, 47–66.

unidirectional ET in $S-A_1-A_2$ can be achieved by the competition of forward ETs from $^*S$ to A_1 and A_1^- to A_2 and the backward ETs from A_1^- and A_2^- to S^+. The question arises how to organize this competition. The theories of ET processes in condensed media give a clear answer. The rate of non-adiabatic ET reactions is given by

$$k_{ET} = (2\pi/\hbar)|H_{rp}|^2 FC , \quad (1)$$

where H_{rp} is the electron exchange matrix element and FC is the Franck-Condon weighted density of states. H_{rp} describes the overlap of the donor and acceptor orbitals and falls off strongly with the distance between the donor and acceptor moieties. FC depends on the thermodynamic driving force for the reaction, ΔG (the energy gap) and on the changes in bond lengths and bond angles of the donor, acceptor and surrounding molecules, *i.e.* on the reorganization of the supramoleculе and the medium. FC increases with an increase of the energy gap (normal region) and then begins to decrease for large energy gaps (inverted region). Such behaviour of FC results in a pronounced "bell-shaped" energy gap dependence for k_{ET}. The decrease of a solvent polarity reduces the solvent reorganization energy and the energy gap, at which k_{ET} attains a maximal value, decreases. Hence in nonpolar media the rate constant of weakly exergonic forward ET reaction can be more than that of highly exergonic backward one.

The bell-shaped energy gap dependence was theoretically predicted by Marcus [1] in 1956 and demonstrated almost 30 years later for rigid glasses and supramolecules in the pioneer works of Miller, Closs *et al.* [2,3]. Wasielewski *et al.* [4] have observed the pronounced bell-shaped energy gap dependence for the rate constants of the forward and backward ETs in a series of the rigid Zn-porphyrin-quinone dyads (P-Q). In non or weakly polar solvents the forward ET from the singlet excited state of the porphyrin to the quinone moiety occurs in the normal region, whereas the backward ET from the quinone radical anion to the porphyrin radical cation occurs in the inverted region, and k_{ET} attains a maximal value at $-\Delta G \approx 1$ eV. The singlet radical ion pair (RIP) formed in a primary photochemical reaction could be stabilized by the subsequent ET from an electron donor (D) to P^+ or from Q^- to a second electron acceptor. The D-P-Q triads in which D is dimethylaniline or a carotenoid polyene have been prepared and examined by Wasielewski *et al.* [5] and Moore *et al.* [6]. The D^+-P-Q^- lifetime is a few microseconds. Charge stabilization in the flexible $P-(CH_2)_n-Q_1-(CH_2)_m-Q_2$ triads is not so efficient and the corresponding singlet RIP decay is in the picosecond time range [7].

The dependence of the ET rate constant from BPh^- to Q_A on changes in the thermodynamic driving force of the reaction by the substitution of the quinone in the bacterial RC has also shown the pronounced bell-shaped energy gap dependence with the maximum at $-\Delta G \approx 0.6$ eV [8].

In polar solvents k_{ET} attains as a rule a maximal value at $-\Delta G >$ 1.3 eV [9,10] and rate constant of the backward ET is greater than that of the forward ET. Charge separation in supramolecules can be highly efficient in oil, but not in water [11].

An alternative way to make charge separation independent of a solvent polarity includes the use of spin dynamics to control the rate of ET. The lifetime of the combined triplet RIP is sufficiently larger than that of the corresponding singlet one. The relatively long-lived (0.1-1 ms) triplet RIPs formed by photoexcitation of the Zn-porphyrin-viologen (P-$(CH_2)_n$-Vi^{2+}) [12-15] and Zn-porphyrin-quinone dyads [16] have been extensively studied in several research groups.

The triplet RIP recombination may be controlled by an external magnetic field [17,18]. Large magnetic field effects in the recombination of the triplet RIP have been observed by the photoexcitation of P-$(CH_2)_n$-Vi^{2+} [14,15]. Recombination of the triplet RIP is the spin-forbidden process and its rate depends on the rate of spin conversion. The triplet RIP spin dynamics depend on the ratio of the values of the effective electron-nuclear hyperfine coupling (hfc) constant (A_{hfc}) and the exchange interaction (J) between the radical centers and is determined by the distance between the individual radicals and the nature of the spacer. The length and flexibility, as well as the chemical nature of the spacer between radical centers, should influence the RIP behaviour in response to an external magnetic field [19-21].

The triplet RIP state of the porphyrin-viologen dyad could be also stabilized by the subsequent ET from Vi^{+} to A_2. In this connection we have recently prepared [22] the new semirigid supramolecules: Zn-porphyrin-viologen-quinone triads (P-Ph-Vi^{2+}-Q and P-BPh-Vi^{2+}-Q) and Zn-porphyrin-viologen dyads (P-Ph-Vi^{2+} and P-BPh-Vi^{2+}) in which the quinone (Q) is tetrahydrodioxoanthrylenetriptycene and the spacers (Ph and BPh) between Zn-meso-4-methoxyphenyl-tris(4-tolyl)porphyrin (P) and methylviologen (Vi^{2+}) are 1,4-phenylene and 4,4'-biphenylene, respectively (Figure 1). The flexible dyad (P-$(CH_2)_{10}$-Vi^{2+}) was also synthesized and was used as a reference.

This paper will discuss the photoinduced ET dynamics of Zn-porphyrin-viologen dyads and Zn-porphyrin-viologen-quinone triads studied by laser and conventional flash photolysis techniques [23-25].

2. PHOTOINDUCED CHARGE SEPARATION IN THE SUPRAMOLECULES [22-25]

Photochemical reactions of the supramolecules have been studied in organic solvents with different polarity, macroscopic viscosity and in organized molecular assemblies. Since the supramolecules contain an ion fragment (viologen moiety) they are completely insoluble in nonpolar solvents, *e.g.*, heptane. Aggregate formation, which results in a drastic broadening and shape change of the Soret band, occurs in benzene, CCl_4 and water. The supramolecules dissolve in the Triton X-100, sodium dodecylsulfate (SDS), cetyltrimethylammonium chloride (CTAC) and bromide (CTAB) aqueous micellar solutions and the Aerosol-OT (AOT) heptane micellar solutions. In micelles aggregates are not formed. The supramolecule absorption spectra in micellar solutions are close to those in polar organic solvents, Triton X-100 and poly(ethylene glycol), PEG (MW 300).

The dyads and triads under consideration are dynamic entities. Space-filling models show that the mutual rotations of the supramolecule fragments are free, and it is very difficult to pick out their

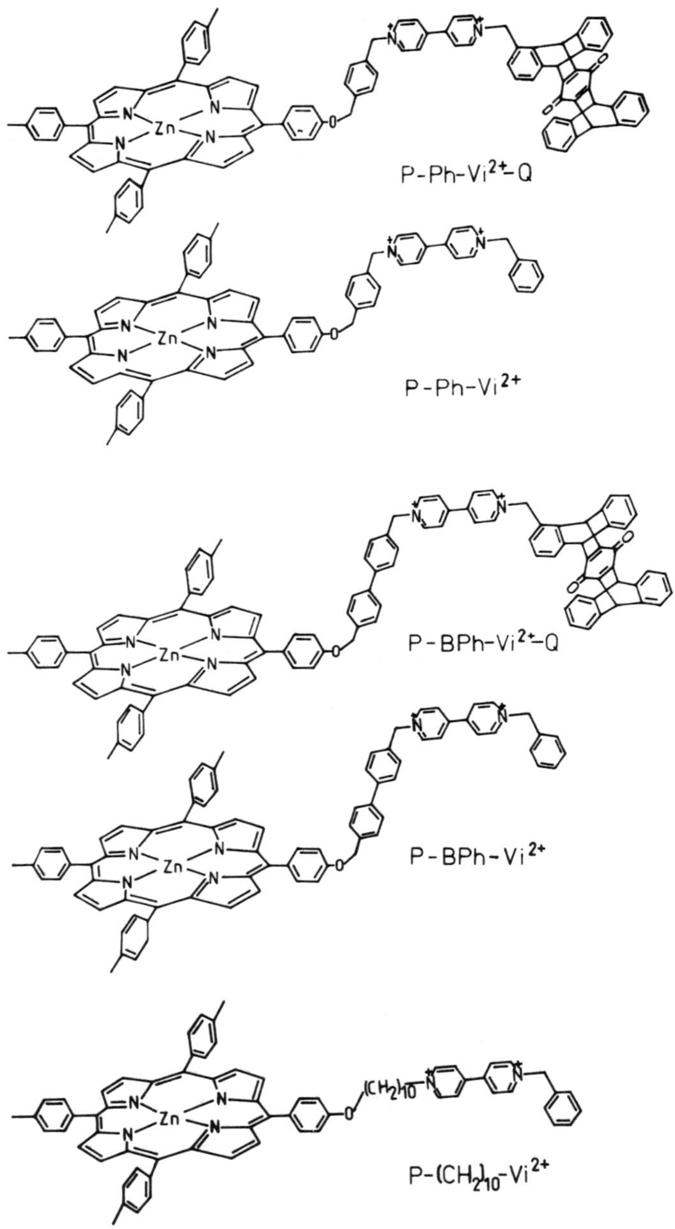

Figure 1. Molecular structures of the dyads and triads.

preferable conformations. In this connection the term "conformer" will be used to denote the multitude of the supramolecule states which possesses the same properties.

The intramolecular ET reaction in a nonrigid combined system should depend on the distances between the supramolecule fragments and their orientations. Two extreme conformers ("opened" and "closed"), with quite different distances between the supramolecule fragments, are responsible for the photochemical behaviour of the dyads and triads. These conformers can interconvert *via* free mutual rotations of the porphyrin, the phenylene spacer, viologen and quinone moieties.

Both the singlet and the triplet excited states of the porphyrin are important in the complex photochemical behaviour of the dyad. It is assumed that the closed and opened conformers do not interconvert within the lifetime of ^1P.

The decrease of the fluorescence yield (ϕ_f) of the porphyrin moiety relative to that of Zn-meso-4-benzoxyphenyl-tris(4-tolyl)porphine, ZnBTP being a model of the supramolecule porphyrin moiety, lends support for the intramolecular quenching of ^1P (Table 1). The value ($1-\phi_f$) represents the contribution of this route.

TABLE 1. Relative fluorescence yields (ϕ_f) of the dyads and triads. Rate constants of the ^3P decay (k_c)

Supramolecule	Solvent	ϕ_f[a]	k_c[b] (10^6 s^{-1})
$P\text{-}Ph\text{-}Vi^{2+}$	Ethanol	0.34	32
" – "	Acetonitrile	0.43	26
" – "	Formamide	0.50	23
" – "	Triton X-100		0.15
$P\text{-}(CH_2)_{10}\text{-}Vi^{2+}$	Acetonitrile		15
$P\text{-}BPh\text{-}Vi^{2+}$	Ethanol	1.0	0.18
$P\text{-}Ph\text{-}Vi^{2+}\text{-}Q$	" – "	0.54	1.0
$P\text{-}Ph\text{-}Vi^{2+}\text{-}QH_2$	" – "	0.54	1.0
$P\text{-}BPh\text{-}Vi^{2+}\text{-}Q$	" – "	0.60	0.28

[a] ϕ_f is measured relative to that of ZnBTP
[b] k_c is determined by a nanosecond laser flash photolysis

The closed conformer of $P\text{-}Ph\text{-}Vi^{2+}$ with a short average distance between the P and Vi^{2+} moieties is responsible for the quenching of ^1P by Vi^{2+}. The intramolecular ET from ^1P to Vi^{2+} results in the formation of the singlet RIP which decays extremely fast ($k_b > 10^8$ s^{-1}). The ^1P quenching by Vi^{2+} competes with the intersystem crossing of ^1P to ^3P. The intramolecular ET from ^3P to Vi^{2+} results in the fast formation of the long-lived triplet RIP within a laser flash (resolution time of a kinetic spectrophotometer is ~5 ns).

In the opened conformer, with a large average distance between the P and Vi^{2+} moieties, ^1P is not quenched by Vi^{2+} and undergoes intersystem crossing to ^3P. The intramolecular ET from ^3P to Vi^{2+} occurs in the nanosecond regime and is directly observed in our experiments.

The rate constant of the ^3P decay (k_c) is independent of solvent polarity (ϵ_s = 25-110) (Table 1). It is quite reasonable to assume that the limiting step of this process is the conformational transition "opened"→"closed" and the subsequent ET from ^3P to Vi^{2+} in the closed conformer is very fast. An alternative explanation may be long-distance ET from ^3P to Vi^{2+} in the opened conformer. However, in this case one should expect the correlation of k_c with solvent polarity [10,11]. In fact this correlation is not observed (see Table 1), and the first assumption seems more probable than the second.

In pure Triton X-100 the k_c value is surprisingly small (Table 1). The opened conformation is possibly stabilized in a solvent consisting of large anisotropic molecules such as Triton X-100, and the conformational transition "opened"→"closed", requiring a sufficient reorganization of the surrounding dyad becomes slow. In PEG, a molecule that is a flexible fragment of the Triton X-100 molecule, the conformational transitions again become fast. Hence the conformational transition in a semirigid combined system can be controlled by the reorganization of the surrounding supramolecule.

The dynamics of the photoinduced ET in the flexible P-$(CH_2)_{10}$-Vi^{2+} are somewhat close to those of P-Ph-Vi^{2+}. However, the kinetics of the transients appearing after a photoexcitation of P-$(CH_2)_{10}$-Vi^{2+}, are multiexponential, and k_c was estimated as the inverse lifetime of ^3P. The participation of several conformers should be considered.

It is surprising that the rate constant of the intramolecular ET from ^3P to Vi^{2+} in P-Ph-Vi^{2+}-Q(QH_2) becomes less than that of P-Ph-Vi^{2+}. The Q(QH_2) attachment to the dyad changes the mutual motion of the supramolecule fragments and the k_c value for P-Ph-Vi^{2+}-Q(QH_2) is less than k_c for P-Ph-Vi^{2+} (Table 1).

Enlargement of the spacer results in a drastic decrease in the value of k_c (Table 1). The opened conformation of P-BPh-Vi^{2+} becomes preferable, and the quantum yield of the intersystem crossing of ^1P to ^3P tends to unity. The limiting step of the ^3P decay is the conformational transition "opened"→"closed".

The relative fluorescence yield of ^3P-BPh-Vi^{2+}-Q is less than that of ^3P-BPh-Vi^{2+} (Table 1). It is quite reasonable to suggest that the closed conformer of the triad is responsible for the ^1P quenching by Q. Space-filling models of P-BPh-Vi^{2+}-Q show that the formation of the closed conformer with a short distance between the P and Q moieties is possible. In the triad opened conformer the rate of ^3P decay is somewhat greater than that of P-BPh-Vi^{2+} (Table 1).

Intramolecular ET in ^3P-Ph-Vi^{2+} results in the formation of the corresponding RIP. There are two routes for RIP formation in the closed and opened conformers, respectively: (*i*) fast ET within a laser flash; (*ii*) a relatively slow process in the nanosecond time range. The total quantum yield of RIP in polar solvents is close to that of ^3P, being in the range of 0.3-0.5. The appearance of P^+-Ph-Vi^+ is not controlled by a macroscopic viscosity of medium. P^+-Ph-Vi^+ is formed in nonviscous and viscous polar solvents, micelles and reversed micelles.

All P^+-Ph-Vi^+ observed after a laser pulse are formed by the intramolecular ETs from ^3P to Vi^{2+}, *i.e.* they are born in the triplet state. The recombination of the triplet RIP (the backward intramolecular ET) is a spin-forbidden process, a mechanism that depends on the ratio of the 2J and A_{hfc} values. Therefore the mechanism of the RIP recombination will be different in closed and opened conformers of the RIP.

The initial conformation of the triplet RIP (in which it was born) is closed. In $^3[P^{+}\text{-Ph-}Vi^{2+}]$ the exchange interaction of the radical centers is efficient ($2J \gg A_{hfc}$) due to a short average distance between P^+ and Vi^+. Hence the hfc cannot induce the T-S transitions. A possible mechanism for the recombination of $^3[P^{+}\text{-Ph-}Vi^{+}]$ to the ground state is one-step spin-forbidden transition (intersystem ET with the rate constant k_{soc}) induced by the spin-orbit coupling (SOC). Inter-system ET induced by SOC is very important for the recombination of the triplet RIP, the main state of which is the contact one. This kind of RIP includes charge transfer triplet exciplexes in nonpolar solvents [10], RIPs in viscous solvents [26] and RIPs formed by the quenching of the luminescent state of the metal complexes by aromatic amines [27,28]. This process is also important for the recombination of several RIPs in micelles in a strong magnetic field [29,30]. The contribution of the k_{soc} value to the observed rate constant (k_r) of the RIP recombination (Table 2) can be estimated from the limit value of k_r at high B, as it

TABLE 2. Rate constants of the triplet RIP recombination (k_r) in zero and high magnetic field

Supramolecule	Medium	k_r^a (10^6 s^{-1}) B = 0	B = 0.24 T
$P\text{-Ph-}Vi^{2+}$	Ethanol	2.1	0.70
" – "	Acetonitrile	1.8	0.80
" – "	Glycerol	3.3	0.31
" – "	Formamide	2.1	0.40
" – "	PEG	1.3	0.14
" – "	CTAC/H_2O	1.8	0.46
" – "	CTAB/H_2O	2.3	1.1
" – "	SDS/H_2O	1.3	0.36
" – "	Triton X-100/H_2O	1.4	0.36
" – "	AOT/H_2O/heptane	1.8	0.46
$P\text{-}(CH_2)_{10}\text{-}Vi^{2+}$	Acetonitrile	8.8	1.4
" – "	Ethanol	2.2	1.3
$P\text{-Ph-}Vi^{2+}\text{-Q}$	" – "	1.0	0.7
$P\text{-Ph-}Vi^{2+}\text{-}QH_2$	" – "	1.0	0.7

[a] k_r is determined by a nanosecond laser flash photolysis

was done for the RIP recombination in micelles [29,30]. It will be shown (see below) that in zero magnetic field, this contribution is small.

However, it dominates strongly at B = 0.24 T. Thus the k_r values observed at B = 0.24 T (Table 2) are close to the k_{soc} ones.

The intersystem ET induced by SOC cannot compete with the conformational transition $^3[P^+\text{-Ph-Vi}^+]_{cl} \rightarrow {}^3[P^+\text{-Ph-Vi}^+]_{op}$ (k_i). The recombination of $^3[P^+\text{-Ph-Vi}^+]_{op}$ is a complex multistep transition that involves the hfc induced T-S transitions (k_{hfc}) in the opened conformer, the conformational RIP transition $^1[P^+\text{-Ph-Vi}^+]_{op} \rightarrow {}^1[P^+\text{-Ph-Vi}^+]_{cl}$ (k_{is}) and the subsequent fast ET in $^1[P^+\text{-Ph-Vi}^+]$ (k_b) without a spin change. It will be shown below that the contribution of this route in the RIP recombination dominates at zero magnetic field.

The limiting step of the $^3[P^+\text{-Ph-Vi}^+]_{op}$ recombination seems to be the conformational transition $^1[P^+\text{-Ph-Vi}^+]_{op} \rightarrow {}^1[P^+\text{-Ph-Vi}^+]_{cl}$. It is reasonable that the observed rate constant of the RIP decay (k_r, see Table 2) is less than k_b ($> 10^8$ s^{-1}) and k_{hfc} ($\sim 4 \cdot 10^7$ s^{-1}) being estimated as $h^{-1}(A^2_{hfc(1)} + A^2_{hfc(2)})^{1/2}$, where $A^2_{hfc(1)}$ and $A^2_{hfc(2)}$ are the effective hfc constants of each radical [31] :

$$A_{hfc} = (\sum_i a_i^2 I_i(I_i+1))^{1/2} \qquad (2)$$

here the square-route expression contains the sum over the hfc constants a_i and spin quantum numbers I_i of all the relevant magnetic nuclei in the radical (a_i of the benzylviologen [32] and Zn-tetraphenylporphine [33] radical cations were used). The condition $k_{hfc} > k_{is}$, k_{soc}, leads to $B \ll A_{hfc}/g\beta_e$ to a stochastic distribution in the RIP spin states [34] and a monoexponential law for the RIP recombination:

$$k_r(B = 0) = k_{is}/4 + 3k_{soc}/4 \qquad (3)$$

The assumption that the limiting step of the RIP recombination is the RIP conformational transition "opened"→"closed", which can proceed *via* free mutual rotations of the dyad fragments, is in accordance with the independence of the k_r value on solvent polarity and its macroscopic viscosity (Table 2).

Photoexcitation of $P\text{-Ph-Vi}^{2+}$ in Triton X-100 fails to form RIPs. In Triton X-100 the rate of the conformational transition $[^3P\text{-Ph-Vi}^{2+}]_{op} \rightarrow [^3P\text{-Ph-Vi}^{2+}]_{cl}$ becomes slow (k_c, Table 1). It is reasonable that the transition $^3[P^+\text{-Ph-Vi}^+]_{cl} \rightarrow {}^3[P^+\text{-Ph-Vi}^+]_{op}$ is also slow and cannot compete with intersystem ET. In PEG the conformational transitions become fast, because they do not require the strong reorganization of the surrounding supramolecule, and the RIP is again formed (k_r, Table 2).

The enlargement of the spacer results in a decrease of the conformational transition rate (k_c, Table 2), and no formation of RIP is observed during the $^3P\text{-BPh-Vi}^{2+}$ decay. The RIP formed by the intramolecular ET from ^3P to Vi^{2+} is born in a closed conformation. The rate constant of the RIP recombination (k_r) in the closed conformer is greater than k_c (rate constant of conformational transition). The steady-state concentration of RIPs is low, and RIPs are not observed directly in the transient absorption spectrum of the dyad. It is possible that the conformational transition $^3[P^+\text{-BPh-Vi}^+]_{cl} \rightarrow {}^3[P^+\text{-BPh-Vi}^+]_{op}$ becomes slow and cannot compete with the intersystem ET induced by SOC.

In the case of P-Ph-Vi^{2+}-Q(QH_2) the rates of the forward and backward ET become comparable, and the kinetics of the ^3P decay and that of the RIP recombination cannot be recorded separately. The effective rate constant of the laser flash transient decay (k_e) is shown in Table 1. The quantum yield of RIPs formed does not exceed 0.05-0.15. The steady-state concentration of P^{+}-Ph-Vi^{+}-Q is low, and this RIP is not observed directly in the transient absorption spectrum of the triad.

3. MAGNETIC FIELD EFFECT ON THE RECOMBINATION OF THE TRIPLET RADICAL ION PAIR STATE OF THE SUPRAMOLECULE [24,25]

The lifetime of $^3[P^{+}$-Ph-$Vi^{+}]$ is drastically increased by an external magnetic field (B = 0.24 T), whereas its yield remains practically the same. Therefore an application of an external magnetic field results in the pronounced retardation of the triplet RIP recombination (Table 2). The magnetic field effect on the k_r value is observed in nonviscous and viscous polar solvents, as well as in micelles and reversed micelles.

The behaviour of $^3[P^{+}$-$(CH_2)_{10}$-$Vi^{+}]$ in an external magnetic field is somewhat close to that of $^3[P^{+}$-Ph-$Vi^{+}]$ (Table 2). However the accuracy of the magnetic field effect measurement is not high on account of the multiexponential kinetics of the RIP decay.

Application of an external magnetic field (B = 0.24 T) results in an increase in the lifetime of the P-Ph-Vi^{2+}-Q(QH_2) laser flash transients. In an external magnetic field the recombination of the triplet RIP is delayed and its steady-state concentration slightly increases. Due to the low steady-state concentration of RIP, an external magnetic field affects the effective rate constant of the laser flash transient decay by only 30% (Table 2).

External magnetic field does not affect the photochemical behaviour of P-BPh-Vi^{2+} and P-BPh-Vi^{2+}-Q. This is not surprising, as there is a very efficient route, unaffected by an external magnetic field, for the $^3[P^{+}$-BPh-$Vi^{+}]$ and $^3[P^{+}$-BPh-Vi^{+}-Q] recombination, namely the intersystem ET induced by SOC.

The lifetimes of $^3[P^{+}$-Ph-$Vi^{+}]$ are strongly perturbed by an external magnetic field. In glycerol and PEG the k_r value decreases monotonically with increase of B (Figure 2). The $B_{1/2}$ value, defined as the value of B at which the magnetic field effect reaches the half of its saturated value, is ~15 mT. The plot of k_r *vs* B becomes flat at B ≥ 0.1 T. In CTAC, CTAB and SDS aqueous micellar solutions the magnetic field dependence curves show a maximum in k_r followed by a decrease to an asymptotic value at B ≥ 0.1 T (Figures 2 and 3). The magnetic field dependence curves which show a maximum (B_{max} = 3-60 mT) followed by a decrease at B > B_{max}, have already been observed in the geminate recombination of a family of flexible triplet biradicals [19,20]. The k_r value remains constant at B ≤ 10 mT and begins to decrease at B ≥ 0.1 T in ethanol, Triton X-100 aqueous and AOT heptane micellar solutions (Figure 3). The magnetic field effect saturates at B ≥ 0.1-0.2 T.

An external magnetic field splits the triplet sublevels (T_+, T_o

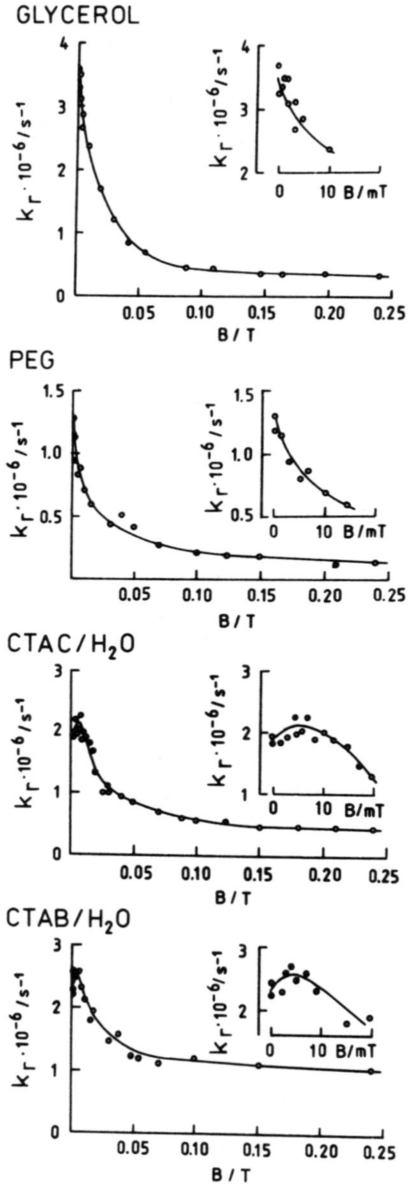

Figure 2. Magnetic field effect on the rate constant of the $^3[P^+\text{-Ph-Vi}^+]$ recombination determined by a nanosecond laser flash photolysis.

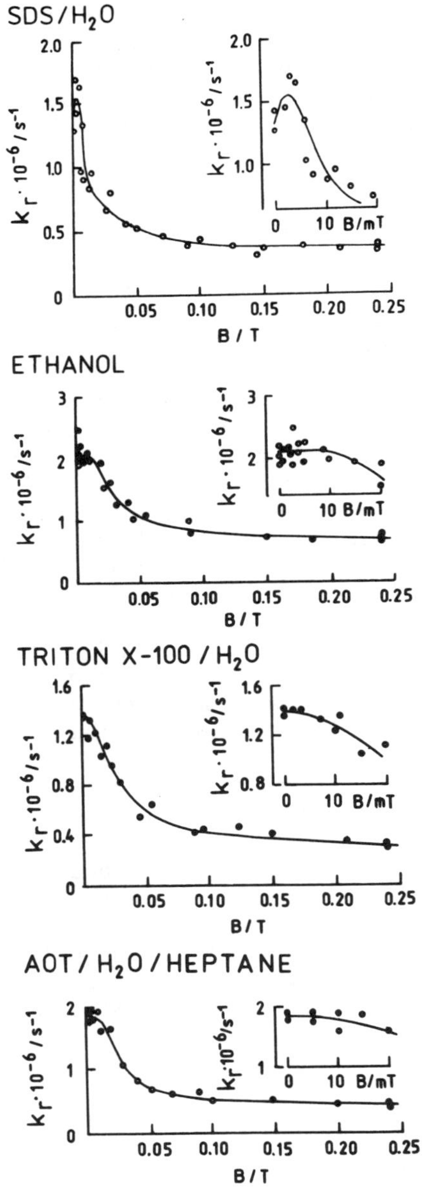

Figure 3. Magnetic field effect on the rate constant of the $^{3}[P^{-}\text{-}Ph\text{-}Vi^{+}]$ recombination by a nanosecond laser flash photolysis.

and T_-) of the RIP due to the Zeeman effect, and the rate constant of the hfc induced $T_\pm$-S transitions decreases. The k_r value begins to decrease when the k_{hfc} value becomes less than that of the limiting step of the RIP recombination (conformational transition $^1[P^+\text{-Ph-Vi}^+]_{op} \rightarrow$ $^1[P^+\text{-Ph-Vi}^+]_{cl}$). The magnetic field effect, which is related to the hfc mechanism, is usually saturated at B < 20 mT [17,18,21]. $B_{1/2}$ can be calculated from the following equation [21] :

$$B_{1/2} = 2(g\beta_e)^{-1}(A_{hfc(1)}{}^2 + A_{hfc(1)}{}^2)/(A_{hfc(1)} + A_{hfc(1)}) \quad (4)$$

According to this approach the $B_{1/2}$ value is 2.3 mT for a given radical pair. In glycerol and PEG the magnetic field effect is particularly tangible for B < 20 mT, typical of the hfc mechanism [17,18,21].

The magnetic field changes the Zeeman splitting between the triplet sublevels T_+, T_o and T_-, but leaves the T_o-S spacing (exchange interaction, 2J) unchanged. Application of the magnetic field leads to a crossing between T_- (or T_+ depending on the sign of J) and S. At B = B_{max}, the efficiency of the intersystem crossing (ISC) reaches a local maximum, because of the T_- (or T_+)-S degeneracy. Thus B_{max} gives a measure of 2J. This mechanism of the ISC is important in CTAC, CTAB and SDS aqueous micellar solutions. In ethanol, Triton X-100 aqueous micellar solutions and AOT heptane micellar solutions the magnetic field dependence curves do not show a maximum. To a first approximation this can be explained in terms of opened and partially closed conformers: $^3[P^+\text{-Ph-Vi}^+]_{op}$ and $^3[P^+\text{-Ph-Vi}^+]_{pcl}$ with 2J ≈ 0 and ~10 mT, respectively. Application of the magnetic field decreases the rate of the ISC in the opened conformer and enhances that in the partially closed one. The compensation of these two effects results in the k_r independence of B in the region of 0-10 mT. The T-S level crossing in the partially closed conformer occurs at B ≈ 10 mT and at B > 10 mT the k_r value begins to decrease.

The magnetic field dependences (Figures 2 and 3) become flat only at B > 0.1 T. Delayed field effect curves have already been observed for the geminate recombination of the radicals in micelles [29,30,35], viscous solvents [26] and for triplet exciplexes [10]. Such character of these curves arises from the relaxation transitions $T_\pm$-T_o, S in the RIP occurring with the rate of paramagnetic relaxation [36,37]:

$$k_{rel} = 1/T_1 = 2\tau_c|V|^2/\hbar^2(1 + \omega^2\tau_c{}^2) \quad (5)$$

Here V is the off-diagonal matrix element of transition due to the hfc anisotropy, anisotropy of the g factor, or still due to the dipole-dipole spin coupling of the radicals; τ_c is the rotational correlation time; the frequency $\omega = g\beta H/\hbar$ (g is the mean value of the isotropic g factors of the radicals). The value of V for relaxation arising from the hfc anisotropy and dipole-dipole interaction between unpaired electrons is independent of B [35]. In this case eq 5 predicts that k_{rel} should decrease with an increase in B in accordance with the form of magnetic field dependences observed (Figures 2 and 3). Hence, relaxation of this kind is important for the RIP under consideration.

In a strong magnetic field only the T_o-S transitions arising from the isotropic hfc and the difference of the g factors of the radicals (the Δg mechanism) are effective in the separated RIP. Therefore, if there had been no alternative T-S evolution routes in RIP, only the RIP generated in the T_o state might be expected to recombine at a strong field. Under these circumstances, one should observe very high magnetic effects. However, as it has already been noted, the magnetic field dependence curves reach a plateau at $B > 0.1$ T (Figures 2 and 3). Hence there should be an alternative channel for the T-S evolution in the RIP, rate of which should be independent of B.

There are two mechanisms for T-S transitions in the separated RIP that are independent of B at high magnetic field: (*i*) relaxation due to g anisotropy and (*ii*) spin-rotation coupling [35,38]. The relaxation due to the g anisotropy is much slower and may be ignored [35]. The spin-rotation relaxation, effective for small radicals with high g anisotropy, can hardly have any effect on the decay of the RIP discussed.

To explain the observed relations at a strong magnetic field (Figures 2 and 3), we have to take into account the spin-orbit coupling in the closed conformer of the RIP where SOC may induce ISC leading to recombination.

In the limit of very high B, when $k_{rel} \ll k_{soc}$, the ensemble of RIP states falls into two independent groups. The RIPs which are produced in T_o and $T_\pm$ disappear with the rate constant $k(T_o)$:

$$k(T_o) = k_{is}/2 + k_{soc} \tag{6}$$

and $k(T_\pm)$:

$$k(T_\pm) = k_{soc} \tag{7}$$

respectively [29,36]. In this approximation, the kinetics of the RIP recombination acquire a two exponential form, which has been observed in the recombination of several triplet RIPs in micelles [29,30,35]. Simple estimates show that a two exponential form of the decay kinetics appears unambiguously when $k_{is} \gg k_{soc}$. For most of the solvents used, this does not take place. One would expect the pronounced two exponential behaviour in glycerol and PEG only. However, it does not take place in glycerol nor in PEG. A monoexponential form of the decay kinetics at high B may be explained in terms of the possible preferable population of $T_\pm$. The initial polarization of ^3P in the molecular axes, due to the anisotropy of SOC ($T_{x,y}$ is mainly produced for porphyrins [39]), may be transformed to the polarization of the opened RIP in the external magnetic axes [17].

According to eq 7, $k_{soc} = k_r(B = 0.24\text{ T})$ (see Table 2). The k_{soc} values observed are 1-3 orders of magnitude less than those for the recombination of the triplet exciplexes formed by ET from aromatic amines to the triplet excited states of quinones [10]. The relation between k_{soc} and the intersystem ET free energy is bell-shaped in combined systems [10,27]:

$$k_{soc} = (2\pi/\hbar)|H_{rp}|^2 FC \tag{8}$$

where H_{rp} is now the SOC coupling matrix element and FC is Franck-Condon weighted density of states. In polar solvents FC reaches a maximum at $-\Delta G$ = 1.2-2.3 eV [9,10]. The k_{soc} for the $^3[P^{+}\text{-Ph-Vi}^{2+}]$ recombination is close to the maximal one ($\Delta G \approx 1.5$ eV [40,41]) and its low value is related to the low value of $|H_{rp}|$. It is not surprising on account of the $|H_{rp}|$ dependence on the conformation and the nature of the radicals [10].

Spin-orbit coupling is strongly enhanced by a heavy-atom environment of the RIP [42]. The RIP under consideration is a cation which can associate with an anion containing a heavy atom. The cation-anion association is possible in cation micelles on account of high local concentration of anions in the double layer of the micelle [43]. The recombination rate constant in aqueous micellar solution of CTAB is greater than k_r in those of CTAC. Figures 2 show that the heavy-atom effect on k_r is the largest at $B \geq 0.1$ T, where the intersystem ET induced by SOC dominates in the RIP recombination, and the smallest in zero magnetic field, where the contribution of this route is not high and the hfc mechanism of the T-S transitions is efficient. SOC enhanced by heavy atoms is very important in the recombination of triplet exciplexes in nonpolar solvents [10] and triplet RIPs in micelles [29,30].

4. QUENCHING OF THE TRIPLET RIP BY MOLECULAR OXYGEN [24]

Charge separation in photosynthesis, occurring in atmosphere of the Earth, is unaffected by molecular oxygen. It is not surprising. Bimolecular reactions controlled by diffusion cannot compete with the very fast ETs in the photosynthetic RC in the picosecond time regime. Charge separation in the supramolecules under consideration is not so fast and molecular oxygen can affect the operation of such a molecular device.

In air saturated ethanol, the lifetime of $^3P\text{-Ph-Vi}^{2+}$ is close to that in the deoxygenated solution, whereas the lifetime of $^3[P^{+}\text{-Ph-Vi}^{+}]$ is shortened. At the $^3O_2$ concentration in air saturated solutions the intermolecular quenching of $^3P\text{-Ph-Vi}^{2+}$ by $^3O_2$ cannot compete with the fast intramolecular ET from ^3P to Vi^{2+}.

The lifetime of $^3[P^{+}\text{-Ph-Vi}^{+}]$ is greater than that of $^3P\text{-Ph-Vi}^{2+}$. Hence a competition between the RIP quenching by $^3O_2$ and the RIP recombination becomes possible. The corresponding second-order rate constant [$k_q(O_2)$] of the RIP quenching by $^3O_2$ is very high [$k_q(O_2)$=(0.8-1.0)$\cdot 10^{10}$ $M^{-1}s^{-1}$] and is close to 1/3 of the diffusion limit (the value of the rate constant for diffusion-controlled reaction with oxygen participation in usual nonviscous organic solvents is (2-3)$\cdot 10^{10}$ $M^{-1}s^{-1}$ [44]). In air saturated ethanol an external magnetic field affects the lifetime of $^3[P^{+}\text{-Ph-Vi}^{+}]$ by only about 20%. Hence, the magnetic field effect on RIP recombination is diminished significantly by the presence of $^3O_2$.

The $^3[P^{+}\text{-Ph-Vi}^{+}]$ quenching by $^3O_2$ results in the RIP transition to the ground state:

$$^{3}[P^{+}\text{-Ph-}Vi^{+}] \xrightarrow{^{3}O_2} {}^{1}[P\text{-Ph-}Vi^{2+}] \qquad (9)$$

No ET from the Vi^+ moiety of the dyad to $^3O_2$ is observed. This is not surprising on account of the $k_q(O_2)$ value being essentially greater than the rate constant of the ET from benzylviologen radical cation to oxygen, $3 \cdot 10^5$ $M^{-1}s^{-1}$.

The rate constant of ^{3}ZnBTP quenching by $^3O_2$, $1.3 \cdot 10^9$ $M^{-1}s^{-1}$ [22], is also less than $k_q(O_2)$, *i.e.* the triplet RIP is more efficiently quenched by $^3O_2$ than is the porphyrin triplet state.

The photoexcitation of $P\text{-Ph-}Vi^{2+}\text{-}Q(QH_2)$ results in the appearance of the excited porphyrin triplet and the triplet RIP states. The efficient quenching rate constant is greater than those for ^{3}ZnBTP and the triplet RIP [$k_q(O_2) = 3.8 \cdot 10^9$ $M^{-1}s^{-1}$].

The photoexcitation of $P\text{-BPh-}Vi^{2+}$ and $P\text{-BPh-}Vi^{2+}\text{-}Q$ results in the appearance of the porphyrin triplet excited state, which is quenched by $^3O_2$ with the same rate constant as ^{3}ZnBTP [$k_q(O_2) = 1.3 \cdot 10^9$ $M^{-1}s^{-1}$]. In air saturated ethanol intramolecular ETs cannot compete with the quenching of ^3P by $^3O_2$.

There are several mechanisms for triplet RIP quenching by the molecular oxygen and other paramagnetic species [44,45]. Molecular oxygen can act as an electron or an energy acceptor and stimulate the T-S transitions by the spin-exchange and in such way enhance the RIP recombination. The ET mechanism is not important as it was mentioned above. Other mechanisms of the RIP quenching by $^3O_2$ appear differently in opened or closed conformers.

In the opened conformer the T-S transitions occur due to the spin-exchange of individual radical centers and $^3O_2$. The efficient rate constant of such process is:

$$k(O_2) = 1/3\, k_{dif}[O_2] \qquad (10)$$

where k_{dif} is the rate constant for a diffusion controlled reaction and 1/3 is the spin-statistical factor.

In zero magnetic field the T-S transitions are very fast, $k_{hfc} > k_{is}$, and the appearance of a new route for them cannot change $k_r(B = 0)$. The T-S transitions induced by $^3O_2$ become important in an external magnetic field. Thus $^3O_2$ switches off the magnetic field effect on the RIP recombination, without acceleration of the RIP decay at B = 0. The separated triplet RIPs in micelles appear to have such behaviour by the quenching by $^3O_2$ and other paramagnetic species [29]. However, in the present case of $^3[P^{+}\text{-Ph-}Vi^{+}]$ the RIP is also quenched at zero magnetic field. This means that the acceleration of the T-S transition in the opened conformer is not the main route of the RIP quenching by $^3O_2$.

In the closed conformer the interaction between the radical centers is very important. Hence the closed RIP should be considered as a kind of the triplet molecule (^3M), which might be quenched by the following mechanisms [44]: energy transfer in singlet encounter complex (formed with spin-statistical factor 1/9):

$$^{3}M + {}^{3}O_2 \longrightarrow {}^{1}(M...O_2) \longrightarrow {}^{1}M + {}^{1}O_2^{*} \qquad (11)$$

or enhancement of ISC in the triplet encounter complex (spin-statistical factor 1/3):

$$^{3}M + {}^{3}O_2 \longrightarrow {}^{3}(M...O_2) \longrightarrow {}^{1}M + {}^{3}O_2 \qquad (12)$$

Analysis of these mechanisms has shown that reaction 11 should not be the main route. The second-order rate constant for RIP quenching by $^{3}O_2$ [$k_q(O_2)$, see Table 1] is significantly greater than 1/9 k_{dif} and the $k_q(O_2)$ value for ^{3}ZnBTP as a reference. This result supports the contention that spin-exchange mechanism of quenching (reaction 12) dominates.

The triplet RIP is born in a closed conformation and the quenching of closed RIP by $^{3}O_2$ [1/3 $k_{dif}[O_2]$] competes with the conformational transition "closed"$\longrightarrow$"opened" (k_1). In air saturated solutions the values of 1/3 $k_{dif}[O_2]$ and of k_1 are of the same order of magnitude ($\sim 10^7$ s^{-1}), so the back ET in the closed RIP in the presence of $^{3}O_2$ may efficiently compete with the corresponding conformational transition.

The mechanism of the RIP quenching by $^{3}O_2$ is quite similar to that of polar triplet exciplexes [46]. The spin-exchange mechanism is effective in these systems under the condition that the triplet state of the contact RIP is higher in energy than the singlet one. Hence, to explain the extremely high $k_q(O_2)$ for $^{3}[P^{+}\text{-Ph-Vi}^{+}]$, we have to suggest with respect to ref 26 that the energy of $^{1}[P^{+}\text{-Ph-Vi}^{+}]$ is lower than that of $^{3}[P^{+}\text{-Ph-Vi}^{+}]$. A lower energy for the triplet RIP state than that of the singlet one is possible. This occurs in the case of the porphyrin-viologen dyad, in polar triplet exciplexes in nonpolar solvents [46] and in flexible biradicals [20].

5. STABILIZATION OF THE TRIPLET RIP BY SUBSEQUENT ET TO A SECOND ELECTRON ACCEPTOR [23]

The radical ion pair state of the supramolecule can be stabilized by the subsequent ET to a second electron acceptor attached to the dyad core. Charge stabilization in P-Ph-Vi^{2+}-Q is not observed, because the subsequent ET from Vi^{+} to Q cannot compete with the fast backward ET from V^{+} to P^{+}.

Photoexcitation of P-BPh-Vi^{2+}-Q results in the appearance of a very long-lived RIP. The transient absorption spectrum of this RIP is attributed to P^{+}-BPh-Vi^{2+}-Q^{-}. RIPs may occur by subsequent intramolecular ET from Vi^{+} to Q or from ^{3}P to Q directly. On account of the low steady-state concentration of P^{+}-BPh-Vi^{+}-Q it is impossible to estimate the relative importance of these routes. The quantum yield of P^{+}-BPh-Vi^{2+}-Q^{-} is 0.10-0.15.

The kinetics of P^{+}-BPh-Vi^{2+}-Q^{-} decay is expressed by the following equation:

$$-d[P^{+}\text{-BPh-Vi}^{2+}\text{-Q}^{-}]/dt = k_{r1}[P^{+}\text{-BPh-Vi}^{2+}\text{-Q}^{-}] + k_{r2}[P^{+}\text{-BPh-Vi}^{2+}\text{-Q}^{-}]^2 \qquad (13)$$

The first term with k_{r1} = 90 s^{-1} relates to the back intramolecular ET whereas the second one relates to the intermolecular reaction of P^{+}-BPh-Vi^{2+}-Q^{-} and another one with k_{r2} = 8•10^{8} $M^{-1}s^{-1}$.

6. CONCLUSIONS

In general, it should be noted that the photochemical behaviour of the new semirigid Zn-porphyrin-viologen dyads and Zn-porphyrin-viologen-quinone triads in many aspects is close to that of flexible ones which possess the conformations with small exchange interactions.

The photoinduced ET reactions of the semirigid dyads and triads are satisfactorily described in terms of the intramolecular processes in the opened and closed conformers. The intramolecular quenching of the singlet, as well as in the triplet states of the porphyrin moiety, occurs in the closed conformer and the triplet-singlet transitions due to the radical-pair mechanisms in the opened one. The conformational transitions between opened and closed conformers are relatively slow being the limiting step of the radical-pair decay at zero magnetic field.

Application of an external magnetic field switches off the triplet-singlet transitions in opened conformer. At high magnetic fields the relatively slow intersystem electron transfer, stimulated by the spin-orbit coupling in closed conformer, is the limiting step of the RIP recombination.

Molecular oxygen quenches the triplet RIP and switches off the magnetic field effect very efficiently. The main route for the RIP quenching by $^{3}O_2$ is the enhancement of the transitions between the triplet and singlet states of closed conformer in which the RIP is born. This mechanism is active on account of the singlet state of closed conformer of the radical pair has a lower energy than the triplet.

The lifetime of P^{+}-BPh-Vi^{2+}-Q^{-} formed by the intramolecular ET in the triad is long enough to use this triad in photobiomimetic devices.

7. ACKNOWLEDGMENTS

We wish to thank Dr. Mary D. Archer and Prof. James R. Bolton for reviewing the manuscript and for many helpful comments.

8. REFERENCES

(1) Marcus R.A. (1956) 'On the theory of oxidation-reduction reactions involving electron transfer', J.Chem.Phys. 24, 966-989.
(2) Miller J.R., Beitz J.V., and Huddleston R.K. (1984) 'Effect of free energy on the rates of electron transfer between molecules', J.Am.Chem.Soc. 106, 5057-5068.
(3) Miller J.R., Calcaterra L.T., and Closs G.L. (1984) 'Intramolecular

long-distance electron transfer in radical anions. The effects of free energy and solvent on the reaction rates', J.Am.Chem.Soc. 106, 3047-3049.

(4) Wasielewski M.R., Niemczyk M.P., Svec W.A., and Pewitt E.B. (1985) 'Dependences of rate constants for photoinduced charge separation and dark charge recombination on the free energy of reaction in restricted-distance porphyrin-quinone molecules', J.Amer.Chem. Soc. 107, 1080-1082.

(5) Wasielewski M.R., Niemczyk M.P., Svec W.A.,and Pewitt E.B. (1985) 'High quantum yield long-lived charge separation in photosynthetic reaction center model', J.Amer.Chem.Soc. 107, 5562-5563.

(6) Gust D., Moore T.A., Liddel P.A., Nemeth G.A., Makins L.R., Moore A.L., Barrett D., Pessiki P.J., Bensasson R.V., Rougee M., Chachaty C., De Schryver F.C., Van der Auweraer M., Holzwarth A.R., and Connolly J.S. (1987) 'Charge separation in carotheno-porphyrin-quinone triads: Synthesis, conformational, and fluorescence studies', J.Amer.Chem.Soc. 109, 846-857.

(7) Mataga N., Karen A., Okada T., Nishitani S., Kurata N., Sakata Y., and Misumi S. (1984) 'Picosecond dynamics of photochemical electron transfer in porphyrin-quinone intramolecular exciplex systems', J.Phys.Chem. 88, 5138-5141.

(8) Gunner M.R., and Dutton P.L. (1989) 'Temperature and $-\Delta G°$ dependence of the electron transfer from BPh^- to Q_A in reaction center protein from *Rhodobackter sphaeroides* with different quinones as Q_A', J.Am.Chem.Soc. 110, 3400-3412.

(9) Mataga N., Asahi T., Kanda Y., and Okada T. (1988) 'The bell-shaped energy gap dependence of the charge recombination reaction on geminate radical ion pairs produced by fluorescence quenching reaction in acetonitrile solution', Chem.Phys. 127, 249-261.

(10) Levin P.P., Pluzhnikov P.F., and Kuzmin V.A. (1989) 'Energy gap dependence for intersystem crossing within charge transfer exciplexes', Chem.Phys. 137, 331-344.

(11) Miller G.R. (1987) 'Controlling charge separation through effects of energy, distance and molecular structure on electron transfer rates', New J.Chem. 11, 83-89.

(12) Harriman A., Porter G., and Wilowska A. (1984) 'Photoredox properties of zinc porphyrin-viologen complexes', J.Chem.Soc., Faraday Trans.2, 80, 191-204.

(13) Kanda Y., Sato H., Okada T., and Mataga N. (1986) 'Formation of a long-lived intramolecular radical ion pair from the singlet excited state of porphyrin-methylviologen combined with flexible chains', Chem.Phys.Lett. 129, 306-309.

(14) Nakamura H., Uehata A., Motonaga A., Ogata T., and Matsuo T. (1987) 'Reaction control of photoinduced electron transfer in porphyrin-viologen linked system by the use of external magnetic fields', Chem.Lett. 543-546.

(15) Saito T., Hirata Y., Sato H., Yoshida T., and Mataga N. (1988) 'Long-lived triplet radical ion pair state of porphyrin-methyl-viologen combined systems as revealed by a large magnetic field effect on its lifetime', Bull.Chem.Soc.Japan 61, 1925-1931.

(16) Lindsey J.S., Delaney Y.K., Mauzerall D.C., and Linschitz H. (1988)

'Photophysics of a cofacial porphyrin-quinone cage molecule and related compounds: Fluorescence properties, flash transients, and electron-transfer reactions', J.Am.Chem.Soc. 110, 3610-3621.
(17) Molin Yu.N. (1984) Spinpolarization and magnetic effects in radical reactions, Elsevier, Amsterdam.
(18) Steiner U.E., and Ulrich T. (1989) 'Magnetic field effects in chemical kinetics and related phenomena', Chem.Rev. 89, 51-147.
(19) Zimmt M.B., Doubleday C.,Jr., and Turro N.J. (1985) 'Magnetic field effect on the intersystem crossing rate constants of biradicals measured by nanosecond transient UV absorption', J.Am.Chem.Soc. 107, 6726-6727.
(20) Doubleday C.,Jr., Turro N.J., and Wang J.-F. (1989) 'Dynamics of flexible triplet biradicals', Acc.Chem.Res. 22, 199-205.
(21) Weller A., Staerk H., and Treichel R. (1984) 'Magnetic-field effects on geminate radical-pair recombination', Faraday Discuss.Chem.Ser. 78, 271.
(22) Batova E.E., and Shafirovich V.Ya. (1989) 'Porphyrin-viologen-quinone triad for unidirectional electron transfer', Dokl.Akad.Nauk SSSR 307, 1131-1134.
(23) Batova E.E., Levin P.P., and Shafirovich V.Ya. (1990) 'Long-lived radical ion pair state of Zn-porphyrin-viologen-quinone triad', New.J.Chem. 4, 269-271.
(24) Levin P.P., Batova E.E., and Shafirovich V.Ya. (1990) 'Magnetic field effect on the recombination of the triplet radical ion pair state of the semirigid Zn-porphyrin-viologen dyad', Chem.Phys. 142, 279-289.
(25) Shafirovich V.Ya., Batova E.E., and Levin P.P. (1990) 'Magnetic field effect on the recombination of the triplet radical ion pair state of the Zn-porphyrin-viologen dyad: The influence of the supramolecule microenvironment', Chem.Phys.Lett., submitted.
(26) Levin P.P., Khudyakov I.V., and Kuzmin V.A. (1989) 'Geminate recombination kinetics of triplet radical pairs in glycerol: Magnetic field effect', J.Phys.Chem. 93, 208-214.
(27) Ohno T., Yoshimura A., and Mataga N. (1986) 'Bell-shaped energy gap dependence of backward electron-transfer rate of geminate radical pairs produced by electron-transfer quenching of Ru(II) complexes by aromatic amines', J.Phys.Chem. 90, 3295-3297.
(28) Steiner U.E., Wolff H.-J., Ulrich T., and Ohno T. (1989) 'Spin-orbit coupling and magnetic field effects in photoredox reactions of ruthenium (II) complexes', J.Phys.Chem. 93, 5147-5154.
(29) Levin P.P., and Kuzmin V.A. (1988) 'Study of the kinetics of geminate recombination of triplet radical pairs including the ketyl radical and radical cation in micelles of sodium dodecyl sulfate by laser photolysis', Bull.Acad.Sci.USSR, Div.Chem. 37, 224-230.
(30) Levin P.P., and Kuzmin V.A. (1989) 'The role of spin-orbit coupling in kinetics of geminate recombination of triplet radical pairs in micelles. Effect of internal heavy atom', Dokl.Akad.Nauk SSSR 292, 134-137.
(31) Schulten K., and Wolynes P.G. (1978) 'Semiclassical description of electron spin motion in radicals including the effect of electron

hopping', J.Chem.Phys. 68, 3292-3297.
(32) Johnson C.S.,Jr., and Gutowski H.S. (1963) 'High resolution ESR spectra of photochemically generated free radicals: The viologens', Chem.Phys. 39, 58-62.
(33) Faier J., Borg D.C., Forman A., Dolphin D., and Felton R.H. (1970) 'π-Cation radicals and dications of metalloporphyrins', J. Am.Chem.Soc. 92, 3451-3459.
(34) Schlenker W., Ulrich T., and Steiner U.E. (1983) 'Magnetic-field-enchanced radical yield from triplet electron-transfer reaction in reversed micelles', Chem.Phys.Lett. 103, 118-123.
(35) Hayashi H., and Nagakura S. (1984) 'Theoretical study of relaxation mechanism in magnetic field effects on chemical reactions', Bull. Chem.Soc.Japan 57, 322-328.
(36) Carrington A., and McLauchlan A.D. (1967) Introduction to magnetic resonance, Harper & Row, New York, Ch.II.
(37) Levin P.P., and Kuzmin V.A. (1990) 'Spin-orbit coupling and paramagnetic relaxation in micellized triplet radical pairs. Determination of relaxation parameters from magnetic field dependences of decay kinetics', Chem.Phys.Lett. 165, 302-308.
(38) Bartels D.M., Lawler R.G., and Trufunal A.D. (1985) 'Electron T_1 measurements in short-lived free radicals by dynamic polarization recovery', J.Phys.Chem. 83, 2686-2707.
(39) Boxer S.G. (1983) 'Model reactions in photosynthesis', Biochem.Biophys.Acta 726, 265-292.
(40) Neumann-Spallart M., and Kalyanasundaram K. (1981) 'On the one and two-electron oxidations of water-soluble zinc porphyrins in aqueous media', Z.Naturforsch. 36b, 596-600.
(41) Bird C.L., and Kuhn A.T. (1981) 'Electrochemistry of viologens', Chem.Soc.Rev. 10, 49-82 .
(42) Ulrich T., and Steiner U.E. (1985) 'Control of photo-electron-transfer induced radical production by micellar cages, heavy-atom substituents and magnetic fields', Tetrahedron 42, 6131-6142.
(43) Mitsui A., Uehata A., Nakamura H., and Matsuo T. (1989) 'Counterion effects on the photoinduced electron-transfer and the reverse process in porphyrin-viologen linked compounds', Chem.Lett. 1445-1448.
(44) Gijzeman O.L., Kaufman K., and Porter G.J. (1973) 'Oxygen quenching of aromatic triplet states in solution', J.Chem.Soc.Faraday Trans 2, 69, 708.
(45) Wang J., Welsh K.M., Waterman K.C., Fehlner P., Doubleday C.,Jr., and Turro N. (1988) 'Dynamics of interaction between a 1,9-bi-radical and lanthanide ions', J.Phys.Chem. 92, 3730-3732.
(46) Levin P.P., Pluzhnikov P.F., and Kuzmin V.A. (1988) 'Mechanism of triplet exciplex quenching by molecular oxygen. Singlet-triplet splitting for the charge transfer state', Chem.Phys.Lett. 152, 409-413.

Electrical Wiring of Redox Enzymes

Adam Heller
Department of Chemical Engineering
The University of Texas at Austin
Austin, Texas

Abstract

In a redox enzyme the functional electron transfer center, normally surrounded by an insulating protein matrix, can be electrically connected to an external source of current through a path of fast electron-relaying redox couples. Such "wiring" is achieved through electrostatically or covalently binding to the enzyme-proteins high molecular weight redox polycations having segments anchored to electrodes. With the wired enzymes subsecond response time amperometric biosensors can be built.

Reprinted with permission from *Accounts of Chemical Research*, **1990,** 23, 128-134. Copyright 1990 American Chemical Society.

E. Pelizzetti and M. Schiavello (eds.), Photochemical Conversion and Storage of Solar Energy, 67–87.

Introduction: Relaying of Electrons in Enzymes

The theme of this account is the chemical modification of regions of large biomolecules from electrical insulators to electrical conductors. Redox enzymes are large molecules of 40,000 (e.g. galactose oxidase) to 850,000 (e.g. choline dehydrogenase) daltons with one or more redox centers. Their average hydrodynamic diameters range from ~55 to ~150Å. In the great majority of enzymes, the redox centers are located sufficiently far from the outermost surface defined by protruding protein or glycoprotein domains to be electrically inaccessible. Consequently, most enzymes do not exchange electrons with electrodes on which they are adsorbed, i.e. their redox centers are neither electrooxidized at positive potentials, nor electroreduced at negative ones. Apparently, part of the protein or glycoprotein shell surrounding the redox centers is there to prevent indiscriminate electron exchange between the different redox macromolecules of living systems. Such exchange would, in the extreme case, lead to an equipotential system, which could not sustain life. Another function of this shell is to stabilize the structure of the enzyme. Because neither function is essential for catalysis, redox enzymes do function when part of the shell is stripped,[1,2] or as we shall see here, when the shell is chemically altered so as to make it electrically conductive.[3-5] Following such alteration, a redox center of an enzyme will directly transfer electrons to an electrode on which the enzyme is adsorbed.[3,4] We call the centers that increase the electron-current flowing through their shells, by accepting and transferring electrons, electron relays.

The distance dependence of the rate of electron transfer in proteins has been the subject of experimental[7-14] and theoretical[15-21] studies during the past decade. Because the process of electron transfer is, in essence, an electron tunneling process,[17] theories conclude and experiments show that the rate of electron transfer decays exponentially with distance, when the distance substantially exceeds atomic dimensions (>3Å). Thus

$$k_{ET} \approx 10^{13} e^{\beta(d-3)} e^{-(\Delta G^{\circ}+\lambda)^2 4RT\lambda} \qquad (1)$$

where k_{ET} is the rate of electron transfer, ß is a constant for a given electron donor/acceptor pair in a defined medium, d is the distance between the donor and the acceptor, $-\Delta G^o$ is the driving energy and λ is the Marcus reorganization energy. Most redox enzymes have sufficiently thick protein or glycoprotein shells to make the product [ßx(d-3)] large enough to decrease the rate of electron-transfer to a negligibly small value for a random encounter between an enzyme and another redox protein or enzyme, or between an enzyme and an electrode.

Function and Design of Electron Relays

Among the several electron-transfer theories, Marcus theory translates the electron transfer rate into simple chemical terms.[7,16] Electron transfer will take place when both the donor and the acceptor assume structural configurations that require no further reorganization upon electron transfer. The energy invested in order to bring the pair to this structure is the Marcus reorganization energy, denoted by λ. At a given reorganization energy, temperature and distance and in a given medium, the rate of electron transfer increases when the process is exoergic ($\Delta G < O$), i.e. the electron hops thermodynamically downhill from a reducing center to an oxidizing center. The difference in the energy of the system is nil for electron transfer between compositionally identical ions in different oxidation states. By measuring the rate of electron transfer within a redox couple one obtains its "self exchange rate". A high self exchange rate is indicative of a small reorganization energy. Couples with small reorganization energies are faster i.e. exchange electrons rapidly with electrodes; thus, the best electron relays in biological macromolecules are also fast redox couples. Furthermore, the faster the couple, the greater the allowed distance at which a given current will flow between an electron donating center of an enzyme and an electron-relay based on this couple, and also between the relay and an electrode. This current will increase when the potential difference between the donor and the relay, or between the relay and the electrode, is increased. Usually, only one of these two electron-hops will be rate controlling.

Of the experimental studies on electron-transfer rates in proteins, those of Gray and colleagues[7,8] are particularly relevant, because they provide quantitative information about the distance dependence at varying reorganization energies and potential differences. For electron transfer between porphyrin-bound Fe^{2+} centers and histidine-bound $[Ru(NH_3)_5]^{3+}$, they find $ß=0.91Å^{-1}$, or a tenfold decrease in the rate of electron transfer upon each 2.1Å increase in distance. Thus for a protein and donor-acceptor pair similar to that studied by Gray et al, insertion of a layer of fast relays at half-thickness across a 30Å thick protein film on an electrode may increase the current, between a sheet of donors on the solution side of the film and the electrode, by a factor as large as 10^7.

The relationship between d, ß, λ, and the potential difference $\Delta V(\Delta G=-e\Delta V)$ is illuminating. Let us consider a monolayer of a redox enzyme covering an electrode and calculate, from equation 1, some d, λ, and ΔV values required for realizing the full enzyme-turnover-limited current density. Glucose oxidase turns over at ambient temperature at a rate of $\sim 10^2 sec^{-1}$ i.e. it produces about 200 transferable electrons per second. Because its radius is ~43Å, there can be up to 1.7×10^{12} enzyme molecules on the electrode surface. The current density, when all redox centers are electrically well connected to the electrode, may thus reach about 3.4×10^{14} electrons $sec^{-1} cm^{-2}$, or $53 \mu A cm^{-2}$. At a 25Å distance between the electron-transferring centers and the electrode, it is possible to reach an electron transfer rate of 200 sec^{-1} when $\lambda \leq 1.0 eV$ and $\Delta V \leq -0.3V$, or when $\lambda \leq 0.4 eV$ and $\Delta V \leq 0V$. For d = 20Å, electron transfer will be effective even at $\Delta V = 0.0V$ for $\lambda \leq 0.9 eV$.[7] Thus with a fast relay, having a redox potential substantially oxidizing with respect to the redox potential of glucose oxidase, electron transfer with a typical reorganization energy of 0.5-1eV may take place across a distance as long as 25Å; and with a fast moderately oxidizing relay an electron transfer distance of 20Å is realizable. We can see why small redox proteins, of 2×10^4 daltons or less, with effective hydrodynamic radii of less than ~21Å can be directly electrooxidized or reduced when adsorbed on electrodes and why glucose oxidase, with 160,000 daltons, and a hydrodynamic radius of 43Å,[1] can not. Nevertheless, even in small proteins, with redox centers that communicate directly with electrodes, protein orientation is important. By selecting electrode surfaces, or by modifying these with orienting promoters, such as positively charged organic or inorganic molecules or ions, the rate of electron transfer can be enhanced.[22-30] We also see from the relationships why a few large enzymes, after complexing small redox proteins (e.g. cytochrome *c*), that are integrated into the external part of their structure, are directly electrooxidized or reduced on electrodes;[31-33] and why most large enzymes, with deep redox centers, are not directly electrooxidized or reduced on electrodes. Glucose oxidase (GO) [E.C. 1.1.3.4] is an example of an enzyme of the latter type. This enzyme catalyzes the transfer of electrons from glucose to oxygen, producing gluconolactone and hydrogen peroxide (equations 2-3). The process involves the $FAD/FADH_2$ redox centers of the enzyme:

$$\text{Glucose + GO-FAD} \xrightarrow[2e^-+2H^+]{} \text{Gluconolactone + GO-FADH}_2 \qquad (2)$$

$$\text{GO-FADH}_2 + O_2 \xrightarrow[2e^-+2H^+]{} \text{GO-FAD} + H_2O_2 \qquad (3)$$

Glucose cannot be selectively electrooxidized at an electrode on which the enzyme has been adsorbed, because the product [ßx(d-3)] is excessive for electron transfer to the electrode. Even at the most anodic potentials that are accessible in aqueous solutions electrode reaction 4 cannot be driven.

$$\text{GO-FADH}_2 \longrightarrow \text{GO-FAD} + 2H^+ + 2e^- \qquad (4)$$

Reaction 3 takes place only because oxygen diffuses into the protein, reducing d sufficiently to allow electron-transfer. In addition to oxygen, other oxidized members of redox couples can diffuse into enzymes to accept electrons from their redox (e.g. $FADH_2$) centers. One of the first studied may have been methylene blue.[34-36]

The physical parameters that define in enzyme electrodes the effectiveness of bound electron relays of diffusing electron carriers are identical: diffusing species are effective when they need not penetrate the enzyme deeply for electron transfer; and relays are effective when they can accept and transfer electrons across substantial distances. In both cases we seek high self-exchange rates, though it is also possible to increase the electron-transfer rates by increasing ΔV. The list of effective diffusional electron acceptors and donors includes ferrocene/ferricinium derivates;[37,38] ruthenium$^{2+/3+}$ ammines[39]; hydroquinones/quinones;[40] reducible and oxidizable components of organic salts;[41,42] and tungsten$^{6+/5+}$ and molybdenum$^{6+/5+}$ octacyanides.[43]

Bonding of Electron Relays to Enzyme Proteins.

Our first study, aimed at establishing that the electrical properties of enzyme proteins can be modified, involved ferrocene/ferricinium carboxylate electron relays (Figure 1).[37,38] Amide links were formed between the carboxylate and part of the 42 lysine amine functions of glucose oxidase through carbodiimide-coupling (equation 5).[3,44]

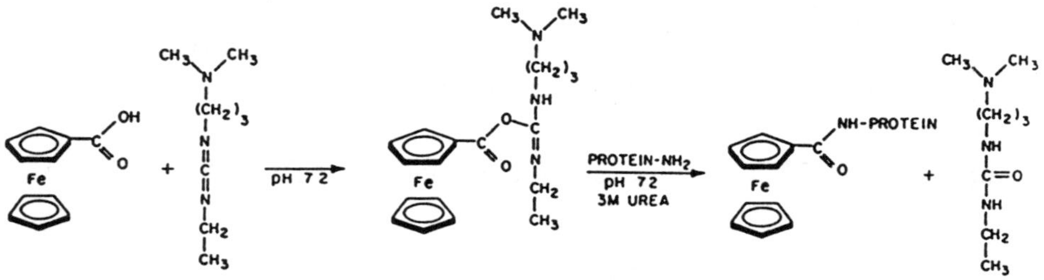

(5)

The reaction proceeds in an aqueous solution at physiological temperature and pH. In the first experiments we were unable to reproduce reliably the (occasionally observed) direct electrical communication between the FAD/$FADH_2$ centers of glucose oxidase and metal or graphite electrodes, via the enzyme-bound relays. But after the enzyme protein was reversibly (partially) unfolded by 2-3M urea, we obtained reproducible results. An average of 12 ferrocene functions could be covalently incorporated in glucose oxidase.[3-5] After removal of the urea by gel permeation chromatography, the relay-modified enzyme retained 60% of its activity and communicated electrically with graphite, glassy carbon, gold and other metals. Communication resulted in a glucose-concentration dependent current at or above the oxidation potential of the relay. (Figures 1,2) The current was produced by a sequence of electron transfer steps, where glucose reduced the enzyme's FAD centers to $FADH_2$; the $FADH_2$ centers were reoxidized to FAD upon transferring electrons to the ferricinium relays, reducing these to ferrocene; ferrocene transferred electrons to the electrode, and was reoxidized to ferricinium. The greater the number of relays we bound to the enzyme, the better its $FADH_2$ centers communicated with electrodes.[5] This was also the case in D-amino acid oxidase.[4,5] The current in relay-modified glucose oxidase electrodes increased when glucose was added to the solution (Figure 2) but did not change when sucrose, other hexoses, or pentoses were added.[5]

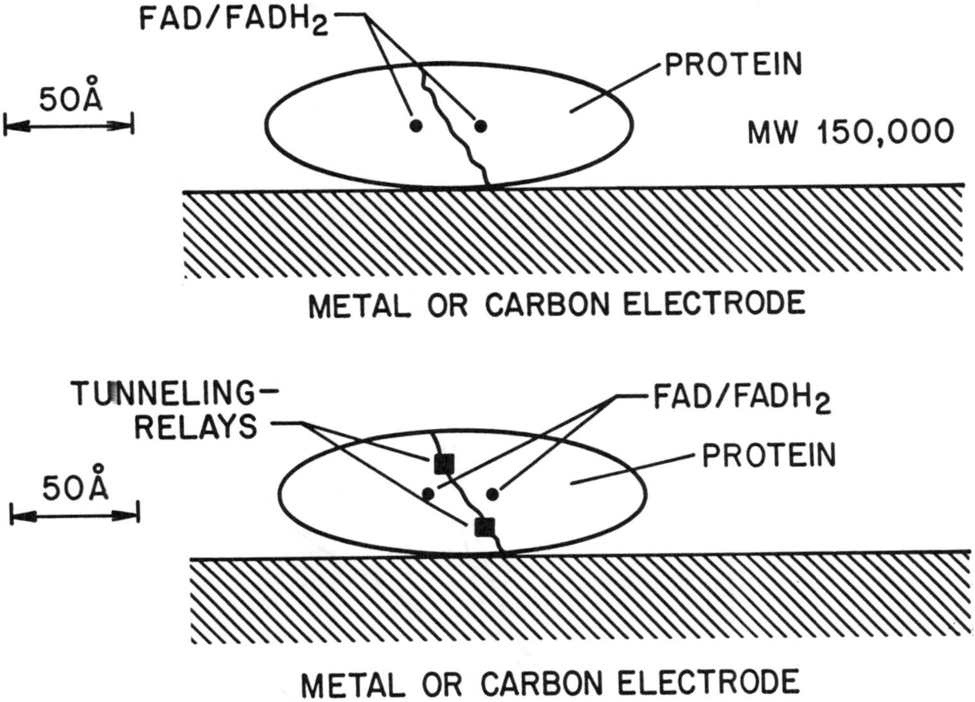

Fig. 1. *Top*: When a native redox enzyme (e.g. glucose oxidase) is adsorbed on an electrode, the electron transfer distances are excessive for electrical communication between the redox centers and the electrode.
Bottom: Flow of electrons through a relay from a redox center of an enzyme (glucose oxidase) to an electrode. A current is observed when the substrate (glucose) transfers a pair of electrons to an FAD center of the enzyme that, in turn, transfers these either to a relay or to a molecular wire, which then transfers the electrons to the electrode.

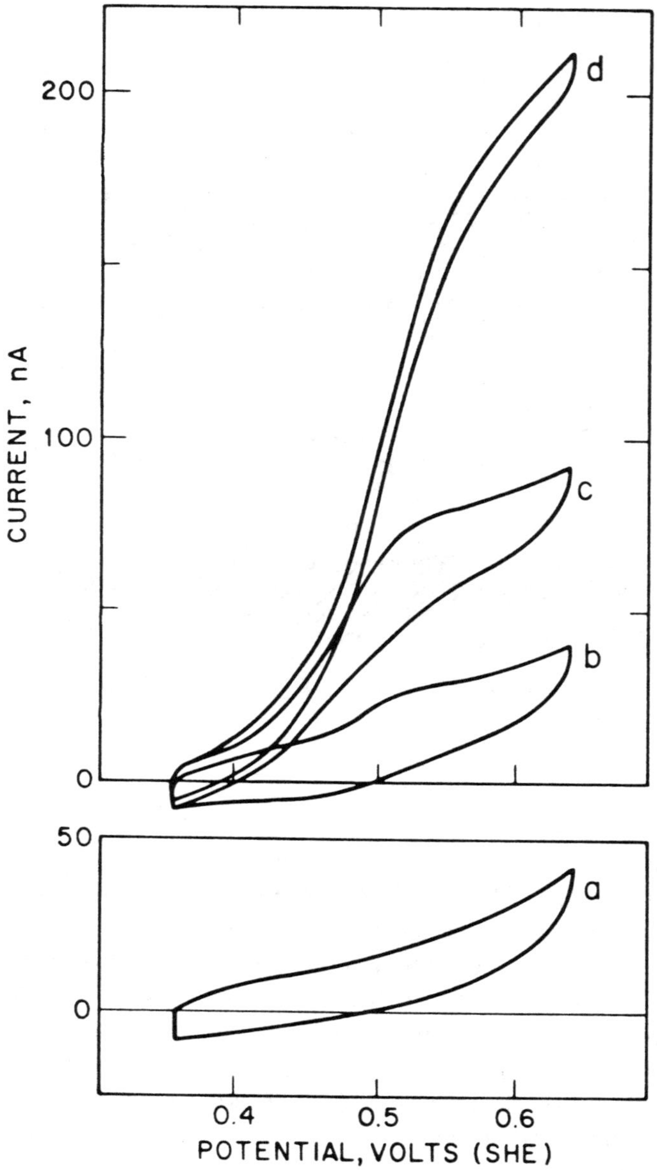

Fig. 2. Cyclic voltammograms (current vs. potential plots) for electrodes made with (a) native glucose oxidase with or without glucose; (b) glucose oxidase modified by covalent bonding of 12 electron-relays per enzyme molecule in the absence of glucose; (c) as in (b), but at 0.8 mM glucose and (d) as in (b) but at 5 mM glucose. Scan rate 2mV/sec.

Another effective glucose oxidase-bound relay that we and Bartlett and Whitaker,[5,45] studied was ferrocene acetic acid, covalently bound to glucose oxidase lysines. In a reaction sequence similar to that shown in equation 5, 21 relays were bound to the enzyme, further enhancing electrical communication.[45] Glucose oxidase was also modified with ruthenium pentammine relays, bound coordinatively to histidine functions of the enzyme, or to pyridine functions that were covalently bound either to tyrosines by azo links, or to lysines through amide links. These relays were less stable than the ferrocenes.[4,5]

Electrical Wiring of Redox Enzymes.

When electron relays are bound to an enzyme, the bulk of the enzyme does not become conductive. Rather, electrons are preferentially conducted along routes defined by the relays. As seen earlier, for relays of low reorganization energy and adequate potential difference (between the FAD/$FADH_2$ center and the relay and between the relay and the electrode) electron transfer distances can be of 20-25Å, and a single relay can be adequate to allow sequential electron hops between the enzyme's redox center and the electrode. Nevertheless, because there are few effective electron transfer paths even if 12 relays are bound to the enzyme,only when the enzyme tumbles on the surface of the electrode is its charge effectively collected: the enzyme must point a specific zone of its surface to the electrode. When a relay modified enzyme is covalently bound to an electrode surface in a random fashion, only a small fraction of the enzymes is properly oriented, and the currents are small. Furthermore, in the presence of oxygen, most of the electrons are transferred to it, to produce peroxide, rather than to the electrode. Our first enzyme electrodes required membranes to contain the freely tumbling enzyme in small volumes near the electrode surface. The membranes needed pores small enough to contain the diffusing enzyme, but large enough to allow in-diffusion of substrate and out-diffusion of product. The membranes required peripheral seals, that increased the manufacturing complexity and cost. Furthermore, in flow systems the membranes limited the rate of substrate transport to, and product transport from, the electrode surface, increasing the time constants.

The need for membranes and seals was obviated by wiring the enzyme with an electron-relaying redox polymer, a segment of

(a)

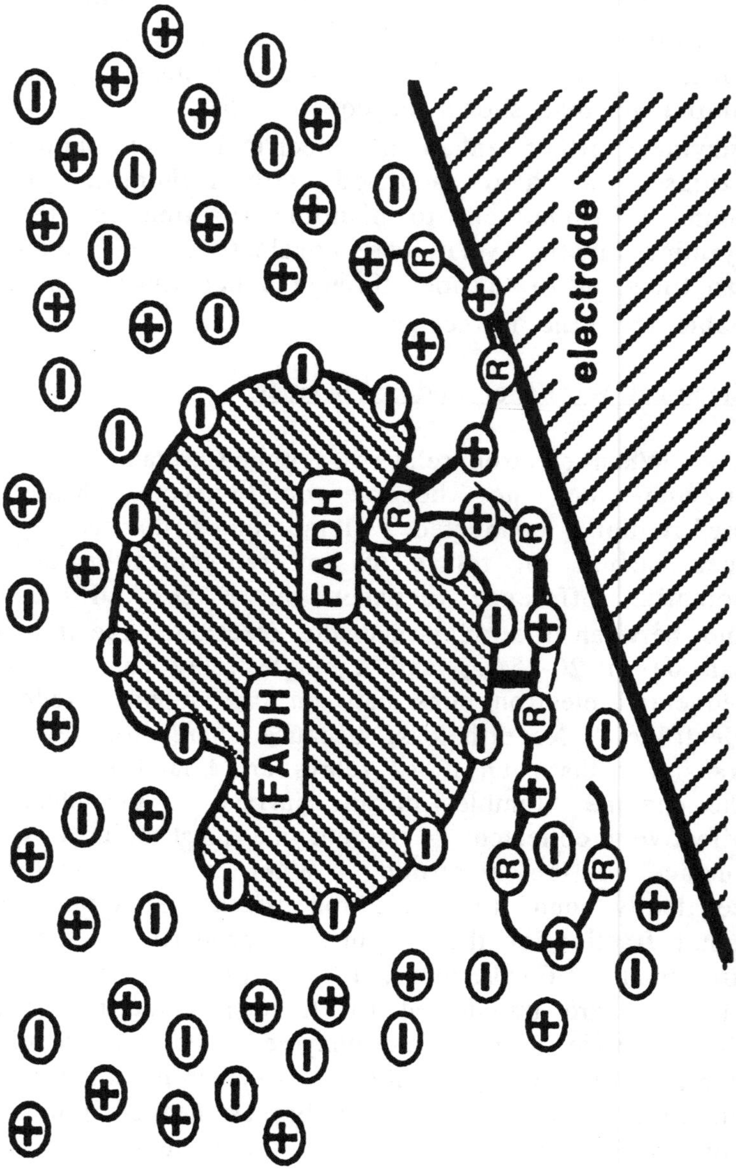

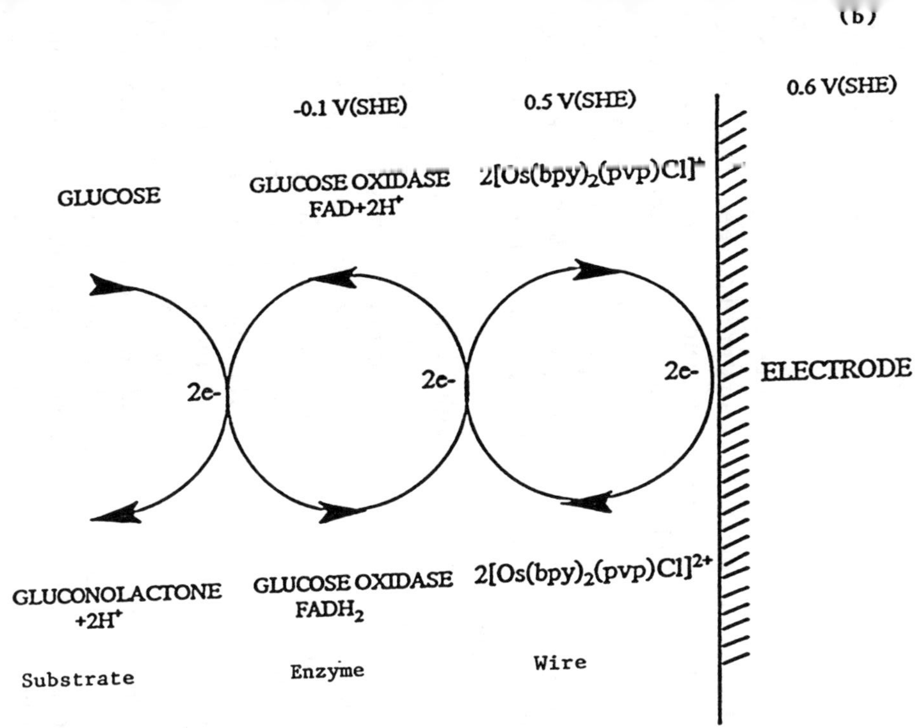

Fig. 3. (a) Wiring of a binary redox enzyme. Fast redox centers (R) of a polycation that is electrostatically and covalently bound to the enzyme relay electrons to the electrode, on which a segment of the polycation is adsorbed. Binding of the polycation to the electrode can be electrostatic when the electrode has a negative surface charge. (b) Electron-transfer steps in the electrooxidation of glucose on a wired glucose-oxidase electrode.

which was bound to the electrode.[6] Through such wiring the enzyme was electrically connected to the electrode, irrespective of its orientation. The resulting enzyme electrodes were then fast and simple to make.

Our method of wiring enzymes followed one of nature's ways of reducing electron transfer distances between redox centers of proteins and enzymes. For electron transfer between the redox centers of cytochrome *c* and enzymes such as cytochrome *c* peroxidase or cytochrome *c* oxidase, or between the ferredoxin and ferredoxin NADP reductase, the center-to-center distances are reduced through complexing the redox protein and the redox enzyme. The complexes are often electrostatic,[46-62] forming between enzymes with negatively charged zones and redox proteins with positively charged ones.[46-48]

We "wired" the enzymes through complexing, but with synthetic, rather than natural, redox macromolecules. The proteins of glucose oxidase, as well as those of other enzymes have at physiological pH a negative surface charge, because of excess of glutamic and aspartic acid over lysine and arginine. The polyanionic enzymes can be electrostatically complexed with polycationic redox polymers of ~10^5 daltons, typically with 100-200 redox centers, and 500-1000 cationic sites. When segments of the redox polymers fold along the enzyme proteins and penetrate these,[6] electron transfer from the $FADH_2$ center of the enzyme to at least one redox center of the polymer, and via this center to the electrode becomes possible. (Fig. 3a) The practicality of such electron transfer was confirmed in our very first (unpublished) experiment, where we complexed glucose oxidase with the copolymer of vinyl-N-methyl pyridinium chloride and vinyl ferrocene, a water-soluble redox polycation. We have preferred, however, osmium based redox polycations over the ferrocenes,[6] because they were shown by Meyer and coworkers to be fast, rugged, and to have potentials in the 0.2-0.5V(SCE) range.[63] Upon complexing commercial, 50,000 dalton poly(vinyl pyridine) with $[Os(2,2'\text{-bipyridine})_2Cl]^{1+/2+}$, and N-methylating the uncomplexed pyridine residues, we obtained an effective electron-relaying, stable polycation; after electrostatic complexing with glucose oxidase, the polycation mediated electron transfer from the substrate-reduced ($FADH_2$) centers of the enzyme to electrodes, cycling as shown in Fig. 3b. Segments of the polycation were found to strongly adsorb on graphite electrodes,[64] as expected from the earlier work of Oyama and Anson.[65] Because the redox polycations both bind and electrically connect the enzymes with electrodes, the enzyme-electrodes can be made simply by dipping abraded, cleaned graphite rods into an aqueous solution of the redox polymer, rinsing, dipping into the enzyme solution, and rinsing again.[64] The current densities are proportional to the concentration of glucose over a broad range, leveling off at $30\mu Acm^{-2}$ at high (30mM) glucose concentration. The electrodes can be kept in dry air, with little or no change, for 2 days. Their current-response is, as expected, rapid: They respond as fast as one can change the fluid in the proximity of electrodes in flow systems. In such systems the current risetimes are as short as 0.25 sec.,

My late teacher of thermodynamics at the Hebrew University, Aharon Katchalsky-Katzir his co-workers,[66,67] followed by more recent workers,[68,69] have shown that the structure of a dissolved polyelectrolyte strongly depends on ionic strength. At low ionic strength, electrostatic repulsion between similar charges on a chain tend to straighten it out even though statistically or entropically the straight chain configuration is unlikely; at high ionic strength the charges are screened by counterions and the macromolecules assume entropically more probable coiled configurations. When the redox polymers are coiled, they do not adequately penetrate crevices in enzymes and the electron transfer distances are not sufficiently reduced for electron transfer. Thus, above 0.5M NaCl concentration, the flow of current from the substrate-reduced enzyme to the electrode stops, even though the enzyme is still substantially active and continues to transfer electrons to small diffusing mediators in the solution.[6,64] Nevertheless, even at 0.5M NaCl the electrostatic redox polycation-enzyme complex does not dissociate, and when an electrode is moved from a glucose solution with 0.5M NaCl to one with 0.15M NaCl, the current flows again.

Because the variation in current with ionic strength limits the usefulness of enzyme electrodes to biosensor applications at fixed ionic strength, we added covalent links to the electrostatic complex in order to keep the enzyme properly wired at high ionic strength. Initially we did so by copolymerizing, at a 20:1 ratio, vinyl pyridine with 4-amino styrene, forming the $[Os(bpy)_2\ Cl]^{1+/2+}$ complex with about one in six pyridine rings, N-methylating the residual rings, diazotizing the amino-styrene functions, and reacting these, at low ionic strength with tyrosine functions of glucose oxidase, to form azo-bonds.[6] Because there are few tyrosines in glucose oxidase the structure was essentially two dimensional. 3-dimensional wired enzyme structures, based on crosslinking the redox polymer chains and binding these to glucose oxidase lysine amines, were subsequently designed and synthesized.[70] An example of a polymerization reaction[72] is shown in Equation 6. In the resulting network, one pyridine ring in about 5 carries an $[Os(bpy)_2Cl]^{1+/2+}$ center.

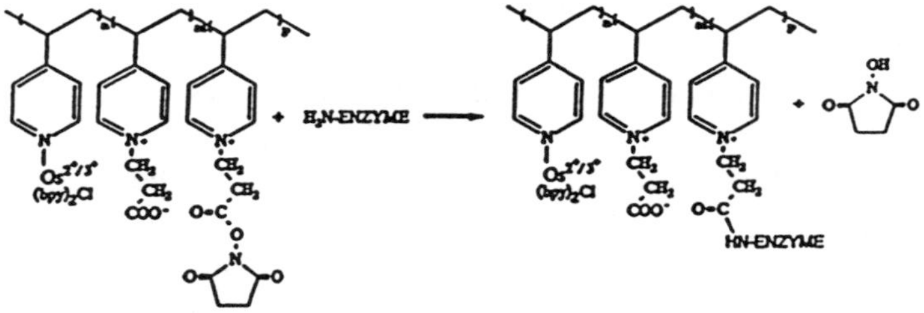

(6)

At a sufficiently high redox polymer to enzyme ratio, the wired enzyme films, of ~1µm thickness, can still be sufficiently conductive to allow remote enzyme molecules to electrically communicate, via the 3-dimensional redox network, with the electrode. We reached in these "thick" films glucose-dependent current densities as high as $\sim 0.5 mA cm^{-2}$, showing that at least 10 equivalent enzyme layers communicated electrically with the electrode (Fig. 4).[70] The superior electrical communication through these films is consistent with the electrical properties of the pure redox polymer.[71]

Amperometric Biosensors Based on Wired Enzymes.

Amperometric biosensors transduce the diffusion controlled flux of a substrate to an enzyme electrode into an electrical current. Such transducers originated in the work of Clark and Lyons.[73] Updike and Hicks subsequently measured amperometrically hydrogen peroxide formation through reaction 3.[74] Substantially advanced *ex vivo* electrodes of these types are now in use and *in vivo* electrodes are under development.[75] Amperometric enzyme electrodes based on diffusing redox mediators, including dyes (e.g. methylene blue),[76] ferrocene derivatives,[77] components of conducting organic metals,[78-79] and quinones[80] have been subject to intensive research. Recently, also electrodes based on conducting polypyrroles with ferrocenes have been reported.[81] Electrodes based on ferrocene derivatives[77] are in commercial production and are used in diverse applications such as monitoring glucose levels in diabetics and food and beverage processing.

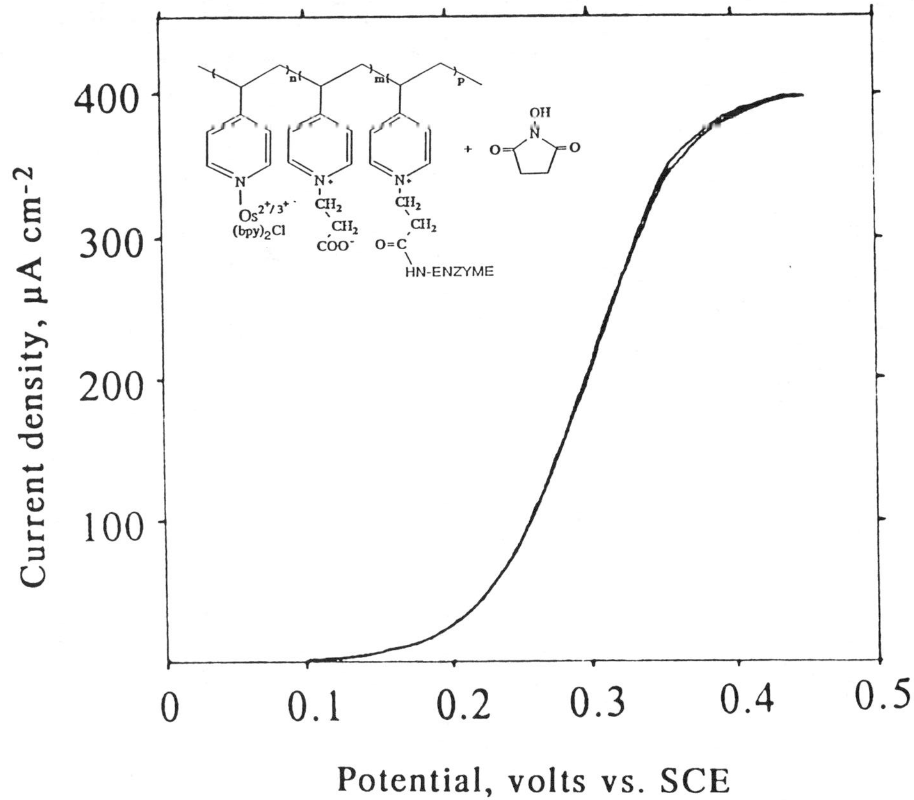

Fig. 4. Cyclic voltammogram, showing a 400μAcm^{-2} glucose-diffusion limited current density reached at 40mM glucose concentration with the wired enzyme network of Equation 6. The scan rate is 5mV/sec.

Amperometric enzyme electrodes can be built also with relay-modified and redox-polymer-wired enzymes. The relevant issues are manufacturability and cost; response time; reproducibility; selectivity; insensitivity to partial pressure of oxygen; detectivity (ratio of signal to noise (S/N)); output current stability *ex vivo*; output current stability *in vivo*; and biocompatibility.

The membraneless glucose electrodes based on 3-dimensional wired structures that we are now building are manufacturable at low cost. They respond to a change in glucose concentration in less than 1 sec. Their reproducibility depends on control of (a) the specific activity of the enzyme that is being wired; (b) the ratio of the wiring-polymer to the enzyme; and (c) the thickness of the wired-enzyme film. Their selectivity depends on the redox potential of the electron relaying centers. The closer this potential is to the redox potential of the enzyme itself, the lesser the likelihood that a potentially interfering substrate will be spuriously oxidized. The electrodes are insensitive to some electrooxidizable, commonly interfering substances such as the analgesic acetaminophen. Fluctuations in current with partial pressure of oxygen, e.g. oxygen concentration in blood, depend on the ratio of the rate of direct electrooxidation of the $FADH_2$ centers to their rate of oxidation by molecular oxygen, and therefore on the rate of electron transfer to, and the electrical resistance of, the 3-dimensional wired enzyme structure. At high osmium-complex concentrations, and in sufficiently thin layers, the competition is won by electron transfer to the electrode via the osmium centers, and the electrodes are relatively insensitive to oxygen. The signal to noise ratio S/N is, in the absence of interfering substrates, proportional to the number of enzyme molecules that are effectively wired to the electrode surface per unit area. At a film thickness of ~1μm, and at typical blood glucose concentrations (~10^{-2}M), a ~10^{-3}Acm^{-2} is achieved. With a low noise potentiostat and only unshielded leads to the biosensor, the noise is less than 10^{-7}Acm^{-2}, i.e. S/N is of 10^4.

The output current stability depends on enzyme durability and on avoiding fouling of the electrodes, primarily by adsorbed proteins. Typical decay rates at 25°C in the absence of proteins are ~5%/day, but are much faster in whole blood. By designing redox polymers that form hydrogels, we are now improving the stability of the bioelectrodes. We are also designing relays that are closer in their potential to those of the enzymes, with the objective of further reducing the residual interference by electrooxidizable species such as urate and ascorbate ions. We are also exploring the range of enzymes that can be electrically wired and are building sensors with these. Currently our list includes, in addition to glucose oxidase, the flavoenzymes D-amino acid oxidase, lactate oxidase and glycerol-3-phosphate oxidase; and lipoamide oxidase, through which NAD^+/NADH requiring enzymes are now coupled to the electrodes.

Acknowledgements

The financial support for this research from the Office of Naval Research, the Texas Advanced Research Program, and the Robert A. Welch Foundation is gratefully acknowledged. Harry B. Gray, Heinz Gerischer and Barry Miller read and improved the manuscript.

References:

1. Nakamura, S.; Hayashi, S.; Koga, K. *Biochem. Biophys. Acta* **1986**, *445*, 294.
2. Yasuda, Y.; Takahashi, N.; Murachi, T. *Biochemistry,* **1971,** *13,* 2624.
3. Degani, Y.; Heller, A. *J. Phys. Chem.*, **1987**, *91*, 1285.
4. Degani, Y.; Heller, A. *J. Am.Chem.Soc.* **1988**, *110*, 2615.
5. Heller, A.; Degani, Y. *"Direct Electrical Communication between Chemically Modified Redox Enzymes and Metal Electrodes: III. Electron Transfer Relay Modified Glucose Oxidase and D-Amino-Acid Oxidase"* in *"Redox Chemistry and Interfacial Behavior of Biological Molecules"*, Dryhurst, G.; Niki, K., Eds.; Plenum Publ. Corp; New York, 1988; p.151-170.
6. Degani, Y.; Heller, A. *J. Am.Chem. Soc.*, **1989**, *111*, 2357.
7. Lieber, C. M.; Karas, J. L.; Mayo, S.L.; Albin, M.; Gray, H.B. *'Long Range Electron Transfer in Proteins"*, Proc. Robert A. Welch Found. Conf. on Chem. Res. XXI. *Design of Enzymes and Enzyme Models*, The Welch Foundation, Houston; 1987; p.9-24.
8. Cowan, J.A.; Gray, H.B. *Chemica Scripta,* **1988**, *28A*, 21.
9. Sykes, A.G. *Chem. Soc. Rev.*, **1985**, *14*, 283.
10. Isied, S.S., *Prog. Inorg. Chem.* **1984**, *32*, 443.
11. Peterson-Kennedy, S.E.; McGourty, J.L.; Ho, P.S.; Sutoris, C.J.; Liang, N.; Zemel, H.; Blough, N.V.; Margoliash, E.; Hoffman, B.M. *Coord. Chem . Rev.* **1985**, *64*, 125.
12. McLendon, G.; Guarr, T.; McGuire, M.; Simolo, K.; Strauch, S.; Taylor, K. *Coord. Chem. Rev.* **1985**, *64*, 113.
13. Takaka, T.; Takenaka, K.; Kawamura, H.; Beppu, Y. *J. Biochem. (Tokyo)* **1986**, *99*, 833.

14. Miller, J.R. in *Antennas and Reaction Centers of Photosynthetic Bacteria,* Michel-Beyerle, M.E., Ed.; Springer Verlag: Berlin, 1985; p.234.
15. Bixon, M.; Jortner, J. *J. Phys. Chem.* **1986**,. *90*, 3795.
16. Marcus, R.A.; Sutin, N. *Biochim. Biophys. Acta*, **1985**, *811*, 265.
17. DeVault, D. *Quantum Mechanical Tunneling in Biological Systems*, 2nd. Ed. Cambridge Univ. Press, Cambridge, 1984.
18. Churg, A.K.; Weiss, R.M.; Warshel, A.; Takano, T.; *J. Phys. Chem.* **1983**, *87*, 1683.
19. Larson, S., *J. Chem. Soc., Faraday Trans. 2*, **1983**, *79*,. 1375.
20. Hopfield, J.J., Proc. Natl. Acad.Sci. USA, **1974**, *71*, 3640.
21. Onuchic, J.N.; Beratan, D.N.; Hopfield, J.J., *J. Phys.Chem.* **1986**, *90*, 3707.
22. Yeh, P.; Kuwana, T. *Chem. Lett.* **1977**, 1145.
23. Armstrong, F. A.; Hill, H.A.O.; Walton, N. *Acc. Chem. Res.* **1988**, *21*, 407.
24. Bowden, E.F.; Hawkridge, F.M.; Blount, H.N. Electrochemical Aspects of Bioenergetics. In *Comprehensive Treatise of Electrochemistry,* Srivanasan, S. et al, Eds.; Plenum, New York 1985, p. 297.
25. Lewis, N.S.; Wrighton, M.S., *Science*, **1981**, *211*, 944.
26. Armstrong, F.A.; Lannon, A.M. *J. Am. Chem. Soc.* **1987**, *109*, 7211
27. Bancroft, F. E.; Blount, H.N.; Hawkridge, F.M. *Adv. Chem. Ser.* **1982**, *201*, 23.
28. Bowden, E.F.; Hawkridge, F.M.; Blount, H.N. *Adv. Chem. Ser.* **1982**, *201*, 159.
29. Castner, J.F.; Hawkridge, F.M. *J. Electroanal. Chem.* **1983**, *143*, 217.
30. Taniguchi, I. *"Interfacial Electrochemistry of Promoter Modified Electrodes for Rapid Electron Transfer of Cytochrome C"* in *"Redox Chemistry and Interfacial Behavior of Biological Molecules"*, Dryhurst, G.; Niki, K. Eds.; Plenum Publ. Corp., New York, 1988; p. 113-124.
31. Assefa, H.; Bowden, E.F. Biochem. *Biophys. Res. Comm.* **1986**, *139*, 1003.
32. Cass,A.E.G.; Davis,G.;Hill,H.A.O.; Noncarrow,D.J. *Biochim. Biophys. Acta* **1985**, *828*,51.
33. Barker, P.D.; Hill, H.A.O. in *Prog. Clin. Biol.Res., 274* (Oxidases Relat.Redox.Syst.), King, T.E.; Mason, H.S.; Morrison, M. Eds., Alan R. Liss Inc., NewYork; 1988, p.419-433.
34. Thunberg, T., *Skand.Arch. Physiol.* **1925**, *46*, 339.
35. Theorell, H., *Biochem. Z.* **1935**, *78*, 263.

36. Theorell, H., in *The Enzymes* Sumner, J.B.; Myrback, K. Eds.; Academic, New York, 1951.
37. Yeh, P.; Kuwana, T. *J. Electrochem. Soc.* **1976**, *123*, 1334.
38. Cass,A.E.G.; Davis,G.; Green, M.J.; Hill, H.A.O.; *J. Electroanal. Chem. Interfacial Electrochem.* **1985**, *190*, 117.
39. Crumbliss, A.L.; Hill, H.A.O.; Page, D.J. *J. Electroanal. Chem. Interfacial Electrochem.* **1986**, *206*, 327.
40. Ikeda, T.; Hiasa, H.; Senda, M. "*Catalytic Oxidation of D-Glucose at an Enzyme-Modified Electrode with Entrapped Mediator*: in "*Redox chemistry and Interfacial Behavior of Biological Molecules*", Dryhurst, G.; Niki, K., Eds., Plenum Publ. Corp.: New York, 1988, p.193-201.
41. Kulys, J.J.; Samalius, A.S.; Svirmickas, G.J.S.; *FEBS Lett.* **1980**, *114*, 7.
42. Albery, W.J.; Bartlett, P.N.; Craston, D.H. *J. Electroanal. Chem. Interfacial Electrochem.* **1985**, *194*, 223.
43. Taniguchi, I.; Miyamoto, S.; Tomimura, S.; Hawkridge,F.M. *J. Electroanal.Chem. Interfacial Electrochem.* **1988**, *240*, 333.
44. Hoare,D.; Koshland,D.E., *J. Biol.Chem.* **1967**, *242*,2247.
45. Bartlett,P.N.; Whitaker,R.G. *J. Chem. Soc. Chem.Comm.* **1987**, 1603.
46. Koppenol, W.H.;Margoliash,E., *J.Biol.Chem.***1982**,*257*,4426.
47. Salemme,F.R., *Annu.Rev.Biochem.* **1978**, *46*, 299.
48. Hazzard,J.T.;Moench,S.J.;Erman,J.E.;Satterlee,J.D.; Tollin,G. *Biochemistry*, **1988**, *27*,2002.
49. Margoliash,E.;Bosshard,H.R., *Trends Biochem.Sci.*, **1983**,*8*,1.
50. Mochan,E. *Biochem.Biophys.Acta* **1970**, *216*,80.
51. Mochan,B.S.; Elliott,W.B.; Nicholls,P.J. *Bioenerg.* **1973**, *4*, 329.
52. Nicholls, P., *Biochim.Biophys.Acta* **1974**, *346*, 261.
53. Miller, W.G.; Cusanovich, M.A. *Biophys.Struct.Mech.* **1975**, *1*, 97.
54. Ng, S.; Smith, M.B.; Smith, H.T.; Millet,F. *Biochemistry* **1977**, *16*, 4975.
55. Koppenol, W.H.; Vroonland,C.A.J.; Braams,R. *Biochim. Biophys. Acta* **1978**, *503*, 499.
56. Poulos,T.L.; Kraut,J.; *J.Biol.Chem.* **1980**, *255*,10322.
57. Simmondsen, R.P.; Weber, P.C.; Salemme,F.R.; Tollin, G. *Biochemistry* **1982**, *21*, 6366.
58. Poulos, T.L.; Mauk, A.G. *J. Biol.Chem.* **1983**, *258*, 7369.
59. Erman, J.E.; Vitello, L.B. *J. Biol.Chem.* **1980**, *255*, 6224.
60. Mauk, M.R.; Reid, L.S.; Mauk, A.G. *Biochemistry*, **1982**, *21*, 1843.

61. Bhattacharrya, A.K.; Meyer, T.E.; Tollin, G. *Biochemistry* **1986,** *25*, 4655.
62. Hazzard, J.T.; McLendon, G.; Cusanovich, M.A.; Tollin, G. *Biochem. Biophys. Res. Commun.* **1988,** *151*, 429.
63. Kober, E.M.; Caspar, J.V.; Sullivan, B.P.; Meyer, T.J. *Inorg. Chem.* **1988,** *27*, 4587.
64. Pishko, M.V.; Katakis, I.; Lindquist,S.-E.; Ye, L.; Gregg, B.A.; Heller, A. *Angew.Chem. Internat.Ed.* **1990,** *29* , 82.
65. Oyama, N.; Anson, F.C.; *J. Am. Chem. Soc.* **1979,** *101*, 3450.
66. Katchalsky, A. *Pure Appl. Chem.* **1971,** *26*, 327.
67. Eisenberg, H. *"Biological Macromolecules and Polyelectrolytes in Solution"*, Clarendon Press: Oxford, 1976.
68. Nagata, I.; Morawetz, H.; *Macromolecules*, **1981,** *14*, 87.
69. Carnie, S.L; Christos, G.A.; Creamer, T.P. *J. Chem. Phys.* **1988,** *89*, 6484.
70. Gregg, B.A.; Heller, A. *Analyt. Chem.* **1990,** 000,000
71. Jernigan, J.D.; Surridge, N.A.; Zvanut, M.E.; Silver, M.; Murray, R. W. *J. Phys. Chem.* **1989,** *93*, 4620.
72. Pollak, A.; Blumenfeld, H.; Max, M.; Baughn, R.L; Whitesides, G.M. *J. Am. Chem. Soc.* **1980,** *102*, 6324.
73. Clark, L.D. Jr.; Lyons, C.; *Ann. NY Acad. Sci.* **1962,** *102*, 29.
74. Updike, S.J.; Hicks, G.P., *Nature*, **1967,** *214*, 986.
75. See, for example (a) Bartlett, P.N.; Whitaker, R.G. *Biosensors*, 1987/8, 3, 359; (b) Scheller, F.; Kirstein, D.; Kirstein, L.; Schubert, F.; Wollenberger, U.; Olsson, B.; Gorton, L.; Johansson, G. *Phil. Trans. R. Soc. London,* **1987,** *B316*, 85. (c) *"Chemical Sensor Technology Vol.1,* Seiyama, T. Ed., Kodansha Ltd.: Tokyo and Elsevier: Amsterdam, copublishers,**1988,** p.195-236. (d) McKean, B.D. and Gough, D. IEEE.Trans.Biomed.Eng. **1988,** *35*, 526. (e) Clark, L.C. ASAIO **1987,** *10*, 837.
76. Silverman, H.P.; Brake, J.M. U.S. Patent 3,506,544,**1970.**
77. Cass,A.E.G.; Davis,G.; Francis,G.D.; Hill,H.A.O.; Aston,W.J.; Higgins,I.J.; Plotkin,E.V.; Scott,L.D.L.; Turner,A.P.F. *Anal. Chem.* **1984,** *56*, 667.
78. Kulys,J.J. *Biosensors,* **1986,** *2*, 3.
79. Albery,W.J.; Bartlett,P.N.; Cass,A.E.G.; *Phil.Trans.Royal Soc.London,* **1987**,*B316*, 107.
80. Ikeda, T.; Hamada, H.; Miki, K.; Senda, M. *Agric. Biol. Chem.* **1985**, *49*, 541.
81. Hale, P.D.; Inagaki, T.; Karan, H.I.; Okamoto, Y.; Skotheim, T.A. *J. Am. Chem. Soc.* **1989**, *111*, 3482.

82. See, for example, Home,P.D.; Alberti, K.G.M.M. *:Biosensors in Medicine: The Clinician's Requirements"* and McCann, J. *"Exploiting Biosensors"* in *"Biosensors: Fundamentals and Applications:,* Turner, A.P.F.; Karube, I.; Wilson, G.S. Eds., Oxford Univ. Press, Oxford, 1987, p. 723-746.

AROMATICS AS ELECTRON TRANSFER SENSITIZERS

A. ALBINI and E.FASANI
Dipartimento di Chimica Organica
Università di Pavia
V le Taramelli 10
I-27100 Pavia, Italy

ABSTRACT. Aromatic molecules are popular as electron transfer sensitizers. On the other hand, they also easily form excited complexes. The interaction between the excited molecule and a quencher goes through different steps, and the observed chemistry can be referred to different species (apolar or polarized exciplex, radical ion pair, free solvated radical ions).

This is exemplified through the results obtained in various systems, viz. aromatics - dienes (cycloaddition and diene dimerization) aromatics - aromatics (4+4 cycloaddition), aromatics - alkylaromatics (benzylation via cleavage of the radical cation), aromatics - arylalkenes (alkene isomerization, cycloaddition), and aromatics σ donor (cleavage of the donor).

1. Electron Transfer Photosensitizers; a Comparison.

1.1 AROMATICS VS TRANSITION METAL COMPLEXES.

Aromatics and transition metal complexes are probably the classes of molecules most largely used as adsorbers in the field of electron transfer photochemistry, a field actively investigated in recent years, [1] and this is because they share some important characteristics, first of all convenient absorption and excited state stability.

$$Ru(bpy)_3^{3+} + A^{\overline{\cdot}} \xleftarrow{+A} Ru(bpy)_3^{2+*} \xrightarrow{+D} Ru(bpy)_3^{1+} + D^{\overset{+}{\cdot}}$$

$$Ru(bpy)_3^{2+*} \longrightarrow Ru(bpy)_3^{2+} + h\nu$$

Ruthenium (II) tris-bipyridine salts, as an example, owe their popularity first of all to the stability of the excited state (contrary to many other complexes monomolecular photodecomposition, viz ligand detachement, is slow) and efficient emission, which makes them suitable for studies of bimolecular photophysics [2,3].

E. Pelizzetti and M. Schiavello (eds.), Photochemical Conversion and Storage of Solar Energy, 89–101.

The large variation of both ionization potential and electron affinity upon electronic excitation makes both oxidation and reduction an easy process. Accordingly, the rate constants for electron transfer in either direction with a great number of donors and acceptors have been determined. These studies have a theoretical relevance and have been compared with current theory for outer-sphere electron transfer.

On the other hand, there are chemical applications, and these are reactions of the oxidized (reduced) partners produced in the original electron transfer, or in an ensuing, secondary transfer, as well as reactions of the strong oxidant $Ru(bipy)_3^{+}$ or, respectively, the strong reductant $Ru(bipy)_3^{+}$ (see the examples in Figure 1, [4,5]. These reactions have been developed for a variety of applications, including, of course, conversion of solar energy into chemical energy, e.g. by producing hydrogen from water (with the intervention of a "sacrificial" donor) [6,7].

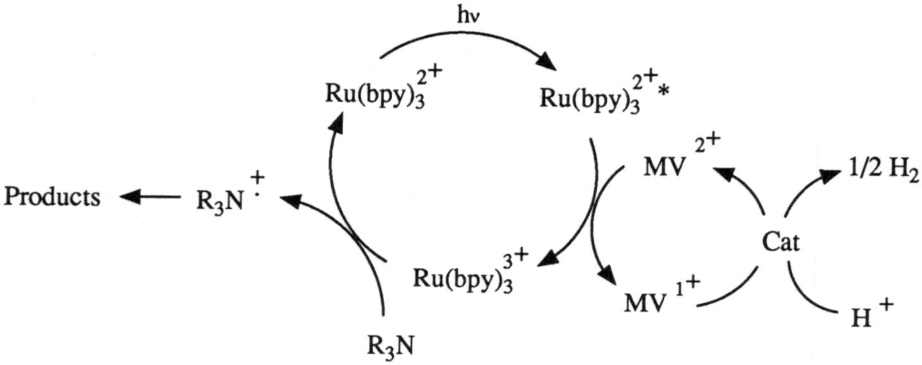

Figure 1

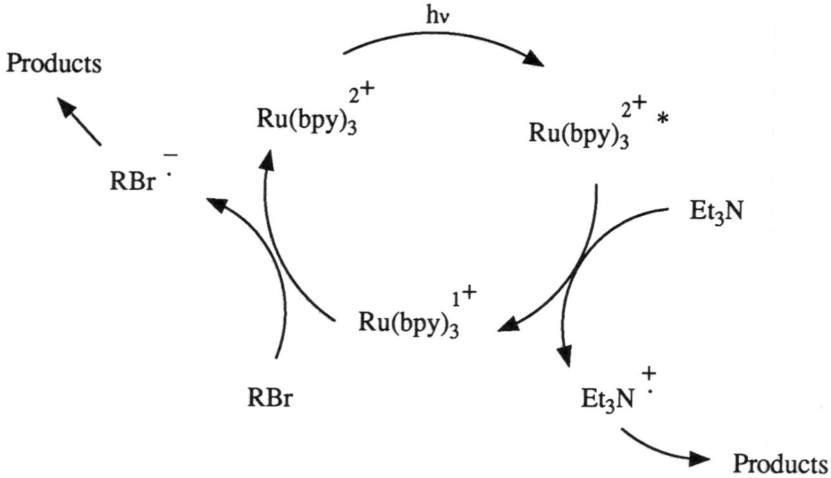

Figure 2

The nearest counterpart of such organo-metallic species in the field of organic chemistry are - as far as photochemical electron transfer is concerned - aromatic compounds. Here again the electronically excited state - or at least the singlet, to which the discussion is limited for the moment - is luminescent and has low monomolecular photoreactivity; furthermore, the redox potentials are not less favourable than in the previous case, and the reduced (radical anion) and oxidized (radical cation) forms are reasonably stable. Therefore, it is not surprising that aromatics are a favourite subject of photophysical studies and available rate constants for electron transfer in literature are again a large number.

Likewise, photochemical reactions following the primary (or a secondary) electron transfer process have been developed and are formally analogous to those discussed above [8], see examples in Figure 3, involving either reactions of the aromatic itself after electron transfer [9,10] or a mere role of for the aromatic of regenerated sensitizer (Figure 4) [11].

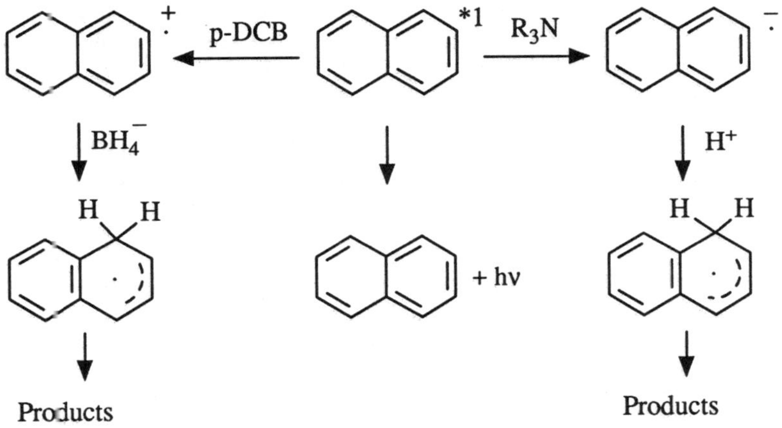

Figure 3

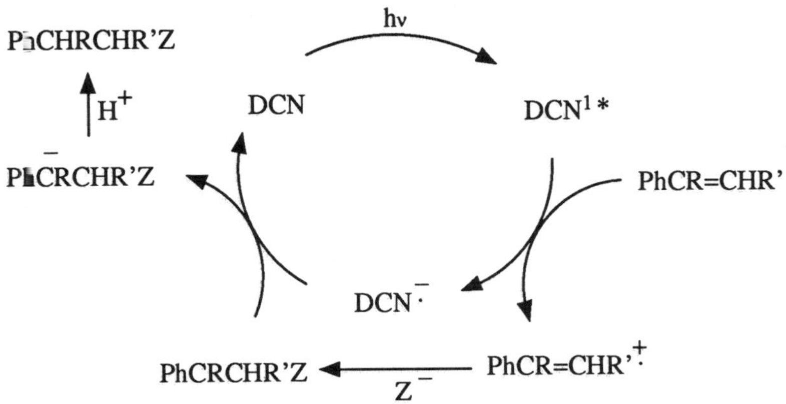

Figure 4

1.2 EXCIPLEXES VS IONS SEPARATION IN THE PHOTOCHEMISTRY OF AROMATICS.

Closer examination, however, reveals more differences than similarities, and this has to do with two questions, viz. i, how full is electron transfer, i.e. is really an electron fully transferred from one species to another, and ii, mode and importance of back electron transfer in preventing irreversible photochemical reaction after the initial electron transfer. The latter is of course a crucial condition for observing any chemistry in these systems, and indeed electron transfer has long been associated by many practitioners with quenching rather than with photochemistry, since both forward and backward electron transfer are usually fast and thus little net chemistry results.

When the two classes of compounds we are considering are compared, it must be noticed that the exchanged electron in metal complexes comes from (or goes to) the metal centre (at least in the end conformation - not necessarily in the exchanging excited state), and indeed a large fraction of recent research is devoted to the isolation of the metal ion by modifying the ligand structure (e.g. in cryptands" or in "sepulchrates"), in order to make electron transfer less easily reversible [12,13]. There are efforts in organic chemistry which many envisaged as going along the same lines, e.g. electron transfer at a fixed dinstance in a bichromophoric molecule, or in an organized medium, but the process remains qualitatively different in the two cases. In the first a more or less shaded (by the ligands) orbitals from within the approximatively spherical complex, in the latter a peripheric π orbitals of the flat aromatic molecules are involved. Therefore, it is not surprising that various types of interaction, besides plain electron transfer and ions separation are possible in the latter case.

This does not mean that electron transfer involving metal derivatives is always just the straightforward collision of two independent species - electron exchange - separation of the new species that is usually assumed to be. As an example, in the ruthenium (II) pyridine - methyl viologen (MV) system in aqueous solution (see Figure 2), a system exaustively investigated in connection with solar energy conversion, it has been recently demonstrated that the efficiency of cage escape for the geminate redox pair is actually a complex function of medium parameters, reflecting self-aggregation of MV^{2+}, aggregation of MV^{2+} and the donor, as well as ionic strength, all factors which affect charge separation [14].

However, the possibility for aromatic molecules of forming complexes with partial or no charge transfer character certainly further complicates the issue. Thus, electron transfer may take place either when the partners are at a certain dinstance (encounter complex, A*---D) or when they are at contact (collision complex, A*D). The last species evolves either towards the ions or towards an exciplex (AD)*, a more or less polarized species where the electronic energy is shared by both partners. Furthermore, also after electron transfer, π interaction as well coulombic attraction between the radical ions is strong, and this gives stability to the various species formed, differing as far as solvation is concerned, viz contact ion pair, ($A^{-\cdot}D^{+\cdot}$), solvent separated (or penetrated) ion pair, ($A^{-\cdot}/D^{+\cdot}$), free solvated radical ions [15]. All this is not merely academic propensity to dissect a phenomenon into the smallest possible parts, but rather answers a synthetic question, since these different species have each a chemistry of their own.

$$A^{*} + D \longrightarrow A^{\overset{-}{\cdot}} + D^{\overset{+}{\cdot}} \longrightarrow A + D$$

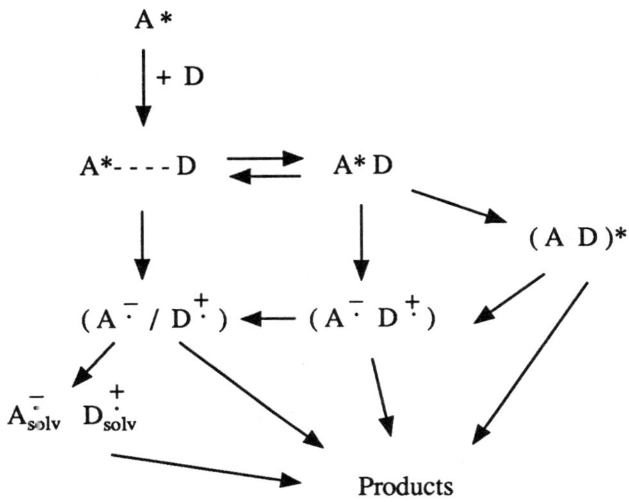

2. Electron Transfer Photochemistry of Aromatics.

Exciplexes were first recognized by their emission, viz through a physical property, and their has been some controversy, several years ago, for the recognizment of their role in photochemical reactions [16]. Likewise electron transfer has been one of the tools of photophysics before entering the field of photochemistry, where is now a common concept, with a tendency to overshadow other mechanisms. [1, 17] Nethertheless, at least when working with aromatics, it is important to take into account that both exciplexes, and thus the stabilizing interaction of π orbitals, and electron exchange are important, and even after initial electron transfer π interaction may be determining.

2.1 REACTION WITH DIENES

A first example is offered by the photoreaction between aromatics and in particular naphthalene and its derivatives, and dienes. The reaction of the parent compound involves no significant charge transfer [18]. The free energy change for electron transfer to the singlet excited state, calculated through the Weller equation, is largely positive, $\Delta G_{et} > 0.3$ eV. The chemical reaction is 4+4 cycloaddition and limiting quantum yield (at infinite quencher concentration) approaches unity.

We found that, when cyano groups are introduced on the naphthalene ring, and thus ΔG_{et} moves toward negative values, the reaction shifts from 4+4 to 2+2 cycloaddition; the limiting quantum yield is now lower, 0.1 to 0.2 and we assume that a different, more polar exciplex is involved [19]. Under the same conditions, as well as at more negative ΔG_{et} values, 4+2 diene dimerization takes place [20-23].

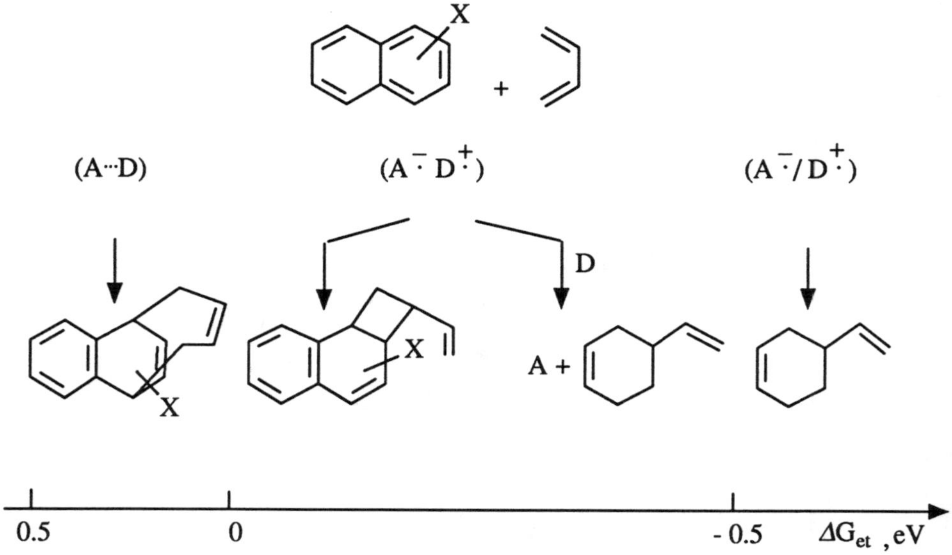

This reaction shows a complex dependence on medium parameters (solvent polarity, presence of salts, concentration of the donor) and we found that it can be explained only by admitting the partecipation of two intermediates in equilibrium. One of these is the above polar exciplex which reacts with a second molecule of the diene to yield a termolecular excited complex (triplex), in turn collapsing to the 4+2 dimer (with a high proportion of the endo isomer). The other one is the radical ion pair, which leads (almost) exclusively to the exo 4+2 dimer. Attribution of the reaction to the pair rather than to the free ions is due to the influence that changing the radical anion has on the reaction quantum yield and selectivity. At very negative ΔG_{et} values, and with triplet acceptors like chloranil the free ions limit is probably reached.

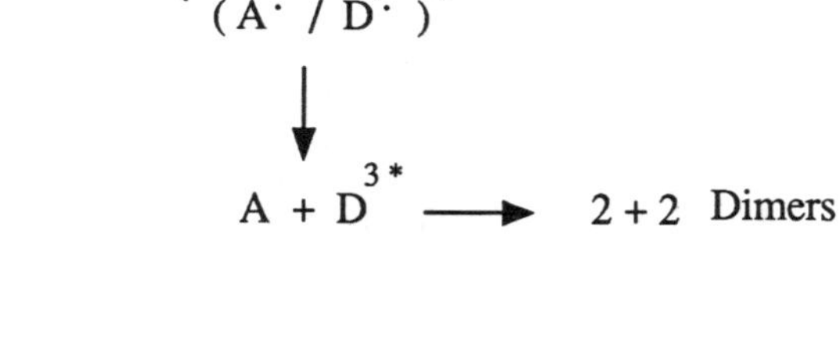

2.2. DIMERIZATION AND CROSS ADDITION OF AROMATICS.

The concept of excimer has been first formulated for the complex between two aromatic molecules, and the role of such a species in aromatic photodimerization has been discussed [24]. Charge-transfer interaction comes into play when two different aromatics and their cross cycloaddition are considered. The excited state surface encounters two minima, one corresponding to the relatively loose exciplex and the other one to the more tight pericyclic minimum leading to reaction.

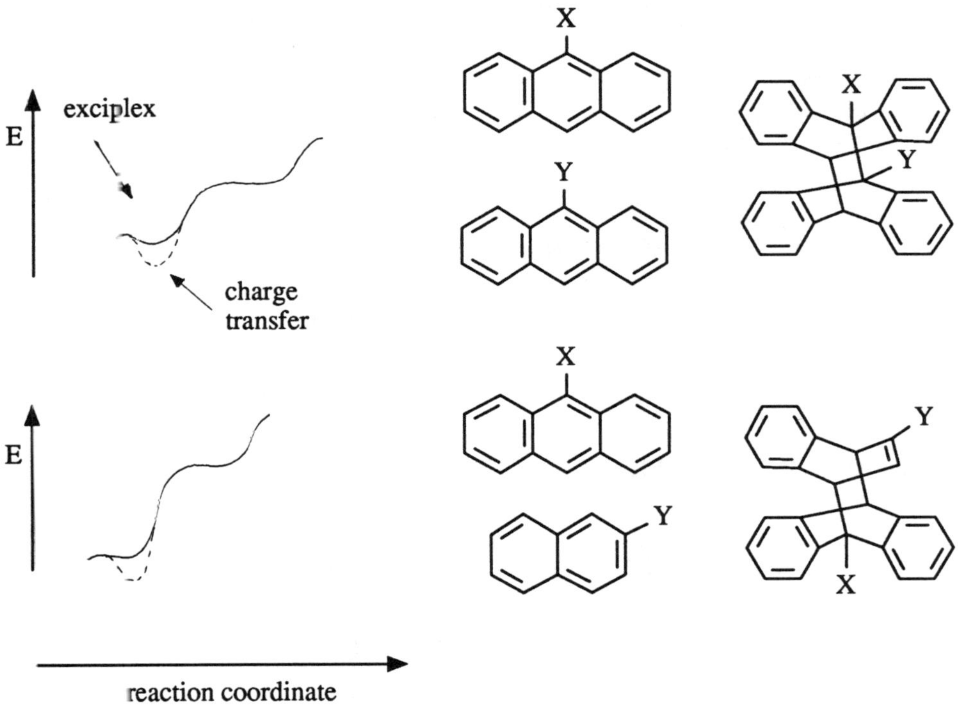

The effect of charge transfer interaction (which lowers the first minimum) is different in the case of the anthracene - anthracene addition, when there is no substantial barrier for the access to the second minimum, and in the case of the anthracene - naphthalene addition, where this barrier is sizeable [25]. Thus, e.g. charge transfer interaction favours preliminary complexation and then cross-addition both in the 9-cyanoanthracene (CNA) - anthracene [25,26] and in the CNA - 9-methoxyanthracene [27] systems. However, while cross-adducts are obtained both from anthracene and 2-cyanonaphthalene, and, with charge transfer in the other direction, from CNA and naphthalene or 2-methylnaphthalene, in the case of CNA and 2-methoxy naphthalene, with ΔG_{et} only slightly more negative, no addition but only quenching (with exciplex emission) is observed [25,26].

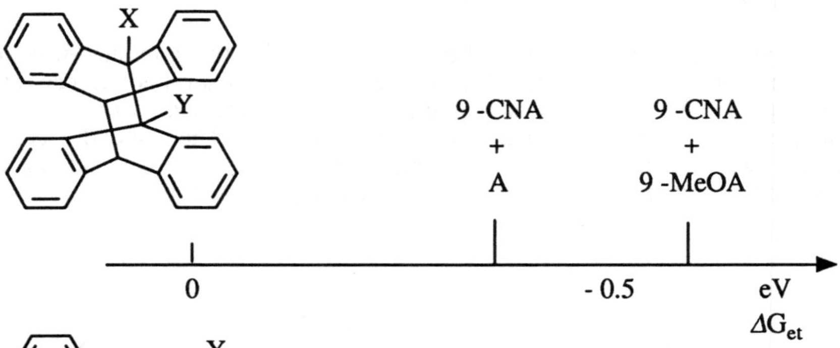

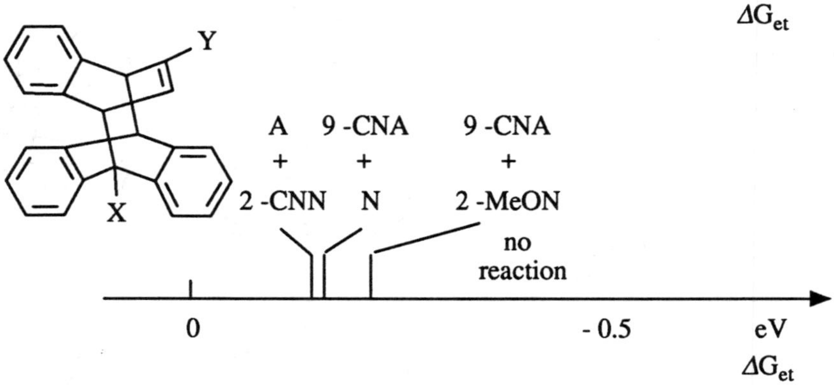

2.3 REACTION WITH ALKYLAROMATICS.

$$A^{*} + ArCHRX \longrightarrow A^{\cdot -} + ArCHRX^{\cdot +}$$

$$ArCHRX^{\cdot +} \longrightarrow \dot{Ar}CRX + H^{+}$$

$$ArCHRX^{\cdot +} \longrightarrow Ar\dot{C}HR + X^{+}$$

$$(X = CMe_3 ,\ CH_2\text{-}CH{=}CH_2 ,\ CHPh_2 ,\ SiR_3 ,\ SnR_3)$$

$$AB \longrightarrow AB^{\cdot +} \longrightarrow A^{\cdot} + B^{+}$$

$$\Delta G_{AB^{\cdot +}} = \Delta G_{AB} + \Delta G_{A\cdot/A^{+}} - \Delta G_{AB/AB^{+\cdot}}$$

$$= \Delta G_{AB} + E_{1/2}(A^{\cdot}/A^{+}) - E_{1/2}(AB/AB^{\cdot +})$$

In the examples discussed above, the chemical reaction competing with back electron transfer is an addition process, either between the two partners in the ion pair or involving a third molecule. Another important process is fragmentation, e.g. ß-fragmentation of benzylic radical cations to yield a benzyl radical and a cation. The thermodynamic feasibility of such reaction can be evaluated if the oxidation potential of the radical is known (or conversily the last parameter can be evaluated from the occurring of the photosensitized cleavage) [28-30].

Indeed, fragmentation may be an efficient process, and both deprotonation (benzylic radical cation are strong acids) and carbon-carbon or carbon-heteroatom bond cleavage have been observed. The reaction occurs either as an independent process of the donor, with the regeneration of the acceptor (e.g. Fig. 5, [30]) or via acceptor-donor addition.

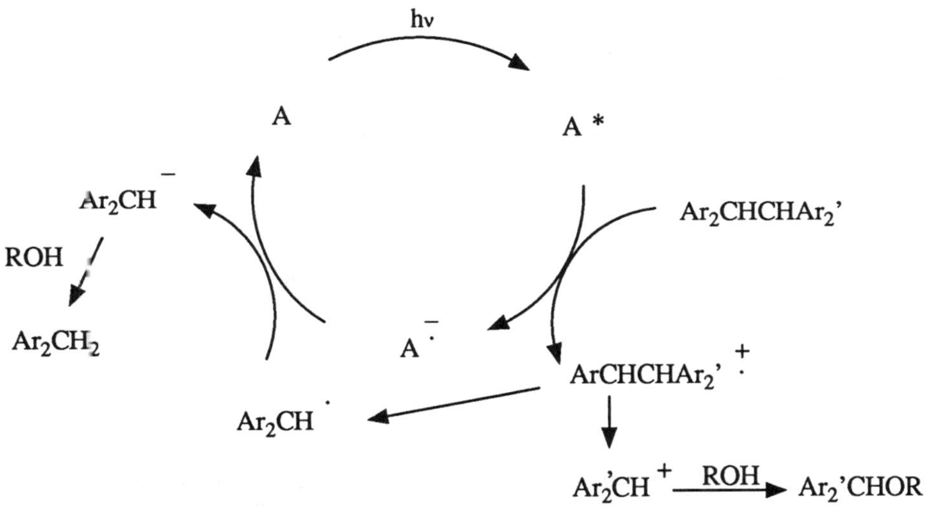

Figure 5

An example of the latter type is the reaction between 1,4-dicyanonaphthalene (DCN) and alkylbenzenes; here proton transfer from the radical cation to the radical anion within the pair is followed by stereoselective addition of the neutral radicals to yield product **1** , and even where the proton is lost to the medium, radical - radical anion combination and reprotonation yield, again stereoselectively, product **2** [31,33]. The difference between reaction via radical ion pair and via free radical ions is traced also in this case: the former intermediate gives a selective reaction with a relatively high quantum yield (up to 0.3), the latter one a non stereoselective, low quantum yield (≪0.1) reaction (cyano group substitution, e.g. to form product **3**, may be observed in this case) [34].

With acceptors different from DCN there is some change, but reaction via free ions remains at low quantum yield [35].

Thus, a species which can be considered a very polar exciplex or a contact ion pair is the intermediate of various reactions, often characterized by moderate to good quantum yield and high (stereo) selectivity.

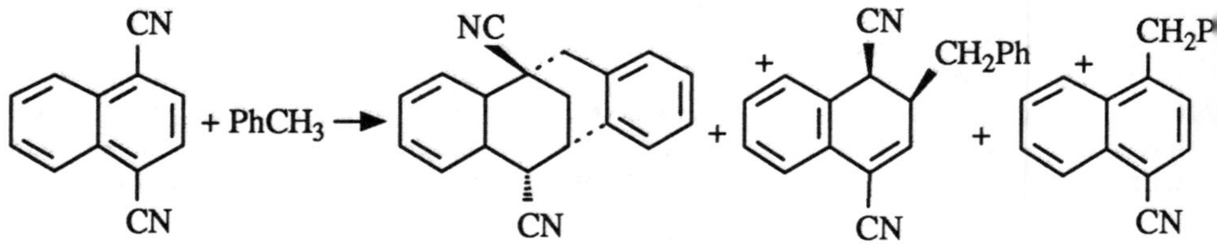

1 2 3

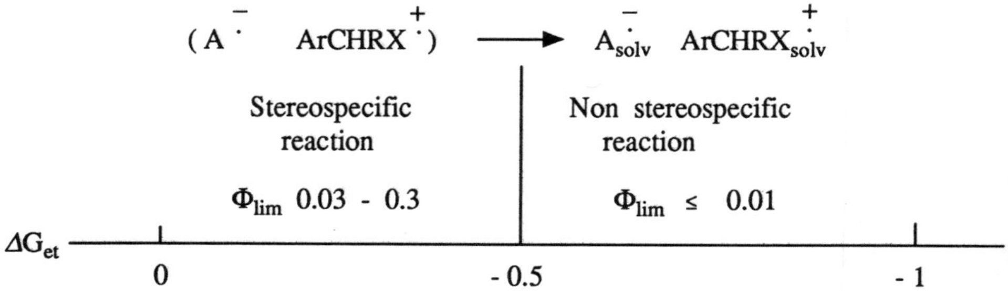

2.4 IONS SEPARATION.

In the previous scheme, separations into free ions is a competing reaction. The quantum yield for the last process has been determined by Gould or Farid for a series of aromatics and found to change over a large range. Assuming that the rate for ion separation, k_s, remains the same in the series, this implies that the rate for back electron transfer, k_{bet}, changes over two order of magnitudes, and correlation with the relative ΔG values evidences an inverse dependence (Marcus inverted region) [36,37].

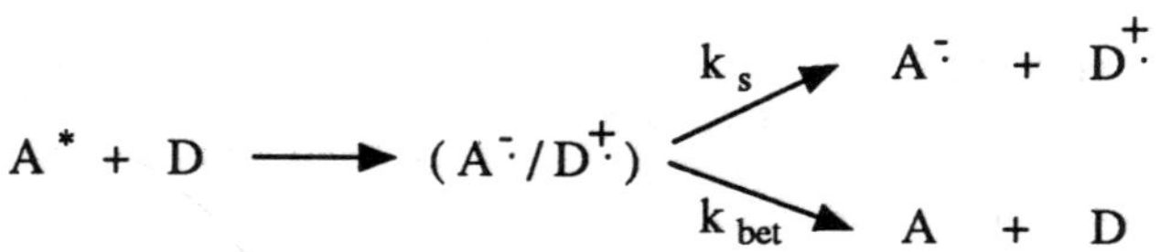

A	D	Φ_{sep}	$-\Delta G_{bet}$	k_{bet}
DCA	Biphenyl	0.83	2.90 eV	$1\times10^8\ s^{-1}$
DCA	Naphthalene	0.57	2.74	3.76×10^8
TCA	Biphenyl	0.24	2.40	1.53×10^9
TCA	Naphthalene	0.125	2.24	3.5×10^9
DCA	Durene	0.041	2.22	1.58×10^{10}

Apart from the theoretical interest, these data explain the function of compounds such as diphenyl as "consensitizers", viz as additives enhancing the quantum yield in electron transfer photosensitized reactions, e.g. oxigenation [38]. Despite the fact that forward electron transfer to diphenyl is not predicted to be fast, the favourable yield of cage escape probably is the winning factor at least for reactions via free radical ions.

Reactions via bona fide free ions typically occur either in the presence of a mechanism for molecular distortion resulting in an important stabilization of one of the radical ions, or when a fast fragmentation is available for a radical ion lacking a chromophore fit for a stabilizing interaction with the counter ion. Under such condition radical ion chain processes may also come in, and lead to overall quantum yield in excess to unity. Examples are the "one way" (Z to E) isomerization of stilbenes with an efficiency for the Z radical cation of 2.4 (the mechanism is not exactly as initially postulated, though, since there is no unimolecular isomerization of the Z-stilbene radical cation [39,40]) as well as reactions of σ donors. In the latter cathegory a $\phi \sim 80$ has been measured for the naphthalene-sensitized isomerization of hexamethyl(Dewar benzene) to hexamethylbenzene [41].

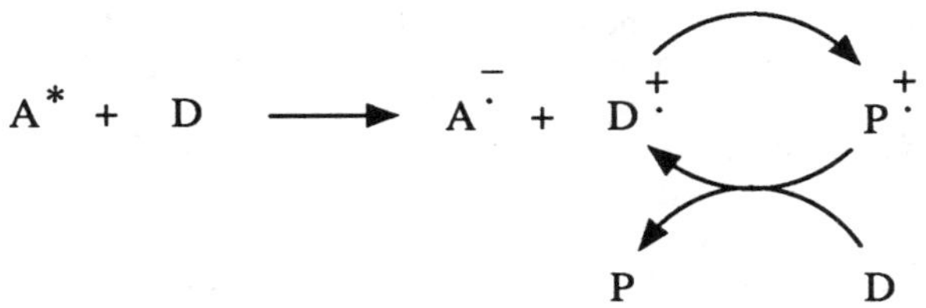

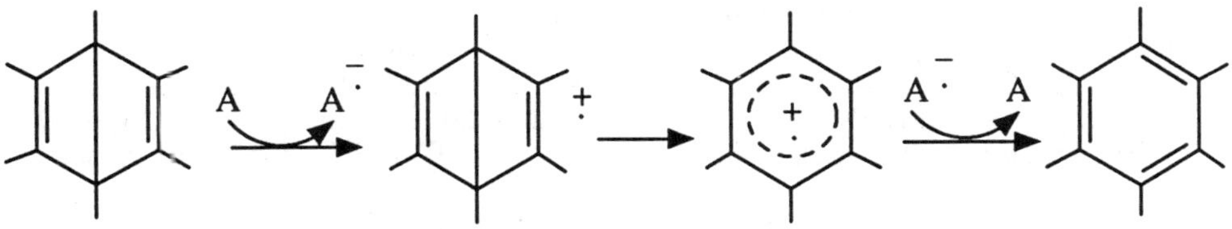

3. Conclusions.

Two conclusions can be obtained at the present stage. On one hand in view of their convenient redox potential, absorption spectrum (extending, if desired, to a part of the visible), stability of the radical ions, high yield (under suitable conditions) of ions separation, possibility of having recourse to secondary transfer for minimizing back electron transfer, aromatics are certainly suitable sensitizers for reactions via (free) radical ions, in some case with high quantum yield owing to the intervention of a chain mechanism. On the other hand, much has to be gained from the opposite situation where the radical ions do not diffuse apart and react in cage, usually with high regio- and stereo-selectivity and satisfactory, albeit of course lower than one, quantum yield to new, high energy products.

Aknowledgement. CNR (Rome) is thanked for continuous support of our research.

4. References.

1 Fox, M.A. and Chanon, M. (eds.) (1988), Photoinduced Electron Transfer, Elsevier, Amsterdam.
2 Juris, A., Balzani, V., Barigelletti, F., Campagna, S., Belser, P., von Zelewsky, A. (1988), Coord.Chem.Rev., 84, 85.
3 Serpone, N. (1988), "Photoinduced electron transfer in hexacoordinate inorganic complexes", in ref. 1, Vol. D, p. 47.
4 Kalyanasundaram, K. (1982) Coord.Chem.Rev., 46, 159.
5 Willner, I., Tsfania, T., and Eichen, Y. (1990), J.Org.Chem., 55, 2656,
6 Harriman, A. and West, M.A. (eds.) (1982), Photogeneration of Hydrogen, Academic Press, London.
7 Gratzel, M. (ed.) (1983) Energy Resources through Photochemistry and Catalysis, Academic Press, New York.
8 Albini, A. and Sulpizio, A. (1988), in ref. 1, Vol. C, p. 88.
9 Barltrop, J,A, (1972), Pure Appl.Chem., 32, 179.
10 Yasuda, M., Pac, C., Sakurai, H. (1981) J.Org.Chem. 46, 788.
11 Maroulis, A.J., Shigemitsu, Y. and Arnold, D.R. (1978) J.Am.Chem.Soc. 100, 535.
12 Balzani V., Sabbatini, N. and Scandola, F. (1986), Chem.Rev., 86, 319.
13 Balzani, V., and Scandola, F. (1988), in ref. 1, Vol. D, p. 148.
14 Hoffman, M.Z. (1988), J.Phys.Chem., 92, 3458.
15 Kavarnos, G.J. and Turro, N.J. (1986), Chem.Rev., 86, 401.
16 Caldwell, R.A. and Creed, D. (1980), Acc.Chem.Res., 13, 45.
17 Mattay, J. (1987), Angew.Chem.Int.Ed.Engl.,26,825
18 Yang, N.C., Libman, J. and Savitzky, M. (1972), J.Am.Chem.Soc., 94, 9226 and 9228.
19 Albini, A., Fasani, E., and Giavarini, F. (1988), J.Org.Chem., 53, 5601.
20 Unpublished date from the authors laboratory.
21 Mattay, J., Trampe, G. and Runsink, J, (1988) Chem.Ber., 121, 1991.
22 Calhoun, G.G., Schuster, G.B. (1984), J.Am.Chem.Soc., 106, 6870.
23 Chung, W.S., Turro, N.J., Mertes, J. and Mattay J. (1989) J.Org.Chem., 54, 4881.
24 Bouas-Laurent, H., Castellan, A., Desvergne, J.P. (1980) Pure Appl.Chem., 52, 2633.
25 Albini, A. and Fasani, E. (1988) J.Am.Chem.Soc., 110, 7760,
26 Albini, A., Fasani, E. and Faiardi, D. (1987) J.Org.Chem., 52, 155.
27 Castellan, A., Lapouyade, R., Bouas Laurent, H. and Lallemand, J. (1975), Tetrahedron Lett., 2467.
28 Wayner, D.D.M., McPhee, D.J. and Griller, D. (1986) Chem.Phys.Let., 131, 189.
29 Wayner, D.D.M., McPhee, D.J. and Griller, D. (1988) J.Am.Chem.Soc., 110, 132.
30 Arnold, D.R. and Popielarz, R. (1990) J.Am.Chem.Soc., 112, 3068.
31 Albini, A., Fasani, E. and Sulpizio, A. (1984) J.Am.Chem.Soc., 106, 3562.
32 Sulpizio, A., Albini, A., d'Alessandro, N., Fasani, E. and Pietra, S. (1989) J.Am.Chem.Soc. 111, 5773.
33 Albini, A., Fasani, E., and Mella, M. (1986) J.Am.Chem.Soc., 108, 4119.
34 Albini, A., Fasani, E., and Montessoro, E. (1984) Z. Naturforsh., 39b, 1409.
35 Lewis, F.D. and Petisce, J.R. (1986) Tetrahedron, 42, 4207.
36 Gould, I.R., Ege, D., Mattes, S.L. and Farid, S. (1987), J.Am.Chem.Soc., 109, 3794.
37 Gould, I.R., Moser. J.E., Armitage, B. and Farid, S. (1989), J.Am.Chem.Soc., III, 1917.
38 Schaap, A.P., Lopez, L. and Gagnon, S.D. (1983) J.Am.Chem.Soc., 105, 663.
39 Lewis, F.D., Dykstra, R.E., Gould, I.R. and Farid, S. (1988) J.Phys.Chem., 92, 7042.

40 Lewis, F.D., Petisce, J.R., Oxman. J.D. and Nepras, M.J. (1985), J.Am.Chem.Soc., 107, 203.
41 Evans, T.R., Wake, R.W. and Sifain, M.M. (1973) Tetrahedron Lett., 701.

Polyelectrolytes: Assessment of their Possible Role in Future Solar Energy Storage

Joseph Rabani
Department of Physical Chemistry and Energy Research Center
The Hebrew University of Jerusalem
Jerusalem 91904, Israel

Abstract

Polyelectrolyte systems have been extensively investigated in recent years, with respect to their potential role in photochemical conversion and storage systems. The inhibition of reactions between oppositely charged ions and the slowness of back reactions between species linked to different polymer molecules, makes them useful for investigations of homogeneous solution photochemical storage. Many polyelectrolytes adsorb to electrodes and to colloids, and modify their surface properties. This feature makes these polymers useful for photoelectrochemical investigations.

In the present manuscript we discuss the properties of polyelectrolytes in relation to photochemical energy storage, in the light of recent developments in the field. This includes pulse radiolysis investigations of model systems, colloid stabilized by polyelectrolytes, electrode mono and bilayers, as well as the role of hydrophobic interactions in the photochemical polyelectrolyte systems. Copolymers with controlled fractions of hydrophobic and electrostatic segments demonstrate the possibility to design polyelectrolytes with specific properties.

Introduction

Polyelectrolytes have been extensively investigated in recent years, with emphasis on their involvement in photochemical conversion and storage systems[1]. These charged polymers have the advantage of possessing well defined chemical structures. They can be engineered to have required charge densities, hydrophobic moieties, as well as being covalently linked to specific segments which can act as electron donors, acceptors, and photosensitizers. They can be designed to posses given molecular weights (and mobilities), as well as be combined with catalytically active materials. The ability of some polyelectrolytes to adsorb to semiconductor colloids and electrodes makes them a potential tool in achieving photoelectrochemical energy storage. Crosslinked polymers, with appropriate redox and photo sensitive segments have potential value in energy fixation, because of the possibility to design rigid systems of this type. Examples for recently used

E. Pelizzetti and M. Schiavello (eds.), Photochemical Conversion and Storage of Solar Energy, 103–119.

polyelectrolytes are given below.

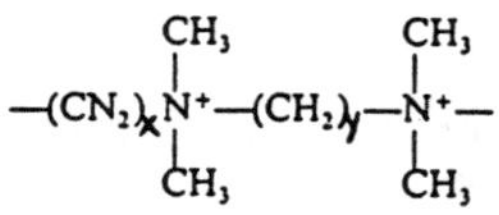

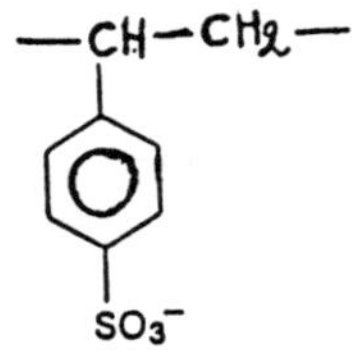

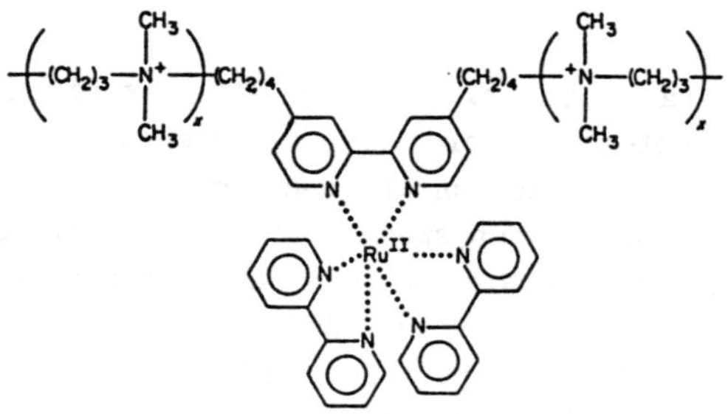

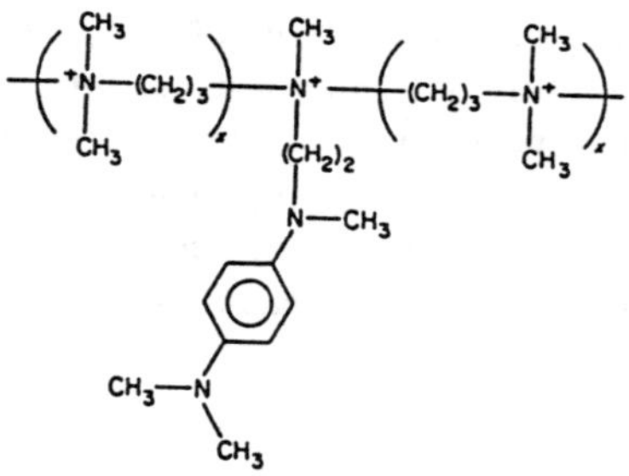

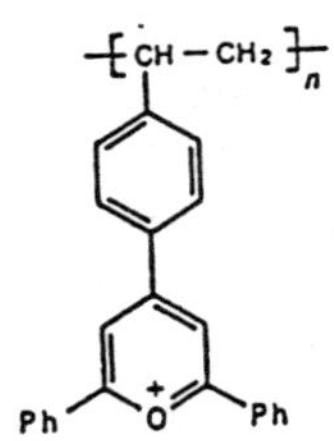

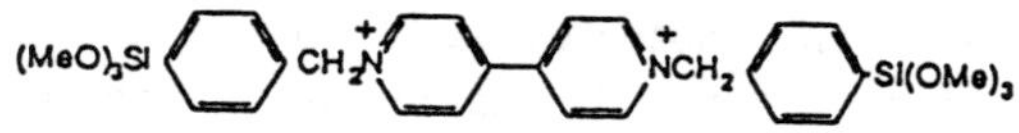

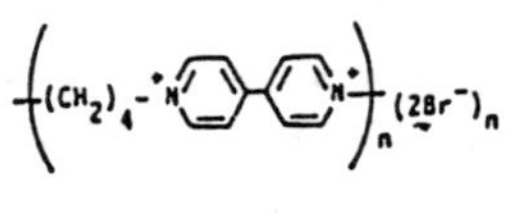

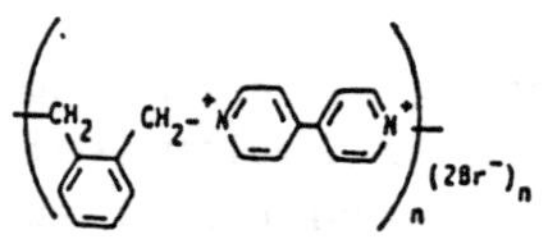

The large variety of potential polyelectrolyte applications calls for periodic assessments of the polyelectrolyte assisted energy storage, in connection with the states of the various related fields. It is well known that often there is a large gap between methods and systems which function "in principle", and actual experimental results. It is the purpose of this manuscript to discuss both principles and experimental results which may suggest directions for future research. Selected examples which will be described in the following serve the purpose of illustration of some of the recent achievements. These are not intended to compose a comprehensive review.

Enhancement of photochemical reactions by polyelectrolytes.

The use of polyelectrolytes for the enhancement of photochemical electron transfer reactions is known for many years. This feature of polyelectrolytes results from the ability of polyelectrolytes to concentrate counter-ions in a small volume near the polymer molecules. Thus, a photochemical reaction between a photosensitizer, S, and a quencher, Q, is enhanced if both reactants are counter-ions of the same polymer. A general reaction scheme of photochemical conversion in the presence of polyelectrolytes is shown below (reactions 1-7).

$S + M \Longleftrightarrow S_P$	Distribution of S (bulk and polymer)	(1)
$Q + M \Longleftrightarrow Q_P$	Distribution of Q (bulk and polymer)	(2)
$S \longrightarrow S^*$	Excitation in the bulk	(3)
$S_P \longrightarrow S_P^*$	Excitation in the micro-volume	(3')
$S^* + Q \longrightarrow S^+ + Q^-$ or $S^- + Q^+$	Bulk electron transfer	(4)
$S_P^* + Q_P^* \longrightarrow S_P^+ + Q_P^-$ or $S_P^- + Q_P^+$	Electron transfer (micro-volume)	(5)
$S^+ + Q^- \longrightarrow S + Q$; $S^- + Q^+ \longrightarrow S + Q$	Bulk back reactions	(6)
$S_P^+ + Q_P^- \longrightarrow S_P + Q_P$; $S_P^- + Q_P^+ \longrightarrow S_P + Q_P$	Micro volume back reactions	(7)

In the above scheme, M stands for an available site on the polymer for the condensation of S and Q. The subscript $_P$ denotes species in the polymer domain. Enhancements by 3-4 orders of magnitude result from concentration of the counterions in a small volume near the polymer molecules[1]. However, very efficient photochemical electron transfer reactions can be obtained without the use of polyelectrolytes, so that the ability to concentrate counter ions may become useful only in relatively rare cases.

The interaction of a photosensitizer with a polymer may increase the stability of the excited state[2]. This was observed in several cases, including the popular photosensitizers $Ru(bpy)_3^{2+}$ and porphyrins. In the presence of aromatic segments in the polymer molecule, it has been attributed to a charge transfer complex with the photosensitizer. The lifetimes of the excited states are usually only slightly increased, although in the case of the water soluble Zn-tetra phenyl porphyrin sulfonate the triplet lifetime increases by several folds upon addition of a pyridinium based polymer[4b]. No example is known where this has been utilized to assist with persisting energy storage.

Inhibition of the photochemical back reactions

The back reactions between the photochemical redox products in homogeneous solutions, result with quick regeneration of the original reactants, accompanied by total loss of the photochemical energy (reactions 6, 7). Such back reactions are always thermodynamically feasible because energy storage must involve the formation of products with a positive free energy change. Thus, the storage of energy must be achieved by providing the system with some sort of kinetic stability. The role of polyelectrolytes in such systems may be the spatial separation between the photochemical products. This can be achieved by the choice of photochemical redox systems where the electron transfer products have opposite charges. The attraction of counter-ions by the polymer, combined with the repulsion of the ions which have the same charge as the polymer is expected to slow the rates of reactions between the photochemical electron transfer products in the presence of polyelectrolytes. Several systems where the primary redox products are oppositely charged ions have been previously reported[1], but none of them shows a pronounced effect in slowing the back reactions. The photochemical systems of this type show retardation of the back reactions by up to one order of magnitude. This is considerably less than the retardation effects (2-3 orders of magnitude) known for some simple inorganic reactions between oppositely charged ions. The reason for the lack of a larger inhibition effect on the photochemical back reactions may be specific to the systems used. These involved either counter ions of the polyelectrolytes which may weaken the polymer electric field at the site of the back reactions, or reactants such as viologens, which are attracted to the polymer by hydrophobic forces. The hydrophobic interactions where suggested to diminish the retardation effect of the electric field[1].

A much more pronounced retardation effect was reported for systems where secondary products linked to different polymer molecules where produced[3]. As an example we'll discuss the $Ru(bpy)_3^{2+}$ - tetramethyl-phenylenediamine (TMPD) system[3a]. Each of the above molecules was covalently linked to a different polyelectrolyte molecule. The polymer pending ruthenium complex serves as an electron acceptor photosensitizer, while the pending TMPD served as an electron donor. Since each of the reactants is linked to a different polymer molecule, there is no direct quenching of the excited photosensitizer by the TMPD. This is due to both the electrostatic repulsion between the two charged polymers, and the slowness of their diffusion. A mediating compound, with redox potential sufficiently low to be able to transfer an electron to the Ru complex, but sufficiently positive to be able to oxidize the TMPD residue afterwards, had to be used. Indeed, preliminary work with this system indicates that energy storage could be achieved, persisting for many seconds, due to the slowness of the back reaction between the polymer pending products, $Ru(bpy)_3^+$ and $TMPD^+$. However, this particular system is not convenient for the purpose of investigations, because of the air sensitivity of the TMPD. In addition, traces of air may catalyze the back reactions, and the system as a whole is not photochemically stable with respect to side reactions. The synthesis of the polymer linked segments is also laborious, as is often the case with this type of systems. However, the principle involved in the above method of retardation of photochemical back reactions has been demonstrated and temporary energy storage for hours or even days, using this method is feasible. An investigation of a system based on the principles described above, in which a three dimensional crosslinked polymer is involved is in progress.

Systems where the photosensitizer and quencher are linked to polymer molecules by means of electrostatic and hydrophobic bindings have not achieved pronounced retardations of the back reactions. Inhibition of back reactions by 1-3 orders of magnitude have been reported[4]. This approach, if successful, could eliminate a great part of the laborious synthesis which is required for covalently linking the active segments to the respective polymer molecules. It is important to understand the reasons for the relatively faster back electron transfer reactions compared to the covalently linked systems. In the latter cases, the photosensitizer and the quencher are linked to different polymer molecules, without being able to change their micro-domain. Reaction between the polymer linked photochemical products involves a close approach of the two polymers, resulting with strong repulsive forces. This is not exactly the case when the reactive species are bound to the polymers by electrostatic or hydrophobic bindings. In the latter cases, electrostatic and hydrophobic crosslinking may occur. The micro-domains for the encapsulation of product molecules are no longer confined to a single polymer molecule, and the retardation of the back reactions becomes less effective because both reactants may exist within the same microvolume. Note that electrostatic crosslinking may be expected if the photochemical active species are multicharged ions. Singly charged ions are unlikely to induce electrostatic crosslinking, but are not suitable for energy storage in polyelectrolyte systems because back reactions via exchange with ions in the bulk are possible since singly charged counter ions are not strictly confined to a small volume near the polymer.

Enlargement of the polymer microvolume by electrostatic and hydrophobic crosslinking was reported for polystyrene sulfonate upon addition of ruthenium or iridium bipyridine compounds[5]. Enhanced bimolecular bipyridine radical reactions take place even in the presence of relatively high concentrations of polystyrene sulfonate. Under the conditions used, there was less than one reacting radical per polymer molecule, and reaction should not have taken place if the micro-volume involved only one polymer molecule. It may be possible to construct a system where two different polymers will be involved, one with hydrophobic segments and the other with covalent or electrostatic binding to a suitable molecule. In such a system, crosslinking will take place only between similar polymer molecules, without decreasing the retardation effect which is expected in a two polymer system of that type. No report on a successful system of that kind has so far been published.

Pulse radiolytic simulations.

Energy storage models can be effectively studied by pulse radiolysis. Recently, we have reported[6] the investigation of polystyrene sulfonate. This water soluble, negatively charged polymer is available commercially at various molecular weights. When an air free, nitrous oxide saturated aqueous solution is irradiated, most of the primary free radicals which are produced radiolytically are quickly converted to OH radicals. The latter react with the polymer predominantly by addition to the aromatic ring of a sulfonatophenyl segment. At near neutral pH, the OH radical adduct which is produced looses a water molecule and is converted to a polymer linked positive radical ion which is stable for several hours at room temperatures (see Fig. 1). The stability of this energy storage system is attributed to several factors. Firstly, a radical radical reaction is retarded by the repulsion between the two negative polymer molecules. In addition, the molecules of the polymer used, with an average molecular weight 70000, diffuse relatively slowly. Finally, there may be a specific effect connected with the particular radical involved here.

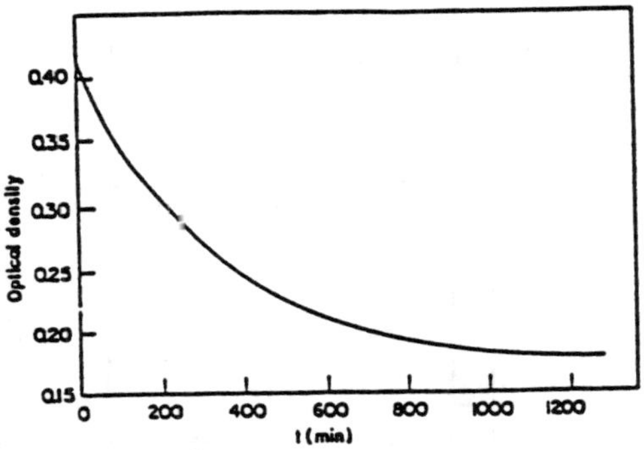

Figure 1. A long lived polymer radical: the decay of the polystyrene-sulfonate cation radical (PSS^+). Initial radical conc.: $2.4x10^{-5}$ M. [PSS] = 0.01 M (in monomer units). From Ref. 6.

For example, it is well known that some radical ions such as the alkoxy benzene radicals posses considerable kinetic stability[7]. The pulse radiolytic work described above shows that the time limit for energy storage in polymer systems may be much longer than that observed with the presently known photochemical systems.

Synthesis by irradiation.

Laborious synthesis is a common obstacle to construction of appropriate photochemical systems involving photosensitizers and quenchers linked to polymers. We have developed a method for linking functional groups to polymers with the aid of ionizing radiation [5], based on radical radical reactions. The reactants are dissolved in water and irradiated under air free nitrous oxide saturated conditions. As discussed above, such conditions favor the conversion of the radiation primary radicals into OH radicals. If, for example, $Ru(bpy)_3^{2+}$ is present, OH radicals react with it predominantly by OH addition to the ligand:

$$OH + Ru(bpy)_3^{2+} \text{ ----> } Ru(bpy)_3^{2+}\text{--}OH \quad (8)$$

The OH adduct decays by radical radical reactions which in this specific case are relatively slow (note that a mixture of radicals is produced. under typical pulse radiolytic conditions, at an initial radical concentration $1x10^{-5}$ M, the lifetimes range from 0.1 to 100 msecs). When a polymer is present, in addition to the Ru complex, the OH radicals are shared between the Ru complex and the polymer. Reaction of OH with a polymer produces a polymeric radical, usually either by abstraction of hydrogen from an aliphatic segment, or by addition to an aromatic ring. The polymer radical then reacts with the

$Ru(bpy)_3^{2+}$--OH adduct, and a covalent link between the Ru complex and the polymer is obtained. The decay of the Ru-bipyridine radicals in the absence and in the presence of a positive polymer radical is shown in Fig. 2a and 2b respectively. Although this method may be useful for practical synthesis, it is not very convenient for research purposes, because a mixture of free radicals and radiation products is usually produced. The method has been successfully applied also to an iridium complex[5].

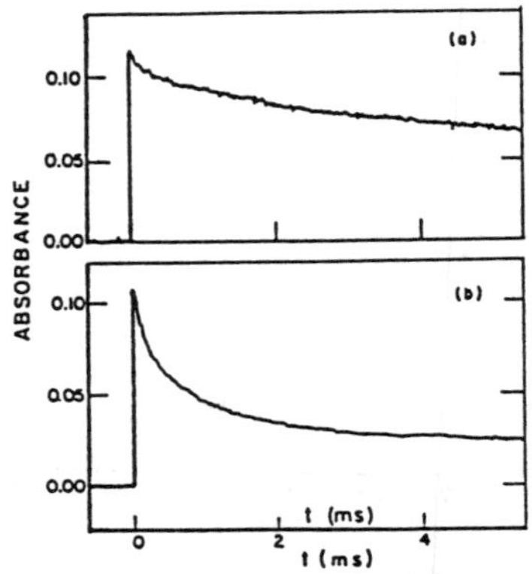

Figure 2. Kinetic traces of the decay of the $Ru(bpy)_3^{2+}$–OH adduct. Monitored at 750 nm with N_2O-saturated solutions of 3×10^{-4} M $Ru(bpy)_3^{2+}$ containing no polymer, 5.3 krd (a), and 1.3×10^{-2} M polybrene, 35 krd (b). Ref. 5a.

Two oppositely charged polymers

When two oppositely charged polyelectrolytes are mixed together, in most cases precipitation occurs. However, when one of the polymers is in excess, or when equivalent concentrations of the two polyelectrolytes are relatively low, the system remains soluble. Under such conditions, each of the polymers retains in its vicinity part of the electric field, even when the other polymer is in excess[8]. Thus, the positively charged water soluble porphyrin, $ZnTMPyP^{4+}$ strongly interacts with the negative polystyrene sulfonate polymer. As a result of this interaction, the initial yield of the porphyrin triplet is decreased by two and a half folds, and the lifetime of the triplet is shortened by about two orders of magnitude. This effect has

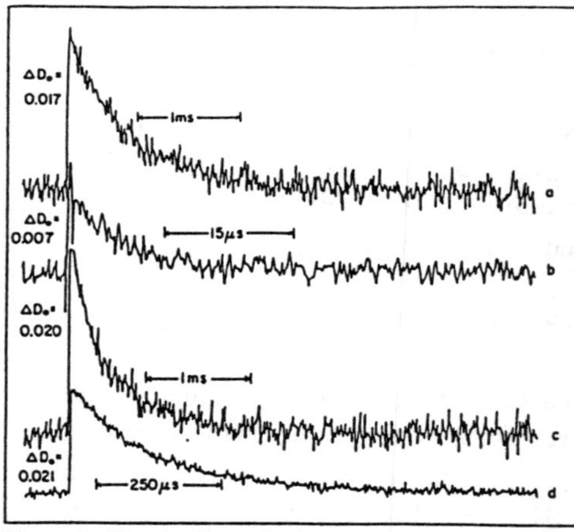

Figure 3. Kinetic effects of polyelectrolytes on the quenching of $^3ZnTMPyP^{4+}$ by Cu^{2+} ions. Laser pulses taken near 435 nm, slightly varied in order to remain at the peak of the Soret band. Detection at 475 nm; 1×10^{-6} M $ZnTMPyP^{4+}$: (a) no additives; (b) 1.50×10^{-4} M PSS added; (c) 1.5×10^{-4} M PSS and 1.9×10^{-4} M PB added; (d) 1.5×10^{-4} M PSS, 1.9×10^{-4} M PB, and 5×10^{-6} M Cu^{2+} added. Ref. 8.

been attributed to triplet triplet annihilation and possible dimer formation because of the concentration effect of the polymer. Addition of excess of the positive polymer "Polybrene" (commercial name), the repeating unit of which includes quaternary ammonium, restores the lifetime and triplet yield which are typical of dilute aqueous solutions. However, the porphyrin is not exchanged under these conditions with the PB, and is not in the bulk of the solution as might be expected. This is obvious from the fact that quenching by Cu^{2+} ions is enhanced by a factor of 30 in the two polymer system, relative to the respective bulk reaction. The latter result shows that both the porphyrin and the copper ions are concentrated by the negatively charged polystyrenesulfonate, despite of the presence of excess polybrene charges. These kinetic effects are demonstrated in Fig. 3.

The oppositely charged polymer systems may open new interesting possibilities for energy storage. For example, it may be possible to use metal complexes with appropriate stability constants so that two accessible oxidation states have opposite ionic charges (e.g. by losing and regaining charged ligands, respectively, in the different oxidation states). Under such conditions, the two oxidation states will become compartmentalyzed in different regions of the two polymer assembly.

Modification of electrodes

Polymer modified electrodes form a fairly new field in which electrochemistry is combined with polymer chemistry. It forms a basis for new approaches in photoelectrochemistry, as well as electrocatalysis and electronics. The technique involves adsorption of a polymer on the electrode surface. The polymer concentrates photoelectrochemically active solutes by its ionic field and by hydrophobic interactions. Conditions may be engineered for photoredox reactions as well as for energy storage. A comprehensive review has been recently published[9].

In order to utilize visible light, narrow band gap semiconductors must be used. The narrow band gap n-type semiconductors are sensitive to corrosion under light illumination. Coating of the electrode with a polymer may be used in some cases for protection against water corrosion, but a more promising use is the application of a polymer layer in order to induce hole transport from the semiconductor surface to the water. This approach has been very recently successfully used by Kaneko, Yao, and Kira[10] in order to achieve sustained water splitting using a CdS electrode modified by the negative polymer Nafion. Nafion is a fluorocarbonated polymer, containing also the ionic sulfonato groups. Because of its highly hydrophobic nature, it can be easily adsorbed to many water insoluble solid materials. When adsorbed to a CdS electrode, it produces a local hydrophobic-electrostatic domain in the vicinity of the electrode. This enables the achievement of high local concentrations of compounds, both ionic and non ionic, at the electrode surface. In the work described above, Ru-red (tri nuclear Ru complex with ammonia as a ligand and oxygen bridges, six positive charges) has been used as an oxidation catalyst. The Ru-red is adsorbed in the Nafion layer. When the CdS electrode is illuminated, the Ru-red accepts corrosive positive holes formed on the CdS surface and leads to water oxidation.

Bilayers.

Bilayers were found to be effective in energy storage[9] via charge trapping. A typical bilayer storage system involves an electrode coated with two different, overlaid films[11]. The redox potentials of the polymers are chosen in relation to the microstructure of the assembly. If the reaction involves reduction of the polymers, and the arrangement is such that polymer A is in contact with the electrode on one side, and with polymer B on its other side, trapping of electrical charge can be achieved when the redox potential of polymer A is more negative than that of polymer B. Polymer A is reduced by the electrode, and subsequently transfers the electrons to polymer B, because of its more negative potential. Electron transfer from polymer B to the electrode does not take place even at high positive electrode potentials because polymer B has no direct contact with the electrode. Thus, an activation barrier is introduced for the back electron transfer reaction. The bilayer assembly is schematically represented in Fig. 4.

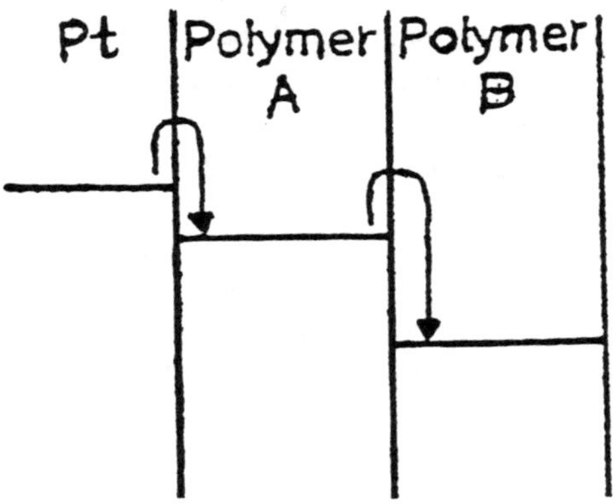

Figure 4. Schematic presentation of a bilayer electrode system.

Recently, Hable, Crooks, and Wrighton[12] have demonstrated charge trapping by quinones, electrostatically bound in a redox active, surface confined, positively charged polymer, containing a viologen derivative as part of the repeating unit. The viologen polymer has the function of polymer A in the above described system, and the quinones, with a more positive potential than the viologen, act instead of polymer B. The quinones apparently remain at the viologen polymer surface, and are unable to penetrate through and reach the electrode. The fact that electrostatic attraction by polyelectrolytes can substitute the laborious preparation and often unstable functionalyzed polymer bilayers opens new possibilities for charge trapping. If the electrode has appropriate photoactive properties, the electrode-polymer system may be a candidate for photochemical energy storage. However, unlike the electrochemical systems, where electrons move only in one direction during electrolysis, a photochemical system is likely to loose a significant fraction of the energy, before trapping occurs, because of the competition with the back reactions between the photochemical primary products. More work is required before the applicability of the principles described above can be known.

Modification of semiconductor colloids.

The positively charged polymer, polybrene, strongly interacts with ZnO colloid particles which are negatively charged at pH 12. The modification of

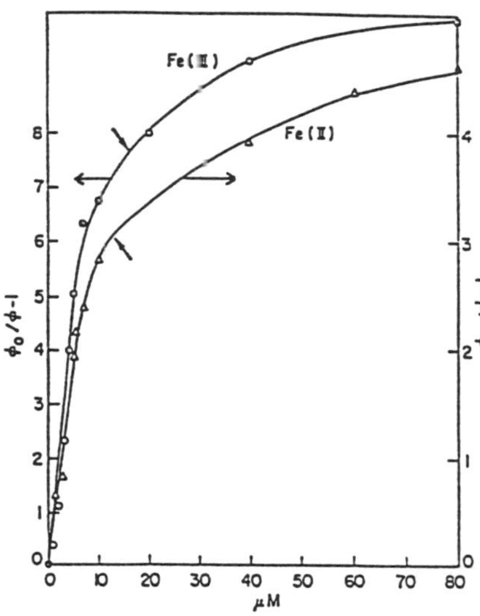

Figure 5. Quenching of ZnO ($2x10^{-7}$ M) emission by ferro and ferricyanide ions in the presence of $2x10^{-5}$ M Polybrene (PB). The arrows pointing towards the curves indicate the PB-colloid- scavenger charge equival- ent points. Ref. 13.

the colloid surface enables electron and hole scavenging by the negatively charged ferricyanide and ferrocyanide ions, respectively[13]. In the absence of the positive polymer, no scavenging takes place by these ions, as can be concluded from the lack of quenching of the known ZnO colloid secondary emission at 560 nm. The efficient quenching of the ZnO emission by ferro and ferricyanide ions is shown in Fig. 5.

An opposite effect, namely, retardation of the scavenging by a polymer stabilizer, is observed when the scavenger and the polymer have similar charges. Methyl viologen is an example. It is known to be an efficient electron scavenger for conduction band CdS electrons[14]. Since the competing electron hole recombination process is very fast, efficient electron scavenging is possible only if the scavenger is initially present at the colloid surface. When the CdS surface is modified by the positive polymer Polybrene, the electron transfer yield decreases several fold as compared with CdS particles stabilized with neutral or negatively charged polymers[15].

In addition to the effect of the polyelectrolytes on the concentrations of scavengers near the surface of the colloid particles, the rates of electron transfer to quenchers, relative to the rates of electron hole combination are also affected by the presence of a polyelectrolyte[15].

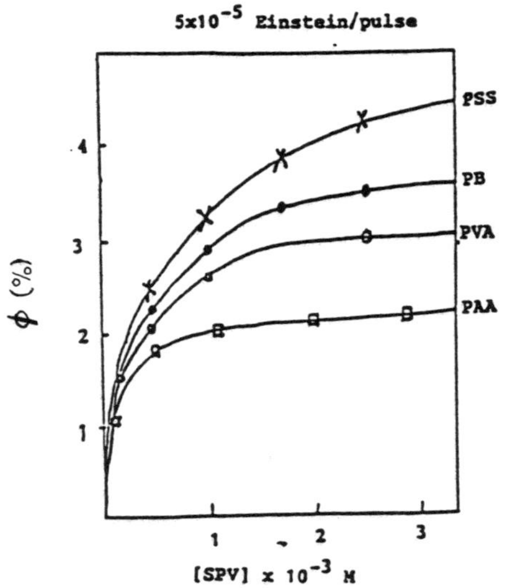

Figure 6. CdS colloid particles stabilized by various polymers: yields of electron scavenging by SPV. Ref. 15.

This conclusion is based on results obtained using electron and hole scavengers with different ionic charges, together with positive, negative, and neutral polymer stabilizers. The polymer field may have an effect on the rates of charge migrations to the colloid surface. In Fig. 6 we demonstrate the concentration effect of the electron acceptor sulfonato-propyl-viologen (SPV) on the yield of electron scavenging in a CdS colloid system stabilized by various polymers. The polymers used were: polystyrenesulfonate (PSS, negative, highly hydrophobic), Polybrene (PB, positive), polyvinyl alchohol (PVA, neutral), and polyacrylate (PAA, negative). Although the increased yield in the PSS system can be attributed to hydrophobic interactions with SPV, the difference between the effects of the negative PAA and positive PB seem to reflect mainly the differences in polymer charges. The results can be qualitatively rationalized if it is assumed that the polymers affect the rates of electron and hole migrations to the colloid surfaces, or the rates of tunneling to the scavengers. These rates may be very sensitive to the electric potential of the semiconductor surface. A remarkable effect of ZnO electrode potential on the rate of electron transfer from the conduction band to a scavenger has been recently observed and attributed to tunneling[16]

The experiments involving the polyelectrolyte effects on colloid photoreactions, and the high stabilities of some colloids, when stabilized with polyelectrolytes point out towards a possibility to construct micro layer structures in colloid solutions, similar to the polymer electrode multilayers, and probably more stable.

The role of hydrophobic interactions.

The possibility to achieve compartmentalization of photochemical products by means of distinct, hydrophobic and electrostatic micro domains existing in the same polymer (will be discussed later), calls for better understanding of the distributions of solutes in the appropriate microenvironments. Vining and Meyer[17] have recently obtained direct evidence for chemically distinct regions within Nafion films on electrodes. Their conclusions are based on the availability of ruthenium tris bipyridine aqua complexes for electrochemical reactions at various pH's. Their results indicate that three phases exist in the Nafion layers, namely: a strictly hydrophobic phase, composed of the fluorocarbon polymer backbone, an electrostatic phase formed by the sulfate groups, and an interfacial phase involving the side chains with fluorinated carbon oxygen links. Distinct phases are possible in Nafion because of the relatively long chains of the various types.

We have recently measured the free energy change of the hydrophobic binding of ruthenium-tris-bipyridine to polystyrene sulfonate[18]. Although compartmentalization does not seem possible in polystyrene sulfonate, because there is only a small distance between the electrostatic and hydrophobic segments, there is evidence for hydrophobic effects on the location of the photosensitizer within the polymer domain. Ferrocyanide ions are known to be efficient quenchers for the ruthenium bipyridine emission in homogeneous aqueous solutions. However, no quenching is observed when a negatively charged polymer (polyvinyl sulfate, polystyrene sulfonate) is added to the solution. The negative polyelectrolytes attract the ruthenium bipyridine and separate it from the ferrocyanide ions which are repelled by the negative fields of the polymers. Quenching is gradually restored by addition of salts. The positive ions added exchange with the ruthenium complex, which is moved, in part, to the bulk, where its excited state is capable to react with the ferrocyanide ions. In the case of polyvinyl sulfate, and in contrast with the polystyrene sulfonate, addition of inert ions make it possible for the ferrocyanide ions to react with the ruthenium complex within the polymer domain. This is observed by the faster decay rate of the ruthenium bipyridine triplet emission in the polymer domain, showing that in addition to exchange between the counter ions from the added salt and the ruthenium bipyridine ions, the inert salt weakens the polymer field. As a result, quenching of the ruthenium-tris-bipyridine within the polymer domain takes place despite of the negative charge of the ferrocyanide ions. The different result in polystyrene sulfonate has been attributed[18] to the hydrophobic binding of the ruthenium complex, which makes it less accessible to the quencher, probably because the ruthenium bipyridine ions are bounded more deeply inside the polymer domain, compared with the polyvinyl sulfate, where the binding is purely electrostatic. The free energy of the hydrophobic binding in this system was measured as -1.4 Kcal/mol. Experiments which were carried out using Nafion layers on SnO_2 electrodes show that the hydrophobic bindings of viologens to Nafion are much stronger than the binding of ruthenium tris bipyridine to polystyrene sulfonate[19]. Ruthenium-tris-bipyridine studied in aqueous Nafion solutions shows much stronger binding than any of the above discussed systems[20]. It seems that hydrophobic interactions may be exploited in future compartmentalization of photochemical reactants and products for optimization of energy storage.

Copolymers with controlled structures, including distinct electrostatic and hydrophobic fractions.

An important group of polymers involves copolymers with controlled fractions of hydrophobic and electrostatic segments. Examples of some of the recently used polymers are shown on the next page. These copolymers may have physical and chemical properties carefully engineered in order to fulfill specific functions. Examples for the use of such copolymers are given below.

Morishima, Furui, Nozakura, Okada and Mataga[21] have prepared a copolymer of styrene (hydrophobic) and 2 acrylamido-2-methyl-propane sulfonate (hydrophillic) which forms micelle like micro structures in water solutions. The formation of the micelles in this system can be controlled by the styrene content. This special microenvironment contains tightly packed styrene residues, with restricted molecular mobility. When a small mole fraction of phenanthroline (chromophor) is incorporated in the polymer, the latter is hydrophobically compartmentalysed. An efficient charge separation is achieved when redox quenchers, e.g. methyl viologen, are added. The reduced viologen product remains in the polymer domain, and reoxidation takes place, although at a rate much slower than the forward electron transfer process. Similar compartmentalization, discussed in terms of "steric protection" has been reported by Delair, Sanquer-Barrie and Webber[22]. These authors used diphenyl anthracene as a photosensitizer, incorporated by covalent bonds to a compact coil of polymethacrylic acid. The hydrophobic interactions were found to enhance charge separation in photoinduced electron transfer from the singlet excited state of diphenyl anthracene segments to viologens.

A copolymer specially engineered for long range energy transfer has been recently reported by Strouse, worl, Younathan, and Meyer[23]. The copolymer used is 1:1 styrene:m,p-(chloromethyl) styrene that contains a polypyridyl complex of ruthenium(II), a polypyridyl complex of osmium(II), and an anthryl derivative which has the triplet energy between the ruthenium and the osmium complexes. Excitation of the ruthenium complex results with energy migration through the anthryl "light pipes" to the osmium complex.

The engineering of copolymers with specific chemical and physical properties seems to be a promising approach for future energy storage systems.

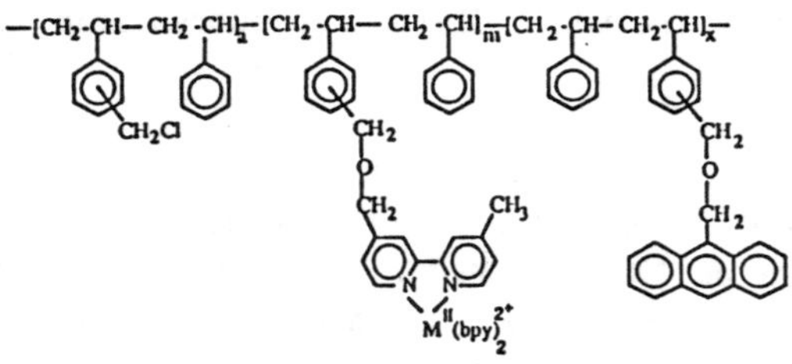

$[PS\text{-}Ru^{II}_3An_{12}Os^{II}_3](PF_6)_{12}$

[a = 12, m = 3 (M = Ru), m = 3 (M = Os), x = 12; a + m + x = 30]

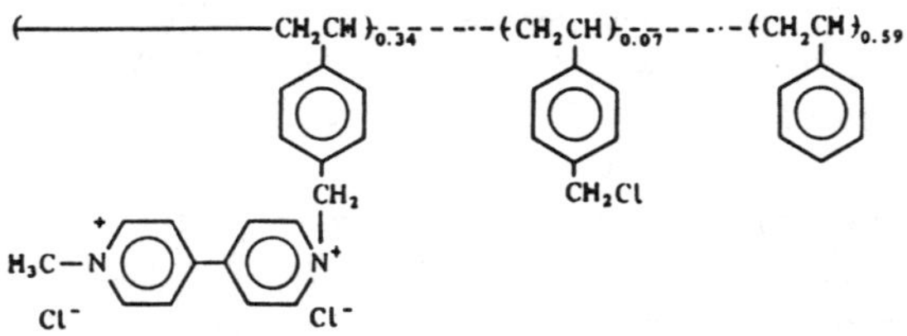

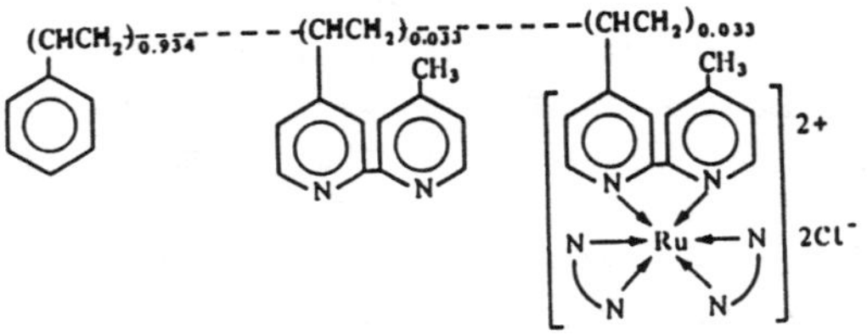

Copolymer-Pendant Ru Complexes

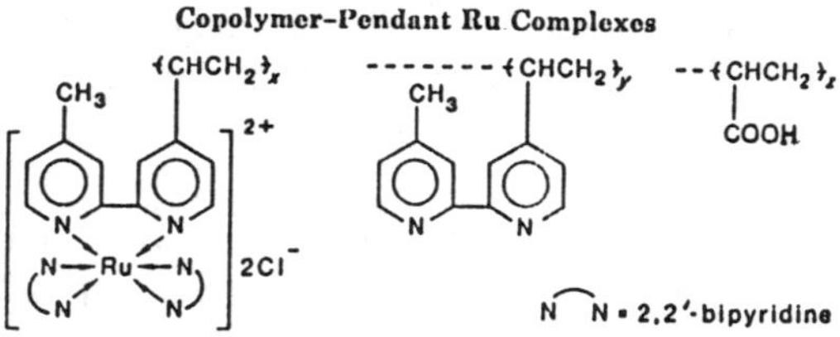

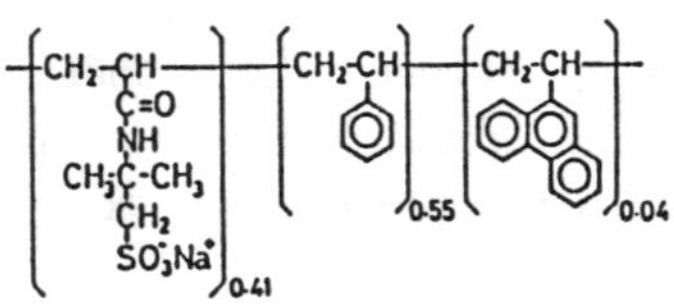

Co-polymers with variable fractions of hydrophobic and electrostatic segments.

References

1. For review: J. Rabani, Photoinduced Electron Transfer, Part B, Ed. M A. Fox and M. Chanon, Elsevier, Amsterdam, (1988).
2. G L. Duveneck, C. V. Kumar, N. J. Turro, and J. K. Barton, J. Phys. Chem. 92 2028 (1988).
3. (a) R. E. Sassoon J. Am. Chem. Soc., 107 6133 (1985).
   (b) J. Olmsted III, S. F. McClanahan, E. Danielson, J. N. Younathan, and T. J. Meyer, J. Am. Chem. Soc.109 3297 (1987).
4. (a) R. E. Sassoon and J. Rabani, J. Phys. Chem. 89 5500 (1985).
   (b) J. Rabani and D. Behar, to be published.
5. (a) P. Neta, J. Silverman, V. Markovic, and J. Rabani, J. Phys. Chem. 90 703 (1986).
   (b) D. Behar, P. Neta, J. Silverman, and J. Rabani Radiat. Phys. Chem. 29 253 (1987).
6. D. Behar and J. Rabani, J. Phys. Chem. 92 5288 (1988).
7. M. Brandys, R. E. Sassoon and J. Rabani, J. Phys. Chem. 91 953 (1987).
8. G. S. Nahor and J. Rabani, Macromolecules 22 2516 (1989).
9. Review: M. Kaneko and D. Wohrle, Advances in Polymer Sci. 84 143 (1988).
10. M. Kaneko, G. J. Yao, and A. Kira, J. C. S. Chem. Commun. 1338 (1989).
11. C. E. D. Chidsey and R. W. Murray, Science 231 25 (1986).
12. C. T. Hable, R. M. Crooks and M. S. Wrighton, J. Phys. Chem. 93 1190 (1989).
13. J. Rabani and D. Behar, J Phys. Chem. 93 2559 (1989).
14. (a) J. Ramsden and M. Graetzel, Chem. Phys. Lett. 132 267 (1986).
    (b) Y. Nosaka and M. A. Fox, J. Phys. Chem. 92 1893 (1988).
    (c) Y. M. Tricot and J. manassen, J. Phys. Chem. 92 5239 (1988).
    (d) Y. Nosaka, H. Miyama, M. Terauchi and T. Kobayashi, J. Phys. Chem. 92 255 (1988).
15. T. Rajh and J. Rabani, to be published.
16. S. Seiichiro, K. Shigehara, and A. Kira, J. Phys. Chem. (to be published.
17. W. J. Vining and T. J. Meyer, J. Electroanal. Chem. 237 191 (1987).
18. A. Slama Schwok and J. Rabani, Macromolecules, 21 764 (1988).
19. O. Johansen, J. W. Loder, A. W. H. Mau, J. Rabani and W. H. F. Sasse, to be published.
20. M. Kaneko, A. Kira, and J. Rabani, to be published.
21. Y. Morishima, T. Furui, S. -I. Nozakura, T. Okada, and N. Mataga J. Phys. Chem. 93 1643 (1989).
22. J. A. Delaire, M. Sonquer-Barrie, and S. E. Webber, J. Phys. Chem. 92 1252 (1988).
23. G. F. Strouse, L. A. Worl, J. N. Younathan, and T. J. Meyer, J. Am. Chem. Soc. 111 9101 (1989).

Photoelectron Transfer Between Molecules Adsorbed in Restricted Spaces

Nicholas J. Turro#, Jacqueline K. Barton*, Donald Tomalia@

#Chemistry Department, Columbia University
New York, New York 10027

*Division of Chemistry and Chemical Engineering,
California Institute of Technology
Padasena, California 91125

@Michigan Molecular Institute
1910 West Andrews Road
Midland, Michigan 48640

Abstract

The photophysics and photoelectron transfer reactions of metal complexes bound to anionic surfaces which provide restricted spaces in aqueous solutions have been investigated by steady state and time resolved luminescence spectroscopy. The electron transfer systems were generated by systematically variation of structures derived from the Ru(II) tris-bipyridyl family as electron donors and methyl viologens and other metal tris-bipyridyl complexes as electron acceptors. The structure of the bipyridyl ligand in the complex was varied with respect to its functionality, size and shape in order to provide electron donors of similar driving force for electron transfer, but of differing binding selectivity and dynamics. In the same vein, metal substitution, e.g., Co(III) for Ru(II), provides a family of acceptors whose structures may be readily varied systematically. In addition, the inherent chirality of the tris-bipyridyl complexes allows opportunities for investigation of enantiomeric selection in binding and in the

E. Pelizzetti and M. Schiavello (eds.), Photochemical Conversion and Storage of Solar Energy, 121–139.

electron transfer step. The anionic surfaces selected for these investigations include those provided by micelles, polyelectrolyte polymers, DNA double helix and starburst dendrimers. However, only the latter two surfaces are considered in detail in this report.

Introduction: Paradigms, Systems, Techniques and Methodologies

The research on photoelectron transfer in restricted spaces described in this manuscript employs the following experimental strtegies:

1. Paradigms: **I.** The supramolecular structural theory of *guest-host* chemistry and dynamics. **II.** The current theories of photoinduced electron transfer.

2. Systems: **I.** Photoinduced electron transfer *guest systems* of metal complex donors and metal complex or viologen acceptor systems. **II.** *Host systems* consisting of interfaces or surfaces that will serve as *nanoscopic reaction spaces* for the electron transfer donor acceptor pairs.

3. Techniques: **I.** Syntheses of families of systematically varying chemical structures. **II.** State of the art steady state and time resolved spectroscopic instrumentation for characterization of products and of transient structure and dynamics.

4. Methodologies: **I.** Development of mechanistic parameters (binding constants, rate constants, etc.) based on the guest-host and electron transfer paradigms. **II.** Measurement of the mechanistic parameters using a variety of techniques (products analysis, time resolved spectroscopic structure characterization, etc.) as a function of the systematic variation of the chemical structure of the (guest and host) systems. **III.** Measurement of mechanistic parameters using a variety of techniques as a function of the systematic variation of external constraints (temperature, viscosity, ionic strength, etc.).

The Photoinduced Electron Transfer Paradigm

The investigation of photoinduced electron transfer reactions has increased spectacularly during the past decade.[1] The intense activity in this area has been driven by at least four factors: (1) the discovery that many photoreactions proceed by mechanisms involving an electron transfer as a key step[2]; (2) the recognition that Marcus's theory[3] of electron transfer provides a powerful paradigm to direct experimental investigations of photoinduced electron transfer; (3) the challenge, both scientific and intellectual, of unravelling of the basic mechanisms of photoinduced electron transfer and the anticipation that such knowledge would assist the harnessing of solar energy and its conversion directly into electricity or its storage as chemical fuels[4]; (4) the goal of elucidating the basic mechanisms of the primary charge separation in photosynthesis[5].

From Marcus theory[3] the structural and thermodynamic factors which influence the rate constant for electron transfer (k_{et}) and the efficiency of electron transfer (Φ_{et}) include the distance of separation (R_{D*A}) between the excited donor D* and the acceptor A, the driving force (ΔG) for the reaction (difference of redox potentials of D* and A), the nature of the intervening medium, orientation factors, the reorganization energy of the pairs D^*/D^+ and A/A^-, diffusional factors, collisional factors and electrostatic factors.

The promise of harnessing solar energy by exploiting photoinduced electron transfer has driven enormous research efforts toward both the development of a fundamental understanding of the mechanisms of photoinduced electron transfer and of the molecular engineering of microheterogeneous reaction spaces to control light induced charge separation by electron transfer.[4] Thus, a desire to effectively harness solar energy has driven both investigations of photoinduced electron transfer processes (and their reversal) and investigations of the nature and dynamics of the restricted reaction spaces of microheterogeneous systems.[6]

Photoinduced Electron Transfer in Restricted Spaces

We have been concerned with photoinduced electron transfer between metal complexes that are non-covalently associated with restricted reaction spaces such as the anionic surfaces provided by polyelectrolytes such as sodium (polystyrene sulfonate), micelles such as sodium dodecyl sulfate, starburst dendrimers such as carboxylate terminated dendrimers and biomolecules such as DNA. These studies are directed towards not only understanding how adsorption of a metal complex to a restricted anionic reaction space influences the efficiency and kinetics of photoinduced electron transfer, but also how the photophysical and photochemical properties of associated metal complex can provide information on binding modes and the structure and dynamics of the interface at which binding occurs. Of particular interest is the fascinating area of environmental control of molecular diffusion and of electron transport from donor and acceptor. Issues such as the relative and absolute spatial location and spatial orientation of the electron donor and acceptor, the spatial dimentionality of molecular diffusion and electron transport, the collision frequencies and collision orientation of the donor and acceptor, etc. are all of interest in these investigations.

From the standpoint of kinetics, the interpretation of rate data of excited states and reactive intermediates embedded in restricted spaces is quite different[4], in general, from that for homogeneous solutions. Decays are commonly non-exponential and the origin of the non-exponentiality must be considered as an unknown in any new system under investigation. Typical sources of non-exponentiality are multiple, non-interacting sites for binding, fractal structures, distributions of sites, statistical considerations, etc. In favorable cases, the decay of a transient can be approximated in terms of two exponentials and tested in terms of reasonableness of fit by systematic variation of decay parameters with transient and environmental structure.

In this report we shall be concerned mainly with **DNA**'s and **dendrimers** as anionic surfaces for exploring photoinduced electron transfer from metal complexes to charged acceptors.

The Photo Systems

Ru(II) complexes were selected as the probe structures to investigate photoinduced electron transfer in restricted spaces. Their selection as probes derives from the extensive backdrop of literature research[7] which has been directed at the measurement of the photophysical and photochemical properties of the electronically excited (triplet) state, *Ru(II), of these complexes. In addition, the tris-bipyridyl complexes of Ru(II) are substitutionally inert, are readily synthesized with a variety of ligand structures and are relatively rigid in the conformational sense. A subtle, but attractive feature of the tris-bipyridyl structures is their chirality, which allows for investigations of enantiomeric selectivity in binding and in the electron transfer step.

Figure 1 shows schematically the photophysical processes and photoinduced electron transfer cycles expected for Ru(II) complexes when the excited state serves as an electron donor.

Figure 2 shows the structures of some of the Ru(II) complexes employed in our investigations and provides a flavor of the strategy of the systematics of the structural variations. In going from $Ru(bpy)_3^{+2}$ to $Ru(phen)_3^{+2}$ to $Ru(DIP)_3^{+2}$, the hydrophobicity and the size of the complexes increases considerably. In the case of binding to DNA, the larger aromatic systems allows better opportunities for intercalation into the major grove of the surface of DNA. These variations in structural features of the complexes are expected to cause a systematic variation in the binding parameters to anionic nanoscopic reactors and to modify the measured electron transfer parameters.

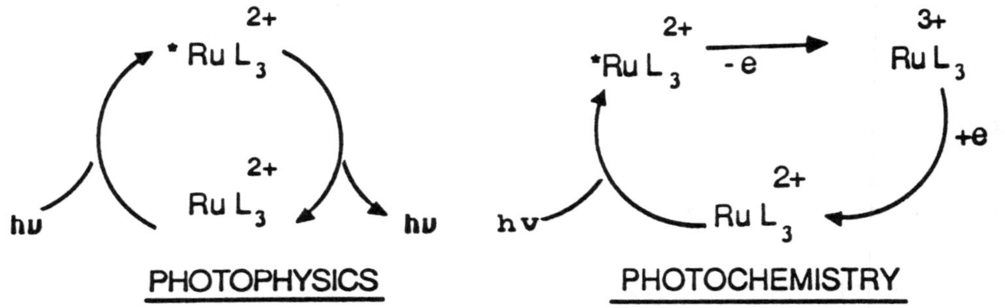

Figure 1. Left: Photophysics of Ru(II)L_3 complexes. Right: Photoelectron electron transfer from the excited (triplet) state of the Ru(II)L_3 complexes.

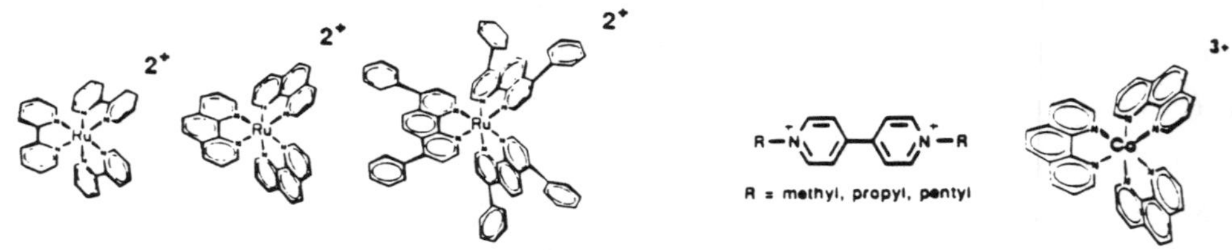

Ru(bpy)$_3^{+2}$ Ru(phen)$_3^{+2}$ Ru(DIP)$_3^{+2}$ MV^{2+} Co(phen)$_3^{2+}$

Figure 2, Structure of the electron donor [Ru(II) complexes, left] and electron acceptor [viologens and Co(III) complexes, right] employed in electron transfer research.

Techniques for determining kinetic information

The photoelectron transfers between Ru(II) complexes and acceptors were investigated by steady state and time resolved single photon counting emission techniques[10]. The rate constant for the quenching of the Ru(II) excited state was extracted from Stern-Volmer analysis of the data. In some cases two distinct decays, assigned to different binding modes of the complex to different sites of the restricted space, were observed by time resolved single photon counting. In these cases rate constants for the quenching of the complex at each site were determined by Stern-Volmer analysis of each decay.

Examples of Restricted Spaces Employed for Investigation of Photoinduced Electron Transfer at Interfaces: DNA and Starburst Dendrimers.

DNA is a generic term for a family of synthetic and natural deoxyribonucleic acid structures, both small molecules and enormous polymers. Even for a fixed composition of a DNA primary structure, a number of additional comformational structural themes are present. For example, the direction of the helical twist structure of the DNA (right handed or left handed) may be different for a given composition and primary structure. In addition local conformations, driven by a variety of interacting factors, may cause the surface of the helix to be rough and possess distinct shapes such as grooves (major or minor) or may cause the surface of the double helix to be smooth and have no significant shape characteristics.

Binding of Metal Complexes to DNA

Investigation of the binding of small molecules, such as metal complexes, to DNA is a field of interse current interest and importance because of the potential benefit of knowledge of binding modes in cancer research and biotechnology[8]. Of special interest are the observations that molecules bound to DNA can cause strand damage, possibly through redox chemistry, and that eventual cleavage occurs when the bound molecules absorb light. Metal complexes of bipyridyls and related ligands provide a particularly versatile class of structures for the investigation of binding to DNA because the ligands provide a rigid scaffolding to orient functional groups that can make specific contacts with the double helix.[8] The metal center itself does not "touch" the DNA and the redox behavior of many complexes of the same metal can be "tuned" with differing shapes, sizes and periferal functionality. Ru(II) complexes possess the further advantage of excellent photophysical properties (strong excited state absorption and intense excited state emission), which may change and be characteristic of binding modes. In addition, these complexes possess well established excited state redox chemistry.[7]

Figure 3 shows schematically the structures of several conformers of DNA.[9] These materials provide very different interfaces for the adsorption of metal complexes so that the

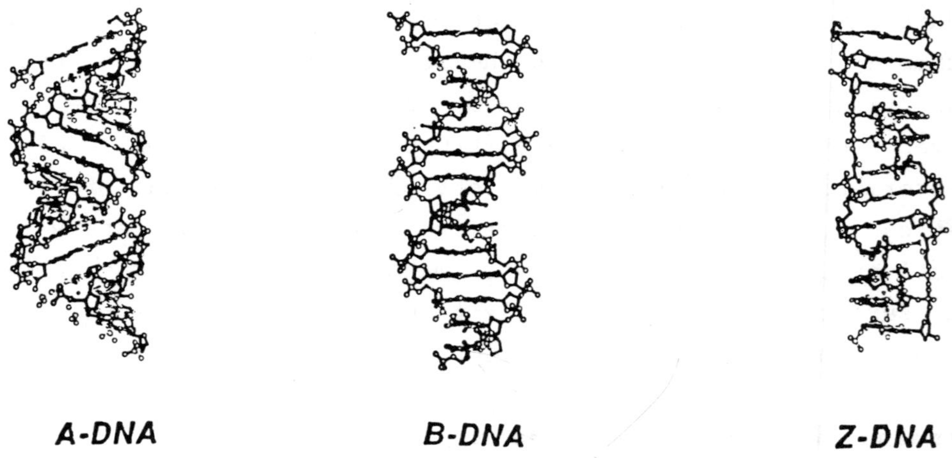

Figure 3. Schematic representation of DNA conformers. Note the differences in the shapes along the double helix.

results of our investigations are expected to provide information that will be important in elucidating the factors that determine affinity and selectivity in non-covalent binding of small molecules to DNAs of differing conformations[10]. In this regard, it is important to understand both qualitatively and quantitatively the contributions from different non-covalent binding modes (electrostatic, hydrophobic, hydrogen bonding and intercalative) to the stabilization of the bound complex. In addition, the shape of the anionic surface of the DNA helix will be an important factor in determining the efficiency of binding of metal complexes with varying ligands. For example, the deep groves of B type DNAs are ideal for intercalative binding, but the more compact structure of A and Z DNA are more conducive to hydrophobic surface binding.

Figure 4 shows schematically the structures of a novel class of materials,termed **starburst dendrimers**, which possess a well-characterized and novel macromolecular structure. and which been recently synthesized by Tomalia, et al.[11] The unique structures of these materials result from systematic and controlled synthesis of molecular branches emanating from a central initator core, as shown schematically in **Figure 4**, which depicts a planar cross section of dendrimer structures of increasing size.

As a specific example of an anionic starburst dendrimer, consider the syntheses of the poly(amidoamine) or **PAMAM** starburst dendrimers (**Figure 5**). Starting with an initiator core of NH_3, the subunits $-CH_2CH_2CO$ and $-NHCH_2CH_2NH_2$ are connected in a branching manner to create "half" (n.5) and "full" (n.0) generations of repeating units. Thus, the resulting dendrimer molecules may terminate with either $-CH_2CH_2NH_2$ (termed n.0 generations) surface groups or with $-CH_2CH_2CO_2CH_3$ groups (termed n.5 generations). The former may be transformed into cationic surfaces ($-CH_2CH_2NH_3^+$, $PAMAM^+$) by addition of a strong acid, or may be transformed into anionic surfaces ($-CH_2CH_2CO_2^-$, $PAMAM^-$) by hydrolysis with strong base. The n.0 generation $PAMAM^+$ dendrimers and the n.5 generation

PAMAM⁻ dendrimers are analogous to the micelles formed by aggregation of cationic and anionic surfactants, respectively.

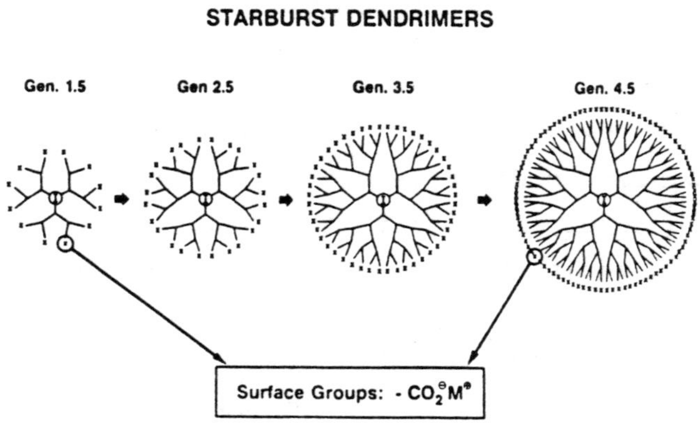

Figure 4. Schematic representation of the branching structure of generations of starburst dendrimers from an initiator core, I, proceeding through branching and terminal groups, Z.

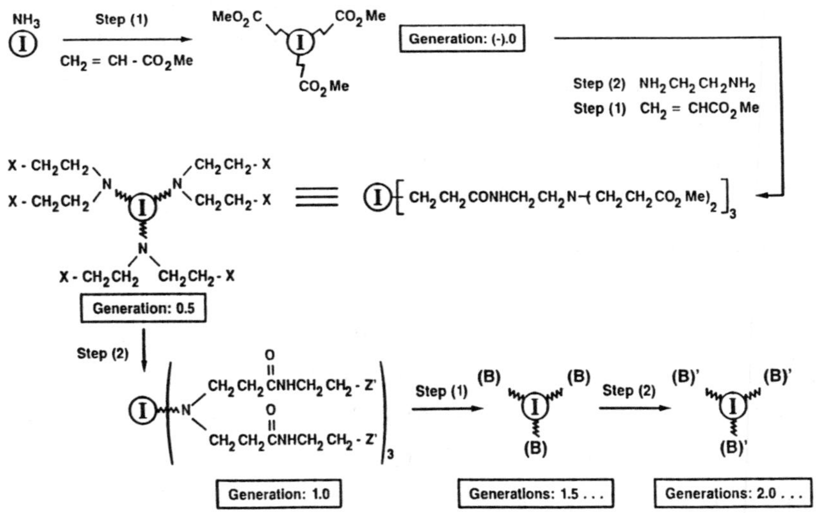

Figure 5. Schematic representation of the synthesis of n.5 carboxylate terminated starburst dendrimers. The sodium salt form of these materials possesses an anionic surface.

We have employed photophysical and photochemical probes to elucidate certain structural details of starburst dendrimers that are not revealed by knowledge of their composition and constitution.[12] For example, the issue of "surface congestion" upon increasing dendrimer size is obvious immediately upon inspection of models (**Figure** 4) or computer simulations of dendrimers.[13] Based on models, the (approximate) diameter of hypothetically spherical dendrimers is expected to increase ca. 10 Å per generation. Because of the exponential increase in surface groups as the dendrimers grow, at some point the surface becomes sterically crowded. One of the questions we wish to probe is how such congestion influences the surface properties of the anionic dendrimers as a function of dendrimer generation.

Another issue to be resolved is the morphology of the dendrimer shape and surface and how it varies with generation size and external variables. Computer simulations of the PAMAM series suggest that dramatic morphological changes occur as one advances from the lower to the higher generations.[13] For example, the simulations for the NH_2 terminated PAMAM dendrimers (n.0 generations) show the earliest generations (1-3) to be very open structures with disk-like shapes; the next generations (4-5) begin to obtain a closed structure and to take on oblate spheroidal shapes; the later generations (6 and higher) are closed structures with nearly symmetrical spherical shapes. These results agree with the results of hydrodynamic measurements of the areas of the surface groups.[11] For the initial generations (1.0-3.0) the area of a surface group is relatively constant at ca. 120 $Å^2$. Beyond generation 3.0, however, there is a steady and linear decrease in the effective area of the surface groups. An interesting issue for investigation is to determine how the change in surface area per surface group, a variable connected to generation size, influences the macromolecular properties of the dendrimers as nanoscopic reactors and the photoinduced electron transfer parameters of metal complexes adsorbed on the dendrimer surface.

Examples of Photoinduced Electron Transfer on DNA.

Among the most salient recent results of the results of our research[10] involving the photophysics and photoinduced electron transfer of metal complexes bound to DNA are the following:

1. The binding of chiral metal complexes (shown in **Figure 2**) to calf thymus DNA (a B type DNA in **Figure** 3) was examined[10a] by changes in the photophysical properties of the complexes monitored with steady state and time resolved techniques. The results were interpreted in terms of several potential binding modes: a dynamic (on the excited state complex lifetime) electrostatic binding to the anionic phosphate atmosphere about the DNA backbone, a relatively loose hydrophobic surface binding within the grooves of the DNA double helix and a relatively rigid intercalative binding within the major groove of the double helix.

2. From Stern-Volmer quenching studies,[10a,b] it was shown that *chiral discrimination occurs for intercalative binding* as manifested by different Stern-Volmer quenching constants for the enantiomeric Δ-Ru(II) complexes compared to the Λ-Ru(II) complexes. The chiral discrimination for intercalative binding was interpreted in terms of the helical asymmetry of a right handed DNA structure which is matched by the asymmetry of the Δ-isomer and leads to better binding, but which is not matched by the Λ-isomer and leads to poorer binding.

3. It was shown[10b] from photophysical investigations employing Ru(II) complexes that the binding modes and base specificity of the complexes with DNAs could be "tuned" to the helix topology.

4. The photoinduced electron transfer from Ru(II) complexes to Co(III) complexes bound to DNA was shown[10c] to be enantiomerically selective and to be sensitive to the length of the DNA. Furthermore, the efficiency of this photoelectron transfer for a given donor-acceptor pair was found[10d] to be greater for surface bound complexes than for intercalated complexes.

5. The use of photophysics of mixed ligand complexes of Ru(II), which allow variation in geometry, size, hydrophobicity and hydrogen-bonding ability by systematic variation of the complex ligands, was employed[10e] to

investigate the factors controlling the binding of the complexes to DNA.

Future Investigations of Metal Complexes as Probes of Binding to DNA, of DNA Structure, and of Photoelectron Transfer on DNA.

We are currently involved in the synthesis of metal tris(phenanthroline) complexes which are covalently bound in a tethered fashion to oligonucleotides, which, in turn, can be employed as hybridization probes and for the detailed investigation of electron transfer along the DNA double helix. **Figure** 6 shows the two ideas schematically: the binding of an oligonucleotide fragment to a complementary region along a single strand of DNA is expected to provide a segment of a duplex structure into which the covalently bound and tethered metal

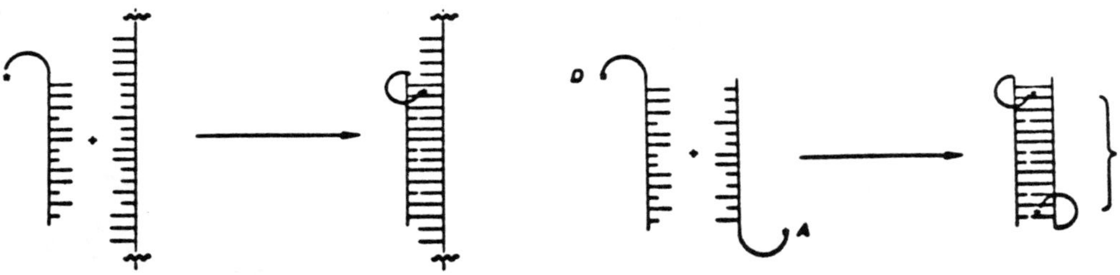

Figure 6. Schematic representation of the covalent tethering of probes to DNA to investigate hybridization (left) and to probe the mechanism of electron transfer (right).

complex can intercalate. Based on the studies described above, the altered photophysical properties (e.g., intensity and lifetime enhancements, Stern-Volmer parameters, etc.) of the intercalated metal complex will signal that hybridization has occurred (**Figure** 6, left).The hybridization concept for investigation of the DNA structure may be extended to examine photoinduced electron transfer between donor and acceptor pairs bound to DNA. For example, the binding of two complementary oligonucleotide metal complex conjugates such as those illustrated in **Figure 7**, left, provides a system which is well suited for probing photoinduced electron transfer reactions along the DNA duplex (**Figure 6**, right). Alteration of the nature of the oligonucleotide fragments will allow study of the efficiency of photoinduced electron transfer as a function of systematic variation of the distance between the metal centers and of the base sequence. The metal $(phen)_3$ complexes shown in **Figure 7** (left) have been synthesized and the attachment of the complex to the 5'phosphate of selected oligonucleotide is currently in progress.

As an example of the development of novel DNA probes, the structure $Ru(bpy)_2(dppz)^{2+}$ shown in **Figure** 7 (right) has been found to be a sensitive probe of DNA, *since it shows no detectable photoluminescence in aqueous solutions at ambient temperatures, but which binds strongly to DNA and in doing so exhibits intense photoluminescence (**Figure** 8).* The complex may be viewed as a "molecular light switch"

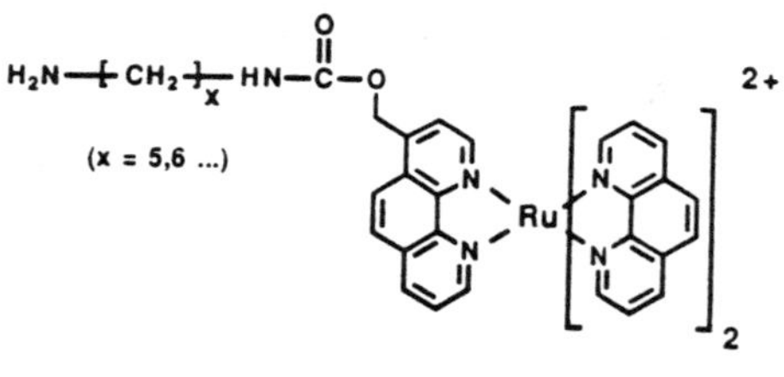

Tethering complexes

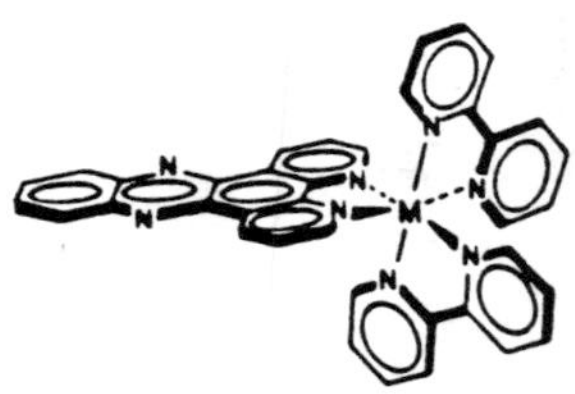

Molecular Light Switch for DNA

Figure 7. Structures of the metal complexes to be employed in the tethering experiments (left) and to serve as a "light switch" for binding to DNA.

for binding to DNA. We are currently exploring the use of $Ru(bpy)_2(dppz)^{2+}$ as a sensitive, non-radioactive probe of DNA structure which is appropriate for both heterogeneous and homogeneous assays. For example, with B-form polynucleotides, emission is centered at 628 nm, whereas with Z-form polynucleotides emission is centered at 640 nm and with A-form DNA only weak emission is observed and is centered at 650 nm.

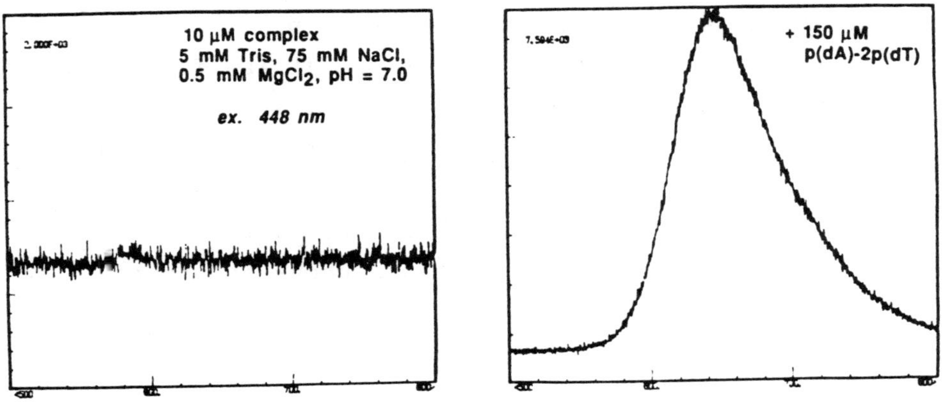

Emission in aqueous solution, no DNA

Emission in the presence of DNA

Figure 8. Emission of the "light switch" for DNA in aqueous solution (left) and in the presence of a synthetic DNA (similar effects are observed with calf-thymus DNA).

Electron Transfer on Starburst Dendrimers as Nanoscopic Reactors.

We have also investigated whether starburst dendrimers can serve as nanoscopic reactors to control photoinduced electron flow between donor and acceptor pairs bound to the dendrimer surfaces. Anionic micelles and anionic polyelectrolytes have been shown to be effective ultrastructures to mediate electron transfer processes between metal complexes.[15] In these investigations, the dendrimer size will be varied over a range of generations (e.g. 0.5 to 9.5 generations for the PAMAM⁻ dendrimers) and sizes (ca 30 Å to 200 Å). The rate constants and efficiencies of photoinduced electron transfer provide important information concerning the mechanism of electron transfer.

Our initial results demonstrated the ability of PAMAM$^-$ dendrimers to mediate photoinduced electron transfer from photoexcited ruthenium complexes [e.g., $Ru(bpy)_3^{2+}$ and related complexes, **Figure** 9) to electron acceptors [e.g., alkyl viologens cations and cobalt complexes, **Figure** 9]. From steady state measurements of the quenching of the photoluminescence of the Ru complexes, it was found that the probe was efficiently quenched in the presence of the higher generations (3.5 to 9.5 G PAMAM$^-$ dendrimers), but that in the presence of the lower generations (0.5 to 2.5 G PAMAM$^-$ dendrimers) the quenching efficiency was very low and essentially equal to that for bulk aqueous solutions.

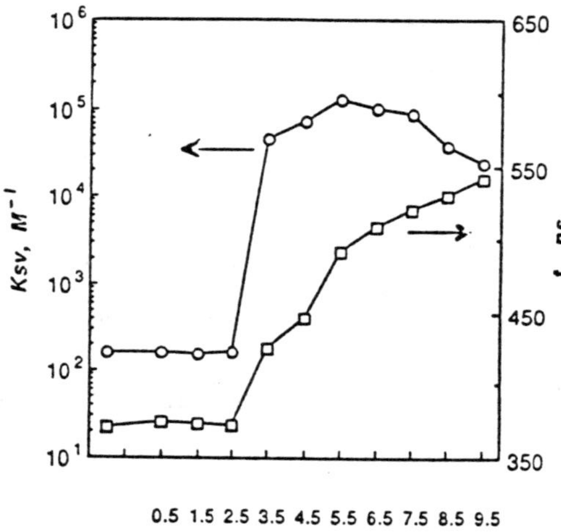

Starburst Dendrimers

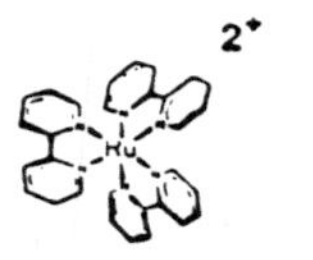

$Ru(bpy)_3^{2+}$

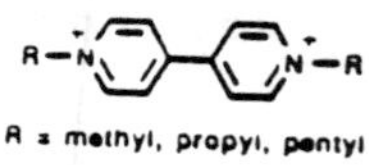

Alkyl viologens

Figure 9. Stern-Volmer constants (left vertical axis) for quenching of the electron transfer quenching of $Ru(bpy)_3^{2+}$ by methyl viologen (MV^{+2}) and emission lifetimes (right vertical axis) of $Ru(bpy)_3^{2+}$ measured by time resolved single photon counting.

The quenching results summarized in **Figure** 9 (left vertical axis) show the high efficiency (large values of the Stern-Volmer constant K_{SV}) for the higher generation dendrimers and provide support for the concept that the dendrimers behave as self-organizing polymers for the donor and acceptor complexes only after the 3.5 generation. Further support for a significant interaction between the higher generation dendrimers and the donor complexes is found in the enhanced luminescence and lengthening of lifetimes (**Figure** 9, right vertical axis) of the Ru complexes as higher generation dendrimer is added to aqueous solutions containing the Ru complexes. That the anionic dendrimer surfaces of higher generation dendrimers behave in a manner similar to that of polyelectrolytes is confirmed by the following observations:

1. The addition of 1 M of a strong electrolyte (NaCl) produces a dramatic decrease in the quenching efficiency due to displacement of the cationic complex and cationic quenchers into the aqueous phase;

2. A plot of luminescence quenching efficiency *versus* the *ratio* of quencher per dendrimer molecule for different concentrations of dendrimer fall on a single line, indicating that it is the number of quenchers associated with a single dendrimer molecule that determines the quenching efficiency;

3. The break in efficiency of photoinduced electron transfer with generation parallels the breaks found for the interactions of cationic surfactants via the pyrene probe method described above.

These results demonstrate that photoinduced electron transfer can be mediated by adsorption of a donor (*Ru complex) and acceptor which reside on the same dendrimer molecule and that the efficiency of the electron transfer may be controlled by varying the structural features of the dendrimer. The profile of results demonstrates that a change in dendrimer structural features occurs about generation 3.5, in agreement with the predicted change in the shape of PAMAM dendrimers predicted by computer simulation and surface group area measurements.

Conclusions

Metal complexes of the tris-bipyridyl family have proven to be very versatile and useful probes of the surface structure and dynamics of microheterogeneous anionic structures such as micelles, globular polymers, starburst dendrimers and DNAs. Each of these restricted spaces in turn have served as "nanoscopic reaction spaces" for the investigation of photoinduced electron transfer. There is a great deal more to be done at both a qualitative and a quantitative level to exploit the science exposed by these studies.

Acknowledgements: The authors thank the NSF, the NIH and the AFOSR for their generous support of this research. The intellectual contributions and scientific skills of the following students and postdoctorals were responsible for the realization of the research reported here: V. J. Kumar, M. D. Purugganan, G. Jaycox, M. C. Moreno-Bondi, G. Orellana, G. Caminati, A. Kirsch-DeMesmaeker and A. Friedman.

References

1. For example see: G.J. Kavarnos and N.J. Turro, **Chem. Rev.**, *86*, 401 (1986); "Photoinduced Electron Transfer: eds. M. A. Fox and M. Chanon, Elsevier, New York, 1988;

2. S. L. Mattes and S. Farid in "Organic Photochemistry", vol. 6, A. Padwa, Ed., Marcel Dekker, New York, 1983, p. 233.

3. R. A. Marcus, **Annu. Rev. Phys. Chem.**, *15*, 155 (1966); R. A. Marcus and N. Sutin, **Biochim. Biophys. Acta**, *811*, 265 (1985).

4. (a) M. Graetzel, "Heterogeneous Photochemical Electron Transfer", CRC Press, Boca Raton, 1989; (b) T. Watanabe, A. Fujishima and K. Honda, "Energy Resources Through Photochemistry and Catalysis", ed. M. Graetzel, Academic Press, New York, 1983; (c) M. Chanon and L. Eberson in "Photoinduced Electron Transfer," eds. M. A. Fox and M. Chanon, Elsevier, New York, 1988, Chapter 1.11, p. 409.

5. J. Ulstrup, "Charge Transfer processes in Condensed Media, Springer-Verlag, West Berlin, 1979; R. D. Cannon, "Elecron Transfer Processes", Butterworth, London, 1980.

6. K. Kalyanasundaram, "Photochemistry in Microheterogeneous Systems",Academic, New York, 1987; J. Rabini, in "Photoinduced Electron Transfer," eds. M. A. Fox and M. Chanon, Elsevier, New York, 1988, Chapter 3.4, p. 642.

7. A. Juris, V. Balzani, F. Barigelletti, S. Campagna, P. Besler and A. Von Zelewsky, **Coor. Chem. Rev.**, *84*, 85 (1988).

8. "Metal Ions in Biology", Ed. T. G. Spiro, vol. 5, 1983; J. K. Barton, **Science**, *233*, 727 (1986); M. B. Fleisher, H. Y. Mei and J. K. Barton, "Nucleic Acids and Molecular Biology", eds. F. Eckstein and B. M. J. Lilley, vol. 2, springer-Verlag, Berlin, 1988.

9. W. Saenger, "Principles of Nucleic Acid Structure", Springer-Verlag, New York, 1984.

10. (a) C.V. Kumar, J.K.Barton, and N.J. Turro, **J. Am. Chem. Soc.**, *107*, 5518 (1985); (b) J.K. Barton, J.M. Goldberg, C.V. Kumar, and N.J. Turro, **J. Am. Chem. Soc.**, *108*, 2081 (1986); (c) J.K.Barton, C.V. Kumar, and N.J. Turro, **J. Am. Chem. Soc.**, *108*, 6391 (1986); (d) M.D. Purugganan, C.V. Kumar, N.J. Turro and J.K. Barton, **Science**, *241*, 1645 (1988); (e) A.M. Pyle, J.P. Rehmann, R. Meshoyrer, C.V. Kumar, N.J. Turro and J.K. Barton, **J. Am. Chem. Soc.**, *111*, 3051 (1989).

11. D. A. Tomalia, H. Baker, J. Dewald, H. Hall, G. Dallos, S. Martin, J. Roeck, J. Ryder and P. Smith, **Macromolecules**, *19*. 2466 (1986); D. A. Tomalia, A. M. Naylor and W. A. Goddard, **Angawante Chem.**, *29*, 139 (1990).

12. M. C. Moreno-Bondi, G. Orellana, N. J. Turro and D. A. Tomalia, **Macromolecules**, *23*, 912 (1990).

13. A. M. Naylor, W. A. Goddard, G. E. Kiefer and D. A. Tomalia, **J. Am. Chem. Soc.**, *111*, 2339 (1989).

14. A. E. Friedman, J-C. Chambron, J.-P. Sauvage, N. J. Turro and J. K. Barton, **J. Am. Chem. Soc.**, *112*, 4960 (1990).

15. (a) Micelles: T. Miyashita, T. Murakata and M. Matsuda, **J. Phys. Chem.**, *87*, 4629 (1983); (b) Polyelectrolytes: G.L. Duveneck, C.V. Kumar, N.J. Turro and J.K. Barton, **J. Phys. Chem.**, *92*, 2028 (1988).

STRATEGIES FOR MIMICKING THE PRIMARY EVENTS OF BACTERIAL PHOTOSYNTHESIS: STRUCTURE, FUNCTION, AND MECHANISM

JAMES R. NORRIS[*†] and MALA RAGHAVAN[†]
[*]*Chemistry Division*
Argonne National Laboratory
Argonne, Illinois 60439

[†]*Department of Chemistry*
The University of Chicago
Chicago, Illinois 60637

ABSTRACT. Some of the possible implications of the photochemistry of photosynthesis for artificial systems are presented. A major aspect of this paper involves the special conditions required to prevent the undesirable charge recombination via the excited triplet state of the primary donor. If a multi-jump model is operative in the primary events of photosynthesis, then artificial model systems are relatively easy to synthesize. Such systems prevent back reactions via a series of downhill chemical reactions. Each downhill step in energy results in an increase in charge separation distance. The annihilation reactions, even to the less energetic triplet state, involve intermediate states uphill in energy and consequently are greatly diminished. However, if superexchange is a correct explanation of photosynthesis, then model systems have not been developed that properly mimic the natural process. In particular, the triplet back reaction can occur due to the lack of any thermal activation barrier. In nature this downhill back reaction appears to be prevented by carefully balancing the energetics using four molecules in the electron transport system. If this is the case, artificial photosynthesis would also require fine tuning of the coupling and energetics with three or four molecules making duplication more difficult to achieve.

Introduction

Photosynthesis is the natural process of light induced charge separation which occurs in a reaction center (RC) protein within about three picoseconds of singlet state excitation. The resulting electron-hole pair lasts for at least ten nanoseconds in the form of a donor-

E. Pelizzetti and M. Schiavello (eds.), Photochemical Conversion and Storage of Solar Energy, 141–150.

acceptor radical pair separated edge-to-edge by about 11 Å. As a result of this long-lived chemistry, nature has a means of producing efficient charge separation with negligible back reactions. This paper in general describes some of the possible implications of this RC chemistry for artificial photosynthesis with emphasis on special conditions required to prevent charge recombination annihilation via the indirect back reaction to the excited triplet state of the primary donor. The basis for such discussions has evolved from a knowledge of the purple photosynthetic bacteria.

The photosynthetic process is so efficient that every photon absorbed in a RC produces a chemical reaction, and therefore, the initial energy conversion act of photosynthesis proceeds with a quantum yield of nearly unity (>0.95) [1-3]. Consequently, an artificial chemical system utilizing the fundamental energy conversion processes of photosynthesis should be able to duplicate the biological process of energy conversion and storage. Not surprisingly, one goal of RC research is to optimize the use of solar energy in chemical systems for the production of useful compounds and fuels.

The bacterial RC can be described in terms of two main systems, the protein matrix and the donor-acceptor complex. The protein part of the *Rb. sphaeroides* RC is further comprised of three separate protein subunits labeled L-, M- and H-. The L- and M-parts contain the actual donor-acceptor pigments and UQs which perform the electron transfer chemistry. The arrangement of the various pigments in the complex without the protein matrix is shown in the X-ray structure of *Rb. sphaeroides* R26 [4], Figure 1. There are four bacteriochlorophylls (Bchls), two bacteriopheophytins (Bpheos), two ubiquinones (UQs), and a high spin ferrous(II) ion (Fe). Two of the four Bchls form the primary donor known as the special pair with one half of the special pair associated with the L-subunit (P_L) and the other half associated with the M-subunit (P_M). The remaining two Bchls will be referred to as the bridging Bchls (B_L), (B_M) and the two Bpheos as (H_L), (H_M) where the subscripts L and M refer to the corresponding protein subunits. The UQs are represented as Q_A and Q_B.

The pigments and the UQs occur in pairs as shown in Figure 1. One half (the right half of Figure 1) of the pigment-UQ complex is primarily associated with the L-subunit and the other half is primarily associated with the M-subunit. Only the Fe does not have a corresponding pair. Moreover, the L-subunit, the M-subunit, and the donor-acceptor complex including the Fe, exhibits an approximate two-fold axis of symmetry, C_2. As a consequence of this two fold symmetry of the pigments and UQ-Fe complex, potentially there exists two pathways for the electron transfer chemistry to occur. If rigorous C_2 symmetry were in operation for the components of electron transfer, both pathways would necessarily be functioning. Interestingly, only the L-branch seems to be photochemically active, implying that the C_2 symmetry must be broken. Despite the knowledge of many symmetry breaking residues of the protein, no mechanism has yet been established for this preferred pathway. The origin of the L-side activity will not be discussed further and any reference to the electron transfer process can be assumed to be associated with the L-branch. The L-bridging Bchl is generally believed necessary for electron transfer between the primary donor special pair and the primary acceptor Bpheo. We will now focus on the chemistry involved with the first step of charge separation; this restricts the discussion primarily to the Bchls and the Bpheos.

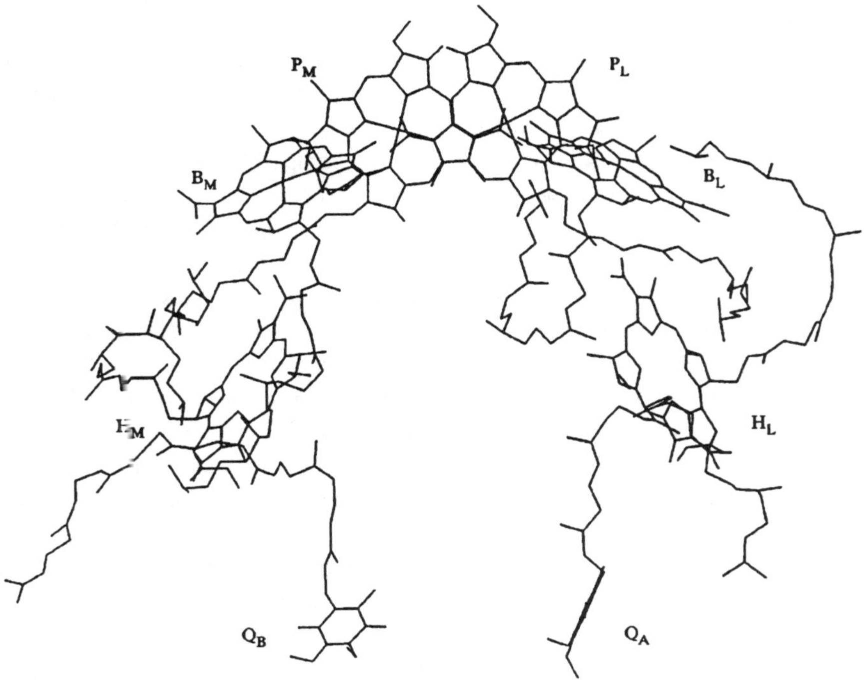

Figure 1. Structure of chromophores in *Rb. sphaeroides* R26 reaction center complex. P_L and P_M are the two Bchls constituting the special pair (P). B_L, H_L and Q_A are the Bchl, Bpheo and the ubiquinone in the photoactive branch L. B_M, H_M and Q_B are the same on the inactive branch M.

Chemical Principles of Photoinduced Charge Separation

Within approximately three picoseconds of optical excitation an electron is transferred from the primary donor to the primary acceptor forming an anion on the Bpheo and a cation on the special pair. The existence and speed of this initial electron transfer reaction is established through transient optical absorption spectroscopy with sub-picosecond time resolution. Whether or not electron transfer involves the bridging Bchl molecule remains to be proven. Moreover, because of the difficulty in experimentally establishing a clear participation of the bridging molecule in the electron transfer act, two basic mechanisms of electron transfer have been predominant. One mechanism involves a short lived, discreet intermediate state where the transferred electron resides for a brief time on the bridging Bchl. The lifetime of the electron on the bridging Bchl is too small for easy experimental verification. The other mechanism invokes superexchange, where participation of the bridging molecule is quantum mechanical. In this latter case, experimental observation of a discrete oxidized or reduced state of the intermediate bridging molecule is intrinsically impossible.

In any event, as soon as the electron has been pumped from the special pair to the L-side Bpheo, within about 200 picoseconds the electron moves from the L-side Bpheo to the UQ of the L-side . Shortly thereafter it moves on to the UQ of the M-side (Figure 1). Eventually, the entire process is repeated so that the UQ on the M-side is ultimately reduced twice. Since the reaction center spans a membrane, this initial process of photosynthesis pumps electrons across the membrane via a single-sided pathway even though the RC structurally exhibits two potential pathways.

However, if the ubiquinone Q_A has been chemically removed or prereduced (e.g., by dithionite treatment in the dark), then the normal forward electron transfer chemistry is blocked in these RCs and charge recombination begins. This annihilation process requires about 10 ns [5]. Two distinct charge recombination routes exist simultaneously. One pathway returns the electron to the primary donor ground state via the singlet manifold. If this pathway alone were active then charge recombination would require about 20 ns [5]. The other pathway returns the electron to the singlet ground state via the first excited triplet state of the special pair. Actually when the triplet state of the primary donor special pair is formed in this process, charge recombination is complete since the hole and the electron have annihilated each other. This triplet annihilation process takes only about 2 ns [5]. Eventually, the excited triplet state of the special pair relaxes to the original ground state. We focus attention on this triplet annihilation process in this paper.

The chemistry that takes place in a RC is rather amazing. We shall proceed by describing such chemistry based on the RC structure and function in terms of artificial photosynthesis. In all our examples, neither the donor nor the acceptor is reactive if both are in their neutral ground state. The electron transfer requires at least one of the participants to be in an excited electronic state that can be generated by the absorption of light. For convenience, throughout this discussion we assume that the light energy is absorbed by the donor and that the donor is promoted into its first excited singlet state.

First we illustrate a hypothetical donor-acceptor pair where the distance of separation between donor and acceptor is relatively large, 25 Å (e.g., the distance between the special pair and Q_A). In this example the average lifetime of the lowest excited singlet state of the typical donor is so short (10 to 100 ns) that electron transfer cannot proceed from the donor to the acceptor in this time and no radical pair formation occurs. In general, each 1.5 Å of separation decreases the electron transfer rate by roughly ten [6]. As a point of reference electron transfer over 10 Å (edge-to-edge) should occur in about 10 ns for molecules like the Bchls. Consequently, the long-distance 25 Å model system is nonfunctional in the excited singlet state since completion of electron transfer would require that the lifetime of the excited state be longer than a second.

In contrast, the longer lived but less energetic triplet state of the donor may survive long enough for the electron transfer event to occur. For a number of years chemists have been able to generate charge separation in model systems based on excited triplet states In general, the triplet state is formed by loss of energy from an excited singlet state. Thus, for maximum conversion the energy available in the singlet state should be used directly, not indirectly via a triplet state. As a consequence, the energy of the charge separated radical pair state should be below the energy of the excited singlet state but well above the energy of the triplet state. In any event, the goal of RC research is to learn how to use the more energetic singlet state.

We therefore proceed to the next example with a hypothetical donor-acceptor pair separated by a very short distance of less than ~4 Å (e.g., the special pair to bridging Bchl distance). In this example, electron transfer occurs during the lifetime of the singlet state of the donor. However, a new problem arises; any charge separation is quickly destroyed by a very efficient and rapid back reaction. The back reaction is the self-annihilation of charge separation and concomitant collapse of the donor and acceptor radical pair. In this case the self-annihilation proceeds so rapidly that the lifetime of the charge separation exists on the order of a few hundred picoseconds. Since the goal of artificial photosynthesis is the production of millisecond charge separation, such a short lifetime is an unsatisfactory solution to the problem. As we move the donor-acceptor further apart (for example to the 11 Å distance of separation between the special pair and Bpheo), then both the reaction rate and the reaction yield decreases precipitously and very quickly the photochemistry does not proceed during the lifetime of the excited singlet state.

One conceivable route to longer-lived charge separation in the next example is via the well known three-site arrangement, where electron transfer proceeds from the initial donor to the terminal acceptor by some mechanism involving the intermediate bridging molecule. This example is similar to the special pair donor, the bridging Bchl molecule and the primary acceptor, Bpheo, of the RC as shown in Figure 2. The technical difference between our hypothetical three site model and the RC is that nature uses four molecules for three sites. The arrangement of molecules is such that electron transfer proceeds within a few picoseconds after light excitation resulting in a donor cation and the third molecule as an anion. In this three molecule system the charge separation can remain stable for longer than 10 nanoseconds. As before, one of the reaction requirements is that the distant, terminal molecule be a good acceptor relative to the excited state of the initial donor molecule and the bridging molecule. In general, the pheophytins are better acceptors than the corresponding chlorophylls, or in other words, bacteriopheophytin is easier to reduce than the corresponding bacteriochlorophyll; thus, the electron would rather be on the Bpheo than on the Bchl.

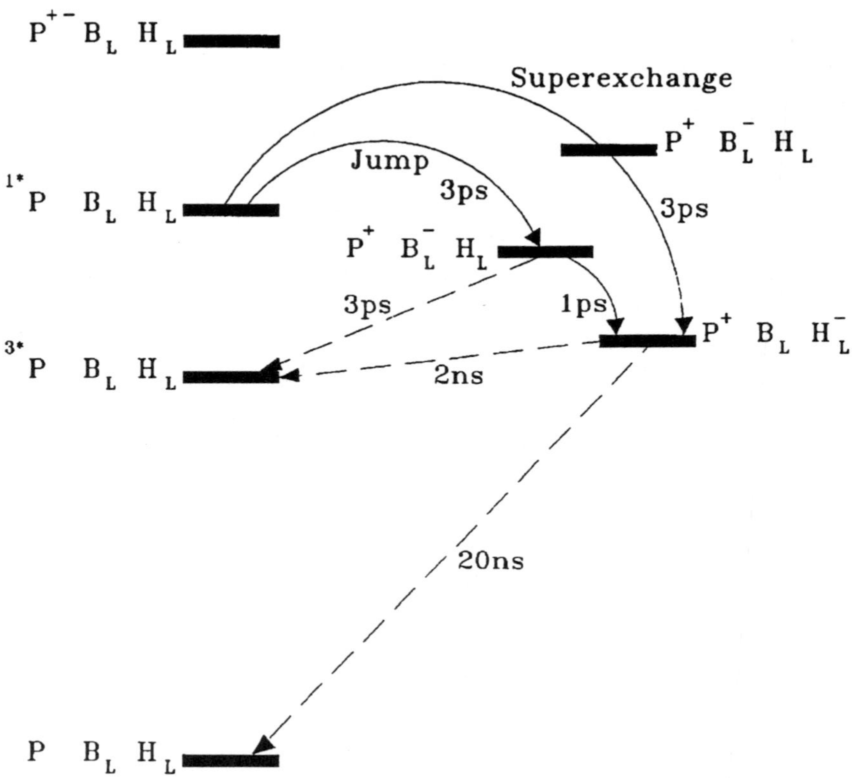

Figure 2. Schematic representation of the primary charge transfer process in the reaction center. P, B_L and H_L are the three different sites involved in electron transfer reaction. The different states are shown as the thick solid lines. The energies of the states are scaled with reference to the ground state energy of the primary donor (P). $P^{+-}B_L$ H_L is the charge resonant state of $^{1*}P$ believed to participate in the super exchange mechanism.

The speed at which the undesirable back reaction occurs depends critically on the distance of charge separation. By having a three-site reaction, the distance of charge separation can be much larger than in a two-site reaction, thus minimizing the direct back reaction rates to the ground state. Suppose that the three-site reaction occurs as it might in photosynthetic bacterial RCs as illustrated in Figure 2. In this mechanism the first electron transfer step, from the donor of site 1, (P), to the intermediate acceptor of site 2, (B_L), may take less than a few picoseconds. If the second forward reaction, i.e., from B_L to H_L proceeds in a time much less than one picosecond, then the overall forward reaction leading to the distant charge separated state $P^+B_LH_L^-$ occurs in about three picoseconds. Because of the large distance between site 1 and site 3, the direct back reaction (i.e., the annihilation of charge separation without participation of site 2) is slow. If the reaction scheme were stopped at site 2, then the time for back electron transfer to the ground state or an excited triplet state located in the donor of site 1 would be significantly short, typically less than one nanosecond. This faster back reaction directly results from the shorter distance between site 1 and site 2.

In addition to the distance, the size of the downhill energy gap also controls the reaction via the so called Frank Condon factor, where in general the larger the gap, the slower the reaction rate. Thus, a large energy gap between the charge separated state and the original ground state is desirable for maximum energy storage as well as for the prevention of the back reaction. However, this large energy gap between ground state and radical pair, charge separation state results in a small energy gap between the triplet state of site 1 and the same radical pair state. Thus, the triplet state back reaction has a favorable Frank Condon factor and thus is faster. Such an explanation agrees with the 2 ns for the triplet back reaction and the 20 ns singlet back reaction of RCs mentioned earlier.

How about the indirect back reaction mechanism? The electron of site 3, (H_L), can jump uphill, back to site 2, (B_L), and then directly return to the ground state of site 1. Note that the indirect mechanism involves one or more intermediate uphill states before a direct jump to the original ground state occurs with any efficiency as shown in Figure 2. Therefore, for efficient and long lived charge separation, the forward rate for the formation of $P^+B_LH_L^-$, must be larger than the direct recombination rate of $P^+B_L^-H_L$ to the ground state or the triplet state of site 1.

In general, to prevent the indirect back reaction mechanism from contributing, each forward reaction step should be slightly downhill in energy storage. In essence each forward reaction gives up some energy as "heat" but gains charge separation stability. Arguments based on microscopic reversibility and detailed balance show that, as long as sufficient energy is lost in the forward reactions (ultimately in the form of heat), the reverse uphill reaction rates are slower than the forward reaction rate constant. Thus, each forward step of the photosynthetic process wastes a little energy as heat in order to reduce the indirect back reaction rates via its preceding excited states. At the same time that the back reaction rates for the excited states are minimized by giving up a little energy in the forward reaction steps, the distance between cation and anion increases. As a result, the time for direct electron transfer back to the ground state and even more importantly to the excited triplet state of site 1, ($^{3*}P$), also increases.

Thus, in the photosynthetic RC both direct and indirect back reaction rates are minimized for both the singlet and the triplet manifold. In fact, in the RC all forward reaction steps increase the distance of charge separation except when the electron goes from Q_A to Q_B. By the time the electron is on the first UQ, the distance of charge separation is 25 Å. This distance between the special pair cation and the reduced ubiquinone, Q_A^-, is so large that the back reaction to the original ground state takes longer than the time required for cytochrome to reduce the special pair cation. After reduction of oxidized special pair by cytochrome, the 45 Å distance of charge separation (the distance between Q_B and a heme group of docked cytochrome) is so large that the back reaction is virtually eliminated.

The fact that photosynthesis uses at least five steps to generate stable charge separation is agreed upon, although, the mechanism of photosynthesis as detailed above has not yet been established. Instead, the above mechanism is just one of many that have been proposed. However, the general principles pertaining to kinetics, distance, and stable charge separation are the same for all models proposed to explain RC photochemistry.

Superexchange Mechanism

One of the other mechanisms for the initial charge separation of bacterial photosynthesis is based on the concept of superexchange. The relevance of this superexchange mechanism to artificial photosynthesis is the major purpose of this paper. In this model, regardless of experimental time resolution, detection of an electron transfer intermediate, $P^+B_L^-H_L$, (shown in Figure 2) prior to the reduction of the primary acceptor, bacteriopheophytin, is impossible. The justification of superexchange is to explain how an electron can be transferred at least 11 Å (nearest π edge-to-edge distance) in about three picoseconds without a measurable intermediate. In this view essentially a supermolecular complex exists such that the effective distance of electron transfer is less than 4 Å instead of 11 Å if the intermediate Bchl were not present. In other words, although at least three molecules must be involved in such a mechanism, this represents only a two site case where the edge-to-edge distance is quite short.

If this mechanism proves to be a correct explanation for photosynthesis, then its duplication in model systems appears to be rather difficult. The reason for the complexity is that some special considerations must be given to preventing the back reaction of charge recombination or self annihilation in such a two site case. We point out, that if a distance of less than 4 Å is involved in the forward reaction as a result of superexchange in the charge separation process, then the same distance and the same supermolecule complex may be expected to be involved in the back reaction. Thus, to be compatible with experimental observations on the RC, one would conclude that in nature the supermolecule complex can only be operative in the forward charge separation reaction and not in the back reaction. Especially if the effective distance between the donor supermolecule and the acceptor is 4 Å, then the back reaction to the triplet state of the primary donor is slightly downhill in energy and thus in principle may be expected to occur very rapidly (in ps). Apparently, if this mechanism is operative in nature, this highly favorable downhill back reaction has been prevented. The next question to be addressed is how could nature diminish the back reaction for this "two site" case.

Friesner and Won [7] have expanded on a model proposed by Thurnauer et.al., [8] based on charge separation within the special pair dimer [8]. The detailed mechanism proposed provides a straightforward explanation for fast forward chemistry and slow backward chemistry while the superexchange mechanism is operative. We note that such a model incorporates four molecules, a dimer special pair donor, a superexchange site of the bridging Bchl and the primary acceptor. The occurrence of four molecules instead of three provides a weakly coupled dimeric species essential for the singlet charge resonance state ($P^{+-}B_L H_L$), (Figure 2). The singlet charge resonance states within the special pair quantum mechanically connect the primary donor special pair to the primary acceptor Bpheo via the bridging Bchl. The same analysis shows that the triplet state of the primary donor, i.e., the triplet of site 1, is not significantly coupled to the charge separation state and thus the back reaction to the triplet state is also not significant.

However, the problem of such a mechanism for artificial photosynthesis is severe. One major difficulty is the requirement of four molecules and the complications for their synthesis. Moreover, until the precise details of this four molecule superexchange mechanism have been established, the energetic and geometrical requirements for the synthesis of such a four molecule complex will be unknown.

Summary

If a multiple discrete jump model is operative in the primary events of photosynthesis, then various model systems have been synthesized with many of the properties of bacterial photosynthesis [9,10]. Such a model prevents back reactions by a series of downhill chemical reactions using well established charge separation principles. Each downhill step in energy gains charge separation distance. The back reaction, even to the less energetic triplet state, involves intermediate states uphill in energy and thereby is not favorable. On the other hand, if superexchange is a correct mechanism of photosynthesis, then no model system has been developed that mimics properly the natural process. The back reaction to the triplet state of the primary donor is downhill in energy and thus can occur in principle. In nature this downhill back reaction appears to be prevented by the use of four molecules in the donor-acceptor complex. Thus, we suggest that superexchange artificial photosynthesis requires four molecules to mimic the natural event sufficiently. In this view superexchange artificial photosynthesis is the more difficult challenge.

Acknowledgement

This work was supported by the Division of Chemical Sciences, Office of Basic Energy Sciences, U.S. Department of Energy, under contract W-31-109-Eng-38.

References

[1] Bolton, J. R., Clayton, R. K. and Reed, D. W. (1969) "An identification of the radical giving rise to the light-induced electron spin resonance signal in photosynthetic bacteria", Photochem. Photobiol. 9, 209-218.

[2] Wraight, C. A. and Clayton, R. K. (1974) "The absolute quantum efficiency of bacteriochlorophyll photo-oxidation in reaction centers of Rhodopseudomonas sphaeroides", Biochim. Biophys. Acta 333, 246-260.

[3] Loach, P. A. and Sekura, D. L. (1968) "Primary photochemistry and electron transport in Rhodospirillum rubrum", Biochemistry 7, 2642-2649.

[4] Chang, C. H., Schiffer, M., Tiede, D., Smith, U. and Norris, J. R. (1985) "Characterization of bacterial photosynthetic reaction center crystals from Rhodopseudomonas sphaeroides R-26 by X-ray diffraction", J. Mol. Biol. 186, 201-203.

[5] a. Norris, J. R., Bowman, M. K., Budil, D. E., Tang, J., Wraight, C. A. and Closs, G. L. (1982) "Magnetic characterization of the primary state of bacterial photosynthesis", Proc. Natl. Acad. Sci. 79, 5532-5536.
b. Bowman, M. K., Budil, D. E., Closs, G. L., Kostka, A. G., Wraight, C. A. and Norris, J. R. "Magnetic resonance spectroscopy of the primary state, P^F, of bacterial photosynthesis", Proc. Natl. Acad. Sci. 78, 3305-3307.

[6] Closs, G. L. and Miller, J. R. (1988) "Intramolecular long-distance electron transfer in organic molecules", Science 240, 440-447.

[7] Friesner, R. A. and Won, Y. (1989) "Spectroscopy and electron transfer dynamics of the bacterial reaction center", Biochim. Biophys. Acta 977, 99-122.

[8] Thurnauer, M. C., Katz, J. J. and Norris, J. R. (1975) "The triplet state in bacterial photosynthesis: possible mechanisms of the primary photo-act", Proc. Natl. Acad. Sci. 72, 3270-3274.

[9] Moore, T. A., Gust, D., Hatlevig, S., Moore, A. L., Makings, L. R., Pessiki, P. J., De Schryver, F. C., Auweraer, V. M., Lexa, D., Bensasson, R. A. and Rougee, M. (1988) "Photoinitiated electron transfer in carotenoporphyrin-quinone triads: enhanced quantum yields via control of reaction exergonicity", Israel Journal of Chem. 28, 87-95.

[10] Wasielewski, M. R., Gaines, G. L., O'Neil, M. P., Svec, W. A. and Niemczyk, M. P. (1990) "Photoinduced spin-polarized radical ion pair formation in a fixed-distance photosynthetic model system at 5K", J. Am. Chem. Soc. 112, 4559-4560.

APPLICATION OF BIOCATALYSTS AND ARTIFICIAL CATALYSTS IN ARTIFICIAL PHOTOSYNTHETIC ASSEMBLIES

I. Willner and B. Willner
Institute of Chemistry
The Hebrew University of Jerusalem
Jerusalem, 91904, Israel

ABSTRACT. Biocatalysts, heterogeneous catalysts and homogeneous catalysts are applied as catalytic materials in artificial photosynthetic assemblies. Photoinduced CO_2-fixation to formate proceeds in the presence of a Pd-colloid. A photogenerated hydrido-rhodium complex is applied for the regeneration of NAD(P)H cofactors and nicotinamide model compounds. Photogenerated artificial electron carriers are electrically communicated with redox enzymes.

1. Introduction

Carbon dioxide reduction to organic fuel products and reduction of nitrogen or nitrogen oxide compounds to ammonia provide a useful pathway for the conversion of abundant materials to fuel products and valuable agrochemicals. Solar light induced carbon dioxide reduction and N_2-fixation systems could provide man-made chemical assemblies mimicking photosynthesis in nature. The constant growth in atmospheric CO_2 content and the increase in nitrous oxide pollutants contribute to an increasing environmental concern (the greenhouse effect). The development of artificial photosynthetic systems could remove these pollutants and have a general environmental impact.

E. Pelizzetti and M. Schiavello (eds.), Photochemical Conversion and Storage of Solar Energy, 151–180.

Photosensitized electron-transfer (ET) reactions, (eq. 1), and formation of an "electron-hole" pair of ionic products represent the general pathway for the conversion of solar light energy into chemical potential [1]-[4]. Figure 1 outlines the design of an artificial photosynthetic assembly based on photosensitized

$$S + A \overset{h\nu}{\rightleftharpoons} S^{+} + A^{-} \qquad (1)$$

ET reactions. Light induced ET from the excited photosensitizer, S, to the electron acceptor A, results in conversion of light energy to redox potential. Reduction of the oxidized photosensitizer by an electron donor, D, recycles the light harvesting component, and the intermediate redox products A^{-} and D^{+} can then be utilized in subsequent energy storage chemical transformations. Control of photoinduced ET reactions by means of organized microenvironments [5]-[10], and design of light harvesting materials [11]-[14] have been subjects of extensive research.

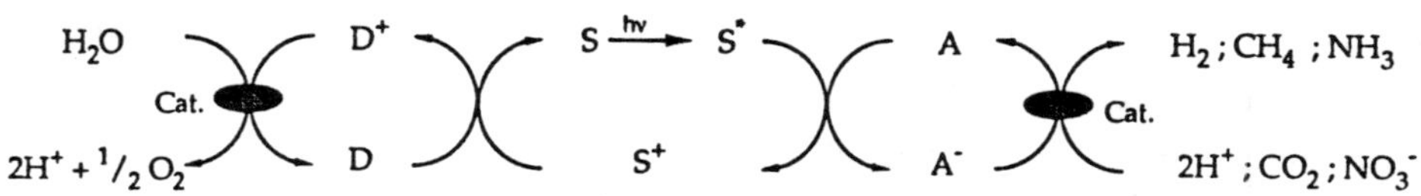

Figure 1: Scheme of an artificial photosynthetic assembly.

One of the major problems in tailoring artificial photosynthetic devices is, however, the linkage of the intermediate photogenerated redox products to the fuel production chemical transformations. For effective utilizatiion of the photoproducts, participation of catalysts in the photosynthetic assemblies is essential [15]. In this account we will highlight the application of heterogeneous catalysts, homogeneous catalysts and enzymes (biocatalysts) as catalytic materials that can be linked to photogenerated redox products. Specific emphasis will be directed towards the development of catalysts for CO_2-fixation and NO_3^- or NO_2^- reduction.

Most of the abundant raw materials that might act as substrates for the fuel products, i.e., water, carbon dioxide, nitrogen, are generated in nature in energy releasing transformation. As a result, these compounds reveal high chemical stability and consequently chemical inertness. Thus, activation of these materials toward chemical reactivity by proper catalysts is essential. A further function of catalysts in artificial photosynthetic assemblies is the need to invoke multi-electron transfer processes. This is best rationalized by following the reduction processes of CO_2 to various C_1-fuel products (eq. 2-eq. 6) and the oxidation of water to various products (eq. 7-eq. 8). Following these examples, we

$$CO_2 + e^- \longrightarrow CO_2^{-\cdot} \qquad E° = -1.9 \text{ Volt} \qquad (2)$$

$$CO_2 + 2e^- + 2H^+ \longrightarrow HCOOH \qquad E° = -0.61 \text{ Volt} \qquad (3)$$

$$CO_2 + 4e^- + 4H^+ \longrightarrow CH_2O + H_2O \qquad E° = -0.48 \text{ Volt} \qquad (4)$$

$$CO_2 + 6e^- + 6H^+ \longrightarrow CH_3OH + H_2O \qquad E° = -0.38 \text{ Volt} \qquad (5)$$

$$CO_2 + 8e^- + 8H^+ \longrightarrow CH_4 + 2H_2O \qquad E° = -0.24 \text{ Volt} \qquad (6)$$

$$H_2O \longrightarrow HO\cdot + H^+ + e^- \qquad E° = 2.38 \text{ Volt} \qquad (7)$$

$$2H_2O \longrightarrow O_2 + 4H^+ + 4e^- \qquad E° = 0.82 \text{ Volt} \qquad (8)$$

realize that reduction of CO_2 or oxidation of water involve multi-electron transfer processes. Furthermore, it is also implied that the oxidation or reduction processes are thermodynamically favoured as the number of electrons involved in the ET processes increases. Thus, catalysts capable of storing electrons or oxidation equivalents, are desirable. Heterogeneous and homogeneous catalysts as well as biocatalysts are capable of acting as multi-electron charge (or hole) carriers. This subject has recently been reviewed, and the mechanisms through which the different catalytic materials maintain these functions were considered in detail [15].

A further aspect involved in using catalysts in photosynthetic assemblies relates to their specificity. One can realize that reduction of carbon dioxide can proceed to different C_1-fuel products, and that oxidation of water leads to different oxidation products. For eliminating chemical reactions between the products and for facilitating their subsequent separation, the specificity in the catalyst performance is desirable. Furthermore, specificity in the catalytic activity is imortant to eliminate the catalyst interactions with the reaction medium. For example, reduction of CO_2 in aqueous environments is accompanied by the thermodynamically feasible H_2-evolution process (eq. 9).

$$2H^+ + 2e^- \longrightarrow H_2 \qquad E° = -0.42 \text{ Volt (pH=7)} \tag{9}$$

Thus, the capability of eliminating H_2-evolution should be an intrinsic feature of a specific CO_2-fixation catalyst.

We conclude that catalytic materials introduced in artificial photosynthetic assemblies should exhibit several complementary functions:

i) The catalyst should activate the substrate towards chemical reactivity.

ii) The catalyst should act as an electron (or hole) sink capable of effecting multi-electron transfer reactions in the catalyst-substrate assembly.

iii) The catalyst should exhibit product specificity and eliminate electrical interaction of the redox products with undesirable ingredients of the system.

2. Photosensitized Reduction of CO_2/HCO_3^- to Formate Using a Pd-β-Cyclodextrin Colloid as Heterogeneous Catalyst.

Carbon dioxide exists in water in three different hydrated forms (eq. 10). The concentration of the different hydrated forms (bicarbonate, carbonate) depends on the pH value of aqueous medium, and thus, control of the catalyst interactions with a certain hydrated species is desirable.

$$CO_2 + H_2O \underset{}{\overset{K_H=2.56\times10^{-3}}{\rightleftharpoons}} H_2CO_3 \overset{pK_a=6.35}{\rightleftharpoons} HCO_3^- + H^+ \overset{pK_a=10.32}{\rightleftharpoons} CO_3^{=} + 2H^+ \tag{10}$$

Carbon dioxide reduction at a catalyst site might proceed by two alternative pathways (Figure 2). ET from the reduced relay to the catalyst site generates a

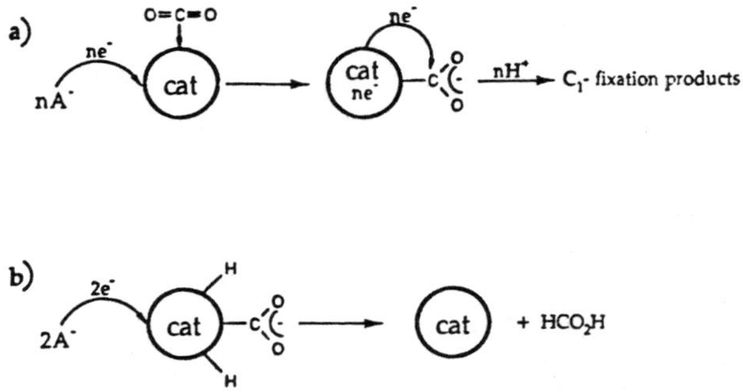

Figure 2: Carbon dioxide fixation at catalytic sites: a) by ET to catalyst activated CO_2 followed by protonation steps; b) by hydrogenation of catalyt activated CO_2.

multi-electron charge relay. Subsequent ET to catalyst activated CO_2 followed by protonation of the reduced intermediates then leads to CO_2-fixation products (Figure 2(a)). A second pathway for CO_2-fixation is outlined in Figure 2(b). This route involves ET from the reduced relay to the catalyst site and subsequent formation of intermediate hydrogen-atom (or hydrido)-catalyst species. Hydrogenation of catalyst activated CO_2 could then lead to carbon dioxide fixation products, provided that H_2-evolution (by dimerization of catalyst associated H-atoms or protonation of hydrido species) is kinetically unfavoured as compared to the hydrogenation route. We have earlier described the application of ruthenium (Ru) and osmium (Os) colloids as heterogeneous catalysts for the photosensitized reduction of CO_2 to methane, (eq. 6) [16,17]. In these photosystems, the $8e^-$ reduction of carbon dioxide to CH_4 proceeds through the direct ET from the metal catalyst (acting as multi-electron relay), to surface activated CO_2 (Figure 2(a)).

In this account the reduction of CO_2/HCO_3^- to formate using a heterogeneous catalyst composed of palladium (Pd) colloid will be discussed [18]. The catalyst's performance proceeds by the hydrogenation route outlined in Figure 2(b). The photosystem is composed of deazariboflavin (**1**) acting as the light harvesting compound, N,N'-dimethyl-4,4'-bipyridinium dichloride, methyl viologen, MV^{2+}, (**2**), as electron acceptor and oxalate as electron donor. The catalyst included in the photosystem is composed of a Pd-colloid stabilized by β-cyclodextrin. β-Cyclodextrin is a macrocyclic polysugar composed of seven glucose units linked in a circular structure by 1, 4 glycoside bonds [19]. Illumination of the photosystem, pH=6.8, in the absence of CO_2/HCO_3^- results in effective H_2-evolution. The sequence of ET processes leading to H_2-evolution are summarized in Figure 3. Oxidative ET quenching of the photosensitizer results in $MV^{+\cdot}$ that acts as intermediate electron carrier. Charging the Pd-colloid by electrons results in metal associated H-atoms (or hydrides), and their dimerization leads to H_2-evolution. Oxalate reduces the oxidized photosensitizer, a process that regenerates the light active compound. In the presence of CO_2/HCO_3^- the photosystem alters its functions. Upon illumination, the major photoproduct is formate, and only trace amounts of H_2 are formed, Figure 4. The quantum efficiency of formate production corresponds to $\phi=1.1$. Labelling experiments, applying $H^{13}CO_3^-$, confirm that formate originates from the reduction of bicarbonate. Introduction of gaseous CO_2 to the photosystem results in only inefficient formation of formate. Addition of carbonic anhydrase, an enzyme catalysing the hydration of CO_2, results in acceleration of formate production, to a value similar to that in the presence of CO_2/HCO_3^- composition. We thus conclude that the photosystem exhibits selectivity towards carbon dioxide fixation, and H_2-evolution is eliminated. Also, the species being reduced in the photocatalytic assembly is HCO_3^- rather than CO_2.

Some insight into the catalytic activity of the Pd-colloid is obtained from kinetic studies accompanying the reduction of HCO_3^-. The rate of formate production as a function of bicarbonate concentration, exhibits a saturation curve (Figure 5(A)). Such saturation curve suggests the existence of HCO_3^- activation sites on the Pd-catalyst leading to maximum catalytic effectiveness at full coverage of these sites. Examination of kinetic isotope effects reveals that the rate of formate formation in a photosystem composed of CO_2/DCO_3^-, D_2O is identical to that observed in the system CO_2/HCO_3^-, H_2O. Yet, the H_2-evolution process in the absence of CO_2/HCO_3^- exhibits a substantial isotope effect, $k_H/k_D \approx 3$. Thus, formation of Pd associated H-atoms by charging the colloid with electrons transferred by $MV^{+\cdot}$ is rapid as compared to the discharge of molecular hydrogen. The slow discharge of Pd-H intermediates allows subsequent hydrogenation of metal activated bicarbonate to formate (eq. 11). The saturation curve of the formate

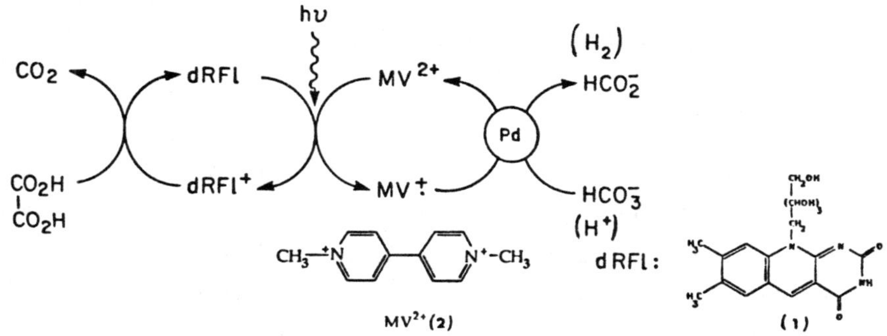

Figure 3: Scheme for H_2-evolution and CO_2/HCO_3^- reduction in the photocatalytic assembly including **Pd-β-CD** catalyst.

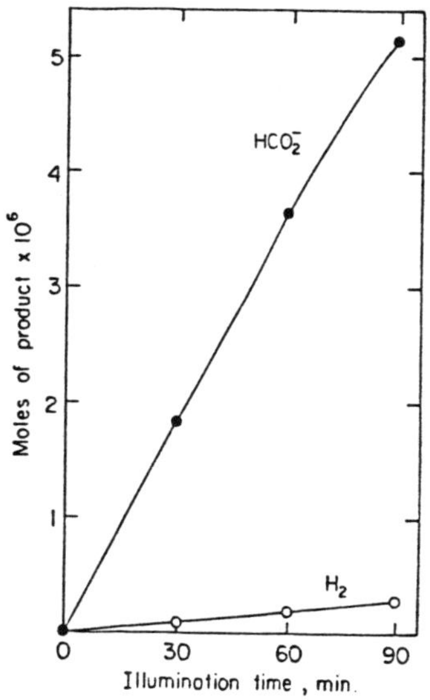

Figure 4: Rate of a) formate production and b) H_2-evolution as a function of illumination time in the photosystem including: 3 ml 0.1 M $NaHCO_3$ solution, pH=6.8, [dRFl]=$3x10^{-5}$ M, [MV^{2+}]=$1x10^{-3}$ M, [oxalate]=$5x10^{-2}$ M, and Pd-β-CD=40 $mgxL^{-1}$.

$$H\text{-}Pd\text{-}H + HCO_3^- \underset{k_{-1}}{\overset{k_1}{\rightleftharpoons}} [H\text{-}Pd\text{-}H(HCO_3^-)] \xrightarrow{k_p} Pd + HCO_2^- + H_2O \quad (11)$$

production rate, Figure 5(A), can be analysed by the active site kinetic model of enzymes, where the rate of the process, v_o, is given by eq. 12. K_m is given by eq. 13 and V_{max} represents the maximum rate of the process. Rearranging eq. 12 to eq. 12(a), suggests a linear relationship between $1/v_o$ and $1/[HCO_3^-]$. Figure 5(B) shows that analysing the saturation curve by such active site model gives the proper linear dependence implying that the Pd-colloid includes specific bicarbonate activation sites. From this curve the values $V_{max} = (1.5 \pm 0.5) \times 10^{-5}$ $M \cdot min^{-1}$ and $K_m = (1.5 \pm 0.6) \times 10^{-2}$ M are derived.

$$v_o = \frac{V_{max}[HCO_3^-]}{K_m + [HCO_3^-]} \quad (12)$$

$$\frac{1}{v_o} = \frac{1}{V_{max}} + \frac{K_m}{V_{max}[HCO_3^-]} \quad (12a)$$

$$K_m = \frac{k_{-1} + k_p}{k_1} \quad (13)$$

The photosensitized reduction of CO_2/HCO_3^- to formate is inhibited by various additives, i.e., amines, aldehydes, thiols. Figure 6(A) shows the decline of the rate of formate production in the photosystem at increasing concentrations of added mercaptoethanol, acting as inhibitor. Added mercaptoethanol not only retards the formation of formate, but restores the activity of the photosystem towards H_2-evolution. When formate production is eliminated by the inhibitor it exhibits an H_2-evolution activity similar to that observed in the absence of CO_2/HCO_3^-. The

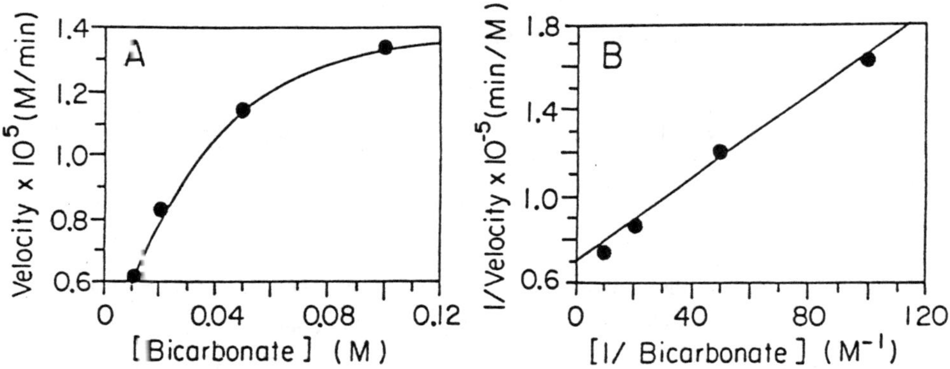

Figure 5: (A) Initial rate of formate production as a function of HCO_3 concentration. Photosystem is composed of $[dRFl]=3x10^{-5}$ M, $[MV^{2+}]=1x10^{-3}$ M, $[oxalate]=5.5x10^{-2}$ M and Pd-β-CD=40 $mgxL^{-1}$.

(B) Kinetic analysis of saturation curve (Figure 5A) using eq. 12a.

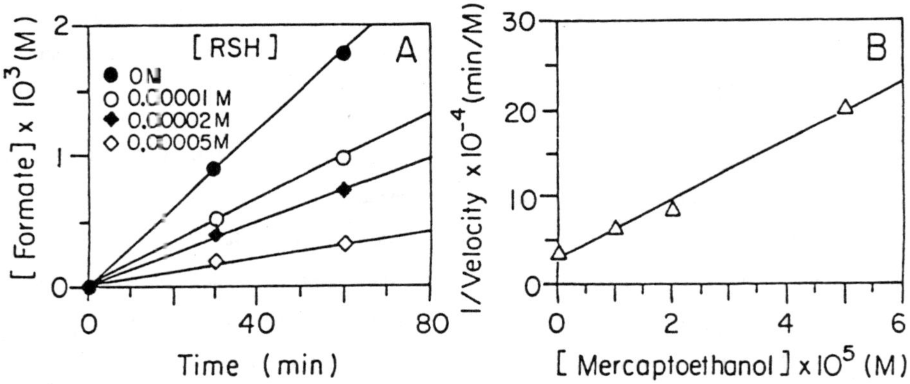

Figure 6: (A) Inhibition of formate production rate at different concentrations of mercaptoethanol. (B) Kinetic analysis of inhibition experiments (Figure 6A) applying eq. 15.

inhibition process can be analysed in terms of a competitive inhibition mechanism, where the active sites are blocked by the inhibitor, I, eq. 14.

$$\begin{array}{l} \text{H-Pd-H} \underset{}{\overset{+HCO_3^-}{\rightleftharpoons}} [\text{H-Pd-H}(HCO_3^-)] \longrightarrow \text{Pd} + HCO_2^- + H_2O \\ \quad +I \Big\Updownarrow \\ [\text{H-Pd-H(I)}] \end{array} \qquad (14)$$

Competitive inhibition of the active sites of Pd implies that the relation between the rate of formate formation and the inhibitor concentration will follow eq. 15, where K_I is given by eq. 16. Figure 6(B) shows the analysis of the inhibited rates of formate production in terms of a competitive inhibition mechanism. The derived linear dependence supports the inhibition mechanism and allows us to estimate the association constant of mercaptoethanol to the bicarbonate activation sites, $K_I = (2.9 \pm 1.5) \times 10^{-6}\ M^{-1}$.

$$1/v_0 = \frac{K_m[RSH]}{V_{max} \cdot K_I[HCO_3^-]} + \frac{1}{V_{max}} \quad 1 + \frac{K_m}{[HCO_3^-]} \qquad (15)$$

$$K_I = \frac{[\text{H-Pd-H}][RSH]}{[\text{H-Pd-H}(RSH)]} \qquad (16)$$

This kinetic analysis allows us to schematically present functions of the Pd-colloid in the reduction of bicarbonate to formate, Figure 7. The Pd-colloid includes specific sites for HCO_3^- activation. Charging of the Pd colloid by electrons supplied by the electron carrier generates surface Pd-H species. As discharge of H-atoms is a slow process, hydrogenation of metal activated HCO_3^- to formate proceeds. The inhibitor, mercaptoethanol, specifically binds to the bicarbonate activation sites. Under such conditions, bicarbonate is desorbed and only the slow H_2-evolution process takes place.

Other Pd-colloids stabilized by glucose, amylose, polyvinyl pyrrolidone exhibit similar catalytic activities towards bicarbonate reduction. Also, immobilization of Pd-β-cyclodextrin onto TiO_2 powders, followed by dialysing off the β-cyclodextrin stabilizer, results in effective photocatalytic semiconductor materials for CO_2/HCO_3^- reduction [20,21].

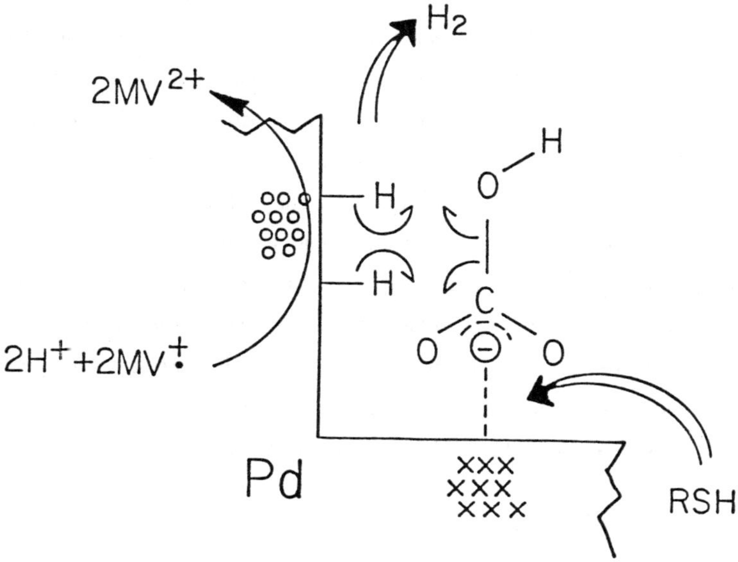

Figure 7: Scheme for the catalytic functions of the Pd-β-CD catalyst in the photosensitized reduction of CO_2/HCO_3^- to formate.

3. Photosensitized Regeneration of NAD(P)H Cofactors by Homogeneous Rhodium Complexes.

Dihydronicotinamide cofactors act as two-electron carriers for many redox enzymes. Photosensitized regeneration of NADH and NADPH cofactors has been reported by us, using artificial photosystems coupled to biocatalysts [22-24]. The photosensitizers composing the photosystems for NADH and NADPH regeneration involve Zn(II)-meso-tetra-methylpyridinium porphyrin and Ru(II)-tris-bipyridinium, respectively. The enzymes lipoamide dehydrogenase and ferredoxin reductase mediate the formation of NADH and NADPH respectively, by utilizing photogenerated N,N'-dimethyl-4,4'-bipyridinium radical cation, $MV^{+\cdot}$, formed in the two photosystems. Thus, the biocatalysts mediate the conversion of a single ET product, to the biologically active two-electron relays. These photosensitized NAD(P)H regeneration systems have been coupled to various endoergic photosynthetic transformations such as reduction of ketones and keto-acids [23], photosynthesis of amino-acids [25], and carbon dioxide fixation through carboxylation of keto-acids [26], (i.e. photosynthesis of malic acid and isocitric acid). Biocatalysed artificial photosynthetic assemblies can be further developed by two conceptual approaches: i) Development of artificial catalysts acting as hydride (two electron) carriers could substitute native enzymes for regeneration of biologically active NAD(P)H cofactors, and ii) Direct electrical communication between artificial electron carriers and the redox active sites in enzymes might exclude the need for the cofactors. These two general approaches have recently been developed in our laboratory and will be discussed in this and the forthcoming sections.

Hydrido-transition metal comlexes are major catalytic species in various chemical transformations. Generation of hydrido-metal species by a photosensitized ET process, eq. 17, could be a central two-electron relay intermediate for various photocatalysed transformations such as hydrogenation (by insertion of olefins), hydroformylation, CO_2-fixation, and NAD(P)H cofactors regeneration, Scheme 1.

$$M^{n+} + 2e^{-} + H^{+} \longrightarrow M^{n+}\text{—}H \qquad (17)$$

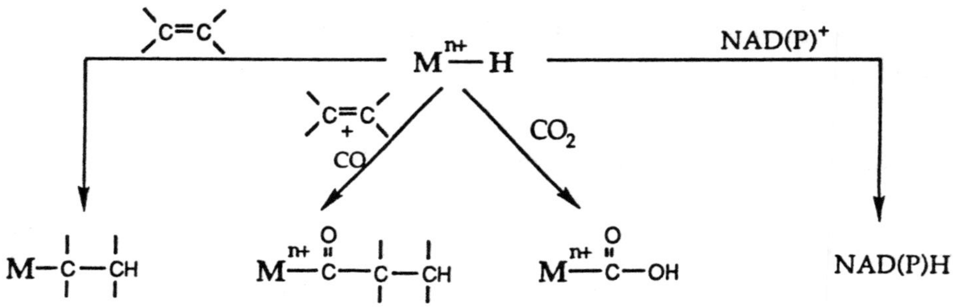

Scheme 1: Catalytic transformations by hydrido-metal intermediates.

A photosystem for the production of the hydrido-rhodium species derived from chloro-tris-(diphenylphosphinobenzene-m-sulfonate)rhodium(I), $RhCl(dpm)_3^{3-}$, (3), has been developed [27,28]. The photosystem is composed of $Ru(bpy)_3^{2+}$ as photosensitizer, ascorbate as electron donor and $RhCl(dpm)_3^{3-}$ as electron carrier. Illumination of this photosystem results in the formation of the hydrido rhodium complex derived from (3). The sequence of reactions leading to the hydrido-rhodium photoproduct is outlined in eq.18-eq.20. Reductive ET quenching of the photosensitizer results in the reduced photosensitizer, $Ru(bpy)_3^+$. This photoproduct reduces (3),

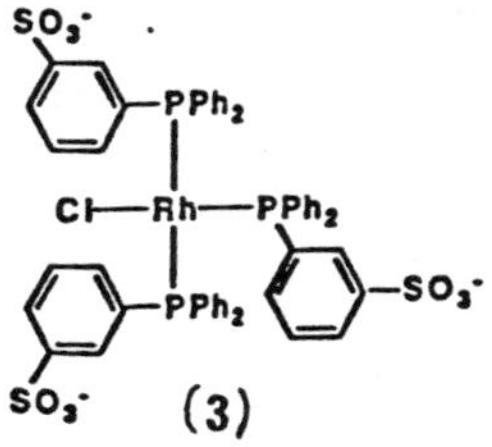

(3)

that undergoes disproportionation to the hydrido-Rh(I) photoproduct. The latter product is a versatile catalytic intermediate for different transformations. Protonation of the hydrido complex leads to H_2-evolution [27], while olefins or acetylenes undergo hydrogenation [28] and a mixture of alkenes and carbon-monoxide leads to hydroformylation products [28].

$$*Ru(bpy)_3^{2+} + HA^- \longrightarrow Ru(bpy)_3^+ + HA\cdot \quad (HA^- = ascorbate) \qquad (18)$$

$$Ru(bpy)_3^+ + Rh(I)Cl(dpm)_3^{3-} + H^+ \longrightarrow Ru(bpy)_3^{2+} + Rh(II)HCl(dpm)_3^{3-} \qquad (19)$$

$$2Rh(II)HCl(dpm)_3^{3-} \longrightarrow Rh(I)HCl(dpm)_3^{4-} + Rh(I)Cl(dpm)_3^{3-} + H^+ \qquad (20)$$

The hydrido-rhodium photoproduct acts also as a two-electron carrier for the reduction of the biologically active NAD^+ and $NADP^+$ cofactors [29]. Illumination of the photosystem that includes $RhCl(dpm)_3^{3-}$ as electron carrier in the presence of $NAD(P)^+$, results in the formation of NADH and NADPH. The quantum efficiencies for the formation of NAD(P)H cofactors are similar and correspond to $\phi = 5x10^{-3}$. It should be noted that nicotinamide cofactors are reducible at positions 2 and 4. Chromatographic analyses reveal that the reduced NAD(P)H cofactors are the biologically active 1,4-dihydronicotinamides with a purity exceeding 93%. Thus, the photogenerated hydrido-rhodium complex, eq. 20, mediates the regioselective formation of NAD(P)H cofactors, eq. 21.

$$Rh(I)HCl(dpm)_3^{4-} + NAD(P)^+ \longrightarrow Rh(I)Cl(dpm)_3^{3-} + NAD(P)H \qquad (21)$$

Synthetic nicotinamide analogues can be similarly reduced in the photosystem composed of $Ru(bpy)_3^{2+}$ as photosensitizer, ascorbate as electron donor and

$RhCl(dpm)_3^{3-}$ as catalyst. Illumination of the photosystem that includes N-benzyl nicotinamide, BNA^+, (4), results in the formation of N-benzyl-1,4-dihydro-nicotinamide, BNAH, with a quantum efficiency corresponding to $\phi=2.0x10^{-2}$. Figure 8 shows the build-up of BNAH in the system at time intervals of illuminations. It will be shown that the synthetic analogue, BNAH, provides a model compound for biocatalysed NAD(P)H systems and can duplicate several functions of native systems.

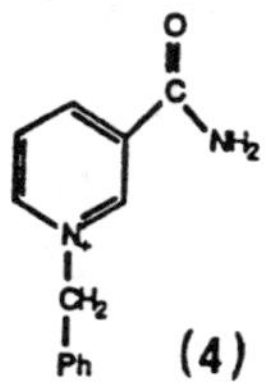

(4)

In the presence of the enzyme alcohol dehydrogenase, AlcDH, photoregenerated NADH mediates the reduction of acetaldehyde to ethanol, eq. 22. The rate of ethanol formation as a function of illumination time is displayed in Figure 9(a). The conversion yield of the ketone to alcohol and the turnover numbers of the biocatalyst and artificial catalyst are displayed in Table 1. Thus, photosensitized regeneration of NADH with the artificial catalyst allows subsequent coupling of the reduced cofactor to biocatalysed transformations.

$$CH_3\overset{O}{C}H + NADH \xrightarrow{AlcDH} CH_3CH_2OH + NAD^+ \qquad (22)$$

Interestingly, the reduced synthetic cofactor analogue, BNAH, can be coupled to a chemical reaction mimicking the activity of the enzyme alcohol dehydrogenase. Biocatalysed reduction of ketones requires the presence of metal ions, i.e. Zn^{++} or Mg^{++} in the active site structure, in addition to the reduced cofactor. The artificial assembly mimicking the biocatalytic system is schematically displayed in Figure 10. The assembly is composed of two phases. In the aqueous phase, the photosystem composed of $Ru(bpy)_3^{2+}$, ascorbate, and $RhCl(dpm)_3^{3-}$ is localized, and the electron mediator BNA^+ and Mg^{2+} are also included. The organic phase includes the substrate, benzil, (5). Illumination of the stirred two-phase system results in the reduction of (5) to benzoin (6), eq. 23. Figure 9(b) shows the rate of benzoin formation as a function of illumination time. Table 1 includes the turnover numbers of the components included in the system. Mechanistic studies

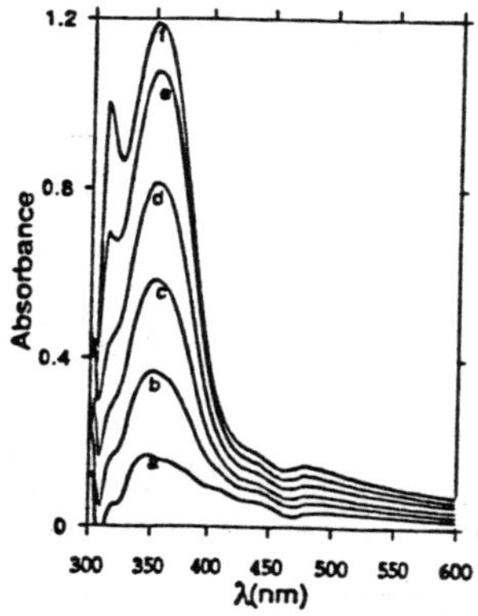

Figure 8: Photoinduced formation of BNAH at time intervals of illumination, (a) 1 min; (b) 2 min; (c) 3 min; (d) 4 min; (e) 6 min; (f) 10 min. $[Ru(bpy)_3^{2+}]=1.4x10^{-3}$ M, pH=6.8 (0.1 M phosphate buffer).

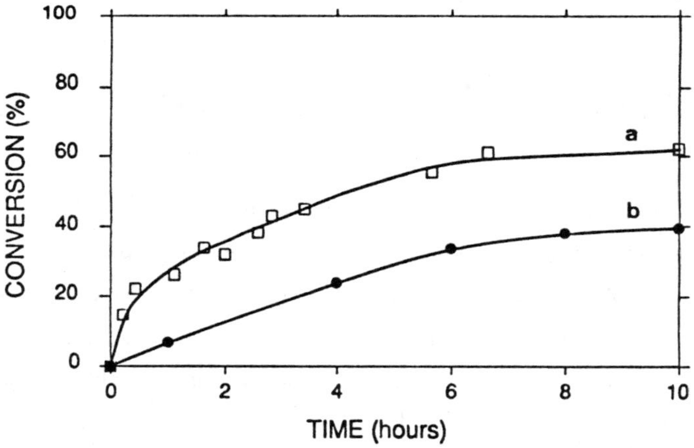

Figure 9: Rates of photoinduced conversion of acetaldehyde into ethanol (a) (□) and of benzil into benzoin (b) (●) using the NADH cofactor and BNAH model-cofactor regeneration systems. The ethanol producing system was composed of 2.5 ml: $[Ru(bpy)_3^{2+}]=1.4x10^{-3}$ M, $[ascorbate]=5.0x10^{-2}$ M, $[(\mathbf{1})]=6.5x10^{-5}$ M, $[NAD^+]=3.0x10^{-4}$ M, alcohol dehydrogenase 2 units, $[CH_3CHO]=1.8x10^{-2}$ M, pH=6.8 (0.1 M phosphate buffer). The benzoin-producing system contained an aqueous phase (2 ml): $[Ru(bpy)_3^{2+}]=1.4x10^{-3}$ M, $[ascorbate]=1.9x10^{-2}$ M, $[(\mathbf{1})]=2.0x10^{-4}$ M, $[BNA^+]=5.0x10^{-3}$ M, $[Mg(ClO_4)_2]=5.0x10^{-5}$ M, pH=6.8 (0.1 M phosphate buffer) and an organic phase (methylene chloride 1 ml):$[benzil]=5.0x10^{-3}$ M.

Table 1: Conversion yields and turnover numbers of the components involved in the reduction of acetaldehyde and benzil by photosensitized regeneration of NADH and BNAH, respectively.

System Product	Illumination time (hours)	Conversion (%)	Turnover number[(a)] $RhCl(dpm)_3^{3-}$	NAD^+	BNA^+	Catalyst[(b)]
Ethanol	10	65	180	3	-	1740
Benzoin	30	40	5	-	20	20

(a) Turnover number is defined as $\left(\frac{\text{number of moles of product}}{\text{number of moles of component}}\right)$

(b) catalyst is the enzyme AlcDH for ethanol production and Mg^{2+}-ions for benzoin formation.

$$\underset{(5)}{Ph\text{-}\overset{O}{\overset{\|}{C}}\text{-}\overset{O}{\overset{\|}{C}}\text{-}Ph} + BNAH \xrightarrow{Mg^{++}} \underset{(6)}{Ph\text{-}\overset{OH}{\overset{|}{C}H}\text{-}\overset{O}{\overset{\|}{C}}\text{-}Ph} + BNA^+ \qquad (23)$$

reveal that Mg^{2+} ions are essential to effect the reduction of (5). In the aqueous phase, the photosystem generates the reduced photoproduct, BNAH. The latter photoproduct binds to Mg^{2+} ions, and due to its hydrophobic character is extracted into the organic phase. The phase-transferred Mg^{2+}-ion, acts as Lewis acid and binds the ketone function group to form a supramolecular assembly composed of $[BNAH\text{-}Mg^{2+}\text{---}O{=}C{<}]$. In the supramolecular structure the carbonyl function is activated towards reduction, and hydride transfer producing the resulting alcohol, (6), is induced [30].

The principle of regenerating NAD(P)H cofactors by means of homogeneous hydrido-transition metal complexes has general implications. The photosystem regenerating the cofactors could be substituted by hydride donors in dark processes, i.e., formate, alcohols. Indeed, with formate as hydride donor, regeneration of NAD(P)H takes place, using $RhCl(dpm)_3^{3-}$ as intermediary hydride carrier [29]. Recently, a mini-reactor for the regeneration of biologically active NADH cofactor has been assembled based on this concept [31].

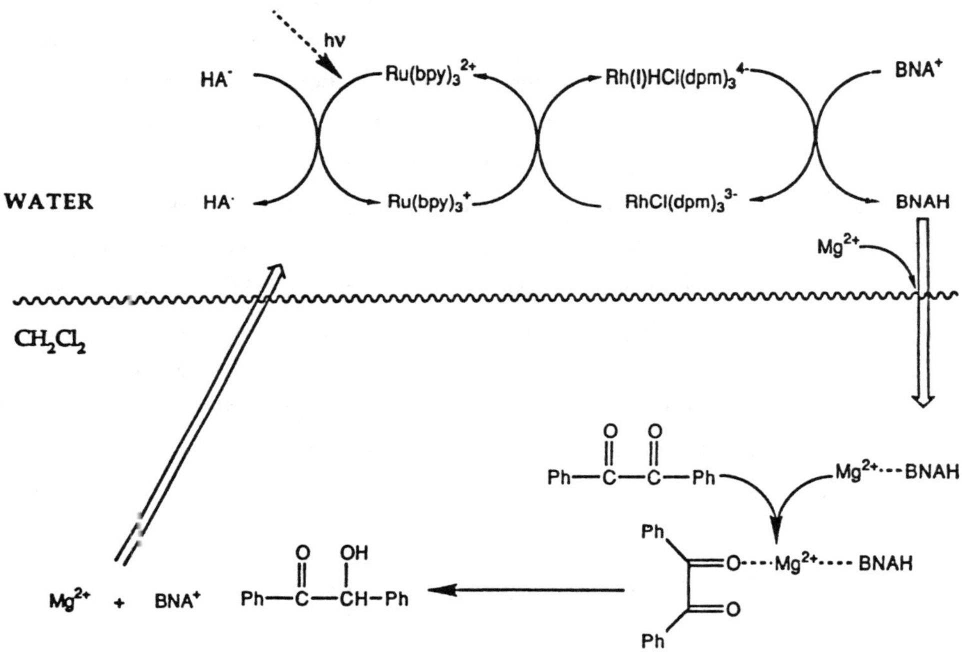

Figure 10: Scheme for the photosensitized reduction of benzil (5) in a water-oil two phase system using BNAH-Mg^{2+} model.

The principle of regenerating NAD(P)H cofactors by means of homogeneous hydrido-transition metal complexes has general implications. The photosystem regenerating the cofactors could be substituted by hydride donors in dark processes, i.e., formate, alcohols. Indeed, with formate as hydride donor, regeneration of NAD(P)H takes place, using $RhCl(dpm)_3^{3-}$ as intermediary hydride carrier [29]. Recently, a mini-reactor for the regeneration of biologically active NADH cofactor has been assembled based on this concept [31].

4. Electrical Communication between Photogenerated Electron Carriers and Redox Enzymes.

Substitution of native cofactors by artificial electron carriers could provide a novel approach to drive biotransformations, otherwise mediated by the cofactors.

The active sites in redox enzymes, are, however, shielded by an insulating protein shell. Natural cofactors are recognized by the protein backbone and their penetration into the protein assembly allows electrical communication. Thus, electrical communication between artificial electron carriers and redox sites requires the permeability of the protein backbone towards the electron relay.

Photogenerated N,N'-dimethyl-4,4'-bipyridinium radical cation, $MV^{+\cdot}$, exhibits permeability into various proton backbones, and electrical communication could be established with different redox enzymes. Photosystems including a photosensitizer, i.e. $Ru(bpy)_3^{2+}$ or Zn-porphyrins, sacrificial electron donors and MV^{2+} as electron acceptor, have been coupled to the enzymes formate dehydrogenase, ForDH [32], and hydrogenase, Hyd [33]. In these systems, photogenerated $MV^{+\cdot}$ mediates carbon dioxide fixation to formate (eq. 24) and hydrogen evolution (eq. 25) in the presence of the biocatalysts.

$$CO_2 + 2MV^{+\cdot} + 2H^+ \xrightarrow{\text{ForDH}} HCO_2H + 2MV^{2+} \qquad (24)$$

$$2H^+ + 2MV^{+\cdot} \xrightarrow{\text{Hyd}} H_2 + 2MV^{2+} \qquad (25)$$

The photosensitized reduction of nitrate to ammonia, eq. 26, has recently been accomplished in our laboratory by communicating the artificial electron carrier, $MV^{+\cdot}$, with the two enzymes: nitrate reductase, NitraR, and nitrite reductase, NitriR, operating in a series [34].

$$NO_3^- + 8e^- + 10H^+ \longrightarrow NH_4^+ + 3H_2O \qquad (26)$$

The photosystem is composed of $Ru(bpy)_3^{2+}$ as photosensitizer, N,N'-dimethyl-4,4'-bipyridinium, MV^{2+}, as electron acceptor and Na_2EDTA as sacrificial electron donor. Illumination of the photosystem in the presence of the enzyme NitraR and NO_3^- as substrate results in nitrite as photoproduct. Similarly, illumination of the photosystem in the presence of NitriR and NO_2^- as substrate, results in the formation of ammonia. Figure 11(A) shows the rates of NO_2^- and ammonia formation in the respective photosystems. Introduction of the two enzymes, NitraR and NitriR into the photosystem that includes NO_3^- as substrate, results in the multi-electron reduction of nitrate to ammonia. Figure 11(B) shows the rate of NO_2^- and ammonia formation in this photosystem. It is evident that at primary stages of illumination, NO_2^- is accumulated and as illumination proceeds, ammonia becomes the major photoproduct of the system. The mechanism leading to the formation of ammonia is schematically outlined in Figure 12. Oxidative ET quenching of the excited photosensitizer results in $MV^{+\cdot}$, and the light active material is regenerated by oxidation of Na_2EDTA. The reduced photoproduct, $MV^{+\cdot}$, is recognized by the two enzymes, NitraR and NitriR and acts as electron carrier for the biocatalysts. The first enzyme, NitraR, mediates the two electron reduction of NO_3^- to nitrite, eq. 27, while the second enzyme, NitriR, mediates the multi-electron reduction of NO_2^- to ammonia, eq. 28. Table 2 provides a summary of the quantum efficiencies and turnover numbers of the different components included in the biocatalytic photosynthetic assemblies for the reduction of CO_2 to formate

Table 2: Quantum yields and turnover numbers of components in photosystems for NO_3^--fixation.

Photosystem	ϕ	$Ru(bpy)_3^{2+}$	MV^{2+}	biocatalyst
NO_3^- reduction to NO_2^- [(a)]	0.08	80	18.5	$6.2x10^4$ (NitraR)
NO_2^- reduction to NH_4^+ [(b)]	0.06	38.5	7	$2.4x10^4$ (NitriR)
NO_3^- fixation to NH_4^+ [(c)]	0.06	32	6	$9x10^4$ (NitraR)
				$2.1x10^4$ (NitriR)

[(a)] Composition of photosystem: $[Ru(bpy)_3^{2+}]=7.4x10^{-5}$ M, $[MV^{2+}]=3.2x10^{-4}$ M, $[Na_2EDTA]=2x10^{-2}$ M, $[NO_3^-]=9.9x10^{-3}$ M, NitraR (E.C.1.9.6.1) 0.2 units.
[(b)] Composition of photosystem: $[Ru(bpy)_3^{2+}]=7.4x10^{-5}$ M, $[MV^{2+}]=3.2x10^{-4}$ M, $[Na_2EDTA]=2x10^{-2}$ M, $[NO_2^-]=1x10^{-2}$ M, NitriR (E.C.1.6.6.4) 0.06 units.
[(c)] Composition of photosystem: $[Ru(bpy)_3^{2+}]=7.4x10^{-5}$ M, $[MV^{2+}]=4.2x10^{-4}$ M, $[Na_2EDTA]=2x10^{-2}$ M, NitraR 1 unit and NitriR 0.35 units. F.W. of NitraR and NitriR ≈ 200,000, cf. M.W.W. Adams and L.E. Mortenston in "Molybdenum Enzymes", T.S. Spiro, Ed., John Wiley, 1985.

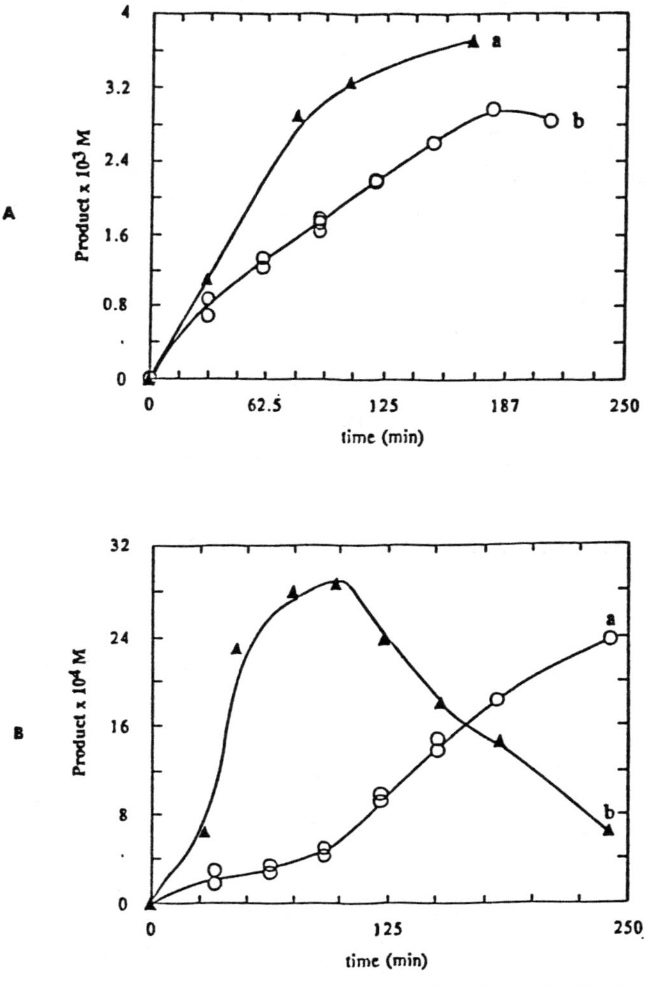

Figure 11: (A) Rates of products formation as a function of illumination time. In all systems $[Ru(bpy)_3{}^{2+}]=7.4x10^{-5}$ M, $[Na_2EDTA]=0.02$ M. (a) (▲) NO_2^--formation, pH=7.0, Tris buffer 0.1 M, $[MV^{2+}]=3.2x10^{-4}$ M, $[NO_3^-]=9.9x10^{-3}$ M, nitrate reductase=0.2 units. (b) (o) NH_4^+ formation, pH=8.0, Tris buffer 0.1 M, $[MV^{2+}]=4.2x10^{-4}$ M, $[NO_2^-]=0.01$ M, nitrite reductase 0.06 units.

(B) NO_2^- and NH_4^+ concentrations in the combined system, as a function of illumination time. (a) (o) NH_4^+. (b) (▲) NO_2^-. pH=8.0, Tris buffer 0.1 M, $[Ru(bpy)_3{}^{2+}]=7.4x10^{-5}$ M, $[Na_2EDTA]=0.02$ M, $[MV^{2+}]=4.2x10^{-4}$ M, $[NO_3^-]=0.1$ M, nitrate reductase 1.0 units, nitrite reductase 0.35 units.

$$NO_3^- + 2e^- + 2H^+ \longrightarrow NO_2^- + H_2O \qquad (27)$$

$$NO_2^- + 6e^- + 6H^+ \longrightarrow NH_4^+ + 2H_2O \qquad (28)$$

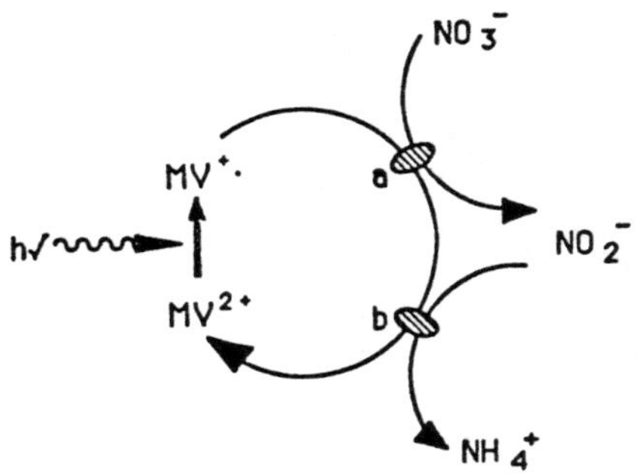

Figure 12: Scheme for biocatalysed sequential reduction of NO_3^- to NH_4^+. a = nitrate reductase, b = nitrite reductase.

and the reduction of NO_3^- and NO_2^- to ammonia. We realize that the photosystem for the fixation of NO_3^- to ammonia exhibits substantial stability and high quantum efficiency. Practical implications of such a photosystem could be solar production of fertilizers, eventually through decontamination of atmospheric nitrous oxides.

In the previously discussed systems, electrical communication between the electron carrier and the active sites is maintained through diffusion of the relay component into the protein backbone. Conversion of biocatalytic photosynthetic systems into practical devices, requires the immobiliztion of the assembly into a rigid-matrice form, allowing long-term recycling of the components. Thus, one should consider means to establish electrical communication of biocatalytic redox sites with their environment in organized rigid assemblies, where diffusion of the electron carrier is eliminted. We find that ET communication between redox enzymes and their macroscopic environment can be established by immobilization of the enzymes into a polymer matrice containing pendant redox groups [35]. The enzyme nitrate reductase (E.C. 1.9.6.1), NitraR, has been entrapped in an acrylamide copolymer composed of the monomers acrylamide and N-methyl-N'-propyl-4,4'-bipyridinium acrylamide (7). The average loading of the acrylamide copolymer by the

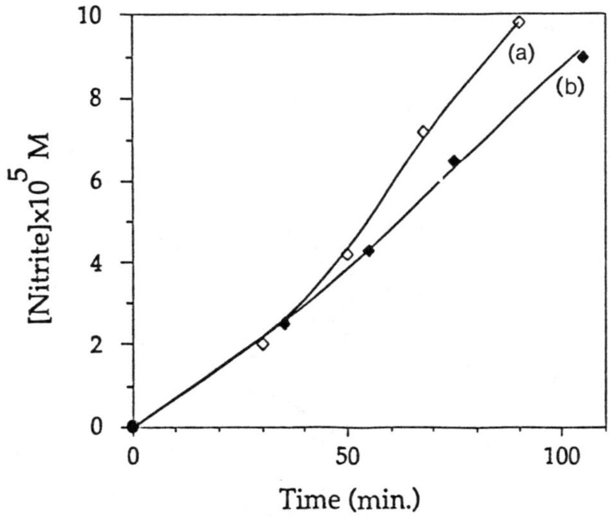

Figure 13: Rate of NO_3^- reduction in the assembly composed of nitrate reductase, 0.23 units (based on native enzyme) immobilized in the acrylamide: (7) copolymer. In all experiments, $[NO_3^-]=4.4\times10^{-3}$ M in 2.3 ml Tris buffer, pH=7.44: (a) through illumination in the presence of $Ru(bpy)_3^{2+}$, 5.5×10^{-5} M, and Na_2EDTA, 4.4×10^{-3} M; (b) Dark reduction in the presence of sodium dithionite, 9.5×10^{-3} M.

bipyridinium redox groups corresponds to 2.7%. Introduction of the enzyme NitraR, immobilized in the redox polymer, into an aqueous photosystem that includes $Ru(bpy)_3^{2+}$ as photosensitizer, Na_2EDTA as electron donor, and nitrate as substrate results upon illumination, in the photoreduction of NO_3^- to nitrite. The rate of NO_2^- formation upon illumination of the system is shown in Figure 13(a). The reduction of NO_3^- to nitrite is also driven in the dark, by adding dithionite to the enzyme being immobilized in the redox matrice. The rate of NO_2^- formation in the dark chemical reaction is displayed in Figure 13(b). It is evident that the rates of NO_2^- formation in the light induced process and in the dark chemical reduction reaction are very similar, implying that the rate limiting process of NO_3^- reduction is independent of the primary reducing component of the systems. Figure 14 outlines schematically the sequence of reactions that occur in the photochemical assembly that includes the redox polymer. Excitation of the photosensitizer results in oxidative ET quenching by the bipyridinium polymer, and the reduced bipyridinium radical cation pendant polymer is formed. The oxidized

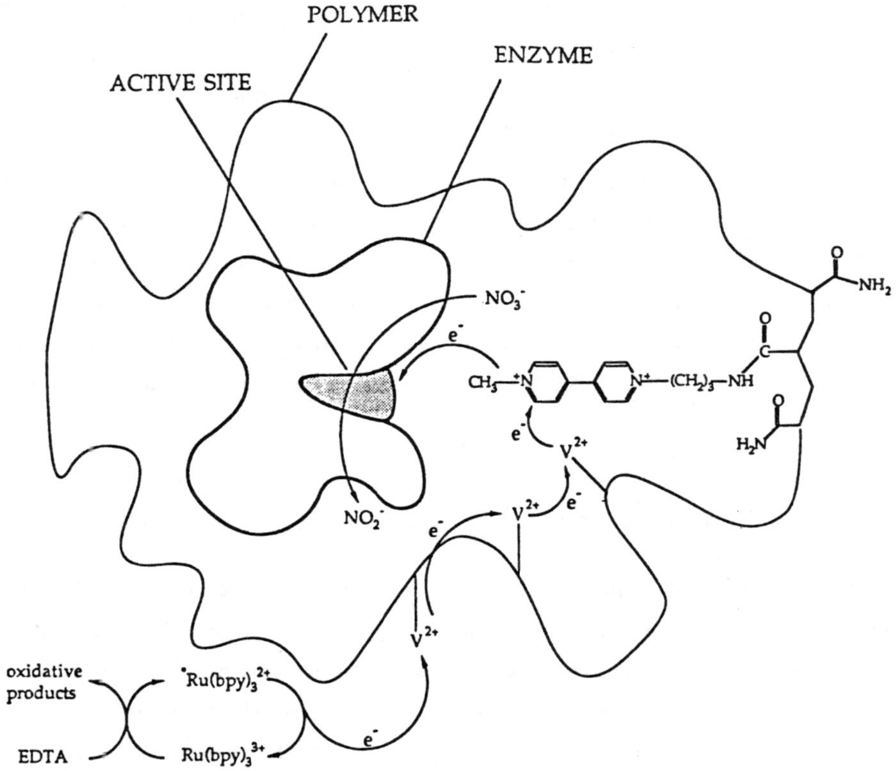

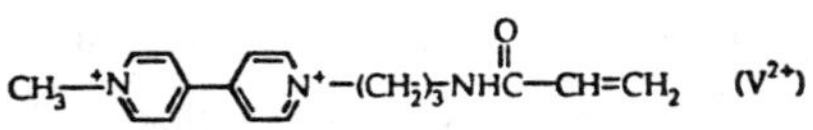

(7)

Figure 14: Photoinduced reduction of NO_3^- to NO_2^- in an organized assembly composed of NitraR immobilized in an acryalmide-bipyridinium redox polymer. ET communication between the redox polymer and the enzyme active site takes place in the redox polymer-enzyme assembly.

photoproduct, $Ru(bpy)_3^{3+}$, oxidizes Na_2EDTA, a process that recycles the photosensitizer. Electron exchange between polymer bound oxidized bipyridinium components and the reduced radical cation site occurs, a process leading to charge distribution across the polymer backbone. It is assumed that immobilization of the enzyme into the redox matrice forces some of the pendant bipyridinium groups into proper distance allignment with the biocatalyst active site, through which ET communication proceeds. Thus, the bipyridinium polymer sites in proximity to the enzyme active site act as electron sinks for collecting polymer distributed electrons and their supply to the biocatalyst active site.

The rate of ET from the reduced redox polymer to the biocatalyst's active site can be determined by means of time-resolved laser flash photolysis experiments, Figure 15. Upon flashing the redox polymer in the presence of the photosensitizer, $Ru(bpy)_3^{2+}$, but without including the NitraR enzyme, the reduced polymer is formed, and its decay, Figure 15(a), by recombination with the oxidized photoproduct, (eq. 29), is followed. The recombination rate constant corresponds to $k_b = (9.5 \pm 2) \times 10^5\ M^{-1} \cdot sec^{-1}$.

$$Ru(bpy)_3^{3+} + V^{+\cdot}\text{—Polymer} \xrightarrow{k_b} Ru(bpy)_3^{2+} + V^{2+}\text{—Polymer} \qquad (29)$$

A similar time-resolved experiment is performed with the enzyme NitraR being immobilized in the redox polymer, Figure 15(b). It is evident that the reduced polymer decays faster in the presence of the enzyme, implying that reduction of the enzyme provides a competitive route to the recombination rate, eq. 30. Deconvolution of the decay curve in the presence of the enzyme, by the decay pattern observed in its absence allows the calculation of the ET rate constant, k_{et}, from the redox polymer to the enzyme active site. The value of k_{et} corresponds to $(9 \pm 3) \times 10^5\ sec^{-1}$. We thus realize that ET communication with biocatalytic sites can be tailored in rigid matrices by the immobilization of catalysts in redox matrices. The value of the ET rate constant, observed in this organized system, implies that the electron carrier pendant on the polymer and the enzyme redox sites exist at an average distance of 2-3 Å through which ET takes place.

$$V^{+\cdot}\text{-Polymer} \begin{cases} \xrightarrow{+Ru(bpy)_3^{3+}} Ru(bpy)_3^{2+} + V^{2+}\text{-Polymer} \\ \xrightarrow[k_{et}]{+E(Enzyme)} V^{2+}\text{-Polymer} + E^{-\cdot} \end{cases} \qquad (30)$$

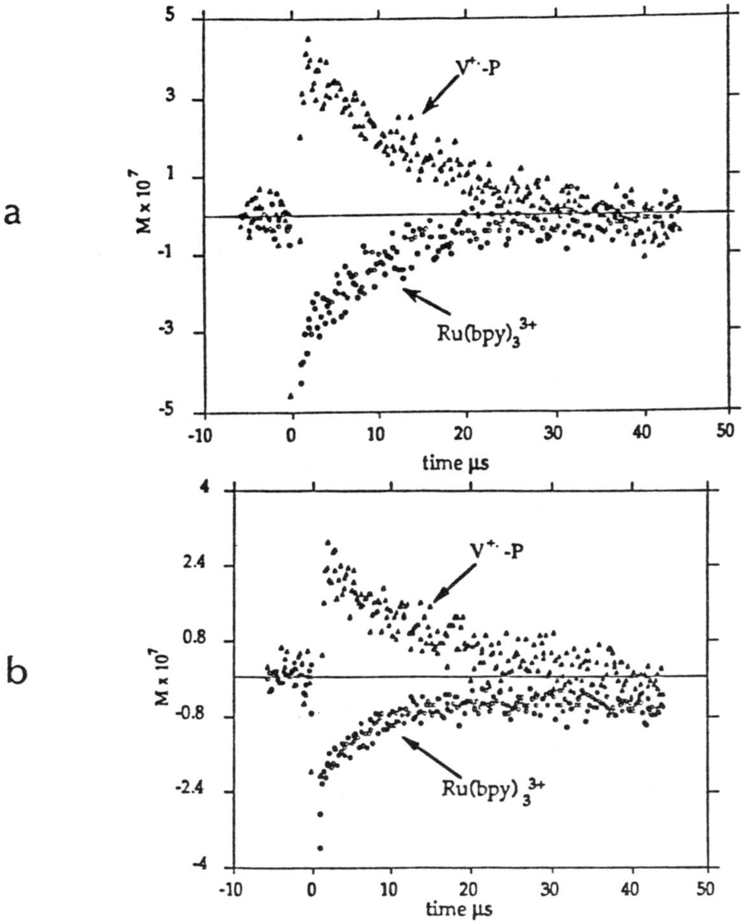

Figure 15: (a) Transient absorption spectra of photoproducts formed upon illumination of $Ru(bpy)_3^{2+}$ and bipyridinium-acrylamide redox copolymer. The system is illuminated with a laser flash, $\lambda = 430$ nm. Transient absorption corresponds to bipyridinium radical cation-copolymer (followed at $\lambda = 602$ nm). Bleach spectrum corresponds to formation of $Ru(bpy)_3^{3+}$ (followed at $\lambda = 465$ nm). Decay curves correspond to the recombination of electron transfer products.

(b) Transient absorption spectra of redox products formed in a photosystem composed of $Ru(bpy)_3^{2+}$ and NitraR immobilized in the bipyridinium-acrylamide redox copolymer. Experimental conditions as in (a). Decay curves correspond to the back electron transfer reaction of intermediate photoproducts and competitive electron transfer process from the reduced polymer to the enzyme.

5. Conclusions and Perspectives

Recent developments in the organization of artificial photosynthetic assemblies using homogeneous catalysts, heterogeneous catalysts and biocatalysts have been reviewed. Such organized photosynthetic systems could provide universal fuel generation systems and man-made devices for the removal of environmental pollutants. Specific emphasis of this report is directed towards the development of CO_2-fixation and NO_3^--reduction processes. Artificial homogeneous and heterogeneous catalysts mimic various functions of enzymes and resemble the kinetic characteristics of natural enzymes. Artificial catalysts can also be coupled for the regeneration of biologically active materials, i.e. NAD(P)H cofactor. At present, it is impossible to point to a superior class of catalytic materials. Enzymes exhibit high substrate-product selectivities and unique chiral and steric specificities. In turn, homogeneous and heterogeneous catalysts are expected to reveal high stabilities. We believe that future efforts should be directed towards the development of all three classes of catalytic materials. Organization of integrated systems, where photoinduced ET reactions are coupled to new redox catalysts, seems to be a viable approach to harness solar light energy for the production of fuels and valuable chemicals.

The progress in tailoring artificial photosynthetic systems has substantial implications to other scientific challenges. Regeneration of biologically active cofactors by means of artificial homogeneous catalysts, and the design of electrical communication between artificial electron carriers and redox enzymes could have broad applicability in biotechnological processes [36]. Chemically and electrochemically induced biocatalysed synthesis could be designed utilizing these concepts. Electrical communication of redox enzymes with their macroscopic environment, through immobilization of the biocatalysts in redox matrices, has broad interest in developing biosensor devices [37,38] and in the design of electrocatalysed biotransformations.

Acknowledgement: The support of the Kernforschung Anlage, Jülich, FRG, is gratefully acknowledged. The names of the coworkers appear in the list of references. Their skills and enthusiasm enabled the scientific progress.

References:

1. Bard, A.J. (1980) "Photoelectrochemistry", Science (Washington D.C.), 207, 139-144.
2. Meyer, T.J. (1989) "Chemical approaches to artificial photosynthesis", Acc. Chem. Res., 22, 163-170.
3. Grätzel M. (ed.) (1983) Energy Resources through Photochemistry and Catalysis, Academic Press, New York.
4. Willner, I., Mandler, D. and Maidan, R. (1987) "Bio-models and artificial models for photosynthesis", Nouv. J. Chim., **11**, 109-121.
5. Willner, I. (1985) "Photosensitized electron-transfer reactions in organized systems. The role of synthetic catalysts and natural enzymes in fixation processes", in M.A. Fox (ed.), Organic Phototransformations in Nonhomogeneous Media, A.C.S. Symposium Series, American Chemical Society, Washington D.C., No. 278, pp. 191-209.
6. Turro, N., Grätzel, M. and Braun, A.M. (1980) "Photophysical and photochemical processes in micellar systems", Angew. Chem., Int. Ed. Engl., **19**, 675-696.
7. Matsuo, T., Takuma, K., Tsutsui, Y. and Nishijima, T. (1980) "Electron transfer across interfaces between water and hydrophobic region as catalyzed by photoexcited amphipathic ruthenium complex", J. Coord. Chem., **10**, 187-193.
8. Fendler, J.H. (1987) "Atomic and molecular clusters in membrane mimetic chemistry", Chem. Rev., **87**, 877-899.
9. Thomas, J.K. (1980) "Radiation induced reactions in oganized assemblies", Chem. Rev., **80**, 283-299.
10. Kalyanasundaram, K. (1987) Photochemistry in Microheterogeneous Systems, Academic Press, London.
11. Balzani, V., Bolletta, F., Gandolfi, M.T., Maestri, M. (1987) "Bimolecular electron transfer reactions of the excited states of transition metal complexes", Top. Curr. Chem., **75**, 1-64.
12. Kalyanasundaram, K. (1982) "Photophysics, photochemistry and solar energy conversion with <u>tris</u>(bipyridyl)ruthenium(II) and its analogues", Coord. Chem. Rev., **46**, 159-244.
13. Juris, A., Balzani, V., Barigelletti, F., Campagna, S., Belser, P., von Zelewsky, A. (1988) "Ru(II) polypyridine complexes: photophysics, photochemistry, electrochemistry, and chemiluminescence", Coord. Chem. Rev., **84**, 85-277.

14. Darwent, J.R., Douglas, P., Harriman, A., Porter, G. and Richoux, M.-C. (1982) "Metal phthalocyanines and porphyrins as photosensitizers for reduction of water to hydrogen", Coord. Chem. Rev., **44**, 83-126.

15. Willner, I. and Willner, B. (1990) "Artificial photosynthetic model systems using light induced electron transfer reactions in catalytic and biocatalytic assemblies" in J. Mattay (ed.), Topics in Current Chemistry - Photoinduced Electron Transfer in Chemistry and Biology, Springer-Verlag, Heidelberg, in press.

16. Maidan, R. and Willner, I. (1986) "Photoreduction of CO_2 to CH_4 in aqueous solutions using visible light", J. Am. Chem. Soc., **108**, 8100-8101.

17. Willner, I., Maidan, R., Mandler, D., Dürr, H., Dörr, G. and Zengerle, K. (1987) "Photosensitized reduction of CO_2 to CH_4 and H_2-evolution in the presence of Ru and Os-colloids: strategies to design selectivity of products distribution", J. Am. Chem. Soc., **109**, 6080-6086.

18. Willner, I. and Mandler, D. (1989) "Characterization of Pd-β-cyclodextrin colloids as catalysts in the photosensitized reduction of bicarbonate to formate", J. Am. Chem. Soc., **111**, 1330-1336.

19. a) Bender, M. and Kamiyama, L. (eds.), (1978) Cyclodextrin Chemistry, Springer, Berlin.

b) Saenger, A.Q. (1980) "Cyclodextrin inclusion compounds in research and industry", Angew. Chem., Int. Ed. Engl., **19**, 344-362.

20. Goren, Z., Willner, I., Nelson, A.J. and Frank, A.J. (1990) "Selective photoreduction of CO_2/HCO_3^- to formate by aqueous suspensions and colloids of Pd-TiO_2", J. Phys. Chem., **94**, 3784-3790.

21. Albers, P. and Kiwi, J. (1990) "Photochemical generation of formate via bicarbonate/carbon dioxide reduction on palladium dispersions", New J. Chem., **14**, 135-139.

22. Mandler, D. and Willner, I. (1984) "Solar light induced formation of chiral 2-butanol in an enzyme catalyzed chemical system", J. Am. Chem. Soc., **106**, 5352-5353.

23. Mandler, D. and Willner, I. (1986) "Photosensitized NAD(P)H regeneration systems: applications in the reduction of 2-butanone, pyruvic and acetoacetic acids and in the reductive amination of pyruvic and ketoglutaric acids to amino acids", J. Chem. Soc., Perkin Trans. 2, 805-811.

24. Willner, I. and Mandler, D. (1989) "Enzyme catalysed biotransformations through photochemical regeneration of nicotinamide cofactors", Enzyme Microb. Technol., **11**, 467-483.

25. Mandler, D. and Willner, I. (1986) "Photoinduced enzyme catalysed synthesis of amino acids by visible light", J. Chem. Soc., Chem. Commun., 851-853.

26. Willner, I., Mandler, D. and Riklin, A. (1986) "Photoinduced CO_2-fixation forming malic and isocitric acids", J. Chem. Soc., Chem. Commun., 1022-1024.

27. Oishi, S. (1987) "A water-soluble Wilkinson's complex as homogeneous catalyst for the photochemical reduction of water", J. Mol. Catal., **39**, 225-232.

28. Willner, I. and Maidan, R. (1988) "Visible light photoinduced hydrogenation and hydroformylation processes using water soluble rhodium triphenylphosphine complexes", J. Chem. Soc., Chem. Commun., 876-877.

29. Willner, I., Maidan, R. and Shapira, M. (1990) "Thermal and photochemical regeneration of nicotinamide cofactors and a nicotinamide model compound using a water-soluble rhodium phosphine catalyst", J. Chem. Soc., Perkin Trans. 2, 559-564.

30. Dan, P. and Willner, I. (1984) "Cyclic $NAD^+/NADH$ models: participation of Zn^{++} in the reduction of ethyl-benzoylformate", Nouv. J. Chim., **8**, 719-721.

31. Steckhan, E., Herrmann, S., Ruppert, R., Thömmes, J. and Wandrey, C. (1990) "Continuous generation of NADH from NAD^+ and formate using a homogeneous catalyst with enhanced molecular weight in a membrane reactor", Angew. Chem., Int. Ed. Engl., **29**, 388-390.

32. Mandler, D. and Willner, I. (1988) "Photochemical fixation of carbon dioxide: enzymatic photosynthesis of malic, aspartic, isocitric and formic acids in artificial media", J. Chem. Soc., Perkin Trans. 2, 997-1003.

33. Okura, I. (1985) "Hydrogenase and its application for photoinduced hydrogen evolution", Coord. Chem. Rev., **68**, 53-99.

34. Willner, I., Lapidot, N. and Riklin, A. (1989) "Photoinduced enzyme catalyzed reduction of nitrate (NO_3^-) and nitrite (NO_2^-) to ammonia (NH_3)", J. Am. Chem. Soc., **111**, 1883-1884.

35. Willner, I., Lapidot, N. and Riklin, A. (1990) "Electron transfer communication between a redox polymer matrice and immobilized enzyme. Activity of nitrate reductase in a viologen-acrylamide copolymer", J. Am. Chem. Soc., in press.

36. Chenault, H.K., Whitesides, G.M. (1987) "Regeneration of nicotinamide cofactors for use in organic synthesis", Appl. Biochem. Biotechnol., **14**, 147-197.

37. Degani, Y., Heller, A. (1989) "Electrical communication between redox centers of glucose oxidase and electrodes via electrostatically and covalently bound redox polymers", J. Am. Chem. Soc., **111**, 2357-2358.

38. Heller, A. (1990) "Electrical wiring of redox enzymes", Acc. Chem. Res., **23**, 128-134.

IMMOBILIZED PHOTOSYNTHETIC SYSTEMS FOR THE PRODUCTION OF FUELS AND CHEMICALS

D.O.HALL, K.K.RAO and I.H.PARK
Division of Biosphere Sciences
King's College London
London W8 7AH, U.K.

ABSTRACT

The efficiency of solar energy conversion by photosynthesis in the natural ecosystem seldom exceeds 1% on an annual basis, but this represents stored energy. Biomimetic systems with isolated components of the photosynthetic light energy harvesting and conversion apparatus have been designed with the aim of achieving higher solar energy conversion efficiencies. Chloroplasts, thylakoid membranes and PSI particles from plants and cyanobacteria have been used together with redox catalysts and enzymes for the photoproduction of H_2, H_2O_2 and $NADPH_2$. Thylakoids and PS II membranes have also been used for the generation of H_2 and photocurrents respectively with water as the electron source. Cyanobacteria are good "biocatalysts" for photoproduction of H_2, NH_3, H_2O_2, organic acids etc. Other microalgae are potential sources for glycerol, β carotene and other plant pigments. Microalgae are also prime potential organisms for the absorption of CO_2 from the environment since they require only inorganic nutrients and water for growth. Immobilization of the microalgae in gels, polyvinyl and polyurethane foams, hollow fibre membranes etc. stabilises their photosynthetic activity and facilitates their manipulation in photobioreactors.

1. Introduction

1.1 Importance of photosynthesis

All our fossil (oil, coal, gas) and biological (biomass) fuels are derived via the process of photosynthesis. The amount of carbon fixed annually by photosynthetic processes is approximately 10^{11} tonnes. The energy content of this fixed carbon is $3x10^{21}$ J per annum, which is about 10 times the world's present annual consumption of energy. Photosynthetic carbon fixation using solar radiation is a significant contributor to our energy resource: it has been and will continue to be one of the major sources of energy on Earth. At present photosynthesis is the only process which removes vast amounts of CO_2 from the atmosphere; this unique role played by photosynthesis is vital in reducing the atmospheric CO_2 content and thus preventing excessive global warming. Photosynthesis and CO_2 recycling has recently attracted the attention of policy makers concerned with energy planning, global warning and environmental depollution [1].

In addition to CO_2 fixation to carbohydrates, photosynthetic processes can also be used to produce lipids (fats and oils), ammonia, amino acids, proteins, hydrogen, hydrogen peroxide, etc.

The theoretical maximum efficiency for photosynthetic solar energy conversion, defined as the ratio of stored chemical energy to the incident solar radiation on the photosynthesizer, is calculated to be about 13%; the practical maximum conversion efficiency under optimal conditions is 9% (J.R. Bolton and D.O. Hall, unpublished). However, this maximum conversion efficiency is never achieved in the field; for most plant species it seldom exceeds 1% on an annual basis. There is therefore a great interest in developing photobiological/photochemical systems which would mimic the energy conversion process of natural photosynthesis, hopefully with better efficiency.

1.2 Photomimetic systems

There are two types of natural photosynthesis viz oxygenic (water-splitting) carried out by plants, algae and cyanobacteria and anoxygenic (where water is not used as electron source) operating in photosynthetic bacteria. In both types the essential features are: light energy absorption by antenna pigments, transfer of absorbed energy to a reaction centre

E. Pelizzetti and M. Schiavello (eds.), Photochemical Conversion and Storage of Solar Energy, 181–192.

with special chlorophylls, charge separation and transfer to quinones via pheophytins and finally energy conversion to ATP and NAD(P)H through redox proteins. All these processes occur at a fast rate through components located in membranes. Photobiologists and photochemists are involved in understanding and mimicking various segments of this photosynthetic energy conversion process. These include:

(a) synthesis and structure analysis of Mn complexes mimicking the PS II oxygen evolution complex [2],

(b) synthesis and study of reaction rates and kinetics of multicomponent carotenoid-porphyrin-quinone complexes mimicking light absorption and energy transfer in photosynthetic bacteria. Molecular triads, tetrads, and recently a pentad have been assembled [3],

(c) synthesis of electron donor-acceptor compounds to understand the mechanism of electron transfer [4],

(d) design of proteo-lipid molecular assemblies (e.g. with photosensitizer dyes entrapped in lipid vesicles) to study electron migration and charge separation in vesicles [5],

(e) building photochemical and photoelectrochemical arrays comprising photosensitizer, electron relay, and redoxcatalysts for water splitting or for CO_2 reduction [6],

(f) isolation of thylakoid membranes, individual photosystem particles (PSI and PS II) or bacterial chromatophores and designing "in vitro" systems for the photoproduction of energy rich chemicals, and

(g) culturing cyanobacteria or algae under "stress" for the excretion of specific metabolites [7].

Our laboratory has mainly been interested in the last two (f and g) topics and emphasis will be placed on these processes.

2. Immobilization

Immobilization is the process of attaching cells or their constituent biocatalysts to a solid matrix so that they do not move independently when placed in a fluid environment. Immobilization can be carried out by physical means, such as adsorption or entrapment in a gel or foam matrix, or by chemical methods such as covalent binding. The advantages of immobilization are stabilization of catalytic activity, often enhanced reduction rates due to concentration of catalyst surface, increased cell metabolism and cell wall permeability, ease of separation and reuse of catalysts from fluid products, and process control in a bioreactor. Main disadvantages are loss of activity during immobilization, the enzyme activities are extracellular and limitation of reaction rates due to diffusion resistance. Gels (acrylamide, agar, alginate, carrageenan, chitosan), glutaraldehyde-albumin sponges, reticulate foams (polyvinyl and polyurethane) polypropylene and polysulfone hollow fibres and ceramics (silica) are the usual materials used as transparent or reflecting supports for immobilization.[8,9]

Entrapment in porous gels or foam matrix is the favoured technique for the immobilization of chloroplasts, chromatophores, plant and algal cells, etc as it does not modify or cause any damage to the organelle or cell. Since light is a direct substrate for photosynthesizers the immobilization matrix must be reasonably transparent as well as being porous for penetration of the cells. In our immobilization work with chloroplasts and cyanobacteria we have used agar and alginate gels, polyurethane and polyvinyl foams, and membrane hollow fibres.

Alginate is a polymer of mannuronic acid and guluronic acid and is isolated from marine algae. Alginate bead immobilization is performed by mixing a suspension of chloroplasts or algal cells with an equal volume of 2% sodium alginate in buffer and dispensing the mixture in drops to a 0.1 M $CaCl_2$ solution kept stirred in an ice bath. Calcium alginate gel beads with entrapped photosynthetic material are obtained. [10]

Urethane is formed by the reaction between an isocyanate and a hydroxyl group:

$RNCO + R'\text{-}OH \rightarrow RNHCOOR'$ where R is an alkyl or aryl group. Condensation with other isocyanates, alcohols or amines produces a cross-linked urethane polymer. Principal isocyanates used are toluene 2,4 diisocyanate (TDI) and diphenyl methane diisocyanate (MDI). The hydroxyl compounds generally used are polyether or polyester based. During hydrolytic condensations of urethanes CO_2 is evolved. The overall texture of polyurethane matrix is dependent on the chemical nature of the polyol, the temperature of condensation and the presence of added catalyst.

Cell immobilization can be performed either by entrapment, i.e. by stirring an aqueous suspension of cyanobacterial cells into a urethane prepolymer (eg Hypol 2002, W.R.Grace Limited, London) and generating the polymer by condensation or by adsorption of the cells on to small pieces of polymer foams by culturing. Entrapment results in a homogeneous distribution of the cells within the matrix; however loss of cell viability has been reported as a result of entrapment.

Polyvinyl foams are usually prepared by the condensation of polyvinyl alcohol. These foams which retain reactive OH groups, are transparent and quite stable. Immobilization of cells is carried out mainly by adsorption.

Membrane hollow fibre reactors have been used for a long time for the immobilization and production of animal cells; they are now finding application in the immobilization of plant, algal and bacterial cells. These membranes are made from hydrophobic (polysulfone, polypropylene, etc.) or hydrophillic (rayon, cellulose acetate) polymers. They provide a high surface area and can act as a molecular sieve. Potential advantages of hollow-fibre bioreactors are high cell retention, more efficient continuous processing, automatic separation of products, recovery and reuse of biocatalysts, and amenability to scale up. Major limitations are the requirement of a particulate-free media, fibre rupture, maintenance of sterile conditions, and difficulty in the supply of nutrients, especially gases. [11,12] A polysulfone hollow fibre membrane bioreactor used for the production of NH_4^+, with immobilized *Anabaena azollae*, is shown in Fig. 1 (Wang and Hall, Biomass, in press).

Serum albumin polymers cross linked with glutaraldehyde is a favoured technique for chemical immobilization of chloroplasts, chromatophores adapted by Thomas et al [8]. Criteria for successful immobilization are maintenance pf photosynthetic activity measured as oxygen exchange or fluorescence, and in the case of immobilized cells, the maintenance of viability and growth as determined by chlorophyll concentration or enzyme activity (eg nitrogenase or hydrogenase).

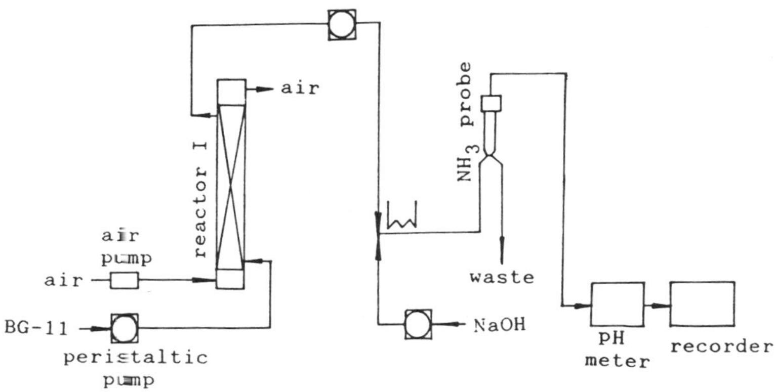

Fig. 1. Schematic illustration of a membrane hollow fibre reactor for the continuous production and monitoring of NH_4^+ from immobilized *A. azollae*.

Table 1. Products from immobilized Photosynthetic Systems

Matrix used and Material Immobilized	Biotechnological Application
AGAR	
Mastigocladus, Phormidium	H_2 production
Anabaena 7363	H_2, fuel cells
Chromatium and *Rhodospirillum rubrum*	H_2 production
chloroplasts + catalysts	N_2 + $NADPH_2$
Photosystem I + catalysts	H_2
thylakoid membranes	H_2O_2
ALBUMIN-GLUTARALDEHYDE	
thylakoid membranes	H_2, ATP
Scenedesmus	biochemical studies
chromatophores	ATP
PSII membranes	ATP
ALGINATE GEL	
Botryococcus	hydrocarbons (oils)
Chlorella emersoni	glycollate
Dunaliella sp	glycerol
Anabaena 27893	ammonia
Mastigocladus laminosus	ammonia
Synechococcus 40607	glutamate
Rhodobacter capsulata (Ba)	H_2
thylakoid membranes	H_2O_2
POLYVINYL AND POLYURETHANE FOAMS	
Porphyridium and *Chlamydomonas* sp	exocellular polysaccharides
Anabaena azollae	ammonia and H_2
Phormidium	H_2O_2
Various cyanobacteria	H_2 and $NADPH_2$
Capsicum frutescens	Capsaicin (alkaloid)
MEMBRANE HOLLOW FIBRES	
A. azollae	H_2 and ammonia

3. Applications

Typical examples of the use of immobilized systems for the photoproduction of fuels and chemicals are shown in Table 1. Details can be found in recent reviews and monographs. [13-16]

3.1 Hydrogen

Hydrogen is the ideal "green" fuel, its combustion product (water) is non-polluting and is renewable. Besides the feed stock (water) is cheap and abundant. There is presently renewed interest in the development of solar hydrogen production technologies.

3.1a Photosynthetic Bacteria

Hydrogen is a metabolic product of photosynthetic bacteria, H_2 evolution being catalysed by the nitrogenase enzyme. Feasibility of H_2 production has been demonstrated with immobilized cells of *Rhodospirillum rubrum* (agar), *Chromatium* sp (agar) and *Rhodobacter capsulata* (Ba alginate). [17] Laboratory-scale bioreactors packed with immobilized cells produced H_2 as efficiently as free-living cells. Substrates for carbon and electron source (malate, lactate, etc.) are available in organic wastes from agroindustry; simultaneous depollution of industrial wastes can be achieved together with H_2 production. If energy input from substrates is disregarded the efficiency of H_2 production could be near unity. The disadvantage of the process is that the photosynthetic bacteria cannot use water as substrate and that the photobioreactor has to be maintained anaerobic to prevent oxygen-inhibition of nitrogenase.

3.1b Cyanobacteria

Nitrogen fixing cyanobacteria (blue-green algae) have been successfully immobilized and used for the photoproduction of energy. H_2 evolution is catalysed mainly by nitrogenase and to a lesser extent by hydrogenase. Cyanobacteria are good harvesters of sunlight and are easy and economical to grow. Most H_2 evolution studies have been performed with *Anabaena* spp; *A. cylindrica* and *A. azollae*. Several marine cyanobacteria have also been screened for their potential as solar H_2 producers [18]. Cyanobacteria are quite amenable to immobilization. An outdoor biophotolytic system using *A. cylindrica* B629 suspended in glass beads was shown to produce H_2 continuously for over three weeks. Continuous H_2 evolution lasting for over 5 months was observed from *A. azollae* immobilized in polyvinyl or polyurethane foams [19]. The viability of the enzymes and photocatalysts is preserved for months after immobilization. At the present time, immobilized cyanobacterial cells appear to be the most promising source for H_2 generation in photobioreactors.

3.1c Green Algae

Hydrogenase activity can be induced in green algae by growing them anaerobically; the algae will then evolve H_2 and O_2 stoichiometrically by water photolysis. Maximum light energy conversion efficiencies (3 to 12%) were reported with thin films of algae entrapped on filter papers [20]. However the extreme oxygen sensitivity of hydrogenase limits the use of these organisms for biophotolytic H_2 production.

3.1d Chloroplast membranes (thylakoids)

Electrons and protons generated by light-induced water splitting at Photosystem II can be transported to PSI where they can reduce an exogenous electron mediator. The reduced mediator can be coupled to hydrogenase or Pt to produce H_2. Several laboratories have studied H_2 production by this route with thylakoids and catalysts immobilized in various types of matrices. Although immobilization resulted in some stabilization of the system, problems of photoinhibition of thylakoid components and oxygen inhibition of catalysts could not be solved completely [10]. The photobioreactor could not function for more than a few hours - we have to learn more about photoinhibition before a viable system can be developed. An alternative would be to bypass Photosystem II and use only Photosystem I activity for H_2 production. This, of course, would be more expensive since it would necessitate the consumption of PSI electron donors, as in photosynthetic bacteria.

3.2 HYDROGEN PEROXIDE

Hydrogen peroxide is an energy-rich compound; it liberates more than 100 kJ per mol when it decomposes to water and O_2. Isolated chloroplasts are able to photoreduce O_2 to H_2O_2 by the so called "Mehler" reaction [21]. The rate of H_2O_2 formation can be enhanced by the addition of PSI electron acceptors such as methyl viologen, flavin nucleotides, quinones, etc., all of which can catalyze the reduction of O_2 to H_2O_2 via superoxide. Light-induced formation of H_2O_2 mediated by riboflavin, has been shown with spinach thylakoid immobilized in different gels and foams - best yield was from agar-entrapped chloroplasts [22]. The reaction was 'short lived' with water as electron source (possibly due to light and O_2 inhibition of PSII); however, with PSI donors H_2O_2 production continued for hours. Production of H_2O_2 with initial rates of 500 μmol H_2O_2 per mg chlorophyll.h, from microalgae illuminated in the presence of methyl viologen has been reported [23].

We have recently succeeded in designing a photobioreactor for continuous H_2O_2 production with polyvinyl foam immobilized *Phormidium laminosum* cells as photosynthesizer and methyl viologen as PSI electron acceptor. The arrangement of the 'reactor' is shown in Fig 2 (Park, Rao & Hall, unpublished).

3.3 ATP and NAD(P)H_2

These molecules are formed during photosynthetic electron transport and phosphorylation and are high value, energy-rich, biochemicals. *In vitro* systems consisting of chloroplasts or bacterial chromatophores immobilised in matrices such as agar, polyacrylamide, serum albumin foam, barium alginate or polyurethane have all been used for photophosphorylation yielding ATP. In all of the reported studies immobilization resulted in stabilization of photosynthetic activities of the isolated organelle [24].

Cyanobacterial cells immobilized on polyurethane foams photoreduced NADP to $NADPH_2$ in the presence of catalytic amounts of the enzyme ferredoxin-NADP dehydrogenase [25]. Photochemical reduction of NAD to NADH can be accomplished by band gap illumination of TiO_2 colloids in solutions of NAD in the presence of rhodium tris bipyridyl complex, a nonenzymatic reduction of NAD [26].

4. Ammonia

Nitrogen fertilisers (nitrate, ammonium, urea, etc.) contribute up to 50% of the energy input in modern agriculture. Development of photobiological or photochemical N_2 fixation processes could significantly reduce the burning of fossil fuels or wood (thus incidentally reducing CO_2 emission) in the chemical plants presently producing nitrogen fertilisers. Ammonia in addition is a valuable chemical fuel.

It is estimated that as much as 10^8 tonnes of N_2 per annum is converted to ammonia by biological processes (cyanobacteria and rhizobia-legume symbiosis). The energy demand for this N_2 fixation is ultimately met by the sun via photosynthesis. Although large amounts of N_2 are fixed by cyanobacterial nitrogenase only in a few cases can this fixed N_2 be directly utilisable as a crop fertiliser. One such special case is the *Anabaena-Azolla* symbiotic association which is extensively used as a natural fertiliser in rice fields and which is capable of high rates of N_2 fixation (up to 3 kg of N fixed per hectare per day). In this association the N_2 fixed by the cyanobacteria is exchanged for photosynthates produced by the *Azolla* fern - any unused ammonia is excreted into the medium [27].

We have chosen *Anabaena azollae* to develop an immobilized photosynthetic reactor for the continuous production of ammonia. *A. azollae* cells were immobilized in hydrophilic polyurethane or polyvinyl foam pieces or in polysulphone hollow fibres. A photobioreactor was constructed with the foam-immobilized cyanobacteria packed into a fluidised bed column with constant circulation of growth medium. The medium was occasionally given pulses of L-methionine-D-L-sulfoximine, an inhibitor of glutamine synthetase, which minimised the conversion of ammonia to glutamine and other metabolites. The ammonia in the effluent was determined, either colorimetrically or polarographically. A photobioreactor with *A. azollae* immobilized in polyvinyl foam generated one mmol NH_3 per litre effluent per hour which is equivalent on an area basis to the fixation of 900 kg of N per hectare per annum. This rate should be compared to the typical N fixation rates of 400 kg per ha per annum in legumes and 110 kg per ha per annum in the natural *Anabaena-Azolla* symbiosis. The only limitation in this immobilized system is that the ammonia in the effluent is rather dilute (0.1 mM); however, this will not be a drawback if the effluent is channelled to irrigation water for the crop.

5. Other Fuels and Chemicals

Hydrocarbons (oils and lipids) are produced as major metabolic products of several green algae. For example, *Botryococcus brauni* can synthesize hydrocarbons up to 80% of its dry mass. The hydrocarbons can be extracted by treatment of the cells with hexane without affecting their photosynthetic activity. Hydrocarbon recovery and yields were substantially increased, relative to free cells, by immobilization of the alga in polyurethane foams [28].

Polyurethane-immobilized *Porphyridium cruentum* (a red alga) have produced sulphated polysaccharides for long periods in a tubular photobioreactor at the rate of 20 g per sq metre per day. Immobilized green algae (*Chlorella* and *Chlamydomonas*) secretes extracellular polysaccharides used in food and agriculture. The green alga *Dunaliella*, immobilized in Ca alginate, can be used for the production of glycerol and β carotene. Immobilized microalgal systems is a rapidly expanding field of biotechnology [29].

6. Environmental Depollution

Two of the challenging problems facing scientists concerned with the environment are the accumulation of "greenhouse" gases in the atmosphere and the accumulation of inorganic fertilisers (nitrates and phosphates) in the biosphere causing eutrophication in water reservoirs, lakes and rivers. What contribution can photobiologists make towards the alleviation, if not complete abolition, of these pollution problems?

6.1 Tertiary Treatment of Water

The effluents from primary and secondary urban waste water treatments may still contain unacceptable levels of N (NO_3^-, NO_2^-, NH_4^+) and P (PO_4^{3-}) compounds for use in domestic water supply. These inorganics are responsible for the eutrophication of water bodies producing microalgal blooms which are very difficult to remove from the water. Tertiary treatment of water with immobilized microalgae may provide a solution to this pollution problem. Preliminary studies have shown that passage of secondary treated water through chitosan-immobilized *Phormidium laminosum* (a cyanobacterium) removed 98% N and 60% P compounds in 24 h. The microalgae consumes these nutrients during their growth in the chitosan matrix[30]. Periodically the algae can be stripped out of the matrix and stored or used as biomass.

We have now successfully immobilized *Phormidium laminosum* in polyvinyl foams and utilised the immobilized cells for photosynthetic nitrate uptake (Gil, Garbisu, Hall, Serra and Bazin, unpublished). Nitrate uptake was maintained for 3 months with very little evidence of cell leakage from the foams.

6.2 Abatement of the CO_2-induced "greenhouse effect".

The build up of CO_2 in the atmosphere (27% increase to date, with half the rise occurring over the last 30 years) and the possible contribution of the other so called 'greenhouse gases' to the overall warming of the earth (about a 0.5° increase over the last century) is now widely recognised. The 'greenhouse effect' with its accompanying sea-level rises, changing rainfall patterns and possible increasing plant productivity are subjects of much study in trying to understand the present situation and where the increasing atmospheric emissions may lead us in the future. The short-term future (next 100 years) appears to depend on how we use fossil fuels and how the changing atmosphere affects climate and the photosynthetic productivity of plants and microorganisms, especially on a regional basis.

Unfortunately, it is the rate of change which will determine whether we can adapt to changing circumstances. Humans presently determine the rate of build up of greenhouse gases in the atmosphere because they control fossil fuel use and most disturbances to the biota.

Over the past two centuries up until the 1920s-1940s, biospheric emissions and ocean buffering have been the major controls of atmospheric CO_2 levels. About a third to half way through this century (the timing is still debated) fossil fuel CO_2 emissions became the predominant source of atmospheric CO_2; the extent of the relative contribution from fossil fuel and biosphere sources is also contentious. Notwithstanding this debate, it must be appreciated that the rate of increase in the use of fossil fuels over the last 15 years has fluctuated greatly, thus making it especially difficult to forecast future energy trends and hence atmospheric CO_2 concentrations. Thus in the period 1967-73 commercial fuel use increased by about 6.4% per annum, from 1973-83 the rate of increase was only 0.1% per annum, due to the two 'energy crises', but since 1983 the increase has been 2.9% per annum as the world adjusts to higher energy prices and learns to use energy more efficiently. A brief summation of carbon fluxes and sinks is compiled in Table 2 [31].

If the CO_2 content of the atmosphere were to double over the next century to reach about 540 ppm (pre-industrial level about 270 ppm) it is predicted that the Earth will experience an average warming up of about 3° over preindustrial temperatures. This 'greenhouse effect' could be produced solely as a result of CO_2 emissions or a combination of CO_2 and other 'greenhouse gases' such as CH_4, CFCs (chlorofluorocarbons), N_2O and O_3.

6.3 Algae and Cyanobacteria in CO_2 abatement

One possible strategy for large-scale CO_2 fixation would involve the application of present photobiological knowledge in algal and cyanobacterial systems. Photobioreactors could be attached to CO_2-emitting sources, for direct conversion of CO_2 to fuels or storage compounds. For example, production of hydrocarbon fuels and other oils by *Botryococcus, Nannochloropsis*, etc. can be envisaged in large open ponds into which the CO_2 output stream is directly injected. Alternatively, closed photobioreactors which exclude outside air could be used for production of hydrogen, ammonium and other compounds. Such algal systems would require only simple growth medium (brackish water or sea water could be used in many cases), need little or no nitrogen fertiliser and could be set up on desert or sterile land, where they would not compete with agriculture.

If the aim were to permanently store carbon, the algal biomass could be fixed with aldehydes, fixed as organic carbonates or stored dry in desert areas. Section of algal strains (a research area which shows considerable potential) would be required for maximum CO_2-scavenging and photosynthetic efficiency, as well as stability of carbon content after harvesting. Although more technological barriers would need to be overcome, algal systems offer certain advantages over carbon storage as wood. Strain selection and genetic manipulation would be greatly enhanced by the short generation time, offering the potential of a wide range of products and characteristics. Efficiency of light energy conversion (and hence CO_2 fixation rate) is also potentially much higher than for forests, which would reduce the requirement for land.

Table 2. *Carbon fluxes and sinks—a synthesis of data from Barnola et al.* (1987), *Bolin et al.* (1986), *Gifford* (1980), *Houghton et al.* (1987), *Kohlmaier et al.* (1987), *Trabalka* (1985), *Lieth* (1985), *Siegenthaler & Oeschger* (1987) *and Mintzer* (1987) [31].

CO_2 in atmosphere (1986)	346 ppm (= 730 Pg carbon)
CO_2 in atmosphere (present annual increase)	1.5 ppm (= 3.2 Pg carbon)
CO_2 in atmosphere (preindustrial)	272 ppm (= 575 Pg carbon)
CO_2 in atmosphere (change from preindustrial to 1986)	74 ppm (= 27%; 155 Pg carbon remained in atmosphere)
CO_2 fossil-fuel emissions (1986)	5.3 Pg carbon per year
CO_2 fossil-fuel emissions (cumulative to date)	183 ± 15 Pg carbon
CO_2 biospheric emissions (present)	1.8 ± 0.9 (1.0–2.6) Pg carbon per year
CO_2 biospheric emission (cumulative to date)	150 ± 50 Pg carbon
Fossil fuels carbon content (reserves)	770 Pg carbon
Fossil fuels carbon content (resources)	6500 Pg carbon
Biomass (terrestrial; 80% in trees)	560 Pg carbon
Biomass (aquatic)	3 Pg carbon
Gross primary production (terrestrial)	120 Pg carbon per year
Net primary production (terrestrial)	60 Pg carbon per year
Net primary production (aquatic)	46 Pg carbon per year
Net primary production (total)	106 Pg carbon per year
Soil carbon content	1515 Pg carbon
Ocean-surface (75 m) carbon content	725 Pg carbon
Ocean-intermediate and deep carbon content	38000 Pg carbon
CO_2 uptake by ocean	1.9 Pg carbon per year } 2.7 aquatic
CO_2 uptake by freshwater	0.8 Pg carbon per year } 2.7 aquatic

$1\ Pg = 10^{15}\ g$

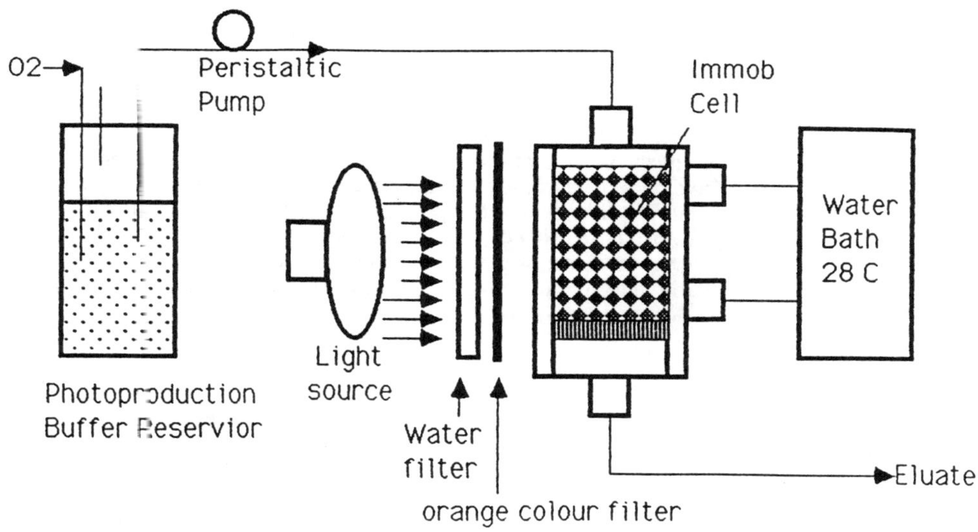

Figure. 2 . Column reactor for the photoproduction of H2O2 using PV-immobilized Phormidium laminosum cells.

Another possible response to the build-up of atmospheric CO_2 would be the large-scale reforestation (or other revegetation) of previously degraded lands. At present, it is estimated that about 1.8 billion tonnes of carbon are released annually by forest destruction and also large amounts from other forms of devegetation such as grassland burnings, compared with about 5.2 billion tonnes per annum from fossil fuel burning. The first step would therefore be to halt the net removal of trees and other plant cover, and then to reverse it by tree planting combined with good forest management. This obviously must be combined on a global scale with increased energy use efficiency and other measures to decrease the emissions of greenhouse gases.

7. Electricity Generation via Immobilized Photosystems

Whole cells of cyanobacteria, isolated thylakoids, individual photosystems and even chlorophyll deposited on semiconductor electrodes can be used for the generation of photocurrents and photovoltages. *Mastigocladus laminosus* immobilized in Ca alginate and deposited on the surface of an optically transparent SnO_2 electrode when illuiminated in an electrochemical cell not only generated a photocurrent, but also continued to grow on the electrode surface [32]. The "living algal electrode" functioned for more than 20 days, generating a steady photocurrent. In a much more simplified system, Photosystem II membrane particles immobilized on a dye-derivatised TiO_2 electrode generated photocurrents of the order of 35μ A cm^{-2} of electrode surface and 10 m A mg^{-1} chlorophyll. Maximum incident photon-to-current conversion efficiency was 12% [33]. Biosolar batteries with photosynthetic pigments sandwiched between electrode surfaces may soon be available.

8. Conclusion

We have outlined some processes whereby immobilized photosynthetic systems can be applied for the production of fuels and chemicals and for the alleviation of environmental pollution. Most of these techniques are still in a laboratory stage and need scaling up before extensive field trials. The problem inherent in solar energy conversion technology is that the incident energy is diffuse with wide fluctuations in insolation over both daily and annual cycles, this disadvantage in solar energy utilisation whereby large collecting surfaces are required should be weighed against its unique advantage that solar energy is both cost and pollution free. We do not venture at present to compare the economics of solar energy technology against those using convential energy derived from fossil fuels and nuclear fission. However, there is now a renewed interest, all over the world, in solar conversion, particularly in the production of H_2 from water. Interest and support for other processes, we hope, would follow.

9. References

1. Summaries of Invited Presentations, 'Solar Photoconversion Processes for Recycling CO_2 from the atmosphere', Colarado Springs, USA, March 1990. SERI, Golden, Colorado.

2. Rutherford, A.W. (1989) 'Photosystem II, the Water-splitting enzyme', TIBS, 14, 227-232.

3. Gust, D. and Moore, T.A. (1989) 'Mimicking Photosynthesis', Science, 244, 35-41. See also: New Scientist, London. 26 May 1990, p.34.

4. Verhoeven, J.W.; Kroon, J., Paddon-Row, M.N. and Oliver, A.M. (1989) 'Electron and energy transfer across rigid, saturated hydrocarbon bridges': molecular engineering of the bridge configuration' in D.O.Hall and G. Grassi (eds.) Photoconversion Processes for Energy and Chemicals, Elsevier Applied Science, London pp 100-108.

5. Haussling, L. Mertesdorf, C. and Ringsdorf,H. (1989) 'Photochemically induced structural changes in multilayers and columnar systems' ibid, pp.53-67.

6. Willner, I. Mandler, D. and Maidaw, R. (1987) 'Bio-models and artificial models for photosynthesis'. New J. of Chemistry, 11, 109-122.

7. Hall, D.O. and Rao, K.K. (1988) 'Immobilized photosynthetic membranes and cells for the production of fuels and chemicals', in Graber, B.P., Schmur, J.M. and Chapman, D. (eds.) Biotechnological Applications of Lipid Micro-structures, Plenum Press, New York pp. 225-249.

8. Mosbach, K., ed. (1987) Methods in Enzymology. Vol. 135, Immobilized Enzymes and Cells. Academic Press, New York.

9. Brouers, M., Shi, D-J., and Hall, D.O. (1988) "Immobilization methods for cyanobacteria in solid matrices', Methods in Enzymology, 167, 629-636.

10. Gisby, P.E. and Hall, D.O. (1980) 'Biophotolytic H_2 production using alginate-immobilized chloroplasts, enzymes and synthetic catalysts', Nature 287, 251-253.

11. Bunch A.W. (1988) The uses and future potential of microbial hollow-fibre bioreactors, J. Microb. Methods, 8, 103-119.

12. Cheryan, M. and Mehaia, M.A. (1986) 'Membrane bioreactors' in W.C.McGregor (ed.) Membrane Separations in Biotechnology, Marcel Dekker, New York pp. 252-301.

13. Hall, D.O. and Rao, K.K. (1989) 'Immobilized photosynthetic membranes and cells for the production of fuels and chemicals' Chimica oggi, 7, (3) 41-47.

14. Borowitzka, M.A. and Borowitzka, L.J. eds. (1988) Microalgal Biotechnology. Cambridge University Press.

15. Cresswell, R.C.; Rees, T.A.V. and Shah, N. eds. (1989) Algal and Cyanobacterial Biotechnology. Longman Scientific & Technical, London.

16. Trevan, M.D. and Mak, A.L. (1988) 'Immobilized algae and their potential for use as biocatalysts'. TIBTECH, 6, 68-73.

17. Vignais, P.M., Colbeau, A., Wilson, J.C. and Jouanneau, Y. (1985) 'Hydrogenase, nitrogenase and hydrogen metabolism in the photosynthetic bacteria', Advances Microb. Physiol. 26, 156-232.

18. Matsunaga, T. and Mitsui, A. (1982) 'Seawater based hydrogen production by immobilized marine photosynthetic bacteria' Biotech. Bioeng. Symp. 12, 441-450

19. Shi, D-J., Brouers, M. Hall, D.O. and Robins, R.J. (1987) 'The effects of immobilization on the biochemical, physiological and morphological features of *Anabaena azollae*', Planta, 172, 293-308.

20. Greenbaum, E.C. (1985) 'Platinized chloroplasts : a novel photocatalytic material', Science 230, 1373-1375.

21. Foyer, C.H. and Hall, D.O. (1980) 'Oxygen metabolism in the active chloroplast' TIBS 5, 188-191.

22. de la Rosa, M.A., Rao, K.K. and Hall, D.O. (1986) 'Hydrogen peroxide photoproduction by free and immobilized spinach thylakoids, Photobiochem. Photobiophys.11, 173-187.

23. de la Rosa, F.F. and Morales, I. (1989) 'Photobiological production of hydrogen peroxide : A model system'. Solar Energy 43, 373-377.

24. Coquempot, M-F. and Thomas, D. (1984) 'Immobilized chloroplast membranes : Kinetic aspects of their continuous use in batch and chemostat', Enzyme Microb. Technol. 6, 321-324.

25. Muallem, A. Bruce, D. and Hall, D.O. (1983) 'Photoproduction of H_2 and $NADPH_2$ by polyurethane-immobilized cyanobacteria', Biotech. Lett 5, 365-368.

26. Cuendet, P. and Graetzel, M. (1984) 'Photoreduction of NAD to NADH in sermiconductor dispersions', Photochem. Photobiol. 39, 609-612.

27. Shi, D.J. and Hall, D.O. (1988) '*Azolla* and immobilized cyanobacteria : from traditional agriculture to biotechnology, Plants Today, 1, 5-12.

28. Frenz, J., Largeau, C. and Casadevall, E. (1989) 'Hydrocarbon recovery by extraction with a biocompatible solvent from free and immobilized cultures of *Botryococcus braunii*, Enzyme Microb. Technol. 11, 717-724.

29. Oren-Shamir, M., Avron, M. and Degani, H. (1988) FEBS Lett., 233, 124-128.

30. de la Noue, J. and Proulx, D. (1988) 'Biological tertiary treatment of urban waste waters with chitosan-immobilized *Phormidium*', Appl. Microbiol. Biotechnol. 29, 292-297.

31. Hall, D.O. (1989) 'Carbon flows in the biosphere : present and future', J. Geol. Soc. London. 146, 175-181.

32. Ochiai, H., Tanaka, A. and Fukui, S. (1983) 'Immobilized Organelles', Appl. Biochem. Bioeng. 4, 153-187.

33. Rao, K.K., Hall, D.O., Vlachopoulos, N. Grätzel, M., Evans, M.C.W. and Seibert, M. (1990) 'Photoelectrochemical responses of photosystem II particles immobilized on dye-derivatised TiO_2 films' J. Photochem. Photobiol. B. Biol. 5, 379-389.

IMPROVEMENTS IN SOLAR ENERGY CONVERSION

R. MEMMING

Institut für Solarenergieforschung GmbH

Sokelantstr. 5, 3000 Hannover 1, FRG

ABSTRACT. In the present paper various aspects of photoelectrochemical utilization of solar energy and the progress in this field during the last couple of years are discussed.

1. Introduction

During the last decade there have been many efforts in several laboratories around the world to develop photoelectrochemical systems based on semiconductor materials for the utilization of solar energy During this period, hundreds of papers on photoelectrochemical systems and reactions have been published [1] - [17]. In the present paper an overview of the progress in the last few years is given. As will be shown below , the main progress has been made in the field of regenerative photoelectrochemical cells. On the other hand, the progress concerning the photoelectrochemical production of chemical fuels is rather limited. Various research groups concentrated on fundamental problems in order to get a better understanding of effects and parameters being essential in photoelectrochemical cells. In addition, semiconductor electrochemistry had a considerable impetus on the fabrication of pure solid state solar cells, as will be illustrated in few examples.

2. Regenerative Photoelectrochemical Cells

2.1. RELATION TO SOLID STATE PHOTOVOLTAIC CELLS

Besides photoelectrochemical cells there are two different types of pure solid state devices such as pn-photocells and Schottky type cells. All three types of cells can be described by the same current-potential equation, namely by

$$j = -j_o \left[\exp\left(\frac{-eU}{nkT}\right) - 1 \right] + j_{ph} \qquad (1)$$

in which j_o is the saturation current of a reversely polarized diode or the exchange current at equilibrium, and U is the applied voltage, j_{ph} the photocurrent density and n the so-called "ideality factor", which is n=1 for ideal junctions, but is usually n>1.

A typical current-potential curve is given in Fig. 1

E. Pelizzetti and M. Schiavello (eds.), Photochemical Conversion and Storage of Solar Energy, 193–212.

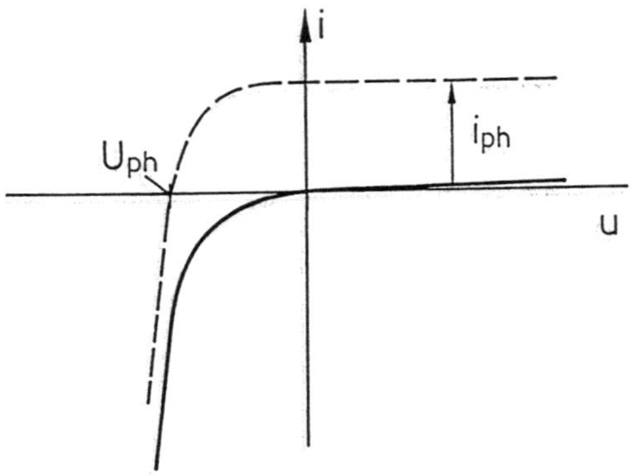

Figure 1. j-U characteristic of a solar cell.

The photopotential U_{ph} as measured under open circuit conditions occurs at a potential at which the photocurrent is equal to the corresponding dark current, i.e. where the total current is zero. According to eq. (1) one obtains:

$$U_{ph} = \frac{nkT}{e} \ln \left[\frac{j_{ph}}{j_o} + 1\right] \tag{2}$$

It is evident that the photovoltage is essentially determined by j_o. It is essential to note that j_o which determines the dark forward current depends on the mechanism of charge transfer at the interface. In the case of a pn-junction the forward current is determined by the injection and recombination of minority carriers, and j_o is given by the Shockley equation [18], i.e.

$$j_o = en_i^2 \left(\frac{D_p}{N_D L_p} + \frac{D_n}{N_V L_n}\right) \quad \text{(minority carrier device)} \tag{3}$$

in which D_n and D_p are the diffusion constants and L_n and L_p the diffusion lengths of electrons and holes, respectively; N_D and N_V are corresponding donor and acceptor densities. In the case of a Schottky-diode the forward current is usually described by the thermionic emission of majority carriers from the semiconductor to the metal (e.g. electrons from the conduction band for an n-type semiconductor) (Figure 2a). Then we have [18]

$$j_o = A \frac{m^*}{m_e} T^2 \exp\left(\frac{-e\Phi_B}{kT}\right) \quad \text{(majority carrier device)} \tag{4}$$

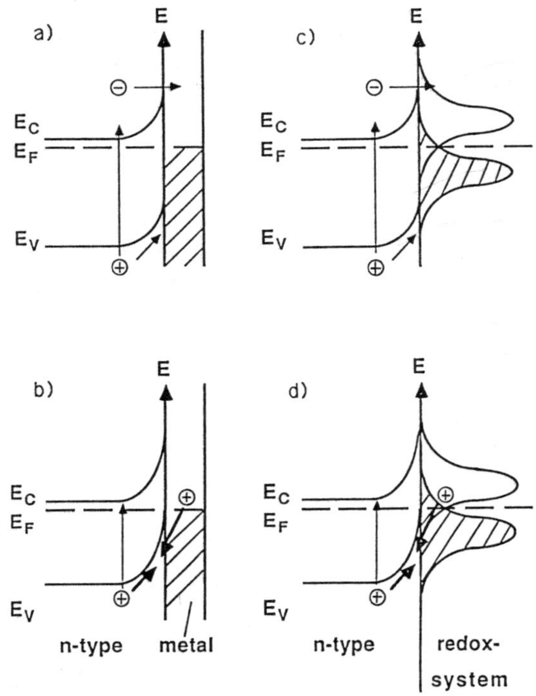

Figure 2. Energy scheme of a Schottky junction (a and b) and of semiconductor electrolyte interface (c and d).

in which $e\phi_B$ is the barrier height of the junction, m* the effective mass and m_e the electron mass in free space. A is given by

$$A = \frac{4 \pi m_e kT}{h^3} = 120 \frac{A}{cm^2} \tag{5}$$

For large barrier heights, $e\phi_B \sim E_g$, the forward current can also be determined by injection of minority carriers as illustrated in Fig. 2b for an n-type semiconductor. In the latter case one has a pure valence band process and j_o is given by

$$j_o = j_{v,o} = e n_i^2 \frac{D_P}{N_D L_P} \quad \text{(minority carrier device)} \tag{6}$$

The forward current of a majority carrier device depends only on the barrier height (eq. (4)), whereas that of a minority carrier device is mainly determined by recombination and diffusion parameters of the minority carriers. Using, for instance, $D_p = 10\ cm^2 s^{-1}$, $L_P = 1 \times 10^{-4}$ cm, $n_i^2 = 3 \times 10^{13}$ cm and $N_D = 3 \times 10^{16}\ cm^{-3}$ (good quality n-GaAs), one obtains $j_{v,o} = 10^{-17} A\ cm^{-2}$.

Applying the thermionic emission model, such a low exchange current is only obtained for a barrier height of $e\phi_B = 1.35$ eV, i.e. a value which corresponds to the bandgap of GaAs ($E_G = 1.4$ eV). In most cases, however, the barrier height is much smaller (one half to 2/3 of the bandgap [19], so that in general the exchange current j_o is relatively large, leading to rather small photovoltages.

On the other hand, the injection current in a minority carrier device can be considerably higher if the recombination occurs mainly within the space charge layer. Then we have [18]

$$j_{o,rec} = \frac{1}{2}\ e\ d_{sc}\ \sigma\ v_h\ N_t\ n_i \tag{7}$$

in which d_{sc} is the thickness of the space charge layer, N_t the density of recombination centers, s the cross section of the electrons and holes and v_{th} the thermal velocity of the charge carriers. Mostly it is difficult to distinguish between recombination in the bulk and within the space charge region. Usually it is decided from measurements of the ideality factor n (see equation (1)) whether the space charge recombination is involved, since n=1 for bulk recombination and n=2 for space charge recombination. In the case of an ideal majority carrier device n should always be unity.

In principle the same process should occur in regenerative photoelectrochemical cells, as illutstrated in Fig. 2c and d. However, the charge transfer can essentially be determined by the kinetics at the semiconductor-liquid interface. For instance, in the case of a majority carrier device consisting of an n-type semiconductor and a proper redox couple in the liquid, the cathodic forward current is given by [17]

$$j_c^- = ek_c^- c_{ox} n_s = k_c^- c_{ox} n_o \exp\left(-\frac{e\Phi_{sc}}{kT}\right) \tag{8}$$

in which c_{ox} is the concentration of the oxidized species of the redox system, n_S and n_O the electron density at the surface and in the bulk and ϕ_{SC} the potential across the space charge region. The corresponding exchange current, which can be used in eq. (1), is given by (majority carrier device)

$$j_{c,o} = e\,k_c^- \, C_{ox}\, n_o \exp\left(-\frac{e\Phi_{sc}{}^{o}}{kt}\right) \tag{9}$$

in which $\phi_{SC}{}^{o}$ is the band bending at equilibrium. Although we have here a majority carrier device, this exchange current $j_{c,o}$ is usually much smaller than that expected from the thermionic emission model. In the latter model, which has been proved quantitatively for many solid state Schottky junctions, it is assumed that all electrons reaching the surface are transferred to the metal. In the case of semiconductor/liquid junctions, the kinetics can be much slower because less empty energy states of the redox couple may be present at $E=E_C$. Accordingly, the thermionic emission model is only an upper limit in electrochemical systems, as proved for few systems [20]. This small exchange current $j_{c,o}$, usually observed in electrochemical systems, can be very advantageous because higher photovoltages can be reached.

In principle, a minority carrier device can also be made with semiconductor junctions by using a redox couple of a sufficiently positive standard potential (Fig. 2d). In this case, if the kinetics is fast, the current is determined by the recombination in the bulk or within the space charge region, and $j_{v,o}$ is given by eqs. (6) and (7).

2.2. CONVERSION EFFICIENCIES

A photocurrent can flow through a solar cell if a band bending below the semiconductor surface exists at equilibrium. Following the thermodynamic theories developed by Ross et al. [21-24] and Bolton et al. [25], an upper limit of the difference of quasi-Fermi levels of electrons and holes can be obtained for a certain solar radiation, i.e.

$$\mu_{max} = E_{F,n} - E_{F,p} = eU_{ph} \tag{10}$$

This difference of quasi Fermi levels is identical to the maximum photovoltage, which can be obtained with a solar cell, as illustrated in Fig. 3, for a photoelectrochemical system. It is assumed herethat the hole transfer is fast, so that the quasi-Fermi level of holes is pinned to

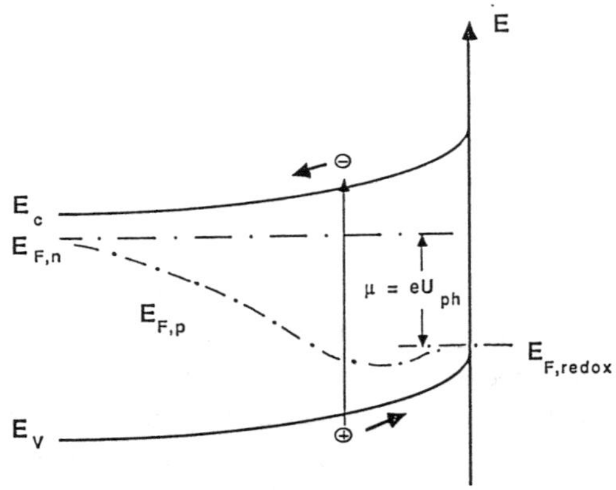

Figure 3. Position of quasi-Fermi levels during illumination

the redox potential. Using the theory mentioned above, μ_{max} and the maximum photovoltage can be calculated for semiconductors as shown for AMO in Fig. 4 [26]. Acording to this result, the upper thermodynamic limit for a photovoltage is smaller than the bandgap. Further, it should

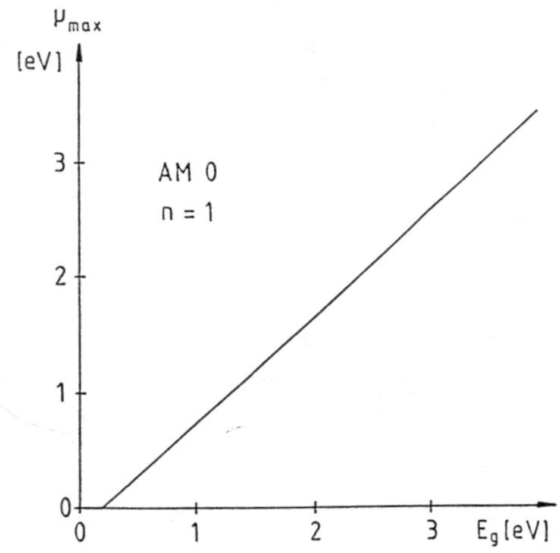

Figure 4. Maximum difference between quasi-Fermi-levels (photovoltage) vs. bandgap

be emphasized that the output power P of a solar cell should be maximized. According to Ross et al. [21], the photovoltage at the maximum power point is given by

$$\mu_p \approx \mu_{max} - kT \ln\left(\frac{\mu_{max}}{kT}\right) \qquad (11)$$

The application of this theory to regenerative photoelectrochemical cells leads to some consequences. In principle, the photocurrent onset should occur close to the flatband potential U_{fb}. Selecting a redox couple, the standard potential of which is close to the valence band then $U^o_{redox}-U_{fb} > \mu_{max}/e$. A good rectangular like current-potential curve of a

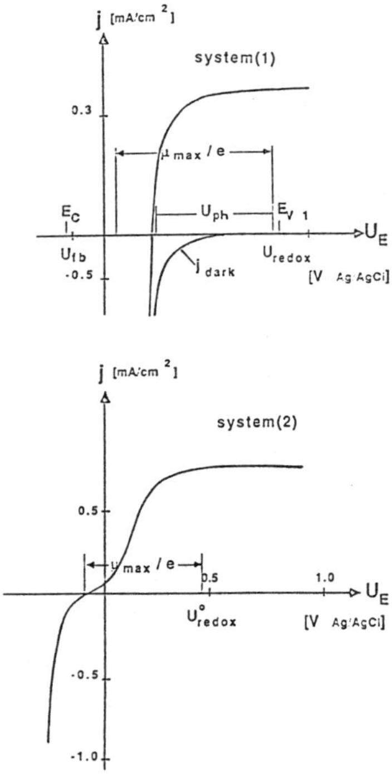

Figure 5. j-U_E characteristic of a semiconductor-liquid junction.

(a) U^o_{redox} at E_V; $U_{ph} < \mu_{max}/e$
(n-WSe_2/Fe(phen)$_3^{2+/3+}$; minority carrier device [43])

(b) U^o_{redox} above E_V; U_{ph} limited by μ_{max}
(n-WSe_2/$Fe^{2+/3+}$; minority carrier device, small k_c [43])

photoelectrochemical solar cell can only be obtained if the forward dark current (cathodic current at n-type) reaches sufficiently high values at electrode potentials $U_E > U^o_{redox} - \mu_{max}/e$, as shown in Fig. 5a.This is valid for a majority as well as for a minority carrier device. Choosing, however, another redox system where $U^o_{redox} - U_{fb} < E_G/e$ and the cathodic dark corresponds to a conduction band process, the situation may be different. Assuming that the kinetics of the electron transfer from the conduction band is rather slow, then large cathodic currents are only reached at relatively negative cathodic potentials (Fig. 5b). In this case, the photovoltage U_{ph} may be equal to μ_{max}/e, however, the current-potential characteristics is very poor (Fig . 5b), i.e. the fill factor is expected to be low.

In conclusion it can be stated that photovoltage and j-U characteristics depend strongly on the forward dark current. In a minority carrier device, this current is mainly determined by the kinetics of majority transfer at the interface, whereas in a minority carrier device the dark current depends on the diffusion constant and length, i.e. on bulk properties of the semiconductor (see also previous chapter).

2.3. EVALUATION OF RECENT SYSTEMS

During the last couple of years, research concentrated mainly on semiconductors which have a suitable bandgap with respect to a high conversion efficiency. Examples are CdSe (1.7 eV); GaAs (1.4 eV); WSe_2 (1.2 eV); Si (1.1 eV); $CuInSe_2$ (1.0 eV) and FeS_2 (0.95 eV). Many investigations have been performed by using sulfide/polysulfide (S^{2-}/S_n^{2-}) as a redox couple. Especially the group of the Weizmann Institute has studied corresponding systems in detail (see ref. [27] and [28] and literature cited therein). One example is n-CdSe, the current-potential

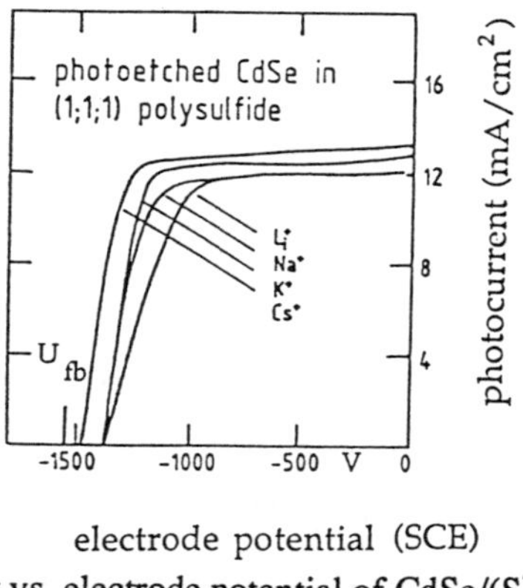

Figure 6. Photocurrent vs. electrode potential of CdSe/(S^{2-}/S_n^{2-}) in aqueous solution [29]

curve of which is given in Fig 6. Since the redox potential occurs about in the middle of the gap (U^o_{redox} -U_{fb} = 0.80 V; U_{fb} and U^o_{redox} are indicated in Fig. 6), the forward current of the cell must be due to an electron (majority carrier) transfer from the conduction band to the redox system. A photovoltage of U_{ph} = 0.7V was obtained. Assuming for a moment that this cell would behave like an ideal Schottky junction and the thermionic emission model can be applied ther - according to eqs. (4) and (2) - only U_{ph} = 0.3 V would be expected. Accordingly, the exchange current must be much smaller here and is entirely controlled by the kinetics of electron transfer at the interface. This conclusion is supported by the result that during illumination the flatband potential is reached (see Fig. 6). The Weizmann group also found an influence of cations such as various alkali [29] which was related to ion pairing for strongly hydrogenated cations, resulting in a decreased activity of the active polysulfide at the electrode. Since the potential dependence of the dark current was not measured, it is not clear whether the cations have also an effect on the kinetics of the electron transfer. Very recently, Licht et al. reported a photovoltage of 1.2 V for a CdSe/$[Fe(CN)_6]^{4-/3-}$-cell [30]. The photodecomposition of the redox couple was avoided by adding CN^--ions to the electrolyte.

During the last couple of years, $CnInSe_2$ elctrodes were studied extensively. In iodine solutions, a high fill factor and a good conversion efficiency were determined, but $CuInSe_2$ was found to be unstable in I_3^- solutions [31, 32]. Surprisingly, the stability could be greatly increased by adding Cu^{1+}-ions to the electrolyte [31, 32]. It could be shown that this stabilization is caused by the formation of a passivating layer of p-$CuISe_3$-Se^o [33]. Accordingly, a pn-heterojunction is formed which is actually the active part of the solar cell.

Rather high photovoltages and conversion efficiencies have been obtained with Si- and GaAs-electrodes, combined with suitable redox systems. In the case of Si, corresponding n-type electrodes were contacted with acetonitrile or methanol containing ferrocene derivatives as a redox couple [34, 35, 36]. Using dimethyl ferrocene (Me_2Fe), a photovoltage of 0.68 V at i_{ph} = 20 mA cm^{-2} has been obtained. The barrier height E_c^s-eU^o_{redox} was determined by measuring the temperature dependence of the open circuit voltage. Since an extrapolation to T = O yielded U_{ph}-E_G/e, it was concluded that the forward current is determined by injection of minority carriers (minority carrier device), i.e. the exchange current j_o is determined by eq. (6). This assumption was supported by the result that U_{ph} varied in the right direction if Si-electrodes of different dopings, diffusion-constants and -lengths were used. Such a calculation implies, however, that the injection current is governed by bulk recombination of the minority carriers, i.e. the quality factor must be n = 1. Unfortunately, the authors did not publish the dark current-potential characteristics which may prove this result.

Similar results have recently been obtained in our laboratory with n-GaAs in contact with aqueous HCl-solutions containing Cu^{1+}/Cu^{2+} as a redox couple [37, 26]. This cell shows a surprisingly high stability (see under section 2,5). As indicated in Fig. 7, the flatband potential occurs at U_{fb} = -1.0 V, whereas the redox potential is U^{o}_{redox} = +0.34 V, i.e. the latter is very close to the valence band. Detailed investigations of n- and p-type electrodes have shown that the cathodic current corresponds to a pure valence band process. This redox process is a reversible reaction, i.e. it is diffusion controlled. Accordingly, the kinetics of a hole injection is a very fast process. The total current can be split into a kinetically and a diffusion controlled current according to the equation [27].

$$\frac{1}{j} = \frac{1}{j_{kin}} + \frac{1}{j_{diff}} \qquad (12)$$

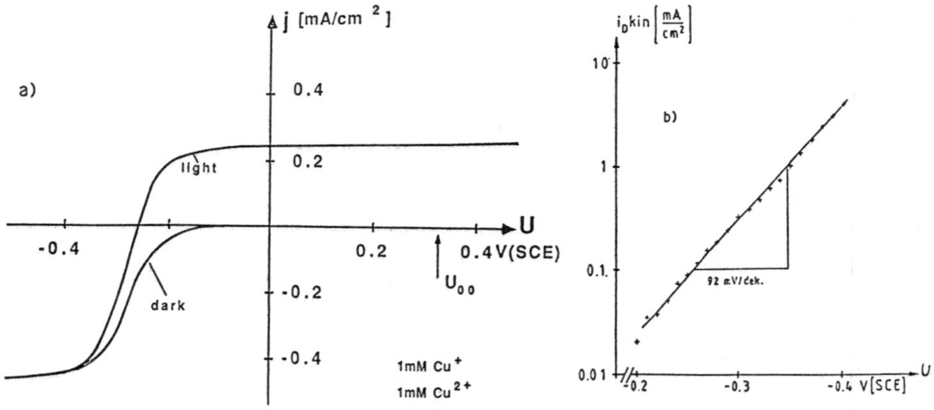

Figure 7. a) Photo- and dark current vs. potential of n-GaAs/(Cu^{1+}/Cu^{2+}) in 6 M HCl [26]
b) semilog plot of j_{kin} vs. U_E (see text) [26]

Experimentally, j_{kin} and j_{diff} were obtained at different electrode potentials from measurements with a rotating GaAs disc electrode, and the resulting potential dependence of j_{kin} is given in Fig. 7b. The slope of the semilogarithmic plot is 92 mV/decade, i.e. the quality factor must be about n = 1.5. Accordingly, - as already discussed in section 2.1. - the injection of minority carriers is not only followed by recombination in the bulk, but also within the space charge layer of the GaAs-electrode. Such a space charge or even surface recombination leads to a relatively high dark current, depending on the density of recombination centers. This explains the relatively low photovoltage of 0.5 V, compared to the theoretical value of 0.9 V given by

Fig. 4. Since here the photovoltage is only determined by the semiconductor properties, higher values can only be realized by improving the material.

It should be mentioned that similar or even higher values of Uph have been obtained with the S^{2-}/S_n^{2-}-couple at n-GaAs, although the barrier height is much smaller [38] [39]. In this case the forward current is determined by electron transfer, and obviously the kinetics is rather slow, so that the exchange current j_o is small.

With respect to mechanistic aspects, n-WSe_2 is a very interesting electrode material. Tenne et al. [40 - 42] have reported a conversion efficiency of 14% and a photovoltage of 0.6 V, using I_3^-/I^- as a redox system in aqueous solutions. This redox couple is rather favorable with respect to the electrode stability against corrosion, because it is strongly adsorbed at the surface. The authors emphasized that the high efficiency was only obtained by prolonged anodic photoetching in KOH. They explain this result by a reduction of surface recombination upon etching. Accordingly, the photocurrent rises more rapidly with increasing potential , as illustrated in Fig. 8. It is interesting to note that the photocurrent and consequently the quantum

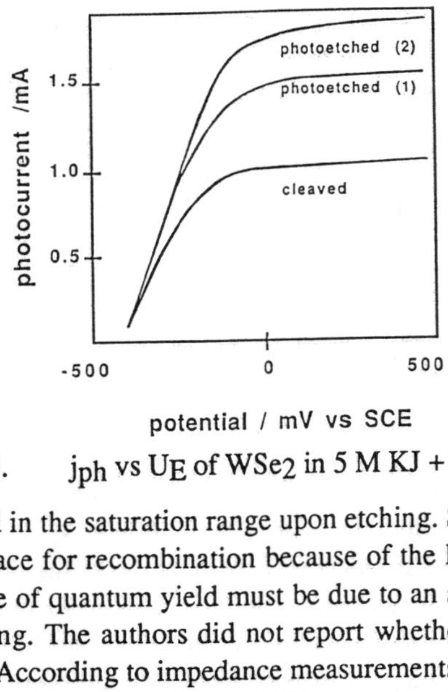

Figure 8. j_{ph} vs U_E of WSe_2 in 5 M KJ + 0.01 M I_2 [41]

efficiency also increased in the saturation range upon etching. Since in this range no electrons are available at the surface for recombination because of the large upward band bending we assume that this increase of quantum yield must be due to an activation of additional surface areas during photoetching. The authors did not report whether this system is a minority or majority carrier device. According to impedance measurements in our laboratory, the flatband potential is very cathodic in iodine solutions [43], a result which may indicate that this is a minority carrier device.

It is well known that the density of steps at the surface of the "van der Waal" type of layer compounds plays an important role [44]. Accordingly the crystal growth is of great importance. Since research groups used different crystal sources, the results are sometimes difficult to compare. The density of steps has a great influence on the stability, because the anodic decomposition occurs preferably at steps [44]. In our studies with WSe_2 we only used crystals where U_{fb} was independent of pH, which is an indication of good stability [45]. We investigated various redox reactions at WSe_2 [46]. In the case of the redox couple $Fe(phen)_3^{2+/3+}$ we found that the standard potential occurs at the valence band at the surface and that the cathodic dark current is due to injection of holes. Accordingly, this cell is a minority carrier device, and the corresponding j-U characteristics is given in Fig. 5a (see section 2.2). The ideality factor is around 1.5 so that recombination at the surface and within the space charge region dominates [43]. In the case of other redox systems, such as e.g. Fe^{2+}/Fe^{3+}, the standard potential of which occurs in the middle of the gap. Here the cathodic current is governed by the transfer of majority carriers. Since it attains higher values only near the flatband potential, the kinetics must be slow (Fig. 5b). It is interesting to note that the U_{ph} reaches here the thermodynamic limit leading to an S-shaped j-U_E-curve, as discussed in section 2.2.

Another promising n-type material seems to be FeS_2 (pyrite), having a bandgap of 0.95 eV. Since this material exhibits only a direct bandgap, the absorption is high in the long wavelength range ($\alpha > 6x10^5 cm^{-1}$ for $E_{ph} > 1.3$ eV) [47] [48]. Although theoretically a photovoltage of 0.55 V is possible, only $U_{ph} \approx 0.2V$ has been obtained with cells containing I_3^-/I^- as a redox system. According to XPS and UPS measurements, strong Fermi level pinning was observed due to the existance of FeS moities at the surface [49]. Various other redox couples, such as e.g. Br^- or $[Fe(CN)_6]^{3-/4-}$, have been successfully tested. Since with positively charged redox ions no photocurrent was observed, Tributsch et al. [48] interpreted this effect by a coordination of the negatively charged redox ions to the positively charged ion center.

A very interesting approach has been used by Nakato and Tsubomura [50]. They have developed a model as illustrated in Fig . 9.If big metal clusters (>10 mm) are deposited, the energy bands are modulated (Fig. 9a), The barrier height $e\phi_B$ underneath the metal is determined entirely by the semiconductor/metal contact. In this case, the photovoltage is independent of the redox potential as proved with cells of n-Si/Pt/redox couple [51] (see lower curve in Fig. 10). In the case of much smaller islands, i.e. ≤ 5 nm the quantitized levels of the electrons in the potential well in the conduction band should become considerably higher, so that the electrons have a higher kinetic energy. Since the potential well is very narrow, the band modulation diminishes rapidly toward the interior of the crystal. These two effects make the effective barrier height little different from that of a naked surface (Fig. 9b); i.e. $e\phi_B$ determines the photovoltage. Since $e\phi_B$ depends on the redox potential, the photovoltage can be changed by varying the standard potential as proved with Si-electrodes (upper curve in Fig. 10). High photovoltages of up to 0.68 V at j_{ph} = 20 mAcm^{-2} were obtained. This system is certainly a majority carrier device, and the electrons are channelled through the small Pt-islands. The authors describe the forward current by the thermionic emission model (eq. (4)). They assume, however, that the actual surface area covered by Pt-islands is very small, leading to a correspondingly small forward current and therefore to a high U_{ph} [51].

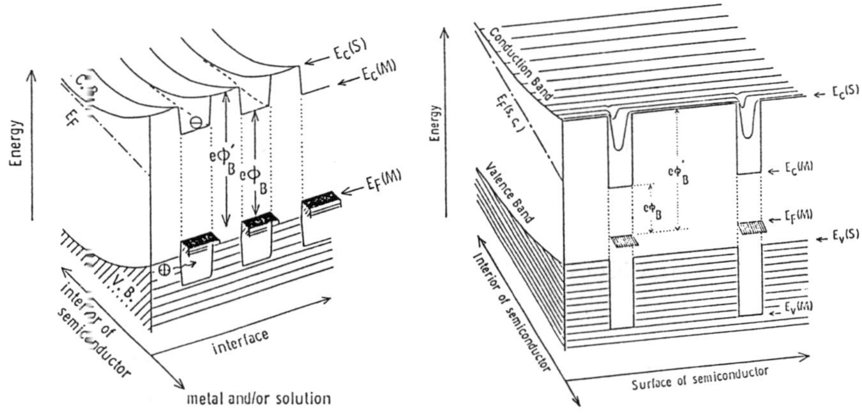

a) d ≫ 10 nm b) d < 10 nm

Figure 9. Schematic energy diagram of n-type semiconductor electronic bands at the solid/liquid interface modulated by discontineous metal coating [50].

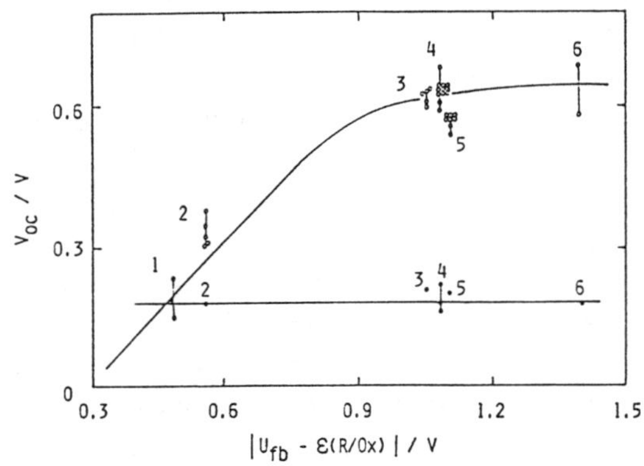

Figure 10. V_{oc} versus $[U_{fb}-\varepsilon(R/Ox)]$ for Pt-coated alkali-etched n-Si (o) and for Pt-coated n-Si (•). Redox couples used are as follows: 1,$Fe(C_2O_4)_3^{4-}/Fe(C_2O_4)_3^{3-}$ (pH 6.2); 2,I^-/I_3^- (7.6MH+); 3,Fe^{2+}/Fe^{3+} (pH 0.5); 4,Br^-/Br_2 (8.6 MH+); 5,$Fe(CN)_6^{4-}/Fe(CN)_6^{3-}$ (pH 7.6); 6,Br^-/Br_2 (pH 2.9) [51].

Very recently, sensitization effects have been applied in regenerative cells using TiO_2-electrodes covered by a sensitizer [52]. In this case, the excitation occurs not in the semiconductor, but in the adsorbed dye. Although only a dye monolayer is usually effective [55], conversion efficiencies of 12% at $\lambda = 470$ nm have been reported by Grätzel [52]. This efficiency was achieved by using a novel structurated TiO_2-surface of an extremely large surface area.

Various attempts have been made with new semiconductors. In most cases, the material is still not sufficiently characterized. One example are ternary compounds, such as $Cd_x Hg_{1-x} Te$. The bandgap is a function of composition. Quantum yields of 0.4 in polysulfide solutions have been reported recently [54].

2.4. STABILITY OF SEMICONDUCTOR ELECTRODES

The photostability of semiconductor electrodes is a severe problem especially in photoelectrolysis cells. The stability can be thermodynamically or kinetically controlled. The first type of stability has been found with various oxide semiconductors, the bandgap of which is rather high. A reasonable or good stability of regenerative cells was also found with semiconductors absorbing light in the visible range and with redox systems of a standard potential being located within the gap of the semiconductor. In several cases it has been reported that the stability decreased with increasing light intensity. Rather complex kinetic models have been derived to explain this effect (for references see [17]). In this context it is very interesting that a very good stability of GaAs in HCl was found with Cu^{1+}/Cu^{2+} as a redox couple, although the standard potential occurs at the upper edge of the valence band [55]. As already mentioned in section 2.3., the charge transfer between the semiconductor and this redox system is extremely fast, so that the redox process competes sufficiently well with the anodic decomposition. It is further interesting to note that the stability was independent of light intensity. The two competing valence band processes, decomposition and redox reaction, could be interpreted without assuming any surface intermediates. Studies with a variety of redox systems have shown that the ratio of decomposition and redox currents at illuminated n-GaAs electrodes can be predicted from corresponding dark currents at p-electrodes [26]. In connection with this problem, a new concept concerning the compatibility of redox reactions at n- and p-electrodes has been developed [26, 37] which cannot be discussed here.

Another interesting case is WSe_2 and other layer compounds. Polarizing for instance an n-WSe_2-electrode (low density of steps at the surface) anodically in an aqueous solution of low pH without any redox system, then the photocurrent is entirely due to anodic decompositon.

However, a high overvoltage (~0.6 V) for the onset of photocurrent with respect to the flatband potential in the dark, has been observed. Capacity measurements have shown that U_{fb} is shifted upon illumination to the onset potential of the photocurrent [56]. The corresponding downward shift of the energy bands in given in Fig. 11. This result has been interpreted by

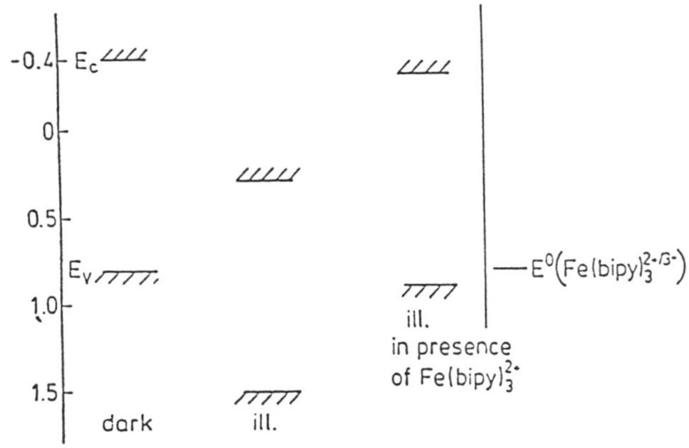

Figure 11. Position of energy bands of WSe_2 in the dark under illumination [46].

assuming that the decomposition potential is located below the dark position of the valence band, i.e. the reaction rate must be very small. Upon addition of a suitable redox couple, such as $Fe(phen)_3^{2+}$, to the electrolyte the flatband potential was not shifted during illumination. Accoringly, the energy bands remain pinned to their dark position as shown in Fig. 11 (right side); i.e. at a position at which no decomposition occurs. Experiments with a rotating ring-disc assembly has proved that the stability factor is unity for photocurrents below the diffusion saturation current [56].

3. Production Of Chemical Fuels

Since the corrosion problem is very severe in the photoelectrolysis of H_2O, only large bandgap oxide semiconductors have been applied in the early stage of this research. The most difficult task is here the oxidation of H_2O to oxygen. More recently, some promising attempts have been made with d-band transition metal chalcogenides, such as e.g. RuS_2 (E_g = 1.3 eV), at which O_2 was produced at anodic polarization during illumination. It is interesting to note that the flatband potential is shifted upon illumination by 1.8 V, so that the valence band occurred at +2.2 eV, whereas the photocurrent onset was found at +1.2 V (NHE) [57]. Since an XPS-analysis has proved the formation of few layers of RuO_2 on the surface [58], and since RuO_2 is a metalic conductor, one must conclude that a semiconductor/metal Schottky is formed. Because of the large anodic shift of the bands and the low bandgap, the cell can only be operated under an additional external bias.

Also with n-$MoSe_2$ O_2-formation has been observed under external bias [43]. In this case, a relay compound ($Rubipy_3^{2+}$) has been used, which pinned the energy bands at a position (see also section 2.4.) that corrosion was avoided. Here the relay compound was oxidized by hole transfer, and the Ru (III) formed in this process oxidized H_2O in the solution in the presence of a catalyst.

Frequently it has been attempted to avoid corrosion by depositing a conducting polymer on the semiconductor surface. This technique has been successfully applied by Horowitz et al. [59], who reported H_2O-photocleavage in a self-driven cell consisting of CdS, coated with a conducting polymer and a Pt-counter electrode. The question arises, however, whether it is reasonable to use a semiconductor on which a semiconductor-metal (conducting polymer) Schottky junction is formed. Since this is a solar cell by itself, it would be safer to separate the Schottky junction from the electrolyte in order to avoid any corrosion problem.

Various attempts have been made to produce alcohol or methane by electrochemical reduction of CO_2. Besides the methanol production at GaAs-electrodes (As III-face) [60], more recently the formation of CH_4 and C_2H_6 at Cu-electrodes have been reported [61, 62, 63]. New attempts to reduce CO_2 photoelectrochemically were made by using p-CuO and p-SiC electrodes [64]. In the case of CuO, the current efficiency for the formation of CH_4 was initially 60%. The electrode, however, was not stable. At microcrystalline p-SiC the main product was methanol or ethanol besides H_2.

4. Application of Semiconductor Electrochemistry in Fabricating Solid State Photocells

Properties of Si- and GaAs-Schottky junctions produced by metal evaporation in vacuum have been extensively studied during the last 20 years (see e.g. [18], [19]). According to these investigations, the barrier height never exceeded above about 2/3 of the bandgap, even not with metals of a high work function. This was explained by Fermi level pinning due to surface states. Few years ago, n-GaAs/metal Schottky junctions have been made by electrochemical deposition of the metal (Au, Cu, Pt) [65]. Detailed measurements of the space capacity and j-U-curves yielded a barrier height of about 1.2 eV (Fig. 12b), i.e. a value being much higher than that obtained after evaporating the metal in UHV. It was concluded that the density of surface states is obviously much smaller at a cathodically polarized GaAs-electrode, probably due to the formation of a hydride monolayer on the GaAs-surface. Although ϕ_B is very large, the forward current (ideality factor between 1.1 and 1.2) could be quantitatively explained by the thermionic emission model [65]. A photovoltage of U_{ph} = 0.5 V at a photocurrent of 0.3 $mAcm^{-2}$ has been obtained (Fig 12a).

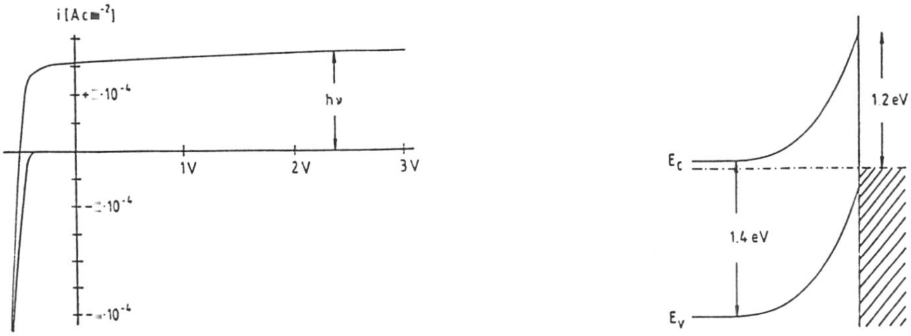

Figure 12.: n-GaAs/Au Schottky junction
a) current vs. voltage
b) energy diagram at equilibrium [65]

Similar observations have recently been made with Si on which a film of poly-(trimethylsilylcyclooctatetraene) was electrochemically deposited. From measurements of the photovoltage vs. temperature a barrier height of about the bandgap has been obtained [66]. It was concluded that the dark current is determined by injection of minority carriers.

Another example is the WSe_2/Au diode, the gold being deposited by vacuum evaporation. If Au was deposited on a freshly cleaved surface, the j-U-curve indicates ohmic behaviour of the contact showing a small photoeffect (dot-dashed and dotted curves in Fig. 13) [67]. On the other hand, treating the surface by running the WSe_2 as a photoanode under short circuit conditions in a polyiodide electrolyte for about 30 min, before the deposition of Au, then a good diode behaviour was found (solid and dashed curves in Fig. 13). A photovoltage of 0.73 V was measured under intense illumination (~ 20 suns). Hodes interpreted the effect of iodine treatment by assuming a passivation of defects [67]. The defects could act as short circuits in parallel with the good diode behaviour of the defect free region.

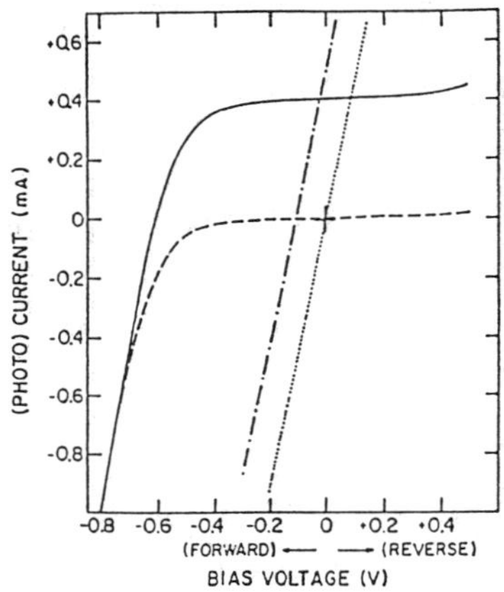

Figure 13. (Photo)current-voltage characteristics of WSe_2/Au diodes.
WSe_2 area = 4.5 mm^2. (Dotted curve) untreated, dark; (dot-dashed curve) untreated, illuminated; (dashed curve) treated, dark; (solid curve) treated, illuminated [67]

5. References

[1] Harrris, L.A. and Wilson, R.H., Annu. Rev. Mater. Sci., p.99, 1978

[2] Gerischer. H., in Solar Energy Conversion, (Topics in Applied Physics, Vol. 31), Seraphin, B.O., Ed., Springer-Verlag, Berlin, 1979, p.115

[3] Memming, R., Phil. Tech. Rev., 38, p.160, 1978/1979

[4] Nozik, A.J., Annu. Rev. Phys. Chem., 29, p.189, 1978

[5] Rajeshwar, K., Singh, P., and Dubow, J., Electrochim. Acta, 23, p. 1117, 1978

[6] Butler, M.A. and Ginley, D.S., J. Mater. Sci., 15, p.1, 1980

[7] Memming, R., Electrochim. Acta, 25, p.77, 1980

[8] Heller, A., Accts. Chem. Res., 14, p.154, 1981

[9] Tributsch, H., in Solar Engergy Materials, (Structure and Bonding Ser., Vol. 49), Clarke, M.J. et al., Eds., Springer-Verlag, Berlin, 1982, p.127

[10] Bard, A.J., Science, 207, p.139, 1980

[11] Grätzel, M., in Solar Energy Materials, (Structure and Bonding Ser., Vol 49), Clarke, M.J. et al., Eds., Springer-Verlag, Berlin, 1982, p.37

[12] Lewis, N.S., Annu. Rev. Mater. Sci., 14, p.95, 1984

[13] Pleskow, Y.U.V., Progr. Surf. Sci., 15, p.401, 1984

[14] Hodes, G., Fonasch, S.F., Heller, A. and Miller, B., in Advances in Electrochemistry and Electrochemical Engineering, Vol. 13, H. Gerischer, H., Ed., John Wiley & Sons Inc., New York, 1985, p.113

[15] Tributsch, H., in New York Trends and Applications of Photocatalysis and Photoelectrochemistry for Environmental Problems, Schiavello, M., Ed., D. Reidel, Dordrecht, 1988, p.297
[16] Memming, R., Top. Curr. Chem., 143, p.81, 1987
[17] Memming, R., in Photochemistry and Photophysics, Vol.2, Rabek, J.F., Ed., CRC-Press, Boca Raton (USA), 1990, p.143
[18] Sze, S.M., Physics of Semicondurctor Devices, 2nd. ed., John Wiley & Sons, New York 1981
[19] Brillson, L.J., Surf. Sci. Rep., 2, p.123, 1980
[20] Memming, R., Ber. Bunsenges., 91, p.353, 1987
[21] Ross, R.T. and Hsiao, T.-L., J. Appl. Phys., 48, p. 4783, 1977
[22] Ross, R.T. and Collins, J.M., J. Appl. Phys., 51, p.4504, 1980
[23] Ross, R.T., J. Chem. Phys., 45, p.1, 1966
[24] Ross, R.T., J. Chem. Phys., 46, p.4590, 1967
[25] Bolton, J.R., Haught, A.F. and Ross, R.T., in Photochemial Conversion and Storage of Solar Energy, Connolly, J.S., Ed., Academic Press, 1980, chap. 11
[26] Reineke, R., PhD-thesis, University of Hamburg, 1988
[27] Hodes, G., J. Photochem., 29, p.243, 1985
[28] Hodes, G., in Energy Resources through Photochemistry and Catalysis, Grätzel, M., Ed., Academic Press, New York, 1983, p.521
[29] Licht, S. and Manassen, J., J. Electrochem., 132, p.1076, 1985
[30] Licht, S. and Peramunage, D., Am. Chem. Soc. Meeting, Boston, 1990, abstract 94
[31] Cahen, D. and Chen, Y.W., Appl. Phys. Lett., 45, p.746, 1984
[32] Menezes, S., Lewerenz, H.J. and Bachmann, K.J., Nature, 305, p.615, 1983
[33] Lewerenz, H.J. and Kötz, E.R., J. Appl. Phys., 60, p.1430, 1986
[34] Rosenbluth, M. and Lewis, N.S., J. Am. Chem. Soc., 108, p.4689, 1986
[35] Abrahams, J.L., Casagrande, L.G., Rosenblum, M.D., Rosenbluth, M.L., Santangelo, P.G., Tufts, B.J. and Lewis, N.S., Nouv. J. Chim., 11, p.157, 1987
[36] Lewis N.S., Rosenbluth, M.L., Casagrande, L.G. and Tufts, B.J., in Homogeneous and Heterogeneous Photocatalysis, Pelizzetti, E. and Serpone, N., Eds., D. Reidel Publ. Comp., Dordrecht, 1986, p.343
[37] Reineke, R. and Memming, R., submitted to J. Phys. Chem.
[38] Chang, K.C., Heller, A., Schwartz, B., Menezes, A. and Miller, B., Science, 196, p.1097, 1977
[39] Parkinson, B.A., Heller, A. and Miller, B., J. Electrochem. Soc., 126. p.954, 1979
[40] Tenne R. and Wold, A., Appl. Phys. Lett., 47, p.707, 1985
[41] Tenne R. Spahni, W., Calzaferri, G. and Wold, A., J. Electroanal. Chem., 189, p.247, 1985
[42] Tenne, R. and Wold, A., Ber. Bunsenges., 90, p.545, 1986
[43] Sinn, Ch., PhD-thesis, University of Hamburg, 1989
[44] Kautek, W., Gerischer, H. and Tributsch, H., Ber. Bunsenges. Phys. Chem., 83, p.1000, 1979
[45] Lewerenz, H.J., Gerischer, H. and Lübke, M., J. Electrochem. Soc., 131, p.100, 1984
[46] Sinn, Ch., Meissner, D. and Memming, R., J. Electrochem. Soc., 137, p.168, 1990
[47] Ennaoui, A. and Tributsch, H., J. Electroanal. Chem., 204, p.185, 1986

[48] Li, X.P., Alonso Vante, N. and Tributsch, H., J. Electroanal. Chem., 242, p.255, 1988

[49] Pettenkofer, C. and Jaegermann, W., Internat. Conf. on Photochem. Conversion and Storage of Solar Energy, Palermo, abstract 122, 1990

[50] Tsubomura, H. and Nakato, Y., J. Nouv. Chim., 11, p.167, 1987

[51] Nakato, Y., Ueda, K., Yano, H. and Tsubomura, H., J. Phys., 92, p.2316, 1988

[52] Grätzel, M., in Photochemical Energy Conversion, Norris, J.R. and Meisel, D., Eds., Elsevier, New York, 1989, p.284

[53] Memming, R., Progr. Surface Sci., 17, p.7, 1984

[54] Neumann-Spallart, M., Tamizhmani, G. and Levy-Clement, C., Internat. Conf. on Photochem. Conversion and Storage of Solar Energy, Palermo, abstract 121, 1990

[55] Reineke, R. and Memming, R., in Photoconversion Processes for Energy and Chemicals, Hall, D.O. and Grassi, G., Eds., Elsevier Appl. Science, New York, 1989, p. 129

[56] McEvoy, A.J., Etman, M. and Memming, R., J. Electroanal. Chem., 190, p.225, 1985

[57] Kühne, H.M. and Tributsch, H., J. Electroanal. Chem., 201, p.263, 1986

[58] Jaegermann, W. and Tributsch, H., Progr. Surf. Sci., 29, p.1, 1988

[59] Gnigue, D., Horowitz, G., Roncali, J. and Garnier, F., J. Electroanal. Chem., 269, p.337, 1989

[60] Frese, K.W. and Canfield, D., J. Electrochem. Soc., 130, p.1772, 1983; 131, p.2518, 1984

[61] Hori, Y., Kikuchi, K., Murata, A. and Suzuki, S., Chem. Lett., 897, 1986

[62] Cooke, R.L., McDuff, R.C. and Sammels, A.F., J. Electrochem. Soc., 134, p.1873, 1987; 134, p.2375, 1987

[63] Kim, J.J., Summers, D.P. and Frese, K.W., Electroanal. Chem.245, p.223, 1988

[64] Hashimoto, K., Luo, J. and Fujishima, A., Internat. Conf. on Photochem. Conversion and Storage of Solar Energy, Palermo, abstract , 1990

[65] Reineke, R. and Memming, R., Surf. Sci., 192, p.66, 1987

[66] Ginsberg, E.J., Sailor, M.J., Goman, C.B., Lewis, N.S. and Grubbs, R.H., Am. Chem. Soc. Meeting, Boston, 1990, abstract 212

[67] Hodes, G., Appl. Phys. Lett., 54, p.2085, 1989

MULTIELECTRON TRANSFER AND COOPERATIVE INTERFACIAL MECHANISMS IN PHOTOELECTROCHEMICAL ENERGY CONVERSION

Helmut TRIBUTSCH
Hahn-Meitner-Institut
Dept. Solare Energetik
D-1000 Berlin 39, Germany

ABSTRACT. In order to explore the possibilities of autocatalysis and nonlinear cooperative mechanisms for photoelectrochemical energy conversion, two model systems have been investigated: (a) illuminated p-type $CuInS_2$ and $CuInSe_2$ in contact with hydrogen peroxide containing electrolytes can undergo sustained photoelectrochemical oscillations/3/. (b) the homogeneous oscillating Belousov-Zhabotinskii reaction, which occurs between bromate ions and an organic substrate (malonic acid) and can be made photosensitive when $Ru(bipy)_3^{2+}$ is used as metal ion redox catalyst /4/.

The following possibilities for photoelectrochemical energy conversion via nonlinear cooperative mechanisms will be discussed:

- photoinduced interfacial electron transfer can be synchronized by autocatalysis even across a macroscopic interface (shown with model system (a)).
- the presence of autocatalytic steps and cooperative behaviour influences interfacial electrode stability (shown with model system (a)). New strategies for kinetic electrode stabilization may be imaginable.
- photoelectrodes for sensor applications with digital and all or non responses are feasible shown with model system (a)).
- periodic photoelectrochemical energy conversion dissipates less energy than a nonperiodic stationary one.
- interfacial autocatalytic steps may enable complicated entropy reducing mechnanisms of photochemical synthesis (analogy with biological synthesis of macromolocules)
- a homogeneous oscillating reaction which dissipates chemical energy may, when made photosensitive and periodically illuminated, be reversed to produce chemical energy from light (in exploration with model system (b)).

Autocatalytic steps in interfacial reactions will not be possible with electron transfer mechanism based on tunnelling processes. Strong interfacial chemical interaction will be necessary such as that provided by semiconducting materials and photocatalysts with energy bands derived from d-states.

Introduction

The conversion of photon energy into electrical and chemical energy involves energy conversion processes which are proceeding away from equilibrium and involve mechanisms which are basically irreversible since the process of generation of (photoelectro)chemical energy from light energy cannot easily be reversed. In spite of this well defined thermodynamic situation concepts from irreversible thermodynamics have not yet found access to photoelectrochemical reasoning. Quasi-thermodynamic descriptions of the illuminated semiconductor/electrolyte interface such as the quasi-Fermi level concept /1/ are widely dominating discussions and photoelectrochemical theory. This is in part understandable since they have proved to be very helpful in the

E. Pelizzetti and M. Schiavello (eds.), Photochemical Conversion and Storage of Solar Energy, 213–233.

qualitative description of photoelectrochemical phenomena proceeding at semiconductor interfaces. On the other hand, using quasi-equilibrium approaches we cannot hope to predict or explore new phenomena of photon energy conversion.

As known from the development of irreversible thermodynamics during the last decades with its significant consequences for the understanding of evolution, biological control und structural organization, as well as for a large variety of physical chemical mechanisms/2-5/, new phenomena can be expected far from equilibrium under conditions of autocatalysis. This situation of autocatalysis is thermodynamically equivalent to a negative excess entropy production (entropy production at constant thermodynamic fluxes) and introduces a nonlinearity and instability which gives rise to cooperative phenomena (e.g. oscillations, dissipative structures, multi-steady state phenomena). Under such conditions, the distance from equilibrium and nonlinearity turn out to be sources of order and reduction of entropy . This occurs, of course, at the expense of the energy input and the entropy production of the overall irreversible reaction. It is in the meanwhile generally accepted among specialists in this field that many biological phenomena including bioenergetic ones involve mechanisms of irreversible thermodynamics. It is already for this reason that an exploration of the potential of autocatalysis for photoelectrochemical energy conversion appears to be justified.

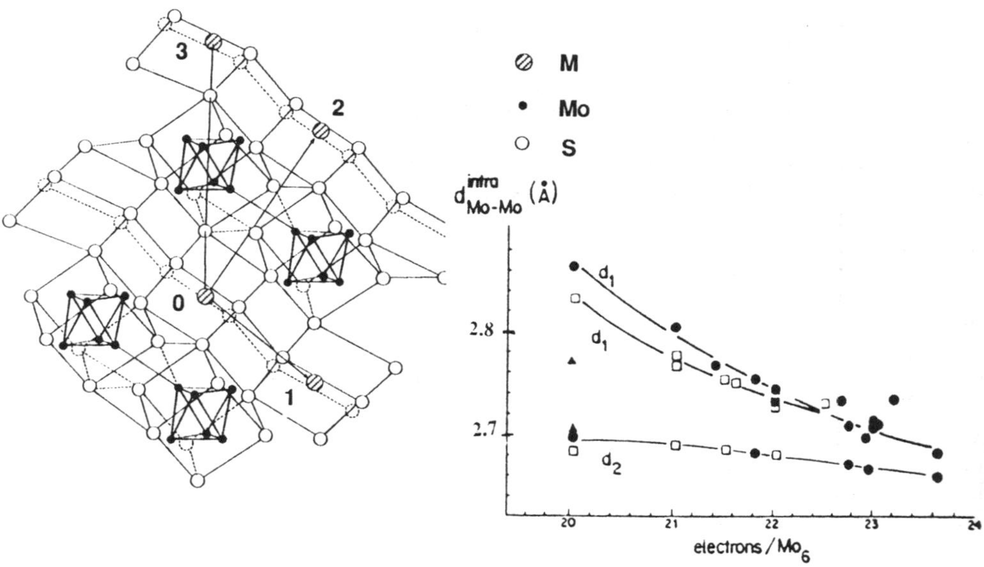

Fig.:1 The crystal structure of M Mo_6X_8 is shown (M = inserted species) together with the dependence of bonding distances within the cluster on the formal number of electrons within the cluster/22/. Interfacial electron transfer from these clusters should increase their size.

Experimental Evidence

Experimental studies on multielectron transfer catalysis for energy conversion using semiconducting cluster compounds ($M_x Mo_{6-x}X_8$ with M=metal,X=S,Se) /6-8/ indicate a favourable effect of autocatalytic steps. Molybdenum metal, for example, is a much less favourable catalyst for the reduction of oxygen to water in an acid environment than the cluster compound Mo_6Se_8, which contains molybdenum actahedra in regular distances within the crystal lattice. Catalytic properties are much improved when mixed cluster compounds of the type $M_x Mo_{6-x}X_8$ with M=metal,X=S,Se are used. One advantage of semiconducting cluster compounds as compared to metals is a higher d-state density in the valence and conduction band which has been shown to exert a very favourable effect on the catalytical properties of photoelctrodes /9/. However, a very interesting additional factor has been taken into consideration: the octahedric clusters change their volume and their electron density distribution depending on the number of electrons stored, (Fig.: 1) and thus show a dynamic behaviour during interfacial electron transfer: the first electron(s) transferred exert(s) a feedback on the catalytic properties of the cluster.

Theoretical arguments confirm the favourable effect of autocatalysis. While rate and energetic efficiency of ordinary multistep reactions drastically decrease with the number of steps, the same reaction can proceed at a high rate and can efficiently propagate free energy when autocatalytic steps are involved. Such mechanisms have long been known to play a key role in biological systems /5/. Among the well known examples is the uptake of oxygen by hemoglobin where the uptake of the first oxygen molecule exerts an autocatalytic effect on the uptake of the following molecules which significantly accelerates the kinetics as compared to myoglobin where autocatalysis is absent. Another example may be multiheme proteins such as cytochrome c_3, which are interesting models for intramolecular electron transfer.

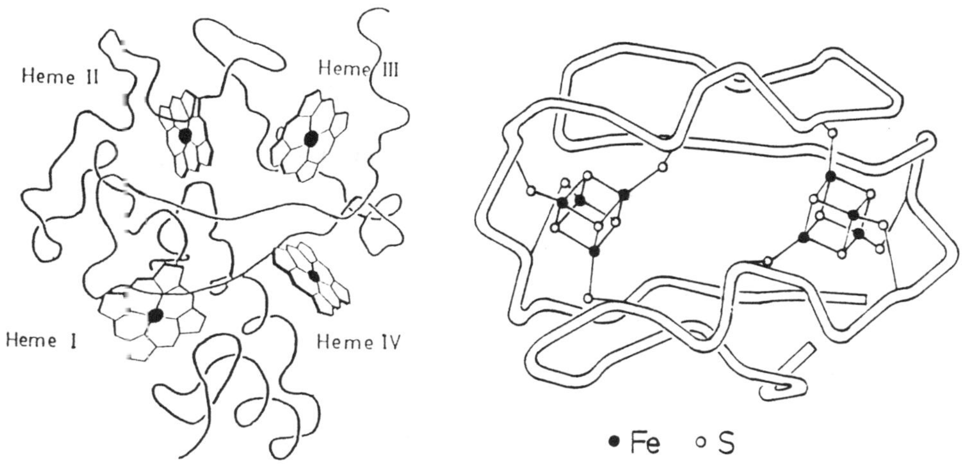

Fig.:2 Structure of cytochrome c_3 which allows strong cooperative heme-heme interaction, which is the basis for efficient electron transfer and a remarkable dynamics of macroscopic and microscopic redox potentials. On the right side ferredoxin is shown which contains iron-sulfur clusters which are also favourable systems for electron transfer.

In cytochrome c_3 from Desulfovibrio vulgaris a strong cooperative heme-heme interaction is observed, which is partially related to its unique heme orientation and

heme environment. Some structural changes occur reversibly during oxidation reduction reactions inducing exceptional electron transfer and redox behaviour/10,11/. Feedback or autocatalysis must be involved.
It is well established that in mechanisms far from equilibrium, autocatalysis may (under conditions of negative excess entropy production) produce cooperative behaviour (oscillations, all or non responses, multiple steady states, dissipative structures, thermodynamic chaos).

Photo-Autocatalytic Mechanisms: A Challenge for Solar Energy Conversion

Let us start with a relation describing entropy production density σ in an open system/5/:

$$\sigma = -\Sigma_i j_i \text{ grad } \frac{\mu_i}{T} + \Sigma_\rho w_\rho \frac{A_\rho}{T} = \Sigma_k J_k X_k \qquad (1)$$

(σ = entropy density, j = generalized flux, μ_i = chemical potential of constituent i, w_ρ = rate of reaction ρ, A_ρ = affinity of reaction ρ, J_k = generalized thermodynamic fluxes, X_k = generalized thermodynamic forces). To calculate the entropy production in an illuminated, energy converting liquid junction which generates photocurrents and photopotentials we have to consider that the photocurrent (I_{ph}) is related to the reaction rates w_p through $I_{ph} = z_pFw_p$ (F = Faraday constant, z_p= charge equivalent). Analogous to the situation in dark electrochemistry, it can be shown that the presence of a photopotential only manifests itself by altering the affinity in the above relation (1) :

$$A_p \to A_p^* = A_p + z_pF(\varphi_p^{*I} - \varphi_p^{II}) = (\mu_p^I + z_pF\varphi_p^{*I}) + (\mu_p^{II} + z_pF\varphi_p^{II}) \qquad (2)$$

(I: solid phase, II: liquid phase). φ_p^*= applied potential, which during illumination includes the difference between the quasi-Fermi level potential for electrons and holes (photoassisted reaction).

The entropy production of a photoelectrochemical interface can thus be written:

$$\sigma = \left(\frac{1}{T^I} - \frac{1}{T^{II}}\right)\frac{d\Phi}{dt} - \Sigma p\left(\frac{\mu_p^I}{T} - \frac{\mu_p^{II}}{T}\right) + A_p^{I*}\frac{v^I}{T} + A_p^{II*}\frac{v^{II}}{T} \qquad (3)$$

Assuming a negligible temperature difference due to illumination ($T^I = T^{II}$) and a negligible change in mole number passing from I to II, we find for the product of entropy production and temperature, which is equivalent to the energy dissipation E_{diss} by the system:

$$\sigma T = E_{diss} = A_p^{I*}w_p^I + A_p^{II*}w_p^{II} > 0 \qquad (4)$$

In a generalized form for simultaneous reactions the energy dissipation can be written

$$E_{diss} = \Sigma A_p^* w_p > 0 \qquad (5)$$

$$\text{where} \quad A_p^* = -\Sigma v_p^* (\mu_p + z_pF\varphi_p^*) = -\frac{\partial G^*}{\partial \xi}_{p,T} \qquad (6)$$

which is the partial derivative of the Gibbs free energy of the photogenerated species with respect to the extent ξ of the reaction.

Interesting new properties of photoelectrochemical energy conversion can only be expected far from equilibrium under conditions of autocatalysis as to be shown in the following: The excess entropy production density, the time derivative of $(1/2)(\partial^2\sigma)$ which can cause an instability of a steady state (thermodynamic branch) distant from equilibrium can be written as follows/5/

$$\frac{d}{dt}\frac{1}{2}(\partial^2\sigma) = -\Sigma_i\, \partial j_i\, \partial(\mathrm{grad}\, \frac{\mu_i}{T}) + \Sigma_\rho\, \partial w_\rho\, \partial\frac{A_\rho}{T} = \Sigma_k \partial J_k \partial X_k \quad (7)$$

(∂J_k = variation of generalized thermodynamic flux, ∂X_k = variation of generalized thermodynamic forces). The first part of this relation originates from vectorial phenomena (charge transport, chemical diffusion), the second from scalar phenomena (interfacial chemical reaction). In the linear range of irreversible thermodynamics relation (7) always has a positive value as a consequence of the second law of thermodynamics. Variations of forces are accompanied by variations of fluxes in the same direction. It is, however, easy to show that autocatalysis or, more generally, chemical reactions involving nonlinear steps have the tendency to destabilize the thermodynamic system far from equilibrium by making the excess entropy production negative: the signs of variations of forces and fluxes could then become opposite. During dissipation of photoelectrochemical energy, cooperative ordered behaviour would become possible as demonstrated for irreversible chemical reactions/5/.

This conclusion can be demonstrated in the following simple way: When comparing the following two interfacial photoelectrochemical reaction steps involving photogenerated electrons n(hv)

$$n(hv) \;+\; X \;\underset{k_2<--}{\overset{k_1-->}{\rule{2em}{0.4pt}}}\; A \quad (8)$$

$$n\,hv) \;+ A \;+\; X \;\underset{k_2<--}{\overset{k_1-->}{\rule{2em}{0.4pt}}}\; 2X \quad (9)$$

and calculating the excess entropy production term $\partial w\, \partial\frac{A^*}{T}$

we find (for T, A, k_1,k_2 , n = const and a constant photopotential φ_p^*):

$$\partial w\, \partial\frac{A^*}{T} = k_1 k_B\, n(\frac{\partial X^2}{X^o}) \;>\; 0 \quad \text{for (8) ,which will always be positive}$$

and

$$\partial w\, \partial\frac{A^*}{T} = -\frac{k_B}{X^o}(k_1 A n - 2k_2 X^o)(\partial X)^2 \quad \text{for (9) , which can not only be}$$

positive but can also become negative.

Autocatalytic, or more general, nonlinear photochemical reactions of the type (9) will therefore tend to destabilize the system , but additional nonlinear steps will be needed to produce instabilities, which will cause cooperative behaviour such as oscillations, multi steady state behaviour or dissipative structures.

Preconditions for Autocatalytic Steps and Distance from Equilibrium

How is it possible to keep an interfacial reaction sufficiently far from equilibrium to accomplish cooperative behaviour? In the simplest case it is necessary to maintain an energy flux, which interconnects an autocatalytic reaction with an energy supplying recovery process. The autocatalytic step can be of the type described by reaction (9) followed by

$$X \frac{k_3\text{-->}}{k_4\text{<--}} B \qquad (10)$$

and $$n(h\nu) + B \quad k_5\text{--->} \quad A \qquad (11)$$

In order to maintain the reaction within the region of a negative thermodynamic force - flux relation (the flux decreases with increasing force) it is also necessary that $k_1 >> k_2$ in relation (9).

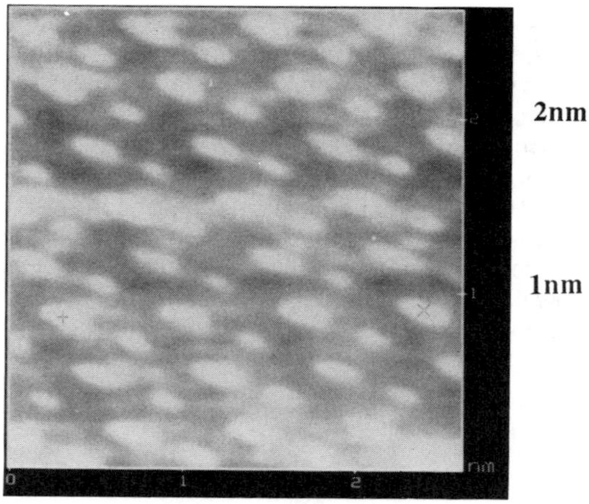

Fig.:3. Scanning Tunnelling Electron Microscope measurements taken of d-band semiconductors make the distribution of transition metals visible. Here, the Re_4 cluster chains parallel to the van der Waals surface of ReS_2 within the S-Re-S sandwich structure are shown (Photoelectrochemistry of ReS_2 described in /33/).

The question arises as to what properties a photoelectrode should have to allow autocatalysis and thus cooperative mechanisms. There is no doubt that chemical interaction and bonding must be involved for this type of mechanisms to occur. Ideal tunnelling processes, in which photoexcited electrons are exchanged with redox systems in the electrolyte under negligible bonding interaction cannot be expected to be suitable for autocatalysis considering that they are not even suitable for ordinary multielectron transfer catalysis. Strong photogenerated interfacial bonding must be involved which, for example, is possible with d-band semiconductors. They enable the transfer of photogenerated charges via d-states to redox species which can be involved in coordination type interfacial bonding/9,12/. A larger number of photoactive transition metal compounds, especially sulfides and selenides, have been identified, in

which individual, paired or clustered transition metals account for the interfacial electron exchange and chemistry (e.g. sulfides and selenides of Mo, W, Pt, Ru, Fe, Re; RuP_3, $M_x Mo_{6-x}X_8$ with M=metal,X=S,Se or $Re_6Se_8Br_2$). Although still not sufficiently explored, such systems may, in principle, be suitable for an interfacial chemistry involving feedback during interfacial electron transfer. It may, however, take a lot of effort to explore such photoinduced interfacial coordination chemistry by electrochemical and spectroscopic means.
The possible advantage of autocatalytic reaction steps in nonlinear and cooperative mechanisms for photoelectrochemical energy conversion has not yet been investigated. The entire field of irreversible thermodynamics is not yet sufficiently explored to provide a reasonable general basis for a systematic treatment of photoelectrochemical energy conversion. However, the following advantages can be demonstrated on the basis of simple arguments:

Reduction of Dissipation of Photoelectrochemical Energy in Oscillating Reactions.

The dissipation of photoelectrochemical energy, which is described by relations (5) and (6), is equivalent to the product of reaction rate w_p (the flux) and the affinity A_p^* (the force) , which involves the photopotential φ_p^* and the Gibbs free energy change - $(\partial G^*/\partial \xi)_{p,T}$ of the photoelectrochemical reaction (6). When nonlinearities are present, these two quantities can be phase shifted in such a way that the rate is highest when the affinity is smallest. As already shown for the oscillating biochemical mechanism of glycolysis/5/, dissipation can thus become less than for steady state situations without oscillations. In other words, for the same rate of consumption of photon energy the energy transduction for the generation of electricity or fuels is larger in the oscillating mode than in the non oscillating steady state mode.

The possibility of phase shifting the rate w_p (the flux) and the affinity A_p^* (the force) of photochemical reactions makes it in fact possible to control dissipation, to either increase or decrease it. A small change of affinity $(\varphi_p^*, -(\partial G^*/\partial \xi)_{p,T})$, equivalent to a small change of photon density can trigger a large change of the rate w_p of the photoelectrochemical reaction.

It is characteristic of such oscillations that they can be entrained by external periodic perturbations /5/. This means that the frequency of a photoelectrochemical oscillation would in the vicinity of the systems natural frequency adjust itself to the external perturbation. Entrainment bands can in fact be found at subharmonics of the oscillator, and the higher the amplitude, the broader the bands of perturbation. Within these entrainment bands thermodynamic force and flux of the oscillating system can optimally counteract to form a dissipation minimum.

In order to study cooperative mechanisms and demonstrate properties such as frequency entrainment for a photoelectrochemical process, a photoelectrochemical oscillation has been identified in the following way: It was known that copper containing metal sulfides (e.g. CuS,Cu_2S,$CuFeS_2$,Cu_4PS_5) show electrochemical oscillations during potentiostatic or galvanostatic reduction of H_2O_2 to water/14/. Only copper containing materials showed these types of oscillations. This indicates that peculiar properties of the oxidized copper species, especially the instability of the oxidation state Cu^{3+}, leading to the equilibrium

$$(Cu^{3+})(O^{2-}) \longleftrightarrow (Cu^{2+})(O^{-}) \qquad (12)$$

is involved. The O^- state (hole on oxygen p-state) is instable and triggers a reaction sequence which involves CuOH, CuO and Cu surface sites and intermediates and radicals of the H_2O_2 decomposition which are linked by autocatalytic reaction steps/15-17/.

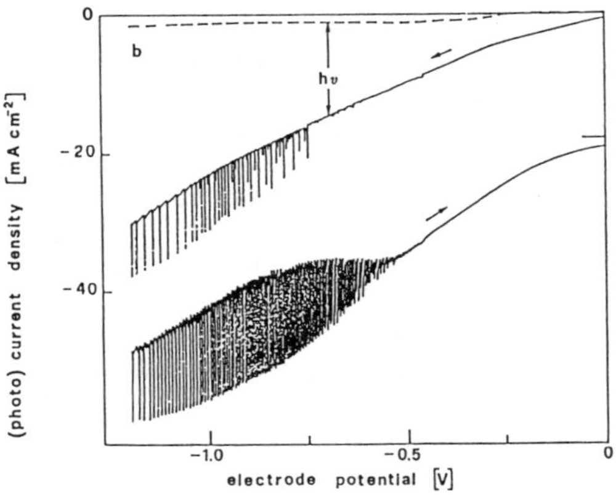

Fig,: 4 Dynamic photocurrent-voltage characteristics of p-$CuInS_2$ in aqueous 0.1 M KCl solution containing 0.4 M H_2O_2 (borax buffer, pf 9); 10mV/s, in dark and during illumination. Photoinduced oscillations are observed

In order to control such an oscillating mechanism with light energy, it appeared to be sufficient to identify copper containing p- type semiconductor electrodes which require photoexitation of electrons for the reduction of H_2O_2. It could indeed be demonstrated that illumination of p-$CuInSe_2$ and p-$CuInS_2$ produced the desired oscillating photoelectrochemical reduction /15/. Fig.: 4 shows a photocurrent voltage diagram of $CuInSe_2$ in contact with a H_2O_2 containing aqueous electrolyte. Photoelectrochemical reduction of H_2O_2 is only observed during illumination and oscillations start beyond a critical onset potential which contributes to the affinity of the reaction. The presence of oscillations depends on several parameters such as the composition of the electrolyte, the mass transport conditions near the interface and the pretreatment of the electrode. It was possible to determine experimentally that it is the quasi-Fermi potential for electrons, that is the electrochemical potential of excited electrons, which controls the affinity A_p^* and thus the onset of cooperative oscillating behaviour. It could be shown that individual oscillations involve the entire electrode surface, which means that the excitation, during which electrons are transferred to the electron acceptor, is affecting the entire semiconductor/electrolyte interface. This explains the magnitude of photoinduced oscillations which are of the order of 10 $mAcm^{-2}$.

The fact that light can be used to produce oscillations or to suppress them, to trigger them individually or to modulate their frequency or amplitude is remarkable since such non linear photoinduced responses could be developed for use in new types of photosensors. In principle they would work lige the eye with a photoreceptor and an excitation emitting nerve in series. Such systems could have remarkable advantages with respect to signal processing. Of cours, it would be necessary to develope stable non linear photosensitive interfaces, which most probably will be of a solid state type.

As discussed above and according to relation (5), energy dissipation E_{diss} will be reduced for the same turnover of reacting species, as compared to the non oscillating mode, due to the phase shift between thermodynamic flux and force. This peculiar property of energy economy of oscillating energy conversion mechanisms is also evident in the above mentioned frequency entrainment. As seen in Fig.: 5, application of an external perturbation frequency (light or potential modulation) causes frequency entrainment of photoelectrochemical oscillations, which in certain frequency intervalls is equivalent to a reduction of energy dissipation.

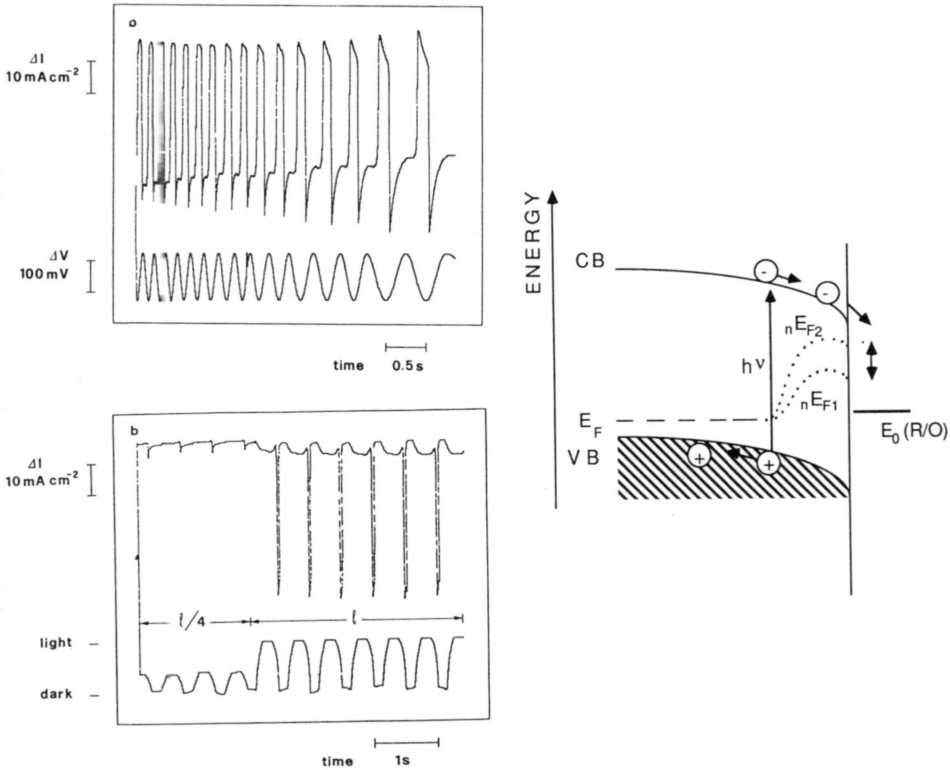

Fig.: 5. Frequency entrainment of oscillations observed with illuminated p-type (above) and intrinsic (below) $CuInSe_2$ when perturbing the oscillating system by a potential and light modulation respectively/. The energy scheme on the right side visualizes the critical role of the quasi-Fermi level.

Efficient Build-Up and Propagation of Photoelectrochemical Energy along a Reaction Chain

Another example demonstrating the relevance of irreversible thermodynamics and photoautocatalysis for solar energy conversion concerns photoelectrochemical energy losses and kinetical obstacles occuring during multistep reactions. Such reactions have typically to be handled during solar generation of fuels (e.g. reduction of CO_2 , or N_2 , photoelectrolysis of water). Typically for a multi-electron transfer the thermodynamic energy (per electron) needed for the reaction decreases with increasing number of electrons transferred in the process. At the same time, however, the kinetic requirements become increasingly demanding. For a linear photoelectrochemical reaction sequence with N steps as shown in Fig.: 6 , for which, for simplicity, we assume identical reaction constants k and a constant supply of electronic charge carriers n, the final product Z is produced at a rate /5/

$$\frac{dZ}{dt} = \frac{kn(A+NZ)}{(N+1)} \qquad (13)$$

In the presence of a large affinity (A>>Z) this means that the rate of synthesis of the product Z decreases with increasing length N of the reaction sequence. An increase of N from 0 to 1 reduces the rate to 1/7 th. Since the concentrations gradually decrease from A to Z in the form $A>X_1>X_2>.....X_N > Z$, there is a parallel degradation of electrochemical potentials: $\mu_A>\mu_{X1}>\mu_{X2} >...>\mu_Z$ (Fig.: 6). A poor functional performance of the reaction mechanisms is the consequence.

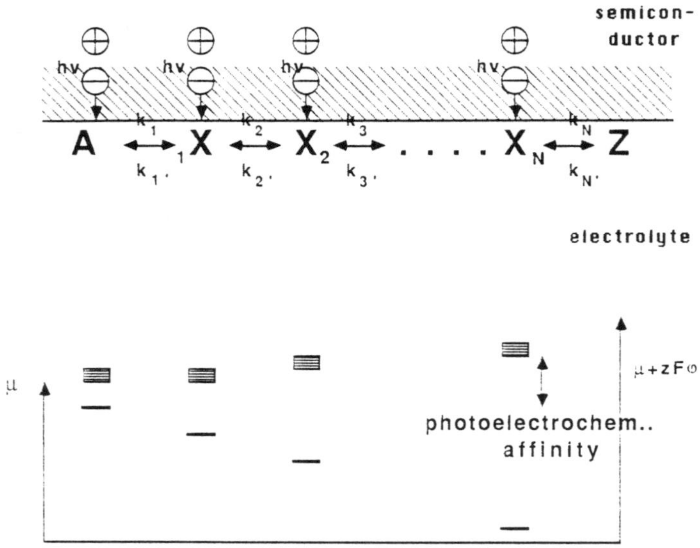

Fig.: 6 Schematic drawing describing a photoelectrochemical reaction sequence (multi-electron transfer reaction) and the poor propagation of the photoelectrochemical potential.

Let us now assume a multistep photoelectrochemical mechanism during which a catalyst C is produced having an influence on the rate w_p of individual steps:

$$w_p \text{ ---> } (1 + aC)w_p \qquad (14)$$

This determines the extent of catalytic activity. If the kinetics of the photoelectrochemical reaction chain is calculated now in analogy to calculations performed for dark biochemical reactions /5/ it can be found that for a --> ∞ the thermodynamic condition, that is, the concentration and the corresponding chemical potential is not any more degradating gradually along the chain but occurs abruptly. This means that the catalyst action is to transfer the photoelectrochemical potential without degradation towards the end product (Fig.: 7). We find the situation: $\mu_A \sim \mu_{X1} \sim \mu_{X2} \sim ...\mu_{XN} > \mu_Z$ and for the propagation of the affinity $A_p^* = -\Sigma \nu_F^* (\mu_p + z_p F \phi_p^*)$, since every reaction step is receiving photoelectrochemical energy by reacting with a photoexcited charge carrier n, a gradual increase towards the final intermediate N occurs. Since its affinity $A_p^* = -\partial G^* / \partial \xi)_{p,T}$ is the partial derivative of the Gibbs free energy of the photogenerated species with respect to the extent ξ of the reaction, more photoelectrochemical energy is available for the synthesis of the final energy storing product.

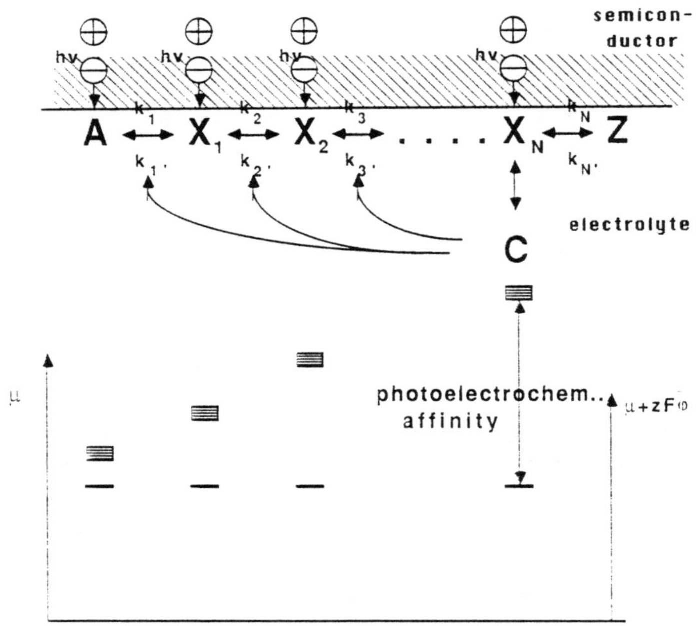

Fig.: 7 Schematic drawing describing a photoelectrochemical reaction sequence (multi-electron transfer reaction) involving autocatalysis. The kinetics and the propagation of the photoelectrochemical potential is much more efficient.

In principle, the autocatalytic transduction of electrochemical potential explained in Fig.:7 can also be applied to the transfer of an individual electron along a suitable molecular pathway. The electron density increases towards the final acceptor thus inducing a feedback (via electronic interaction) which would significantly improve electron transfer. This consideration supports our efforts to develope photoelectrodes which facilitate photoinduced interfacial electron transfer vial strong chemical interaction (coordination chemistry) which may allow autocatalysis/9,12/. It also may help understanding why, in natural electron transfer chains, electron transfer is so remarkably efficient (Fig.: 2)

Improvement of Reaction Efficiency

In biological systems, complicated molecules are found to be present in steady state concentrations, which are by orders of magnitude larger than the expected equilibrium concentration. This is possible because far from equilibrium kinetic factors like catalytic activity may outplay and compensate thermodynamic probability. While near equilibrium catalytic action would not be able to shift appreciably the steady state concentration of a product, a significant increase would be possible far from equilibrium. It is important to remember that distance from equilibrium can be designed by arranging suitable molecular pathways and autocatalytic steps as shown in Fig.: 7 or formulated via the chemical steps (9),(10) and (11).
We have seen that catalytic activity (14) sustains, between the last intermediate product N and the final product Z, a large affinity of reaction A_p^* and according to relation (1) a high rate of entropy production. If entropy production is calculated for the presence of catalytic activity (relation (14)), a minimum of entropy production, which is approached by the system in the steady state, is found at higher values of total entropy production. There is consequently a higher turnover of species and energy during the reaction. In other words: photoelectrochemical energy conversion becomes more efficient by keeping, through autocatalytic steps, one or more of the intermediate reactions involved far from equilibrium.

Why is such an increased reaction efficiency needed in photoelectrocatalysis? Practically all materials which have proven to be useful for photoelectrocatalysis up to now have energy gaps which are too large to permit reasonably efficient energy conversion according to thermodynamics (e.g. TiO_2, $SrTiO_3$). It can be shown that this is due to the fact that the photogenerated overvoltage η_H increases linearly with the energy gap.

For an n-type photoelectrode material in an electrochemical solar cell, we can write /13/ (Δp^* = concentration of minority carriers (holes), E_G = energy gap, ΔG_{ph}= free energy needed for fuel generation (e.g. photoelectrolysis), η_{cat}= overpotential at the cathode, $i_{ph}R$ = potential drop in external circuit, $\phi_H{}^D$ = potential loss in double layer, N_cN_v = product of electronic state density in conduction band and valence band respectively, n_0 = electron concentration in conduction band)

$$\eta_H = \frac{kT}{e}\ln\Delta p^* + \frac{E_G}{e} + \frac{kT}{e}\ln\frac{n_0}{N_cN_v} - i_{ph}R - \frac{\Delta G_{ph}}{e} - \eta_{cat} - \phi_H{}^D \qquad (15)$$

This relation is a consequence of properties of the quasi-Fermi level. When inserting numerical values into relation (15) assuming $\Delta p^* = 10^y$, N_c , $N_v = 10^{19}\ cm^{-3}$, and $n_0 = 10^{18} cm^{-3}$, setting internal resistance losses and the overpotential loss at the counter electrode both to 0.1 V we arrive at the following quantitative relation (in Volt):

$$\eta_H = y \times 0.059 + \frac{E_G}{e} - 1.18 - 0.1 - 0.1 - \phi_H^D \qquad (16)$$

The overvoltage dependent activation energy for electron or hole transfer reactions can thus be reduced to much smaller values for large gap than for small gap semiconductors since E_G gives a very high contribution (1-3.5 eV)to the overpotential. In order to provide efficient photoelectrocatalysis with semiconductors of lower energy gap, they have to provide electron transfer pathways across the electrode/electrolyte interface which are characterized by a low activation energy and a high interfacial minority carrier concentration Δp^*. These are exactly two necessary preconditions for driving an interfacial reaction far from equilibrium. A high concentration of photoinduced minority carriers in the electrode surface (a large $\ln\Delta p^*$) would provide a high photoelectrochemical affinity, while a low activation energy would guarantee a high reaction rate. Combined with an autocatalytic reaction step (as in equation (9)), this would be sufficient to guarantee a large distance from equilibrium and thus a high reaction rate of the overall reaction due to autocatalysis.

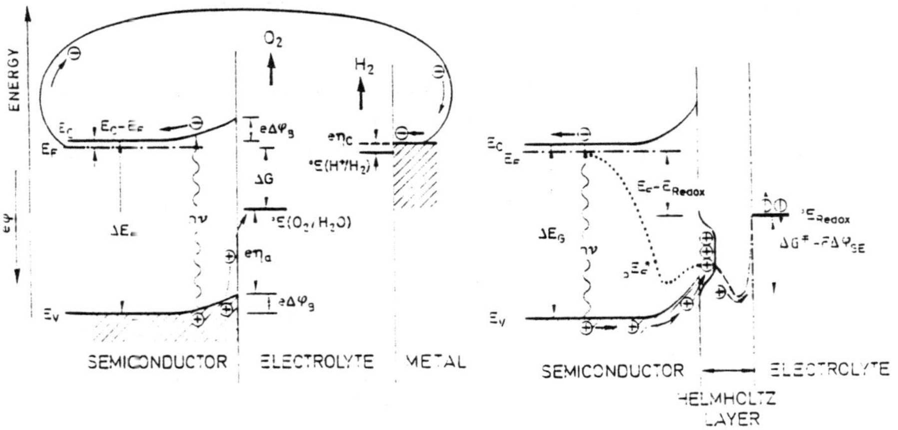

Fig.:8 Energy scheme of photoelectrochemical cell used to derive factors determining the size of the overpotential (relation 15). On the right side, the surface concentration of minority carriers (quasi-Fermi level) as well as the activation energy for hole transfer are identified as the most crucial factors determining the rate of hole transfer. But autocatalysis would be needed for efficient multi-electron transfer.

Can the Stability of Electrode-Electrolyte Interfaces be Controlled by Autocatalysis?

In the linear range of irreversible thermodynamics, entropy production σ can be decomposed into two uncoupled parts:

$$\sigma = \sigma_{ch} + \sigma_d = -\Sigma_i \, j_i \, \mathrm{grad}\frac{\mu_i}{T} + \Sigma_\rho \, w_\rho \frac{A_\rho}{T} = \frac{\delta S}{\delta t} > 0 \qquad (17)$$

The scalar quantity σ_{ch} , defined as

$$\sigma_{ch} = \Sigma_\rho w_\rho \frac{A_\rho}{T} \geq 0 \qquad (18)$$

where w_ρ is the rate of the ρ-th chemical reaction and A_ρ the relevant affinity, describes the contribution of the chemical reaction to the entropy production. The vectorial quantity σ_d, defined as

$$\sigma_d = -\Sigma_i j_i \, \mathrm{grad} \frac{\mu_i}{T} \qquad (18)$$

where j_i is the flux of the i-th component with chemical potential μ_i, considers entropy production due to vectorial processes. They include transport processes and corrosion. The independence of σ_{ch} and σ_d in the linear range of irreversible processes simply expresses the experimental knowledge that chemical reactions in a homogeneous medium do not induce vectorial processes such as an oriented diffusion flux. However, in the non linear range of irreversible thermodynamics, where cooperative phenomena are possible, σ_{ch} and σ_d can be coupled since the property of isotropy may be changed independent of the structure of the medium at equilibrium. This is the reason why in a homogeneous medium, spacial structures can develope (e.g. in the Belousov-Zhabotinski reaction). This means that when the chemical contribution σ_{ch} to the entropy production is reduced due to the cooperative behaviour of the interfacial reaction (oscillation), the vectorial contribution to the entropy production σ_d could increase. When a symmetry-breaking instability is involved, the degree of inhomogenicity of a system far from equilibrium may therefore spontaneously change. Such an instability may be induced by the spontaneous rise of vectorial mass transport (diffusion).

The initiation of a cooperative interfacial process (e.g. an electrochemical oscillation) may therefore be coupled with a change of interfacial corrosion which has a vectorial component. It is imaginable that with appropriate feedback mechanisms either an enhancement or a reduction of corrosion may be possible.

In order to explore the possibility of controlling and modifying electrode corrosion by autocatalytic pathways, we have investigated the semiconducting material $CuFeS_2$ ($\Delta E_F = 0.54$ eV) in contact with an aqueous electrolyte during H_2O_2 reduction. /16,17/. By slightly changing the external conditions this reaction can either proceed in a stationary non-oscillatory or in a purely oscillatory way. Since the initial reactants and the final products are basically identical, simple reversible thermodynamic arguments /1/ would not suggest a difference in electrode stability. However, when a total, potentiostatically passed charge of 10 Ccm^{-2} is transferred through the electrode/electrolyte interface during sustained current oscillations, a significantly higher visible electrode corrosion is observed. The electrode surface is morphologically much more disrupted. EDAX measurements of the element composition of the electrode surface demonstrate a much higher oxygen content and a lower sulfur content of the interface. In the potential regions immediately adjacent to the oscillatory region, both visible corrosion and oxide formation is significantly reduced. These results may support the conclusion that cooperative interfacial processes involving autocatalysis can indeed influence the corrosion mechanism. It may be speculated that other cooperative mechanisms (chaos ?) may reduce interfacial corrosion. However, this matter will require further investigation.

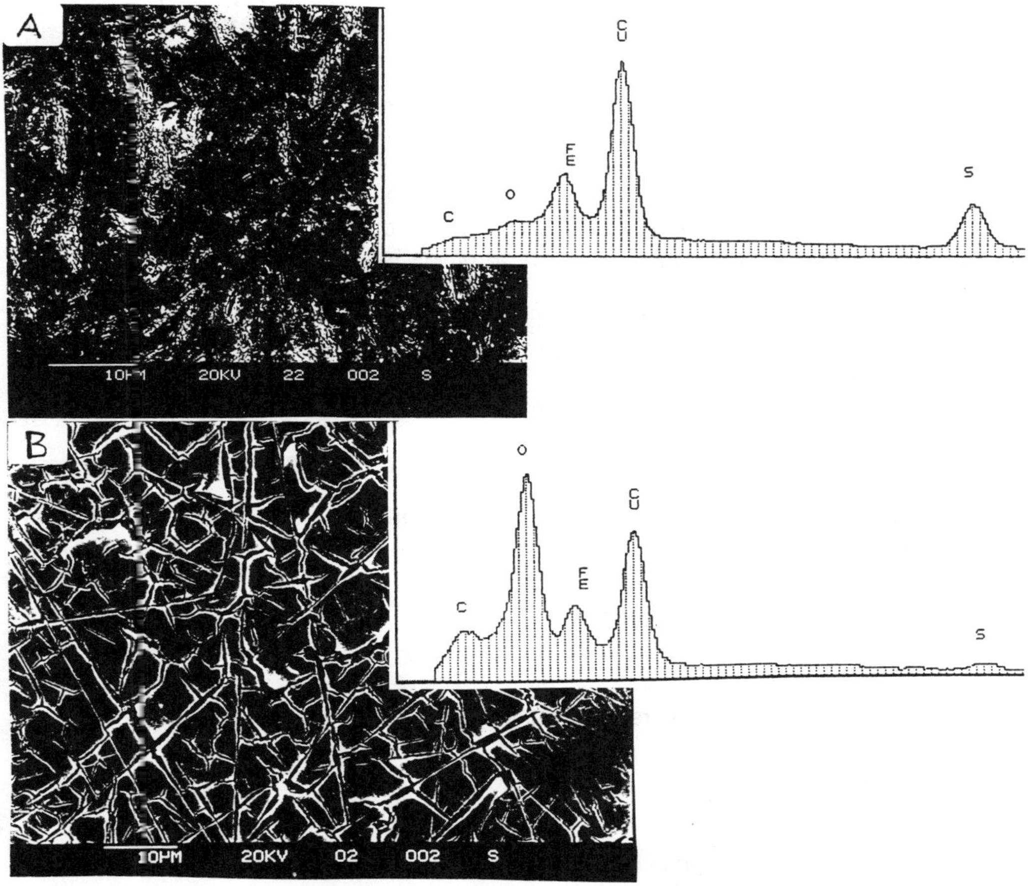

Fig.:9. SEM pictures and EDAX (energy dispersed analysis of X-rays) measurements demonstrate that for the same electrical charge passed through a $CuFeS_2$ / electrolyte interface (10 C cm^{-2}) oscillatory reduction of H_2O_2 on $CuFeS_2$ (ΔEG=0.53eV) is more corrosive (B) than non-oscillatory reduction (A). EDAX measurements taken after polarization at -0.9 V (B)and -1.0 V (A) respectively (vs SCE).

Photochemical Energy Conversion in Oscillating Homogeneous Solutions

Reversible thermodynamics does not forbid conversion of light into chemical energy with reasonably high efficiency, but practical experience shows that reverse reactions are leading to a dissipation of most of this energy. For this reason only neglibible energy conversion efficiencies are observed even with selected systems (e.g. light induced cis-trans isomerisation). Inhomogeneities and oriented transport of charge carriers and reaction products respecively are needed to obtain reasonably high energy conversion efficiencies.

Chemical solutions in which autocatalytic steps are pushing the system sufficiently far away from equilibrium to produce self organisation offer new opportunities: In the Belousov-Zhabotinskii reaction/2,4,18,19/, for example, oscillations, bistability or deterministic chaos can be observed. Chemical waves are exhibiting sharp profiles which have also been confirmed with computer models (Fig.:: 10a). Typically, the reaction proceeds between bromate ions and an organic substrate such as malonic acid. The added metal ion catalyst can be selected to be photoactive /20/, such as the ruthenium bipyridyl complex $Ru(bipy)_3^{2+}$. Crucial in the complicated reaction scheme is the autocatalytic oxidation of the $Ru(bipy)_3^{2+}$ and its reduction by brominated organic substrate. Bromide is the most important intermediate since it acts as an inhibitor to autocatalytic oxidation. By excitation of $Ru(bipy)_3^{2+}$ the rate of bromide production can be controlled photochemically. In this way, the reaction regime can be selected and the formation and behaviour of chemical waves controlled/20/.

The overall reaction can be approximated by :

$$2BrO_3^- + 3CH_2(COOH)_2 + 2H^+ \dashrightarrow 2BrCH(COOH)_2 + 3CO_2 + 4H_2O \qquad (19)$$

When decribing the intermediate $HBrO_2 = X$ and Ru(III) = Z, their time dependent formation can be approximately written as

$$\frac{dX}{dt} = X(1-X) - (\beta(h\nu) + \gamma \frac{Z}{(Z_c - Z)}(\frac{X-\mu}{x+\mu})) \qquad (20)$$

where the photon energy enters via the the function $\beta(h\nu)$

and $$\frac{dZ}{dt} = X - \alpha \frac{Z}{Z_c - Z} \qquad (21)$$

The following differential equation for $X = HBrO_2$ with periodical solutions results:

$$\frac{d^2X}{dt^2} + b(h\nu)\frac{dX}{dt} + c(h\nu)X = 0 \qquad (22)$$

Here $b(h\nu)$ and $c(h\nu)$ are complicated functions depending on photon energy. An interesting situation may arise when the oscillating Belousov-Zhabotinskii reaction is periodically illuminated. In this case relation (22) is a differential equation with periodically variable coefficients. Such equations describe parametric energy conversion (e.g. electrical circuit with periodically modulated capacity, stretched electrical wire which is periodically heated with an alternating current and thus excited to mechanical oscillations, postulated biological mechanisms/30/) which describes gradually increasing amplitudes of X. It can therefore be concluded that under appropriate conditions periodical illumination may lead to an periodical build-up of $HBrO_2$ (which could be extracted from the system). In other words: chemical energy can be formed at the expence of light energy/31/. We are dealing with a new kind of photosynthesis aided by cooperative mechanisms.

It cannot be expected that the postulated mechanism is easy to be demonstrated experimentally in a straightforward way without detailled analysis of the complicated photosensitive Belousov-Zhabotinskii reaction. We also cannot say whether a mechanism of this type will ever be of some practical value. However, such a mathematically demonstrated possibility of a photosynthesis, assisted by cooperation emphasizes the opportunities involved in the study and developement of autocatalytic mechanisms for solar energy conversion.

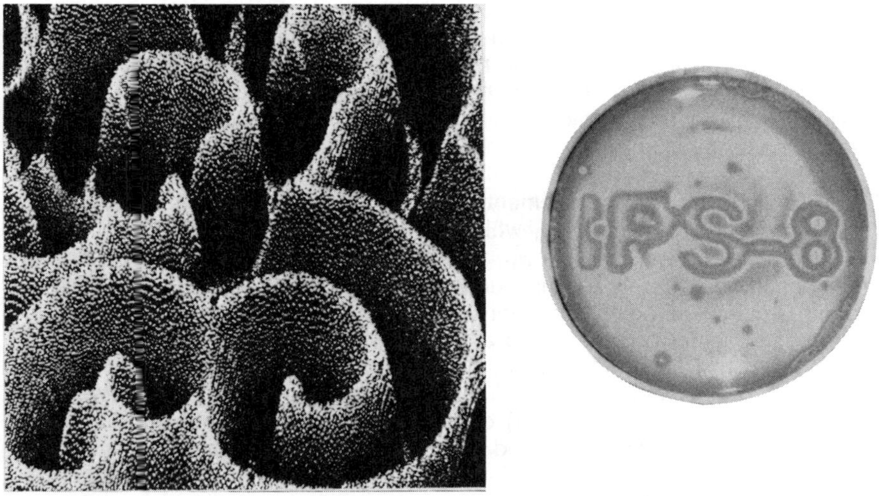

Fig.: 10 Structure of propagating chemical waves in the Belousov-Zhabotinski reaction from simulations showing anisotropy (a)/29/ and (b) projection of a photo-image (IPS-8) to demonstrate how the dissipative structures can be controlled by light.

In the discussed oscillating mechanism excitation of $Ru(bipy)_3^{2+}$ which leads to the formation of bromide, occurs within a pattern of sharply structured propagating waves. Photon energy conversion thus occurs under asymmetric, vectorial conditions generated by autocatalytic processes.

In biological structures, synthesis of complicated molecules, which involves a considerable reduction of entropy, typically occurs via autocatalytic pathways far from equilibrium/5/. There is no reason why photosynthetic reactions at photoactive electrodes should not take advantage of such powerful mechanisms

Discussion

Most photoelectrocatalytic research up to now has been aimed at photoelectrocatalysis using large gap semiconductors since they proved to be experimentally more successful. This advantage can easily be explained since the photoinduced overpotential generated at such electrodes is much larger and the rate of photoelectrochemical mechanisms correspondingly higher(compare equations (15 and(16)). In the case of TiO_2 and $SrTiO_3$, photoelectrochemically generated radicals open a wide spectrum of chemical reaction possibilities. However, the photoelectrocatalytical mechanisms developed do not support a main thrust of solar energy utilization. For such applications, photoelectrocatalysts with lower energy gap (1-2 eV) would have to be used for the purpose of achieving higher energy conversion efficiencies. In order to approach this goal, it will be necessary to significantly improve catalytic efficiency of photoelectrochemical mechanisms. The most straightforward way to accomplish this would be by increasing both the rate constant (decreasing activation energy) and the photochemical affinity (optimizing the position of the quasi Fermi-level by maximizing the concentration of interfacial minority carriers Δp^*). A chemical precondition for accomplishing this goal would be the developement of photoelectrodes which are able to involve an appropriate interfacial photoelectrochemistry permiting chemical bonding for the formation of suitable intermediates. d-band semiconductors with mononuclear and polynuclear transition metal centers, which have been developed in our laboratory (e.g./9/13/6/), have shown that complicated multi-electron transfer processes (e.g. photoevolution of oxygen from water, reduction of oxygen to water in acid medium) can favourably be catalyzed with such interfacial coordination-chemical mechanisms.

Theoretically, a further improvement could be obtained by making the interfacial photoelectrochemical reaction, which already proceeds distant from equilibrium, cooperative. As we have shown, this could be accomplished by introducing one or more autocatalytic steps. As a consequence photoelectrochemical energy would efficiently be propagated and complicated multistep electron transfer mechanisms of relevance for energy transfer could proceed at reasonably high rates.

It has, however, to be emphasized that the high thermodynamic energy conversion efficiency calculated for equilibrium conditions (e.g. 27 % for for water photoelectrolysis via a tandem system/21/) cannot be reached far from equilibrium. Some loss in thermodynamic efficiency would however be acceptable provided a reasonable reaction rate could be achieved for the photogeneration of chemical fuel using visible solar energy._Since multi-electron transfer tends to become less energy and rate efficient with an increasing number of electron transfer steps, the introduction of autocatalytic mechanisms appears to be necessary. During the electron transfer reaction the catalytic center should change its structure so as to to modify electron transfer rates. Such a feedback would lead to a more efficient collective eletron exchange with the reacting substance.

In order to take advantage of the described possibilities offered by irreversible thermodynamics and autocatalysis a broad basis of fundamental knowledge will have to be elaborated. Photoinduced oscillations have occasionally been observed with semiconductor electrodes (Table 1). Passivation phenomena leading to a negative differential resistance in combination with a concentration polarization have been invoked to explain the phenomena (compare /28/ for a general theory for electrochemical oscillations). Many different mechanisms for the generation of oscillations are imaginable. Most of them may not be interesting for energy conversion. It will be necessary to search for autocatalytic mechanisms supporting photoinduced electron transfer.

At the moment, we have little experience with the developement of photoelectrodes which provide reaction pathways far from equilibrium and which are suitable for autocatalysis. It is, however, as already mentioned, difficult to immagine such mechanisms without interfacial mechanisms involving chemical bonding interaction between redox species and the photoelectrode.

Table 1.: Oscillation phenomena observed with illuminated semiconductor electrodes:

Material	Electrolyte	Potential	Observations	Ref.
n-GaAs	0.01M $Na_2B_4O_7$ + 0.02M NaOH + 1 M KCl (pH=10)	-0.375VvsSCE	sus. oscill.	/23/
n-CdTe	1 M CsOH 1 M. Cs_2S	+0,446V_{vsPt}	sus. or unst. oscill.	/24/
n-CdS	0.3 M p-mercap-to benzoic acid/ (pH=14)	0.V_{vsSCE}	damped oscill.	/25/
n-Si	0.1 M NH_4F at pH 4	3.5-4.5V_{vsSCE}	damped oscill.	/26/
n-Bi_2S_3	1 M Na_2S	galvano-static	sus. oscill.	/27/

The following strategies have been shown to be useful for the identification of catalytically reactive semiconductor electrode interfaces:

The search for photoactive materials with adequate electronic structure. Photoactive materials have to be chosen in such a way as to initiate an appropriate type of interfacial reactions. Semiconductors with a high contribution of d-states to valence and conduction bands will facilitate photoinduced interfacial coordination chemical mechanisms.

The development of interfacial coordination photoelectrochemistry. The interfacial complexes formed may serve to efficiently transfer electrons via bridges and inner sphere electron transfer mechanisms. Molecular bridges may intentionally be built to channel photoinduced charge carriers efficiently through the semiconducter/electrolyte interface to avoid recombination and corrosion processes.

Interfacial reactions via transition metal clusters. Multielectron transfer mechanisms need suitable transition metal complexes with the capacity to serve as an electron reservoir and which is able to provide adjacent bonding sites. Transition metal clusters which form part of the crystal structure and which participate in electron conduction, or which may be attached to the semiconductor interface via electron transferring bridges have been shown to be promising photocatalytic centers.

Relevance for photosynthetic oxygen evolution

T. Watanabe et al./32/ have discussed a remarkable paradoxon when analyzing photosynthetic oxygen evolution as a photoelectrochemical mechanisms. They estimated an effective photocurrent density through the reaction centers of i_{ox} = 10 $mAcm^{-2}$ with practically negligible overpotential, while the best electrochemically investigated systems would require for similiar current densities overpotentials of the order of 300 -350 mV. The authors assume a step by step charging of the manganese complex prior to oxygen evolution. It is suggested here that also the catalytic manganese center for oxygen evolution from water in photosynthesis reacts as a system far from equilibrium, aided by autocatalysis, and would thus only be understandable on the basis of nonlinear irreversible thermodynamics: The capture of positive charges in the manganese center of oxygen evolution leads to a structural and electronic change of the center which may have a feedback on the rate of transfer of electrons from water into the manganese complex. According to the above discussed principles light-generated free energy may thus efficiently be transferred into oxygen evolution in a cooperative way.

We have succeeded in developing photoelectrodes (p-type $CuInS_2$ and $CuInSe_2$ during H_2O_2 reduction), which exhibit lightinduced cooperative behaviour. It could be shown that photoinduced oscillations can be controlled and synchronized either by electrode potential or light pulses. This means that the position of the quasi Fermi level, which affects the electrochemical affinity, and controls the distance from equilibrium, also controls the occurrence of oscillations which involve macroscopic cooperative electron transfer. Synchronization of oscillations with external perturbations also demonstrates that periodical photoelectrochemical processes can adjust to a minimum of energy dissipation.

The experimental results obtained with illuminated p-type $CuInS_2$ and $CuInSe_2$ during H_2O_2 reduction as a model system demonstrate that photoinduced electron transfer can be synchronized even over a macroscopic electrode. This suggests that, in presence of autocatalysis, cooperative electron transfer should also be possible on a microscopic molecular scale.

Acknowledgements

The author thanks Dr. H. Jungblut for taking the scanning tunnelling microscope picture of ReS_2 and Drs. L. Kuhnert, L. Pohlmann and J. Enderlein for a helpful discussion of the lightinduced Belousov-Zhabotinskii - Reaction. Stimulating discussions with collaborators, especially with S. Cattarin, N. Alonso-Vante and G. Smestad are gratefully acknowledged.

References

/1/ H. Gerischer in " Solar Power and Fuels", J. Bolton ed. Academic Press, New York (1977)

/2/ A.N. Zaiken and A.M. Zhabotinsky, Nature,225,235(1970

/3/ H. Haken, Synergetics, (1978) Springer, Berlin

/4/ R.J. Field and M. Burger (eds) Oscillations and Travelling Waves in Chemical Systems (Wiley, New York,1977)

/5/G. Nicolis and I. Prigogine, Selforganization in Nonequilibrium Systems (Wiley, New York 1977), chapter 15; M. Sanglier, PhD Dissertation, University of Brussels (1976); M. Sanglier and G. Nicholis (1976), Biophys. Chem. 4 (1976) 113

/6/ N. Alonso-Vante and H. Tributsch, Nature 323 (1986) 431

/7/ N. Alonso Vante, B. Schubert and H. Tributsch, J. Catalysis 112(1988)384

/8/ N. Alonso-Vante, W. Jaegermann, H. Tributsch, W. Hönle and k. Yvon, J. Am. chem. Soc. 109(1987) 3251

/9/ H. Tributsch in Photocatalysis and Environment-Trends and Applications, NATO ASI Series, Kluwer Academ, Pub. Darmstadt, Boston, N.Y. (1988)

/10/K. Niki, Kawasaki, K. Hinnen, C., Higuchi, Y. and Yasuaka, N (1986) in Frontiers in Bioorganics

/11/ K. Fan, Dissertation, Yokohama National University (1990)
/12/K. Büker, H. Collel, N. Alonso-Vante and H. Tributsch, poster presented at IPS-8 meeting, Palermo (1990)
/13/ H. Tributsch, in "Photoelectrocatalysis, N. Serpone, E. Polizetti, eds. J. Wily & Sons, Inc. N.Y. (1989)
/14/ H. Tributsch, Ber. Bunsengel. Phys. Chem. 79 (1975) 570; 79 (1975) 580
/15/ S. Cattarin and H. Tributsch, Chem. Phys. Lett. 148 (1988) 221; Ber. Bunsenges. Phys. Chem. 92 (1988) 1345
/16/ S. Cattarin and H. Tributsch, J. Electrochem. Soc. In press
/17/ S. Cattarin, S. Fiechter, C. Pettenkofer and H. Tributsch J. Electrochem. Soc. in press
/18/ P.M. Wood and J. Ross, J. Chem. Phys. 82 (4) 1985
/19/ J. Ross, S.C. Müller, C. Vidoch, Science (1988), 240, 460
/20/ L. Kühnert, K.J. Agladze and V.I. Krinsky, Nature 337, 244 (1989)
/21/ J.R. Bolton, S.J. Strickler and J.S. Connolly, Nature 316 (1985) 452
/22/ K. Yvon in Superconducativity in Ternary Compounds, Fischer, M.B. Maple, eds.Springer Verlag, Berlin, N.Y. (1982) Vol. 1
/23/R.L. Van Meirhaeghe, F. Cardon and W.P. Gomes Electrochim. Acta, 24 (1979) 1074
/24/ V. Marcu and R. Tenne, J. Chem. Phys. 92 (1988) 7089
/25/ P. Josseaux, J.C. Micheau and A. Kirsch-De Mesmaeker, Electrochim. Acta 30 (1985) 1093
/26/ H. Gerischer and M. Lübke, Ber. Bunsenges. Phys. Chem. 92 (1988) 573
/27/ L.M. Peter, J. Electroanal. Chem.98 (1979) 40
/28/ H. Degn, Trans. Faraday Soc. 64 (1968) 1348
/29/ adapted from a computergraphic obtained by S.C. Müller, T. Blesser and B. Hess
/30/ H. Tributsch J. Theor. Biol. 52 (1975)17
/31/ H. Wetzel and H. Tributsch, Proc.of the 2ndInt.Conf.on Unconventional Photoactive Solids,(1986).Plenum Publ. Corp.
/32/ T. Watanabe, M. Kobayashi and T. Sagara, Abstract, VIIIth International Congress on Photosynthesis, 1989 Stockholm; Physiolgia Plantarum 76 (3,part2) A147 and poster presented
/33/ B. Schubert and H. Tributsch, J. Appl. Electrochem. in press.

DYNAMICS AND EFFICIENCY OF PHOTOVOLTAIC PROCESSES IN SUPRAMOLECULAR STRUCTURES AND SEMICONDUCTORS

F. WILLIG
Fritz-Haber-Institut der Max-Planck-Gesellschaft,
Faradayweg 4-6,
D-1000 Berlin 33
W. Germany

ABSTRACT. Light harvesting and electron-hole pair separation was studied with picosecond time resolution in supramolecular structures and in semiconductors. Slow singlet exciton diffusion requires an absorption coefficient $>10^5$ cm^{-1} throughout the solar specturm for efficient light harvesting in the supramolecular antennae system or the assembly of several such less efficient systems in series to facilitate complete light harvesting. The at least 100 times faster minority carrier diffusion in semiconductors imposes a much weaker requirement on the absorption coefficient for complete light harvesting in semiconductors. Optimized electron-hole pair separation along a molecular chain requires a very fast initial hop (ps) and a standard free energy difference smaller than 0.25 eV. Thus the reorganisation energy must be kept small with rigid molecules incorporated into a rigid environment with not too different values for the optical and static dielectric constant. A desirable fuel producing solar energy converter could consist of an efficient photovoltaic semiconductor device protected against corrosion and coupled directly to molecular fuel producing catalyzers.

1. Introduction

In this paper we present time resolved signals of the important processes that contribute to the photovoltaic effect firstly in supramolecular structures, secondly in mixed molecular/semiconductor systems, and thirdly in simple semiconductor devices for comparison. Essential requirements for an efficient photovoltaic effect will be compared with the material properties of supramolecular structures and of semiconductors. It will be shown that these differences in material properties make the construction of an efficient photovoltaic machine more difficult in the case of a supramolecular structure than in the case of a semiconductor device. Supramolecular structures offer the possibility of extreme miniaturisation and may provide easier ways for direct coupling of fuel developing catalyzers to the photovoltaic machine.

E. Pelizzetti and M. Schiavello (eds.), Photochemical Conversion and Storage of Solar Energy, 235–249.

2. Time-resolved photovoltaic processes in supramolecular structures

2.1 LINEAR REACTION CENTER WITH 3-DIMENSIONAL ANTENNAE SYSTEM

A linear chain of molecules facilitating spatial electron-hole pair separation will be called a reaction center (RC) in accordance with the nomenclature introduced by the work on photosynthesis. The energy of the impinging light quanta has to arrive at the entrance of the linear reaction center. Efficient light harvesting can be achieved in a 3-dimensional molecular antennae system within a sphere of radius L, the exciton diffusion length, as illustrated in Fig. 1.

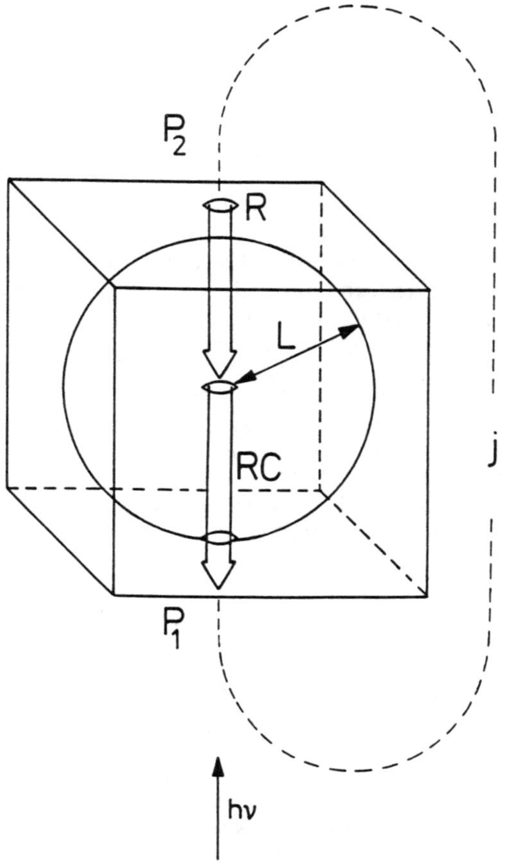

Figure 1. Scheme illustrating a 3-dimensional molecular energy transfer (antennae) system with exciton diffusion length L coupled to a linear molecular electron-hole separating chain (RC). R represents a slow charge conserving process, i.e. the formation of a second fuel product P_2 or a slow current path in the photovoltaic case (j).

The simplest design of a molecular reaction center is illustrated in Fig. 2. The forward hopping rate constant has to compete effectively with the reaction channel leading to the ground state. Typical time constants for these compet-

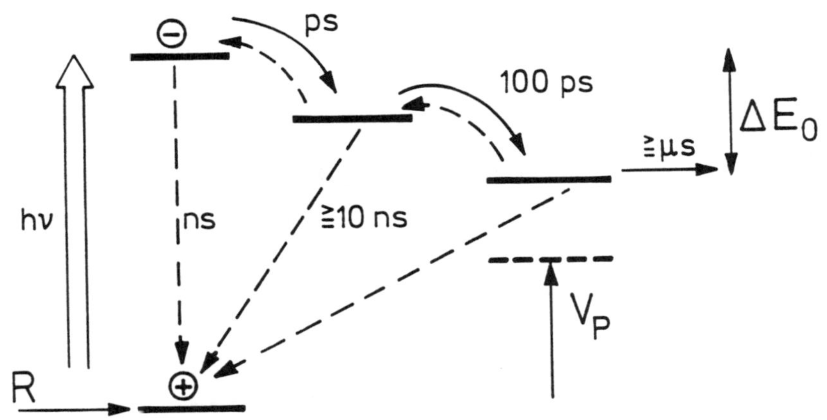

Figure 2. Scheme illustrating efficient electron-hole pair separation in a linear molecular chain. Typical time constants for downhill hopping are given that compete effectively with recombination. ΔE_o represents the minimum downhill standard free energy difference of about 0.25 eV for the optimized product of current flow and photovoltage /1/. V_p is the stationary change in free energy for a given light flux at the entrance. R stands for the charge conserving process as in Fig. 1.

ing channels are indicated in Fig. 2. For such a given set of forward and competing rate constants one can determine the minimum standard free energy difference required for the first hop and for the total standard free energy difference $\Delta E°$, i.e. for all the hops, by linking the backward rate constants (uphill) via the Boltzmann factor to the values of the forward rate constants and by optimizing for the greatest product of current flow times photovoltage /1/. The optimized standard free energy difference $\Delta E°$ is about 0.25 eV for a set of forward and competing rate constants as shown in Fig. 2. The dimensions of such an electron-hole pair separating molecular chain can be as small as 30 Å. It is not difficult to achieve electron hopping between molecular donor-acceptor pairs on the required time scale. A recent experimental example is given in Fig. 3. Fast electron transfer in the range of a few ps in a molecular donor-acceptor pair requires firstly that the standard free energy difference of the reaction is of the same order of magnitude as the reorganisation energy of the reaction to minimize slow-down due to Franck-Condon factors /2/, and secondly that the molecular bridge (cyclohexyl in Fig. 3) is sufficiently transparent, or in case of free space between donor and acceptor, that the distance is short enough to ensure an electron transfer integral of the order of $\geqq 50$ cm^{-1}. The latter is about 10 cm^{-1} in the system shown in Fig. 3 /3/. To keep the standard free energy difference close to the optimum 0.25 eV in the reaction center (Fig. 2) it is necessary to avoid a large reorganisation energy for the reaction. The latter increases with the difference between the static and optic or infrared dielectric constants /4/, and it is particularly large in a polarizable medium like water. Thus, rigid reactant molecules in a rigid (solid) environment are advantageous

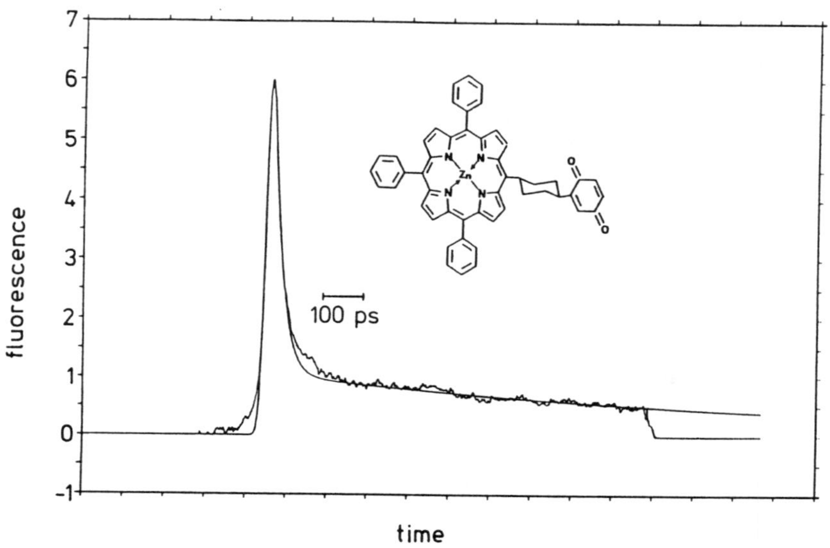

Figure 3. Fluorescence decay indicating fast electron transfer (25 ps) through a rigid cyclohexane bridge between the excited Zn-porphyrine singlet state donor and the quinone acceptor molecule. The slow fluorescence component (almost horizontal) arises from 9 percent Zn-porphyrine that is linked to hydroquinone in place of quinone. The solid curve is the calculated fit.

to achieve a fast electron transfer reaction requiring a minimum standard free energy difference of about 0.25 eV. The properties of a molecular antennae system will be discussed in the following section.

2.2 PLANAR REACTION ZONE WITH 3-DIMENSIONAL ANTENNAE SYSTEM

a) *Light harvesting.* The case where a 3-dimensional molecular antennae system delivers the excitation energy to a planar array of aliged reaction centers is depicted in Fig. 4. Light harvesting can be studied for such a system as light absorption and transfer of excitation energy to the surface of a molecular single crystal, where the excitation energy is quenched via electron transfer to acceptor molecules at the crystal surface. The corresponding exciton diffusion transients will be shown below. It is important to realize the requirements regarding the structural perfection of such a molecular antennae system. This is illustrated in Fig. 5. The dashed curves show the vibrational structures and the Stokes' shift between the absorption and fluorescence spectra (uncorrected) of an anthracene single crystal. The solid curves in Fig. 5 represent the absorption and fluorescence spectra of a multilayer system of the same anthracene chromophores assembled with the Langmuir-Blodgett technique /5/ after a suitable modification of the molecule has been carried out as shown on the left hand side as insert in Fig.5. The spectra of the multilayer system contain characteristic features betrying geometrical disorder, i.e. the presence of many structural traps. The spectral dependence of the initial fluorescence decay shown in the

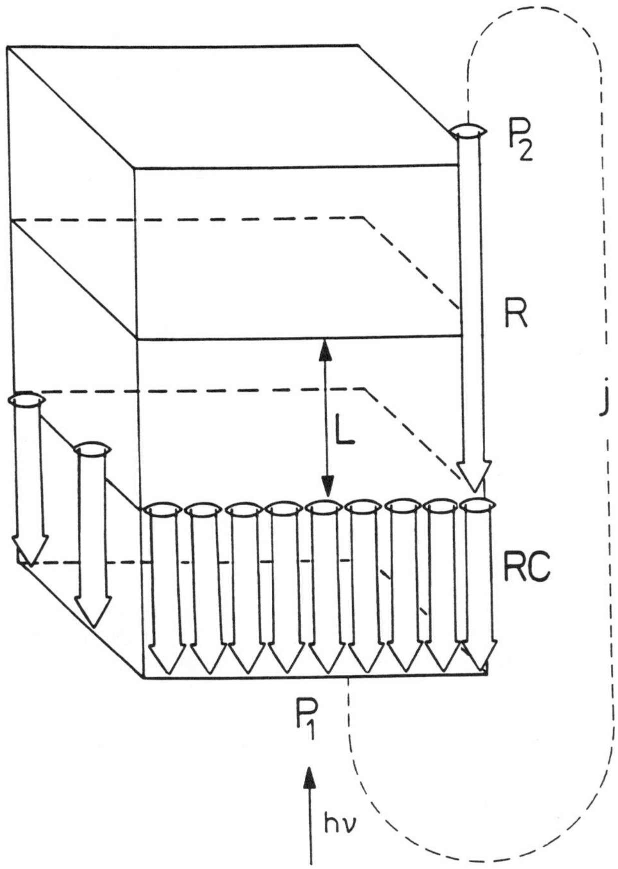

Figure 4. Scheme illustrating a 3-dimensional molecular antennae system with exciton diffusion length L coupled to a planar molecular system of aligned electron-hole pair separating chains RC. R represents the slow charge conserving process, i.e. either formation of a second fuel P_2 or a current loop in the photovoltaic case (j).

insert on the right hand side indicates that higher energy levels are depopulated during the lifetime of the excited state (exciton). Vapor grown single crystals with the same structure as the LB multilayer system /6/ exhibit similar disorder features as the LB multilayer system, though to a lesser degree /7/. Structural traps slow-down significantly the exciton transport and decrease the flux of excitons in the molecular system since the radiative and non-radiative decay channels compete with the detrapping process. The latter requires some ns at room temperature for a typical trap depth of a few tenths of an eV.

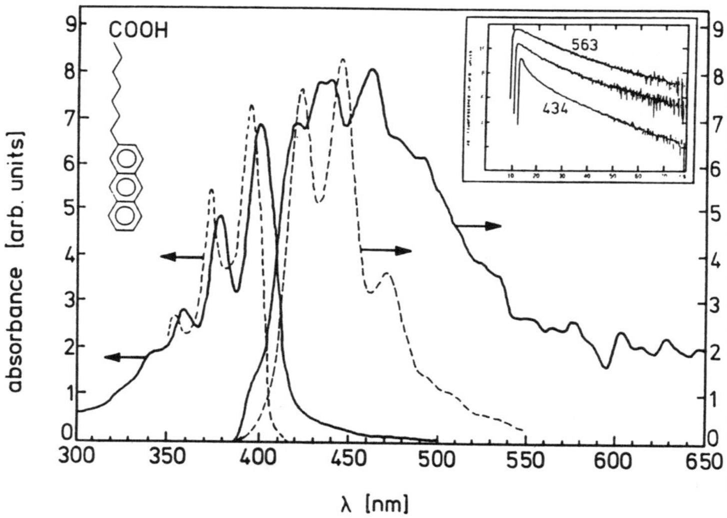

Figure 5. Comparison of the absorption and fluorescence spectra (uncorrected) of an almost perfectly ordered molecular system, i.e. a vapor grown ultrapure single crystal (dashed curves), with those of an ordered Langmuir-Blodgett multilayer system (solid curves) /7/ of the same chromophore anthracene. Three typical features indicate significant disorder in the LB multilayer system: Firstly, the long foot extending to the red of the absorption edge; secondly, the broad almost structureless fluorescence spectrum reaching further into the red; thirdly, non-exponential decay curves (insert) with faster initial decay at the short wavelength side and slower initial rise at the long wavelength side of the fluorescence spectrum.

The time resolved measurement of light harvesting is shown in Fig. 6 for a vapor grown ultrapure anthracene crystal illuminated through the front surface, where fast electron transfer quenching of the singlet excitons occurred at acceptor molecules /8/. The singlet exciton transients were measured as picosecond time resolved injection photocurrent transients. They showed the expected delayed arrival for increasing penetration depths of the incident light. The insert in Fig. 6 shows that the penetration depth of the incident light was varied from 0.06 µm at $\lambda = 396$ nm to 6 µm at 410 nm. The calculated transients (not shown) agreed with a singlet exciton lifetime of 7 ns and light harvesting over a diffusion length $L = \sqrt{D\tau} \approx 0.1$ µm /8/. This is a typical case since the exciton diffusion coefficient is $\leqq 10^{-1}$ cm^{2}s^{-1} in a molecular crystal at room temperature. Thus, only very strongly absorbed light ($\alpha L \geqq 1$) can be harvested efficiently in well ordered molecular system. It appears very difficult to realize a well ordered molecular system with an absorption coefficient $\geqq 10^5$ cm^{-1} throughout the solar spectrum. Moreover, the system must be extremely well ordered and free of impurities to obtain a diffusion coefficient $\geqq 10^{-2}$ cm^2 s^{-1}. Many structural traps have been detected in the multilayer system assembled with the Langmuir-Blodgett technique for the compound shown in the insert of Fig. 5, even though the anthracene chromophore has almost the same space

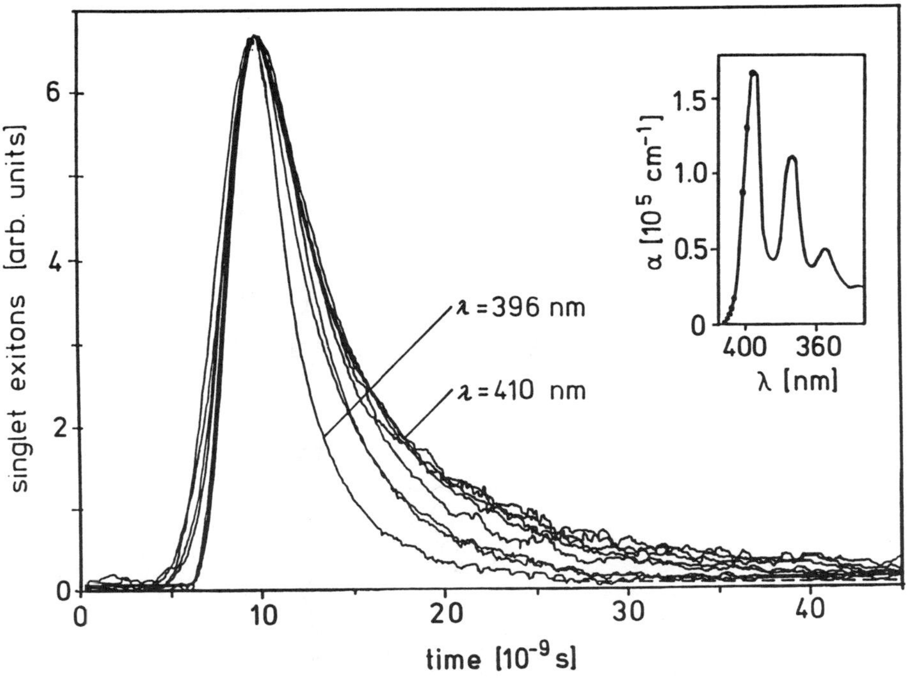

Figure 6. Transients of singlet excitons (normalized to the same maximum height) arriving at the illuminated surface of an anthracene single crystal where they were quenched via electron transfer. The wavelength of the incident light was varied along the absorption edge (insert) to realize different penetration depths. Light harvesting in this 3-dimensional molecular antennae system is limited by the diffusion length L of the excitons as illustrated in Fig. 3.

filling as the alkane chain /6/. For most chromophores there will be a larger difference in space filling and the multilayer will be even less ordered.

b) Experimental example. Fig. 7 shows a recent attempt to assemble an experimental system with some of the properties illustrated in Fig. 4. Light is absorbed by the aligned thiapyrillium molecules in the Langmuir-Blodgett film, by the perylene moeities in the LB film, and in the perylene single crystal. The excitation spectrum of the photocurrent (top) shows that energy transfer occurred from the perylene moeities to the thiapyrillium moeties. It is not clear whether this involved exciton diffusion or direct energy transfer. The behavior of the antennae system is complicated in this case by the very efficient formation of perylene excimers both in the single crystal and in the LB film /7/. The perylene single crystal serves here also as the conducting element R in Fig. 4. The holes hop through the crystal to the rear end that is connected to the solu-

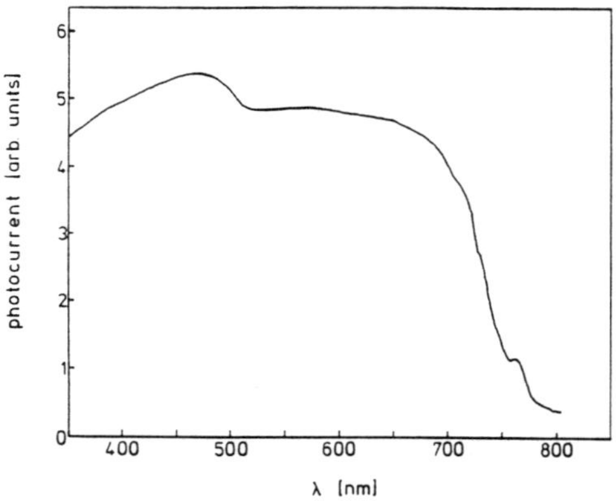

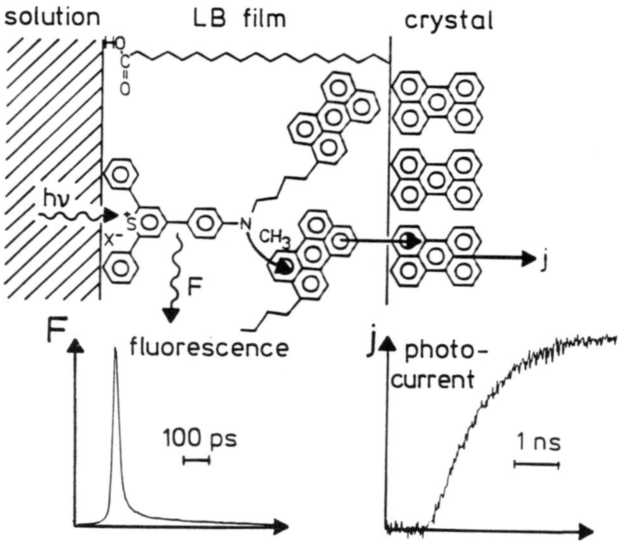

Figure 7. Electron-hole pair separating molecular chains aligned in a Langmuir-Blodgett film /10/ and coupled to a perylene single crystal that serves as antennae system and hopping path for the current loop of the holes. The rear of the crystal is connected to the solution in front of the LB film. The excitation spectrum of the photocurrent is shown on top (uncorrected). The time resolved signals at the bottom show fast fluorescence quenching due to electron-hole pair separation (ps) and the slower rise (ns) of the photocurrent that is controlled by hole traps.

tion in front of the LB film, thereby providing for a closed current loop. In the absence of an applied electric field hole transport through the α-perylene insulator crytsal occurs via diffusion and is opposed by a corresponding Coulombic barrier, set-up by the counter and image charges at the contacts. For a molecular insulator crystal of finite thickness, e.g. $\geqq 1\ \mu m$, hole transport by diffusion against a barrier is very slow and inefficient. Therefore, the system shown in

Fig. 7 cannot be utilized for a photovoltaic application. However, when an external electric field $\geqq 10^5$ Vcm^{-1} is applied one can collect the injected holes and measure the time dependence of hole transport from the thiapyrillium dye to the surface layer of the perylene crystal /9/. The corresponding rise time of the photocurrent was found in the ns range (Fig. 7, bottom). Fluorescence quenching of the excited singlet state of the thiapyrillium donor molecule of the hole was much faster and occurred with a time constant $\leqq$ 1 ps /10/ (Fig. 7 bottom). The slower rise time of the photocurrent was attributed to hole traps formed by the perylene moeities in the LB film. The supramolecular structure shown in Fig. 7 is certainly very inefficient as solar energy converter but it permits picosecond time resolved measurements of necessary reaction steps that have to be realized in a supramolecular photovoltaic device.

3. Time resolved photovoltaic processes in mixed molecular/semiconductor structures

Injection of electrons from the excited singlet state of adsorbed dye molecules into the conduction band of a semiconductor (Fig. 8) has been studied since

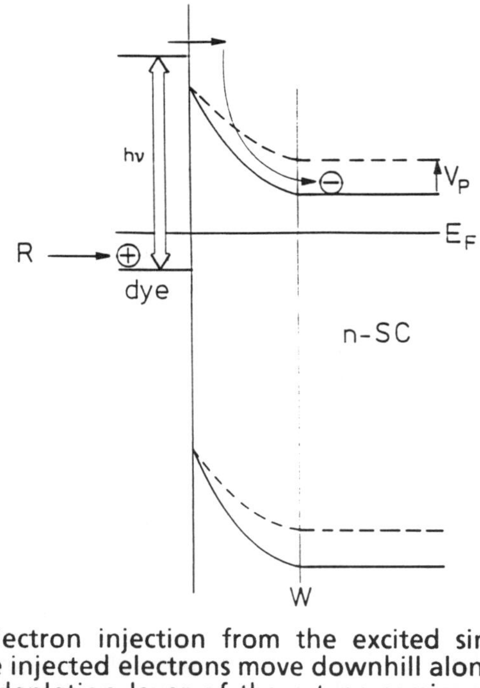

Figure 8. Electron injection from the excited singlet state of adsorbed molecules. The injected electrons move downhill along the potential gradient set-up in the depletion layer of the n-type semiconductor. V_p indicates the photopotential and R the connection to the external circuit. Light harvesting is limited by light absorption in the adsorbed dye layer.

many years. It is generally accepted as the primary process in photography. Recently Grätzel et al. /11/ described an interesting modification of the experimental system where the active surface area of the semiconductor and thus the number of light absorbing molecules was increased by up to a factor of 100 compared to a planar electrode surface (Fig. 9). A very high probability close to 80 percent of electron injection from the excited dye molecules was achieved with the polycrystalline porous TiO_2 electrodes. A chelate-type bonding of dye molecules to the TiO_2 surface appears to be of significance. It is not clear whether this bonding prevents an in-between adsorption layer of e.g. water molecules and contaminants or whether it helps in the attachment to favorable sites for electron injection. The injection process was found faster than 10 ps with the appropriate dye molecules /12/. The specific method of preparing the polycrystalline porous TiO_2 electrodes /11/ suggests that a high density of trap states should be formed at the grain boundaries.

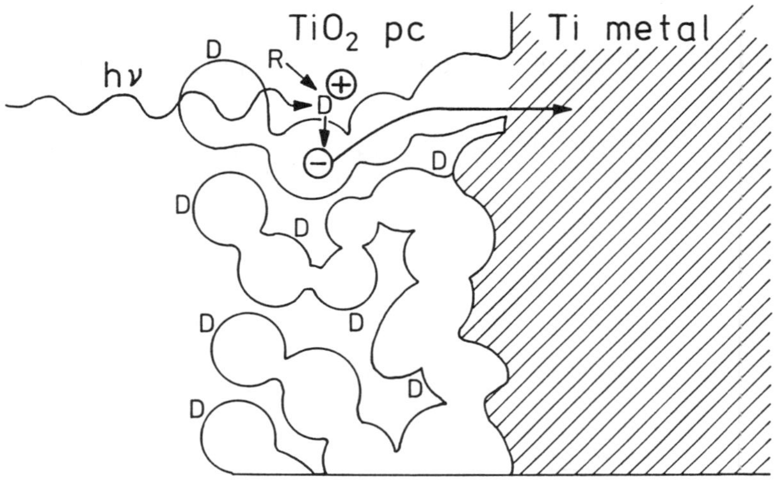

Figure 9. Illustration of significantly improved light harvesting in a system of the type shown in Fig. 8, achieved with a grossly enlarged surface area (about 200 times the geometrical area). The current yield, i.e. collected electrons per incident photons reached over 80 percent at the maximum of the dye absorption with an extremely porous polycrystalline TiO_2 electrode /11/.

Indeed, the characteristic non-exponential delayed rise of the photocurrent to its stationary value and the corresponding delayed decay (Fig. 10) is caused by complete filling and emptying of deep traps /13/. It should be noticed that these deep traps cannot fucntion as recombination centers since only majority carriers are injected into the depletion layer. The grossly enlarged surface area leads to almost 100 percent light absorption at the peak of a strong transition /13/. It has not been checked whether the complete solar spectrum can be covered with a suitable combination of different adsorbed dye molecules to obtain close to 100 percent light absorption. The maximum open circuit photovoltage could then approach the difference betwen the conduction band edge at the surface and the Fermi energy in the dark. This would correspond to the efficiency of a corresponding small band gap semiconductor device, provided the injected electrons can leave the polycrystalline semiconductor only via the conduction band and not via states in the band gap.

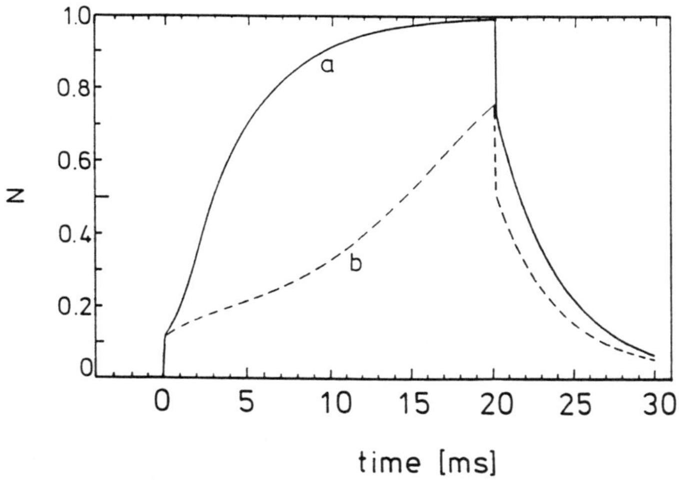

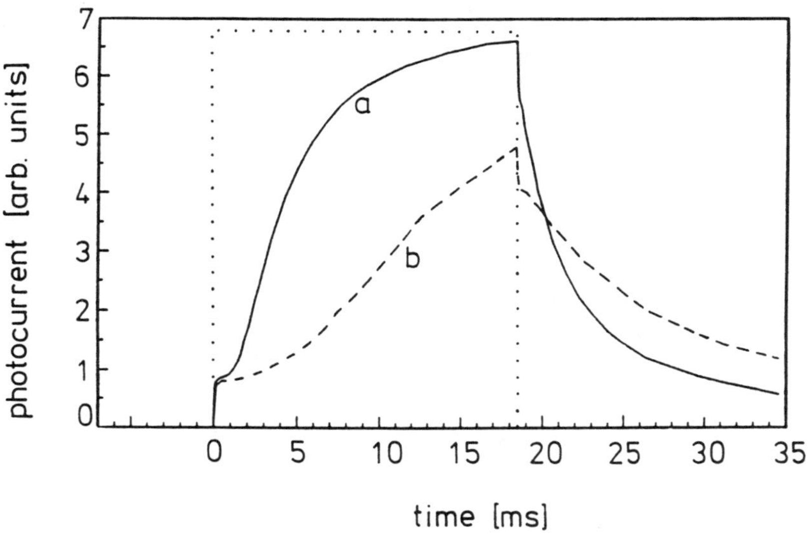

Figure 10. Rise of the injected photocurrent in the system illustrated in Fig. 9 /13/ to its stationary value. The dotted area indicates the illumination period. The two experimental curves at the bottom correspond to 20 m Watt/cm2 (curve a) and 2 m Watt/cm2 (curve b) of 600 nm light. The two curves at the top are numerical simulations of the free charge carrier concentration N in a three level system, i.e. a free carrier level with entrance and exit and two different trap levels. The lower trap level becomes completely filled when N attains its stationary value. Curves a correspond to a 10 times higher injection rate than curves b and show a much faster rise due to faster filling of the lower trap level.

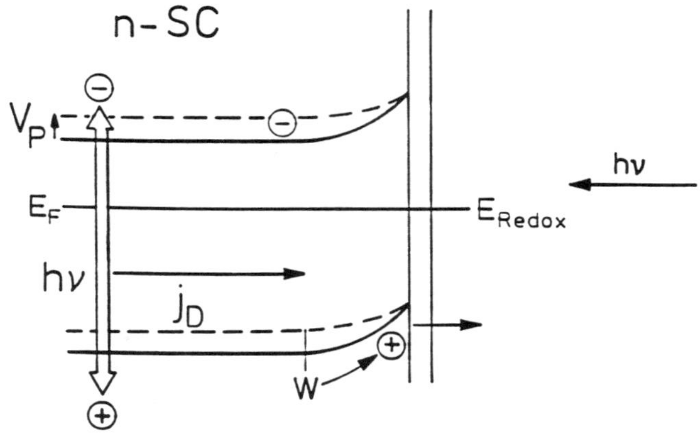

Figure 11. Light-harvesting and electron-hole pair separation at a semiconductor electrode. Long diffusion lengths of screened minority carriers, due to a high diffusion coefficient and mobility, and fast electron-hole pair separation in potential gradients set-up in doped semiconductors are the basic features of the most efficient photovoltaic devices.

4. Time resolved photovoltaic processes in a semiconductor/electrolyte cell

Semiconductors with a high minority carrier mobility (several 100 cm^2/Vs) and a correspondingly large diffusion length (> 100 μm) combine efficient light harvesting via minority carrier diffusion with very fast and thus efficient electron-hole pair separation in a potential gradient set-up in the depletion layer at a contact or a p-n junction /14/. Such a system is illustrated in Fig. 11. The picosecond time resolved build-up of the photovoltage in such a system is shown in Fig. 12 /15/. There are two different contributions to the experimental photovoltage (top), the first arises from the drift current of electron-hole pairs formed in the depletion layer (center) and the second from the diffusion current of screened minority carrier generated in the bulk that are moving towards the edge of the depletion layer (bottom), where they contribute to the measured signal. Integration over these current contributions in the measured signal across the external load resistance is due to RC shaping /16/. For front illumination as in Fig. 11 and a wavelength of 600 nm the drift and diffusion current contributions are of comparable magnitude in GaAs /15/. Photovoltaic solar energy conversion efficiencies greater than 20 percent have been reached with one band gap semiconductor devices and higher efficiencies with e.g. tandem cells.

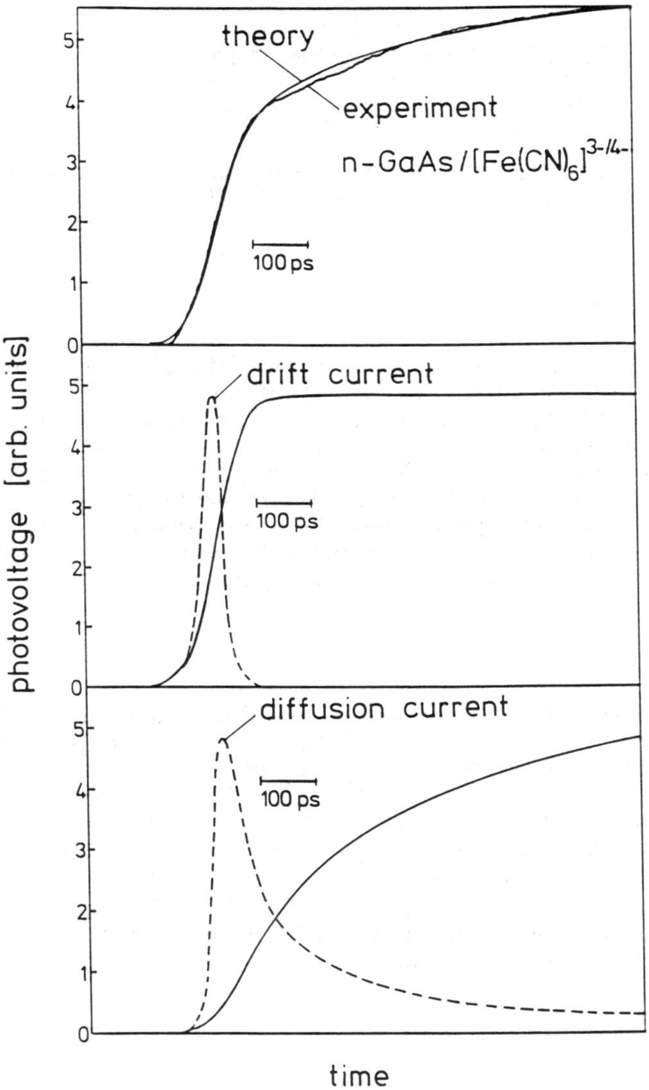

Figure 12. Picosecond time-resolved rise of the photovoltage in an n-GaAs electrode with a depletion layer at the surface (upper curves) /15/. Light absorption inside of the depletion layer and outside of the depletion layer contribute as drift current (central curves) and diffusion current (lower curves) to the time dependence of the photovoltage. The integration over the current signals arises in this short time window due to the RC constant (several 10 ns) of the circuit /16/.

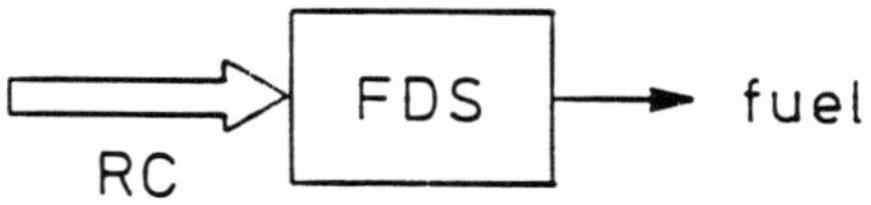

Figure 13. Scheme illustrating the goal of developing catalyzer systems (FDS) that are coupled directly to the photovoltaic device (RC) and generate suitable fuel compounds.

Summary of 1, 2 and 3

The advantage of the semiconductor devices for photovoltaic solar energy conversion over the supramolecular structures arises due to the several 100 times larger diffusion coefficients and minority carrier mobilities in the semiconductors as compared to the exciton diffusion coefficients and charge carrier mobilities in the supramolecular structures. Overall high photovoltaic solar energy utilization could be achieved with many supramolecular structures placed on top of each other and with separate reaction chambers for the production of fuels P_1 and P_2, much like in the many thylakoid membranes assembled in a leaf. This is of course a very demanding construction.

5. Fuel production

The most elegant solution of solar energy driven fuel production is direct coupling of the catalyzer to the photovoltaic machine (Fig. 13). In the case of semiconductors in contact with water this would require a stable protection against corrosion that does not disturb the functioning of the catalyzer. This has been achieved for H_2 evolution /17/. Catalyzers for multi-electron reactions, such as O_2 evolution from water, are beginning to take shape as complex molecules whose synthesis is inspired by biological catalyzers /18/. It may turn out that, following nature, such molecular catalyzers can be coupled easier to molecular photovoltaic machines than to semiconductor surfaces. However, the latter approach promises the highest conversion yields.

Acknowledgement

The author is grateful to the DFG, SFB 337 and to the Fonds der Chemischen Industrie for financial support

References

/1/ M. Van der Auweraer, F. Willig, and K.-P. Charlé, Chem. Phys. Lett., 128, 214 (1986);
K.-P. Charlé and F. Willig, to be published.

/2/ S. Efrima and M. Bixon, Chem. Phys. Lett., 25, 34 (1974);
J. Ulstrup and J. Jortner, J. Chem. Phys., 63, 4358 (1975).

/3/ J. von Gersdorff, M. Huber, H. Schubert, D. Niethammer, B. Kirste, M. Plato, M. Möbius, H. Kurreck, R. Eichberger, R. Kietzmann, and F. Willig, Angew. Chemie, 102, 690 (1990).
/4/ R. A. Marcus, J. Chem. Phys., 43, 679 (1965).
/5/ W. S. Durfee, W. Storck, F. Willig, and M. von Frieling, J. Am. Chem. Soc., 109, 1297 (1987).
/6/ W. S. Durfee, M. Gruszecka, T. A. Olszak, W.Storck, H. Bradaczek, and F. Willig, J. Am. Chem. Soc., 111, 3699 (1989).
/7/ D. Weiß, R. Kietzmann, J. Lehnert, and F. Willig, to be published.
/8/ K. Bitterling and F. Willig, unpublished results.
/9/ M. Eichhorn, F. Willig, K.-P. Charlé, and K. Bitterling, J. Chem. Phys., 76, 4648 (1982).
/10/ R. Eichberger, F. Willig, and W. Storck, Mol. Cryst., Liqu. Cryst., 175, 19 (1989).
/11/ N. Vlachopoulos, P. Liska, J. Augustynski, and M. Grätzel, J. Am. Chem. Soc., 110, 1216 (1988);
P. Liska, N. Vlachopoulos, M. K. Nazeeruddin, P.Comte, and M. Grätzel, J. Electroanal. Chem. 259, 59 (1989).
/12/ R. Eichberger and F. Willig, Chem. Phys. 141, 159 (1990).
/13/ K. Schwarzburg, F. Willig, and P. Liska, to be published.
/14/ S. M. Sze, Physics of Semiconductor Devices, 2nd ed., chapter 14, Wiley, New York, 1981.
/15/ K. Bitterling and F. Willig, J. Electroanal. Chem. 204, 211 (1986).
/16/ F. Willig, Ber. Bunsenges. Phys. Chem. 92, 1312 (1988).
/17/ A. Heller, Science, 223, 1141 (1984).
/18/ K. Wieghardt, Angew. Chemie, 101, 1179 (1989).

MECHANISMS OF ORGANIC TRANSFORMATIONS ON SEMICONDUCTOR PARTICLES

D. W. BAHNEMANN
Institut für Solarenergieforschung
Sokelantstraße 5
D 3000 Hannover 1
West Germany

ABSTRACT. Highlights of the current state-of-the-art of research activities in photocatalytic transformations of organic molecules on semiconductor particles are presented. These include the specific reactivity of such molecules due to the stabilization of intermediate free radicals at the particle surface. The effect of the size of the catalyst grain on the product distribution is expounded. Examples from time-resolved studies are given to further elucidate mechanistic details. The influence of surface charge and zeta-potential as well as the possible intermediacy of hydroxyl radicals are discussed. It is shown that variations of the particle morphology or its environment (solvent effect) often result in drastic differences in the photocatalytic reactivity. Similar observations are made when different materials are tested for the same reaction. Results from recent publications are given to illustrate areas of common knowledge, but also to draw the attention to obvious contradictions.

1. Introduction

Photocatalysis has been an area of rapidly growing interest over the last ten years starting with the initial stimulants of the direct photoelectrochemical conversion of radiant into chemical energy. A number of excellent review articles has been published recently [1-5] and an enormous part of the current knowledge in the field has been covered in several books [6-9]. It is the aim of this status report to highlight a few areas of current interest concerning the understanding of mechanisms involved in the transformation of organic molecules during photocatalytic processes on semiconductor particles. No attempt can and will be made to fully cover the area since the author is limited by both, time and space. Any omission of important contributions by collegues around the world was certainly unintentional and will hopefully not offend anybody.

1.1 A DEFINITION OF PHOTOCATALYSIS

The proper definition of " photocatalysis " has been published elsewhere [10], thus only a brief clarification should be given here. Semiconductor particles are ideal photocatalysts for a desired reaction

$$A \longrightarrow B \qquad (1)$$

E. Pelizzetti and M. Schiavello (eds.), Photochemical Conversion and Storage of Solar Energy, 251–276.

if - products B are formed with high specificity (often the product distribution will differ from the homogeneous case),
- they remain unchanged at the end of the reaction,
- electrons and/or holes generated by the absorption of photons with an energy exceeding the semiconductor bandgap energy are required for the process, and
- no net photon energy is stored in the final products, i.e. reaction (1) is exothermic and in principle only kinetically retarded.

Especially the first point of this definition differs considerably from that of conventional catalysis. As the forthcoming short tour through photocatalysis will demonstrate, the product distribution is often completely altered in photocatalytic systems as compared with the homogeneous situation. This can

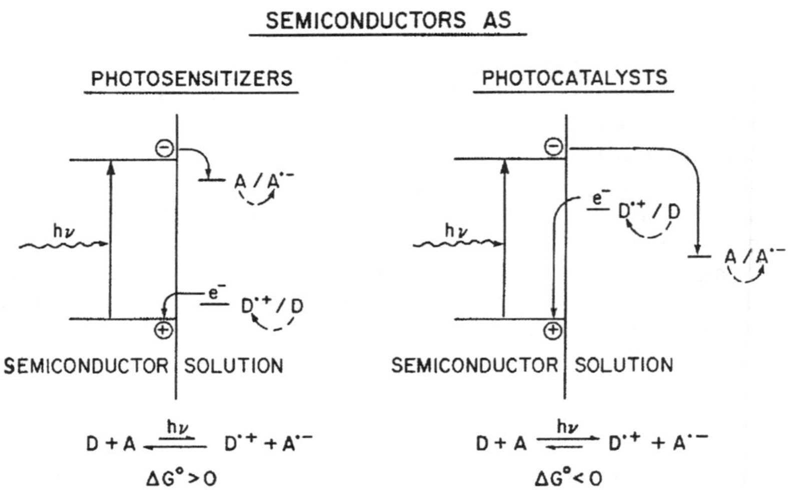

Figure 1. Energetic considerations for the definition of semiconductor particles as photosensitizers or photocatalysts, respectively.

be advantageous especially where specific products are desired.

1.2 ENERGETIC CONSIDERATIONS

The energetic requirements (cf. section 1.1, point 4) of a photocatalytic process are compared with those of photosynthesis in Figure 1. While it is evident that no net storage of light energy into the final products is achieved in photocatalysis, it has be shown by many authors that free radical intermediates play an important role [11-19]. Indeed, the formation of these radical species from the original educts is generally highly endothermic and requires a considerable amount of the incident photon energy [20-25]. The energetic requirements for such a free radical mechanism are illustrated on the left side of Fig. 1. The nature of these radical intermediates, their interaction with the semiconductor surface, their lifetime and their reactivity are among the most crucial factors determining the yield of photocatalytic conversions as well as product distribution and specificity.

Fig. 2 shows a cartoon which is frequently used to illustrate photocatalytic processes. It consists of a superposition of the energy bands of a semiconductor (valence band VB, conduction band CB, band gap energy E_g) and the geometrical image of a spherical particle. Electron / hole pairs are generated

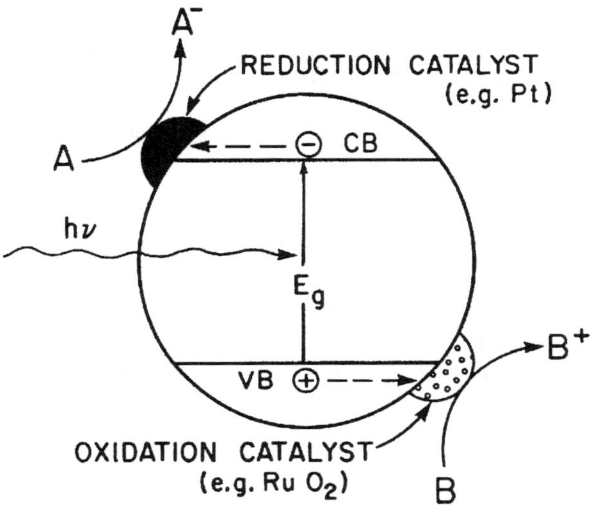

Figure 2. Schematic illustration of a semiconductor particle used as a redox photocatalyst (energetic and geometric representations are superimposed).

within the particle following the absorption of photons with an energy hυ exceeding E_g. If these charge carriers reach the particle surface before they recombine they can be transferred to electron acceptors (A) or donors (B), respectively, thus initiating the desired redox chemistry. It can sometimes be useful to deposit additional catalysts such as platinum (Pt) or ruthenium dioxide (RuO_2) onto the particles to lower overpotentials and thereby enable a more efficient charge transfer [2, 4].

1.3 PARTICLE SIZE EFFECT

The semiconductor / electrolyte interface has been studied extensively using photoelectrochemical systems [26-30]. Thermodynamic equilibration between the electrochemical potential of the solid and the chemical potential of the surrounding (mostly aqueous) solution leads to a depletion of majority carriers near the semiconductor surface and thus to a bending of its energy bands. The resulting electric field facilitates the separation of photogenerated electron / hole pairs within this space-charge layer (SCL) with the minority carriers being drawn towards the semiconductor / electrolyte interface.

Typical values for the thickness of the SCL(d_{sc}) are calculated to be around 100 nm (for a semiconductor with dielectric constant $\varepsilon = 10$, dopant concentration $n_0 = 10^{17} cm^{-3}$, potential drop across SCL: $U_{SC} = 1$ V). As particle diameters become considerably smaller than d_{SC} the band bending within the particle will become negligable and the description given in Fig. 2 will be correct.

In a recent review Memming has calculated the transit time necessary for charge carriers formed within the bulk to reach the surface of the semiconductor particles[31]. While in a particle with a diameter d of 100 Å it takes only 1 ps before electrons and holes can be trapped in energetically favorable sites at the semiconductor / electrolyte interface, a transit time of 10 ns is computed for d = 1 μm. As lifetimes have been reported to be in the ns time regime for electron / hole pairs in semiconductor particles [32], recombination within the bulk will be a competitive process in μm - sized particles. These calculations also evince that the band-bending described above should not pose a problem on the photocatalytic action of larger semiconductor particles [31]. While flat - band conditions are a priori fulfilled in e. g. a 100 Å particle, illumination times in the order of a few milliseconds (using ordinary laboratory conditions) are sufficient to reach flat - band in 1 μm particles, a prerequisite for the transfer of majority carriers via the particle surface.

2. Specific Reactivity of Semiconductor Particles

It has already been pointed out that driving redox reactions with products different from those obtained through other methods is one of the main reasons which lead to the widespread utilization of suspensions of semiconductor particles for photocatalytic applications. As early as 1977 Kraeutler

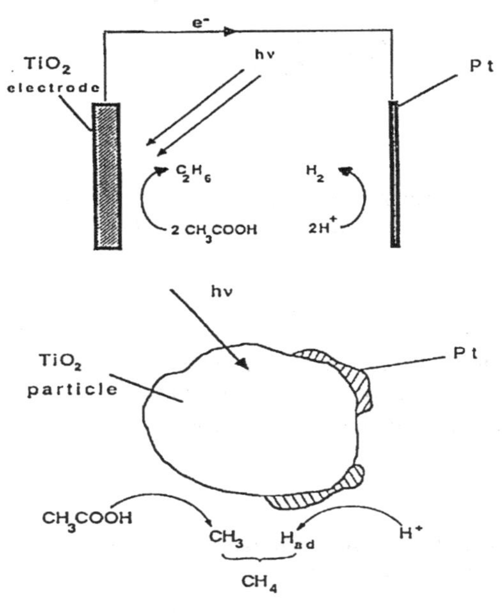

Figure 3. Photo-Kolbe decarboxylation of acetic acid at TiO_2 electrodes (upper part) and TiO_2 particles (lower part), taken from [33].

and Bard observed that while ethane is the main product when acetic acid is oxidized on illuminated TiO_2 electrodes with Pt counter electrodes, methane is formed when the same reaction is performed using platinized TiO_2 particles [34]. The mechanistic model which accounts for these at first sight surprising results is illustrated in Fig. 3 (taken from [33]). The photoelectrochemical system shown in the upper part of this figure ensures complete separation of anodic and cathodic processes. Methyl radicals formed in the initial one-electron oxidation step (Photo-Kolbe type decarboxylation)

$$CH_3COOH \quad + \quad h^+_{VB} \longrightarrow {}^{\bullet}CH_3 \; + \; CO_2 \; + \; H^+_{aq} \qquad (2)$$

combine in a subsequent reaction to form ethane:

$${}^{\bullet}CH_3 \quad + \quad {}^{\bullet}CH_3 \longrightarrow C_2H_6 \qquad (3)$$

Conduction band electrons, e^-_{CB}, induce the formation of molecular hydrogen at the platinum counter electrode.

The high yield of methane formation observed when aqueous suspensions of platinized TiO_2 particles are illuminated with $h\upsilon > E_g$ is explained by the simultaneous generation of methyl radicals (cf. reaction (2)) and hydrogen atoms (reaction (4))

$$H^+ \quad + \quad e^-_{CB} \xrightarrow[Pt]{} {}^{\bullet}H_{ads} \qquad (4)$$

on the surface of the same particle (cf. lower part of Fig. 3), where they subsequently combine:

$${}^{\bullet}CH_{3\,ads} \quad + \quad {}^{\bullet}H_{ads} \longrightarrow CH_4 \qquad (5)$$

Following these pioneering studies various groups have directed their efforts towards a more detailed investigation of fundamental differences in the reactivity of particulate systems as compared with bulk electrodes or homogeneous redox chemistry. A few examples taken from the research activities of M. A. Fox and co-workers illustrate the specific selectivity of photocatalytic organic transformations on illuminated TiO_2 particles (Scheme 1) [35]. It is evident that the type of products obtained in these oxidations depends largely on the nature of the oxidant. No general rule can be deduced from the examples given in the scheme. While the two-electron transfer is favored in example (1), a one electron oxidation appears to be the main path in example (2). Even ring opening reactions are sometimes observed (example (3)).

Recently Muzyka and Fox studied the mechanism of the photocatalytic oxidation of vicinal cyclohexane-dicarboxylic acids in suspensions of TiO_2 powders [36]. They showed that the one-hole oxidation process is the predominant path of both stereo isomers and Photo-Kolbe type products are formed in close similarity with reactions (2), (4), and (5). However, since molecular oxygen is present in most of their systems as an electron scavenger,

$$O_2 \quad + \quad 2e^-_{CB} \quad + \quad 2H^+ \longrightarrow H_2O_2 \qquad (6)$$

the authors also observed the formation of products following the energetically favored reaction of triplet O_2 with the carbon centered radical generated in the initial decarboxylation step (reactions (7a)-(7c)).

$$R-COOH + h^{+}_{VB} \longrightarrow {}^{\bullet}R + CO_2 + H^{+}_{aq} \tag{7a}$$

$$^{\bullet}R + O_2 \longrightarrow R-O_2^{\bullet} \tag{7b}$$

$$R-O_2^{\bullet} \longrightarrow \longrightarrow \longrightarrow R'=0 \tag{7c}$$

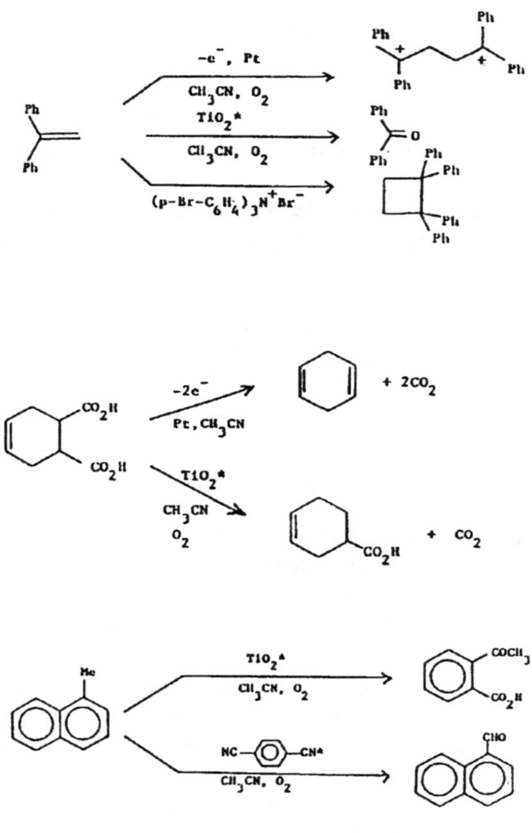

(1)

(2)

(3)

Scheme 1. Reaction mechanisms observed during the oxidation of various organic compounds using different redox systems, taken from [35].

It is interesting to note that the latter path is favored in case of the cis-isomer of the dicarboxylic acid. This can be explained by a different adsorption geometry of the two isomers with the cyclohexane ring system of the cis-form lying parallel to the particle surface [36] thus favoring the reaction with molecular oxygen which is known to be strongly adsorbed onto metal oxide particles [37].

Another example for the specific photocatalytic reactivity of metal oxide particles is the irreversible two-electron oxidation of methylviologen (1,1'-dimethyl-4,4'-bipyridylium chloride, MV^{2+}) on TiO_2[38]. Two strongly fluorescing products are formed with comparatively high quantum yields upon ultra-bandgap irradiation of aqueous colloidal suspensions of TiO_2 containing MV^{2+}: 1',2'-dihydro-1, 1'-dimethyl-2'-oxo-4, 4'-bipyridylium (upper structure) ($\phi = 0.032$) and 3,4-dihydro-1,1'-dimethyl-3-oxo-4,4'-bipyridylium (lower structure) ($\phi = 0.01$).

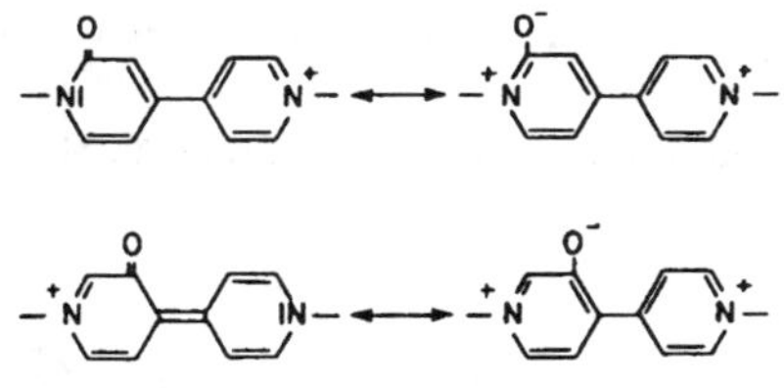

Two photons are required for the formation of each of the two products, thus an overall quantum yield $\phi = 0.084$ is calculated for these oxidation processes. Since these products are only by-products in the

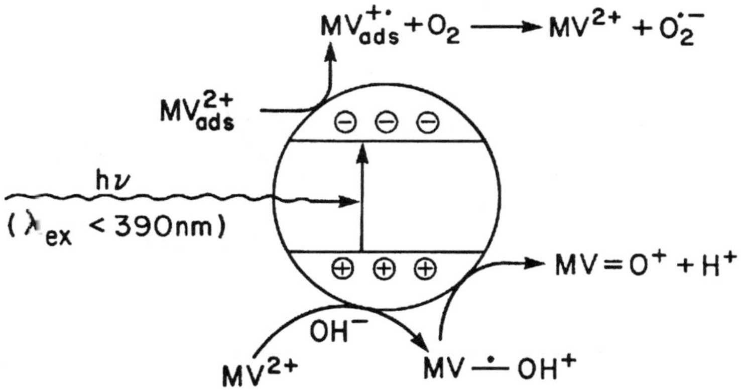

Figure 4. Mechanism of the photocatalytic oxidation of methylviologen on TiO_2 particles.

homogeneous free radical oxidation of MV^{2+} a simultaneous or consecutive two-hole transfer mechanism on the particle surface is suggested. The proposed reaction mechanism is given in Fig. 4 . The oxidation of methylviologen by h^{+}_{VB} photogenerated in TiO_2 particles will irreversibly remove this compound from the reaction mixture. Since MV^{2+} is used as an electron relay compound for various homogeneous and heterogeneous systems [39-50], it must be ascertained that oxidizing equivalents capable of attacking this organic molecule (e. g. $^{\bullet}OH$, $h^{+}_{VB}(TiO_2)$) will not be formed during its operation. Similar considerations are valid for light harvesting system incorperating organic materials since they cannot be regenerated like in living organisms. There are certainly many reason to employ such probe molecules for model studies, however, the author cannot envisage technologically feasible approaches to collect and convert solar energy where organic molecules play a crucial role.

It has been pointed out throughout this chapter that the specificity of particulate semiconductor suspensions in directing the reaction paths into desired channels is often well explained by the stabilization of free radical intermediates on the particle surface. This point is further illustrated by the investigation of the photocatalytic reduction of halothane (2-bromo-2-chloro-1,1,1-trifluoroethane) using platinized TiO_2 particles [51]. The overall mechanism of this process is given in Figure 5. The initial one-electron reduction of halothane via

$$CF_3CHBrCl \quad + \quad e^-_{CB} \quad \longrightarrow \quad CF_3{}^{\bullet}CHCl \quad + \quad Br^- \qquad (8)$$

results in the formation of bromide and a carbon centered radical. Studies of the latters homogeneous

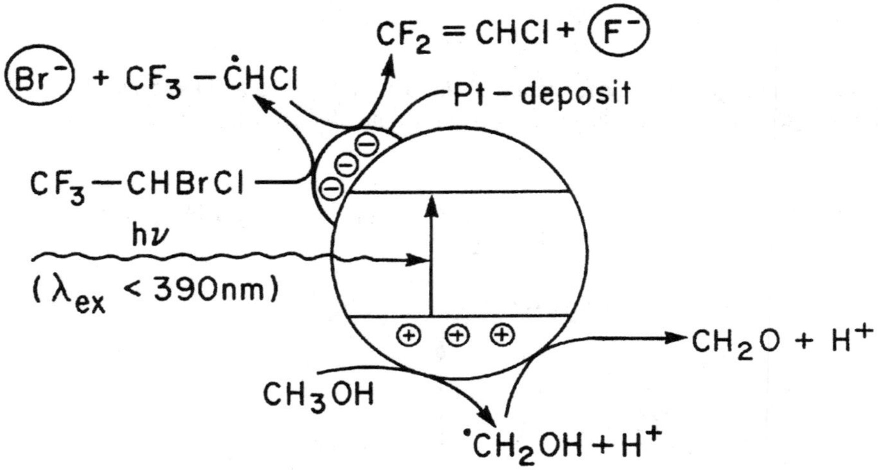

Figure 5. Mechanism of the photocatalytic reduction of halothane on platinized TiO_2 particles.

phase chemistry have shown that it reacts rapidly with molecular oxygen yielding trifluoroacetate as final product [52, 53] :

$$CF_3{}^{\bullet}CHCl \quad + \quad O_2 \quad \longrightarrow \quad CF_3CHClO_2{}^{\bullet} \qquad (9a)$$

$$CF_3CHClO_2{}^{\bullet} \quad \longrightarrow \quad \longrightarrow \quad CF_3CO_2{}^- \qquad (9b)$$

Further reduction of the radical formed in reaction (8) can in homogeneous solution only be observed under anoxic conditions in the presence of appropriate reducing agents, e.g. ascorbate[54]:

$$CF_3{}^{\bullet}CHCl \quad + \quad red. \quad \longrightarrow \quad CF_2CHCl \quad + \quad F^- \quad + \quad ox. \qquad (10)$$

Using the heterogeneous system described in Figure 5 it could be demonstrated that high yields of fluoride ions are also found in air-saturated suspensions evincing that the intermediate $CF_3{}^{\bullet}CHCl$ radical resides on the particle surface where it is reduced either by a subsequent e^-_{CB}

$$CF_3{}^{\bullet}CHCl + e^-_{CB} \longrightarrow CF_2CHCl + F^- \qquad (11)$$

or by the hydroxymethyl radical which is formed upon the one-electron oxidation of the hole scavenger methanol (cf. Fig. 5) :

$$CF_3{}^{\bullet}CHCl + {}^{\bullet}CH_2OH \longrightarrow CF_2CHCl + CH_2O + H^+ + F^- \qquad (12)$$

Thus, two electrons are transferred in subsequent steps to one halothane molecule before it has a chance to leave the TiO_2 surface. Stabilization of otherwise highly reactive free radical intermediates appears to be a key feature of the surface of metal oxide particles such as TiO_2.

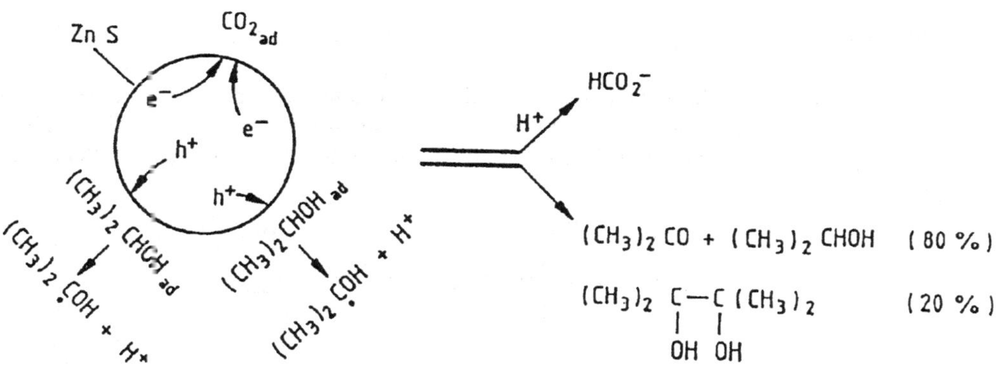

Figure 6. Illustration of the two-electron plus one-hole redox mechanism in the illumination of a ZnS sol containing CO_2 and 2-propanol (The two 2-propanol molecules generally are semioxidized at different colloidal particles), taken from [55].

3. Mechanistic Implications of the Particle Size

In contrast to the examples discussed in the previous chapter radical species formed in the initial one-electron redox step sometimes behave as if they were generated in homogeneous solution. In a study of the photocatalytic oxidation of 2-propanol in aqueous suspensions containing small colloidal zinc sulfide particles, Henglein and co-workers observed a product distribution typical for free radical intermediates [55]. A schematic presentation of their results is given in Fig. 6. The fact that pinalcol ($(CH_3)_2(COH)_2(CH_3)_2$) is detected as one of the products shows that the reaction proceeds via $(CH_3)_2{}^{\bullet}COH$ radicals. The observed product ratio of 80 % acetone and 20 % pinalcol is well known from homogeneous phase chemistry. Therefore it seems rather improbable to postulate mainly surface chemical reaction paths to explain these results. It is worth noting that the simultaneously occurring reduction of CO_2 proceeds via a two-electron transfer in the employed ZnS suspensions

(cf. Fig. 6).

These studies reported by Henglein et al were performed using extremely small ZnS particles with diameters between 2-3 nm. Assuming the photon flux of typical laboratory illumination conditions (i.e. the light of 450 W Xe-lamp is absorbed in 5 to 100 ml reactant suspensions), the time interval between the successive absorption of two photons by the same semiconductor particle is about 1ms. This apparently exceeds the residence time of the radical intermediate on the ZnS surface andradicals formed at different particles react with each other in the bulk solution. Recently, analogous experiments were performed with suspensions of larger particles with grain sizes of 2-3μm [56]. Under these conditions, where a photon is absorbed by one particle almost every nanosecond, no formation of pinalcol was observed. Initially formed radicals (reaction (13)) are further oxidized on the same ZnS particle

$$RCH_2OH \quad + \quad h^+_{VB} \longrightarrow R^{\bullet}CHOH \quad + \quad H^+ \qquad (13)$$

by a subsequently formed hole (reaction (14)) before their diffusion into the surrounding solu-

$$R^{\bullet}CHOH \quad + \quad h^+_{VB} \longrightarrow RCHO \quad + \quad H^+ \qquad (14)$$

tion becomes relevant. While this model can be readily used to describe the particle size effect upon the oxidation of aliphatic alcohols, it fails to explain observations by Yanagida et al. who studied photocatalytic reduction processes on different sized CdS-grains [57]. The investigated systems are

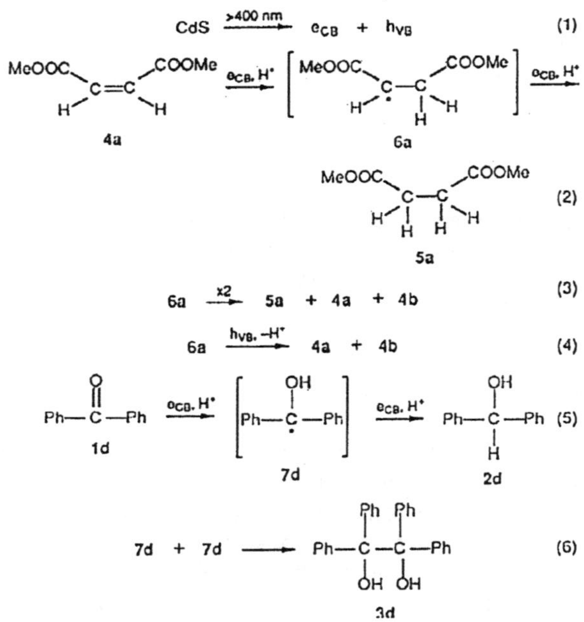

Scheme 2. Reaction mechanisms observed during the reduction of organic molecules using CdS particles of different sizes as photocatalysts, taken from [57].

summarized in Scheme 2. When small colloidal CdS particles (q-CdS) were employed, two-electron reduction processes prevailed (reactions (2) and (5) in the scheme). If, however, larger CdS grains were used, the free radical intermediates 6a and 7d reacted with each other (reactions (3) and(6)) to form products like those observed in homogeneous free radical reduction processes, α-hydroxy alkyl radicals have been postulated as intermediates in both studies (cf. reaction (13) and (7d) in Scheme 2), therefore the stability of a radical on the catalyst surface cannot be taken as a relevant explanation for these observations. A unified model capable to account for particle size effects on both, reduction and oxidation processes, is currently not available in the literature. In this context , one should remember that a two-electron reduction process was also observed by Henglein et al. using small ZnS particles as catalysts [55].

Another research area where systematic investigations of the particle size effect have lately been initiated is the so- called "q-size" regime. The latter term has been given to ultrasmall colloidal semiconductor particles or clusters which exhibit different photophysical properties than the corresponding bulk material, i.e. the apparent band gap energy E_g is shifted to higher values as compared with the solid state value when particle diameters become smaller than those of the corresponding exciton [58]. In a recent puplication, Nosaka et al. have found higher quantum yields for the photocatalytic reduction of methylviologen (reaction (15)) on q-size CdS particles [59]. This can be explained by a shift

$$MV^{2+} \quad + \quad e^-_{CB} \longrightarrow MV^{+\bullet} \qquad (15)$$

of the Fermi level or the conduction band edge, respectively, to more negative values as E_g increases with decreasing cluster size. Similar results are obtained by the same authors when sulfopropylviologen (SPV) is used as an electron acceptor with q-size In_2S_3 particles being the photocatalyst. As the intensity of the incident light is increased, however, higher quantum yields for SPV-reduction are observed with larger In_2S_3 agglomerates. A kinetic model is proposed to account for these observations[59].

4. Detection of Intermediates by Time-resolved Studies

Most of the free radical intermediates postulated in the mechanisms given above are too short-lived to be detected by conventional spectroscopic (e.g. absorption or fluorescence spectroscopy, esr) or analytical techniques (e.g. GC/MS, HPLC). On the other hand, light scattering in suspensions or on powder surfaces imposes considerable experimental problems on time-resolved studies. Diffuse reflectance spectroscopy as introduced by Wilkinson offers an elegant method to circumvant the latter problem and has been applied successfully to study processes on semiconductor powders [60]. Another way to avoid artefacts due to light scattering lies in the utilization of colloidal suspensions of small particles (a rule-of-thumb tells us that scattering is minimal when particle diameters $d < \lambda/20$, with λ: observation wavelength of light) which can be prepared to appear like clear solutions [61, 62]. Only studies of the latter kind will be reported here.

A limited number of examples exist in the literature where the charge carriers initially formed upon light absorption in a semiconductor particle (i.e. electrons and holes) exhibit a characteristic absorption spectrum after they have been trapped in surface sites. Figure 7 shows the absorption spectrum of an electron trapped via

$$Ti^{IV} \quad - \quad e^-_{CB} \longrightarrow Ti^{III} \qquad (16)$$

in colloidal TiO_2 particles [19]. The absorption vs time trace (given as the inset of Fig. 7) shows that the absorption change is found to be stable over several milliseconds. Steady-state illuminations evince that this absorption does not decay in the absence of molecular oxygen or other oxidants provided a hole scavenger is present in the solution [62]. Polyvinylalcohol (PVA) has been used as electron donor in the experiments shown in Fig. 7. It reacts with h^+_{VB} to form an α-hydroxy radical :

$$PVA + h^+_{VB} \longrightarrow PVA(\text{-}H)^{\bullet} + H^+_{aq} \qquad (17)$$

which has reducing properties and thus can lead to the formation of another Ti^{III}- site (current-doubling effect) via injection of an electron into the conducting band of TiO_2 :

$$PVA(\text{-}H)^{\bullet} + Ti^{IV} \longrightarrow PVA(\text{-}H_2) + Ti^{III} + H^+_{aq} \qquad (18)$$

The extinction coefficient of Ti^{III}- sites on TiO_2 has also been determined : $\varepsilon = 800\,M^{-1}\,cm^{-1}$ at 780 nm [63].

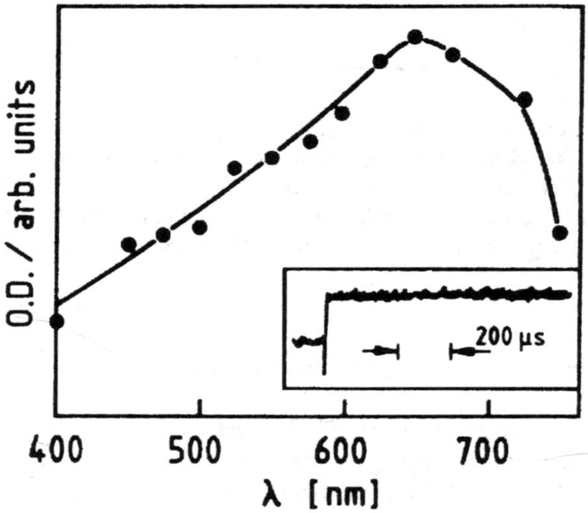

Figure 7. Absorption spectrum of a laser-flashed solution of 6.3 x 10^{-3} mol dm^{-3} TiO_2 containing 5x 10^{-3} mol dm^{-3} PVA and time profile of the absorption signal (inset) at pH 10, taken from [19].

Trapping of e^-_{CB} is an extremely fast process. Figure 8 shows the spectra of Ti^{III} in colloidal TiO_2 following a 20 ps-laser flash [32]. The transient spectrum is already fully developed immediately after the flash. Apparently, trapping occurs instantaneously as the electrons reach the particle surface. The experiments shown in Fig. 8 were performed in the absence of hole scavengers, thus the transient absorption of Ti^{III} disappears quickly due to charge carrier recombination

$$Ti^{III} + h^+_{VB} \longrightarrow Ti^{IV} + \text{heat} \qquad (19)$$

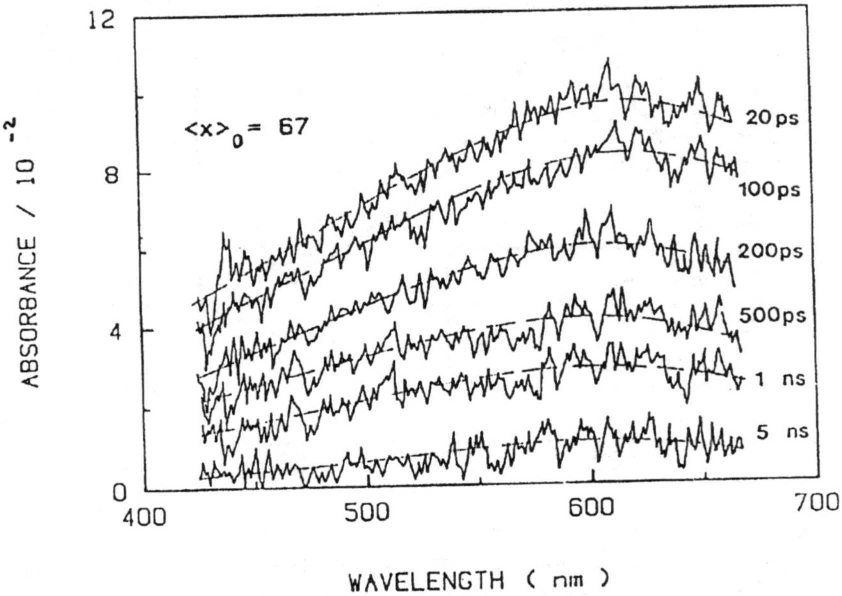

Figure 8. Transient spectra observed at various time intervals after picosecond excitation of colloidal TiO_2. [TiO_2] = 17 g/L, pH 2.7, solution deaerated with Ar, optical path length 0.2 cm. Average number of electron-hole pairs present initially in one TiO_2 particle is 67, taken from [32].

It could be shown that the lifetime of a single e^-/h^+ pair on a TiO_2 particle is $\tau = (30 \pm 15)$ns [32.].The detection of transient absorption spectra due to trapped e^-_{CB} has not been reported in colloidal suspensions of other semiconductors. However, when extremely small (q-size) particles of transition metal sulfides (e.g. CdS, FeS_2, ZnS) are illuminated in the presence of hole scavengers a blue shift of the absorption spectrum is observed [64-67]. This so-called Burstein shift is caused by excess charge on the small particle and has been explained by a band filling model [61] or alternatively by the polarization of subsequently formed excitons [64]. Similar results have been obtained with q-ZnO and q-TiO_2 [61, 62, 68]. In the latter case, it could be shown that excess negative charges due to surface hydroxyl groups are sufficient to observe the Burstein shift [62]. A detailed review of these photophysical phenomena has been published recently [58].

When electrons are scavenged efficiently during the band-gap illumination of a semiconductor particle, i.e. by oxidants adsorbed on its surface, valence band holes (h^+_{VB}) are left behind which get trapped on surface sites (SS) via

$$h^+_{VB} + SS \longrightarrow SS^{+\bullet} \qquad (20)$$

Figure 9 shows the transient absorption spectrum attributed to such an oxidized surface site ($SS^{+\bullet}$) in colloidal TiO_2 particles when Pt deposits are present as electron scavengers [19].

$$e^-_{CB} + (Pt)_x \longrightarrow (Pt)_x^- \qquad (21)$$

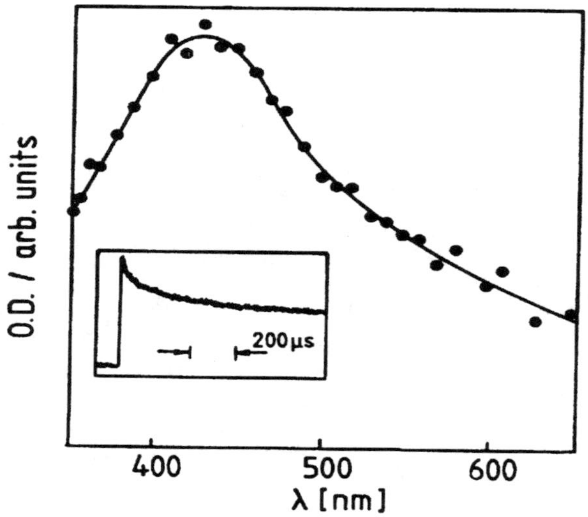

Figure 9. Absorption spectrum of a laser-flashed TiO_2-Pt colloid (3.8×10^{-3} mol dm^{-3} TiO_2 and 1.6×10^{-5} mol dm^{-3} Pt) and decay of the absorption (inset) at pH 2.5, taken from [19].

The absorption vs. time trace given as inset in this figure shows that the transient decays even in the absence of additional electron donors probably due to combination reactions

$$SS^{+\bullet} \quad + \quad SS^{+\bullet} \longrightarrow SS^{2+} \quad + \quad SS \qquad (22)$$

leading to the formation of doubly oxidized surface sites. Anodic corrosion is normally observed as a consequence of these processes [69]. Comparatively weak and broad transient absorption bands peaking around 650 nm have been observed in colloidal CdS in the presence of acetone as electron scavenger [70]. They were attributed to holes trapped in sulfur-centered surface states :

$$h^+_{VB} \quad + \quad S^{2-} \longrightarrow S^{-\bullet} \qquad (23)$$

Following the disappearence of the transients described so far, it is possible to study mechanistic details of one-electron redox processes occuring on semiconductor particles upon band-gap illumination.Subsequently formed free radical intermediates often possess a characteristic absorption spectrum of their own thus yielding further information about the respective process. Detailed kinetic studies of the associated electron transfer mechanisms have been published [71].

Charge injection into semiconductors can also be studied conveniently by following the resulting transient absorption or Burstein shift, respectively [72, 68]. For example, it could be shown that free radicals generated in solution can inject electrons [64]

$$(CdS)_x \; + \; (CH_3)_2C^{\bullet}OH \longrightarrow (CdS)_x^- \; + \; (CH_3)_2CO \; + \; H_{aq}^+ \qquad (24)$$

or holes [70]

$$(CdS)_x \quad + \quad {}^{\bullet}OH \longrightarrow (CdS)_x^{+} \quad + \quad OH^{-} \qquad (25)$$

into small semiconductor clusters.

Finally, it should be mentioned that some controversy exists concerning the transient absorption of trapped h^{+}_{VB} in TiO_2. Langford et al. recently published the spectrum shown in Fig. 10 which they attributed to this intermediate [72].The comparison with Fig. 9 clearly shows the associated problem. While adsorbed molecules were used as electron scavengers by Langford and co-workers, metal islands served this purpose in the other experiments by Bahnemann et al [19]. Further studies are certainly necessary to elucidate the reason for the obvious discrepancy.

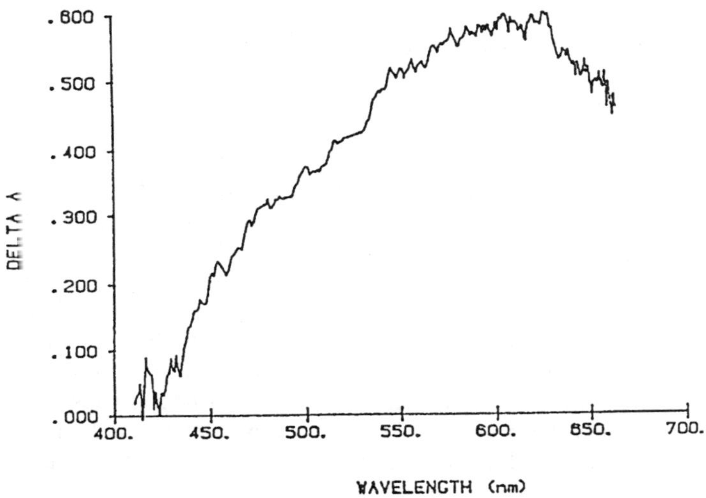

Figure 10. Difference of transient spectrum at 10 ns minus that at 500 ps for the $CuPcTs^{4-}$- TiO_2 system, taken from [72].

5. The Nature of the Oxidizing Intermediate

It has been shown in the previous chapter that electrons and holes trapped in surface states of semiconductors often exhibit characteristic absorption spectra. While trapped e^{-}_{CB} can almost always be regarded as reduced metal centers (cf. eq. (19)), the chemical nature of a hole localized at such a surface state still poses an unresolved question. Multiple attempts have been made to decide whether hydroxy radicals are indeed formed in aqueous suspensions of e.g. TiO_2 particles. Due to the extremely short life-time of $^{\bullet}OH$ in aqueous solution most of the experimental approaches employed had to be indirect. For example, illuminations were carried out in the presence of spin-trapping agents so that products could be identified and quantified by e s r [73].Another approach was to use probe molecules the $^{\bullet}OH$-adducts of which are well known and could be analyzed qualitatively and quantitatively

[74, 75]. Often there are, however, two possible routes to arrive at the same reaction products. Valence band holes could form hydroxy radicals upon trapping at surface hydroxy groups of a metal oxide semiconductor (cf. eq. (21)).

$$h^{+}_{VB} \quad + \quad OH_S^- \longrightarrow \ ^{\bullet}OH_S \qquad (26)$$

The reaction of a probe molecule P will then result in the observed product distribution

$$^{\bullet}OH_S \quad + \quad P \longrightarrow P^{\bullet}\text{-}OH \longrightarrow \text{products} \qquad (27)$$

Alternatively, P could be oxidized directly by valence band holes

$$P \quad + \quad h^{+}_{VB} \longrightarrow P^{+\bullet} \qquad (28)$$

followed by the hydrolysis of the intermediate radical cation $P^{+\bullet}$

$$P^{+\bullet} \quad + \quad H_2O \longrightarrow P^{\bullet}\text{-}OH \quad + \quad H^{+}_{aq} \qquad (29)$$

thereby leading to the same products.

Recently, Fox et al. have attempted to identify the hydroxy adducts ($P^{\bullet}$- OH) directly by their characteristic transient absorption spectrum using a laser flash photolysis technique combined with diffuse reflectance measurements [76]. Employing different probe molecules, such as Pt^{2+}, phenothiazine or methylviologen, they were not able to detect any transient signals which could be explained by $^{\bullet}$OH-adduct formation. These observations do, however, not prove that hydroxy radicals are not envolved in redox processes on metal oxide particles since acid-base catalysis favored on the highly hydroxylated surfaces (see also next chapter) might result in the rapid depletion of the $^{\bullet}$OH-adducts before they could be detected within the time-resolution of the kinetic apparatus.

Another attempt to answer the question of the intermediacy of $^{\bullet}$OH radicals in photocatalytic systems is illustrated in Scheme 3. It has been established in detailed radiation chemical investigations that hydroxy radicals attack acetate ions mainly at the methyl group [77]:

$$^{\bullet}OH \quad + \quad CH_3COO^- \longrightarrow \ ^{\bullet}CH_2COO^- \quad + \quad H_2O \qquad (30)$$

In the presence of air the radicals thus formed react quickly with molecular oxygen leading to the product distribution given in the lower part of Scheme 3 [78]. Direct oxidation of acetate, e.g. on Pt electrodes, results in the well-known Kolbe decarboxylation with the formation of methyl radicals.

$$h^{+}_{VB} \quad + \quad CH_3COO^- \longrightarrow CH_3COO^{\bullet} \longrightarrow \ ^{\bullet}CH_3 \quad + \quad CO_2 \qquad (31)$$

A considerably different product distribution results when these methyl radicals react with O_2[79]. The evolution of products observed upon the photocatalytic oxidation of acetate in aqueous suspensions containing TiO_2 as the catalyst is shown in Figure 11 [80]. It is obvious that glycolate and glyoxylate are both formed as primary products of the photocatalytic process. Since the intermediacy of $^{\bullet}CH_2COO^-$ radicals appears to be a prerequisite for the formation of both products (cf. Scheme 3) these results strongly suggest that $^{\bullet}$OH radicals are formed on TiO_2 surfaces upon band-gap illumination. These radicals are most likely strongly adsorbed on the oxide particles and should not be regard-

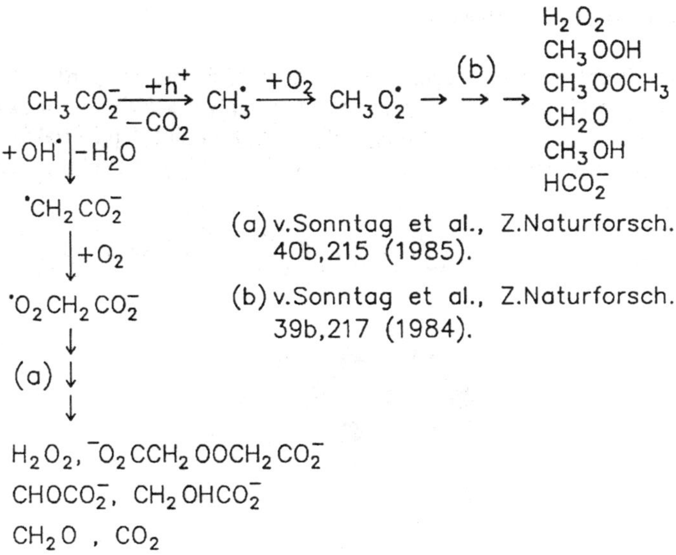

Scheme 3. Proposed reaction mechanism for the oxidation of acetate by h^+_{VB} or $^\bullet OH_S$, respectively.

ed as freely diffusing species. Further results of this study including a detailed analysis of the very complex pH-dependence involved will be published elsewhere [ref].In the case of transition metal chalcogenide semiconductors the intermediacy of chalcogenide radicals has been proposed(cf. eq. (23))[70]. Yet, much more experimental evidence is required to support this suggestion.

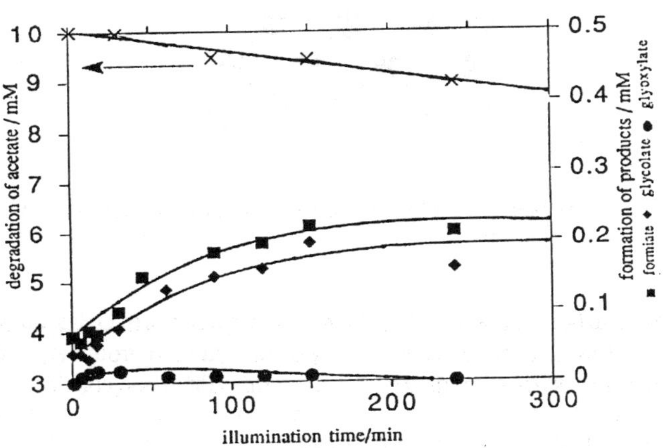

Figure 11. Photocatalytic oxidation of acetate at pH 10.6; 0.5 g/L TiO_2 (P25) in aqueous solution; O_2 saturated; hν ≥ 320 nm; T = 298 K .

6. The Role of Surface Chemistry

It has already been pointed out throughout this article that acid-base properties of metal oxide surfaces can have considerable implications upon their photocatalytic activity. Some examples from the recent literature will be given in the following to illustrate this point. A pictorial discription of the processes occuring at a metal oxide surface in contact with water is given in Figure 12 [81]. Adsorption

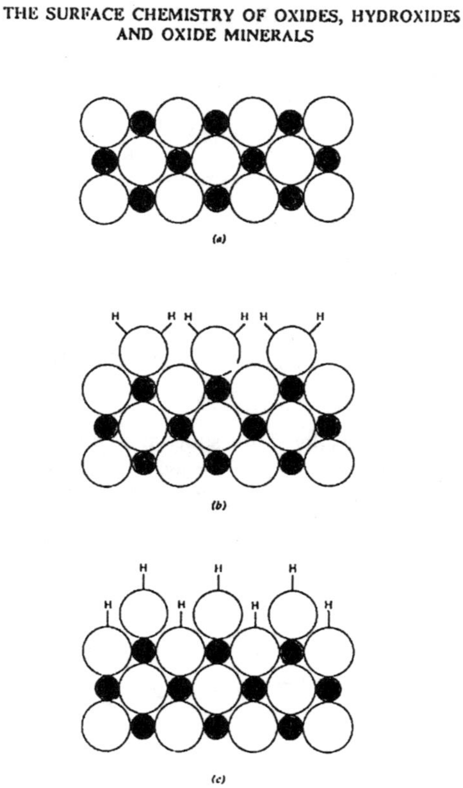

Figure 12. Schematic representation of the cross section of the surface layer of a metal oxide, taken from [81].

of H_2O molecules at surficial metal sites is followed by dissociation of OH-groups leading to a coverage (4 - 10 nm^{-2} [81]) with chemically equivalent metal hydroxyl groups (M-OH). Due to the amphoteric behaviour of most metal hydroxides two equilibria must be considered:

$$\text{M-OH} + H^+ \rightleftharpoons \text{M-OH}_2^+ \qquad K_1 \qquad (32)$$

$$\text{M-OH} \rightleftharpoons \text{M-O}^- + H^+ \qquad K_2 \qquad (33)$$

The zero-point-of-charge (pH_{zpc}) of metal oxide particles is defined as the pH where the concentration of protonated and deprotonated surface groups are equal :

$$pH_{zpc} = 1/2\,(pK_1 + pK_2) \qquad (34)$$

pH_{zpc} values have been determined for a wide variety of metal oxide particles. Examples for the most common materials are given in Table 1 [82]. As a consequence of this amphoteric behaviour the metal oxide surface is predominantly positively charged below pH_{zpc} and negatively charged above. The influence of this surface charge on the photocatalytic activity is demonstrated in Figure 13 [83]. While the oxidation of the anion trichloroacetate is only observed with reasonable yield below the pH_{zpc} of the employed catalyst TiO_2, the cation chloroethylammonium is predominantly oxidized at high pH. The first result could also be explained by the Nernstian shift of the band edges to more ne-

Zero Point of Charge

Material	pH_{zpc}
SiO_2	2.0
TiO_2	3.5 - 6.7
$\alpha - Fe_2O_3$	6.5 - 8.5
ZnO	9.0 ± 0.3
$\alpha - Al_2O_3$	9.1
CuO	9.5
MgO	12.5

Table 1. pH-values at the zero point of charge for various metal oxide powders, taken from [82].

gative values with increasing pH leading to a diminishing oxidation potential of h^+_{VB} at high pH [11]. The latter result, however, is in clear contrast to this model and can only be rationalized by the electrostatic adsorption model where cations are more readily accommodated at negative sites. In fact, computer simulations of this surface charge model using a modified Gouy-Chapman-Stern (GCS) approach have been successfully applied to fit the results shown in Figure 13 [83].

Lateley, Meisel et al have studied the charge transfer from viologen radical ions formed in homogeneous solution to ferric oxide particles as a function of pH [84]. They observed a rapid electron

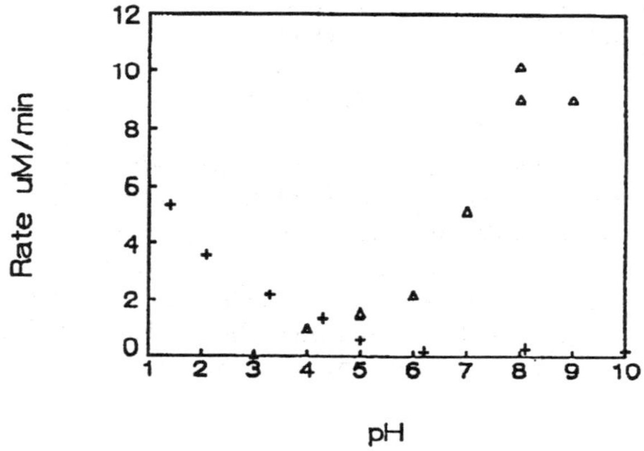

Figure 13. pH dependence of the rate of degradation of trichloroacetate (+) and chloroethylammonium (Δ) ions in illuminated suspensions of TiO_2 (0.5 g/L). Actinometry (300 nm-400nm) I≅10^{-4} Einstein l^{-1} min^{-1}, taken from [83].

injetion from methylviologen radical cations ($MV^{+\bullet}$) above the pH_{zpc} of the material while the anionic species ($ZV^{-\bullet}$)transferred electrons efficiently only at low pH. Using the GCS model to calculate the respective surface potential a reasonable fit of the experimental data was obtained over the whole pH range. The above examples illustrate that a detailed knowledge of the surface chemistry of semiconductor particles is extremely important to predict and understand their photocatalytic activity.

7. Influence of the Material Properties

Titanium dioxide is the material the photocatalytic activity of which has been studied most extensively over the past ten years. This explains why in most of the examples given above TiO_2 has been used as the catalyst. The widespread usage of TiO_2 can be understood by the fact that it is readily prepared, inexpensive and non-toxic. Moreover, it is extremely stable in aqueous suspensions with a diminishing rate of photo-corrosion. Upon band-gap illumination titanium dioxide is found to be an extremely active photocatalyst for various redox processes. However, the applicability of TiO_2 for photocatalytic applications driven by solar light is limited since due to its wide band gap (E_g = 3.2 eV) it absorbs less than 5% of the available photons.

A comparison of the photocatalytic activity of different metal oxide particles is shown in Fig. 14 [85]. While the oxidative photodegradation of chloroacetate is accomplished within 30 minutes with TiO_2 as the catalyst, the same reaction requires 13.5 hours of illumination to attain completion when ZnO is used. Hematite (α-Fe_2O_3) exhibits no photocatalytic activity for the same reaction, i.e. quantum yields smaller than 10^{-5} are encountered. At this point it should be noted that many publications concerning the photocatalytic activity of various semiconducting materials do not report values for quantum efficiencies. If any technical applicability of a process is envisaged quantum yields of 10^{-5} or less will certainly not suffice. Therefore it is of great importance that such values are measured and made available.

The method of preparation of a given semiconductor has an enormous impact on its photocatalytic activity. The anatase modification of TiO_2 is for example found to be more active than rutile for most applications [86, 87]. Amorphous colloidal particles can sometimes possess similar activities as anatase [88]. The implication of the particle size effect has already been discussed (cf. section 3), but depending upon preparation the number of active surface groups and hence the catalytic properties may vary drastically. Systematic studies addressing some of these important questions are currently not available. One of the reasons for these missing informations might be the lack of appropriate analytical tools to fully characterize the real surface of particles in contact with water or other solvents.

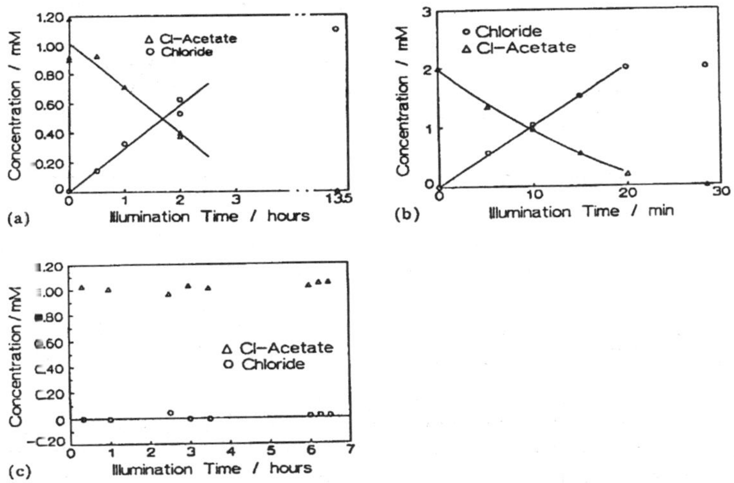

Figure 14. Photocatalytic degradation of chloroacetate with concomitant formation of chloride in aqueous suspensions of metal oxides. (a) 2×10^{-2} M ZnO powder, pH 8, air, UV light (300-400 nm) with $I \approx 5 \times 10^{-3}$ M hν min^{-1}; (b) 6.3×10^{-3} M TiO_2 (P25) powder, pH 5, air, UV light; (c) 1×10^{-3} M α-Fe_2O_3 colloid, pH 5, air, full lamp spectrum, taken from [85].

8. Conclusions

Photocatalysis has been a rapidly growing research area over the last decade. This review has specifically been concerned with data from recent systematic investigations in this field. Especially, various mechanistic aspects of organic transformations during photocatalytic processes on semiconductor particles have been discussed. It has been shown, that the specific reactivity of photocatalytic systems leading to product distributions different from those reported in the respective homogeneous studies can often be explained by a considerable stabilization of free radical intermediates on the particle surface. The surface chemistry of the semiconductor particles in contact with the (mostly aqueous) electrolyte thus plays an important role. Evidence for the intermediacy of $^{\bullet}OH$ radicals during photocatalytic oxidations on metal oxide surfaces has been presented. However, it has also been pointed out that the amount of data which can be unambigously interpreted in this regard is still limited. The effect of grain size and particle morphology upon product distribution and yield, respectively, has been discussed. It has been demonstrated that time-resolved kinetic studies using colloidal suspensions of extremely small semiconductor particles can be conveniently used to investigate mechanistic details of photocatalysis. While areas of common knowledge have been summarized, the attention has also been drawn to obvious contradictions existing in the recent literature.

9. References

[1] Kamat, P. V. and Dimitrijević, N. M. (1990) ' Colloidal semiconductors as photocatalysts for solar energy conversion ' , Solar Energy 44, 83-98.

[2] Bard, A. J. (1988) ' Semiconductor particles and arrays for the photoelectrochemical utilization of solar energy ' , Ber. Bunsenges. Phys. Chem. 92, 1187-1194.

[3] Fox, M. A. (1986) ' Charge injection into semiconductor particles - importance in photocatalysis ' , La Chimica e L'Industria 68, 59-64.

[4] Kalyanasundaram, K., Grätzel, M., and Pelizzetti, E. (1986) ' Interfacial electron transfer in colloidal metal and semiconductor dispersions and photodecomposition of water ' , Coord. Chem. Rev. 69, 57-125.

[5] Anpo, M. (1989) ' Photocatalysis on small particle TiO_2 catalysts. Reaction intermediates and reaction mechanisms. ' , Res. Chem. Intermed. 11, 67-106.

[6] Serpone, N. and Pelizzetti, E.(1989) Photocatalysis: Fundamentals and Applications,Joh n Wiley & Sons, Inc., New York.

[7] Fox, M. A. and Chanon, M. (1988) Photoinduced Electron Transfer Part A-D, Elsevier Science Publishers, Amsterdam.

[8] Pelizzetti, E. and Serpone, N. (1986) Homogeneous and Heterogeneous Photocatalysis (NATO ASI series, Series C: Mathematical and Physical Sciences Vol. 174), Reidel Publishing Company, Dordrecht.

[9] Schiavello, M. (1988) Photocatalysis and Environment : Trends and Applications (NATO ASI series. Series C, Mathematical and physical sciences Vol. 237), Kluwer Academic Publishers,Dordrecht.

[10] Kisch, H. (1989) ' What is photocatalysis ' , in N. Serpone and E. Pelizzetti (eds.), Photocatalysis: Fundamentals and Applications, John Wiley & Sons, Inc., New York, pp. 1-8.

[11] Duonghong, D., Ramsden, J., and Grätzel, M. (1982) ' Dynamics of interfacial electron-transfer processes in colloidal semiconductor systems ' , J. Am. Chem. Soc. 104, 2977-2985.

[12] Moser, J. and Grätzel, M. (1983) ' Light-induced electron transfer in colloidal semiconductor dispersions: single vs. dielectronic reduction of acceptors by conduction band electrons ' , J. Am. Chem. Soc. 105, 6547-6555.

[13] Serpone, N., Sharma, D. K., Jamieson, M. A., Grätzel, M., and Ramsden, J. J. (1985) ' Photophysical and photochemical primary events in semiconductor particulate systems. Colloidal CdS with Methylviologen. ' , Chem. Phys. Lett. 115, 473-476.

[14] Ramsden, J. J. and Grätzel, M. (1986) ' Formation and decay of methylviologen radical cation dimers on the surface of colloidal CdS: separation of two-and three-dimensional relaxation ' , Chem. Phys. Lett. 132, 269-272.

[15] Albery, W. J., Brown, G. T., Darwent, J. R., and Saievar-Iranizad, E. (1985) ' Time-resolved photoredox reactions of colloidal CdS ' , J. Chem. Soc., Faraday Trans.1, 81, 1999-2007.

[16] Kamat, P. V. (1985) ' Photoelectrochemistry in colloidal systems : interfacial electron transfer between colloidal TiO_2 and thionine in acetonitrile ' , J. Photochem. 28, 513-524.

[17] Gopidas, K. R. and Kamat, P. V. (1989) ' Photoelectrochemistry in particulate systems. 11. Reduction of phenosafranin dye in colloidal TiO_2 and CdS suspensions ' , Langmuir 5, 22-26.

[18] Bahnemann, D., Henglein, A., Lilie, J., and Spanhel, L. (1984) ' Flash photolysis observation of the absorption spectra of trapped positive holes and electrons in colloidal TiO_2.' , J.

Phys. Chem. 88, 709-711.

[19] Bahnemann, D., Henglein, A., and Spanhel, L. (1984) ' Detection of the intermediates of colloidal TiO_2-catalysed photoreactions ', Faraday Discuss. Chem. Soc. 78, 151-163.

[20] Wardman, P. (1989) ' Reduction potentials of one-electron couples involving free radicals in aqueous solution ', J. Phys. Chem. Ref. Data 18, 1637-1755.

[21] Stanbury, D. M. (1989) ' Reduction potentials involving inorganic free radicals in aqueous solution ', Advances in Inorganic Chemistry 33, 69-138.

[22] Bielski, B. H. J., Cabelli, D. E., Arudi, R. L., and Ross, A. B. (1985) ' Reactivity of HO_2/O_2^- radicals in aqueous solution ', J. Phys. Chem. Ref. Data 14, 1041-1100.

[23] Wayner, D. D. M., McPhee, D. J., and Griller, D. (1988) ' Oxidation and reduction potentials of transient free radicals ', J. Am. Chem. Soc. 110, 132-137.

[24] Rao, P. S. and Hayon, E. (1974) ' Redox potentials of free radicals. 1. Simple organic radicals ', J. Am. Chem. Soc. 96, 1287-1294.

[25] Steenken, S. and Neta, P. (1982) ' One-electron redox potentials of phenols. Hydroxy - and aminophenols and related compounds of biological interest ', J. Phys. Chem. 86, 3661-3667.

[26] Gerischer, H. (1980) ' The principles of photoelectrochemical energy conversion ', in F. Cardon, W. P. Gomes and W. Dekeyser (eds.) Photovoltaic and Photoelectrochemical Solar Energy Conversion, Plenum Press, New York, pp. 199-263.

[27] Memming, R. (1983) ' Processes at semiconductor electrodes ', in B. E. Conway (ed.), Comprehensive Treatise of Electrochemistry 7, Plenum Press, New York, pp. 529-592.

[28] Bard, A. J. and Faulkner, L. R. (1980) Electrochemical Methods: Fundamentals and Applications, John Wiley & Sons, Inc., New York.

[29] Tributsch, H. (1986) ' Photoelectrolysis and photoelectrochemical catalysis ', in Modern Aspects of Electrochemistry 17, Plenum Press, New York, pp. 303-355.

[30] Pelskov, Y. V. (1990) Solar Energy Conversion : A Photoelectrochemical Approach, Springer Verlag, Berlin Heidelberg.

[31] Memming, R. (1988) ' Photoelectrochemical solar energy conversion ', Topics Curr. Chem. 143, 79-112.

[32] Rothenberger, G., Moser, J., Grätzel, M., Serpone, N., and Sharma, D. K. (1985) ' Charge carrier trapping and recombination dynamics in small semiconductor particles ', J. Am. Chem. Soc. 107, 8054-8059.

[33] Memming, R. (1990) ' Photoelectrochemical utilization of solar energy ', in J. F. Rabek (ed.), Photochemistry and Photophysics II, CRC Press, Inc., pp. 143-189.

[34a] Kraeutler, B. and Bard, A. J. (1977) ' Photoelectrosynthesis of ethane from acetate ion at an n-type TiO_2 electrode. The photo-Kolbe reaction ', J. Am. Chem. Soc. 99, 7729-7731.

[34b] Kraeutler, B. and Bard, A. J. (1978) ' Heterogeneous photocatalytic decomposition of saturated carboxylic acids on TiO_2 powder. Decarboxylative route to alkanes ', J. Am. Chem. Soc. 100, 5985-5992.

[35] Fox, M. A. (1987) ' Photocatalysis on modified semiconductor surfaces and on bipolar photoelectrodes ', New J. Chem. 11, 129-133.

[36] Muzyka, J. L. and Fox, M. A. (1990) ' Selective photoelectrochemical oxidation of vicinal cyclohexanedicarboxylic acids: a mechanistic study ', J. Org. Chem. 55, 209-215.

[37] Heiland, G., Mollwo, E., and Stöckmann, F. (1959) ' Electronic processes in zinc oxide ', Solid State Physics 8, 191-323.

[38] Bahnemann, D. W., Fischer, Ch.-H., Janata, E., and Henglein, A. (1987) 'The two-electron oxidation of methyl viologen', J. Chem. Soc., Faraday Trans.1, 83, 2559-2571.

[39] Miller, D. S. and McLendon, G.(1981) 'Quantitative electrochemical kinetics studies of "microelectrodes" : catalytic water reduction by methyl viologen/colloidal platinum', J. Am. Chem. Soc. 103, 6791-6796.

[40] Kosanić, M. M., Dajić-Jovanović, R. P., and Nikolić, A. D. (1983) 'Temperature effect in the photocatalytic evolution of hydrogen from aqueous solution of proflavine', Bulletin de la Société Chimique Beograd 48, 229-233.

[41] Heyrovský, M. (1983) 'Catalytic and photocatalytic reduction of water by the reduced forms of methylviologen', J. Chem. Soc., Chem. Commun., 1137-1138.

[42] Nenadović, M. T., Mićić, O. I., Rajh, T., and Savić, D. (1983) 'Temperature effect on the photoinduced reduction of methyl viologen with several sensitizers and the evolution of hydrogen from water', J. Photochem. 21, 35-44.

[43] Matheson, M. S., Lee, P. C., Meisel, D., and Pelizzetti, E. (1983) 'Kinetics of hydrogen production from methyl viologen radicals on colloidal platinum', J. Phys. Chem. 87, 394-399.

[44] Brandeis, M., Nahor, G. S., and Rabani, J. (1984) 'Reactions of colloidal platinum in aqueous solutions containing methyl viologen, its cation radical, and hydrogen, studied by pulse radiolysis', J. Phys. Chem. 88, 1615-1623.

[45] Degani, Y. and Willner, I. (1983) 'Photoinduced hydrogen evolution by a zwitterionic diquat electron acceptor. The functions of SiO_2 colloid in controlling the electron-transfer process.

[46] Prasad, D. R., Hessler, D., Hoffman, M. Z., and Serpone, N. 'Quantum yield of formation of methylviologen radical cation from the photolysis of the $Ru(bpz)_3^{2+}$ /methylviologen/EDTA system', Chem. Phys. Lett. 121, 61-64.

[47] Moradpour, A., Amouyal, E., Keller, P., and Kagan, H. (1978) 'Hydrogen production by visible light irradiation of aqueous solutions of RU $(BIPY)_3^{2+}$', Nouv. J. Chim. 2, 547-549.

[48] Grätzel, M. (1981) 'Artificial photosynthesis : water cleavage into hydrogen and oxygen by visible light', Acc. Chem. Res. 14, 376-384.

[49] Ebbesen, T. W., Tembe, B. L., and Kozak, J. J. (1984) 'Factors affecting the efficiency of photochemical water cleavage systems. The reaction between O_2 and the reduced electron acceptor.', J. Phys. Chem.88, 683-688.

[50] Johansen, O., Launikonis, A., Loder, J. W., Mau, A. W.-H., Sasse, W. H. F., Swift, J. D., and Wells, D. (1981) 'Solar reduction of water. 1. Competition between light-induced hydrogen formation and hydrogenation of methylviologen in the system water-tris(2,2'-bipyridine)ruthenium dication-ethylenediaminetetraacetic acid-platinum', Aust. J. Chem. 34, 981-991.

[51] Bahnemann, D. W., Mönig, J., and Chapman, R. (1987) 'Efficient photocatalysis of the irreversible one-electron and two-electron reduction of halothane on platinized colloidal titanium dioxide in aqueous suspension', J. Phys. Chem. 91, 3782-3788.

[52] Mönig, J., Asmus, K.-D., Schaeffer, M., Slater, T. F., and Willson, R. L. (1983) 'Electron transfer reactions of halothane-derived peroxyl free radicals, $CF_3CHClO_2^{\bullet}$: measurement of absolute rate constants by pulse radiolysis', J. Chem. Soc. Perkin Trans. II, 1133-1137.

[53] Mönig, J. and Asmus, K.-D. (1984) 'Formation and reactions of halothane peroxy free radicals, $CF_3CHClO_2^{\bullet}$', in W. Bors, M. Saran, and D. Tait (eds.), Oxygen Radicals in Chem-

istry and Biology, Walter de Gruyter & Co., Berlin, pp.57-63.

[54] Mönig, J., Krischer, K., and Asmus, K.-D. (1983) ' One-electron reduction of halothane and formation of halide ions in aqueous solutions ' , Chem.-Biol. Interactions 45, 43-52.

[55] Henglein, A., Gutiérrez, M., and Fischer, Ch.-H. (1984) ' Photochemistry of colloidal metal sulfides. 6. Kinetics of interfacial reactions at ZnS-particles ' , Ber. Bunsenges. Phys. Chem. 88, 170-175.

[56] Müller, B., Meissner, D., and Bahnemann, D., J. Phys. Chem., manuscript in preparation.

[57] Shiragami, T., Pac, C., and Yanagida, S. (1990) ' Visible-light-induced two-electron-transfer photoreductions on CdS: effects of morphology ' , J. Phys. Chem. 94, 504-506.

[58] Henglein, A. (1989) ' Small-particle research : physicochemical properties of extremely small colloidal metal and semiconductor particles ' , Chem. Rev. 89, 1861-1873.

[59] Nosaka, Y., Ohta, N., and Miyama, H. (1990) ' Photochemical kinetics of ultrasmall semiconductor particles in solution : effect of size on the quantum yield of electron transfer ' , J. Phys. Chem. 94, 3752-3755.

[60] Wilkinson, F., Willsher, C. J., Uhl, S., Honnen, W., and Oelkrug, D. ' Optical detection of a photoinduced thermal transient in titanium dioxide powder by diffuse reflectance laser flash photolysis ', J. Photochem. 33, 273-278.

[61] Bahnemann, D. W., Kormann, C., and Hoffmann, M. R. (1987) ' Preparation and characterization of quantum size zinc oxide : a detailed spectroscopic study ' , J. Phys. Chem. 91, 3789-3798.

[62] Kormann, C., Bahnemann, D. W., and Hoffmann, M. R. (1988) ' Preparation and characterization of quantum-size titanium dioxide ' , J. Phys. Chem. 92, 5196-5201.

[63] Kölle, U., Moser, J., and Grätzel, M. (1985) ' Dynamics of interfacial charge-transfer reactions in semiconductor dispersions. Reduction of cobaltceniumdicarboxylate in colloidal TiO_2 ' , Inorg. Chem. 24, 2253-2258.

[64] Henglein, A., Kumar, A., Janata, E., and Weller, H. (1986) ' Photochemistry and radiation chemistry of semiconductor colloids : reaction of the hydrated electron with CdS and non-linear optical effects ' , Chem. Phys. Lett. 132, 133-136.

[65] Kamat, P. V., Dimitrijevic, N. M., and Nozik, A. J. ' Dynamic Burstein-Moss shift in semiconductor colloids ' , J. Phys. Chem. 93, 2873-2875.

[66] Liu, C.-y. and Bard, A. J. (1989) ' Effect of excess charge on band energetics (optical absorption edge and carrier redox potentials) in small semiconductor particles ' , J. Phys. Chem. 93, 3232-3237.

[67] Liu, C.-y. and Bard, A. J. (1989) ' Irradiation-induced absorption edge shifts in colloidal particles of FeS_2 (pyrite) ' , J. Phys. Chem. 93, 7047-7049.

[68] Haase, M., Weller, H., and Henglein, A. (1988) ' Photochemistry and radiation chemistry of colloidal semiconductors. 23. Electron storage on ZnO particles and size quantization ' , J. Phys. Chem. 92, 482-487.

[69] Gutiérrez, M. and Henglein, A. (1983) ' Photochemistry of colloidal metal sulfides. 4. Cathodic dissolution of CdS and excess Cd^{2+} reduction ' , Ber. Bunsenges. Phys. Chem. 87, 474-478.

[70] Baral, S., Fojtik, A., Weller, H., and Henglein, A. (1986) ' Photochemistry and radiation chemistry of colloidal semiconductors. 12. Intermediates of the oxidation of extremely small particles of CdS, ZnS, and Cd_3P_2 and size quantization effects (a pulse radiolysis study) ', J. Am. Chem. Soc. 108, 375-378.

[71] Meisel, D. and Mulac, W. A., ' Photochemical and radiation chemistry techniques in the study of colloidal systems ' , Colloids and Surfaces 35, 179-199.

[72] Arbour, C., Sharma, D. K., and Langford, C. H. (1990) ' Picosecond flash spectroscopy of TiO_2 colloids with adsorbed dyes ' , J. Phys. Chem. 94, 331-335.

[73] Schindler, M., private communication and poster at the " 11. Vortragstagung der Fachgruppe Photochemie der GDCh ", Duisburg, FRG, Nov. 1989

[74] Izumi, I., Dunn, W. W., Wilbourn, K. O., Fan, F.-R. F., and Bard, A. J. (1980) ' Heterogeneous photocatalytic oxidation of hydrocarbons on platinized TiO_2 powders ' , J. Phys. Chem. 84, 3207-3210.

[75] Matthews, R. W. (1984) ' Hydroxylation reactions induced by near-ultraviolet photolysis of aqueous titanium dioxide suspensions ' , J. Chem. Soc., Faraday Trans.1, 80, 457-471.

[76] Fox, M. A.(1990)' Control of photocatalytic oxidative selectivity on irradiated TiO_2 powders : a diffuse reflectance kinetic study ' , in E. Pelizzetti and M. Schiavello (eds.), Proceedings of the 8th International Conference on Photochemical Conversion and Storage of Solar Energy.

[77] Neta, P., Simic, M., and Hayon, E. (1969) ' Pulse radiolysis of aliphatic acids in aqueous solutions. 1. Simple monocarboxylic acids ' , J. Phys. Chem. 73, 4207-4213.

[78] Schuchmann, M. N., Zegota, H., and von Sonntag, C. (1985) ' Acetate peroxyl radicals, $^{\bullet}O_2CH_2CO_2^-$: a study on the γ-radiolysis and pulse radiolysis of acetate in oxygeneted aqueous solutions' , Z. Naturforsch. 40b, 215-221.

[79] Schuchmann, H.-P. and von Sonntag, C. (1984) ' Methylperoxyl radicals : a study of the γ-radiolysis of methane in oxygenated aqueous solutions ' , Z. Naturforsch. 39b, 217-221.

[80] Wolff, K. and Bahnemann, D. (1990) ' Are OH radicals formed during photocatalytic processes on TiO_2 ? A detailed study of the oxidation of acetate ' , J. Phys. Chem., manuscript in preparation.

[81] Stumm, W. and Morgan, J. J. (1981) Aquatic Chemistry, John Wiley & Sons, New York.

[82] Bahnemann, D. (1990) ' Photocatalytic formation of sulfur-centered radicals by one-electron redox processes on semiconductor surfaces ' , in C. Chatgilialoglu and K.-D. Asmus (eds.), Sulfur-centered Reactive Intermediates in Chemistry and Biology (NATO-ASI Series), Plenum Press, London, New York, pp. 103-120.

[83] Kormann, C., Bahnemann, D. W., and Hoffmann, M. R. (1990) ' On the photolysis of aqueous solutions of organic molecules in the presence of TiO_2 suspensions ' , Env. Sci. Technol., in press.

[84] Mulvaney, P., Swayambunathan, V., Grieser, F., and Meisel, D. (1990) ' Effect of the ζ potential on electron transfer to colloidal iron oxides ' , Langmuir 6, 555-559.

[85] Kormann, C., Bahnemann, D. W., and Hoffmann, M. R. (1989) ' Environmental photochemistry : is iron oxide (hematite) an active photocatalyst ? A comparative study : α-Fe_2O_3, ZnO, TiO_2 ' , J. Photochem. Photobiol., A : Chemistry, 48, 161-169.

[86] Sclafani, A., Palmisano, L., and Schiavello, M. (1990) ' Influence of the preparation methods of TiO_2 on the photocatalytic degradation of phenol in aqueous dispersion ' , J. Phys. Chem. 94, 829-832.

[87] Sclafani, A., Palmisano, L., and Davì, E. (1990) ' Photocatalytic degradation of phenol by TiO_2 aqueous dispersions : rutile and anatase activity ' , New J. Chem. 14, 265-268.

[88] Chandrasekaran, K. and Thomas, J.K. (1984) ' Photochemical reactions of amorphous and crystalline titanium dioxide powder suspensions in water ' , J. Chem. Soc., Faraday Trans.1, 80, 1163-1172.

PHOTOCATALYSIS : HETEROGENEOUS REGIME. CATALYSTS, ADSORPTION AND NEW TECHNIQUES

P. PICHAT
Unité de Recherche Associée au CNRS
"Photocatalyse, Catalyse et Environnement"
Ecole Centrale de Lyon
B.P. 163, 69131 Ecully Cedex
France

ABSTRACT. This report attempts to present the state of the research in photocatalysis on powder semiconductors by considering the studies that allow one to determine the properties of the solids and to obtain information on surface sites and adsorbed species. First, the importance of surface features and associated phenomena, the dependence of adsorption equilibria and surface entities mobilities upon band-gap illumination, and the information that can be derived from kinetic measurements and some spectroscopic methods, are briefly discussed. Then, the objectives and results of recent studies attempting to prepare particulate photocatalysts with controlled texture or higher stability, as well as semiconductors deposited on or incorporated in inorganic supports, are reviewed. Finally, the capabilities of contactless time-resolved microwave conductivity and photoacoustic spectroscopy are shown.

1. Introduction

This report deals only with heterogeneous photocatalysis on semiconductors that are used in the form of powders, which means that related studies on microcrystals or colloids are not included. It focuses on new aspects of the preparation and studies of photocatalysts, particularly those concerning their surface properties. The term "heterogeneous photocatalysis" is employed here to denote the phenomena by which the optical excitation of a solid, which remains unaltered, is utilised to increase the rate of a chemical reaction.

In order not to overlap with previous reviews more than necessary for the understanding, we have preferentially cited publications appeared since the last International Conference on Photochemical Conversion and Storage of Solar Energy (I.P.S.-7, Evanston, 1988). Also, as recommended by the organisers of I.P.S.-8, studies communicated in Palermo are mentioned whenever they correspond to the topics of this report.

In the first section, the importance of surface features and surface phenomena in heterogeneous photocatalysis is emphasised by comparison with investigations that tend to regard semiconductors almost exclusively from the energetic viewpoint. The effects of band-gap illumination on the adsorption equilibria and the lability of adsorbed/surface oxygen atoms (in the case of semiconductor oxides) are briefly recalled. The information that can be drawn from kinetic studies of photocatalytic reactions is critically examined.

E. Pelizzetti and M. Schiavello (eds.), Photochemical Conversion and Storage of Solar Energy, 277–293.

Finally, recent progresses in the applications of some spectroscopic methods to investigate adsorbed species on powder semiconductors are cited.

The second section enlightens some special efforts made to prepare particulate photocatalysts that possess controlled texture and/or are supported on inorganic materials. The objectives and results are discussed.

The last section deals with examples of physical techniques that have recently been applied to study the effects of some modifications of particulate photocatalysts on the charge carriers. These techniques are contactless, time-resolved microwave conductivity using nanosecond pulse-radiolysis, and photoacoustic spectroscopy using a dual-beam configuration and continuous UV-illumination to excite the semiconductor.

2. Importance of surface features and surface phenomena

2.1. COLLECTIVE ELECTRON MODEL AND LOCALISED MODEL

Because of the wide development of research in photoelectrochemistry on semiconductors which has followed the report by A. Fujishima, K. Honda and S. Kikuchi [1], most interpretations on the interactions between light-excited semiconductors and the surrounding medium have originally been based on the **collective electron model** of semiconductors. This model refers to thermodynamics, since it considers the energy levels which can be occupied by electrons in the solid, the energy levels of the so-called surface states and finally redox potentials of the species present in the external medium. The surface states can be intrinsic ; this term designates defects due to the termination of the crystal lattice. The extrinsic surface states include impurities, various surface defects such as vacancies, surface groups and adsorbed species. Whereas the energy levels of the valence band and the conduction band, as well as those of redox compounds in solution, are generally known, data yielding the positions of defined surface states on the energy scale are rare. In addition, this collective electron model, however useful to indicate whether a given type of electron transfer is possible or not, has the disadvantage of considering adsorbed species and surface features from the energetic viewpoint only.

Therefore, the **localised model** , which is founded on the concept of surface sites, should equally be considered. This model not only refers to the nature of the semiconductor, which provides the energy levels of its bands, but also tries to take into account the identity of the particular sample used. For powder samples the preparation determines the texture, i.e. the mean grain size, the morphology (spheres, polyhedra, needles, etc.) and the degree of crystallinity. As a result, the exposed crystal planes should differ and the number of surface irregularities, such as steps, kinks, etc., varies for a given powder semiconductor. Also the pretreatments play a major role in the nature of the surface sites. For example, pretreatments at high temperature of semiconductor oxides produce Lewis-acid sites which can chemisorb basic molecules, thereby facilitating electron transfers between these adsorbates and the light-excited solid. The density of surface hydroxyl groups also depends on the preparation and pretreatments.

The **surface sites** of a photocatalyst determine the amount and nature of the **adsorbed species** formed from the reactants. Even in the absence of dissociative adsorption, particular bonds of the reactant can be activated as a result of the interaction with the surface sites. The ensemble constituted by an adsorbed species and the corresponding surface site can induce a stereochemistry which cannot be explained by regarding the surface from the thermodynamic viewpoint only. Also, this ensemble determines the mobility of surface entities. In turn the importance of reverse reactions depends on that mobility. The strength of the bonds between the surface sites and the intermediates conditions the distribution of the products, i.e. the selectivity of the photocatalytic reaction. Scheme I illustrates that point. (A is a reactant, B and C are products). Because of the pattern made by the vertical arrows and the horizontal line representing the solid surface, this scheme has been named "rake mechanism" (J.E. Germain in "Catalytic Conversion of Hydrocarbons" (1969) Academic Press, p. 259).

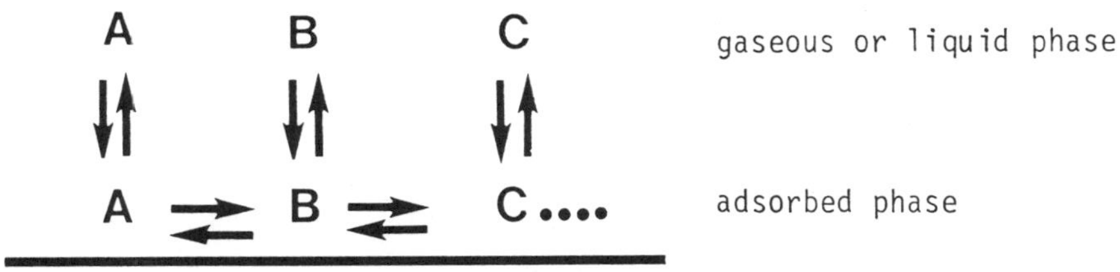

Scheme I

2.2. PHOTOADSORPTION AND PHOTODESORPTION

Because of the creation of electron-hole pairs, it is obvious that band-gap illumination can modify the adsorption equilibrium of charged adsorbed species (sometimes called "ionosorbed" species). In the case of large entities, the adsorption equilibrium of configurations with highly localised partial charge can also be affected. Direct reactions corresponding to photodesorption or photoadsorption are symbolised by equations (1-3) :

$$A^{-}_{ads} + h^{+}_{VB} \rightleftarrows A_{ads} \rightleftarrows A_{gas} \quad (1)$$

$$B_{gas} \rightleftarrows B_{ads} \quad (2)$$

$$B_{ads} + e^{-}_{CB} \rightleftarrows B^{-}_{ads} \quad (3)$$

We have previously reviewed these phenomena [2]. Their study present technical difficulties and does not give rise to many publications presently. In addition, it has been showed that for the adsorption of dioxygen on semiconductor oxides - in fact the most often investigated case - the photoadsorbed amounts are low as compared with the amounts previously adsorbed in the dark [2,3]. In other words, the coverage of the surface by adsorbed species forms essentially in the dark before illumination.

However, photosorption phenomena should not be ignored when studying a photocatalytic reaction, since the band-gap illumination is susceptible to influence the lability of the adsorbed species. Although that principally concerns the ionosorbed species, perturbations for adsorbed neutral molecules can also be envisaged ; for example some photodesorption of water from TiO_2 was observed and attributed to the weakening of the electron-donating bond between the oxygen atom of H_2O and a Ti^{4+} cation whose charge can be decreased by one unit under band-gap illumination [4].

Also, measurements of the amounts of photoadsorbed dioxygen have been used as a means of showing the effects of modifications upon the density of electrons available at the surface of a semiconductor oxide [3]. See Figure 1.

To our knowledge, there exists only one report dealing with photodesorption or photoadsorption in liquid phase : in that case the fluorescence of dissolved perylene was used as an indirect probe of the concentration of dissolved dioxygen in a suspension of TiO_2 in ethanol [5].

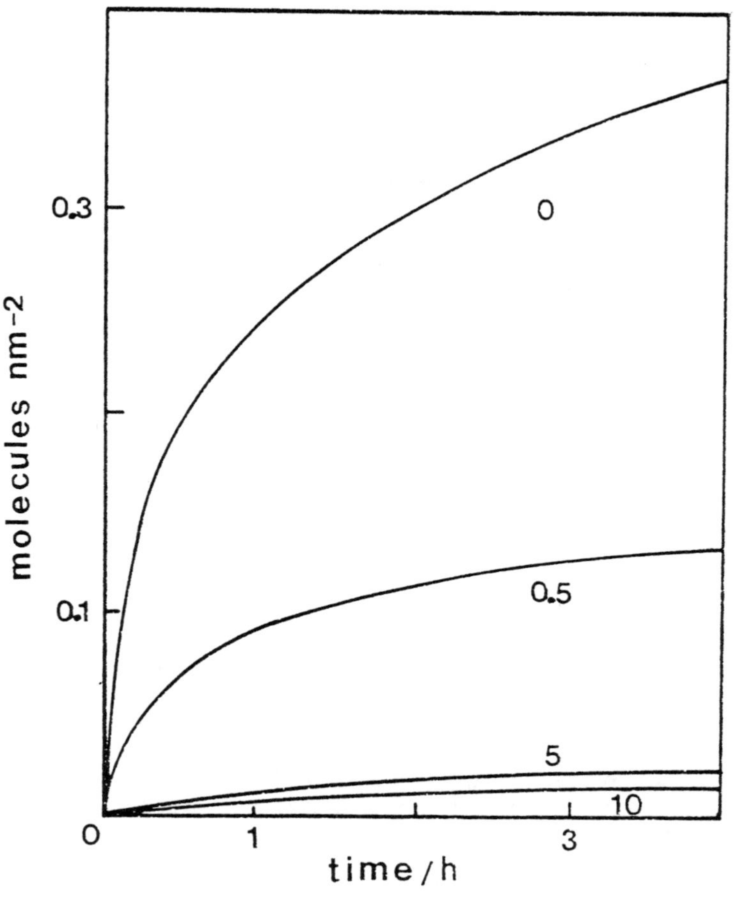

Figure 1. Dioxygen amounts adsorbed per surface unit of titanium dioxide (Degussa P-25) as a function of illumination time (after equilibration in the dark). Conditions : initial pressure 0.266 kPa ; $300 < \lambda/nm < 400$ (Corning 7.60 filter) ; radiant flux ca. 3.75 mW cm^{-2} ; Pt contents (in wt %) as indicated ; sample pretreatment : evacuation at 423 K for 15 h [Reprinted with permission from J. Phys. Chem. (1984), 86, 5210-5215, H. Courbon, J.-M. Herrmann and P. Pichat, Copyright 1984, American Chemical Society].

2.3. LIGHT-INDUCED OXYGEN LABILITY OF SEMICONDUCTOR OXIDES

The effect of band-gap illumination on the lability of oxygen atoms in the case of semiconductor oxides is best illustrated by studies of oxygen isotope exchange between gaseous dioxygen and adsorbed/surface oxygen atoms.

Band-gap illumination results in the removal of valence-band electrons from their normal location on the 2p orbitals of oxygen anions and thereby weakens the binding energy between these anions and the metal cations of the oxide. Added to the effect of illumination upon the equilibrium between negatively-charged adsorbed oxygen species and gaseous dioxygen, this phenomenon explains the light-induced lability of adsorbed/surface oxygen.

Indeed, for a number of semiconductor oxides, oxygen isotope exchange involving adsorbed/surface oxygen atoms has been found to occur at room temperature [2,3,6], whereas, in the dark, high temperatures are required (Figure 2). The rate of isotopic exchange is even faster with $N^{18}0$ over UV-illuminated titanium dioxide (N_2 and partially labelled N_20 are formed simultaneously) [6].

Light-induced oxygen lability is an important phenomenon which should be remembered when studying photocatalytic reactions involving oxygen. In fact, correlations have been observed between this phenomenon and photocatalytic oxidations over various semiconductor oxides. For example, the same orders of activities have been found for oxygen isotope exchange oxidations under UV-illumination over several oxides or over various specimens of the same oxide. More directly, the oxygen isotope exchange over titanium dioxide has been shown to be inhibited by a hydrocarbon, such as isobutane, within the duration needed to oxidise this compound [2,6].

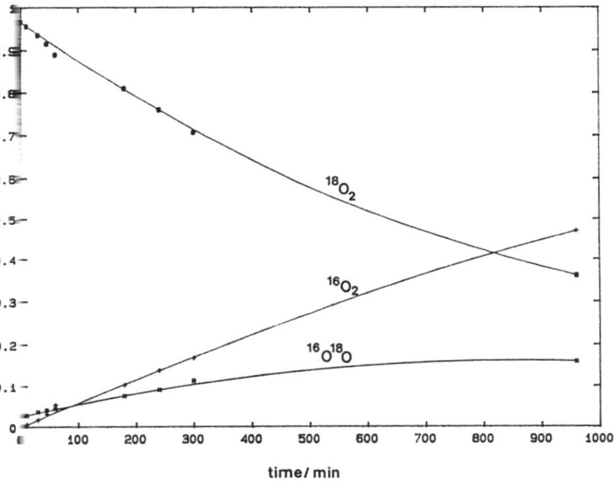

Figure 2. Changes in the isotopic fractions of gaseous dioxygen contacted with $KNb0_3$ (surface area 0.4 m^2 g^{-1}) as a function of illumination time. The points refer to the experimental results, the lines correspond to a kinetic model. The simultaneous growths of the $^{16}0_2$ and $^{16}0\,^{18}0$ fractions show the existence of two types of heteroexchange. Conditions : room temperature ; initial pressure ca. 6 Pa ; $\lambda \geq 290$ nm ; radiant flux ca. 3.7 mW cm^{-2} ; sample pretreatment : calcination in 13.3 k Pa dioxygen at 723 K for 3 h and subsequent evacuation at 293 K for 1 h. [G. Métrat, H. Courbon and P. Pichat, unpublished results].

2.4. CHEMICAL KINETICS AND INFORMATION ON ADSORPTION

From simple measurements of the rate of a photocatalytic reaction as a function of the concentration of a given reactant or product, valuable information can be derived in principle. For example, these measurements should allow one to know whether the active species of an adsorbed reactant are dissociated or not, whether the various reactants are adsorbed on the same surface sites or on different sites, and whether a given product inhibits the reaction by adsorbing on the same sites as one of the reactants. However, that implies referring to kinetic models. The Langmuir-Hinshelwood model, which indicates that the reaction takes place between both reactants at their equilibrium of adsorption, has often been used to interpret kinetic results of photocatalytic reactions in gaseous or liquid phase [6]. For example, the steady-state photocatalytic rates of hydrogen peroxide decomposition over titanium dioxide (calculated by substracting the photochemical decomposition) were numerically fitted to this model by iterative variations of the reaction rate constant, the adsorption constant and a parameter γ which accounts for a possible decrease in the initial concentration C_0 occurring before the illumination was started ($C = C_0 - \gamma$). This parameter was found to be very important for the fitting [7].A contribution of the Eley-Rideal mechanism (reaction between one non-adsorbed and one adsorbed reactant) has sometimes been proposed (Figure 3) [8].

Very recent considerations [9] have nonetheless shown that conclusions should be drawn with caution from the kinetic results. The example chosen was that of the degradation of an organic pollutant in an aqueous suspension of a semiconductor. It was assumed that 0H° radicals formed at the surface of the solid were the active species and

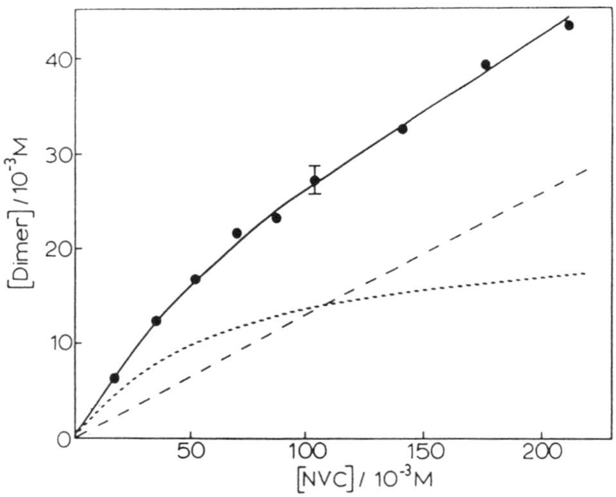

Figure 3. Effect of the concentration of N-vinylcarbazole (NVC) on the rates of dimer formation (filled circles). The overall rates are separated into Langmuir-Hinshelwood (...) and Eley-Rideal (---.) components. The cadmium sulphide sample was illuminated at $\lambda > 430$ nm for 30 min with a 150-W xenon lamp. [Reprinted with permission from Tetrahedron, 42, 6277 (1986), H. Al-Ekabi and P. de Mayo, Copyright 1986, Pergamon Journals Ltd.].

the question was whether these radicals predominantly react in the adsorbed phase or in the solution at some distance of the solid surface. Four possibilities were considered : the Langmuir-Hinshelwood model (equation 4), the Eley-Rideal model (with either the pollutant or the 0H° radicals reacting when adsorbed ; equations (5,6) and finally the case where the organic compound R and the active species react in the homogeneous phase (equation 7) :

$$0\overset{\circ}{H}_{s} + R_{ads} \rightarrow P_{ads} \tag{4}$$

$$0\overset{\circ}{H}_{aq} + R_{ads} \rightarrow P_{ads} \tag{5}$$

$$0\overset{\circ}{H}_{s} + R_{aq} \rightarrow P_{aq} \tag{6}$$

$$0\overset{\circ}{H}_{aq} + R_{aq} \rightarrow P_{aq} \tag{7}$$

The authors demonstrated that in any case the expression of the rate r is

$$r = \frac{k_{obs}\ \kappa\ C}{1 + \kappa\ C + \kappa_i C_i} \tag{8}$$

where k_{obs} is the observed reaction rate constant, C is the concentration, and the subscript i indicates an intermediate. This equation has the form expected for the Langmuir-Hinshelwood mechanism if κ is the adsorption constant. But if κ is regarded as a mere kinetic parameter, the three other mechanisms can be considered as valid [9]. In other words, the kinetic experiments do not allow one to discriminate between the possibilities symbolised by equations (4-7). In line with these considerations, two I.P.S.-8 posters [10-12], dealing with the degradation of organic pollutants in aqueous titanium dioxide suspensions, raise the problem that the experimental results are difficult to interpret if some reaction steps are not supposed to proceed in the aqueous phase. This issue is a very important one, not only for fundamental reasons, but above all to implement the photocatalytic method in the degradation of water organic pollutants.

From the uncertainties on the conclusions that can be drawn in some cases from the kinetic measurements, it must not be deduced that the adsorption modes of a given molecule cannot play a major role in photocatalytic reactions. A recent example concerning the oxidation of 1,4-pentanediol, using titanium dioxide as main photocatalyst and acetonitrile as solvent, illustrates that role [13]. For durations corresponding to the almost complete consumption of the diol (0.02M), 4-hydroxypentanal was formed in amounts exceeding those of 4-oxopentanol by a great factor (up to ca. 7). This high selectivity was found to correlate with the ratio of ca. 5.6 of the adsorption constants on the same photocatalyst, using the same solvent in the dark, for 1-pentanol and 2-pentanol. This ratio shows that adsorption effects are very likely to be at the origin of the selectivity observed. The adsorption bond of 1,4-pentanediol with titanium dioxide preferentially involves the region of the molecule containing the primary alcohol function which accordingly undergoes the more rapid oxidation [13].

2.5. SPECTROSCOPIC STUDIES AND HETEROGENEOUS PHOTOCATALYSIS

Spectroscopic methods are required to determine the nature of the adsorbed species and surface sites. In fact, they have not yet been used or developed in the field of heterogeneous photocatalysis as fully as might be desired. Most often they have been employed in the absence of illumination. Even in that case, the information obtained may allow one to say which of the adsorbed species is involved in photocatalysis. A monodentate acetate species was thus considered as responsible for the decomposition of ethanoic acid over UV-illuminated $MgO\text{-}TiO_2$ samples [14]. Very recently, infrared spectra of 2-nitrophenol adsorbed on titanium dioxide were scanned after various durations of *in situ* UV-illumination in the presence of dioxygen [15]. This type of experiments can bring useful knowledge on the photodegradation pathways of organic pollutants, despite the fact that IR spectroscopy cannot be used for studies in aqueous media.

Laser-flash excitation of semiconductors coupled with fast optical analysis of the transmitted light has been very precious to study colloids. For powders, information can be derived from the diffusely reflected light. Results obtained with alumina onto which aromatic hydrocarbons had been adsorbed were reported in 1981 [16]. Progress in the time scale from milliseconds to nanoseconds has then considerably increased the capabilities of this technique [17]. The difficulty resides in the kinetic analysis of the transient signals and in the spectral interpretation, which limits the systems that can be investigated. This book contains a chapter by M.A. Fox and collab. [18] presenting such an analysis and showing that diffuse-reflectance flash photolysis can provide information as to whether some organic and inorganic molecules are oxidised in aqueous TiO_2 suspensions by direct electron transfer. Unlike IR spectroscopy, this technique allows the study of aqueous solutions.

3. Special preparations of particulate photocatalysts

3.1. PARTICULATES WITH CONTROLLED TEXTURE

One of the objectives of the preparation of photocatalyts is to obtain particles with controlled sizes for two purposes : one is academic, a better understanding of heterogeneous photocatalysis is expected from better-defined solids ; the other one is related to the use of particles that can be more easily separated from a liquid medium for applications such as water purification.

In Lyon, a flame reactor has been frequently used for the preparation of pure or mixed oxides in the form of nonporous particles with controlled morphologies (spheres, polyhedra) [19]. The absence of porosity warrants the homogeneous illumination of the particles. The morphologies and the particle sizes (usually with a narrow distribution) can be mastered by adjusting the temperature, the flow rates (H_2 and O_2) and the concentration(s) of the compound employed to generate the oxide. Figure 4 shows an example of an iron oxide sample thus prepared.Recently, MoO_3 [20] and WO_3 with surface area comprised between 20 and 90 $m^2\ g^{-1}$ have thus been prepared. One of the MoO_3 samples has successfully been used under UV-illumination at room temperature to prepare the $H_{0.34}\ MoO_3$ bronze by use of an alcohol as a hydrogen source [21].

At I.P.S.-8, the preparation of fine, spherical and uniform ZnO (doped or undoped) particles by a somewhat similar method has been presented [22]. The powders are obtained by pyrolysis of a spray. The crucial point is the control of the size distribution of the droplets by the specially-designed nozzle atomizer. A sol-gel method has been implemented in a manner that allows one to obtain TiO_2 spheres, ca. 1 mm in diameter, having a surface area of 84 m^2 g^{-1} owing to their porosity [23]. These big particles minimise filtration problems.

3.2. PHOTOCATALYSTS WITH HIGHER STABILITY

Another problem which new types of preparations attempt to answer is the production of stable forms of photocatalysts that are known to corrode in water under band-gap illumination. At I.P.S.-8, it was reported that the apparently simple fusing of sodium nitrate with ruthenium trichloride at 773 K for >30 minutes gives rise to a form of ruthenium (IV) oxide (Adams ruthenium) with not only a great surface area, but also a high degree of crystallinity and a low water content [24]. These two latter properties increase the stability of the oxide. On the other hand, it would be interesting to render iron oxides photocatalytically active because of their low price and their absorption in the near-visible region. An attempt was presented at I.P.S.-8. The preparation is based on the formation of beta-iron (III)-oxide-hydroxide containing foreign ions [25].

TiO_2 P-25 produced by Degussa (Germany) is certainly the specimen of titanium dioxide that is the most commonly employed in heterogeneous photocatalysis. That is due to its great photocatalytic activity with respect to those of other commercial samples. According to a poster presented at I.P.S.-8 the structural and allotropic heterogeneities of Degussa P-25, which has been investigated in detail, might well explain its activity [26]. The qualitative interpretation is based on potential changes occurring in zones where anatase and rutile are in contact.

3.3. SEMICONDUCTORS SUPPORTED ON OR INCORPORATED IN POWDER INORGANIC MATERIALS

Deposition of semiconductors on photoinsensitive inorganic supports has been explored as a method of improving the possibilities of photocatalysts. Various objectives can thus be sought for : (i) production of smaller semiconductor particles as a result of the dispersion, (ii) inclusion of these particles in a geometrically restricted environment such as that formed by zeolites and clays, and (iii) usage of a structuring matrix to form epitaxial semiconductors. Also, effects can be expected from the acidic/basic or hydrophobic/hydrophylic properties of the support.

Such methods of preparation have principally been used for titanium dioxide. Recent reports deals with TiO_2 on SiO_2 [27], and TiO_2 [28] or $Fe_2 O_3$ [29] incorporated in montmorillonite. These solids were prepared by impregnating the support either with a titanium (or iron) compound [27,29] or with a titania sol [28]. Chemical vapour deposition, a more unusual technique in this field, has been applied to form thin films of TiO_2 particles on plates of fused silica [30]. This technique could also be envisaged to attach titanium dioxide to suitable supporting materials in view of water decontamination (see R.W. Matthews's chapter in this book). Both types of preparation allow one to obtain tiny particles whose absorption is displaced to shorter wavelengths as compared with larger particles (the so-called "blue shift effect"). Accordingly the reducing power of these particles is increased, which can be useful to reduce protons for instance.

To our knowledge, no study has determined whether or not the selectivity can be modified by the dispersion and/or support effects. Moreover, although calculations have

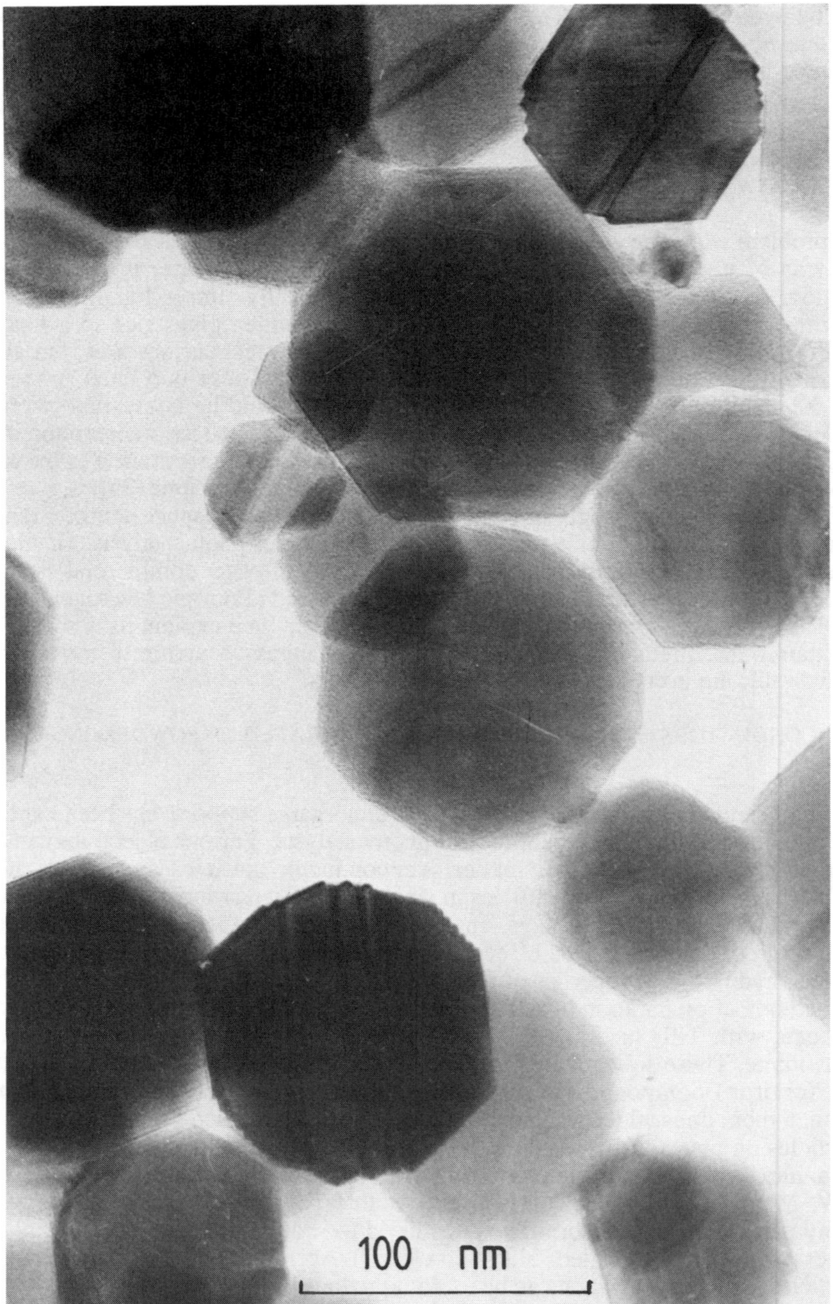

Figure 4. Transmission elecron micrograph of a γ-Fe_2O_3 sample prepared by the flame reactor method.

shown that the activity per mass unit of semiconductor is improved as expected because of the dispersion, the choice of the reference unsupported material obviously affects any estimation of the improvement. Besides, the photons that are absorbed in the photoinsensitive support decrease the overall efficiency and these losses are not indicated generally. Despite this latter drawback, these supported semiconductors might present various advantages as pointed out at the beginning of this section. These potential gains have not been fully explored and, in particular, no research in this field was presented at I.P.S.-8.

4. Examples of new usages of physical techniques to study powder photocatalysts

4.1. TIME-RESOLVED MICROWAVE CONDUCTIVITY (TRMC)

This contactless method, developed by J.M. Warman and M.P. de Haas [31,32] employs nanosecond pulses of high energy (3 MeV electron) from a Van de Graff accelerator. Figure 5 shows the sample-holder.Changes in the conductivity of the semiconductor that are due to the electronic charge carriers generated by these pulses are monitored by TRMC in the frequency range 26.5 to 38 GHz. The sum of the intrinsic mobilities of the electron and holes are estimated from the magnitude of the end-of-pulse conductivity.

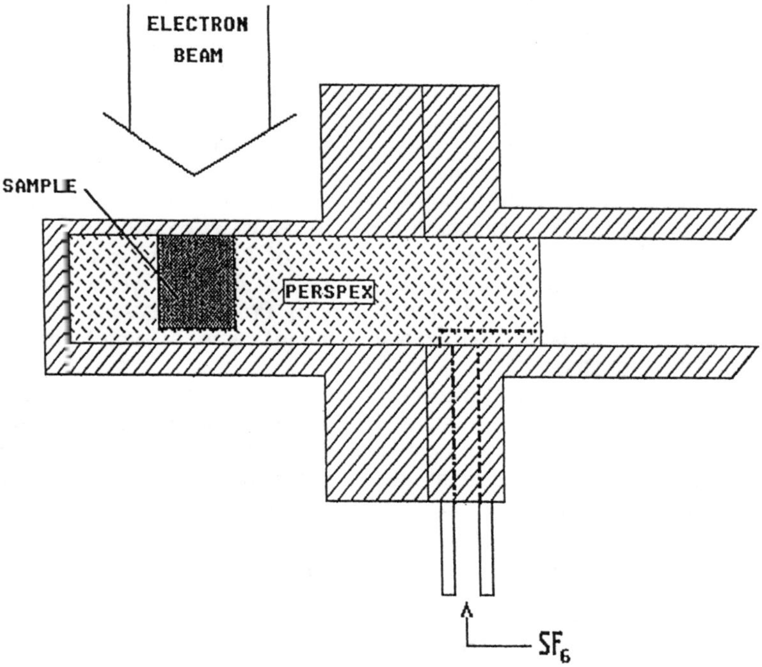

Figure 5. Perspex block used for the time-resolved microwave conductivity measurements. The cavity, into which the sample powder was compressed by hand, was cylindrical (1.7 mm in radius, 3.0 mm deep). The metal waveguide sections are shown diagonally shaded. [Reprinted with permission from ref. 33, Copyright 1990].

The sensitivity of the method was demonstrated by investigating the TRMC of an alumina sample [33]. A conductivity value of 3 x 10^{-8} m^2/Vs was determined for this insulating material. Although this is an extremely small mobility by semiconductor standards (*vide infra*), it corresponds to a diffusion time of a charge carrier from the bulk to the surface of a few tens of nanoseconds for an alumina powder whose average particle diameter is 20 nm. This explains why adsorption of oxygen [34] and decomposition of methanol [35] have been observed upon illumination, in the short-wavelength UV-region, of high-surface-area alumina.

For several nonporous anatase samples prepared in a flame reactor (*vide supra*) and formed by particles with a narrow size distribution, the mobility values were found to increase with increasing mean particle sizes [33]. Thus values of 5 x 10^{-6} m^2/Vs and 140 x 10^{-6} m^2/Vs were determined for average radii of 5 nm and 47 nm, respectively. This effect has been thought to be related to shallow trapping of electrons at localised surface sites whose density becomes greater as the particle size decreases because of the expected higher number of irregularities. Photoacoustic spectra (*vide infra*) of similar anatase samples have recently shown that the photochromism, attributed to Ti^{3+} localised sites, also increases with decreasing particle size, however up to a maximum value reached for an average radius of ca. 25 nm [36].

While chemical studies probe particulate semiconductors basically from the outside, TRMC may be seen as giving an electron's eye view of changes at the interface from within the particles. Two effects, that of the adsorption of an alcohol and that of platinum particles on the surface of titanium dioxide, are briefly presented here to illustrate the capabilities of TRMC.

The absolute magnitude of the end-of-pulse conductivity is apparently not significantly influenced by the introduction of 2-propanol in the measurement cell, at least for low total radiation doses, although a quantitative analysis is made difficult because of the background dielectric loss of the alcohol [33]. But the conductivity decay is markedly slowed down in the presence of 2-propanol, i.e. the neutralisation of trapped electrons is delayed (Figure 6). This is understandable since alcohols are capable of trapping holes either directly or indirectly. As expected, the effect of 2-propanol is suppressed when using a titanium dioxide pigment that is coated by a silica-alumina layer which hinders the access of the alcohol to the semiconductor (Figure 6). Provided its sensitivity proves to be sufficient, TRMC could be a sensitive test of the efficacity of such layers which are widely used in paint industry.

TRMC measurements [33] have also been performed for a series of Degussa P-25 samples that have been impregnated with a H_2PtCl_6 solution and thermally reduced in dihydrogen in such a way that platinum particles of an average diameter of ca. 2nm were formed with a narrow size distribution almost independent of the platinum content [37]. The end-of-pulse conductivity is decreased in the presence of the platinum particles as compared with a naked sample that has similarly been treated except that H_2PtCl_6 was replaced by HCl. This result is qualitatively in agreement with earlier direct-current photoconductance measurements carried out under vacuum for the same types of samples [38]. It confirms the attraction of electrons to the platinum particles and further demonstrates the interest of TRMC to determine the effect of surface modifications on the fate of charge carriers.

Finally, to exemplify the potential capability of the TRMC experimental approach to observe the influence of bulk lattice impurities, measurements were performed on a titanium dioxide specimen which was homogeneously doped with chromium III cations (the flame reactor allows such a regular distribution). The conductivity found in the

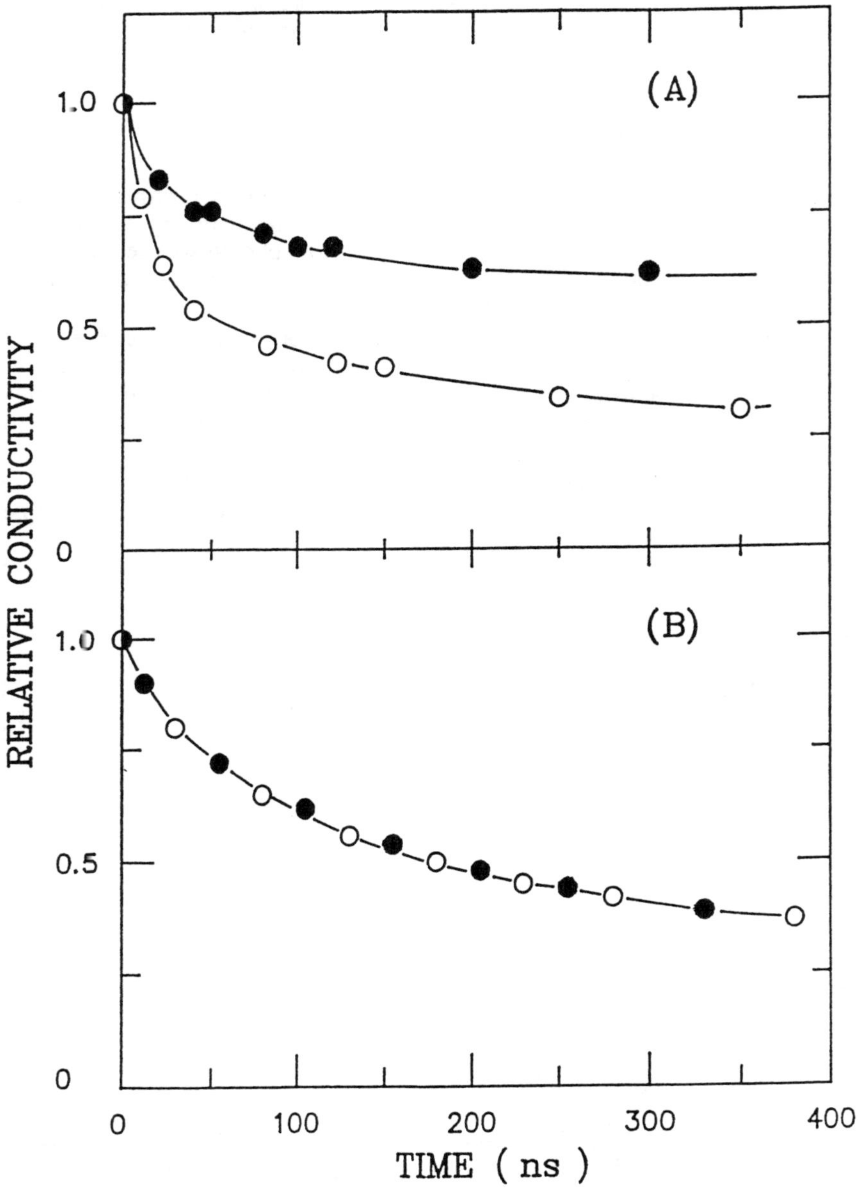

Figure 6. The effect of adding 2-propanol (filled circles) on the radiation induced conductivity transients for A) an uncoated rutile sample from Tioxide and B) a Tioxide sample silica-alumina coated ("R-TC4"). The signals have been normalised to the same end-of-pulse heights in both cases [Reprinted with permission from ref. 33, Copyright 1990].

absence of substitutional cations was completely suppressed by the chromium doping [33]. That was interpreted in terms of the extremely rapid diffusion of the charge carriers to the bulk impurity centres which would only be a few nanometres apart at the high concentration used (0.85 ion %).

4.2. PHOTOACOUSTIC SPECTROSCOPY

Photoacoustic spectroscopy is also a technique that can be very informative. However it is limited to semiconductors that gives rise to a photoacoustic signal upon band-gap excitation. Titanium dioxide belongs to this category. Its photoresponse has the form of a quasi continuum absorption with a shallow maximum at 1100 nm. It has been attributed to small-polaron excitation (probably bulk and surface Ti^{3+} ions) [39].

To show the possibilities of this experimental tool, photoacoustic spectra were recorded in a dinitrogen/water vapour atmosphere (dinitrogen was used to remove dioxygen that suppresses the photochromism ; the presence of water was found to enhance the response) for Cr^{3+}-doped TiO_2 and Pt/TiO_2 samples [40] identical or similar to those also examined by TRMC. The absence of photoresponse in the Cr^{3+} doped sample can readily be interpreted by efficient charge recombination at the substituted cations in accord with DC photoconductance and photocatalytic results [41]. For the Pt/TiO_2 samples, the

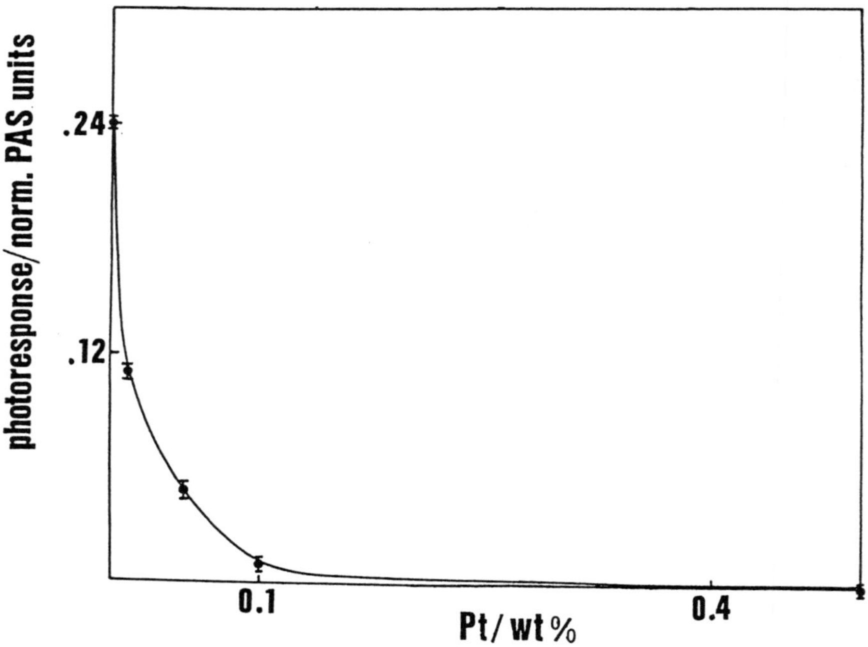

Figure 7. Photoresponse (normalised at 900 nm) of Degussa P-25 titanium dioxide in N_2-H_2O vapour atmosphere against Pt content. The platinum particles were formed by impregnation with H_2PtCl_6 and subsequent reduction by H_2 [Reprinted with permission from ref. 40, Copyright 1988, CNRS-Gauthiers-Villars].

photoacoustic signal of titanium dioxide decreases progressively as a function of the Pt content and is suppressed for a content as low as 0.5 wt % corresponding to one Pt crystallite per TiO_2 particle (mean diameter 30 nm) (Figure 7). This tends to show that the interparticle charge transfer is remarkably efficient. On the other hand, the photochromism is more sensitive to the presence of Pt particles than TRMC is [33]. This difference in sensitivity cannot easily be explained because of very distinct experimental conditions. However, qualitatively, results obtained by both techniques are in agreement.

The photochromism of titanium dioxide was also used to investigate the possible transfer of electrons between this material and another powder semiconductor photosensitive in the visible region, viz. CdS [42]. Upon illumination at wavelengths > 450 nm, no response was detected for TiO_2 (absence of photosensitisation) or for CdS (absence of photochromism), but for mechanical mixtures of both materials a photoacoustic spectrum was recorded (Figure 8). From its form and behaviour, it was inferred that it was due to Ti^{3+} ions. In other words, at least in the case chosen, photoacoustic spectroscopy is able to bring direct evidence for interparticle electron transfer from band-gap excited CdS.

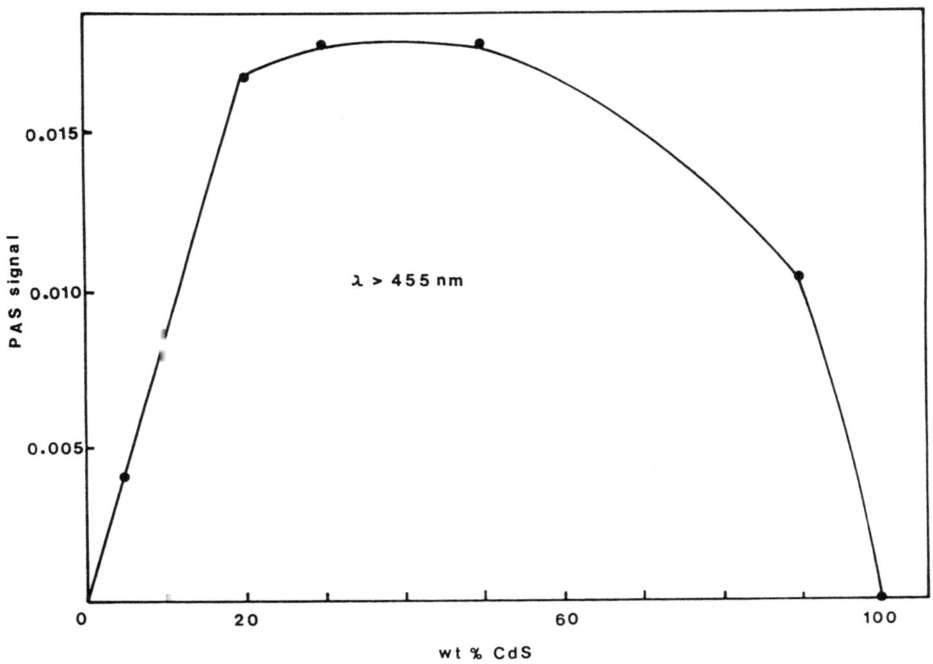

Figure 8. Photoresponses (normalised at 900 nm) of mechanical mixtures of titanium dioxide (Degussa P-25) and cadmium sulphide (Fluka, yellow, purissimum) in N_2 - H_2O vapour atmosphere [42].

5. Conclusion

Surface phenomena obviously play a major role in photocatalysis on semiconductors even though recent considerations and experimental facts may imply that some active species created on the solid surface react in fact in homogeneous medium rather than in the adsorbed phase. Therefore research directed to a better characterisation of these surface phenomena is still needed as are required studies aiming at a better appraisal of the effects of photosorption and light-induced mobility of surface atoms upon photocatalytic reactions. Preparation of specially designed photocatalysts is obviously necessary both for basic and practical objectives. It does not generally involve an expensive equipment. However, securing the material possessing the expected properties is not an easy task and can be time-consuming as often are the characterisations. Nevertheless we think that progress can be achieved in that direction. Recent reports have shown that novel applications of physical techniques to powder semiconductors bring invaluable information - and, at the same time, as it is frequently the case, raise new questions. A more intense use of these techniques under relevant conditions requires an increased funding of research in heterogeneous photocatalysis. Such an increase can be expected as a consequence of the growing interest due to the involvement of photocatalysis and related phenomena in the chemical transformation of various substances in our environment.

Acknowledgements.
The author is indebted to his CNRS collaborators whose names appear in the references, as well as to visiting scientists (Drs. B. Jenny and J.G. Highfield) for their contributions. He expresses his gratitude to Drs. M.A. Fox (Austin), J.M. Warman and M.P. de Haas (Delft) for fruitful collaborations.

References

1. Fujishima, A., Honda, K. and Kikuchi, S. (1969) J. Chem. Soc. Jpn. 72, 108-113.
2. Pichat, P. (1986) in Pelizzetti, E. and Serpone, N. (eds.), Homogeneous and Heterogeneous Photocatalysis, D. Reidel Publ. Co, pp. 533-553.
3. Courbon, H., Herrmann, J.-M. and Pichat, P. (1984), J. Phys. Chem. 88, 5210-5214.
4. Munuera, G., Rives-Arnau, V. and Saucedo, A. (1979) J. Chem. Soc., Faraday Trans. I 75, 736-746.
5. Oelkrug, D., Flemming, W., Füllemann, R., Günther, R., Honnen, W., Krabichler, G., Schäfer, M. and Uhl, S. (1986) Pure Appl. Chem. 58, 1207-1218.
6. See : Pichat, P. and Herrmann, J.-M. (1989) in Serpone, N. and Pelizzetti, E. (eds.) Photocatalysis, John Wiley, pp. 217-250 and references therein.
7. Jenny, B. and Pichat, P. (in press) Langmuir.
8. Al-Ekabi, H. and Serpone, N. (1989) in Serpone, N. and Pelizzetti, E. (eds.) Photocatalysis, John Wiley, pp. 457-488.
9. Turchi, C.S. and Ollis, D.F. (1990) J. Catal. 122, 178-192.
10. Book of Abstracts, 8 th Int. Conf. on Photochemical Conversion and Storage of Solar Energy (I.P.S.-8), (1990) Palermo, Italy.
11. Cunningham, J. and Al-Sayyed, G. in ref. 9., p. 210.
12. Percherancier, J.-P., Perichet, G., Meallier, P. and Pouyet, B. in ref. 9, p. 247.
13. Fox, M.A., Ogawa, H. and Pichat, P. (1989) J. Org. Chem. 54, 3847-3852.

14. Palmisano, L., Schiavello, M., Sclafani, A., Coluccia, S. and Marchese, L. (1988) New J. Chem. 12, 847-852.
15. Augugliaro, V., Palmisano, L., Schiavello, M., Sclafani, A., Marchese, L., Martra, G. and Miano, F. (in press) J. Phys. Chem. and ref. 9, p. 243.
16. Kessler, R.W., Oelkrug, D. and Wilkinson, F. (1981) Appl. Spectrosc. 36, 673.
17. Wilkinson, F. and Willsher, C.J. (1987) Tetrahedron 43, 1197-1209.
18. Fox, M.A., Draper, R.B., Dulay, M. and O'Shea, K., this book ; Draper, R.B. and Fox, M.A. ; (1990) J. Phys. Chem. 94, 4628-4634 ; (in press) Langmuir.
19. See for example : Formenti, M., Juillet, F. Meriaudeau, P., Teichner, S.J. and Vergnon, P. (1972) J. Colloid Interf. Sci. 39, 79-89.
20. Hoang-Van, C., Zegaoui O., Pommier, B. and Pichat, P. (in press) in Delmon, B. et al. (eds.), Preparation of Catalysts V, Elsevier.
21. Pichat, P., Mozzanega, M.-N. and Hoang-Van, C. (1988) J. Phys. Chem. 92, 467-470.
22. Gonzalez-Carreno, T., Bickley, R.I. and Serna, C.J. in ref. 9, p. 222.
23. Pelizzetti, E., Minero, C., Carlin, V., Borgarello, E., Miano, F. and Serpone, N. in ref. 9, p. 261.
24. Mills, A. and Davies, H. in ref. 9, p. 220.
25. Manassen, J. and Hirsch, J. in ref. 9 p. 199.
26. Bickley, R.I., Gonzalez-Carreno, T., Lees, J.S., Palmisano, L. and Tilley, J.D. in ref. 9, p. 197.
27. Domen, K., Sakata, Y., Kudo, A., Maruya, K. and Onishi, T. (1988) Bull. Chem. Soc. Jpn. 61, 359-362.
28. Yoneyama, H., Haga, S. and Yamanaka, S. (1989) J. Phys. Chem. 93, 4833-4837.
29. Miyoshi, H., Yoneyama, H. (1989) J. Chem. Soc., Faraday Trans. I 85, 1873-1880.
30. Sato, S. Sobczynski, A., White, J.M., Bard, A.J., Campion, A., Fox, M.A., Mallouk, T.E. and Webber, S.E. (1989) J. Photochem. Photobiol. A : Chem. 50, 283-290.
31. Warman, J.M. and de Haas, M.P. (in press) in Tabata Y. (ed.) Pulse Radiolysis of Irradiated Systems, CRC Press Inc.
32. Warman, J.M., de Haas, M.P. and Wentinck, H.M. (1989) Radiat. Phys. Chem. 34, 581.
33. Warman, J.M., de Haas, M.P., Pichat, P., Koster, T.P.M., van der Zouwen-Assink, E.A., Mackor, A. and Cooper, R. (in press) Radiat. Phys. Chem.
34. Balard, H., Mansour, A., Papirer, E. and Pichat, P. (1985) J. Chim. Phys. 82, 1051-1058.
35. Lafrance, C.-P., Kaliaguine, S., Roberge, P.C. and Pichat, P. (1984) in Kaliaguine, S. and Mahay, A. (eds.) Catalysis on the Energy Scene, pp. 309-317.
36. Highfield, J.G. and Pichat, P., unpublished results.
37. Pichat, P., Mozzanega, M.-N., Disdier, J. and Herrmann, J.-M. (1982) Nouveau J. Chim. 6, 559-564.
38. Disdier, J., Herrmann, J.-M. and Pichat, P. (1983) J. Chem. Soc. Faraday Trans I, 79, 651-660.
39. Highfield, J.G. and Grätzel, M. (1988) J. Phys. Chem. 92, 464-467.
40. Highfield, J.G. and Pichat, P. (1989) New J. Chem. 13, 61-66.
41. Herrmann, J.-M., Disdier, J. and Pichat, P. (1984) Chem. Phys. Lett. 108, 618-622.
42. Highfield, J.G., Grätzel, M. and Pichat, P., unpublished results.

Photophysics of small semiconductor clusters: the effects of quantum confinement and dielectric confinement

Ying Wang
Central Research and Development Department
E. I. DuPont de Nemours & Co.
P. O. Box 80356
Wilmington, DE 19880-0356

Abstract. Small semiconductor cluster in the size regime of tens to hundreds of Å represents a fascinating new class of materials. The advancement in cluster syntheses now allows us to study their size-dependent properties systematically. In this paper we report recent progress towards the understanding of the quantum size effect and the dielectric confinement effect. The effects of surface-trapped electron-hole pair on the photophysical and nonlinear optical properties of CdS clusters as well as the exciton luminescent properties of ammonia-passivated CdS clusters will also be discussed.

1. Introduction:

Recent success in synthesizing small semiconductor clusters has opened the doorway for systematic studies of clusters in the condensed phase[1-3]. Because of the delocalized nature of the electrons and holes in the semiconductors, the cluster properties are sensitive to size. This is the so-called quantum size effect, which becomes important as the diameter of the particle approaches to that of the exciton. It is manifested by a blue-shift in the band gap and an enhancement in the volume-normalized exciton oscillator strength. Furthermore, because of the boundary established by the different refractive indices of the semiconductors and the surrounding media, the interaction between the light field and the particle becomes size-dependent. The field intensity near, at, and inside the particle surface can be enhanced considerably compared to the incident intensity. This enhancement in local field should be important in the study of spectroscopy, photophysics and photochemistry of these semiconductor particles.

E. Pelizzetti and M. Schiavello (eds.), Photochemical Conversion and Storage of Solar Energy, 295–305.

Several review articles on this subject are available in the literature[1-3]. Here we report a few recent progress made in this area from our laboratories.

2. Quantum size effect- exciton energy vs. cluster size.

In our previous study of the quantum size effect on the optical band-gap of PbS clusters,[4] we have demonstrated experimentally that the simple particle-in-a-box model based on effective mass approximation cannot explain the data. Instead, we developed a tight-binding model which successfully reproduced the observed size-dependence. We have now extended the study to CdS. For II-VI clusters such as CdS, reliable data covering a large enough size range has been difficult to obtain. This is because the shift in the exciton energy becomes pronounce only for CdS clusters smaller than ~60Å, a region in which the necessary characterization by x-ray diffraction becomes difficult unless suitable samples are available. This problem is now alleviated with the recent advances in the syntheses of quantum-confined semiconductor clusters.[5-8]

Figure 1 shows the size-dependence of the exciton energy of CdS clusters ranging in diameters from 60 to 10 Å. Three types of samples have to be synthesized for this study: CdS clusters in Nafion[5,6], thiophenolate-capped CdS clusters[7], and 10-Å single size CdS clusters[8].

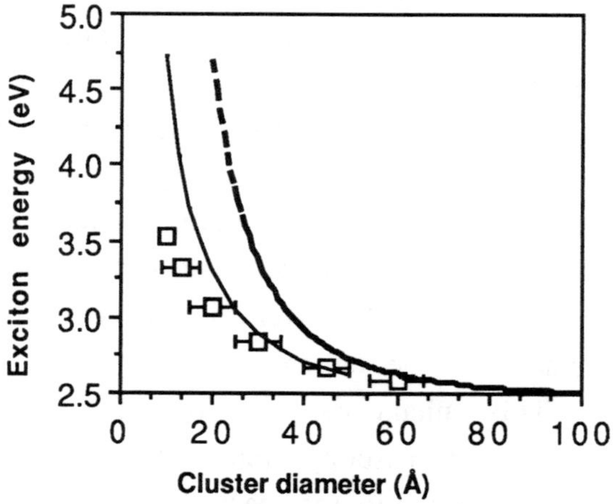

Figure 1. The size dependence of the exciton energy of CdS clusters (taken from reference 8)

In Figure 1, the dashed line represents the simplest three-dimensional confinement model based on the effective-mass approximation[9-11], which predicts the energy shift, ΔE, as

$$\Delta E = \frac{\hbar^2 \pi^2}{2 R^2}\left[\frac{1}{m_e} + \frac{1}{m_h}\right] - \frac{1.786\, e^2}{\varepsilon R} - 0.248\, E^*_{Ry} \qquad (1)$$

where R is cluster radius, m_e is the electron effective mass, m_h is the hole effective mass, ε is the dielectric constant, and E^*_{Ry} is the effective Rydberg energy, $e^4/2\varepsilon^2\hbar^2(m_e^{-1} + m_h^{-1})$. The first term in eq. (1) represents the kinetic energy[9], the second term the Coulomb energy[10] and the third term is a result of the correlation effect[11] The dashed line is calculated using m_e=0.19, m_h=0.53, and ε=5.5. Clearly, the calculation cannot reproduce the experimental results, indicating the failure of the effective-mass approximation.

The solid line in Figure 1 represents the results of a 13 parameters tight-binding calculation by Lippens et al[12]. The agreement with the experimental data is quite good for clusters larger than ~20Å. Below 20 Å, the tight-binding calculation starts to deviate from the experimental value. One possibility for the discrepancy between the calculation and the experiment is the change in the crystal structure for clusters smaller than ~20Å. From the x-ray data we found that there is a ~3% contraction in the (111) lattice constant for 15 Å CdS clusters[8]. Since the tight-binding parameters are optimized to the band structure of bulk sphalerite-type CdS, they are not suitable for clusters whose structure deviate from the bulk.

Based on this study on CdS and previous study on PbS[4], it is clear that one has to go beyond the effective mass approximation to obtain a quantitative description of the quantum size effect. We have shown that semi-empirical tight-binding method provides an effective tool if the cluster structure stays the same as that of the bulk.

3. The effects of surface-trapped electron-hole pairs.

A convenient way to study the effect of surface trapped e-h pairs on the absorption spectrum of a semiconductor cluster is to generate them with a short laser pulse and then probe the spectrum with a second pulse. We have performed such pump-probe experiments[5,6] and found that the exciton absorption is bleached during the presence of the trapped electron hole pairs (this is a third order nonlinear optical process) and recovers as the trapped electron-hole pair decays away. Furthermore, there is no appreciable change in the position of the exciton peak for timescales longer than tens of picoseconds. By studying the laser power dependence of the bleaching efficiency, we found that the bleaching saturates when there is about one trapped electron-hole pair per cluster[5,6], that is, one trapped electron-hole pair can bleach the exciton absorption of the whole cluster.

The efficient bleaching of the exciton absorption is due to the strong interaction between the exciton and the trapped e-h pair. Since the exciton diameter is comparable to or smaller than the cluster diameter, the exciton "senses" the presence of the trapped e-h pair immediately upon generation. Theoretical modelling confirms the experimental results.[6] A variational calculation shows that in the presence of a surface trapped e-h pair, the exciton energy shifts to the red by only ~50 meV but its oscillator strength is reduced by 90% for a 25Å-radius CdS cluster.[6] The basic physics can be easily understood by considering the overlap of the electron and hole wavefunction. In the absence of the trapped e-h pair the hole wavefunction is located at the center of the cluster and has good overlap with the electron wavefunction. By introducing trapped electron and hole to the cluster surface, the hole wavefunction is now localized by the trapped electron because of the heavier effective mass of the hole, while the electron wavefunction is still delocalized. This reduces the spatial overlap of the electron and hole wavefunction and thus the oscillator strength of the exciton.

This example illustrates the very large effect surface defects can have on the exciton oscillator strength, a possible reason why the exciton absorption peak is either weak or absent in many of the semiconductor clusters synthesized so far, in addition to the size dispersion problem.

4. Surface passivation of CdS clusters with ammonia

Typical semiconductor clusters as prepared contain surface defects. As a result, the initial photo-generated electron-hole pairs (excitons) get trapped very rapidly. The photoluminescence spectrum usually show a broad trap emission, red-shifted from the exciton absorption peak. Several methods have been reported for passivating the surfaces [13-15]. Recently, we developed a method of passivating the CdS cluster surfaces with ammonia[6]. As a result of this passivation procedure, the photoluminescence spectrum now exhibits two bands, a sharp high energy band with peak at 464 nm and a broad low energy band in the 500-750 nm region (Figure 2). The broad low energy luminescence is due to the surface trap states commonly observed for small CdS clusters. The sharp band-edge luminescence overlaps almost exactly with the exciton absorption peak at 462 nm. The Stokes-shift is only ~20 meV, corresponding to very shallow defects if they exist. Based on this almost mirror image relationship of the spectral peak positions, we attribute the sharp luminescence to the bound exciton luminescence.

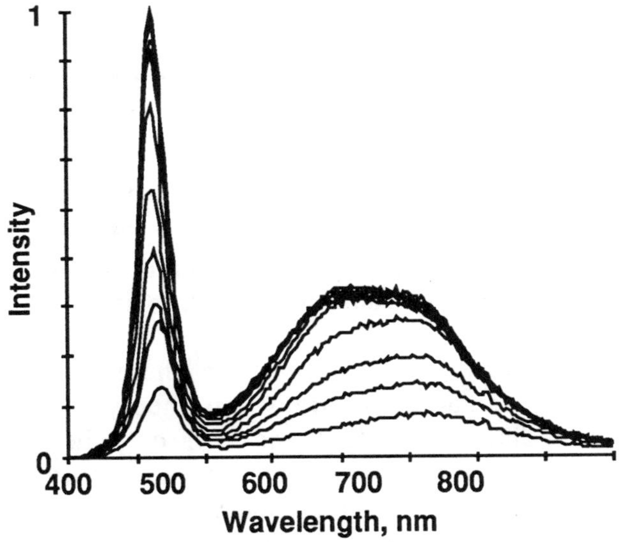

Figure 2. The temperature-dependent luminescence spectra of ~40-Å ammonia-passivated CdS clusters in Nafion® film.

The temperature dependence of the luminescence intensity (Figure 3) provides further insight into the mechanism of exciton relaxation dynamics. From Figure 3 we see that the bound exciton luminescence intensity at 464 nm is rapidly quenched after warming up to ~50 K, due to multiphonon-induced radiationless decay processes.

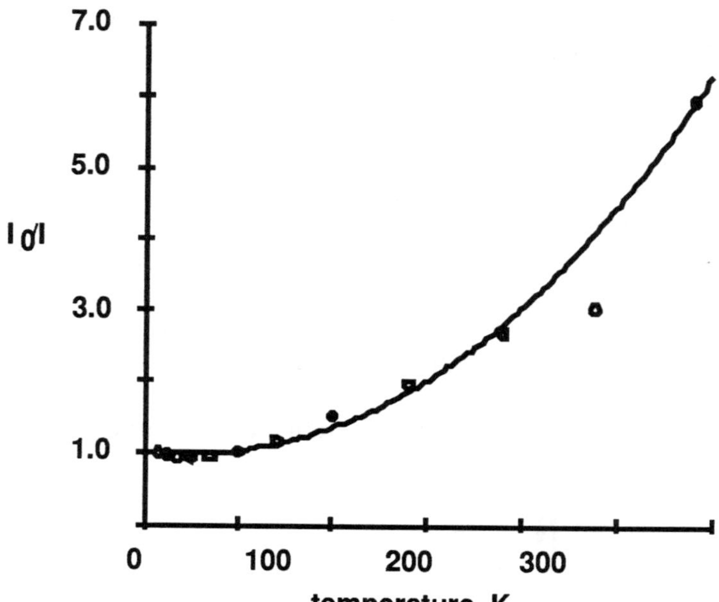

Figure 3. Temperature dependence of the relative luminescenc intensity at 462 nm. I_0 represents the luminescence intensity at 7.6 K. Solid curves represent theoretical fits based on multiphonon-induced radiationless transition theory.

A quantitative analysis was performed on these data based on the multiphonon-induced radiationless transition theory. Our data analysis[15] have been discussed before and will only be briefly described here. There are three parameters in the theory describing the temperature dependence of the nonradiative relaxation of an excited state stimulated by the excitation and emission of the mediating phonons: the effective mediating phonon frequency ($\hbar\omega$), the Franck-Condon displacement factor, G, and the electronic coupling parameter, C. When the radiative rate at 0 K is unknown, as in the present case, the data can still be fitted with three parameters but the last parameter, C'(=C/k_r), contains the radiative rate at 0 K and therefore has no physical meaning.[15]

Initially we found that the data in Figure 3, can be fitted with the following parameters: $\hbar\omega$=195 cm^{-1}, G=105, C'=9.8x10^{-38}, if we assume the bound exciton relaxes directly to the ground state (i.e. energy gap=464 nm). However, this large displacement factor of 105 is too large to be consistent with the observed spectral shift of ~20 meV, which suggests the real gap of the radiationless relaxation is much smaller than the apparent 21550 cm^{-1} (464 nm) value. To maintain a small G value, the data can only be fitted with a small energy gap. The best fit (solid line in Figure 3) are obtained with the following parameters: $\hbar\omega$=147 cm^{-1}, G=0.06, ΔE=501 cm^{-1}, and C'=1.0x10^{-38}. Two important conclusions can be drawn from this data analysis: (1) the radiationless relaxation of the bound exciton is mediated by the CdS lattice phonons, rather than by the high frequency host phonons as observed for trap states luminescence (6,15). (2) The exciton decays to a relatively shallow defect states, spaced ~500 cm^{-1} below, rather than directly to the ground state. The nature of this shallow trap is not clear at present.

The controlled modification of the cluster surfaces is important for the fundamental understanding of these clusters and their photophysical and photochemical properties. These data show initial success towards the realization of this important goal.

5. The dielectric confinement effect

Since the semiconductor clusters are always immersed in certain medium, such as polymers, glasses, solutions, or air, there exists a difference in index of refraction at the interface. When illuminated by light, the field intensity near, at, and inside the particle surface can be enhanced considerably compared to the incident intensity. This enhancement in local field may be considered as a result of the "dielectric confinement" effect and can be an important factor in the study of spectroscopy, photophysics and photochemistry of these semiconductor clusters and particles.

The size and wavelength dependence of the local field intensity on the surface of a dielectric sphere can be computed with the Lorenz-Mie theory. The Lorenz-Mie theory is usually used to calculate the far-field scattering efficiency, Q_{sca}, of a dielectric sphere.[16] Q_{sca} is defined as the ratio of the scattered intensity (at a distance far from the sphere) and the incident intensity per geometrical cross section, πa^2:

$$Q_{sca} = \lim_{R \gg a} \frac{R^2}{\pi a^2} \int_0^{2\pi} \int_0^{\pi} \vec{E}_s \bullet \vec{E}_s^* \sin\theta d\theta d\varnothing \qquad (2)$$

where a is the particle radius, $\vec{E}_s$ is the scattered electric field at distance R from the center of the sphere. Since experimentally the observation is always carried out at far-field, the integration is carried out at $R \gg a$.

The same theory can be used to calculate the field intensity at the surface of a sphere by evaluating the integral at R=a. Messinger et al[17] introduced a new parameter, Q_{nf}, which gives a measure of the local field intensity at the surface.

$$Q_{nf} = \frac{R^2}{\pi a^2} \int_0^{2\pi} \int_0^{\pi} \vec{E}_s \bullet \vec{E}_s^* \sin\theta d\theta d\varnothing \Big|_{R=a} \qquad (3)$$

In terms of the scattering coefficients,

$$Q_{nf} = 2\sum_{n=1}^{\infty} \{|a_n|^2[(n+1)|h_{n-1}^{(2)}(ka)|^2 + n|h_{n+1}^{(2)}(ka)|^2] + (2n+1)|b_n|^2|h_n^{(2)}(ka)|^2\} \qquad (4)$$

where $h_n^{(2)}$ is the Hankel function of the second kind. Equation (4) provides the basis for calculating the local field on the surface of a dielectric sphere.

We have performed this local field calculation on many semiconductors. Here we only show the results for TiO_2, a commonly studied photocatalyst. Figure 4 shows the dependence of Q_{nf} on the TiO_2 particle radius at three wavelengths: 300, 400, and 1060 nm. The refractive index of the medium is 1.33 (water).

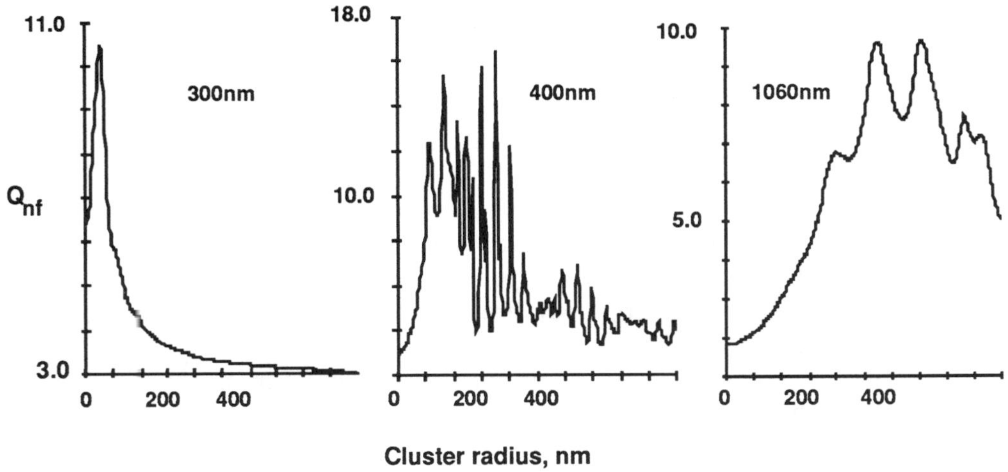

Figure 4. The near-field enhancement factor, Q_{nf}, plotted as a function of TiO_2 particle radius at light wavelengths of 300, 400, and 1060 nm. Refractive index of the surrounding medium is 1.33.

From figure 4 we see that the field intensity at the particle surface is enhanced compared to the incident intensity. For very small clusters the intensity enhancement factor is relatively small (2-3). As the cluster size increases, the enhancement factor increases monotonically in the Rayleigh size regime (< ~1000Å). For larger particles, oscillations in Q_{nf} occur. Many sharp peaks, corresponding to the condition when the light field is in resonant with the sphere, can be observed. Under these structurally resonant conditions, an enhancement factor close to 18 can be achieved (Figure 4).

When the wavelength of light is in the absorbing region (e.g. 300nm, Figure 4), the largest enhancement occurs in the small size regime (<500 Å for TiO_2). When the wavelength of light is in the non-absorbing region (e.g. 1061 nm), the enhancement factor is larger for large particles due to the structural resonances. These structural resonances can only be observed with samples containing very monodisperse particles. We have found[18] that a size dispersion of ~9% (full-width-half-maximum) can totally wash out these structures. It is important to note that even with non-monodisperse particles, the local field intensity enhancement effect still exists.

In a recent third harmonic generation study of small CdS clusters[19], we found that the third order nonlinear optical susceptibility, $\chi^{(3)}$, increases with increasing cluster size. This is attributed to the local field enhancement effects with the additional contribution from the quantum confinement effect. Third harmonic generation is a particularly sensitive technique to sense the local field effect since it is a third order nonlinear optical process.

Recently we have also observed structural resonances in the photoluminescence spectrum of monodisperse micron-size spherical CdS particles[18]. These resonance peaks are absent for very small CdS clusters and also for large but non-monodisperse CdS particles. It is clear that the effect of local field enhancement on the spectroscopic and photochemical properties of semiconductor particles is a subject needs to be explored in the future.

6. Acknowledgement. I thank my collaborators: N. Herron, A. Suna, L.-T. Cheng and E. Hilinski for their contributions to some of the work discussed here.

7. References

[1] Y. Wang, N. Herron, W. Mahler, and A. Suna, J. Opt. Soc. Am. B, 6, 808(1989).
[2] A. Henglein, Topics in Current Chemistry, **143**,113(1988).
[3] M. L. Steigerwald and L. E. Brus, Annu. Rev. Mater. Sci. 19, 471(1989).
[4] Y. Wang, A. Suna, W. Mahler, and R. Kasowski, J. Chem. Phys. **87**,7315, (1987).
[5] E. Hilinski, P. Lucas, and Y. Wang, J. Chem. Phys. **89**, 3435(1988).
[6] Y. Wang, A. Suna, J. McHugh, E. Hilinski, P. Lucas, and R. D. Johnson, 92,6927(1990).
[7] N. Herron, Y. Wang, and H. Eckert, J. Am. Chem. Soc. 112,1322(1990).
[8] Y. Wang and N. Herron, accepted for publication in Phys. Rev. B.
[9] Al. L. Efros and A. L. Efros, Sov. Phys. Semicond. **16**, 772 (1982).
[10] L. E. Brus, J. Chem. Phys. **80**, 4403(1984).
[11] Y. Kayanuma, Phys. Rev. B 38,9797(1988).
[12] P. E. Lippens and M. Lannoo, Phys. Rev. B 39, 10935(1989).
[13] L. Spanhel, M. Haase, H. Weller, and A. Henglein, J. Am. Chem. Soc., **109**, 5649(1987).
[14] T. Dannhauser, M. O'Neil, K. Johansson, D. Whitten, and G. McLendon, J. Phys. Chem., **90**, 6074(1986).
[15] Y. Wang and N. Herron, J. Phys. Chem. **92**,4988 (1988).

[16] " The scattering of Light and Other Electromagnetic Radiation" by M. Kerker, Academic Press, New York, 1969.
[17] J. Messinger, K. Ulrich von Raben, R. K. Chang, and P. W. Barber, Phys. Rev. B, 24, 649(1981).
[18] Y. Wang, unpublished results.
[19] L.-T. Cheng, N. Herron, and Y. Wang, J. Appl. Phys. 66, 3417(1989) and unpublished results.

PHOTOCATALYSIS AND PHOTOINDUCED REACTIONS ON WELL-DEFINED SUPPORTED METAL OXIDES -MOLECULAR SCALE REACTION MECHANISMS-

M. ANPO
Department of Applied Chemistry,
College of Engineering,
University of Osaka Prefecture
Sakai, Osaka 591
Japan

ABSTRACT. Various photocatalysis and photoinduced reactions have been investigated on well-defined anchored V, Mo, and Ti/SiO_2 (or /Vycor glass), and ion-exchanged Cu/SiO_2 (or /Zeolite) catalysts, in order to obtain much detailed information on the photocatalytic reaction mechanisms on a molecular scale.

1. INTRODUCTION

For the complete understanding of photocatalysis, it is requisite to elucidate the chemical natures and reactivities of photo-generated electrons and holes, as well as the excited states of catalysts and the primary reaction processes, including their dynamics (Anpo et al. [1-3]). With well-defined chemically supported "anchored" catalysts, one can expect not only to obtain much detailed information about these problems on a molecular scale but also to achieve more active and selective photocatalytic systems, because it is known that when a metal oxide is highly dispersed on a support in a molecular scale both physical and chemical natures are seriously modified, often resulting in a high catalytic activity and selectivity (Iwasawa [4]). In addition to these problems, with the "anchored" catalysts one can design the active surface species which span the range from discrete molecules to extended semiconductor structures. Such structure changes from discrete molecules, to aggregated clusters, and finally, to bulk semiconductors are of special interest in connection with the recent topics, particle size effect and/or size quantization effect on photocatalysis.

Along these lines, various photocatalysis and photoinduced reactions have been performed on the well-characterized anchored catalysts. The special attention is focused on the understanding of molecular scale reaction mechanisms on the basis of the dynamic studies of the excited states of catalysts and direct detection of reaction intermediate species, together with the analysis of reaction products. The specific photoactivities of anchored catalysts as the model

E. Pelizzetti and M. Schiavello (eds.), Photochemical Conversion and Storage of Solar Energy, 307–321.

of top surface layers of powdered bulk photocatalysts are also given attention.

2. MATERIALS AND METHOD

2.1. Preparation of Catalysts and Their Characterization

Anchoring of $VOCl_3$ onto the supports, i. e., SiO_2 and porous Vycor glass (PVG), was performed using a facile reaction of photoactivated $VOCl_3$ with the surface OH groups on the dehydrated supports at 273 K. Photoactivation of adsorbed $VOCl_3$ molecules was carried out by uv excitation at around 300-350 nm beams (Anpo et al. [5]). Anchoring of $MoCl_5$ and $TiCl_4$ onto the supports was performed using the facile reaction of $MoCl_5$ (Anpo et al. [6]) and $TiCl_4$ molecules (Anpo et al. [7]) with the surface OH groups at 473 K, respectively. These anchoring by chemical vapor deposition (CVD) followed by the treatments with H_2O vapor to replace Cl atoms, and then these samples were finally calcined in oxygen at 750 K. Cu/SiO_2 (and Cu/Zeolite) catalysts were prepared by the ion-exchange method (Anpo et al. [8]).

The structures of anchored V, Mo, and Ti/SiO_2 (and / PVG) has been examined by a number of techniques. On the basis of photoluminescence, Raman, FT-IR, UV, and ESR spectroscopies, together with the recent extended X-ray absorption fine structure (EXAFS) and ^{51}V NMR measurements, the following is established. At low surface coverages, anchored V, Mo, and Ti-oxides form four-coordinated species (tetrahedral structures), whereas at higher coverages they become increasingly six-coordinated (octahedral structures) (Iwasawa [4], Bell et al.[9], Kazansky et al. [10]). Especially, it is well established that with anchored V/SiO_2 catalysts with lower V contents the monoxo-vanadyl groups bound to the SiO_2 surface via three V-O-Si bonds (Bell et al. [9], Yoshida et al. [11]) and that with anchored Mo/SiO_2 catalysts the dioxo-molybdenyl groups bound to the surface via two Mo-O-Si bonds (Iwasawa [4]).

2.2. Apparatus and Procedure

Dynamic photoluminescence studies (i.e., measurements of the spectral shapes and lifetimes in the absence and the presence of reactant molecule) were carried out at 77 and 298 K with a Shimadzu RF-501 spectrofluorophotometer equipped with a personal computer and an apparatus for the lifetime measurement. The quartz cell with windows and a furnace parts was connected to a conventional vacuum system. The V, Mo, and Ti-oxide catalysts of 0.1-0.2 g were heated in air at 773 K, degassed at 453 K. The icn-exchanged Cu^{2+} catalysts were used after drying at 333 K. Uv irradiation of the catalyst in the presence of reactants was carried out at 273-298 K using a high-pressure mercury lamp through a color filter and water filter ($\lambda>$280 nm). Aralysis of reaction products was made by gas chromatography. ESR spectra were measured in the temperature of 77-298 K with Varian (E-3) and/or Jeol (JES-ME-1) spectrometers (X-band). Uv irradiation of the samples was carried out in ESR cavity at 77 K. UV and IR spectra were recorded at 298 K on a Shimadzu UV-210A and a Shimadzu IR-460 spectrometer, respectively (Anpo et al. [1-3, 6,7,12-14]).

3. RESULTS AND DISCUSSION

3.1. Photocatalytic Reactions of Alkenes on Anchored Vanadium Oxides

Figure 1a shows the typical phosphorescence (a) and its excitation spectrum (b) at 77 K, and the UV absorption spectrum (c) at 298 K of anchored vanadium oxide catalyst. These phosphorescence and absorption (i. e., excitation) spectrum are attributed to the following charge transfer processes on the surface monoxo-vanadyl groups in a tetrahedral VO_4 unit structure, involving an electron transfer from O^{2-} to V^{5+} and its reverse radiative decay (Anpo et al. [1-3]). Figure

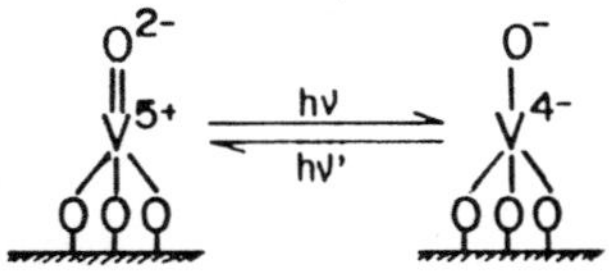

1b shows the second derivative spectrum of the phosphorescence shown in Fig. 1a. The spectra indicate that the energy gap between the (0 →0) and (0 →1) transition bands is to be 1030 cm^{-1}, being in agreement with the vibrational energy of the surface V = 0 bond (vanadyl group) observed by FT-IR and Raman measurements. This shows that the photon energy absorbed by the catalyst is mainly localized on the V = 0 surface bonds.

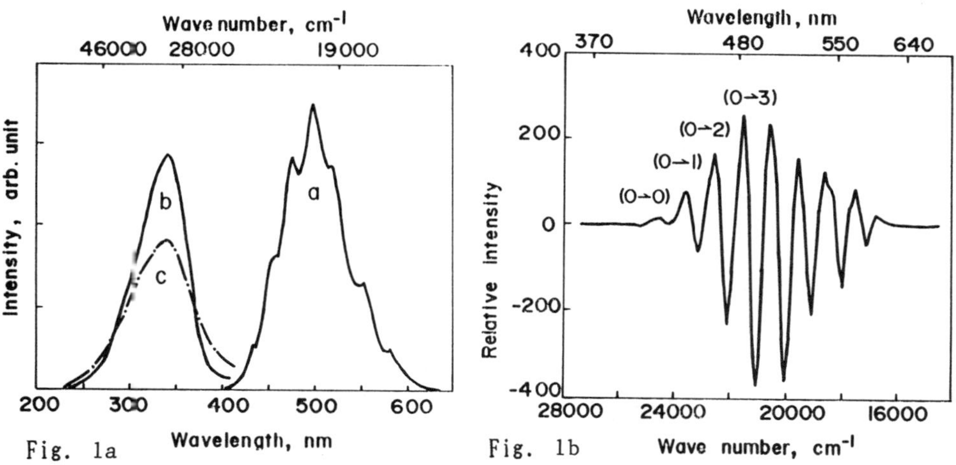

Figure 1a. Phosphorescence spectrum (a), its excitation spectrum (b) at 77 K, and uv absorption spectrum (c) of anchored vanadium oxide at 298 K.

Figure 1b. Second derivative phosphorescence spectrum of Fig. 1a-(a).

Phosphorescence of anchored vanadium oxide catalysts, which increased linearly in intensity with the content of anchored V ions, decayed as a beautiful single exponential. The lifetime was determined to be 8.2 msec at 77 K and 450 μsec at 298 K, respectively. These values scarcely changed with increasing the contents of anchored V ions at least up to 1 V wt%. These values were found to be much longer than those of the impregnated vanadium oxide catalyst, in which the lifetimes become shorter with the contents of V ions. These results suggest that V ions are highly dispersed on anchored vanadium oxides prepared by CVD method, even at higher V contents, as it was expected from the preparation steps.

The addition of O_2, C_2H_4, C_3H_6, and C_4H_8 onto the anchored vanadium oxide led to the quenching of the phosphorescence with different efficiencies, i. e., in the order of $O_2 > C_2H_4 > C_3H_6 > C_4H_8$. The Stern-Volmer plots for the yields and lifetimes of the phosphorescence in the presence of these molecules showed a linear function of the concentration of added molecules, clearly indicate that the quenching of the phosphorescence occurs by the interaction of the added molecules with the charge transfer excited triplet state of the vanadyl species to give nonradiative deactivation pathways.

Uv irradiation of anchored vanadium oxide in the presence of trans-2-C_4H_8 was found to lead to trans → cis (geometrical) and 2-C_4H_8 → 1-C_4H_8 (double bond shift) isomerizations. The yields of these reactions increased linearly with the uv irradiation time. As shown in Fig. 2, the rates of photocatalysed isomerizations are in good linear relationship with the yield of phosphorescence at the reaction temperature, directly indicating that the charge transfer excited triplet state of the vanadyl species plays a significant role in the photocatalysed isomerizations (Anpo et al. [5]).

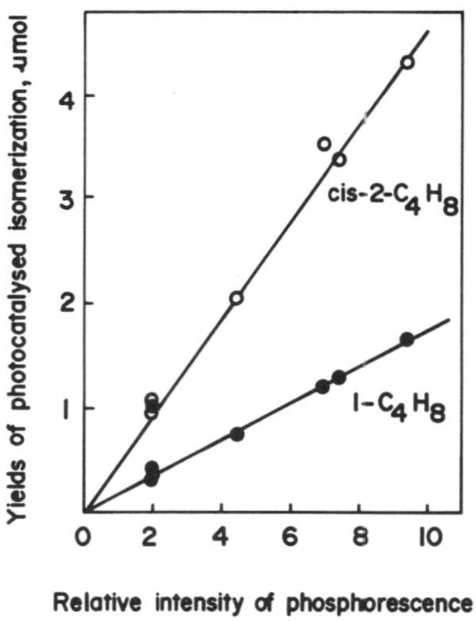

Figure 2. Relationship between the yields of photocatalysed isomerizations of trans-2-C_4H_8 at 280 K and the intensity of phosphorescence of anchored vanadium oxide at 298 K.

The addition of O_2 into the system led to the suppression of the photocatalysed isomerizations, its extent depending on the pressure of O_2. Simultaneously, photooxidation of butenes was found in place of the isomerizations. As shown in Table 1, uv irradiation of anchored vanadium oxide in the presence of a mixture of C_3H_6 and O_2 at 298 K leads to the photooxidation reaction of C_3H_6 (experiment: d and e) (Anpo et al. [12]). Under such photooxidation conditions, the photo-induced uptake of oxygen was enhanced as compared with that in O_2 alone, its extent increasing with the pressure of C_3H_6. The major oxygen-containing products of the photooxidation of C_3H_6 are CH_3CHO, C_2H_5CHO, and $CH_2{=}CHCHO$. The formation of CH_4 and C_2H_4, as well as CO and CO_2 were also observed as minor products. Uv irradiation of anchored vanadium oxide in the presence of O_2 alone led to the formation of O_2^- species adsorbed on the V ions. The O_2^- species were thermally stable at least up to 453 K. After the formation of O_2^- species by uv irradiation of the oxide in the presence of O_2 alone, C_3H_6 was introduced onto the catalyst without the evacuation of O_2 in the system. In this system, the formation of CH_3CHO, C_2H_5CHO, and $CH_2{=}CHCHO$ are observed (experiment: b), though their yields are much lower than those of the photooxidation by less than one order of magnitude. As seen in Fig. 3, ESR mesurements show that O_2^- species adsorbed on V^{5+} ions did not react with C_3H_6 in the temperature ranges of 77 - 200 K, but react with C_3H_6 above 273 K, and the concentration of O_2^- gradually decreasing with the reaction time (Anpo et al. [13]).

TABLE 1
Yields of the reaction products of C_3H_6 over V/PVG oxides under various reaction conditions at 298 K.

No.	Experiments Reaction conditions	10^{-4} ml S. T. P. CH_3CHO	 C_2H_5CHO	 $CH_2{=}CHCHO$
a)	Dark reaction of C_3H_6 and O_2	0	0	0
b)	UV; in O_2 alone, and then C_3H_6 introduced onto the catalyst	0.81	0.42	0.058
c)	UV; in C_3H_6 alone	0.54	0	0
d)	UV; in C_3H_6 and O_2	5.36	4.95	2.42
e)	UV; in C_3H_6 and O_2	20.4	12.8	7.01
f)	Quantum yields	0.00275	0.0017	0.0009

a): thermal reaction of C_3H_6 and O_2 over V/PVG oxide (0.031 V wt%) at 298 K for 1 h.
b): UV-irradiation of V/PVG oxide (0.031 V wt%) was carried out in the presence of O_2 alone, and then C_3H_6 (0.26 Torr) was introduced onto the catalyst and kept for 1 h at 298 K.
c): UV-irradiation of V/PVG oxide (0.031 V wt%) was carried out in the presence of C_3H_6 alone (0.26 Torr) at 298 K for 1 h.
d): UV-irradiation of V/PVG oxide (0.031 V wt%) was carried out in the presence of C_3H_6 (0.26 Torr) and O_2 (0.65 Torr) for 30 min at 298 K.
e): UV-irradiation of V/PVG oxide (0.143 V wt%) was carried out in the presence of C_3H_6 (0.46 Torr) and O_2 (0.33 Torr) for 30 min at 298 K.
f): Quantum yields were determined for the reaction condition d) at 330 nm excitation at 298 K. sample weight: 0.35 g.

From these results, it is likely that the interaction of the charge transfer excited triplet state of $(V^{4+} - O^{-})^{*}$ paired state (or O^{-} species of the excited complexes) with $2\text{-}C_4H_8$ results in the opening of its C = C bond and resulting intermediate species participate in the geometrical isomerization of trans-$2\text{-}C_4H_8$, in a manner similar to that proposed for the photocatalysed isomerization on TiO_2 (Anpo et al. [15]) and MgO (Anpo et al. [16]). The fact that in the presence of O_2 the photooxidation of butenes took place in place of the isomerizations, suggesting that alkene molecules can interact with the photoformed O^{-} hole center of the excited vanadyl groups but no longer react with the trapped electron center, V^{4+} ions, because oxygen can react with V^{4+} center with much higher reactivity than alkenes to form the $V^{5+} - O_2^{-}$ species, as mentioned above.

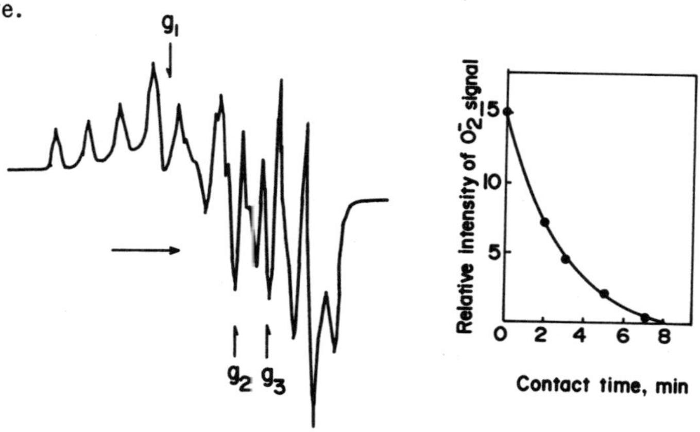

Figure 3. ESR signal of O_2^{-} anion radicals photo-formed on anchored vanadium oxide (recorde at 77 K) and a decrease of the signal by the reaction with C_3H_6 (recorded at 298 K).

As a result, following reaction schemes (schemes I and II) are proposed for the photooxidation of propene and photocatalytic isomerizations of trans-2-

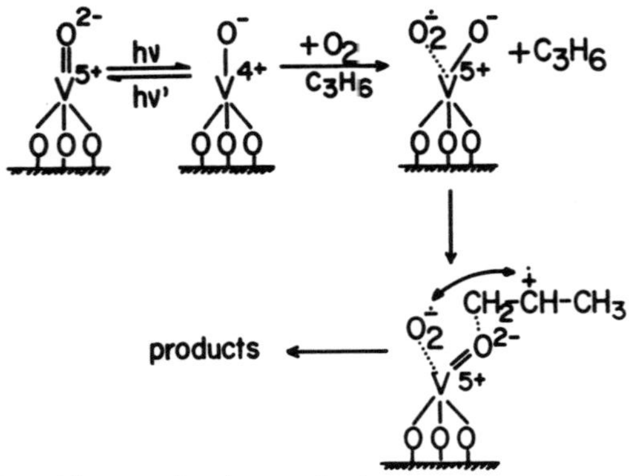

Reaction mechanism of photocatalytic oxidation of C_3H_6 with O_2 on anchored vanadium oxide (scheme I).

butene, respectively. In the reaction scheme I, O_2^- anion radicals are produced on the V^{4+} ion while C_3H_6 cation radicals are expected to be formed on the O^- ion of the excited state of the vanadyl groups. Simultaneously, the V^{4+} and O^- ions go back to the ground state of the vanadyl species. When O_2^- anion radicals and propene cation radicals react the oxidation of C_3H_6 is completed, leading to the formation of oxygen-containing products.

As shown in the scheme II, in the absence of O_2, alkene molecules react with the photo-formed O^- hole centers of the charge transfer excited triplet of the vanadyl species to form bridged-type π complexes shown in the scheme II. There is a source of an electron to recombine with it. Therefore, the energy produced when electron and hole recombine is used for the opening of the C = C bond of butenes.

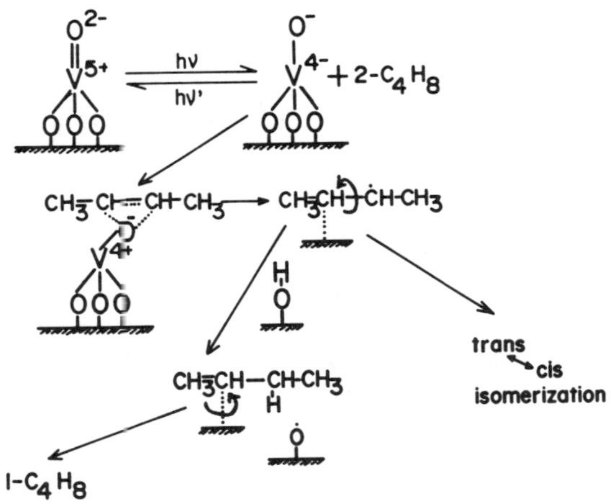

Reaction mechanism of photocatalytic isomerization of trans-2-C_4H_8 on anchored vanadium oxide (scheme II).

3.2. Photoinduced Reactions of Alkenes on Anchored Molybdenum Oxides

Figure 4 shows the photoluminescence spectrum and its excitation spectrum of anchored molybdenum oxide at 77 K. The photoluminescence can be assigned to the phosphorescence induced by excitation of the charge transfer absorption band at around 250-280 nm associated with the four-coordinated (tetrahedral) dioxo-type molybdenyl species. The lifetime of the charge transfer excited triplet state of molybdenyl species is to be ca. 75-78 μsec at 298 K, being nearly constant for Mo contents varying from 0.01-0.5 wt%. It was found that the addition of alkenes such as C_3H_6, C_2H_4, and C_4H_8, as well as O_2 and N_2O onto the oxide led to an efficient quenching of the phosphorescence not only in intensity but also in lifetime with the same effectiveness without any changes in the spectral shapes. These results indicate that the quenching of the phosphorescence involves the dynamic collisional quenching mechanism , whereby added C_3H_6 molecules interact with the molybdenyl species in their charge transfer excited triplet state (Anpo et al. [6,13]), as mentioned in the section 3.1.

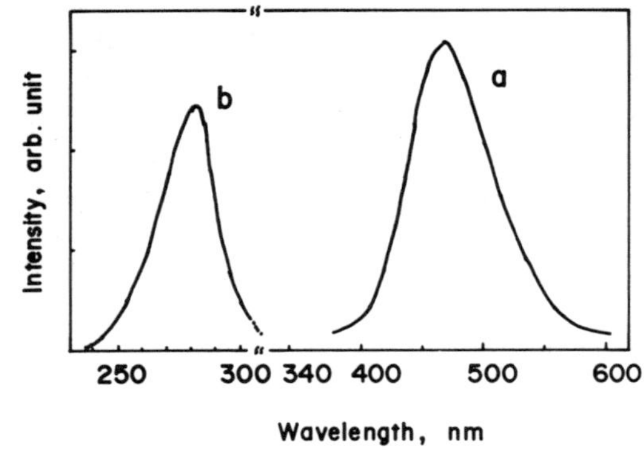

Figure 4. Phosphorescence spectrum (a) and its excitation spectrum (b) at 77 K of anchored molybdenum oxide.

As shown in Fig. 5, uv irradiation at 298 K of anchored molybdenum oxide catalyst in the presence of C_3H_6 induces the formation of C_2H_4 and 2-C_4H_8 with the formation of very samll amounts of CH_3CHO and HCHO. The yields of the photo-induced C_2H_4 and 2-C_4H_8 are almost equal, indicating that the metathesis of C_3H_6 ($2 \bullet C_3H_6 \rightarrow C_2H_4 + 2\text{-}C_4H_8$) takes place under uv irradiation of anchored molybdenum oxide in the presence of C_3H_6. In contrast, no reaction occurred in the dark condition at 298 K. In the case of 2-C_4H_8, the formation of C_3H_6 and C_5H_{10} was observed as products of the photoinduced metathesis of C_4H_8 molecules.

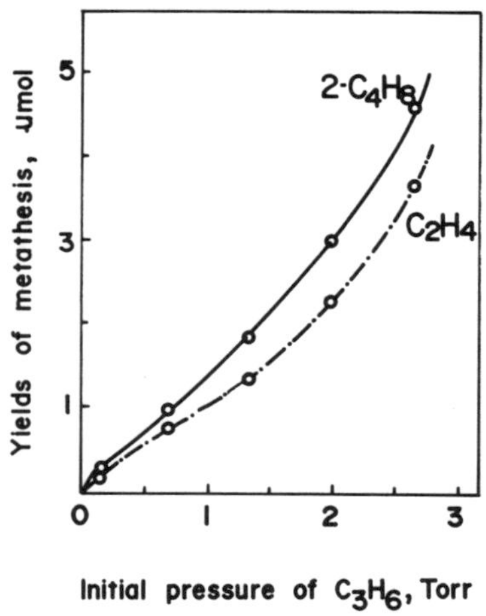

Figure 5. Effect of the initial pressure of C_3H_6 on the yields of 2-C_4H_8 and C_2H_4 in the photoinduced metathesis of C_3H_6 on anchored molybdenum oxide.

On the other hand, uv irradiation of the oxide in the presence of C_2H_4 was found to induce the formation of C_3H_6. It was confirmed that no interaction of C_2H_4 with molybdenum oxide in its ground state, since a complete recovery of the C_2H_4 introduced onto the oxide was observed by desorption at around 290 K. The yields of photoinduced C_3H_6 increased with decreasing the reaction temperature from 298 to 77 K. It was also found that decreasing the temperature of the oxide leads to an increase of intensity of the phosphorescence, its extent being good parallel to that of the yields of photoinduced C_3H_6. These results clearly suggest that the formation of methylene (=CH_2) **formed** from splitting of C_2H_4 occurs in a certain stage of the reaction, resulting in the C_3H_6 formation (Anpo et al. [17]).

Uv irradiation of the oxide at 77 K in the presence of C_2H_4 leads to an appearance of the ESR signal shown in Fig. 6. The signal consists of five lines with their intensities ratio of 1:4:6:4:1, and has a hyperfine splitting of H_a = 24.5 G and g-value of 2.003. This ESR signal is attributed to the bridged-type π complex, $(CH_2 \cdots CH_2)^-$ (bridged through O) species. On ceasing the uv irradiation, the signal decreased in the intensity even at 77 K. After disappearance of the signal, the formation of C_3H_6 was confirmed in the system. These results suggest that on uv irradiated anchored molybdenum oxide the formation of methylene (=CH_2) from C_2H_4 occurs through the formation of the bridged-type π complex intermediates, resulting in the C_3H_6 formation.

It is known that the linear $(CH_2\text{-}CH_2O)^-$ type and bridged-type π complexes, both originating from the O^- species formed thermally from N_2O on the partially reduced molybdenum oxides supported on SiO_2, decompose at higher temperatures as follows: $C_2H_4O^- \rightarrow OH^- + \dot{C}_2H_3$. Therefore, such a feature is completely different from the results observed with the bridged-type π **complexes** originating from the photo-formed O^- species which constitute the charge transfer excited triplet

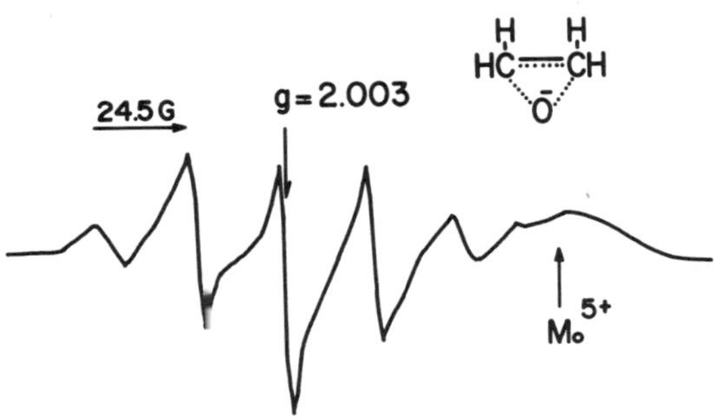

Figure 6. ESR signal of the photo-formed bridged-type π complexes of C_2H_4 on anchored molybdenum oxide (uv irradiation and ESR measurement: at 77 K).

state of molybdenyl groups. In other words, the photo-induced increase in electron density in the Mo ions of molybdenyl species plays a significant role in this specificity, being expected to enhance the back-donation of electron to π^* oribital in the bridged-type π complexes with consequent easier rupture of the

C = C bond, leading to the formation of methylene and HCHO. Thus, the reaction mechanism of the photoinduced metathesis is shown in scheme III. As seen in the scheme, on the anchored molybdenum oxide the energy produced when electrons and holes recombine is used for the C = C bond fission.

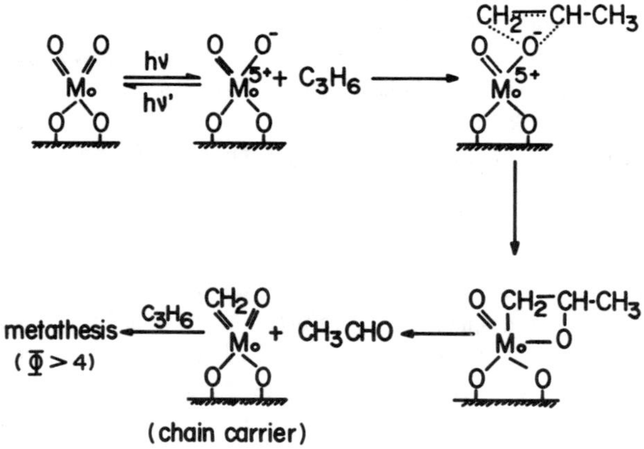

Reaction mechanism of the photoinduced metathesis of C_3H_6 on anchored molybdenum oxide (scheme III).

This scheme suggests that the rate of the photoinduced metathesis depends on both the extent of the reaction of $(Mo^{5+} - O^-)^*$ with the alkene and on the reactivity of the subsequent carbene, which acts as a chain carrier in the metathesis. That is, if a methylene-type carbene ($Mo=CH_2$) reacts with C_3H_6 it results in the formation of a C_2H_4 and a $Mo=CH-CH_3$ type carbene via the four-membered metallacyclobutane intermediate, which is well know as an intermediate species in the metathesis reactions. Thus, the alternative formation of $Mo=CH_2$ and $Mo=CH-CH_3$ type carbenes are induced in the chain reaction. It is also valuable to compare the yield of CH_3CHO (or HCHO) with that of metathesis, i.e., C_2H_4 (or 2-C_4H_8) in order to determine the reactivity of photoproduced carbenes. This comparison indicates that the carbene species formed by uv irradiation is recycled ca. 600 times as a chain carrier, its extent depending on the pressure of alkenes, which leads to the higher quantum yields than 4.0.

3.3. Photocatalytic Decomposition of NOx on Ion-exchanged Cu/SiO_2 Catalyst

How to protect our environment from global air pollution caused by NOx (NO, NO_2, and N_2O), as well as SOx is currently an urgent and serious problem. Ammoxidation using V_2O_5/TiO_2 catalyst has been developed as a de-NOx-ing process. However, it operates only at high temperature and required NH_3. Therefore, what is strongly desired are catalytic systems for direct decomposition of NOx which operate at ambient temperature and ambient pressure conditions without NH_3. Ion-exchanged Cu/SiO_2 and Cu/Zeolite catalysts are given a great deal of attention as potential catalysts for the direct decomposition of NO to N_2 and O_2 (Iwamoto et al. [18]). Utilization of photocatalytic processes in gas-solid systems also seems to be one of the most promising approaches to these environmental problems (Anpo et al. [8]).

Figure 7 shows the typical photoluminescence spectrum (a) and its excitation spectrum (b) at 77 K of ion-exchanged Cu/SiO_2 catalyst which was degassed at 873 K in vacuum. Figure 8 shows the effect of the degassing temperature of the catalyst upon the intensity of the ESR signal due to Cu^{2+} ions (solid curve: a) and of the photoluminescence of the catalyst (solid curve: b). The intensity of the ESR signal decreases with increasing the degassing temperature of the

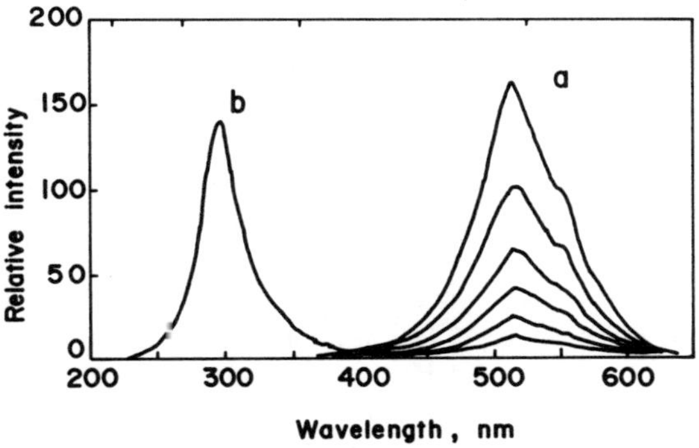

Figure 7. Photoluminescence spectrum (a) and its excitation spectrum (b) of ion-exchanged Cu/SiO_2 catalyst at 77 K. Spectral changes (a) also shows that the addition of NO leads to the quenching of the photoluminescence, its extent depending on the pressure of added NO.

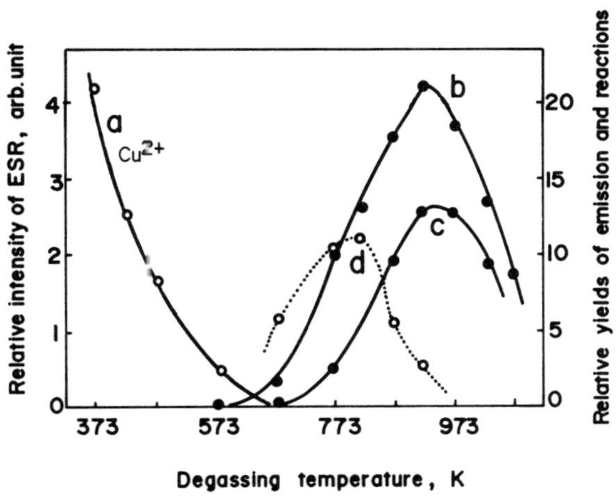

Figure 8. Effects of degassing temperature on the intensity of the ESR signal of Cu^{2+} (a), intensity of the photoluminescence due to Cu^{+} (b), yields of photo-catalysed decomposition of NO at 298 K (c), and yields of thermal decomposition of NO at 673 K (c) on ion-exchanged Cu/SiO_2 catalyst.

catalyst. On the other hand, the photoluminescence of the catalyst is observable only with the catalyst which was degassed at higher temperatures than at where the ESR signal becomes weak and disappears. The intensity of the photoluminescence changes with the degassing temperature, increasing with the degassing temperature, passing through a maximum at 923 K, and then decreases with the degassing temperature (Anpo et al. [8]).

It is known that the absorption band at around 300 nm and photoluminescence spectrum at around 500 nm of Cu^+/SiO_2, Cu^+/Zeolite, and Cu^+/Al_2O_3 are attributed to the absorption of Cu^+ ions due to ($3d^94s^1 \leftarrow 3d^{10}$) and the phosphorescence from the excited state of Cu^+ ions due to ($3d^{10} \leftarrow 3d^94s^1$) (Texter et al. [19], Barrie et al. [20]). Therefore, these results shown in Fig. 8 clearly indicate that with increasing the degassing temperature of the catalyst, the Cu^{2+} ions on SiO_2 are partially reduced to be Cu^+, the concentration of the Cu^+ ions becomes a maximum at the degassing temperature of 923 K.

Figure 7 also shows that the addition of NO on to the catalyst leads to a drastic quenching of the phosphorescence due to Cu^+ ions, its extent depending on the amount of added NO molecules. The lifetime of the phosphorescence became short in the presence of NO, suggesting that added NO molecules interact with the excited state ($3d^94s^1$) of the Cu^+ ions. In fact, uv irradiation of the catalyst in the presence of NO molecules was found to lead to the catalytic decomposition of NO. As shown in Fig. 9, the yields of the photocatalytic decomposition of NO to N_2 and O_2 on the ion-exchanged Cu/SiO_2 catalyst at 275 K increases linearly with uv irradiation time. The major products were N_2 and O_2 with a good mass balance. As shown in Fig. 8 (solid curve: c), the yields of the photocatalytic decomposition of NO increase with the degassing temperature of the catalyst, passing through a maximum at around 923 K, and then decrease. Such changes are parallel to those obtained for the intensity of the phosphorescence due to the Cu^+ ions, indicating that the excited state of the Cu^+ ions,

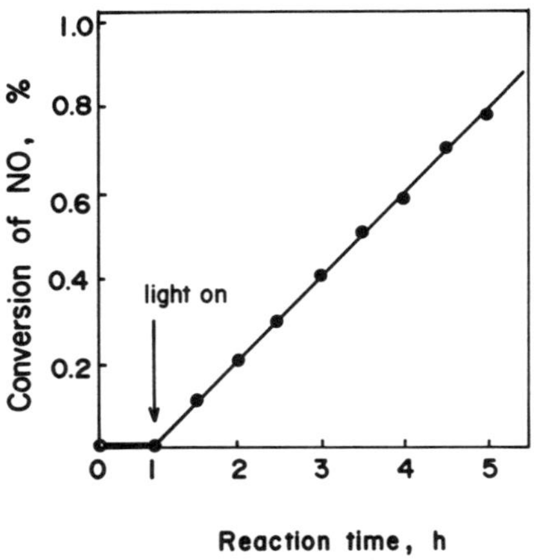

Figure 9. Photocatalysed decomposition of NO molecules to N_2 and O_2 at 275 K on ion-exchanged Cu/SiO_2 catalyst degassed at 873 K.

i. e., $3d^9 4s^1$ state plays a significant role in the photocatalytic decomposition of NO on ion-exchanged Cu/SiO_2 catalyst at normal temperature.

On the other hand, it was found that a thermal decomposition of NO also proceeds on the ion-exchanged Cu catalyst, but only at higher temperatures than 673 K. In contrast to the photocatalytic reaction, major products of the thermal reaction were NO_2, N_2, and O_2 with a minor production of N_2O. The efficiency of the thermal decomposition of NO was strongly depend on the degassing temperature of the catalyst. As shown in Fig. 8 (dotted curve: d), the yields of the thermal reaction increase with degassing temperature of the catalyst, passing through a maximum at around 773 K, and then decrease with temperature. A bad mass balance, i. e., an excess amounts of oxygen (NO_2 and O_2) in the thermal reaction suggests that the decomposition of NO might be accompanied by reduction of the catalyst.

Thus, these results indicate that ion-exchanged Cu catalyst exhibits an activity for both thermal and photocatalytic decomposition of NO molecules. In the thermal reaction, Cu^+ ions and their precursor state's surface species involving lattice O^{2-} ions and/or surface OH groups might associate with the reaction at higher temperatures. On the other hand, as shown in scheme IV, in the case of photocatalytic decomposition of NO on the catalyst at normal temperature, only the excited state of the Cu^+ ions associates with the reaction. It is likely that an electron transfer from the excited state of the Cu^+ ions ($3d^9 4s^1$) into an anti-bonding orbital of NO molecule results in the breaking of the N - O bond of NO molecule. As a result, the activation of NO molecules might be achieved in different ways in the thermal and photocatalytic reactions, resulting in the different reaction products.

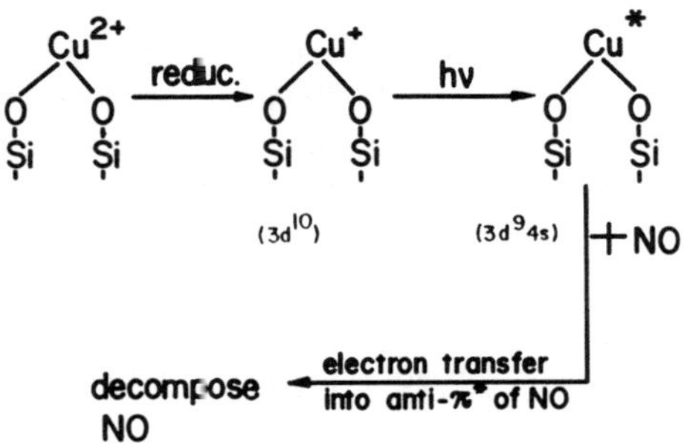

Reaction mechanism of photocatalysed decomposition of NO molecules at 275 K on ion-exchanged Cu/SiO_2 catalyst degassed at 923 K (scheme IV).

4. CONCLUSIONS

The results obtained with well-defined anchored transition metal oxide catalysts are summarised as follows: The charge transfer excited triplet state of V and Mo oxides, i. e., $(V^{4+} - O^-)^*$ and $(Mo^{5+} - O^-)^*$, plays a significant role in the photocatalytic and photoinduced reactions on these anchored V and Mo oxides, respectively. The O^- hole center as a trapped hole and the $Me^{(n-1)+}$ ion as a trapped electron act as key species in the photoreactions. These species are also undoubtedly formed and exist on the surface of unsupported bulk oxide catalysts. Therefore, it is likely that similar situation might be more or less applicable to the photocatalytic reactions on unsupported bulk catalysts, though their existence tends to be hidden by the effects of background bulk ions which are important for achieving real catalytic systems.

The results obtained with the ion-exchanged Cu^{+}/SiO_2 catalyst implied not only the possiblity of its utilization as a new type of photocatalytic materials but also the existence of different reaction mechanisms between the photocatalytic reaction at normal temperature and thermal reaction at higher temperatures.

ACKNOWLEDGEMENTS

These works have been supported by the Ministry of Education of Japan for Grant-in-Aid Scientific Research (Grant No. 01550632) and International Joint Project Research (Grant No. 02044125). Thanks are also due to Osaka Prefecture for Special Project Research, "Functional Dyes and De-NOx-ing Catalysis" (Project chief: Professor T. Kitao).

REFERENCES

1. Anpo, M. and Kubokawa, Y. (1986) 'Photoinduced and photocatalytic reactions on supported metal oxide catalysts', Res. Chem. Intermedi., **8**, 105-124.
2. Anpo, M., Suzuki, T., Yamada, Y., and Che, M. (1988) 'Dynamic studies of excited states and photocatalytic activity of highly dispersed vanadium oxide catalysts', Proc. 9th Intern. Congr. Catal.(Calgary), ed. Phillips, M. J. and Terna, M., **4**, 1513-1520.
3. Anpo, M. (1989) 'Photocatalysis on small particle TiO_2 catalysts. reaction intermediates and reaction mechanism', Res. Chem. Intermedi., 11, 67-106.
4. Iwasawa, Y. (1987) 'Chemical design surfaces for active solid catalysts', Adv. Catal., **35**, 187-264.
5. Anpo, M., Sunamoto, M., and Che, M. (1989) 'Preparation of highly dispersed anchored vanadium oxides by photochemical vapor deposition method and their photocatalytic activity for isomerization of trans-2-butene, J. Phys. Chem., **93**, 1187-1189.
6. Anpo, M., Kondo, M., Coluccia, S., Louis, C., and Che, M. (1989) 'Application of dynamic photoluminescence spectroscopy to the study of the active surface sites on supported Mo/SiO_2 catalysts.-features of anchored and impregnated catalysts-', J. Am. Chem. Soc., 111, 8791-8799.
7. Anpo, M., Aikawa, N., Kubokawa, Y., Che, M., Louis, C., and Giamello, E. (1985) 'Photoluminescence and photocatalytic activity of highly dispersed titanium oxide anchored on porous Vycor glass', J. Phys. Chem., **89**, 5017-5021.

8. Anpo, M., Nomura, T., and Giamello, E. (1990) 'Characteristics of thermal- and photocatalytic decomposition of NO on Cu^+/SiO_2 catalyst', to be published.
9. Went, Gregory T., Oyama, S. Ted, and Bell, Alexis T. (1990) 'Laser Raman spectroscopy of supported vanadium oxide catalysts', J. Phys. Chem., **94**, 4240-4246.
10. Kazansky, V. B. (1977) 'Spectroscopic investigation of oxide surfaces and their interaction with adsorbed molecules', Proc. 6th Intern. Congr. Catal. (London), ed. Bond, G. C., Wells, P. B., and Tompkins, F. C., 1, 50-65.
11. Yoshida, S., Tanaka, T., Nishimura, Y., Mizutani, H., and Funabiki, T. (1988) 'The total structures of vanadium oxide on silica and γ-alumina studied by X-ray absorption (XANES/EXAFS) spectroscopy -the effect of hydration-', Proc. 9th Intern. Congr. Catal.(Calgary), ed. Phillips, M. J. and Terna, M., **3**, 1473-1480.
12. Anpo, M., Suzuki, T., Yamada, Y., Giamello, E., and Che, M. (1990) 'Dynamic studies of photocatalytic oxidation of C_3H_6 with O_2 on vanadium oxide supported on porous Vycor glass', in G. Centi and F. Trifiro (eds.), New Developments in Selective Oxidation, Elsevier, Amsterdam, pp. 683-692.
13. Anpo, M., Kondo, M., Kubokawa, Y., Louis, C., and Che, M. (1988) 'Dynamic studies of the photoinduced metathesis of C_3H_6 and photoreduction of Mo with CO on anchored and impregnated Mo/SiO_2 catalysts', J. C. S., Faraday Trans. I, **84**, 2771-2782.
14. Kubokawa, Y. and Anpo, M. (1985) 'Photocatalytic reactions on metal oxides and their excited states' in M. Che and G. C. Bond (eds.), Adsorption and Catalysis on Oxide Surfaces, Elsevier, Amsterdam, pp. 127-138.
15. Anpo, M., Yabuta, M., Kodama, S., and Kubokawa, Y. (1986) 'Photocatalytic isomerization of butenes over TiO_2 -photo-formed active species and characteristics of reaction-', Bull. Chem. Soc. Jpn., **59**, 259-264.
16. Anpo, M., Yamada, Y., Coluccia, S., Zecchina, A., and Che, M. (1989) 'Photocatalysed isomerization of butenes on MgO powders with coordinatively unsaturated surface ions', J. C. S., Faraday Trans. I, **85**, 609-620.
17. Anpo, M. and Kubokawa, Y. (1982) 'Reaction of C_2H_4 with photoformed O^- hole centers on supported MoO_3', J. Catal., **75**, 204-206.
18. Iwamoto, M. (1990), in M. Misono (ed.), Future Opportunities in Catalytic Industry, Elsevier, Amsterdam, (in press).
19. Texter, J., Strome, D. H., Herman, R. G., and Klier, K. (1977) 'Chemical spectroscopic properties of copper containing zeolites', J. Phys. Chem., **81**, 333-338.
20. Barrie, J. D. and Dunn, B. (1989) 'Optical spectroscopy of copper(I)-doped Na^+-β-alumina', J. Phys. Chem., **93**, 3958-3963.

CONTROL OF PHOTOCATALYTIC OXIDATIVE SELECTIVITY ON IRRADIATED TiO_2 POWDERS: A DIFFUSE REFLECTANCE KINETIC STUDY

MARYE ANNE FOX, R. BARTON DRAPER, MARIA DULAY,
AND KEVIN O'SHEA
Department of Chemistry
University of Texas at Austin
Austin, TX 78712 USA

ABSTRACT. Both direct spectroscopic observation of intermediates formed on irradiated semiconductor surfaces and product analysis from inter- and intramolecular competition experiments with multifunctional organic substrates implicate direct interfacial electron transfer from an adsorbed organic donor as a primary process in photocatalytic oxidation reactions. Time-resolved diffuse reflectance spectroscopy is used to measure kinetic profiles for the decay of photogenerated transients on opaque, photocatalytically active metal oxide powders. The implications of such studies in the mechanistic characterization of photocatalytic detoxification are considered.

1. INTRODUCTION

1.1. Practical Importance of Photocatalysis

Photocatalysis on irradiated semiconductor suspensions provides a method for highly effective, yet spatially controlled, degradation of a wide variety of organic and inorganic substrates.[1-7] Intense investigation within the last decade has shown conclusively that both oxidative and reductive transformations can be initiated by visible or long wavelength ultraviolet irradiation of suspended semiconductor powders brought into contact with a liquid solution containing the molecule of interest. Such reactions are of obvious practical importance in such diverse applications as water purification, controlled detoxification of dangerous materials, and the recovery of precious metals from dilute solutions of their ions. This wavelength responsiveness also renders photocatalysis a potentially significant vehicle for solar energy utilization. Controlling or inhibiting these reactions can also pose a challenge in using such semiconductors for other purposes; for example, the degradation of paint binders in contact with suspended TiO_2 pigments has been recognized as a significant problem in the paints and coatings industry for several decades.

If such important applications are to be realized, it is mandatory that the chemical reactions which describe substrate disappearance be mechanistically understood. This requirement follows not only from the desire to optimize the efficiency of the photocatalytic degradation but also to unambiguously establish that the course of the photodegradation does not involve intermediate formation of species which might be even more toxic, or otherwise environmentally threatening, than the primary substrate. We describe in this article how mechanistic investigations of photocatalysis can provide

E. Pelizzetti and M. Schiavello (eds.), Photochemical Conversion and Storage of Solar Energy, 323–335.

predictive clues to specify those factors which lead to complete mineralization or to partial oxidation products. If the partial oxidation can be controlled by the surface, such catalysts may also find important roles in synthetic transformations. Of particular importance is a determination of whether the photooxidative reactions occurring in aqueous suspensions occur solely by hydroxy radical mediation or by competing direct hole capture via interfacial electron transfer.

1.2. Photocatalysis on Heterogeneously Dispersed Semiconductor Powders

Our research group was among the first to recognize the importance of mechanistic investigations of photocatalytic oxidations of organic substrates on irradiated semiconductor surfaces.[1-7] Consistent with our extensive studies over the last decade, the principles effective in describing interfacial electron transfer at irradiated semiconductor surfaces[8] are also operative in photocatalytic redox reactions of organic molecules. Band-gap photoexcitation of semiconductor powders suspended in polar solvents initiates formation of an electron-hole pair. Trapping of the surface-associated hole by an adsorbed electron donor generates a surface-bound radical cation for which subsequent thermal reactivity is influenced by the proximity of the semiconductor surface. Interfacial electron transfer to attain hole trapping competes with electron-hole recombination, and the quantum efficiency of product formation depends sensitively on the ratio of rates of these two processes. Slow hole trapping can be tolerated if efficient electron trapping (often by adsorbed oxygen) obviates electron-hole recombination.

Similar reactivity is observed on single crystal electrodes, powders, and colloids. The colloids and powders can be considered to be short-circuited electrochemical cells in which charge separation is attained on the same particle in analogy with that observed at a photoanode and photocathode. Photogenerated electron-hole pair formation occurs such that the hole, associated with the valence bonding orbitals, is constrained to a site near the initial light absorption, whereas the electron is poised at a potential near the conduction band edge, population of which allows for facile electron motion throughout the particle. The separated electron and hole can then be trapped via interfacial electron transfer from an adsorbed acceptor and donor, respectively.

Thermodynamic constraints are imposed by the band edges of the semiconductor. In principle, any donor whose oxidation potential lies positive of the valence band edge of the semiconductor can donate an electron, forming a surface-bound cation radical, whereas any acceptor whose reduction potential is less negative than the conduction band edge will take up the photogenerated electron. Because of the very fast recombination of the electron-hole pair, however, efficient charge trapping can occur only if the donor and acceptor are pre-adsorbed on the catalyst before excitation. Either an adsorbed substrate or a redox-active adsorbed solvent can in principle act as the primary electron donor.

1.3. Photooxidation of Organics via Adsorbed Cation Radicals

A semiconductor to be employed in a photocatalytic sequence must also meet several additional requirements: it must be photoactive under accessible illumination wavelengths, photostable under the conditions for the photocatalysis and interfacial electron transfer, non-toxic, and inexpensive. The semiconductors most routinely employed in photocatalytic applications, therefore, are metal oxides and metal chalcogenides: such suspensions (in water at pH 1) often are characterized by valence band edges positive of +2 V vs. SCE and are therefore highly oxidizing toward most organic groups bearing either non-bonded electron pairs or any pi-conjugation. By far, the most thoroughly investigated photocatalyst is TiO_2, a species which has proved to be both highly resilient under the typical conditions for photocatalysis and still highly reactive. In this study, TiO_2 has been examined exclusively.

Observation of primary electron transfer from a specific adsorbent is much easier if the solvent itself is inert to electron transfer to the photogenerated electron-hole pair. Because charged species are usually formed on these irradiated surfaces, a polar solvent is often helpful if appeciable charge trapping is to be realized (in competition with charge recombination with the space-charge region or semiconductor bulk). Acetonitrile or methylene chloride are most commonly employed as redox-inert solvents in which partially oxidized products can be characterized as intermediates in the photocatalysis. In the absence of water, total mineralization of the organic substrate does not occur and products of partial oxidation, often involving photooxygenation, can be isolated. In most cases, the formation of such partially oxidized products can be rationalized by primary formation of surface-bound cation radicals being attacked either by triplet oxygen, adsorbed superoxide, or other accessible nucleophilic traps. This paper will summarize how cation radical reactivity in thioethers leads to detoxification of both diaryl sulfides[9,10] and of aliphatic mustard simulants.[11] A comparison will be made with other photocatalytic conversions involving control of the number of electrons per oxidative event[12] and those specifying regiochemical control in the oxidation of multifunctional organic molecules.[13]

1.4. Photocatalytic Oxidations in Water: Hydroxylation and Mineralization

If the solvent is redox active, however, photocatalysis may occur by an alternate route involving reactive species deriving from solvent oxidation. In water, for example, substantial experimental evidence exists for the intervention of hydroxy radicals as the critical species in photocatalysis. The hydroxy radical's high reactivity is thought to be responsible for the complete detoxification of trace organics found as water impurities. In part, this reasonable surmise derives from the known physical characteristics of aqueous suspensions of TiO_2: the surface of TiO_2 is known to be covered with oxidizable hydroxy groups;[14] the hydroxy radical has been directly detected by electron spin resonance spectroscopy[15] and has been captured by spin traps also characterized by esr;[16] hydroxylated intermediates intermediates,[17,18] formed in these photocatalytic sequences bear close similarity to those attained with Fenton's reagent;[19] and a solvent kinetic isotope effect consistent with the involvement of solvent in the rate determining step in observed on irradiated TiO_2 suspensions.[20] Whether the involvement of the hydroxy radical is ubiquitous or whether direct oxidation of strongly adsorbed substrates can sometimes be competitive (even in water) with solvent-mediated oxidations remains an open question.

1.5. Time-resolved Diffuse Reflectance Spectroscopy in Mechanistic Photocatalysis

A clear description of the role of the hydroxy radical could be attained from spectral and kinetic characterization of transients formed during the photocatalysis. While such investigations are feasible with routine instrumentation on colloidal suspensions, the kinetic characterization of transients on optically opaque semiconductor suspensions typically encountered under heterogeneous photocatalytic conditions dictate that a typical spectroscopic method of characterization must be employed. Diffuse reflectance absorption or emission studies provide just such a tool. Such investigations would also provide very useful information on possible photocatalytic mechanisms which may be operative in water, where much less is known about the reaction order or the rates of reaction of transients formed at the surface. Furthermore, the factors which can influence the rate of electron transfer to a specific adsorbate, which are currently not firmly established, could be clarified. What follows is a description of the use of time-resolved diffuse reflectance spectroscopy to establish the identities of transient organic intermediates involved in oxidative photocatalysis on TiO_2.

A two parameter model developed by Albery to describe kinetics in several heterogeneous systems has been applied to the kinetic characterization by diffuse reflectance methodology of transients formed on irradiated semiconductor powders. The faster decay typically observed on powders than on the analogous colloid can be rationalized by first order kinetics with a distribution of rate constants, implying reduced dimensionality of diffusion on the surface of the powder.[21] The technique has been applied to an analysis of the photocatalytic decomposition of thiocyanate, potassium iodide, methyl viologen, tetrachloroplatinum (II), phenylenediamine, and thianthrene,[22] where direct electron transfer from these strongly adsorbed substrates can be implicated.

2. EVIDENCE FOR DIRECT ADSORBATE OXIDATION

2.1. Photocatalytic Oxidation of Thioethers

We illustrate the direct trapping of photogenerated holes by considering the mechanism of the photocatalytic oxidation of diaryl and dibenzyl thioethers. Band gap irradiation of titanium dioxide suspended as a powder in dry aerated acetonitrile leads to clean sequential conversion first to the sulfoxide and then the sulfone, equation 1, in which the

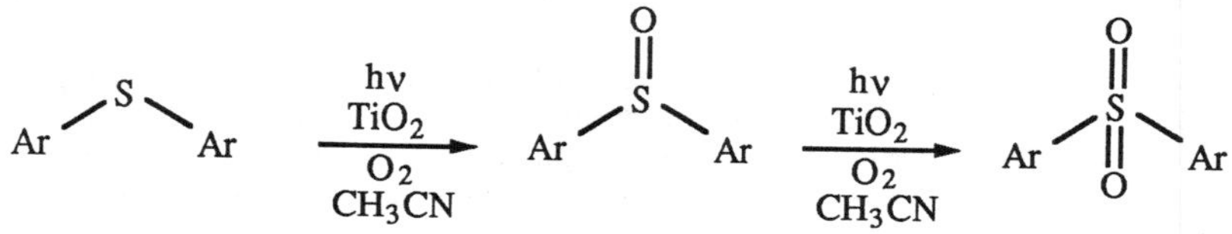

(1)

carbon skeleton remains intact.[9] The analogously para-substituted substrates bearing substituents with variable electron donating and accepting properties are similarly converted to sulfoxides, equation 2. The relative rates observed within this series exactly

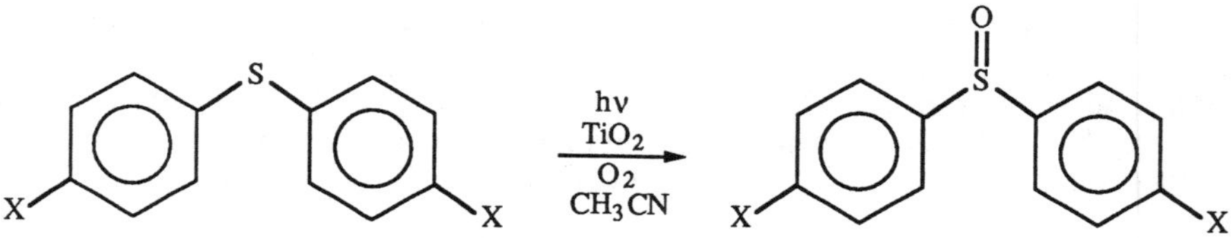

(2)

X = Cl, Br, H, Me, OMe

parallel the oxidation potentials of these compounds, and a linear plot of the logarithm of the reaction rate against σ^+ is observed. When the substituent in equation 2 is methyl, competitive deprotonation occurs at the methyl group, generating a benzylic radical, oxygenation of which gives rise to the thioether aldehyde. Such competitive reactivity is consistent with intermediate formation of a cation radical intermediate.

If the strong aryl-S bond is replaced with a weaker benzyl-S bond, competitive sulfoxidation and C-S cleavage is observed, equation 3.[9] The relative weakness of

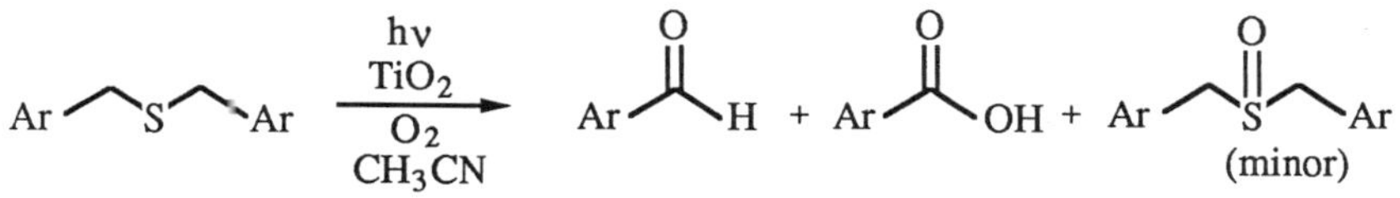
(3)

the C-S bond makes oxidative cleavage the dominant pathway, in contrast to the minor route observed in equation 2, when the additional carbon is present as a p-methyl substituent.

Analogous reactivity is also observed in the mustard simulants shown in equation 4. As with the aryl sulfides, sulfoxidation and subsequent sulfonation occur in high

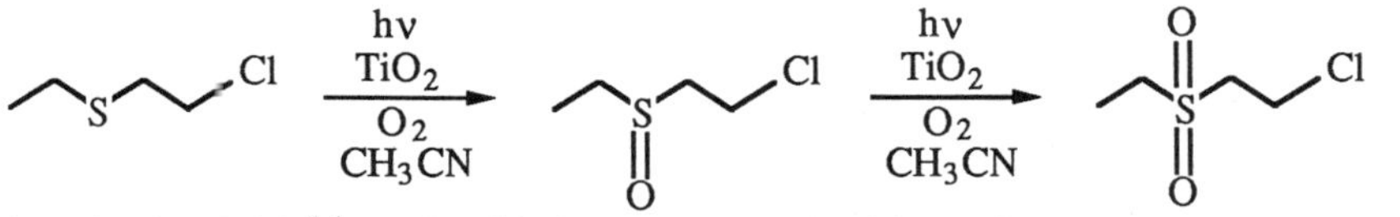
(4)

chemical yield.[11] The high water insolubility of these substrates makes this photocatalytic degradation an exceedingly important route for detoxification of such compounds. Examples are known which show persistence of mustards dispersed during WWI until the mid-1980's.

2.2. Oxygenation of the Thioether Cation Radical

The mechanistic route outlined in equation 5 for sulfur oxidation finds precedent both in

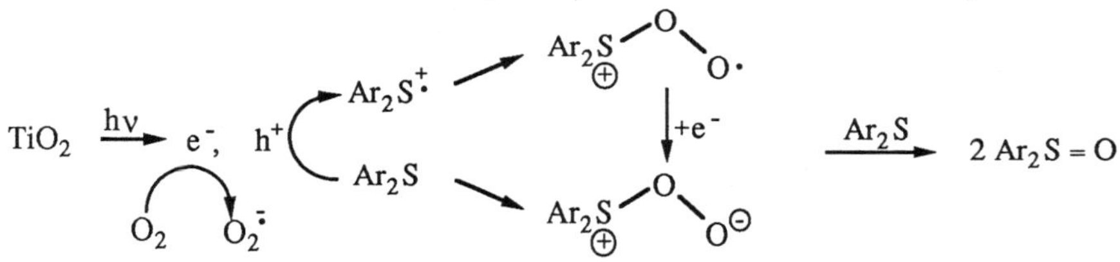
(5)

the electron transfer oxygenation of sulfides and, in subsequent steps, from known characteristics of oxygen reactivity. Hole trapping by interfacial electron transfer to the photogenerated hole on TiO_2 powders produces a diaryl thioether cation radical. This species can be trapped either by adsorbed oxygen or by superoxide formed by electron trapping on a common particle. Attack of the resulting species by diaryl sulfide, either on the surface of the photocatalyst or after dispersive dissociation via mass transfer into homogeneous solution, forms two molecules of the observed product. It is important to note that only the first step in this conversion requires activation by a photon. Subsequent steps are all thermally activated and may be influenced by the metal oxide surface on which the reactive intermediate is formed.

2.3. Cation Radical Trapping by Water

In addition to the slow trapping by oxygen discussed in the preceding section, cation radicals can also be attacked by nucleophiles. This route can significantly complicate mechanistic studies deriving solely from product analysis, since identical products might reasonably be expected in many cases to have formed from either hydroxy radical attack or initial hydration of a primary singly oxidized cation radical. The hydroxylation of phenol, for example, can proceed either by direct attack by a hydroxy radical, producing a characteristic distribution of hydroxylated isomers, equation 6, or by primary electron transfer to generate a cation radical,

$$TiO_2 \xrightarrow{h\nu} e^-, h^+$$

$$h^+ + H_2O \longrightarrow H^+ + {}^\bullet OH$$

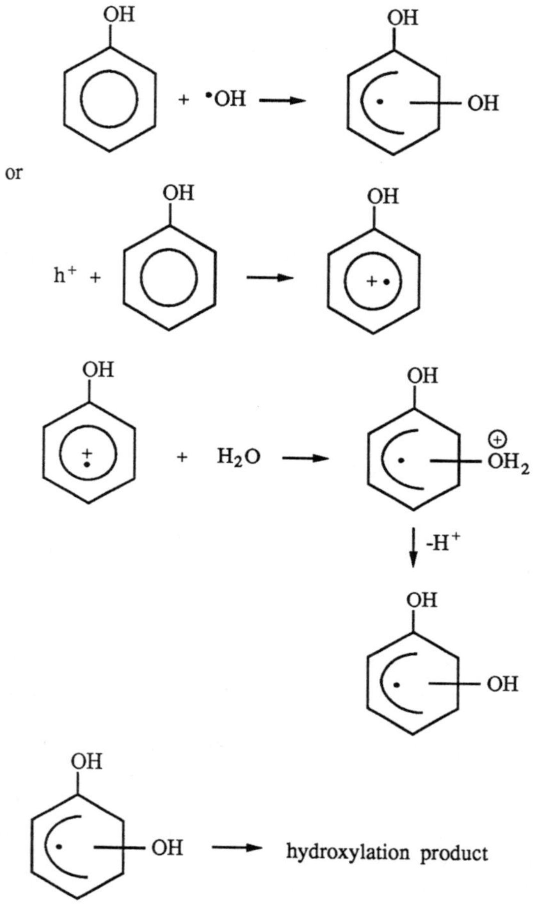

(6)

hydration of which gives a hydronium ion. Deprotonation of this ion then gives the same adduct formed from direct hydroxyl radical attack. Thus, product analysis studies alone may, at least in some cases, be inadequate to unambiguously establish the type of activity involved in the primary photocatalytic event. It is therefore important that any complete mechanistic study admit mechanistic tests which directly characterize charge transfer in the photocatalytic step or include measures for direct spectroscopic characterization of transients formed therein. We consider in Section 3 how time-resolved diffuse reflectance spectroscopy can provide such information.

2.4. Requirement for Prior Adsorption for Active Photocatalysis

It is important to realize that the short lifetime of the photogenerated electron-hole pair on irradiated semiconductor powders requires that a species participating in carrier trapping must be pre-adsorbed before the absorption event occurs. If a molecule is present in solution and must diffuse to the surface of the photocatalyst, efficient trapping of either an electron or hole is kinetically doomed. The requirement for pre-adsorption for efficient charge trapping means that molecules will exhibit reactivity which is ultimately sensitive to pre-adsorption equilibria.[12,13] These pre-association equilibria can be described by a Langmuir adsorption isotherm, equation 7.[5] This model assumes that

$$\Theta = K[C]/(1+K[C]) \qquad (7)$$

surface coverage by adsorbate will be given in terms of solution concentrations [C] and an equilibrium constant K. Catalyzed surface reactions in which a competition between two or more co-adsorbed species for a single active adsorption site are described by an analogous expression, equation 8, in which the substrate of interest i is compared with

$$\Theta = Ki[C_i]/(1+\Sigma_j Ki[C_j]) \qquad (8)$$

the summation of competitively adsorbed species j. Under the conditions in which most photocatalytic oxidations are conducted, this substrate, water, and oxygen are all pre-adsorbed on the surface. Water, however, is in great excess and is often invisible in such kinetic treatments. Since such suspensions are typically bubbled with air or oxygen gas, or are at least in contact with an ambient atmosphere, one can assume that pre-adsorption of oxygen from the gas phase would have occurred before the introduction of substrate.

In any case, the Langmuir equation provides an effective model for understanding pre-adsorption effects in the oxidation of organic molecules. Shown in Figure 1, for example, is the plot obtained from observed rate data at varying concentrations of cyclohexane carboxylic acid in dilute aqueous acid upon photocatalytic oxidation by suspended TiO_2 particles. The observed linearity argues for a strong pre-adsorption of the substrate.

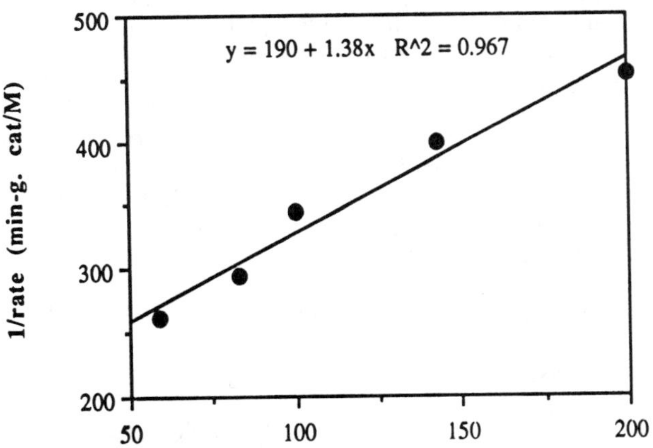

Figure 1. Photodecomposition of cyclohexane carboxylic acid in dilute aqueous acid on irradiated TiO_2 powder (Degusa P-25) under ambient air.

That such pre-adsorption is not limited to aqueous suspension can be easily shown by a comparison of analogous rate measurements conducted in non-aqueous suspension. For example, the plot shown as Figure 2

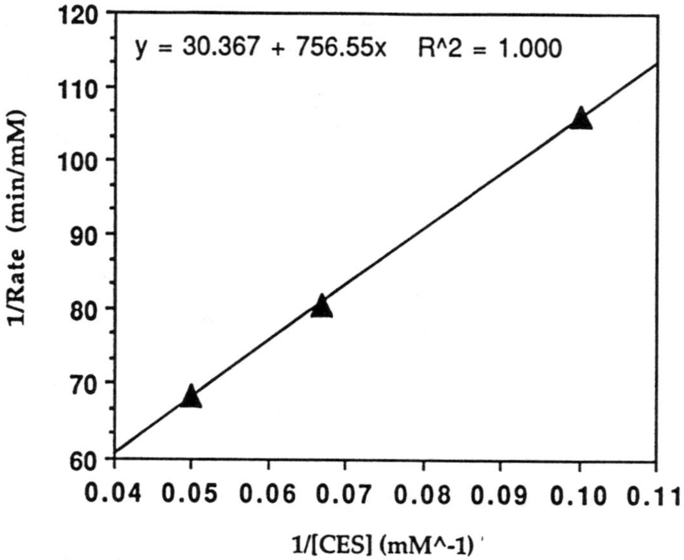

Figure 2. Photodecomposition of 2-chloroethyl ethyl sulfide in dilute acetonitrile on irradiated TiO_2 powder (Degusa P-25) under ambient air.

describes the dependence of the rate of reaction on initial concentration in the TiO_2 photocatalyzed photooxygenation of 2-chloroethyl ethylsulfide as described in equation 4. Even when this reaction is conducted in acetonitrile, clear linearity consistent with pre-adsorption is observed.

One should note that two limiting mechanisms, the Langmuir-Hinshelwood and the Eley-Rideal models, are commonly employed for quantitative analysis of surface reaction kinetics. These differ in that the former model requires a reaction to take place directly between species which are strongly adsorbed (chemisorbed) on the surface, whereas the latter involves a surface reaction taking place between an adsorbed substrate and a non-chemisorbed species. The latter may be either dissolved in solution or physisorbed, i.e., weakly held. From the slopes of the plots shown as Figure 1 and 2, one can surmise that both organic substrates are reasonably strongly adsorbed, and even larger equilibrium constants are typical of the substrates described in the subsequent section as probes for direct oxidation and hydroxy radical interception.

3. TIME-RESOLVED DIFFUSE REFLECTANCE SPECTROSCOPY

3.1. Techniques for Diffuse Reflectance

Although laser flash photolysis has been effectively employed with optically transparent dispersions in the direct spectroscopic characterization of photogenerated transients, such methods are not routinely adaptable to suspensions including opaque materials. In the past, extrapolation of results from colloidal suspensions to powdered photocatalysts have been necessary in order to address the question of spectroscopic characterization of primary photogenerated reactive intermediates. In the absence of a direct method for the measurement of lifetimes and quantum yields for formation of these species, the mechanistic reliability of conclusions extrapolated from measurements on colloids to those occurring on powders (limiting progress in mechanistic characterizations of photocatalytic oxidative degradation of organic substances) relates to the validity of the assumptions on which these experiments are based.[23]

Happily, Wilkinson and coworkers have addressed the question of spectroscopic characterization of intermediates formed on opaque materials by simply changing the geometry of the incident light source and the analyzing optical bench.[24] As shown in the schematic block diagram, Figure 3, this altered geometry allows for discrete collection of diffusely reflected light, to be analyzed in either an absorptive or emissive mode, by collection through an appropriate lens focused onto a monochromator, with the resulting signal intensity being analyzed by a conventional photomultiplier tube/data acquisition system.

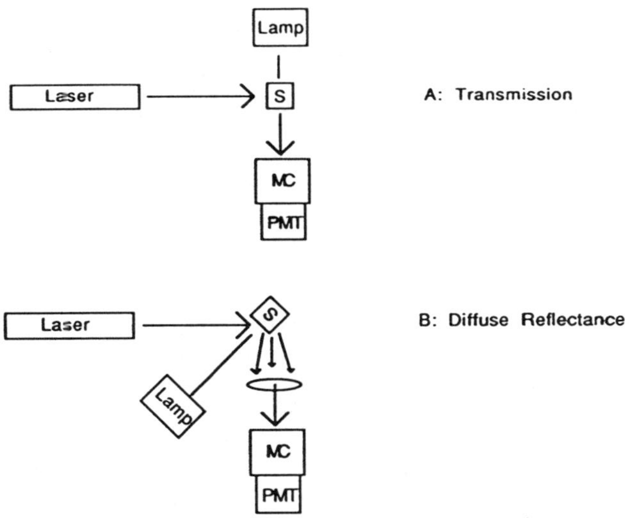

Figure 3. Schematic comparison of conventional laser flash photolysis with a diffuse reflectance measurement.[21]

By differentiating the adsorbed incident light (I_0) from that diffusely reflected (J_0) through a volume element dx, it is possible to treat the observed signal intensity according to the Kubelka-Munk theory,[25] Figure 4.

We have reported elsewhere the use of this technique for the time resolution of the transient absorption spectrum for the oxidative dimerization of thiocyanate, with the rates obtained both on colloidal suspension and on powdered samples being compared with that observed in solution by pulse radiolysis.[21] The time evolution and decay of this signal, however, could not be treated by a simple Kubelka-Munk analysis for either first or second order decay, and the detailed kinetic analysis could best be obtained by employing the two parameter model suggested by Albery for describing heterogeneous reactions occurring in micellar media.[26,27]

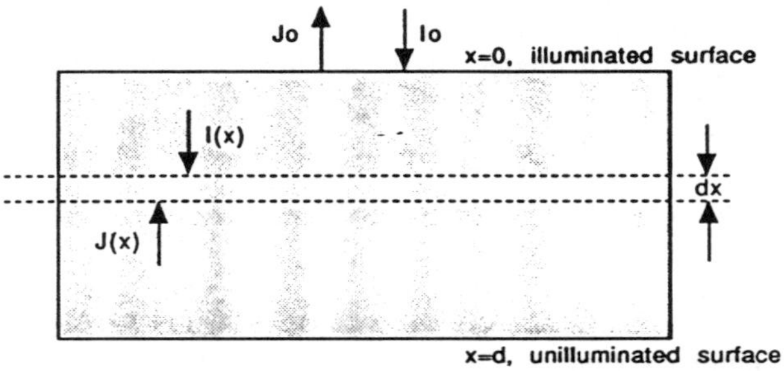

Figure 4. Schematic view of diffusely-reflected light from an infinitesimally thin layer of thickness dx from the negative direction oriented with respect to the incident laser pulse.[25]

This model proved to be useful to investigate not only the absorption characteristics of these transients, but also their dependence on laser intensity, concentration, pH effects in contacting solution, and ionic strength dependence on transient decay kinetics observed on such irradiated titanium dioxide powders.[21] This model employs a cautious adaptive Romberg extrapolation routine taken from a standard sub-routine library to analytically integrate the dependence of observed δJ differences on standard rate constants for formation and decay of the transient, and two fitting parameters describing the effective width of the particle size distribution and a distribution variable which is allowed to take values between plus and minus infinity. By accumulating and averaging roughly 200 measurements of δJ as a function of time, rate dependence could be evaluated by minimization of the sum of the residuals squared to obtain optimal fit and specify the model's two parameters.

Under such conditions, the difficulties in applying appropriate kinetic models disappeared, and the dependence of solution parameters as mentioned above could be established without concern about possible flocculation associated with colloidal suspensions. The average rate constant for decay of the transient on TiO_2 powders was about twenty times shorter than that observed on TiO_2 colloids, thus bringing into sharp focus the question raised earlier regarding the assumed effective modeling of reactions occurring on the surfaces of suspended powders by employing optically transparent colloidal dispersions. This dependence of the rate constant was assigned to a reduction in the dimensionality of the reaction space from three in the colloidal system to two on the powder and, hence, to two dimensional diffusion of the oxidized anion confined on a single particle. No evidence for complications inducing second order kinetic effects via particle-particle encounters could be observed.

3.2. Application of Diffuse Reflectance Methods to Transient Detection on Irradiated Metal Oxide Powders

The applicability of time-resolved diffuse reflectance flash photolysis as a technique of merit for characterization of optically absorptive transients on irradiated semiconductor powders having been established, we undertook a brief survey of several redox-active materials for which either known spectroscopic assignment of the singly oxidized or reduced transients and hydroxy radical adducts have been reported or for which product analysis under steady state conditions suggested hydroxylation as a significant product generating route. Accordingly, diffuse reflectance and transmission flash photolysis methodology has been employed to differentiate the primary reactive intermediates formed upon long wavelength ultraviolet sensitization of titanium dioxide powders (Degusa, P-25), in inducing the photosensitized oxidations of several organic and inorganic redox couples: potassium iodide, 2,4,5-trichlorophenol, methyl viologen dichloride, potassium tetrachloroplatinum (II), tris(1,10-phenanthroline) iron (II) perchlorate, thianthrene, and N,N,N',N'-tetramethyl-p-phenylenediamine, and the photoreduction of methyl viologen dichloride.[22] In most cases, available spectroscopic evidence indicates that the spectroscopic properties of the product of OH radical mediated adduct formation differ from those attributable to the product of direct electron transfer oxidation.

Only iodide and tetrachloroplatinum give identical spectra for direct electron transfer and hydroxy radical mediated reactions. In these systems, spectroscopic properties of transiently detected intermediates cannot differentiate direct electron transfer and OH attack. In the other substrates, however, the assignment of the initially observed products in these systems can demonstrate whether interfacial electron transfer through surface mediated trapping of the electron or whole can effect direct oxidation and reduction in competition with solvent oxidation routes to the highly reactive OH radical.

In all cases, products of direct electron transfer redox reaction are observed. An example is shown as Figure 5 in which the transient absorption spectrum derived from diffuse reflective excitation of TiO_2 with pre-adsorbed thianthrene is shown.[22] The inset shows the decay kinetics for the transient as fit to the Albery model The transient decay kinetics appear to be first order with a Gaussian distribution of reaction rate constants reflecting the distribution of particle radii as expected from independent light scattering measurements and from the application of the Albery model as discussed above.[22]

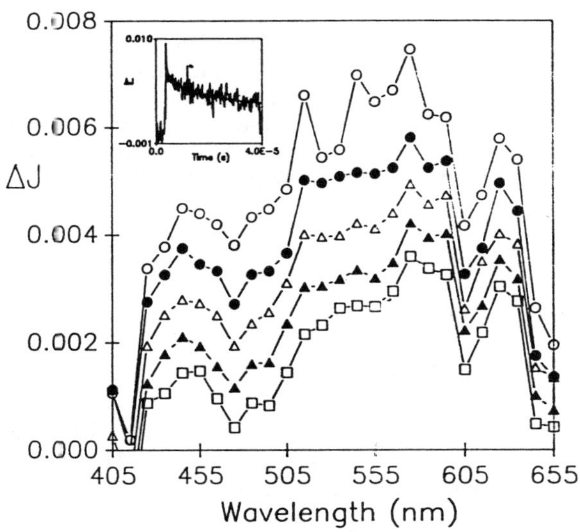

Figure 5. Transient diffuse reflectance absorption spectrum attributed to the cation radical thianthrene generated by flash photolysis of TiO_2 powder in acetonitrile.[22] The respective curves indicate delay time after flash: o = 1.4 ms; • = 4.4 ms; Δ = 11.1 ms; ▲ = 21 ms; □ = 31.2 ms. Inset, delay of transient at 585 nm.

4. CONCLUSIONS

Overwhelming evidence exists that photocatalytic hydroxylation is induced by irradiation of TiO_2 suspended in water containing oxidizable organic molecules. Despite the observation of hydroxylated structures derived in the photocatalytic degradation of aromatic compounds and esr spin trapping of hydroxyl radicals in aqueous solutions of illuminated TiO_2, the diffuse reflectance studies described here clearly indicate that the assumption that direct hole-organic reaction is not significant must at least sometimes be erroneous. The possibility that such single electron redox reactions could be mediated by trapped hole equivalents either on the metal oxide surface (as, for example, by a surface-bound hydroxy radical or by a sub-surface lattice oxygen situated directly beneath an adsorbed hydroxide ion[28]) remains ambiguous from these experiments. What is clear, however, is that electron transfer to an organic adsorbate must be considered as a possible primary reaction step in the conversion of a photogenerated electron hole pair initiating photocatalytic oxidative degradation of strongly adsorbed, easily oxidized substrates such as those employed in this study.

Future experiments will seek to determine whether subsequent thermal reaction occurs on the surface or away from the surface as would be necessary for hydroxy radical initiated attacks on much more weakly adsorbed substrates.[29] Again, spectroscopic methods may be deemed useful in these experiments, since conventional kinetic studies employing either a Langmuir-Hinshelwood or an Eley-Rideal mechanism have recently been shown to be kinetically indistinguishable in addressing this point.[29]

5. ACKNOWLEDGEMENTS

We gratefully acknowledge the support of the US Department of Energy for our work on chemical modification of semiconductor surfaces and the National Science Foundation and the Robert A. Welch Foundation for support for the development of diffuse reflectance flash photolysis methods. The flash photolysis experiments were conducted at the Center for Fast Kinetics Research in Austin which is jointly supported by the National Institutes of Health and by the University of Texas at Austin.

6. REFERENCES

1. Fox, M.A. *Accts. Chem. Res.* **1983**, *16*, 314.
2. Fox, M.A. *Top. Org. Electrochem.* **1985**, *4*, 177.
3. Fox, M.A. *Top. Curr. Chem.* **1987**, *142*, 72.
4. Fox, M.A. *NATO Adv. Sci. Ser.* **1988**, *C 237*, 445.
5. Fox, M.A.; Pichat, P. in "Photoinduced Electron Transfer," Fox, M.A., Chanon, M., Eds., Elsevier, Amsterdam, 1988, Vol. D, Chapter 6.1, p. 241.
6. Fox, M.A. in "Photocatalysis," Serpone, N.; Pelizzetti, E., Eds., Wiley, New York, 1989, Chapter 13, p. 421.
7. Fox, M.A. in "Advances in Electron Transfer Chemistry," Mariano, P., Ed., JAI Press, Greenwich, CT, 1990, in press.
8. Bard, A.J. *Science* **1980**, *207*, 139.
9. Fox, M.A.; Abdel-Wahab, A.A. *Tetrahedron* **1990**, in press.
10. Fox, M.A.; Abdel-Wahab, A.A. *J. Catal.* **1990**, submitted.
11. Fox, M.A.; Kim, Y.S.; Abdel-Wahab, A.A.; Dulay, M. *Catal. Lett.*. **1990**, in press.

12. Muzyka, J.L.; Fox, M.A. *J. Org. Chem.* **1990**, *55*, 209.
13. Fox, M.A.; Ogawa, H.; Pichat, P. *J. Org. Chem.* **1989**, *54*, 3847.
14. Parfitt, G.D. *Prog. Surf. Membr. Sci.* **1976**, *11*, 181.
15. Cereas, E.M.; Burlamacchi, L.; Visca, M. *J. Mater. Sci.* **1983**, *18*, 289.
16. Jaeger, C.D.; Bard, A.J. *J. Phys. Chem.* **1979**, *83*, 3146.
17. Barbeni, M.; Morello, M.; Pramauro, E.; Pelizzetti, E.; Vincenti, M.; Borgarello, E.; Serpone, N. *Chemosphere* **1987**, *16*, 1165.
18. Okamoto, K.; Yamamoto, Y. Tanaka, H.; Tanaka, M.; Itaya, A. *Bull. Chem. Soc. Jpn.* **1985**, *58*, 2015.
19. Barbeni, M.; Minero, C.; Pelizzetti, E.; Borgarello, E.; Serpone, N. *Chemosphere* **1987**, *16*, 225.
20. Cunningham, J.; Srijaranai, S. *J. Photochem. Photobiol. A: Chem.* **1988**, *43*, 329.
21. Draper, R.B.; Fox, M.A. *J. Phys. Chem.* **1990**, *94,* 4628.
22. Draper, R.B.; Fox, M.A. *Langmuir* **1990**, *6*, in press.
23. See for example (a) Fox, M.A.; Chen, C.-C.; Lindig, B. *J. Am. Chem. Soc.* **1982**, *104*, 5823; (b) Brown, G.T.; Darwent, J.R.; Fletcher, P.D.I. *J. Am. Chem. Soc.* **1985**, *107*, 6446; (c) Rotherberger, G.; Moeser, J.; Grätzel, M.; Serpone, N.; Sharma, D.K. *J. Am. Chem. Soc.* **1985**, *107*, 8054.
24. Wilkinson, F.; Willsher, C.J. *Appl. Spectros.* **1984**, *38*, 897.
25. Wendlandt, W.W.; Hecht, H.G. in "Reflectance Spectroscopy", Interscience Publishers, New York, **1966**, 46.
26. Albery, W.J.; Bartlett, P.N.; Wilde, C.P.; Darwent, J.R. *J. Am. Chem. Soc.***1985**, *107*, 1854.
27. Albery, W.J.; Brown, G.T.; Darwent, J.R.; Saievar-Iranizad, E. *J. Chem. Soc., Faraday Trans. 1* **1985**, *81*, 1999.
28. Howe, R.F.; Grätzel, M. *J. Phys. Chem.* **1985**, *89*, 4495.
29. Turchi, C.; Ollis, D. *J. Catal.* **1990**, *122*, 178.

PHOTOCATALYTIC AND PHOTOGRAPHIC HETEROJUNCTIONS

Boris Levy
Department of Chemistry
Boston University
Boston, Massachusetts 02215
U.S.A.

ABSTRACT. The fields of photo-assisted heterogeneous catalysis and silver halide based photographic systems have a number of common features. They both utilize particulate semiconducting materials to absorb the incident radiation and produce photoelectrons and photoholes. Recombination of these charge carriers is counter productive, since they are the reagents relied upon for the initiation of subsequent events crucial to overall process efficiency. From this perspective, it is not unusual that progress in these seemingly diverse fields is proceeding in similar directions. In both fields heterojunction structures are employed to promote the separation of charge carriers on the supposition that this will increase efficiency - as measured by increased product yield in photocatalysis and increased Ag^+ ion reduction in photographic systems. The determination of time resolved charge transport parameters has been recognized as essential for the achievement of a detailed understanding of the photophysics and chemistry of these systems. Their particulate nature places special demands on the techniques employed in the characterization of electronic transport properties and on the interaction of electronic and ionic charge carriers with other chemical species. In the field of silver halide imaging, the time resolved photocharge technique (TRPC) has been applied to the investigation of what have been termed photographic heterojunction diodes and transistors. Descriptions of the formulation and characterization of a number of these heterojunction systems will be given with the intent of illustrating similarities with approaches being taken in the field of photocatalysis. Recent TRPC measurements on TiO_2 and Pt/TiO_2 particulate systems will also be discussed.

1. INTRODUCTION

Heterogeneous photocatalysis and photographic imaging have a number of features in common governing their operation. Current trends in both fields are placing increasing emphasis on the use of multi-phase particulate materials [1,2] to promote the separation of photoholes from photoelectrons, thereby decreasing recombination and also to extend the range of spectral response of

E. Pelizzetti and M. Schiavello (eds.), Photochemical Conversion and Storage of Solar Energy, 337–392.

the primary photosensitive component. In the case of photo-assisted catalysis, one of the objectives is to increase the probability of reaction between adsorbed species and the charge carrier remaining, or having been transported to the surface by diffusion and/or drift. Evidence supporting this approach has been largely based on spectroscopic flash photolysis investigations [3,4]. Correspondence between the intended function of heterojunction catalyst structures and measured chemical reaction rates, has been taken as support for the charge separation mechanism [5].

In both fields attention has been directed toward the need for a more direct electronic means of evaluating the extent of charge carrier separation and of other charge carrier transport parameters [3]. Transient time of flight [6] measurements (TOF) and steady state surface photovoltage [7] (SPV) are frequently used to evaluate the mobility and diffusion length of charge carriers in photoconducting materials of macroscopic dimension where direct attachment of ohmic contacts to the sample is relatively straight forward in principle, although bothersome and time consuming in practice. A problem associated with such contacts is the establishment of depletion layers across the photoconductor/contact interface which can alter the magnitude and under some conditions, the polarity of the photo-induced currents [8,9]. Even without this complication, measurements requiring direct electrical contact with the sample are not practical for particulate photoconductors and certainly not for those of colloidal dimensions. Time resolved microwave photoconductivity (TRMC) [10] avoids this problem and is capable of determining the rise and decay times of photoelectrons in the conduction band. TRMC signal lifetime has been associated with the extent of charge separation, based on the assumption, which may not always be valid, that longer lifetimes can be equated to an increase in charge separation [11].

Another transient technique, not requiring direct electrical contact with the sample is the time resolved photocharge technique (TRPC) [8,9,12]. It can be used to evaluate the following parameters: 1) identify the sign of the electronic photocharge carrier (electron or hole) having the larger velocity; 2) determine the mean velocity of the majority charge carrier; 3) measure the ambipolar mean charge carrier separation distance; 4) determine the kinetics of decay of the TRPC signal as it relates to mechanistic considerations and 5) evaluate the effects of various adsorbed and junction materials on the preceding properties.

Silver halide photographic systems generally utilize micron range silver halide particles. The photographic process depends on the generation of photocharge carriers and on the competition between recombination and charge separation. Trapped photoelectrons, which have become separated from photoholes, attract and reduce interstitial Ag^+ ions to form silver clusters (latent image) [13]. Multi-junction grains have been used to extend the spectral range of the silver halides and also to increase photographic speed for exposures within the silver halide intrinsic absorption band [11,12,14]. Sub-monolayer deposits of adsorbed dyes, metallic deposits and inorganic epitaxial deposits have been used singly and in combination for these purposes. Dyes and inorganic photoconductors, such as PbO, can spectrally sensitize AgBr by electron injection from the sensitizer into the AgBr conduction band. These single junction systems have been considered to function by the photographic diode mechanism [2,12]. Multi-junction systems consisting, for example, of AgBr/Dye/PbO, display increased photographic sensitivity and an increased photoconductive response, for exposures in the spectral region where the dye is the only absorbing component of the heterojunction structure. Such systems have been termed photographic transistors because of their amplified electronic and photographic response [2,12]. Other structures, such as, Dye/Ag_2S/AgBr/Dye have also been analyzed from the viewpoint of a solid state device [14j,k]. Here the energy band parameters suggest a voltage gradient across the AgBr phase capable of drawing photoelectrons toward the Ag_2S/AgBr interface, in agreement with electron micrographs of a partially developed exposed grain which indicate that this interface is indeed the site of silver reduction within the grain [15]. Structures of this type have much in common with photovoltaic cells [16]. Multi-junction silver halide compositions with abrupt AgI, AgBr and AgI_xBr_{1-x} interfaces have very high photographic sensitivity believed to result from an internal field which contributes to the charge carrier separation and greatly extends the lifetime of the photoelectrons as measured by time resolved photocharge (TRPC) as described below [8,9,11,14j,k].

In what follows, is a description a number of particulate heterojunctions designed with the intent of increasing the system response beyond that attainable with a single phase photoconductor. The TRPC technique for characterization of transport properties in single and multi-junction photoconducting materials will also be described, along with illustrative examples from the silver halide, photovoltaic and photocatalytic fields.

2. TIME RESOLVED PHOTOCHARGE (TRPC)

2.1. Description of TRPC Measurement Technique

Front surface absorption of light in a photoconductor generates a concentration gradient of photoelectrons and photoholes. When the charge carriers have unequal velocities, the resulting diffusion current generates a field within the photoconductor, due to the separation of charge types. This, in turn, introduces a drift current, which under equilibrium conditions balances the diffusion current. When a short excitation pulse is used, there is a transient electronic response. The time resolved rise and decay of this photo-induced transient is the essence of the photocharge measurement. In the special case where there are no internal or surface fields present prior to exposure, the technique has been termed the Dember Effect [17]. In the presence of pre-existing fields, the technique has been termed the Photocharge Effect [12]. Pre-existing space charges can be collapsed by use of a CW bias light of sufficiently high intensity, as is commonly practiced in surface photovoltage measurements (SPV) [7]. It is obvious that a number of concurrent processes are occurring during a photocharge transient waveform, i.e. diffusion, drift, trapping, detrapping, recombination and possibly chemical reaction. In these early investigations [12a] and in more recent work [8,9], capacitors in series were used to model a semiconductor/substrate interface in order to extract information on mean charge carrier separation distances. This approach will be discussed in more detail below. Recent investigations by Willig and co-workers [18] on laser induced electronic transients across a semiconductor electrode/electrolyte interface, follow a similar modeling procedure.

2.2. Analysis of TRPC Transient for Determination of Charge Transport Parameters

In analyzing a TRPC signal to obtain a measure of the mean charge carrier separation distance, the procedure we have employed is to observe the instant in time, t_m, at which the absolute value of the TRPC signal amplitude, V_m, reaches a maximum. From the magnitude of V_m and the assumption that the quantum yield of charge carriers is 1, the mean separation distance of photo-induced charge carriers that can account for the experimentally determined V_m is obtained. V_m has been shown to be proportional to L_m, the separation distance of the center of mass of the photoelectrons from the center of mass of the photoholes at the time, t_m, for the reasons similar to those cited above by Tiedje [6], as the basis for time of flight (TOF) measurements. Namely, the photo-induced diffusion current contributes to the charging of the plates of the sample cell in proportion to the number of charges moving and their mean separation distance from charge carriers of opposite sign.

Figure 1 presents a schematic diagram of the TRPC apparatus. It consists of a parallel plate spring loaded sample cell, with 0.5 mil transparent blocking contacts. The sample can be any photoconductor (single crystal, thin film, particulate solid, particulate or colloidal suspension or liquid). When liquid photoconductors or dispersions are used a special sample cell is employed [19]. For simplicity, we show a thin film photoconductor coated on a glass substrate. The photoconducting sample and the substrate or for the case of a particulate coating using an insulating binder, the sample and the binder, can be considered the equivalent of capacitors connected in series (see Fig. 2). For strongly absorbed light, charge carriers produced by a 10ns laser pulse are generated close to the front surface. To insure that the major portion of the charge carrier concentration gradient lies largely beyond the depletion layer a CW light source emitting at weakly absorbed wavelengths has been employed to collapse the depletion layer, thereby minimizing contributions to the photocurrents caused by drift, as in the SPV measurements [7].

For fixed light intensity and constant substrate thickness, it has been shown [14a,9] that L_m = (const.)V_m . The constant term contains experimentally accessible quantities, such as the sample dielectric constant, sample thickness and sample cell calibration parameters, obtained by observing the induced signal across the sample cell electrodes, with sample in place, when a known charge is injected onto one of the electrodes via step voltage charging of a small test capacitor attached to one of the sample cell electrodes. From this we can obtain the correspondence between charge and observed signal magnitude, which

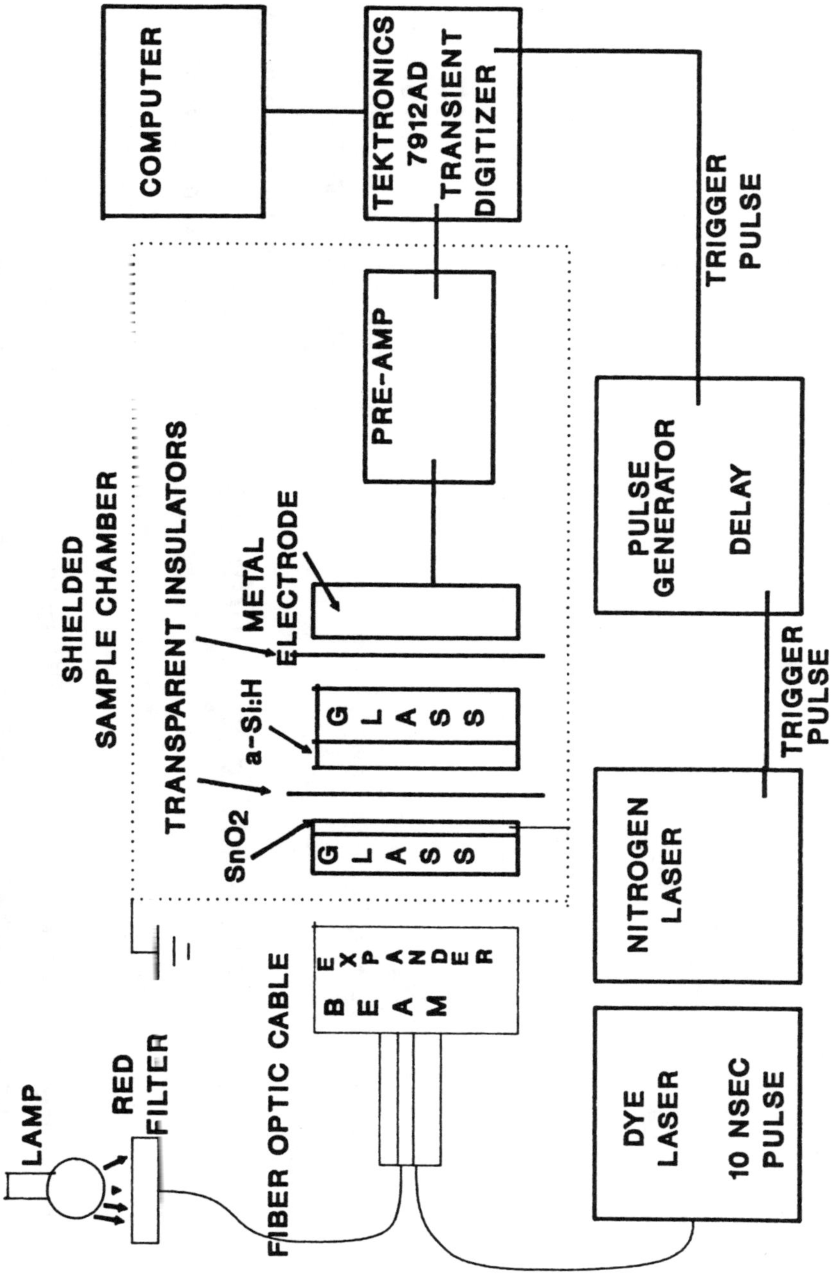

Fig.1 Schematic diagram of photocharge apparatus.

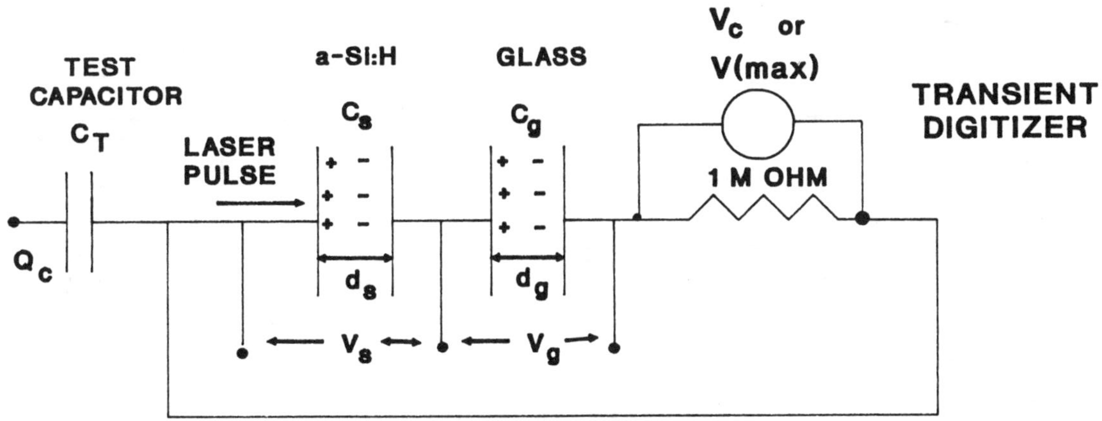

$$V(max) = V_s + V_g$$
$$1/C = 1/C_s + 1/C_g$$
$$C_T \ll C$$
$$L(max) = (Q_c/V_c eN)(d_g k_s/k_g)V(max)$$

$$\boxed{L(max) = KV(max)}$$

V(max) = max TRPC amplitude
K_s = diel. const. of sample
K_g = diel. const. glass
e = electron charge
N = number absorbed quanta
Q_c = charge on C_T
V_c = voltage with Q_c on C_c

Fig.2 Equivalent circuit for TRPC measurement showing basis for determination of charge carrier separation, L(max).

together with knowledge of the number of incident quanta per pulse and the absorption characteristics of the sample, can be used to obtain the mean separation distance of the charge carriers, L_m, that can account for the observed signal. Or from a pragmatic standpoint, the relative values of L_m or V_m, may be used to evaluate trends resulting from variations in material parameters.

The mean ambipolar diffusion velocity of charge separation, v, is given by $v = L_m/t_m = (\text{const.})V_m/t_m$. This determination of velocity, v, of the charge carriers is related to the charge carrier mobility, as obtained by TOF, but unlike the mobility, is independent of the field.

Surface and internal fields can influence the magnitude and direction of the photo-induced currents [14a-c,9]. As mentioned above, experiments indicate that continuous high intensity exposure, at wavelengths sufficiently long to assure uniform absorption throughout the sample thickness, collapses the depletion layer, making it possible to obtain the transport parameters L_m, t_m and v, characteristic of the photoconducting sample, under the same assumptions as have been employed in SPV measurements. On the other hand, should one wish to study interfacial effects due to adsorbed or epitaxial deposits, one can perform experiments with and without the CW high intensity exposure, thereby determining by difference the modulation of these parameters resulting from space charge effects. In this manner, TRPC measurements utilizing blocking contacts consisting of thin transparent insulating layers to electrically isolate the sample, can be effectively used to study space charge effects and other electronic properties of the photoconductor imposed by the environment (i.e. electrodes, passivating layers, photoconducting p-layers and n-layers, spectral sensitizing dyes, etc.) with which it is brought into contact.

For the case of amorphous silicon, α-Si:H, the introduction of defect states by light soaking has been widely studied by such techniques as SPV and ESR. TRPC is also capable of following the degradation effects of light soaking and of providing mechanistic insights by analysis of the kinetics of photocharge decay according to concurrent first and second order processes [14d,e, 14g-i]. These experiments are viewed as support for the interpretation of the photocharge analysis presented above. Similarly, quantitative correlation of photoconductivity and photocharge data have been obtained for α-Si:H samples. Relative mobilities in a series of single crystal Si wafers were compared to the charge carrier velocities, v, as obtained by TRPC and found to be in close agreement. α-Si:H samples, whose ambipolar diffusion lengths were determined by Moore [7] using SPV, were also evaluated by the TRPC technique to obtain the mean separation distance of charge carriers, L_m. The relative values of these parameters were in good agreement [8,9].

A typical TRPC signal for a silver halide photographic emulsion coating is given in Fig. 3. Concurrent first plus second order decay kinetics is frequently found to conform to the experimental data. Correlation of this type of data with photographic data is interpreted on the basis of a competition between the

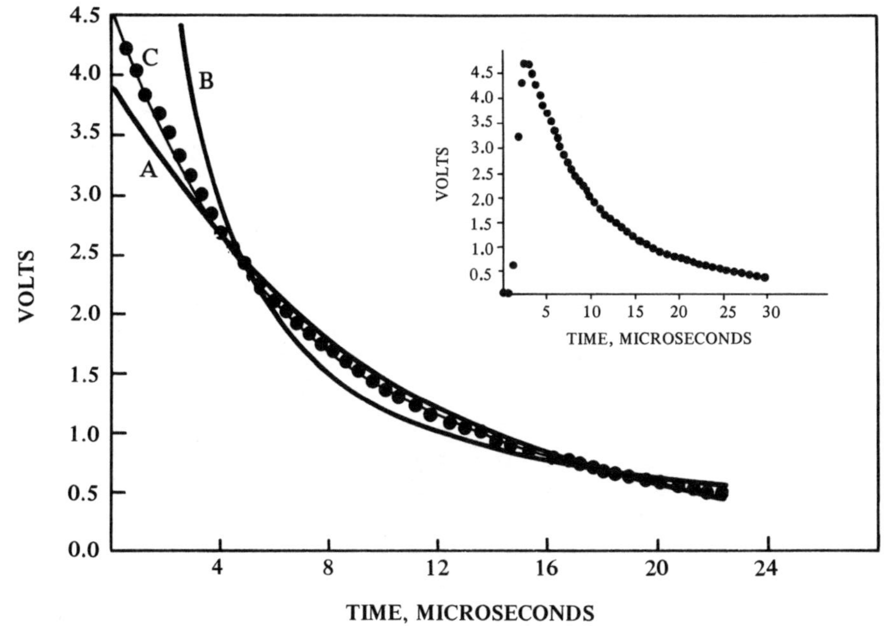

$$-\frac{dV}{dt} = k_1 V + k_2 V^2$$

$$V = 1 \Big/ \left[\left(V_o^{-1} + {k_2}/{k_1} \right) e^{k_1 t} - {k_2}/{k_1} \right]$$

Fig. 3. Insert is photocharge signal generated by a 200 mw/cm^2 light of ≈ 50 ns duration incident on a silver bromide photographic emulsion coating. The signal is amplified by 1000X. The main figure shows the computer fitting of the data points after the onset of signal decay according to a) first order, b) second order and c) concurrent first plus second order decay kinetics [14e].

first order decay process, believed to lead to latent image formation, with the second order recombination process.

3. SILVER HALIDE HETEROJUNCTIONS

3.1 Spectrally Sensitizing Photographic Diodes Formulated From Particulate Dispersions

Spectral sensitization and supersensitization has been reported in the past for heterojunctions formulated by mixing inorganic photoconductor dispersions (inorg. PC) with melted silver halide photographic emulsions [2,12a-c]. Particulate organic sensitizing dye dispersions have also been reported [20]. Photographic diodes, comprising a single junction, AgX/(inorg.PC), provide spectral sensitization for both negative and positive working AgX systems [12a,b]. PbO(orthorhombic)/AgX was one of the early systems investigated [2,12b]. Transmission and scanning electron micrographs of a PbO/AgCl system, formed from a particulate dispersion of PbO combined with a AgCl emulsion, are shown in Figs. 4-5, suggesting intimate contact between the PbO and the AgCl. Electron and hole transport across the junction between sheet crystals of AgBr and AgCl, under light pressure, has also been demonstrated [21], indicating that gelatin is not a necessary component for charge transport across an interface composed of these materials. Systems of this type have a number of interesting features: 1) PbO (orthorhombic) spectrally sensitizes AgCl; 2) PbO (tetragonal) does not spectrally sensitize AgX; 3) Other oxides of lead (Pb_3O_4, PbO_2, Pb_2O_3 do not spectrally sensitize AgX; 4) There is a correspondence between the photographic action and photocharge spectra in the PbO/AgCl system in the spectral region beyond 420nm where only the PbO absorbs a significant amount of light (see Figs. 6-8).

Figure 9 is a proposed energy band diagram which accounts for these diverse experimental observations. An interesting feature is the positive space charge in the thermodynamically equilibrated AgCl phase prior to exposure to light, which can explain the frequently observed desensitization in the region of intrinsic absorption of the AgCl often accompanying spectral sensitization by efficient sensitizers. Under conditions where there is a strong overlap in the absorption spectrum of the AgX with that of the spectral sensitizer, as is found in the AgBr (orthorhombic) system, an enhancement of the photographic response in the AgBr intrinsic absorption region is possible, despite the presence of a positive space charge in the silver halide phase. Other energy band configurations which can produce spectral sensitization, as well as an enhanced response in the intrinsic region of silver halide absorption, have been discussed in earlier papers [2,12a-c]. Similar issues are discussed in a recent study of spectral sensitization of silver halide, formed by evaporation of AgBr on top of evaporated layers of Ge, GaP, PbO, GeS2, SnS or TiO_2 [22]. In confirmation of earlier results, Ge, GaP and PbO sensitized AgBr. GeS_2 and SnS were also shown to be sensitizers for AgBr. TiO_2 functioned as a desensitizer which is not unexpected on the basis of the positions of the conduction band of TiO_2 (-4.2eV) and valence band (-7.2ev) [23]. Energy band

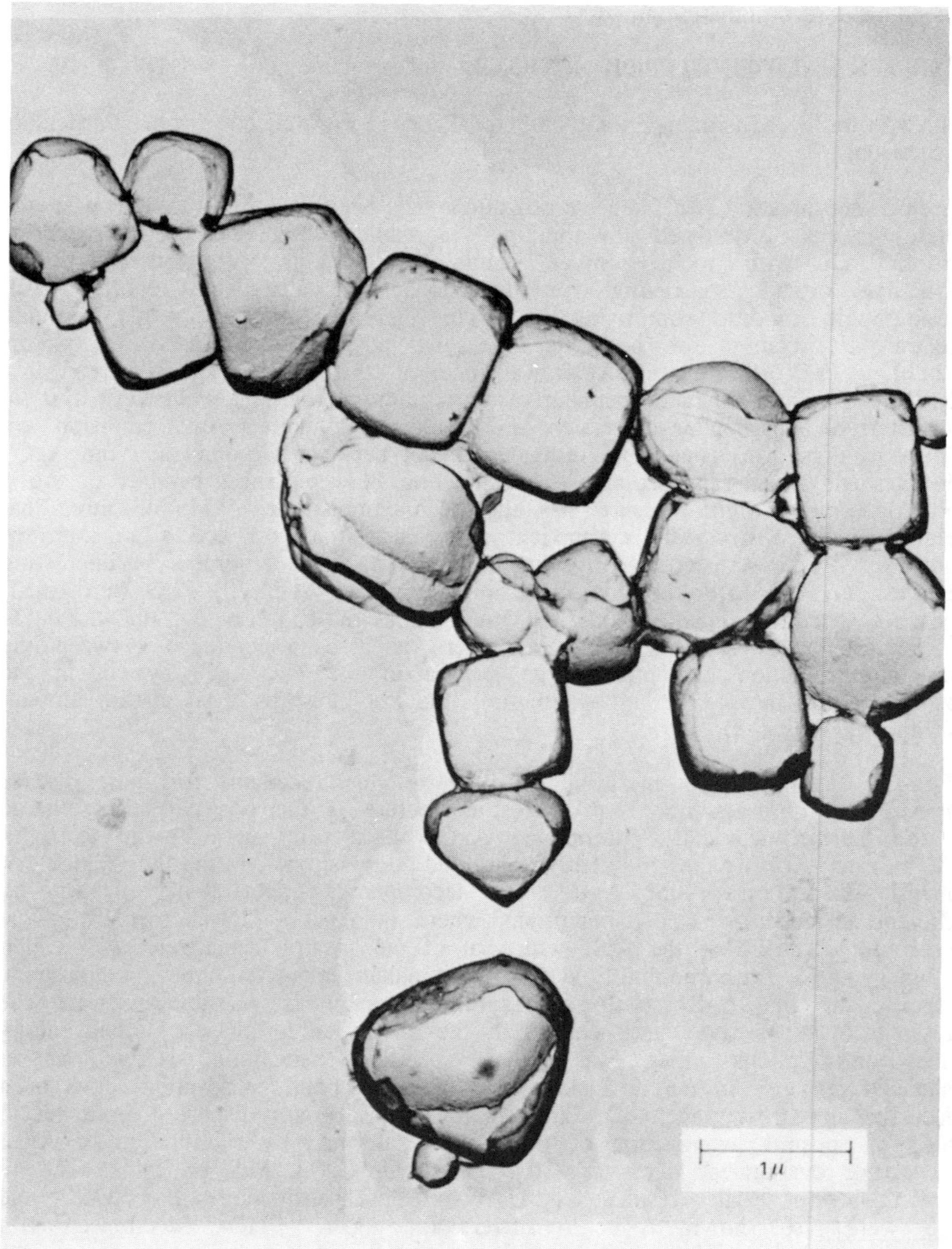
1μ

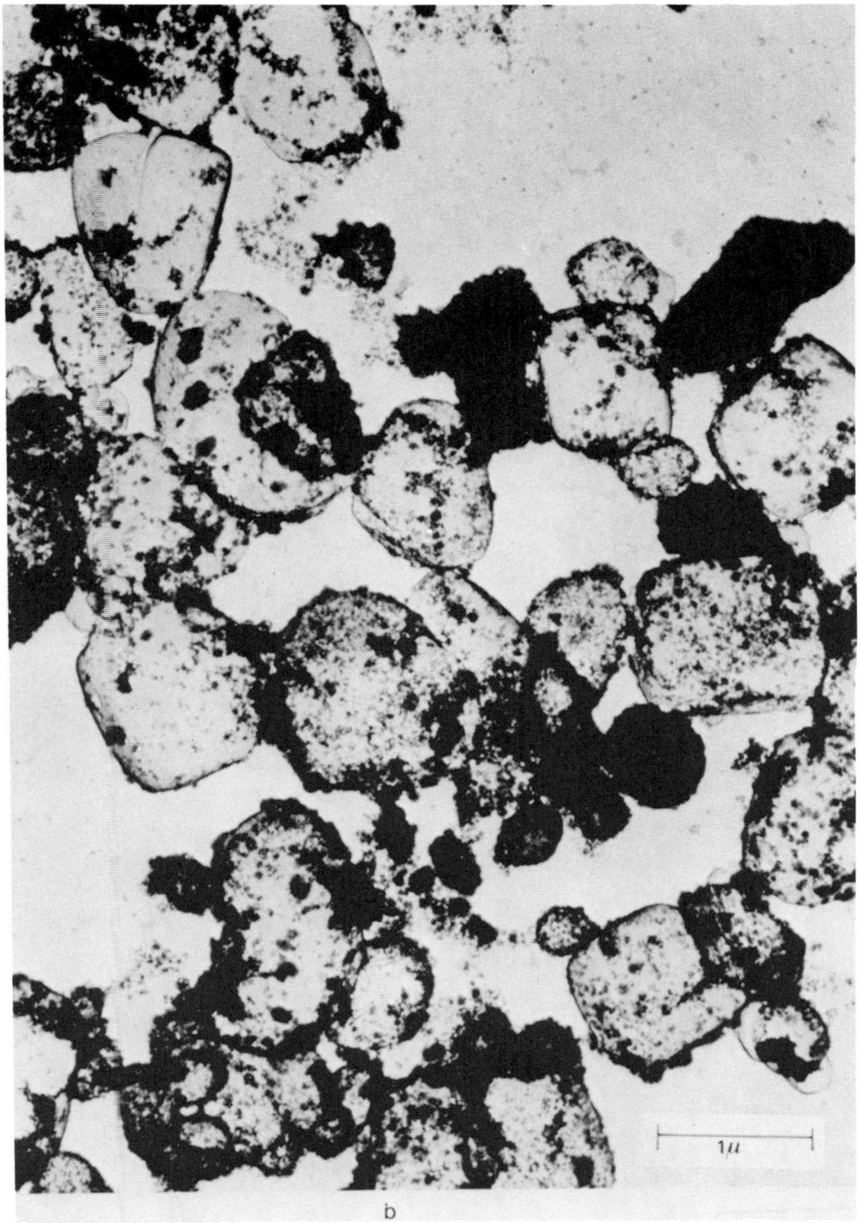

Fig. 4. Transmission electron micrographs of carbon replicas of a) AgCl emulsion grains and b) the same AgCl emulsion in contact with PbO (orthorhombic).

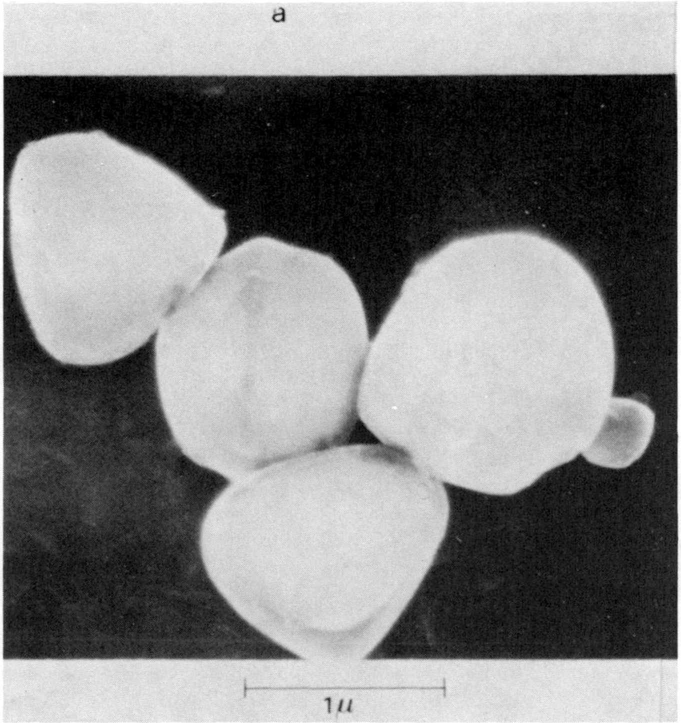
a
1μ

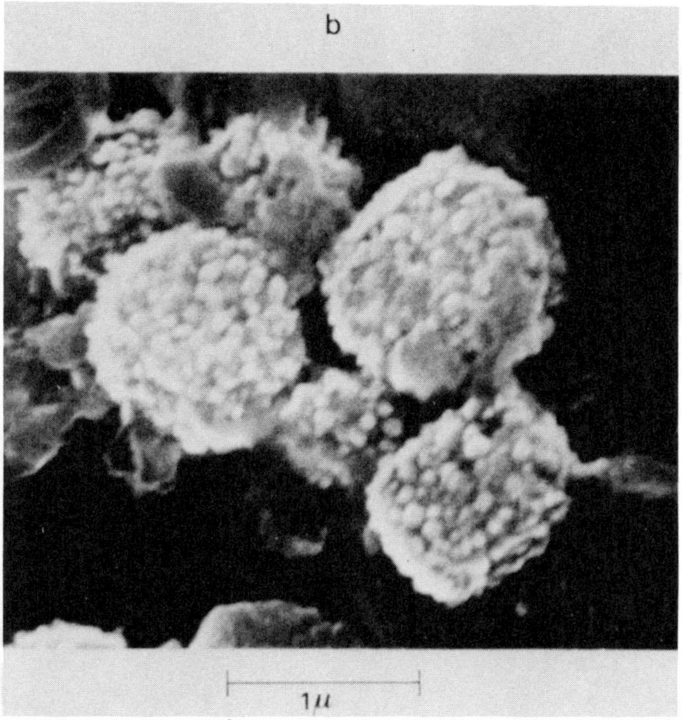

Fig. 5. Scanning electron micrographs of the emulsion grains of Fig. 4. a) AgCl, b) AgCl/PbO (orthorhombic).

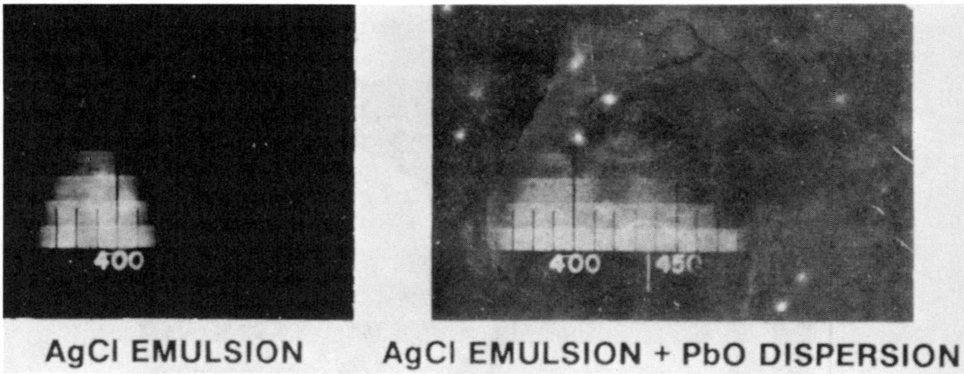

Fig. 6. Action spectra of coatings of emulsions of Figs. 4 and 5. a)AgCl, b) AgCl/PbO.

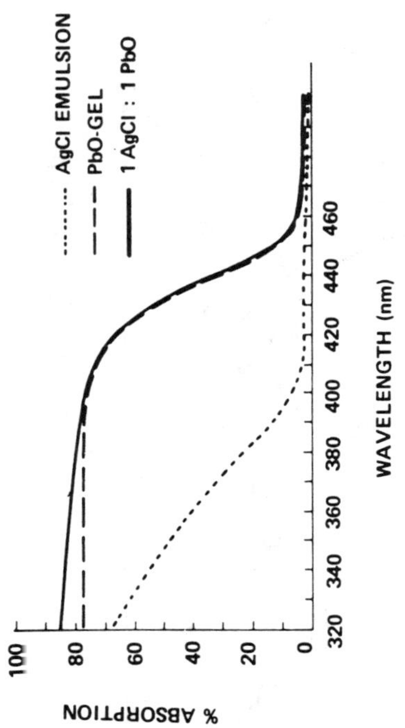

Fig. 7. Absorption spectra of the AgCl and AgCl/PbO coatings of Fig. 6. Also shown is the absorption spectrum of a PbO coating in the absence of AgCl.

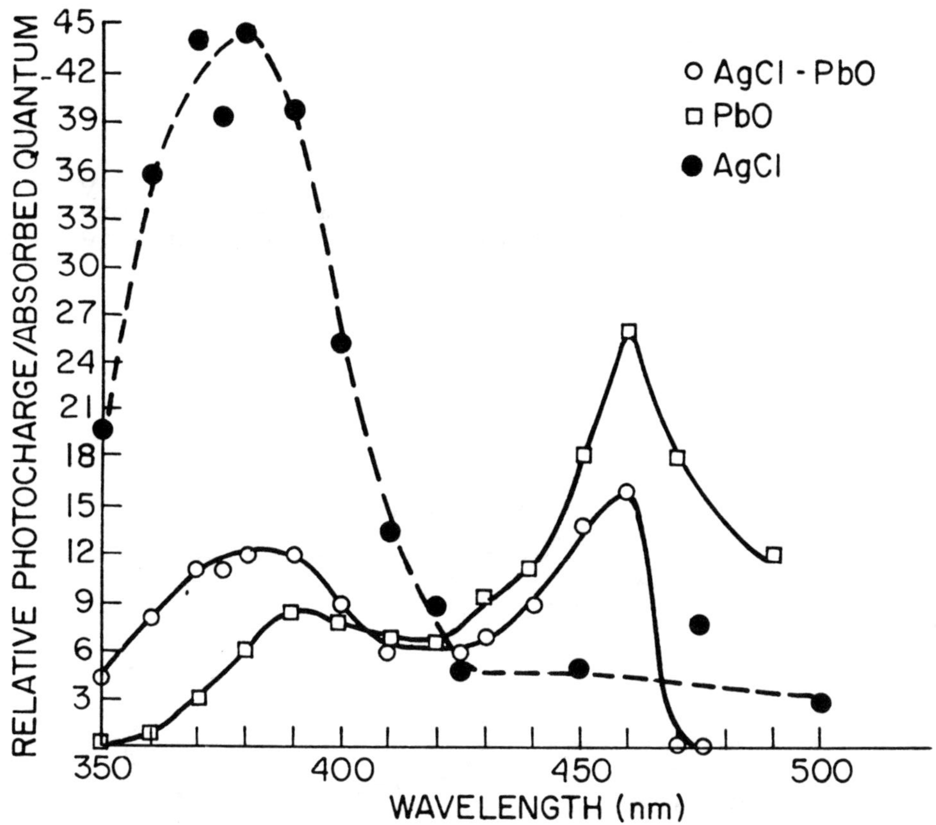

Fig. 8. Relative photocharge response per absorbed quantum of light at the photocharge maximum for the coatings of Figs. 7 and 8. Comparison of Figs. 6-8 shows a correspondence in the new peak at approximately 460 nm where only the PbO absorbs light and in thc desensitization in the intrinsic region of AgCl absorption.

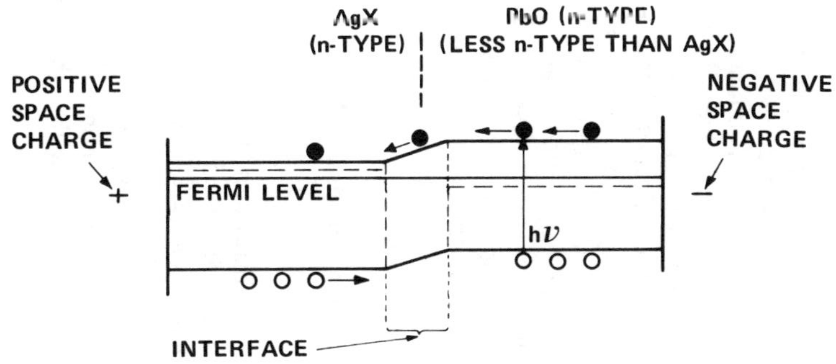

Fig. 9. Proposed energy band structure for the AgCl/PbO photographic diode. After Fermi level equilibration the dye can inject photoelectrons into the AgCl conduction band, accounting for the observed spectral sensitization of the photographic action and of the photocharge spectra. The positive space charge in the AgCl phase accounts for the desensitization of the response in the inherent region of AgCl absorption.

configurations for other photoconducting sensitizers, to be discussed below, can be shown to provide some degree of spectral sensitization accompanied by a dominant sensitization in the region of AgX intrinsic absorption. Phenomenologically these systems are usually described as having been chemically sensitized.

Photoelectron emission investigations of the energy band parameters of PbO (orthorhombic) and PbO (tetragonal) have been performed using the retardation potential technique [2], with the intent of developing an understanding of the failure of the tetragonal form to spectrally sensitize silver halides. These studies indicate that the bottom of the conduction band of PbO (orthorhombic) is >leV higher in energy than the bottom of the conduction band of AgBr and of AgCl, which may in part account for its ability to spectrally sensitize these photoconductors. The situation for PbO (tetragonal) is quite different. Here, the conduction bands of PbO (tetragonal) and of AgBr and AgCl are closer in energy. After Fermi Level equilibration with the silver halides, during formation of the heterojunction diodes, the bottom of the conduction band of PbO(tetragonal) may be below that of AgX, thus creating a barrier for electron injection into the silver halide. Similar studies may be useful for the anitase and rutile forms of TiO_2.

3.2 Chemically Sensitized Photographic Diodes Formulated From Particulate Dispersions

Sensitization within the silver halide intrinsic absorption band, accompanied by a low level of spectral sensitization, has been described for a broad range of elemental and compound inorganic photoconductors [2,24] (see Figs. 10-11). Spectral sensitization of the photographic response is demonstrated by high levels of exposure through a filter transmitting to wavelengths beyond the AgBr intrinsic absorption band. An energy band diagram consisting of a diode structure has been proposed which is consistent with the photographic properties of these systems (see Fig. 12).

Sulfur and sulphur plus gold have traditionally been found to be outstandingly efficient chemical sensitizers, while also providing a low level of spectral sensitization [25]. These later systems will be discussed in the section dealing with epitaxial heterojunctions.

3.3 Photographic Transistors: Double Junction Devices Formulated From Particulate Dispersions

Double junction systems comprising AgX/Dye/(inorg. PC) have been discussed in the literature and the term, "photographic transistor" [2,12c], has been employed to underscore the capability of these systems to provide an amplified photographic, electronic and ionic response over the entire spectral region of light absorption by the system. Action spectra for the AgBr/Dye/PbO system are shown in Fig.13. The dye is a thiocarbocyanine. Order of addition of dye and inorganic sensitizer phases have been shown to be important for photographic

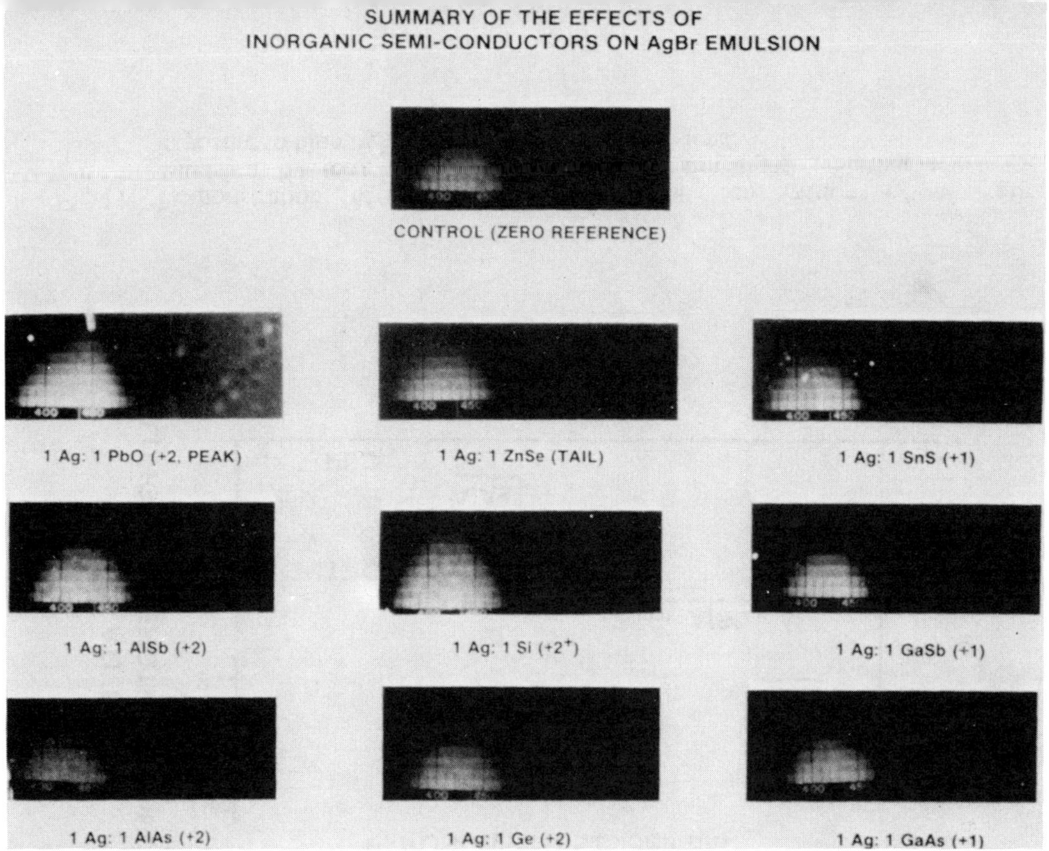

Fig.10. Action spectra of a AgBr emulsion sensitized with dispersions of inorganic semiconductor, demonstrating enhanced response in the intrinsic region of AgBr absorption. The mole ratios of AgBr to semiconductor sensitizer is 1:1. In parenthesis is the relative photographic speed increase in stops (each stop is a factor of 2 in exposure).

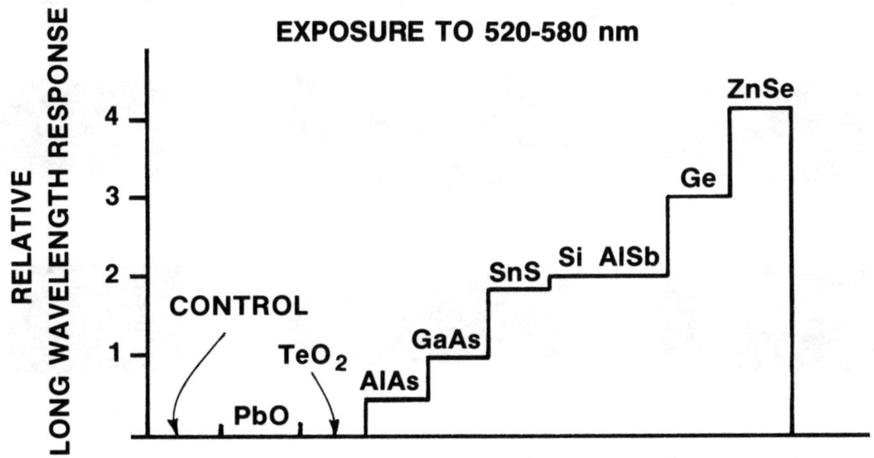

Fig.11. Demonstration of low levels of spectral sensitization of a AgBr emulsion by long exposure to 520 - 580 nm light. Similar levels of exposure to blue light over exposes these emulsions.

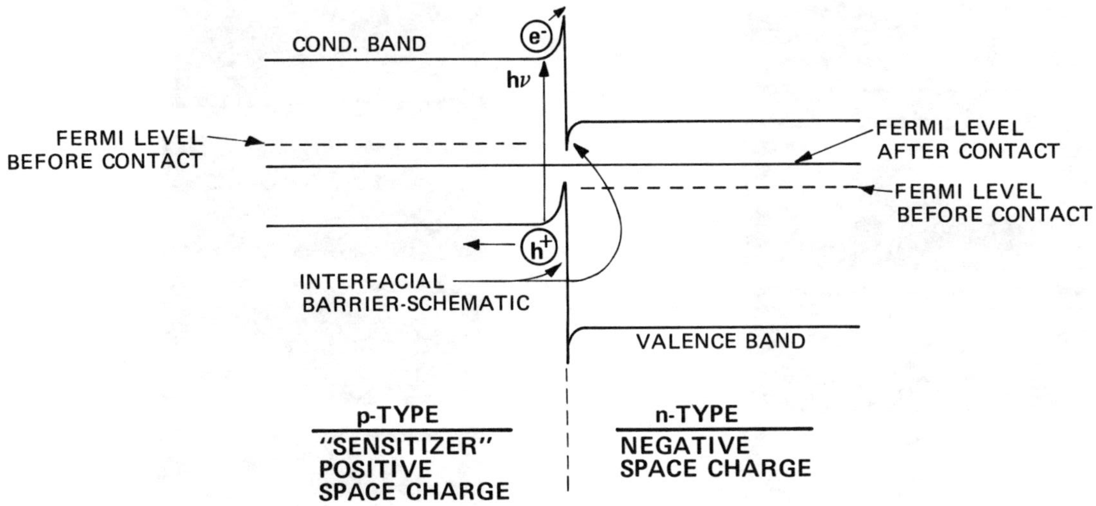

Fig.12. Proposed energy band diagrams to account for the photographic responses of Figs. 10 and 11. Here the sensitizer injects electrons into the AgBr or the AgBr injects positive holes into the sensitizer phase during Fermi level equilibration prior to exposure. This accounts for the increase in response to light within the AgBr absorption band. Low levels of spectral sensitization are also possible at exposure levels of sufficient intensity to remove any barrier that may be present caused by band bending at the sensitizer/AgBr interface.

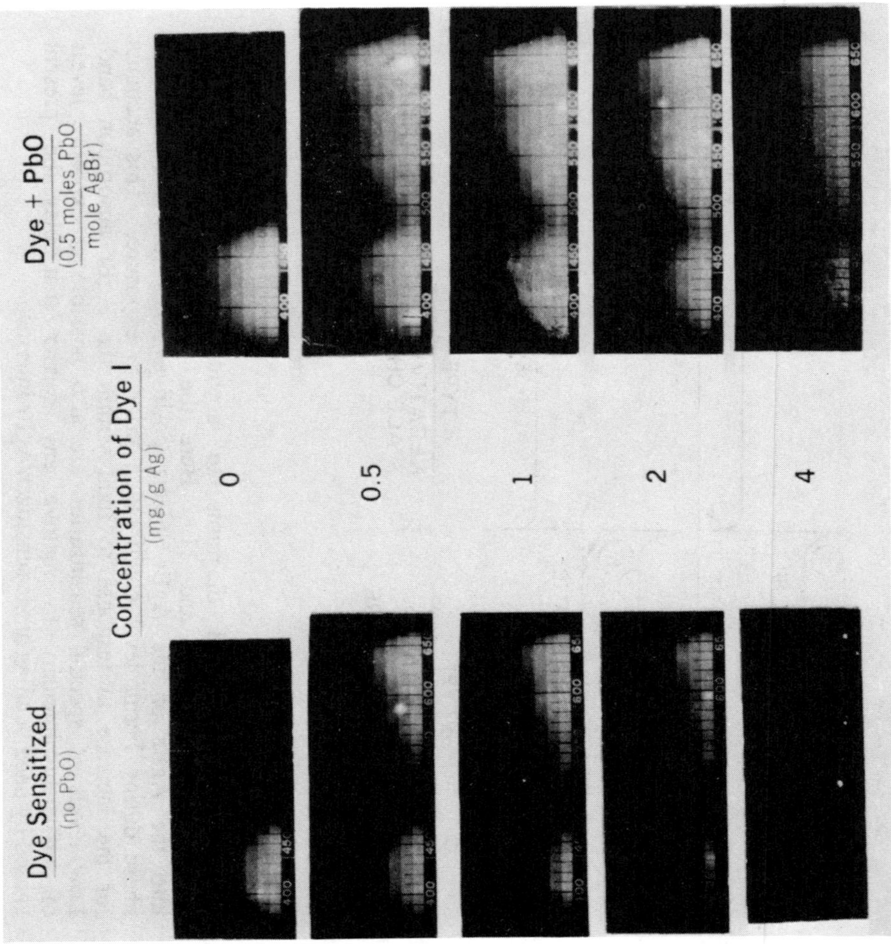
Dye Sensitized
(no PbO)
Concentration of Dye I
(mg/g Ag)
0
0.5
1
2
4
Dye + PbO
(0.5 moles PbO / mole AgBr)

Fig.13. Action spectra of the AgBr/PbO photographic diode and AgBr/Dye/PbO photographic transistors using 0.5 moles PbO per mole AgBr and 0.5 - 4 mg of benzimidazolcarbocyanine dye per g Ag. Also shown are the corresponding action spectra without PbO.

transistor action in certain instances, as shown in Fig. 14 using AgBr, PbO and a benzimidazolcarbocyanine dye, implying that the detailed structure of the heterojunction device is of importance in formulating efficient transistor imaging devices.

Some sensitizers (organic or inorganic) which may be of only marginal practical interest as spectral sensitizers, when used in diode formulations, acquire a substantially more practical aspect when formulated along with a second sensitizing material. A listing of systems in this category is given in reference [2]. Fig. 15 provides sensitometric information for a ZnO/Dye/AgX photographic transistor.

The correspondence of the enhanced photographic action and photocharge spectra, in a spectral region where the addition of PbO to a dye sensitized AgBr emulsion does not increase the total absorption of light by the system, indicates that the photographic as well as the electronic response to light absorption is being amplified (see Fig. 16). Proposed energy level diagrams for a Dye/AgX photographic diode and a PbO/Dye/AgX photographic transistor are given in Fig. 17.

3.4. Comparison of Photographic Diode and Transistor Theory with Other Theories of Chemical and Spectral Sensitization

In the systems described above, each AgX grain forms part of a distinct and separate imaging device, the coating as a whole consisting of the order of 10^6 devices/mm^2. The large compressive forces, provided by the dry gelatin envelope surrounding each of the individual heterojunctions formed in those devices fabricated by mixing an (inorg. PC) dispersion with a melted AgX gelatin emulsion, are believed to be at least partially responsible for insuring intimate contact at the (inorg. PC)/AgX junction [26]. This facilitates equilibration of the system prior to exposure and creates space charge and energy level conditions favoring spectrally sensitized silver cluster formation during and immediately following exposure. Similarly, the chemical sensitizers are considered to acquire properties during dark equilibration of the individual phases that govern the light-induced response of a given system. Proposed mechanisms for chemical and spectral sensitization of AgX which do not consider the consequences of chemical equilibration of the individual phases prior to exposure overlook fundamental thermodynamic principles [2]. Equilibration plays a role in Mein's p-n mechanism for spectral sensitization [27]. However, some problems in interpretation have arisen as a result of Meier's assignment of AgX as a p-type photoconductor, despite the fact that the authors cited by Meier for this assignment have rescinded this original evaluation and have provided evidence that AgX is an n-type photoconductor [28].

Some of the difficulties associated with mechanisms which fail to account for the equilibration of sensitizers with the material to be sensitized, are illustrated schematically in the energy level diagrams of Fig. 18, where the electron

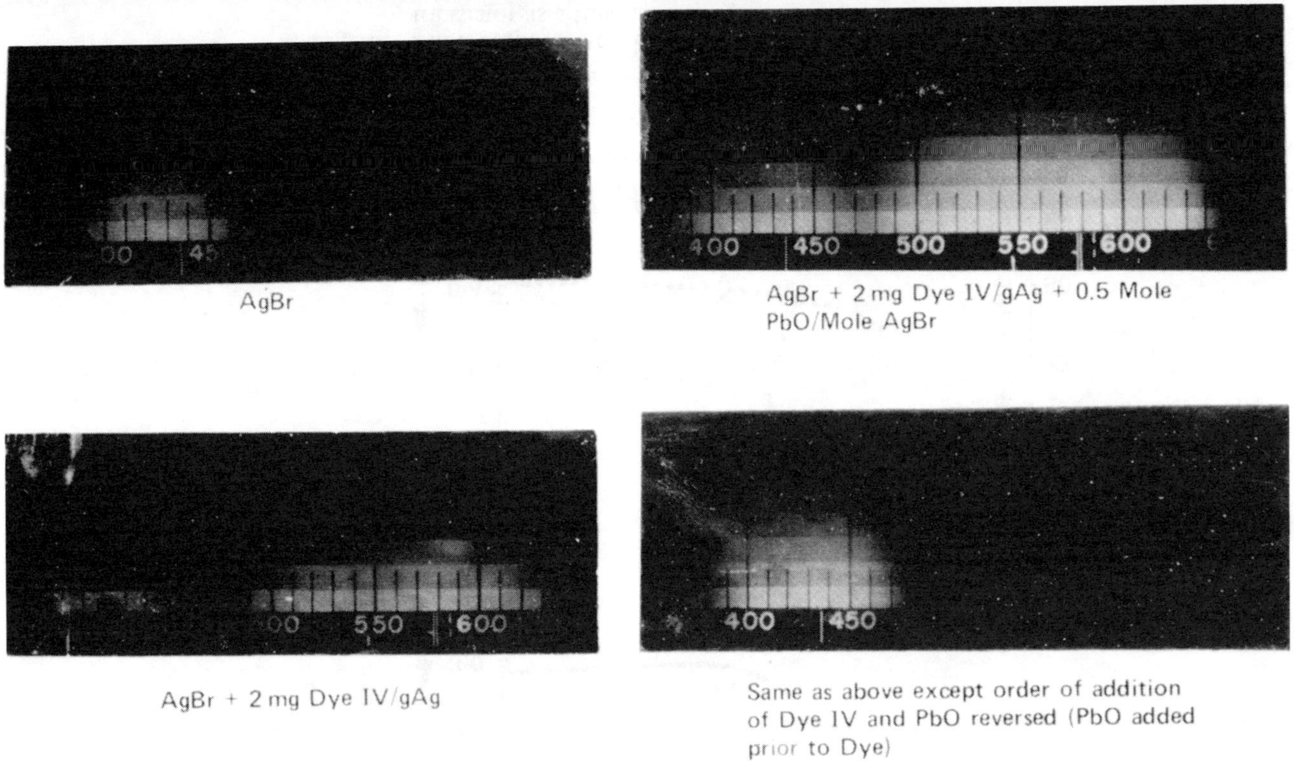

Fig.14. Action spectra of AgBr, AgBr + Dye, AgBr/Dye/PbO and AgBr/PbO/Dye. Only the AgBr/Dye/PbO system displays photographic transistor action.

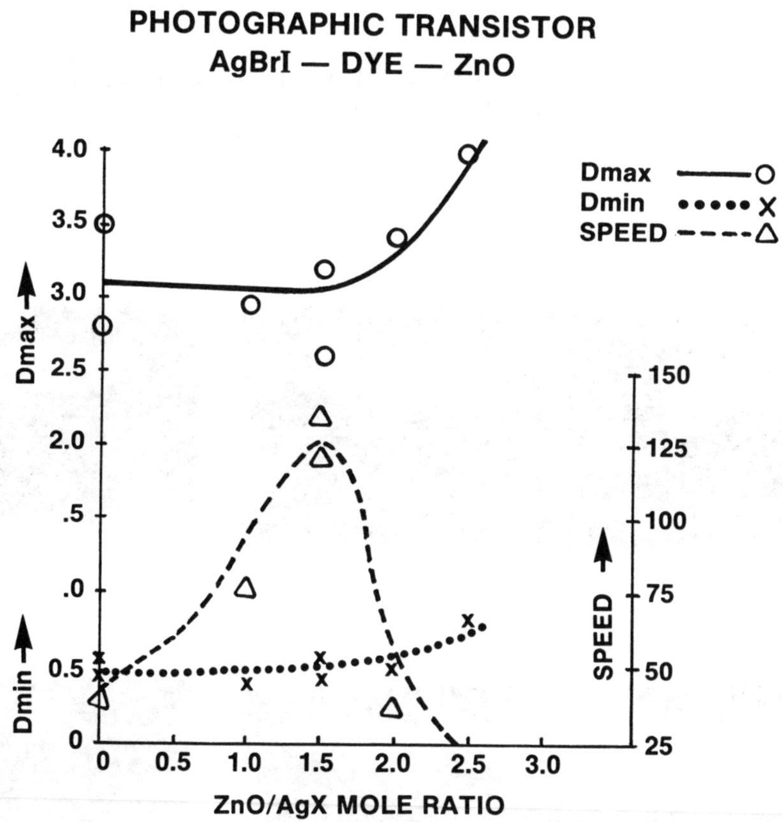

Fig. 15. Photographic sensitometric data for the AgBrI/Dye/ZnO photographic transistor as a function of AgBrI/ZnO mole ratio.

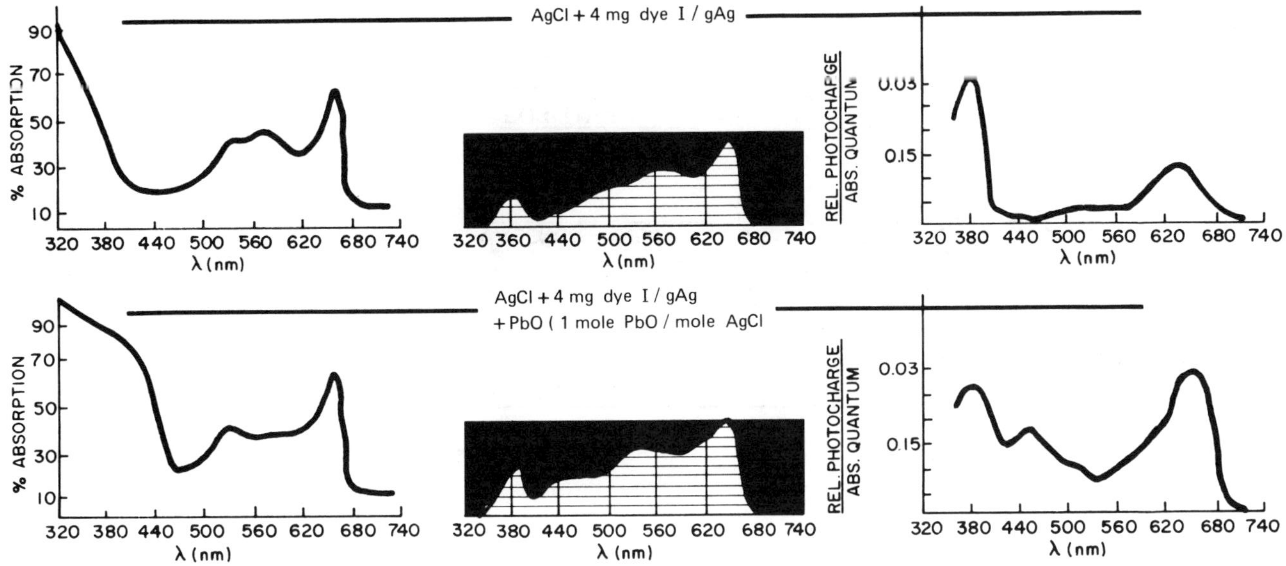

Fig.16. Absorption, photographic action and photocharge spectra of a dye sensitized AgCl emulsion and of the AgCl/Dye/PbO photographic transistor. PbO does not contribute to increased light absorption at approximately 660 nm where the dye J-aggregate has a strong absorption peak. However, with PbO present there is an amplification of the photographic sensitivity and the photocharge response at 660 nm.

a

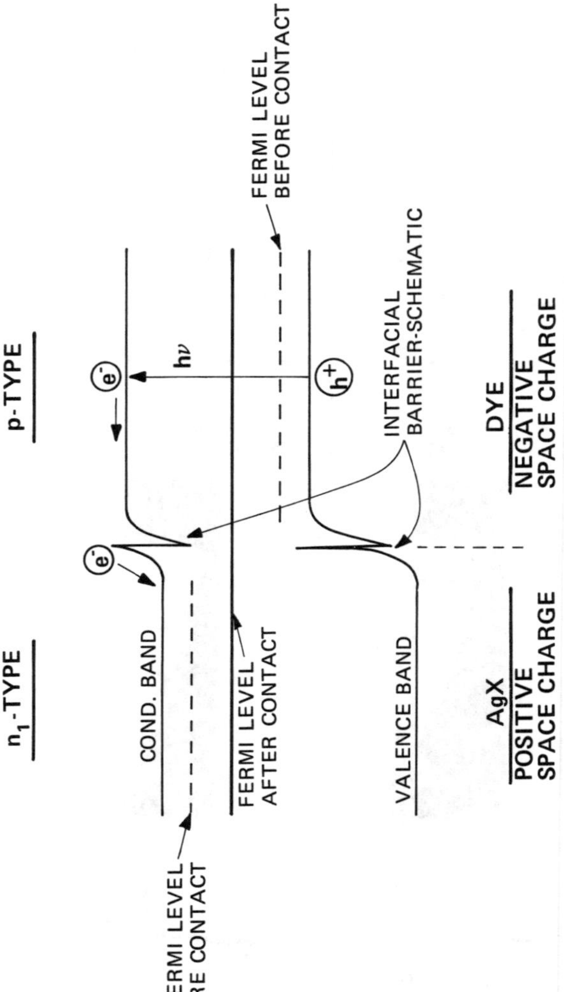
n1-TYPE
p-TYPE
COND. BAND
FERMI LEVEL
AFTER CONTACT
VALENCE BAND
FERMI LEVEL
BEFORE CONTACT
FERMI LEVEL
BEFORE CONTACT
INTERFACIAL
BARRIER-SCHEMATIC
hν
e-
e-
h+
AgX
POSITIVE
SPACE CHARGE
DYE
NEGATIVE
SPACE CHARGE

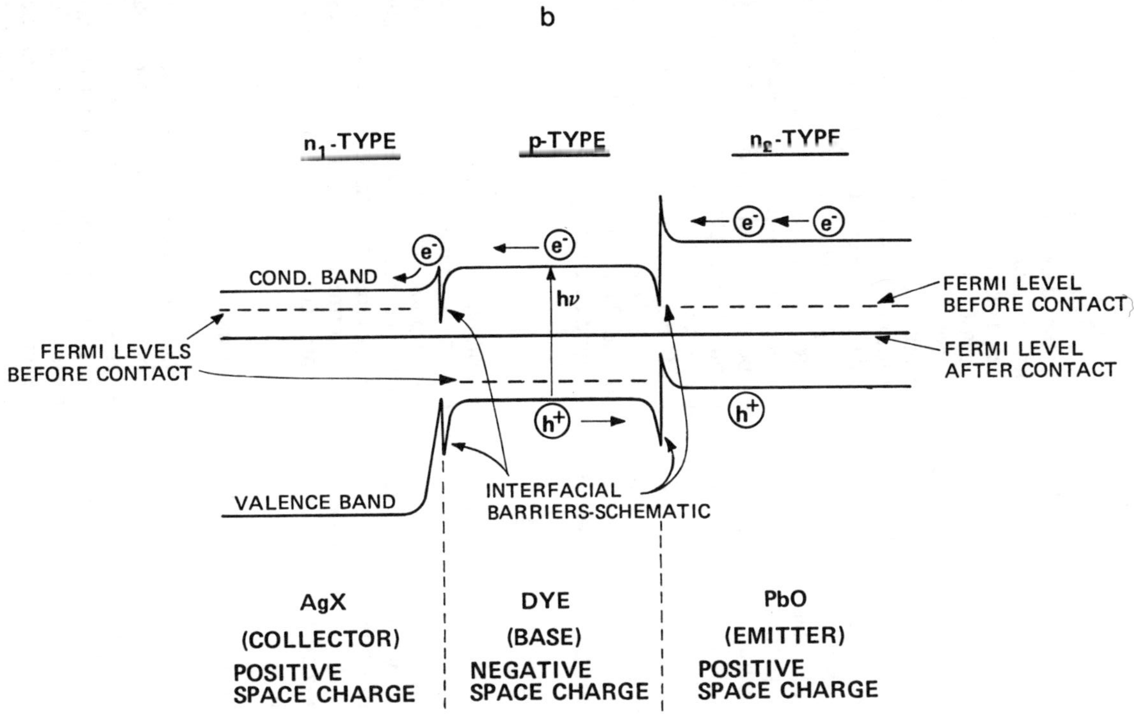

Fig.17. Energy band diagrams for a) AgCl/PbO photographic diode and b) AgCl/Dye/PbO photographic transistor. Light absorption by dye results in electron injection into the AgCl causing the adsorbed dye phase to become more positively charged. This results in a secondary current from the PbO. If the dye layer is thin electron tunneling to the AgBr phase gives rise to the observed photographic and photocharge amplification.

injection and energy transfer mechanisms for spectral sensitization are presented. In Fig. 18, the energy levels of the AgX phase and the adsorbed dye sensitizing phase are depicted as being separated in space and the adjustment of these levels during equilibration in the dark prior to exposure is not considered. Unfortunately, even in more modern versions of these mechanisms, dark equilibration continues to be neglected and the relative positions of the energy levels of the various phases are considered to shift relative to each other only after exposure. In essence, the system is pictured as not having come to equilibrium prior to exposure. It is only as a means of explaining certain specialized situations where spectral sensitization occurs using dyes that would otherwise be incapable of sensitization according to previously formulated theories based on measurements of polarographic oxidation and reduction potentials [29], that the energy levels are considered to be capable of adjusting their relative positions during and immediately subsequent to exposure. In support of the photographic diode model for spectral sensitization of AgBr by adsorbed dye aggregates, a photoelectron emission investigation of AgBr, Dye and of the combined AgBr/Dye system has shown that the Fermi Level of the AgBr/Dye system is intermediate between that of the separate AgBr and Dye phases. This is interpreted as evidence of a dark equilibration of the heterojunction by either: electron injection form the Dye phase into the AgBr phase, or interaction of the Dye dipole with the AgBr band structure [30].

3.5. Epitaxial Heterojunction Imaging Systems

Conditions for charge carrier transport and separation across phase boundaries are expected to be enhanced when the hetrojunction contacts are formed by chemical vapor deposition techniques or by adsorption or precipitation from solution, as compared to more or less random contact between particulate photoconductors of differing composition. To differentiate these two major classes of deposit, the former examples have been termed epitaxial for purposes of this paper. Sensitization by adsorbed dyes are considered to be epitaxial deposits [2,12a-c,14b,c].

3.5.1. Junctions Between Two or More Silver Halide Phases

Epitaxially grown systems comprising junctions between two or more distinct silver halide phases have been reported [11,15], having properties that can best be interpreted according to heterojunction principles. Spectral sensitization of one silver halide phase by another with which it is in epitaxial contact has been observed [11a], as predicted on the basis of the photographic diode mechanism. Energy band diagrams for several of these silver halide heterojunctions have been used to describe their properties in essentially the same manner as has been employed for the AgX/(inorg. PC) systems, which include AgX1/AgX2 systems as a sub-class. Photocharge measurements have been made on the green sensitive silver halide layer of the Konica 3200 ASA color product, verifying the long lifetime of the photoelectrons as obtained by microwave photoconductivity [11d]. The extension of the photoelectron lifetime is believed to be partially responsible for the high speeds obtained with these

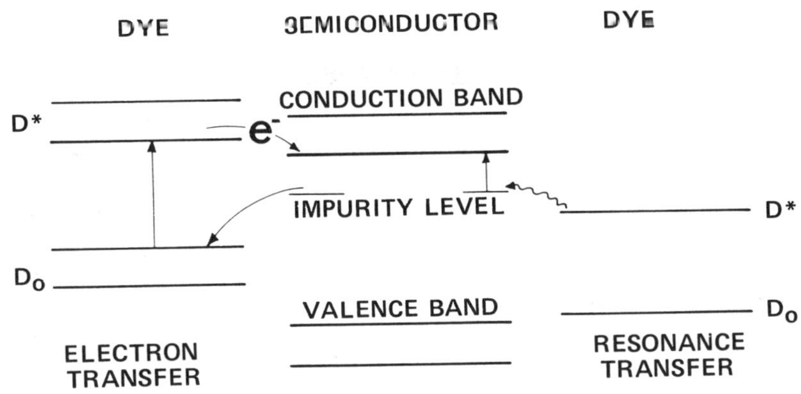

Fig.18. Electron injection and resonance energy transfer mechanisms for spectral sensitization.

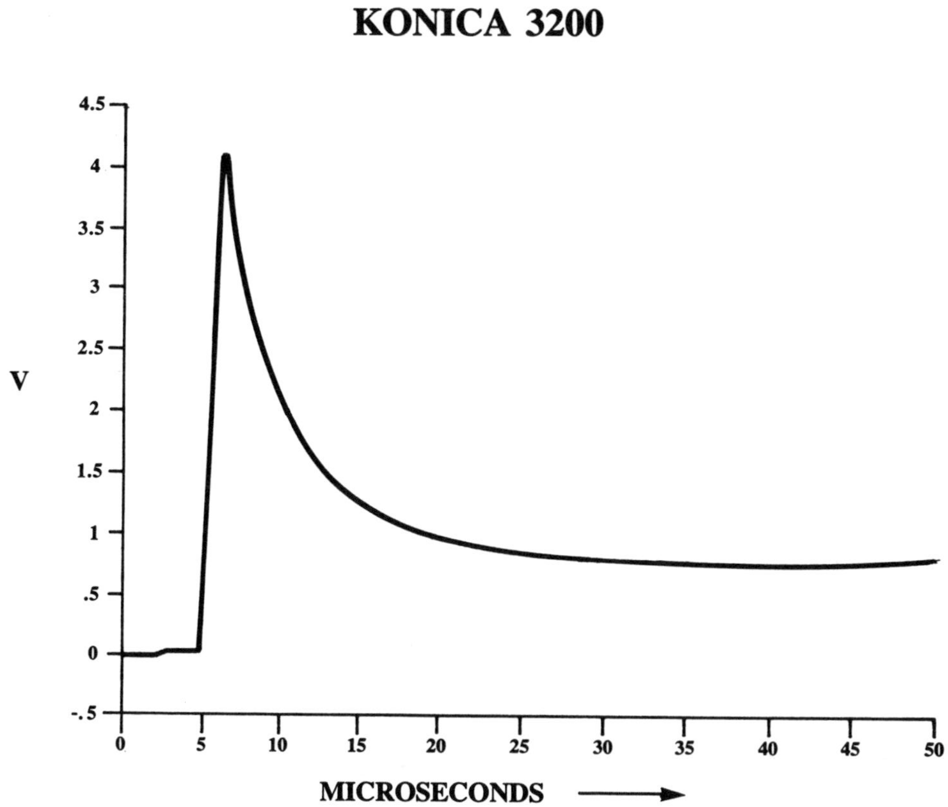

Fig.19. Photocharge transient signal with 560 nm exposure of the green sensitized layer in the Konica 3200 multi-junction photographic product. Persistence of the photocharge response is attributed to interphase electron migration which separates the photoelctrons from the dye positive hole.

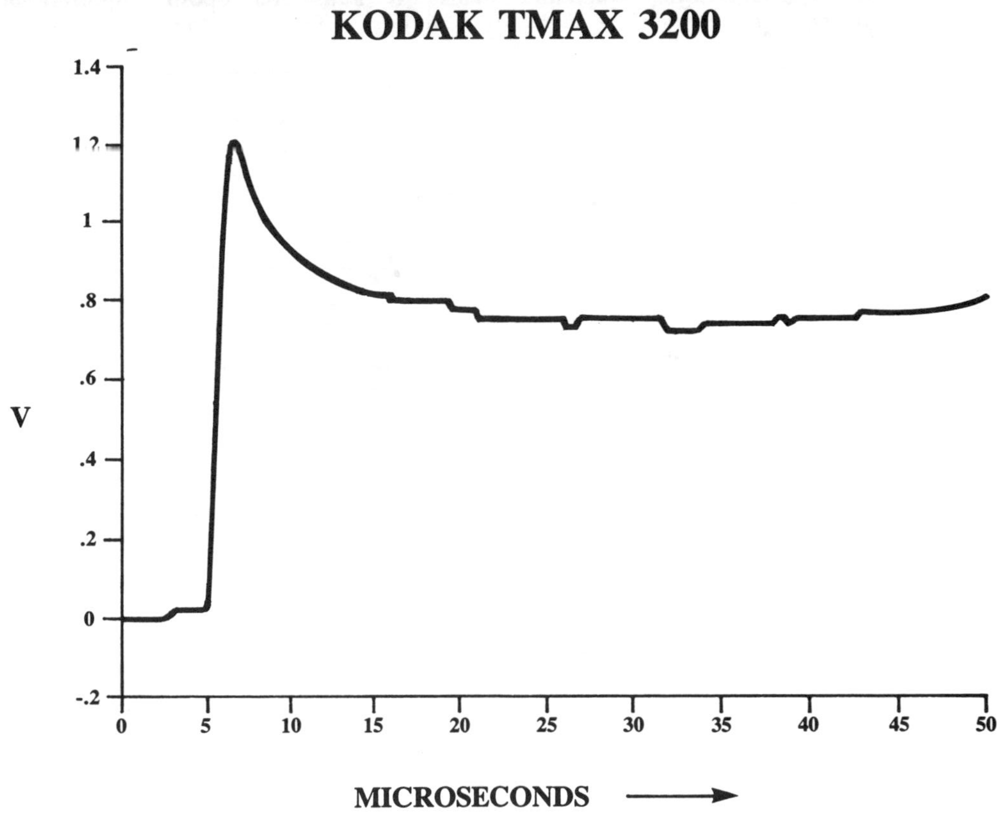

Fig.20. Photocharge transient from 560 nm exposure of the Kodak T Max 3200 product. Speculation concerning the origin of the signal persistence based on tabular AgBr grain with epitaxial AgCl corner deposits is given in text.

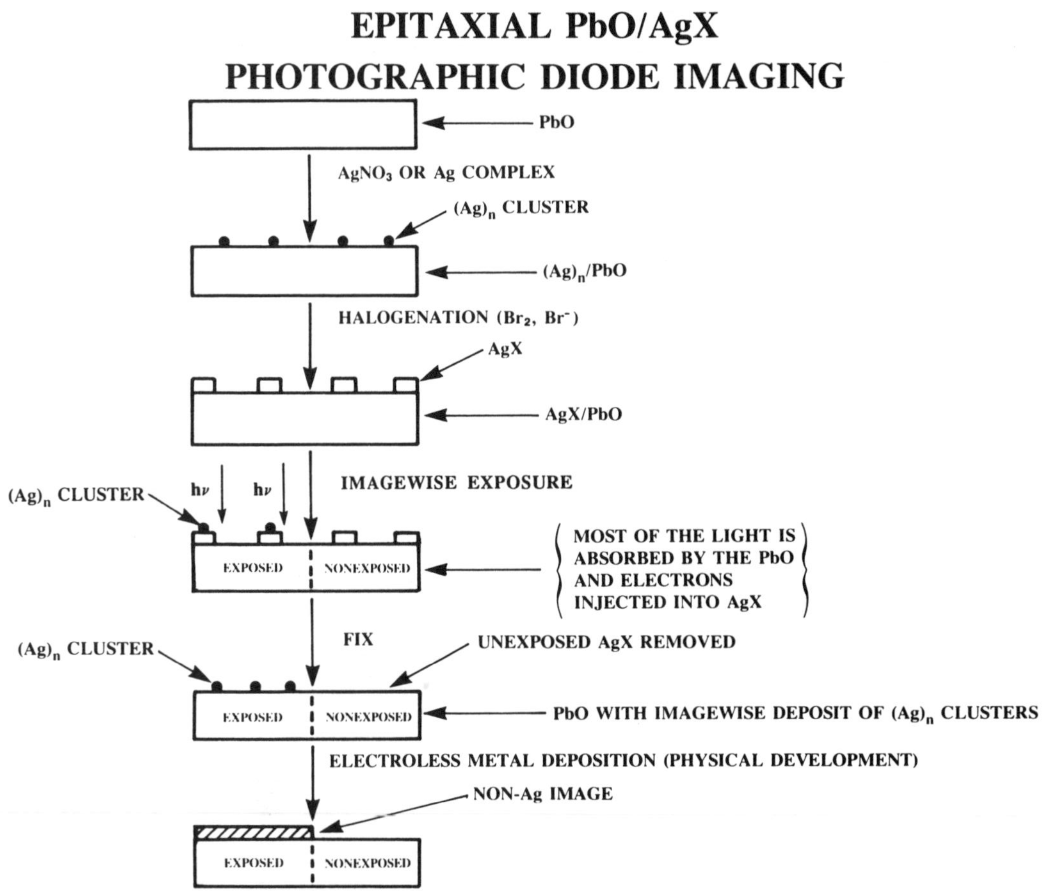

Fig.21. Preparation and imaging scheme for an epitaxial PbO/AgBr photographic diode consisting of small epitaxial AgBr deposits on

TRANSMISSION ELECTRON MICROGRAPH OF EPITAXIAL PbO/AgBr

a

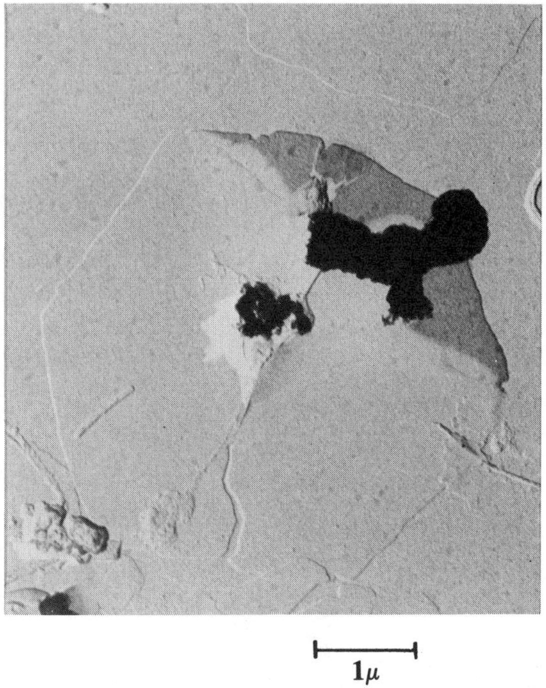

Fig.22. a) Transmission electron micrograph of AgBr epitaxial deposit on a PbO grain using preparation procedure outlined in Fig. 21. b) Microprobe X-ray fluorescence spectrum of dark cubic area attached to the PbO grain identified this region of the micrograph as AgBr.

b

STEM IMAGE

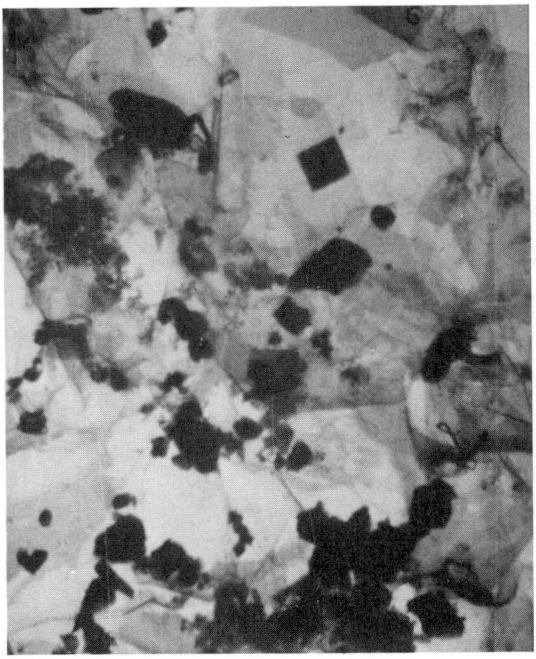

PbO + AgBr Replica

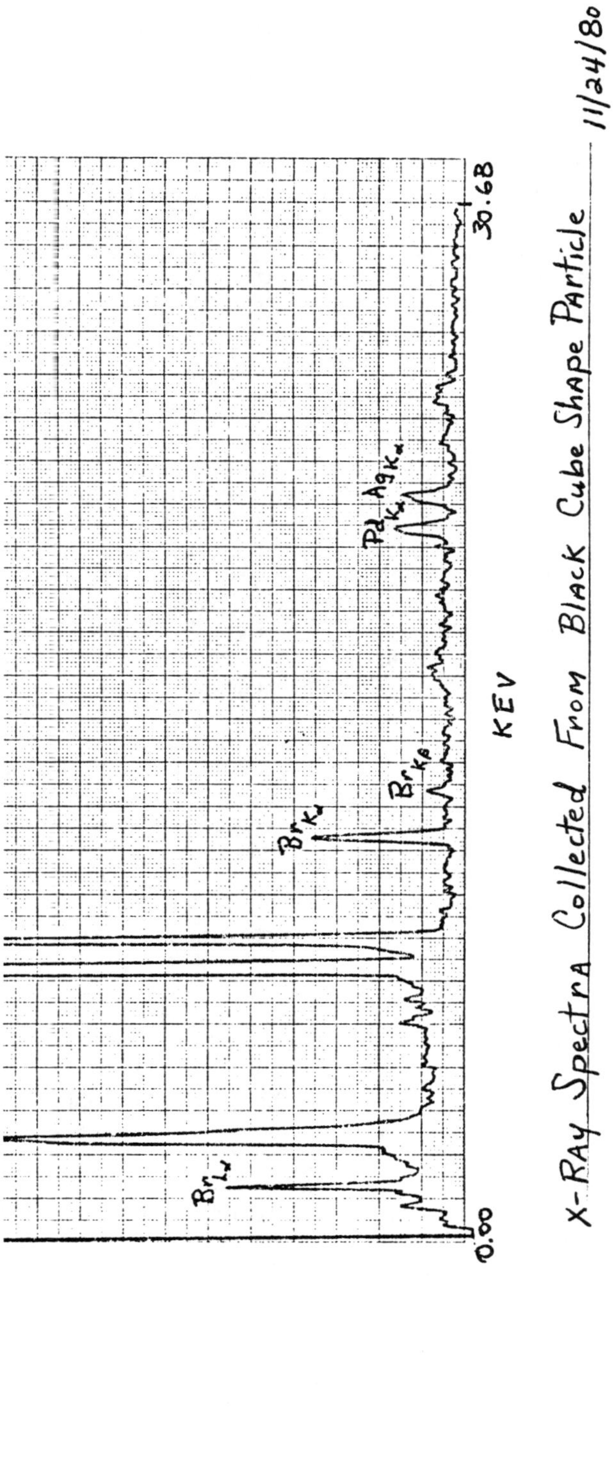
Br Lα
Br Kα
Br Kβ
Pd Kα
Ag Kα
0.00
30.68
KEV
X-RAY Spectra Collected From Black Cube Shape Particle
11/24/80

core-shell multi-structure grains [11d] (see Fig. 19). Photocharge measurements on Kodak TMAX 3200 yields signals of long persistence with only small decay (approximately 25% within 50 microseconds), which we speculate is attributable to a tabular morphology and heterojunction epitaxial edge growths consisting of silver halide with high chloride content [15] (see Fig. 20). In the TMAX 3200 system, we further speculate that latent image formation is directed toward the heterojunction interface, as suggested in a recent paper [11b], aided by a drift assisted electron diffusion. The epitaxial edge growth may supply the silver ions for the formation of the latent image, while the tabular host crystal is the most likely major source of the photoelectrons, which move, to a first approximation, in a lateral direction approximately perpendicular to the incident light on the tabular grain surface. The fields introduced by this geometry may account for the unusual persistence of the photocharge signal and presumably for the high photographic sensitivity.

3.5.2. Epitaxial PbO/AgBr

Another example of an epitaxial imaging system employs microcrystals of PbO (orthorhombic) as the major constituent (weight percent), which are also responsible for absorbing a major fraction of the incident light absorbed by the system. The PbO serves the additional important function of providing the medium upon which controlled quantities of silver ions are adsorbed and spontaneously reduced to form small silver clusters, which remain attached to the PbO surface. The silver clusters are in turn brominated to produce epitaxial PbO/AgBr heterojunctions, comprising the photosensitive elements of the photographic coating. Figure 21 outlines the sequence of steps used in the formulation of this heterojunction.

Fig. 22 presents electron micrographs of these systems showing the nature of the epitaxial growth of the AgBr on the PbO. The dark cubic area in Fig. 22b, in epitaxial contact with the PbO upon which it was grown, was shown by X-ray microprobe analysis to be AgBr. Latent image formation is in the AgBr phase, as evidenced by the visualization of the image using conventional silver halide developing and fixing formulations and procedures, followed by electroless nickel deposition on the imagewise distribution of developed silver clusters.

The range of spectral sensitivity of the system extends beyond the long wavelength absorption edge of the AgBr which together with the low AgBr coverage and consequent low absorptivity, provides clear evidence of spectral sensitization of the AgBr by the PbO. At the AgBr coverages employed (as low as 10mg Ag/m2), light absorption by the AgBr alone is insufficient to provide the required quantity of developed Ag to efficiently catalyze the visualization of the image by electroless metal deposition. Instead, light absorption is primarily by the PbO which injects photoelectrons into the epitaxial AgBr. In addition, Fig. 23 demonstrates that there is sensitivity in the green region of the spectrum, a result found only when the PbO sensitizer is present in combination with AgBr.

3.5.3. Sensitization of Silver Halide by Sulfur and Sulfur Plus Gold

Chemical and spectral sensitization by labile sulfur, S, compounds and by a combination of sulfur plus gold, S + Au, continue to be important components of modern photographic emulsions which have been receiving renewed attention in recent years [31]. Emulsions sensitized by S and by S + Au bear a photographically functional similarity to systems sensitized by inorganic photoconductors such as Si, ZnSe, AlSb, AlAs and others [2]. These sensitizers produce an increased response in the intrinsic region of AgX absorption, along with a low level of spectral sensitization. In addition, significant increases in photographic sensitivity are displayed when sensitizers in this class are combined with a dye sensitized emulsion.

Sulfur sensitization is modeled by the Ag_2S/AgBr epitaxial heterojunction, where second order photocharge decay kinetics has been shown to become increasingly dominant with increase in deposition of an epitaxial Ag_2S phase onto one side of a AgBr single crystal [14i]. The observed increase in second order decay rate constant, (Fig. 23), has been ascribed to reactions of photolytically produced interstitial ions, Ag^+_i, with trapped photoelectrons, e^-_t , where the Ag^+_i ions are the products of reactions with photoholes (Fig. 24).

For S + Au sensitization, modeled by either the AgAuS/AgBr or the Ag_2S/AgAuS/AgBr heterojunctions [32], the second order contribution to the photocharge decay kinetics diminishes with increasingly efficient sensitization [14d]. This suggests that for S + Au sensitization, the primary source of interstitial silver ions is present prior to exposure and is determined by the Frenkel equilibrium between interstitial silver ions and silver ion vacancies. With S + Au sensitization, it appears that photoholes do not react with the product(s) of S + Au sensitization to form Ag^+_i. This difference contributes to the frequently encountered latent image dispersion observed with S sensitization, associated with high intensity reciprocity failure and its less troublesome occurrence with S + Au sensitization.

Photographic evidence for the redistribution of latent image from the interior of a AgBr evaporated layer to the surface, after deposition in the dark of a surface layer of Ag_2S, (see Fig. 25), has been convincingly demonstrated [33], and interpreted as support for the view that a space charge is developed across the Ag_2S/AgBr junction during dark equilibration, such that the Ag_2S phase has a positive space charge and the AgBr phase a negative charge [2,14f,i]. Spectral sensitization by S and S + Au sensitized emulsion systems to wavelengths >620nm, where only the Ag_2S absorbs light, and the enhanced sensitivity in the presence of a sensitizing dye absorbing at wavelengths <620nm, is discussed in reference [31]. In this paper, it is suggested that the dye is a trap for photoholes produced by light absorption in the Ag_2S phase, making possible the elevation of the Ag_2S energy states to a position where

a

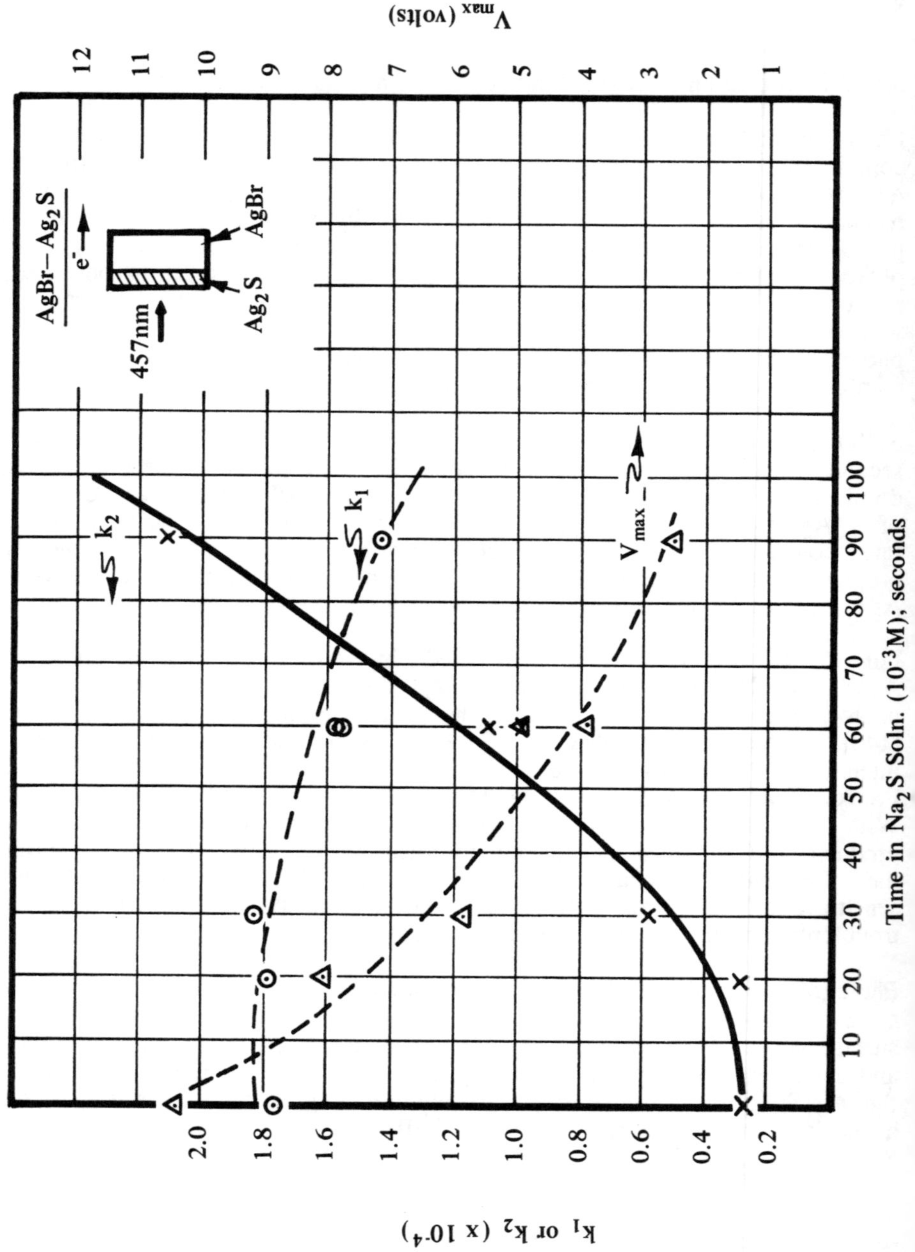
AgBr – Ag2S
e⁻
457nm
Ag2S
AgBr
k2
k1
Vmax
Time in Na2S Soln. (10⁻³M); seconds
k1 or k2 (x 10⁻⁴)
Vmax (volts)
10
20
30
40
50
60
70
80
90
100
0.2
0.4
0.6
0.8
1.0
1.2
1.4
1.6
1.8
2.0
1
2
3
4
5
6
7
8
9
10
11
12

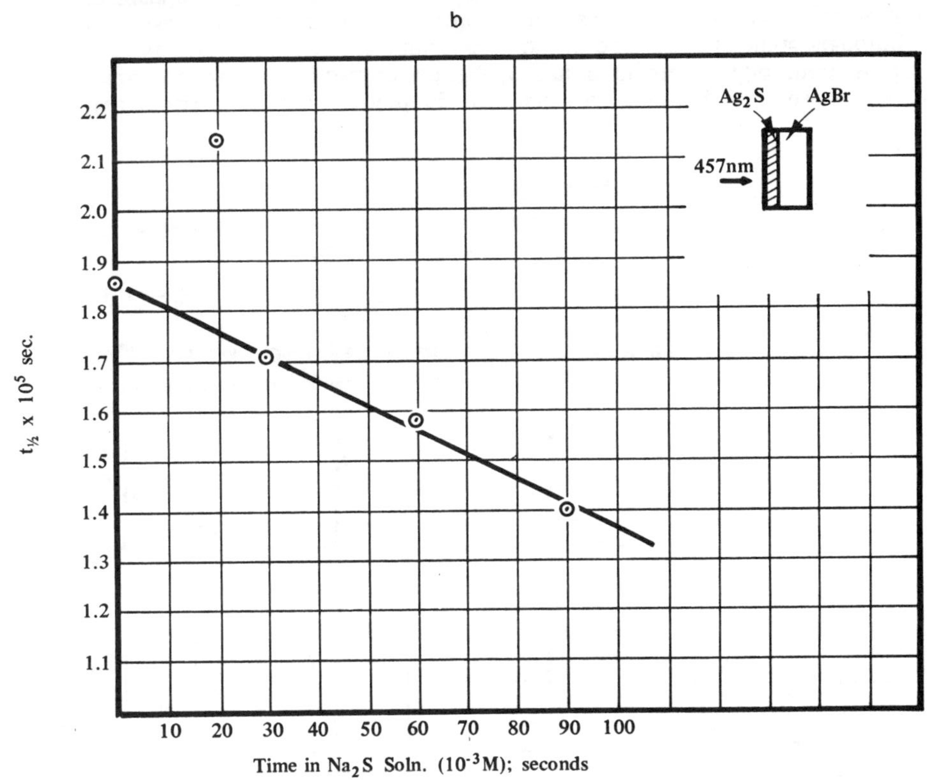

Fig.23. Determination of photocharge decay kinetics of a $AgBr/Ag_2S$ system consisting of a AgBr single crystal with solution deposition of Ag_2S on one surface. a) first order rate constant, k1; second order rate constant, k2 and maximum signal amplitude, Vmax. b) time for photocharge signal to decay to one half Vmax, t1/2, as a function of time in Na_2S solution, which controls the extent of Ag_2S deposition.

PHOTOCHARGE INVESTIGATION OF SULFUR SENSITIZATION

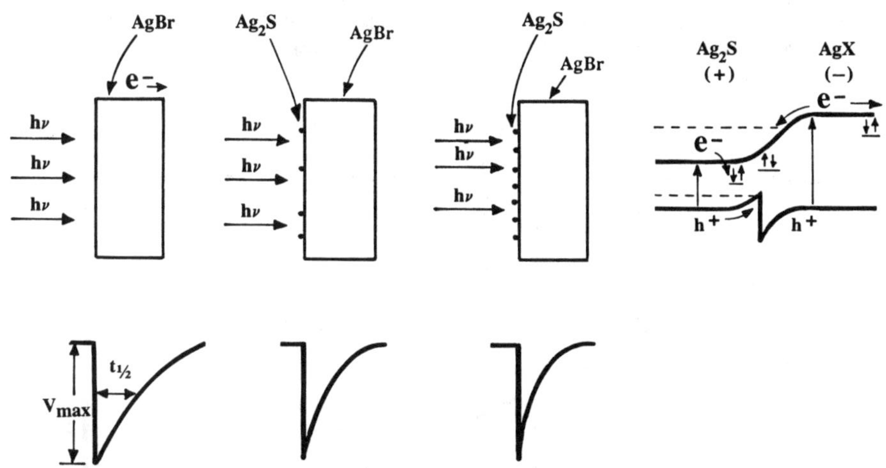

With increasing sulfur sensitization

a) V_{max} Decreases – Due to positive space charge in Ag_2S (e^- drift into Ag_2S)

b) $t_{1/2}$ Decreases }
c) k_2 Increases }   Due to $h^+ + Ag_2S \rightarrow Ag_i^+ + AgS$

$$\downarrow e^-_{\tau}$$

$$Ag^{\circ} \rightarrow \text{---} (Ag)_n$$

Fig.24. Schematic diagram explaining changes in photocharge transients with increased Ag_2S deposition on AgBr single crystal. The increase in the second order rate constant, k2, is important in understanding the increase in reciprocity failure of sulfur sensitized photographic materials.

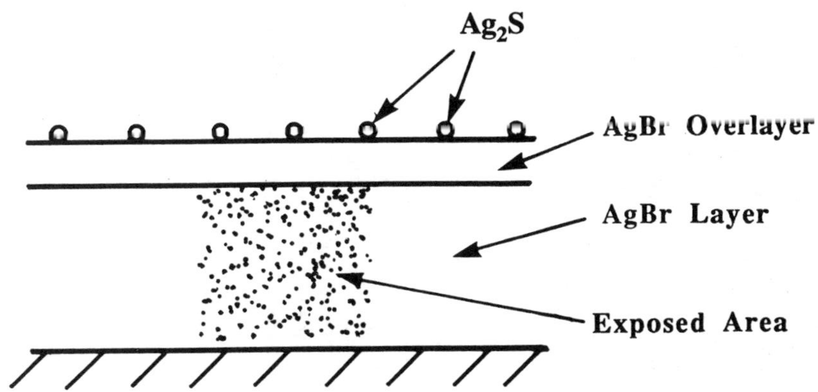

Schematic presentation of a sample with Ag_2S specks separated from the exposed bulk by an overlayer of unexposed AgBr.

Zhelev & Aserov, Photog. Sci. Eng., 1983

Fig.25. Diagram adapted from paper by Zhelev and Aserov showing a configuration which led to a redistribution of internal latent image to surface image after deposition of surface Ag_2S. Positive space charge in the Ag_2S deposit followed by electron and interstitial Ag^+ ion migration is believed to be responsible these observations.

photoelectron injection from the Ag_2S into the AgBr is more favorable.

A dye-inorganic sensitizer interaction has been proposed as an important aspect of the photographic transistor mechanism for supersensitization [2], as discussed above. Here, an initial transfer of electronic and ionic charge carriers between phases in contact occurs prior to exposure during the equilibration of chemical potentials and the establishment of interfacial space charges. In response to light absorption, photo-generated charge carriers are transported across phase boundaries, assisted by diffusion and drift. Secondary currents, originating in phases which need not be absorbing light, can provide amplified currents and photographic action in response to light induced positive fields created by electron injection from an adjacent layer into the AgX phase, as in a phototransistor. The hole trapping mechanism discussed above [31] may also contribute to the amplified photocurrents.

Energy band diagrams have been constructed that explain the diverse photographic, electronic and ionic phenomena for S and S + Au sensitization and for the supersensitization of these systems with dye. These band diagrams are constructed making use of energy band parameters appearing in the literature [31,33-35]. These energy band diagrams have features in common with principles governing the operation of amorphous silicon photovoltaic solar cells, where a field is introduced during the device fabrication which promotes separation of photoelectrons from photoholes, thereby reducing recombination, Fig. 26, [16].

3.5.3.1. The Ag_2S/AgBr Epitaxial Photographic Diode

Following the general features of the energy band diagram of the Ag_2S/AgBr photographic diode provided by Zhelev et al of reference [33], and using the optical transitions from Ag_2S on AgBr appearing in a paper by Ehrlich [34], together with the positions of the top of the valence band (-5.9eV) and bottom of the conduction band (-3.3eV) of AgBr [34,35], one obtains the energy band diagram of Fig. 27. The barrier height of 0.33eV, between the excited state within the Ag_2S conduction band and the bottom of the AgBr conduction band of Fig. 28, is taken from the experimental determination of the temperature dependence of spectral sensitization of AgBr by Ag_2S, by Hamilton et al [31]. This 0.33 eV barrier height corresponds closely with the barrier for electron injection into AgBr from the 1.69eV (730nm) optical transition from the top of the Ag_2S valence band [34]. The ionization potential (top of the valence band) of Ag_2S, (-5.32eV), in contact with AgBr was assumed by Ehrlich to be essentially unaltered from bulk Ag_2S. This is reasonable if the Fermi Level of Ag_2S is close to the bottom of the conduction band, (-4.42eV), as expected for a strongly n-type semiconductor. If the Fermi Level of AgBr is near mid-gap (-4.6eV), as has recently been reported [30], equalization of Fermi levels will produce a positive space charge in the Ag_2S phase and a negative space charge in the AgBr, accounting for the increased sensitivity and commonly observed appearance of surface image in sulfur sensitized emulsions exposed to radiation

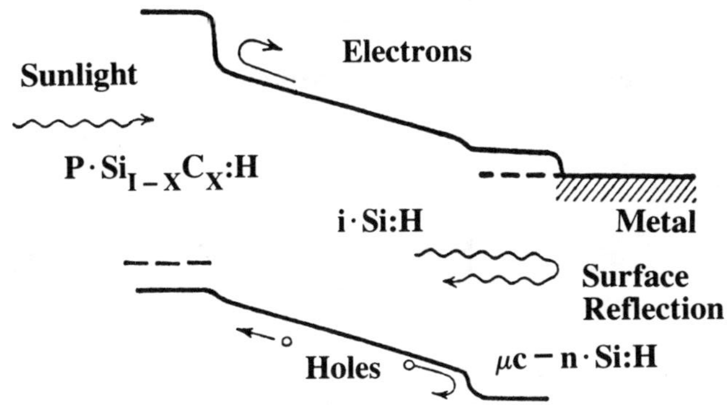

Fig.26. Band structure of a photovoltaic p-i-n device showing potential gradient in the i-layer, which assists charge carrier separation and reduces the extent of recombination. (Adapted from a figure in reference [16].)

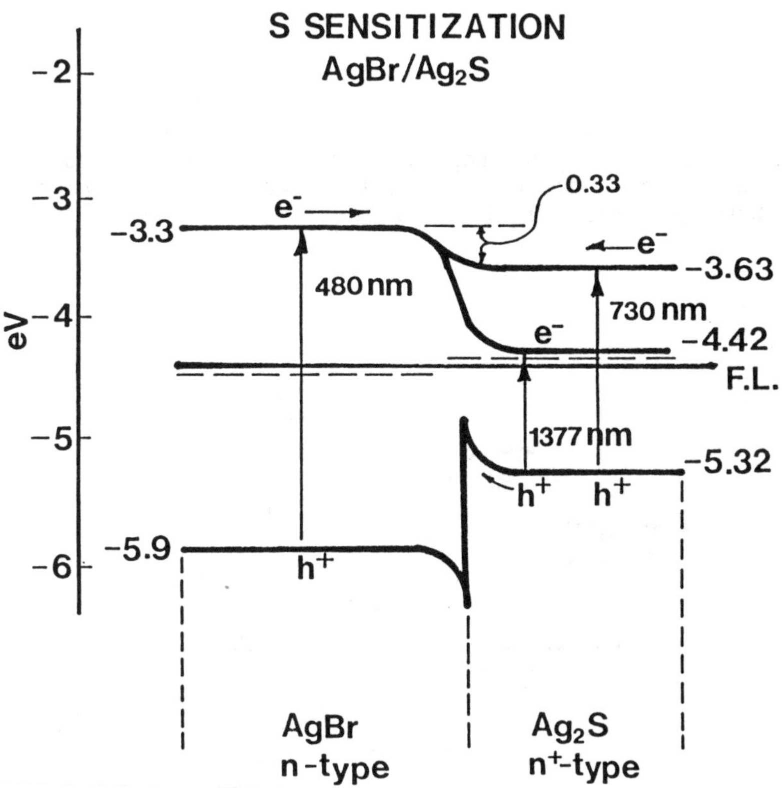

Fig.27. Proposed energy band diagram for the $AgBr/Ag_2S$ photographic diode using data taken from references [31, 33-35].

within the AgBr absorption band.

In the long wavelength region, >620nm, where only the Ag_2S is absorbing, spectral sensitization by the Ag_2S has been observed at high exposure levels [31]. Since the bottom of the conduction band of the Ag_2S is below that of the AgBr phase, the observed inefficiency of spectral sensitization by Ag_2S is understandable. At the onset of >620nm exposure, there is a high probability of recombination of photoelectrons with pre-existing positive holes within the Ag_2S phase, introduced during dark equilibration with the AgBr. High exposure levels create a new situation that results in spectral sensitization [31]. As the population of photoelectrons in the conduction band increases, thereby decreasing the band bending at the interface, the barrier for electron injection into the AgBr is lowered. According to the experimentally determined temperature dependence of spectral sensitization found by Hamilton et al, the (-3.63eV) energy level appears to be the dominant level within the Ag_2S conduction band from which electrons are injected into the AgBr phase in the A_2S/AgBr epitaxial photographic diode.

A fluorescence investigation of AgI and AgI/Ag_2S heterojunction colloids [36] has shown that the formation of Ag_2S on AgI quenches the AgI fluorescence and produces a new fluorescence in the infrared. In agreement with earlier work on sulfur sensitization of AgBr [37], it is postulated that light absorption in AgBr or AgI initiates injection of both electrons and holes into the Ag_2S phase where they recombine with emission of infrared fluorescence. For this to be possible, the bottom of the conduction band and top of the valence band of the Ag_2S must lie within the band gaps of AgBr and AgI. References [36,37] do not consider the possibility of thermodynamic equilibration prior to light exposure of the AgI/Ag_2S and AgBr/Ag_2S heterojunctions, which were shown by Zhelev et al [33] to be responsible for the redistribution of $(Ag)_n$ clusters from the interior of AgBr to the surface upon Ag_2S deposition. We consider that some or all of the infrared fluorescence hole recombination centers are derived from combination of photoelectrons injected from AgBr or AgI into the Ag_2S positive space charge region - where the Ag_2S positive space charge is established during thermodynamic (Fermi Level) equilibration prior to exposure.

3.5.3.2. The DYE/Ag_2S/AgBr/DYE Epitaxial Photographic Transistor

Since optimum levels of sulfur sensitization leave the AgX grains only partially covered with islands of Ag_2S [38], the junctions on either side of a cross section through a given AgBr grain are in general not symmetrically disposed with respect to the surface distribution of Ag_2S. When this system is sensitized with a monolayer of dye, the DYE/Ag_2S/AgBr/DYE structure is created. Using energy band parameters appearing in the papers cited above [31,33-35] and assuming that the Ag_2S/AgBr epitaxial diode has reached thermodynamic equilibrium prior to the addition of the spectrally sensitizing dye (E_{HO} = -4.96eV, E_{LV} = -2.82eV), an energy band structure for the supersensitized system has been constructed consistent with the observed photographic response (see Fig. 28). Following Hamilton et al [31], a lowering

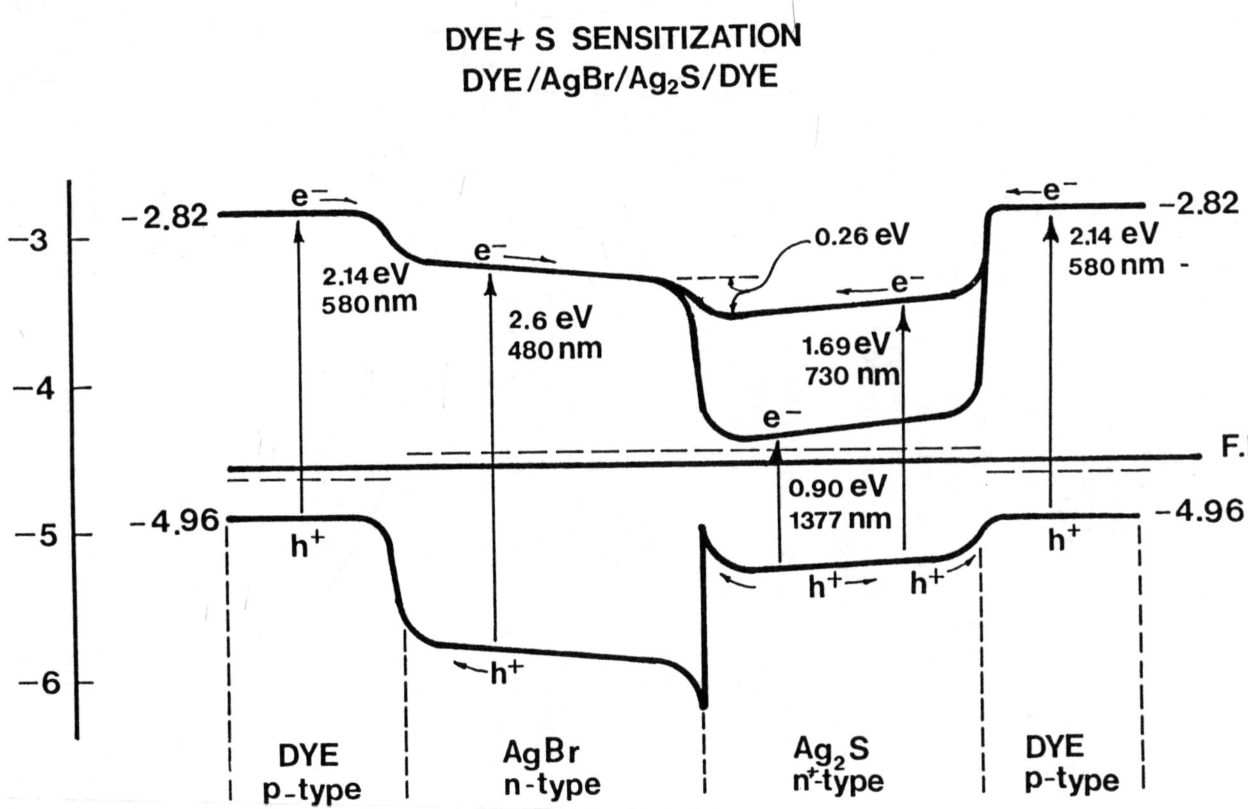

Fig.28. Proposed energy band diagram for the Dye/AgBr/Ag_2S/Dye photographic transistor using data taken from references [31,33-35].

of the barrier for spectral sensitization by Ag_2S to (0.26eV) when dye is present, from (0.33eV) without dye, has been employed.

Charge transfer between the pre-equilibrated AgBr(-)/Ag_2S(+) diode and the adsorbed sensitizing dye, leading to a re-equilibration of the system in the dark, is energetically possible according to the photographic transistor mechanism. Since the top of the valence band of the Ag_2S phase in the AgBr/Ag_2S diode is at (-5.32eV), it can inject a positive hole into the dye valence band (-4.96eV). Hamilton et al provide evidence that this raises the Ag_2S band structure by (0.07eV). Hole injection into the dye can also occur during exposure of the Ag_2S phase to >620nm radiation, as suggested in reference [31].

Based on the above assignments, the band bending and band gaps of the individual phase at the DYE/AgBr and AgBr/Ag_2S junctions of the DYE/Ag_2S/AgBr/DYE device of Fig. 28 are such that a potential gradient is likely to be generated across the AgBr phase. In this respect the AgBr is similar to the amorphous silicon i-layer of a p-i-n photovoltaic solar cell [16]. Due to the Ag^+_i ion conductivity of the AgBr, the potential gradient is expected to be somewhat relaxed, as compared to what it would be if each of the phases comprising the heterojunction were conducting by a purely electronic mechanism [14a]. There is also a potential gradient at the Ag_2S/DYE interface. Absorption of light in each of the components of this imaging device leads to a field assisted flow of electrons toward the AgBr/Ag_2S interface, in agreement with photographic experimental evidence. Positive holes flow in the opposite direction, thereby reducing the probability of recombination and resulting in what can phenomenologically be termed supersensitization. The proposed mechanism describing these events is that of the photographic transistor. The photographic response of this system to exposure in the spectral regions: a) >620nm, where only the Ag_2S absorbs light, b) 500-600 where the dye and the Ag_2S phase absorbs light and c) <500nm, where both AgBr and Ag_2S absorb light, follows from the configuration of the band diagram of Fig. 28.

S +Au sensitization of AgBr and supersensitization by adsorbed dye can be explained by similar procedures for formulation of the energy band diagrams. Space limitations do not make the inclusion of this material possible.

4. EPITAXIAL PHOTOCATALYTIC HETEROJUNCTIONS

Reviews of progress in the field of epitaxial photocatalytic heterojunctions, with special attention to colloidal systems, have appeared in the recent literature [1,4]. In this section several representative examples of single and double junction systems are briefly discussed. The CdS/TiO_2 system, consisting of thin, small diameter CdS deposits on a TiO_2 electrode, provides a demonstration of spectral sensitization of the photocurrent by CdS [39] that is analogous to dye sensitization. In this paper Cd is first deposited on the TiO_2 from $Cd(NO_3)_2$ solutions and then converted to CdS by immersion in a sulfide solution. In agreement with the work of Brus discussed in [4], the onset of

CdS spectral sensitization is shifted to higher energies as the size of the CdS deposit decreases to the quantum well domain (≈ 40 A for CdS). In another paper, inter-particle electron transfer is demonstrated by photoconductivity measurements in a pressed double-pellet of CdS and TiO_2 - with the direction of electron photocurrents dependent on whether the measurements are made in vacuum (electron injection from TiO_2 into CdS) or in the presence of O_2 (electron injection into TiO_2) [40]. Fluorescent quenching of the Cd_3P_2 emission by electron injection into the TiO_2 is described for colloidal samples prepared by wet chemical techniques for attachment of TiO_2 colloids to quantum well sized Cd_3P_2 colloids [41]. Quantum broadening of the Cd_3P_2 band gap with decreasing particle diameter, making electron injection into the TiO_2 possible, is suggested. Double junction systems (i.e. CdS/TiO_2/RuO_2) have also been investigated [1,4,40]. Hydrogen generation by irradiation of H_2S in the presence of CdS or TiO_2/RuO_2 is reported to be inefficient. However, efficiency improves considerably for the double junction system. Particulate semiconductor catalysts with metallic surface deposits have been used extensively to promote a variety of chemical reactions [1,4]. Some aspects of recent TRPC measurements on Pt/TiO_2 are discussed in the following section.

4.1. Feasibility Demonstration of TRPC Measurements on Particulate TiO_2 and Pt/TiO_2 and Preliminary Interpretation

TRPC has been used in a preliminary study of particulate TiO_2 (Degussa P25) and Pt/TiO_2 [8,9], Fig. 29, with the intent of seeking an explanation for the frequently observed improvements in photo-assisted catalytic activity due to platinization [1,4]. In agreement with chemical investigations, TRPC signals generated by a 337nm laser pulse provide evidence of a positive space charge in the Pt phase of the Pt/TiO_2 catalyst (see Fig. 30), which directs a portion of the photoelectron current to the catalyst surface where it can initiate reaction with adsorbed reactants. ESCA measurements of Pt/TiO_2 heterojunctions have been interpreted on the basis of positively charged Pt clusters on the TiO_2 [42]. The ESCA results are discussed in terms of a strong metal-support interaction (SMSI), which has also been the subject of a NMR study of Pt/TiO_2 [43]. The NMR study, although not considered definitive regarding charge transfer, indicates that there is a substantial rearrangement of the electronic structure at the Pt-TiO_2 interface. In related measurements, positive space charges in small $(Ag)_n$ and $(Au)_n$ metallic clusters deposited on AgBr evaporated layers have been detected by TRPC measurements [2]. When TRPC measurements are made with the Pt/TiO_2 catalyst in contact with a gelatin binder, the direction of photoelectron flow toward the Pt deposit is further enhanced, as evidenced by a reversal in signal polarity and an increase in magnitude of the photocharge signal, Fig. 30. In the absence of Pt, the presence of gelatin reduces the signal magnitude while retaining the original signal polarity of the TiO_2 catalyst. Gelatin increases the TRPC signal lifetime of both the platinized and non-platinized TiO_2. A similar extension of the lifetime of optically detected photoelectrons in Pt/TiO_2 in the presence of polyvinyl alcohol has been reported [4]. Multi-phase heterojunction catalysts, with appropriate surface and internal fields, designed to direct charge carriers to the spatially separated

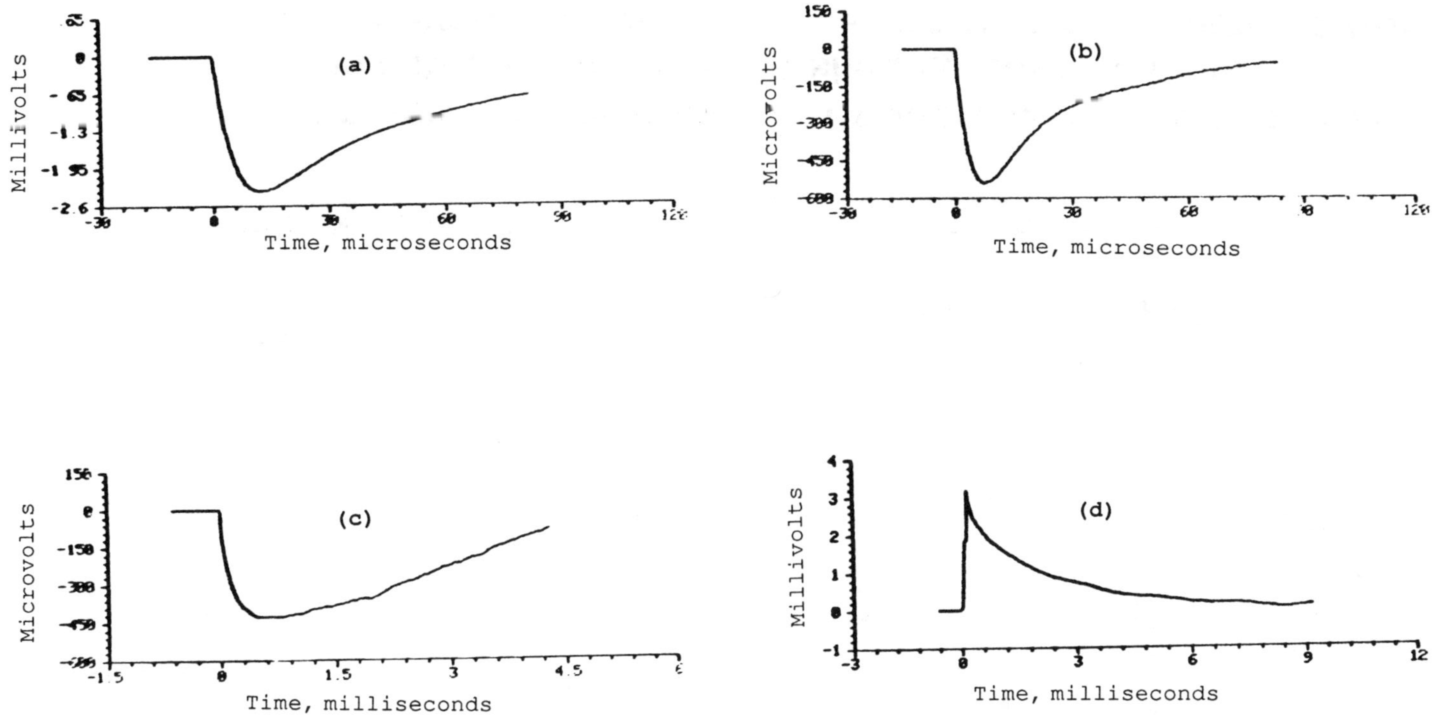

Fig.29. Time resolved photocharge transient of a) TiO_2, Degussa P25, coated from water slurry, b) Pt/TiO_2, Pt deposit on TiO_2, coated from water slurry, c) TiO_2 coated from gelatin suspension and d) Pt/TiO_2 coated from geltin suspension.

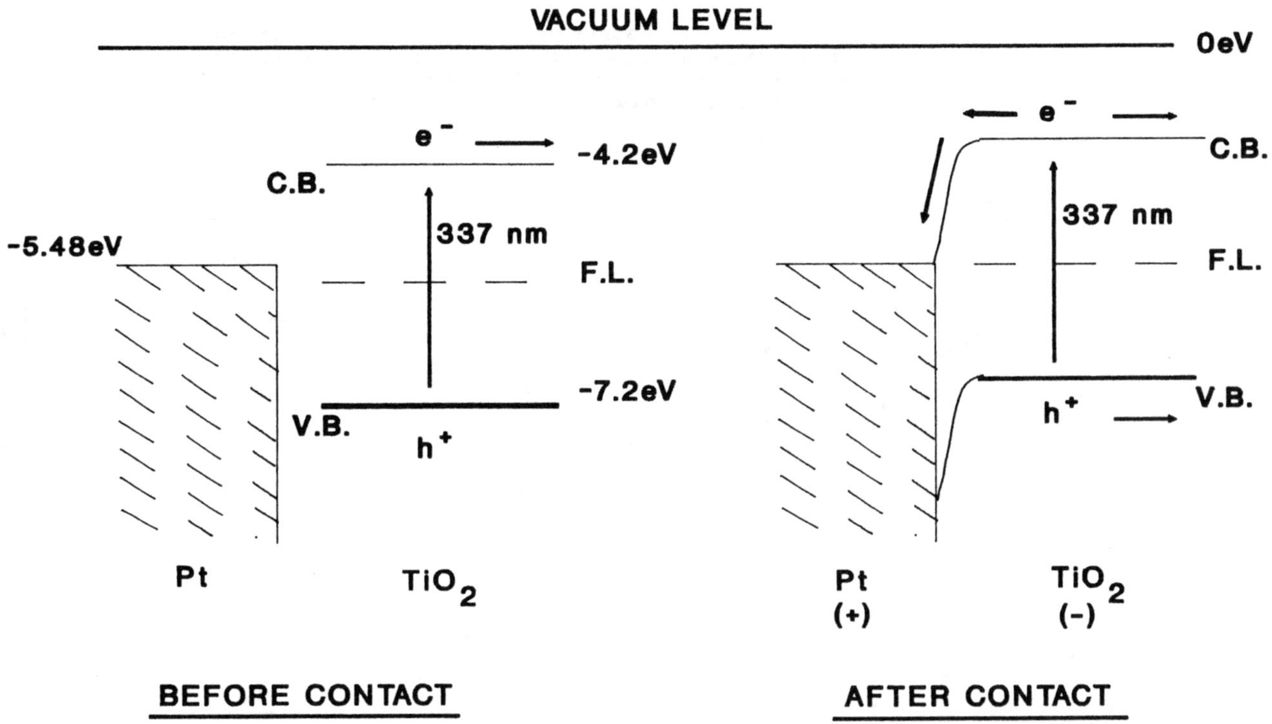

Fig.30. Energy band diagrams of Pt and TiO_2 before and after equilibration in the dark showing space charge and directions of charge carrier transport deduced from TRPC measurements .

regions, provide a viable approach for improving catalyst performance, which can be interpreted according to solid state heterojunction principles. These preliminary measurements suggest that it will be possible to use TRPC techniques for evaluation of catalyst design concepts by monitoring photocatalyst charge carrier dynamics in the absence and presence of adsorbed chemical reactants.

5. Summary

Examples from the photographic literature which illustrate objectives and research directions in common with the field of heterogeneous photocatalysis have been presented. Heterojunctions are now well recognized in both fields for their already demonstrated advantages and for their potential in future applications. A solid state device approach for the fabrication and characterization of photographic and catalyst materials is suggested. For particulate and colloidal systems, time resolved charge carrier transport characterization by direct electronic techniques, such as TRPC, is considered crucial to the efficient development of new and improved materials. Electronic methods should be used to complement fast photochemical spectroscopic techniques that have already been developed to a high degree of sophistication. Since the advantages of following these directions is now well recognized and has been advancing more or less independently in each of these fields, prospects for accelerated progress in the future in both areas is believed to be high.

Acknowledgement: The author would like to express his appreciation to many of his former collaborators and in particular to Marcia Lindsey and Carl Rippon, whose contributions have been invaluable. At Boston University, I would like to thank Kallambella Vijayakumar and Juchang Dong for their preparation of the platinized titanium dioxide sample. Bin Cai's able collaboration in the exploratory photocharge measurements of the photocatalytic systems is much welcomed. I would also like to thank Norman Lichtin for having re-introduced me to the field of heterogeneous photocatalysis and for many helpful discussions.

REFERENCES

1. P.V.Kamat and N.M.Dimitrijevic, Solar Energy **44:** 83(1990).
2. B.Levy, Photogr.Sci.Engr., **27:** 204(1983).
3. J.Cunningham, "Radiation and Photoeffects at Gas/Solid Interfaces", in Comprehensive Chemical Kinetics, eds. C. H. Bramford, C.F.H.Tipper and R.G.Compton, Elsevier, NY, **19:** 291(1984).
4. A.Henglein, "Mechanism of Reactions on Colloidal Microelectrodes and Size Quantization Effects", in Topics in Current Chemistry, **43:** 113(1988).
5. A.Heller, Y.Degani, D.W.Johnson,Jr. and P.K.Gallagher, J.AM.Chem.Soc. **91:** 5987(1987).
6. T.Tiedje, "Information about Band-Tail States from Time-of-Flight Experiments", Semiconductors and Semimetals, **Vol.21**, part C, (1984), ed. J. I. Pankove, Academic Press, Inc., NY, pp 207-238.

7. A.R.Moore, "Diffusion Length in Undoped a-Si:H", ibid., pp. 239-256.
8. B.Levy, "Determination of Photo-Induced Charge Carrier Transport Parameters in Thin Film and Particulate Photoconductors", Paper Summaries, SPSE Annual Meeting, Rochester, NY, May 20-25,1990.
9. B.Levy, "Photocharge: A Time-Resolved Technique for Determination of Charge-Carrier Transport in Photoconducting Micro-Heterojunction Environments", Paper Summaries, Symp. on Electrochemical Surface Science, ACS Annual Meeting, Boston, MA, Apr. 22-27, 1990.
10a. E.V.Baranov and I.A.Akimov, Dokl.Akad.Nauk SSSR **154:** 184(1964).
b. L.M.Kellogg, Photogr.Sci.Engr. **18:** 378(1974).
c. R.J.Deri and J.P.Spoonhower, Photogr.Sci.Engr. **25:** 89(1981).
d. R.A.Bogomolni, H.Tributsch, G.Peterman and M.P.Klein, J.Chem.Phys. **78:**2579(1983).
e. A.Werner and M.Kunst," Influence of Preparation Conditions on Charge Carrier Dynamics in α-Si:H Determined by an In-Situ Technique", in Materials Issues in Amorphous Semiconductor Technology, April 15-18, 1986, Palo Alto, CA, ed. D. Adler, Y. Hamakawa and A. Maden, Mat.Res.Soc.Proc., **70:** 137(1986).
f. D.H.Klosterboer, W.L.Meissner and M.R.V.Sahyun, J.Appl.Phys. **54:** 5161(1983).
g. M.R.V.Sahyun, Chem.Phys.Lettr. **112:** 571(1984).
h. M.R.V.Sahyun, "Time - Resolved Dielectric Loss as a Probe of Interfacial Electron Transfer Dynamics in Unconventional Photoactive Solids", ed. H. Scher, Plenum Publ. Co., 1988, pp.133-143.
i. M.R.V.Sahyun, J.Appl.Phys. **61:** 3044(1987).
j. M.R.V.Sahyun, J.Inorg.Sci. **33:** 94(1989).
k. J.M.Warman, M.P.deHaas, M.Gratzel and P.P.Infelta, Nature **310:** 306(1984).
l. R.W.Fessenden and P.V.Kamat, Chem.Phys.Lettr. **123:** 233(1986).
m. P.V.Kamat, N.M.Dimitrijevic and R.W.Fessenden, J.Phys.Chem. **91:** 396 (1987).
o. C.J.Dobbin, A.R.McIntosh, J.R.Bolton, Z.D.Popovic and J.R.Harbour, J.Chem.Soc. Far. Trans. 1, **82:** 3625(1986).
11a. J.E.Maskasky, Photogr.Sci.Engr. **25:** 96(1981).
b. S.Takada, H.Ayato and S.Ishimaru, Proceedings of the International Congress of Photographic Science, Cologne, 1986, pp. 81-87.
c. S.Bando, Y.Shibahara and S.Ishimaru, J.Imag.Sci. **29:** 193(1985).
d. A.Ezaki, T.Yagi, and T.Iijima, SPSE 40th Annual Conference, May 17-22,1987, Preprints, pp110-111.
12a. B.Levy and M.Lindsey, Photogr.Sci.Engr. **16:** 389(1972).
b. B.Levy and M.Lindsey, Photogr.Sci.Engr. **17:** 135(1973).
c. B.Levy and M.Lindsey, Photogr.Sci.Engr. **17:** 423(1973).
d. B.Levy, Photogr.Sci.Engr. **18:** 347(1974).
e. B.Levy, J.Photogr.Sci. **22:** 200(1974).
f. B.Levy, J.Photogr.Sci., **22:** 251(1974).
g. C.R.Dickson and B.Levy, Photgr.Sci.Engr. **18:** 524(1974).
h. B.Levy, Photogr.Sci.Engr., **19:** 65(1975).
i. B.Levy, Photogr.Sci.Engr. **20:** 273(1976).
j. B.Levy, "Solar Energy Conversion Processes in Photography", Proceedings

of the Progress Review Meeting of the Advanced High-Energy Efficiency Program, March 24-26, 1982, Solar Energy Research Institute, Golden, Colorado, pp. 453-498.

13. J.F.Hamilton, "The Silver Halide Photographic Process", Adv.Phys. **37**:359-441(1988).

14a. B.Levy, Photogr.Sci.Engr. **15**: 279 (1971).

b. B.Levy, M.Lindsey and C.R.Dickson, Photogr.Sci.Engr. **17**: 115(1973).

c. B.Levy and C.R.Dickson, Photogr.Sci.Engr. **19**: 79(1975).

d. K.C.Chang, F.P.Chen and B.Levy, Photogr.Sci.Engr. **24**: 55(1980).

e. B.Levy, K.C.Chang and F.P.Chen, in Growth and Properties of Metal Clusters, J. Bourdon (ed.), Elsevier Publishing Co., 1980,pp. 393-397.

f. K.C.Chang, F.P.Chen and B.Levy, Photogr.Sci.Engr. **25**: 138(1981).

g. F.P.Chen, K.C.Chang, L.Corben, M.Falxa and B.Levy, Photogr.Sci.Engr. **26**: 15(1982).

h. C.Rippon and B.Levy, Photogr.Sci.Engr. **27**: 65(1983).

i. B.Levy, "Photocharge and the Photographic Process", Proc. International East-West Symp.; Oct.28-Nov.2, 1984; Maui, Hawaii, pp. B13-16.

j. B.Levy and C.Rippon, "Epitaxial Photographic Diode and Transistor Heterojunctions: Systems Containing Silver Halide", Proc.International East-West Symp. II; Oct.30-Nov.4, 1988; Kona, Hawaii, pp.F44-48.

k. B.Levy, "An Epitaxial Heterojunction Model for Sulfur and Sulfur Plus Gold Sensitization of AgBr", Paper Summaries, 42nd Annual Conf., SPSE, May 14-19, 1989, pp. 246-250.

15. J.E.Maskasky, "Epitaxial Selective Site Sensitization of Tabular Grain Emulsions", East-West Symposium II, Kona, Hawaii, Oct. 30 - Nov. 4, 1988, pp.C24-29.

16. K.Takahasi and M.Konagai, Amorphous Silicon Solar Cells, John Wiley and Sons, NY, 1986, pp.124-163.

17. H.Dember, Physik. Z. **32**: 554, 856(1931); **33**: 207(1932).

18a. F.Willig, Ber.Bunsenges.Phys.Chem. **92**: 1312(1988).

b. K.Bitterling, F.Willig and F.Decker, J.Electroanal.Chem. **228**: 29(1987); **204**: 211(1986).

c. M.Eichorn, F.Willig, K.P.Charle and K.Bitterling, J.Chem.Phys. **76(9)**: 4648(1982).

19. A.Bergman, C.R.Dickson, S.D.Lidofsky and R.N.Zare, J.Chem.Phys. **65(3)**: 1186(1976).

20. T.L.Penner and P.B.Gilman,Jr., Photogr.Sci.Engr. **20**:97 (1976).

21. M.Kawasaki, H.Hada and H.Uchida, J.Appl.Phys. **60**:3945 (1986).

22. B.Vaupel and H.Bottcher, personal communication, to be published in J.Inform.Rec.Mat., "Model Investigations of the Sensitization of Silver Bromide with Inorganic Semiconductors.

23. H.Gerischer, Angewante Chemie **27**:63(1988).

24. B.Levy, U.S.Pat. 3,976,489, Aug.24, 1976.

25a. S.E.Sheppard, A.P.H.Trivelli and R.P.Loveland, J.Franklin Inst. **200**:51 (1925).

b. E.P.Wrightman, A.P.H.Trivelli and S.E.Sheppard, J.Franklin Inst. **204**:49 (1927).

26. C.R.Berry, M.H.vanHorn and R.L.Griffith, Acta Cryst. **7**:135 (1954).

27. H.Meier, Spectral Sensitization, The Focal Press, London, 1968.
28. A.Terenin and I.Akimov, J.Phys.Chem. **69:**730 (1965).
29. A.A.Muenter, P.B.Gilman,Jr., J.R.Lenard and T.L.Penner, East-West Symposium, Hawaii, 1984, pp. C21-23.
30. T.Tani, private communication; K.Seki, Y.Kobayashi, H.Yanagi, T.Ohta, H.Inokuchi and T.Tani, General Forum of Molecular Structures, Oct. 7-10, 1988, Univ. Electro-Communications, Chofu, Tokyo.
31. J.F.Hamilton, J.M.Harbison and D.L.Jeanmarie, J.Imag.Sci. **32:**17 (1988).
32. K.Kuge, J.Soc.Photogr.Sci.Tech.Japan **50:**108 (1987).
33a. V.Zhelev and Y.Astrov, Photogr.Sci.Engr. **27:**160 (1983).
b. V.Zhelev and J.Malinowski, J.Imag.Sci. **31:**47 (1987).
34. S.H.Ehrlich, Photogr.Sci.Engr. **23:**348 (1979).
35. C.R.Berry, Photogr.Sci.Engr. **19:**93 (1975).
36. A.Henglein, M.Gutierrez, A.Ffojtik and J.Jirkovsky, Ber.Bursenges.Phys.Chem. **93:** 593(1989).
37. V.M.Belous, N.A.Orlovskaya, V.I.Tolstobrov and K.V.Chibisov, Dokl.Akad.Nauk. SSSR **235:**1339(1977).
38. D.Schrijvers, C.Goessens, J.Van Landuyt and R.De Keyzer, "High-Resolution Electron Microscopy Study of NM-Scale Decoration Particles On AgBr Crystals", Preprints, SPSE 43rd Annual Conference, May 20-25, 1990, Rochester, NY.
39. H.Gerischer and M.Lubke, J.Electranal.Chem. **204:** 225(1986).
40a. N.Serpone, E.Borgarello and E.Pelizzetti, J.Electrochem.Soc., **135:** 2760(1988).
b. P.Pichat, E.Borgarello, J.Disdier, J.Herrmann, E.Pelizzetti and N.Serpone, J.Chem.Soc.,Faraday Trans. I, **84:** 261(1988).
41. R.Kietzmann, F. Willig, H.Weller, R.Vogel, D.N.Nath, R.Eichberger, P.Liskas and J.Lehnert, preprint, to appear in Mol.Cryst.Liqu.Cryst (1990).
42. J.Kiwi and P.Italiano, J.Mol.Catal. **50:** 131(1989).
43. J.P.Bucker, J.J. van der Klink and M.Graetzel, J.Phys.Chem. **94:** 1209(1990).

HOMOGENEOUS AND HETEROGENEOUS CATALYTIC SYSTEMS FOR CONVERSION AND STORAGE OF SOLAR ENERGY

Kirill I. Zamaraev and Valentin N. Parmon
Institute of Catalysis, Prospect Akad. Lavrentieva 5, Novosibirsk 630090, USSR

1. INTRODUCTION.

In our lecture at the previous conference I.P.S.-7 [1] we have discussed advances and unsolved problems in the use of catalysis for conversion and storage of solar energy. In this lecture we shall present new results obtained in this area at the Institute of Catalysis (Novosibirsk) within the last two years, i.e. in the period between the I.P.S.-7 and I.P.S.-8.

Two different types of catalytic processes for solar-to-chemical energy conversion will be discussed: (i) photocatalytic ones that are based on photochemical reactions in which solar light quanta serve as energy-rich reagents providing accomplishment of otherwise thermodynamically forbidden catalytic processes, (ii) thermocatalytic ones that are based on endothermic catalytic reactions proceeding upon heating the reagents with solar light [1,2].

2. PHOTOCATALYTIC SYSTEMS FOR SOLAR ENERGY CONVERSION.

At early stages of research the photocatalytic method of solar-to-chemical energy conversion attracted much more attention than the thermocatalytic one. The main reason for this seems to be the well known fact that photocatalytic method is successfully used by living nature to store solar energy via photosynthesis in plants and bacteria.

Among various endoergonic chemical processes, that are discussed nowadays as possible candidates for accomplishing photocatalytic solar energy conversion, photocatalytic cleavage of water into dihydrogen and dioxygen attracts special attention. The development of the photocatalytic systems for water cleavage

E. Pelizzetti and M. Schiavello (eds.), Photochemical Conversion and Storage of Solar Energy, 393–425.

involves the design of catalysts for three types of processes: photocatalysts, PhC, for photochemical charge separation

$$A + D \xrightarrow[PhC]{h\nu} A^- + D^+ \quad (1)$$

which leads to formation of starting reagents for water cleavage, namely strong reductant A^- and oxidant D^+, and catalysts for subsequent multielectron reduction of water to H_2 by the reductant:

$$2A^- + 2H_2O \xrightarrow[cat\ H_2]{} 2A + H_2 + 2OH^- \quad (2)$$

and oxidation of water to O_2 by the oxidant:

$$4D^+ + 2H_2O \xrightarrow[cat\ O_2]{} 4D + O_2 + 4H^+ \quad (3)$$

This scheme (Fig. 1) is similar to the sequence of processes leading to light energy storage during photosynthesis in plants.

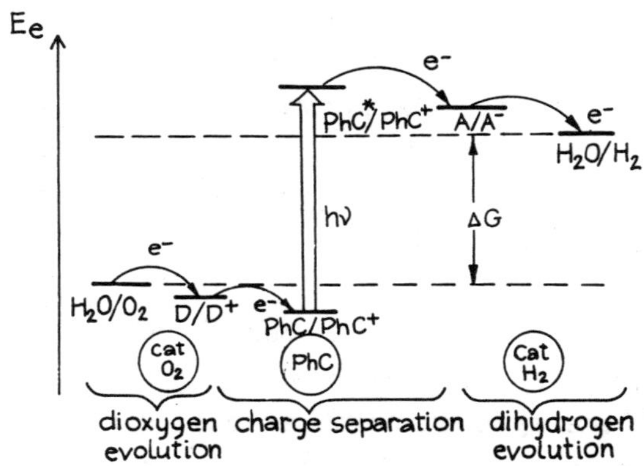

FIG. 1. The scheme of the process of water cleavage in a molecular photocatalytic system. Here and below E_e denotes the energy of the transferred electron; $h\nu$ is the energy of absorbed photon; ΔG is the net stored chemical energy. The curved arrows show paths for electron transfer [1,2].

Depending on the chemical nature of PhC, photocatalytic systems for solar energy conversion are commonly divided into two classes: semiconductor and molecular systems [1-5].

2.1. Molecular photocatalytic systems for H_2 and O_2 evolution from water.

The design of artificial molecular systems for solar-to-chemical energy conversion occurred to be a much more difficult problem than it was expected at the beginning of the work in this field. The direct attack against this problem did not lead to a success, and, thus, the researchers had to start a systematic siege of this problem. The accents in basic research have been shifted to the mechanistic studies of various steps of photocatalytic cleavage of water and search of new catalysts for these steps.

2.1.1. Step of dihydrogen evolution from water.

At present a large set of efficient catalysts for reduction of water to dihydrogen is known (see reviews [1,2,6-8] and references there in). Among these catalysts, those based on highly dispersed noble and non-noble metals seems to be most promising. Interesting data were obtained recently concerning mechanisms of action of such catalysts.

It was found that regularities of H_2 evolution upon reduction of water by one-electron reductants in the presence of finely dispersed metals can be explained in terms of the modified electrochemical "microelectrode" model [9,10a]. According to "microelectrode" model, evolution of H_2 in the presence of colloidal metal particles is assumed to consist of two steps, i.e. (I) - reduction of colloidal particles by the reductant A^-, which leads to the accumulation of a certain total charge by these particles, and (II) - reduction of protons by the reduced colloidal particles, which leads to the formation of H_2. In the original version of "microelectrode" model [9] the situation was discussed when the rate determining step of H_2 evolution was the reaction of proton discharge, while the rate determining step of A^- oxidation was the diffusion of the reductant to the colloid surface. The modified "microelectrode" model [10a] takes into account the facts, that in reality: (i) - step (II) is a complex reaction, which can be limited by either proton discharge (Folmer's mechanism, step (IIa)), or electrochemical desorption of hydrogen atoms (Heyrovsky's mechanism, step (IIb)), or recombination of hydrogen atoms (Tafel's mechanism, step (IIc)), and (ii) - step (I) can be both electrochemically reversible and irreversible, depending on how large is the rate constant of electron transfer from A^- to metal particle [10a].

From the shape of the dependencies of the reaction rates on pH and electrochemical potential of the reductant A^-, it occurs possible to determine, whether step (I) is electrochemically reversible or irreversible and which of the above mentioned reaction steps (I), (IIa,b) or (IIc) are rate-determining one.

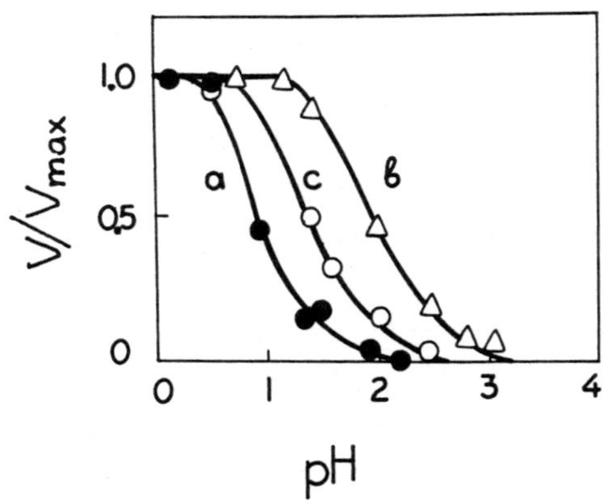

FIG. 2. Dependence between pH and the rate of H_2 evolution from water solutions of $2\cdot10^{-3}$ M VSO_4 in the presence of metal colloids. (a) - copper with the average size of 25 Å; (b) - copper with the average size of 52 Å; (c) - palladium with the average size of 40 Å.

It was shown, that the shape of pH dependence of the rate of the catalytic process of H_2 evolution (Fig. 2) and in particular the values $pH_{1/2}$ and ΔpH for such metals as copper and palladium and such reductants as V^{2+} and MV^+ (MV^+ is cation radical of methylviologen) are quantitatively explained by this model [10a].

The "microelectrode" model was used to explain, e.g., the mechanism of H_2 evolution in the presence of copper colloids. For this metal a procedure was elaborated, which allows one to obtain ultrafine particles with sizes from 25 to 250 Å with narrow size distribution [10b].

It was found that under appropriate conditions, the catalytic activity of colloidal copper catalysts is nearly the same as that for colloidal noble metals. Thus, inexpensive copper can be used as the catalyst of H_2 evolution instead of expensive noble metals. The character of the dependence of the rate of H_2 evolution in the presence of colloidal copper catalysts on pH and electrochemical potential of the reductant suggests that in this case step (I) is electrochemically reversible (for VSO_4 and MV^+ reductants) and step (II) is controlled by Folmer's mechanism. From the analysis

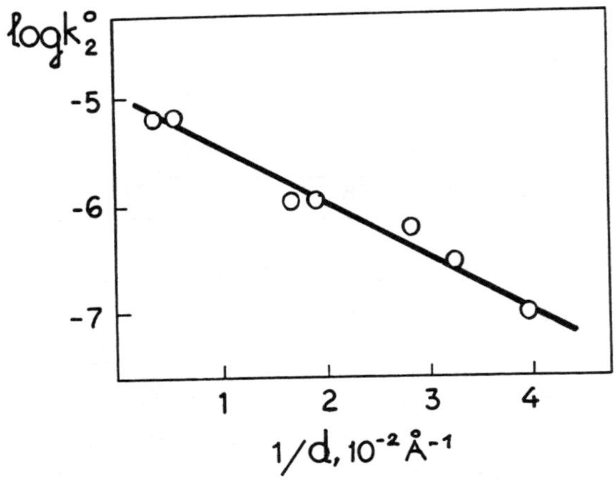

FIG. 3. Dependence of the standard heterogeneous rate constant k_2^o of proton reduction, on the size d of copper colloidal particle - catalyst for dihydrogen evolution [10a]. k_2^o in $cm \cdot s^{-1}$.

of the pH-dependence of the catalytic activity of copper colloids, the standard heterogeneous rate constant for proton reduction on the surface of copper was found. The value of this rate constant depends strongly on the particle size (see Fig. 3) and for the particles with the size of 25-100 Å equals $(0.8\text{-}56) \cdot 10^{-7} cm \cdot s^{-1}$.

From the data of Fig. 3 one can choose the optimal for catalysis size of copper particles. Indeed, as one can easily calculate from the data of Fig. 3, the optimal size of the copper particles providing the maximum activity of the catalyst with respect to the rate of dihydrogen evolution per unit mass of the metal, lies in the rather narrow range of 50-100 Å.

For colloidal Pd ($d \simeq 40$ Å) step (I) was also found to be electrochemically reversible. The elctric potential aquired by Pd particles during the reaction was measured experimentally and found to correspond to accumulation of ca. 30 electrons by a metal particle in steady state conditions. Proton discharge was found to follow Folmer's mechanism.

One can assume, that procedures for kinetic analysis developed in [10a] will be of interest for the studies of the catalysts based not only on copper and palladium, but on other metals as well.

2.1.2. Step of dioxygen evolution from water.

2.1.2.1. Hydroxides of transition metals supported on polysaccharates as catalysts for water oxidation.

The catalysts providing water oxidation to dioxygen are well known as well. These are mainly small free or supported polynuclear particles of oxide- or hydroxide-type compounds of cobalt, iron, manganese, ruthenium and some other transition metals (see review papers [1,2,11-14] and references there in).

For conjugation of the step of O_2 evolution with the step of charge separation in molecular photocatalytic systems for water cleavage, it is important to stabilize catalysts for evolution of O_2 on organic supports. Within the last two years procedures were developed, which allow one to stabilize hydroxide catalysts for O_2

evolution on such organic supports as lipid membranes [13,15] and polysaccharates [13,16,17] .

Note, that the use of organic supports for preparation of the oxygen-evolving catalysts is indeed an important step towards mimicing the oxygen-evolving enzyme operating in the "Photosystem II" of green plants, since this enzyme is known to be stabilized inside the organic matrix.

The first polysaccharate-stabilized oxygen-evolving catalysts had cyclodextrins as the support [16]. In our laboratory we are using as a support for such catalysts much more available polysaccharates like amiloze, which is water-soluble part of starch [17]. Using amiloze, one can prepare both heterogeneous and colloidal catalysts for water oxidation.

Polysaccharates contain electron-donating groups. Thus, they could be expected to undergo rapid oxidation by strong oxidants (like, e.g., trisbipyridyl complex of ruthenium(III), $Ru(bpy)_3^{3+}$), which are necessary to oxidize water. This should result in the dramatic decrease in selectivity of the saccharates-supported oxygen evolving catalysts. However, unexpectedly, it occurred not to be the case. Both selectivity and activity of the catalysts under discussion appeared to be nearly the same as those of the catalysts based on inorganic supports (metal oxides and hydroxides). This observation suggests a rather rigid structuration of the reactive transition metal hydroxides on (or in) the organic matrix, perhaps in the same manner as for the manganese containing reaction center of the oxygen-evolving enzyme in "Photosystem II". The catalytic properties of, e.g., manganese hydroxides supported on polysaccharates are also to some extent similar to those of both natural enzymes and hydroxides supported on inorganic materials. For example, like in the case of the natural enzyme, the pH-region for oxygen evolution by polysaccharates-supported catalysts includes neutral and even slightly acidic region, in contrast to the behavior of "free" colloids of manganese oxides and hydroxides which are more efficient at basic pHs (see Fig. 4).

Polysaccharate supports have a highly hydroxilated surface, which makes them similar to inorganic hydroxides. Note, that the location of the catalytically active centers on the polysaccharates remains still unknown. As suggested in [13,16], they may be located, e.g., inside the "biographical" highly structurated "holes" of cyclodextrins, which determine the size of catalytically active centers . However, amiloze-based catalysts demonstrate the same properties as cyclodextrine-based. In contrast to cyclodextrine, amiloze is not expected to have "biographical" holes since it is a linear slightly branched polymer. Thus, for amiloze the catalytic activity should be, perhaps, attributed to ultradispersed species of transition metal hydroxides chemically (through -O-moieties) bound to the outer surface of polysaccharates globules.

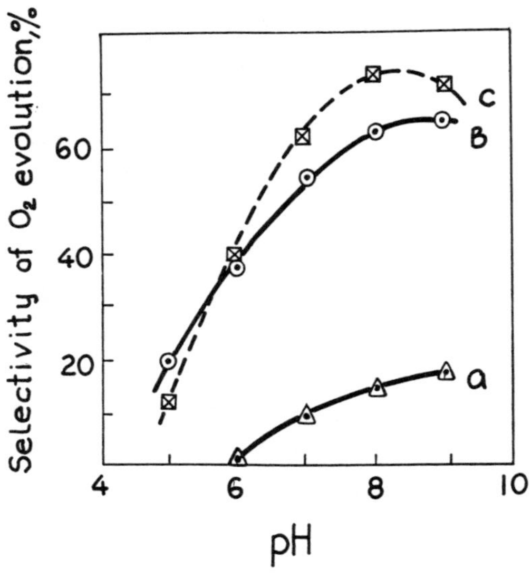

FIG. 4. Comparison of typical pH-dependences of selectivities of O_2 evolution for various manganese-based oxygen-evolving catalysts. (a) - Catalyst obtained as "free" colloid by hydrolysis of $Mn(bpy)Cl_3(H_2O)$; (b) - manganese hydroxide on inorganic NaA-zeolite support; (c) - manganese hydroxide on amiloze. 10^{-3} M $Ru(bpy)_3^{3+}$ is used as the oxidant (according to [17]).

Note, that, when supported on amiloze, hydroxides with the initial oxidation state of manganese atom corresponding to Mn(III) demonstrate much higher catalytic activity and selectivity than hydroxides with the initial state Mn(IV). Moreover, the activity and selectivity of Mn(III) hydroxides decrease upon oxidation of Mn(III) ions in the same catalyst to Mn(IV) state . On the other hand, reduction of Mn(IV) hydroxide to Mn(III) improves dramatically the activity and selectivity.

At present there exist several hypotheses concerning the mechanism of action of oxygen-evolving catalysts and enzymes, especially those containing manganese ions (see, e.g., [13,19,20]). For elucidation of the mechanism it is quite important to know the oxidation state of the catalytically active ion just after release of the dioxygen molecule . For artificial catalysts this state of the catalyst is often considered to coincide with the starting one, e.g., Co(III), Fe(III), Mn(III) (see, e.g.,[13,19,20]). However, our data suggest, that just after release of dioxygen molecule, these catalysts contain metal ions in lower oxidation state than initially, i.e., respectively, Co(II), Fe(II), Mn(II). This suggestion follows from some peculiarities in the chemical behavior of catalytically active species, and seems to be natural when taking into account numerous data concerning dioxygen carriers based on compounds of Co(II), Fe(II), Mn(II) with compositions quite close to those of oxygen-evolving catalysts. Note also, that similarly to oxygen carriers, oxygen-evolving catalysts show sufficient activity not only in water oxidation, but also in reaction of oxidation (by dioxygen) of water soluble organic compounds and hydrazine.

2.1.2.2. Zinc porphyrins as photocatalysts for oxidation of water.

An important goal in design of oxygen-evolving photocatalytic systems is also the search of photocatalysts other than $Ru(bpy)_3^{3+}$ or its close analogs. Attractive candidates for performing this function seem to be metal porphyrins. Note, that the chlorophyll molecules, that play the role of the photocatalyst in the "Photosystem II", are indeed porphyrin-magnesium complexes.

Untill recently only a series of articles by A.Harriman et al. was known (see [21-24] and references therein), where the authors succeeded to couple water-soluble zinc porphyrins (used as the photocatalysts) with the RuO_2, IrO_2 and some other oxygen-evolving catalysts. Unfortunately, it occured difficault to reproduce the evolution of O_2 by these systems. In fact, we succeeded to reproduce practically all the data reported in [21-24] except those concerning dioxygen evolution.

For this reason we decided to look for a new kind of dioxygen evolving photocatalytic systems based on water soluble zinc porphyrins as the photocatalysts (photosensitizers). Our main idea was to introduce into these systems the mediator for electron transfer between the oxidized form of the porphyrin (its cation radical) and the oxygen-evolving catalysts [25]. As such mediator we have used $Fe(bpy)_3^{2+}$ complex. The principle of action of a particular system of this type can be understood from the following example. The irradiation (λ>520 nm) of slightly acidic solution containing zinc tetramethylpiridylporphin ($ZnTMPyP^{4+}$), $Fe(bpy)_3^{2+}$, and $S_2O_8^{2-}$ (irreversible electron acceptor-sacrifice) leads to the formation of the strong one-electron oxidant $Fe(bpy)_3^{3+}$ with quantum yield of ca. 0.2. The reaction mechanism established by means of flash and steady-state photolysis includes photochemical generation of zinc porphyrin cation radical $ZnTMPyP^{5+\cdot}$ via oxidation of the triplet excited porphyrin molecules by $S_2O_8^{2-}$. The porphyrin cation radical oxidizes rapidly mediator $Fe(bpy)_3^{2+}$ to give $Fe(bpy)_3^{3+}$ and $ZnTMPyP^{4+}$. After the accumulation of $Fe(bpy)_3^{3+}$ in the presence of Co^{2+} (the precursor of the oxygen-evolving catalyst) the reaction of dioxygen formation is initiated by adding an alkaline (to shift acidity of the solution to pH12). Dioxygen evolved is detected directly by gas-chromatography.

One should note, that according to our experience neither photochemically, nor chemically generated cation radicals of the water soluble zinc-porphyrins evolve dioxygen by themselves even in the presence of the best artificial oxygen-evolving catalyst. The reason of this remains still unknown.

Note, that in the reported example the accumulation of the oxidized form of the mediator was separated in time with the step of catalytic dioxygen evolution. The background for such procedure is as follows. Inspite of numerous attempts, we have failed so far to find a system, where dioxygen is evolved directly during the action of light in the presence of triplet excited molecules of the metal porphyrins. The reason of this seems to be the extremely high reactivity of triplet porphyrins towards the dioxygen molecules. In order to avoid this difficulty, the Nature uses in the photosynthesis not triplet, but singlet excited porphyrin molecules which are much less active towards dioxygen. But in this situation because of extremely short (ca. 10^{-9} s) lifetime of the singlet excitation of porphyrins, one should have a very rigid and adjusted construction of the chain of electron-transfer. Otherwise, the singlet excitation would dissipate before providing a useful chemical reaction.

In our case, the separation in time of the photochemical step of $Fe(bpy)_3^{3+}$ formation and the non-photochemical step of O_2 evolution from H_2O by the accumulated $Fe(bpy)_3^{3+}$, makes evolution of O_2 possible even though triplet states of the zinc porphyrin are formed during the reaction.

2.1.3. Step of photochemical charge separation and its conjugation with the step of H_2 evolution.

The last years we have focused our efforts on the design of molecular photocatalytic systems based on the suspensions of bilayer lipid vesicles. As the main problems to be solved here, we consider light-induced electron transfer across the lipid membranes and coupling of this electron transfer with the catalysts for water reduction and oxidation.

2.1.3.1. Mechanisms of electron transport across membranes.

As one can see from the analysis of our data and those available in the literature, there are three principal mechanisms of electron transfer through the lipid membrane: (i) migration of the particle of a charge carrier as a whole across the membrane; (ii) electron exchange (presumably, via electron tunneling) between particles located in the inner and outer layers of the bilayer membrane; (iii) "contact" electron transfer through a chain (bridge) of chemical bonds in rigidly structurated submolecular systems like bridge molecules, semiconductor particles, etc. (see review articles [26,27] and references therein).

The electron transfer in a real system based on lipid vesicles, is a combination of these pathways [26,27].

In the last two years we concentrated our attention on the mechanisms (i) and (ii) of photoinduced charge transfer, since the known systems with domination of these types of charge transfer allow to organize rather efficient (with quantum yield of several percent in non-sacrificial systems) electron transportation across the membranes simultaneously with the formation of the reductant-oxidant pair of particles capable of reducing water to H_2 and oxidizing it to O_2 [1,2,25,26]. An additional advantage of these systems is that they can be designed by relatively simple methods of molecular engineering, without the use of too complicated or exotic compounds.

2.1.3.2. Conjugation of charge separation and H_2 evolution.

Most of the works in this direction were made, using lipophilic viologens, V^{2+}, as electron carriers (relays) for electron transfer through the lipid membrane from $Ru(bpy)_3^{2+}$. This process leads to the formation of the oxygen-evolving oxidant $Ru(bpy)_3^{3+}$ [1,2,20,26,27]. However, precise studies of theses systems have shown, that the reducing ability of the viologen cation radicals, V^+, formed upon photoinduced electron transfer from $Ru(bpy)_3^{2+}$ to V^{2+}, is substantially decreased (by ca. 0.2 V) when these cation radicals are embedded into the lipid membranes [26]. This creates an obstacle for simultaneous photoproduction of dihydrogen in such systems, since the resulting reduction potential of the photochemically generated V^+ (ca. -0.25 V vs. N.H.E.) is suitable for dihydrogen evolution only at acidic pH when it is rather difficult to evolve dioxygen. Thus, we had to choose a novel lipophilic electron carrier with more negative redox potential and ability to couple with both $Ru(bpy)_3^{2+}$

photocatalyst (photosensitizer) and the hydrogen-evolving catalyst.

Recently we have found such a mediator [28]. This appears to be 1,4-bis(1,2,5-triphenyl-4-pyridyl)benzene (benzoV^{2+})

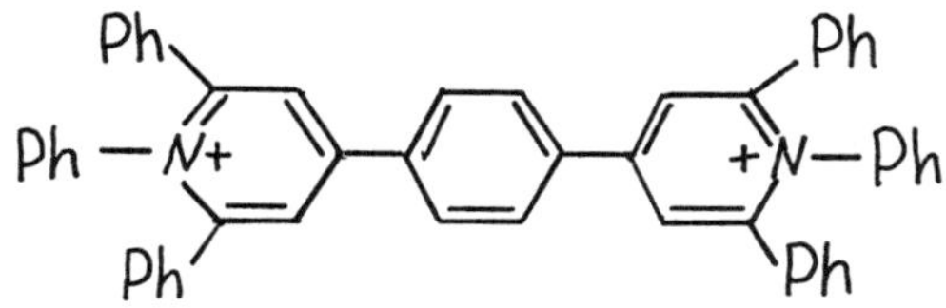

which is an analog to viologens. Fortunately, the potential of its one electron reduction in sodium dodecyl sulfate micellar solutions (which model suspension of vesicles) is equal to ca. -0.52 V vs. N.H.E., i.e. is sufficient to evolve dihydrogen even at slightly basic pHs. There is also a possibility for benzoV^{2+} to be reversibly reduced by two electrons, the difference between two reduction potentials being very small (ca. 0.06 V). Being embedded into the lipid membrane, benzoV^{2+} can serve as the efficient electron acceptor for exited sensitizers, e.g. $^*Ru(bpy)_3^{2+}$, located in the inner or outer water phases of the vesicle suspensions. We have found also, that ultrafine palladium particles (with diameter of ca. 10-20 Å) supported on both inner or outer surfaces, of the same lipid membranes or located only on their outer surface can serve as a hydrogen-evolving catalyst for benzo$V^{+\cdot}$.

All these properties of the benzoV^{2+} mediator have allowed us to design several vesicle-based systems, where photocatalytic dihydrogen evolution occurs even at pH 7.8 with quantum yields exceeding 10^{-2}%. The simplest systems of this type are those evolving dihydrogen either from inner or outer water phases of the vesicle suspension without a transmembrane electron transfer:

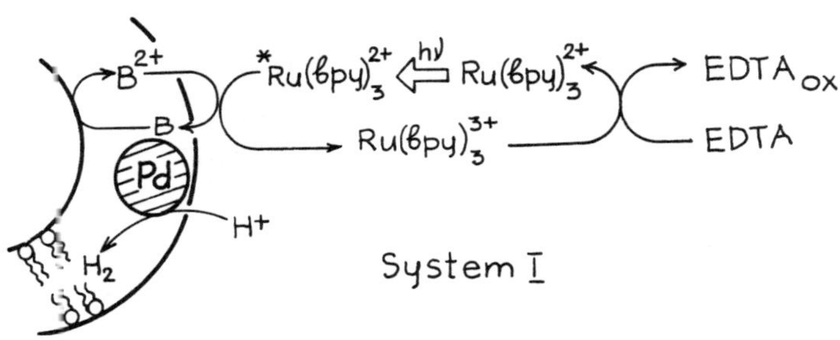

System I

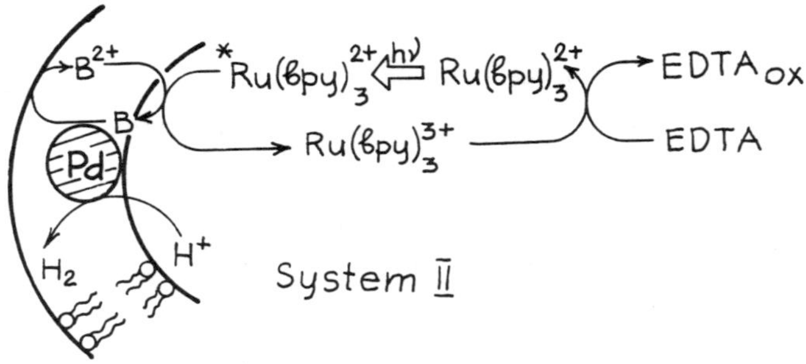

(Here and later B^{2+} stands for benzoV^{2+}.) However, the most interesting system is that allowing to couple dihydrogen evolution with the photoinduced transmembrane electron transfer:

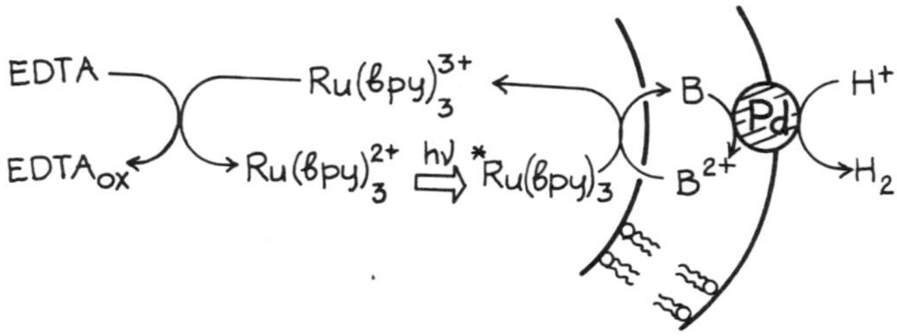

2.1.4. Simultaneous accomplishment of water reduction and oxidation.

The development of appropriate technique for the anchoring of catalysts for water oxidation and reduction on the surfaces of membranes made it possible to construct sacrificial photochemical systems containing vesicles of different types , which are capable to perform the functions of water reduction and oxidation simultaneously. The first systems of this kind, where the simultaneous photocatalytic evolution of dioxygen and dihydrogen was observed, were suggested by Luneva et al [29] (see also [20]). The sacrificial photocatalytic system for dihydrogen evolution was incorporated into the inner volume of lipid vesicles (Fig. 5a) , while the sacrificial photocatalytic system for dioxygen evolution was located in the outer bulk solution (Fig. 5b). Upon illumination of the suspension obtained by mixing of both systems shown in Fig. 5, one could observe a simultaneous evolution of H_2 and O_2. Note, however, that no conjugation of these processes with electron transfer across the membranes was carried out in this system.

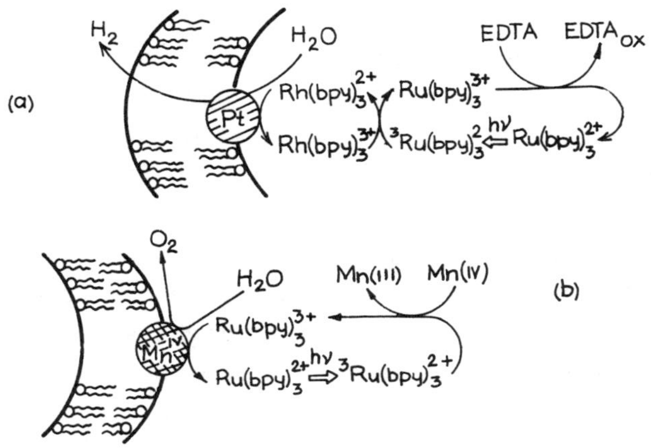

FIG. 5. Application of lipid vesicles for photoinduced cleavage of water in sacrificial systems. (a) - Dihydrogen evolution inside the vesicle cavity in the presence of a Pt metal catalyst anchored to the inner lipid/water interface, (b) - dioxygen evolution in the bulk solution in presence of a manganese catalyst anchored to the outer membrane/water interface (according to [26,27,29]).

In the vesicle suspension of Fig. 5 it occurs possible to isolate the centers for dihydrogen and dioxygen evolution and, thus, to avoid cross reactions of D^+ and A^- with the catalysts for H_2 and O_2 evolution, respectively. However, it turned out that O_2 evolution gradually inhibits the H_2 evolution, because oxygen evolved in the outer volume permeates across the membranes and gradually destroys the apparatus for dihydrogen evolution located inside the vesicles. Note, that such a problem arises also for biological systems adapted to provide simultaneous evolution of H_2 and O_2 [30,31].

The difficulties arising from the penetration of O_2 through the membrane, which separates the oxidative and reductive parts of the photocatalytic systems, are difficult to be solved with the use of microscopic mono- or bilayer membranes for photoinduced charge separation. In fact, it seems rather difficult to prevent the diffusion of a small non-polar oxygen molecule across a lipid or surfactant membrane with the thickness of 30-40 Å. Better isolation between the oxidative and reductive parts of the photochemical system, perhaps, may be provided by macroscopic porous or ion-exchange membranes separating the solutions with systems (a) and (b) of the type presented in Fig. 5 (the way it is usually done in water electrolysis), or in solid layered structures, where two different interlayer compartments containing the catalysts for evolution of H_2 and O_2, can be separated by a rigid layer of framework atoms [39]. It might also occur possible to avoid the inhibiting influence of O_2 on the molecular apparatus for H_2 production, by means of evolving dihydrogen and dioxygen from water consecutively rather than simultaneously. In this case one may benefit from the technique suggested by Gerasimov et al. [25] (see section 2.1.2.2.), which provides the separation in time of the two steps: (i) photogeneration of a strong oxidant, which in this case must be conjugated with the simultaneous process of dihydrogen evolution, and (ii) subsequent oxidation of water with this oxidant.

Thus, the techniques have already been developed which allow to introduce the catalyst for H_2 and O_2 evolution into appropriate places of the membrane system: inside or outside vesicles, on the inner or outer surface of the membranes. However, the problem of efficient conjugation of the catalytic processes of H_2 and O_2 evolution with the photoinduced electron transfer across the membranes is still to be solved.

2.2. Semiconductor photocatalytic systems for production of dihydrogen from solutions of electron donating compounds.

The development of photocatalytic systems mimicing the bacterial photosynthesis, which use as substrates substances that can be oxidized more easily than water, was found to be a much simpler problem than photocleavage of water. These systems can function without membranes which provide the spatial separation of particles A^- and D^+. By the moment a large number of both homogeneous and heterogeneous photocatalytic systems are already known, which mimic bacterial photosynthesis. These are the so-called "sacrificial" photocatalytic systems. They provide a rather efficient evolution of dihydrogen owing to irreversible oxidation of a donor-"sacrifice" under the action of visible light.

The summary of the data obtained so far in this area can be found, e.g., in refs. [3-5,32,33].

From a practical standpoint, the most interesting for providing dihydrogen evolution from the solutions of electron donating compounds are photocatalysts based on colloidal or suspended semiconductors (see, e.g., [3,5,32,33]). The preparation of such semiconductor photocatalysts is, as a rule, a much more simple problem than preparation of molecular photocatalysts, since it does not reqire any complicated synthesis of organic chemicals. The life time of many semiconductor photocatalysts also exceeds notably that for photocatalysts based on organic dyes or metal complexes. However, an important problem to be solved is how to increase the efficiency of the photocatalytic action of the semiconductors under illumination by solar light. In order to solve this problem, more detailed mechanistic studies of the photocatalytic processes over semiconductor particles seem to be necessary.

2.2.1. Photosensitization of semiconductor photocatalysts to visible light.

The most studied semiconductor photocatalysts are based on wide-gap semiconductors like TiO_2, ZnO, etc. However, wide-gap semiconductors are able to absorb only UV-, but not visible, light. Hence, their photocatalytic activity under the action of solar light is rather poor. In order to overcome this drawback, one should elaborate the ways of photosensitizing the wide-gap semiconductor photocatalysts to visible light.

The most known are the following four ways of providing photocatalytic reactions over semiconductor catalysts under the action of visible light: (i) the use of narrow-gap semiconductors; (ii) modification of wide-gap semiconductors with special dopants; (iii) the use of semiconductor particles with heterojunctions; (iv) sensitization of wide-gap semiconductors by organic or metal complex dyes.

The ways (i) and (ii) have certain fundamental restrictions because of certain peculiarities of thermodynamic and electrochemical properties of simple semiconductor materials. For example, simple oxides and simple sulfides are able to evolve dihydrogen from water solutions only, if the width, E_g, of their band-gap is greater than ca. 3.0 and 2.2 eV, respectively (see, e.g. [3,5,32-35]). For this reason, at present we consider as more promising the ways (iii) and (iv).

2.2.1.1. Photosensitization via creation of a heterojunction.

Heterojunctions can be formed in the place of the electrical contact of two different semiconductor phases. In the case of finely dispersed semiconductor photocatalysts a procedure of consecutive precipitation was recently developed in our laboratory to obtain highly dispersed species with the electrical contact between two metal sulfides [35]. The photocatalysts prepared were promoted for water reduction by small metal particles and used for dihydrogen evolution from water solutions containing hydrogen sulfide or sulfide-ions:

$$H_2S \xrightarrow[PhC]{h\nu} H_2 + S \qquad (4)$$

The best results were obtained for microheterojunctions between a wide-gap cadmium or zinc sulfide (or a mixed sulfide of these metals) and a narrow-gap copper or silver sulfide (see Fig. 6).

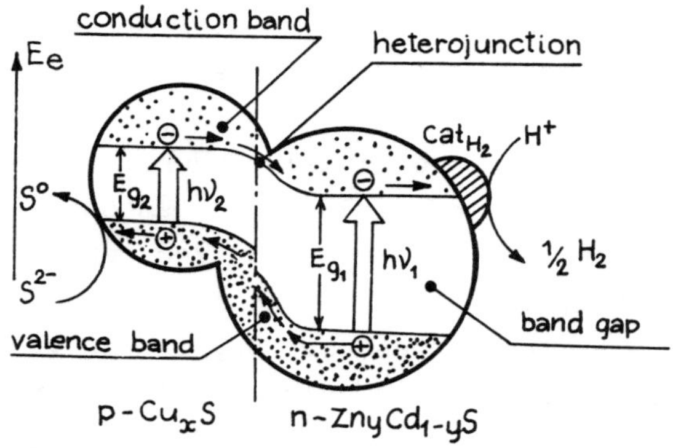

FIG. 6. Idealized energy diagram of electron transfer and the scheme of the photocatalytic dihydrogen formation for the photocatalyst particle with microheterojunction n-$Zn_yCd_{1-y}S$/p-Cu_xS (according to [1,2]).

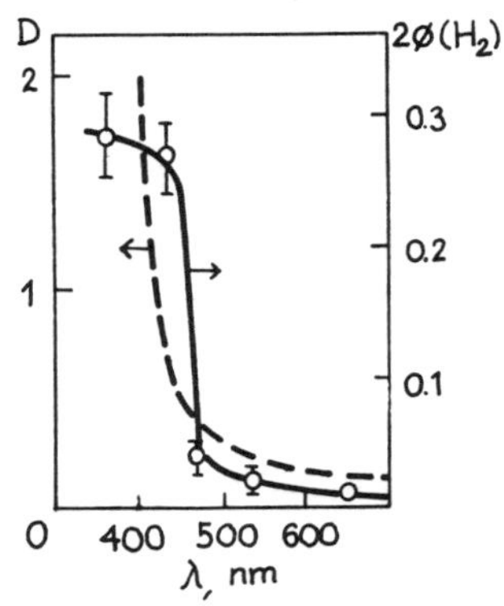

FIG. 7. Absorption spectrum (dashed line) for the Pt-$Zn_{0.17}Cd_{0.83}S/Cu_xS$ suspension and typical spectrum of the photocatalytic action for the same suspension. D is optical density, Φ(H_2) is effective quantum yield of H_2 molecules (according to [35]).

A typical spectrum of photocatalytic action of the Pt-$Zn_{0.17}Cd_{0.83}S/Cu_xS$ catalyst that has a microheterojunction, is shown in Fig. 7.

The presence of an appropriately chosen heterojunction is quite useful because of the following reasons. First, it is equivalent to the creation of an organized chain of vectorial electron transfer and, thus, significantly improves the efficiency of the charge separation process. Second, it provides a notable shift of threshold energy $h\nu_o$ of the photocatalytically active quanta towards smaller values due to the fact that light quanta are now absorbed not only by the wide-band semiconductor, but also by the narrow-band one (see Fig. 6).

It can be seen from Fig. 7, that the photocatalyst Pt-Zn_yCd_{1-y}/Cu_xS provides a noticeable quantum yield for reaction (4) even in the spectral region with $\lambda>500$ nm, i.e. beyond the fundamental absorption band of $Zn_{0.17}Cd_{0.83}S$ semiconductor. Note, that the Cu_xS semiconductor with $E_g<1$ eV ($\lambda_o>1000$ nm), when taken alone, does not show any photocatalytic activity, [35]. In the optimal situation, upon illumination in the absorption band of the wide-band phase, for the Pt-Zn_yCd_{1-y}/Cu_xS photocatalyst with the particle size of ca. 2-4 μm, the effective quantum yield of ca. 50% was obtained. This value exceeds considerably the quantum yield achieved so far for "simple" single-phase semiconductors, the high quantum yields being observed even for very low (ca. 10^{-4} M) concentrations of the sulfide anion in solutions. This allows one to evolve dihydrogen efficiently even from quite diluted solutions of hydrogen sulfide. For illumination at $\lambda>500$ nm (up to near infrared), a rather high (up to 10%) quantum yield was registered as well.

2.2.1.2. Photosensitization with immobilized dyes and their complexes with noble metals.

As it was shown above, photosensitizing of the semiconductor photocatalysts via creation of heterojunction structures, has lead to certain success. Nevertheless, hybrid molecular-semiconductor systems providing photosensitization by means of adsorption or chemical bonding of dye molecules to semiconductor surface, still attracts attention. A driving force for the interest to such systems is an expected possibility to combine within them the advantages of both semiconductor and molecular photocatalytic systems (such, e.g., as higher stability, which is characteristic of semiconductor systems, and greater flexibility for tuning the chain of electron transfer, which is characteristic of molecular systems, etc.).

Sensitization of Pd-TiO_2 with Rhodamine B.

A typical example of sensitization of a semiconductor with a dye is shown in Fig. 8. One can see, that the adsorption of organic dye Rhodamine B on the platinized particles of TiO_2 leads to appearance of a novel, with respect to "pure" TiO_2, band in the spectrum of photocatalytic action of TiO_2 in the sacrificial

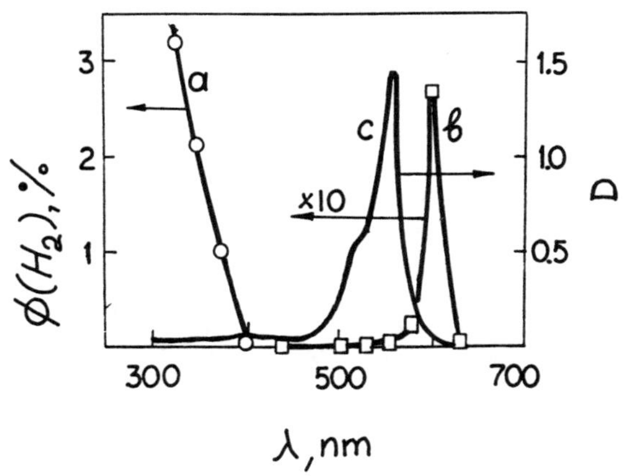

FIG. 8. Typical spectra of the photocatalytic action in dihydrogen evolution from water solutions of oxalic acids, for dispersed platinized TiO_2: (a) - non-sensitized and (b) - sensitized by Rhodamine B. (c) - Optical density D of $1.3 \cdot 10^{-5}$M homogeneous water solutions of Rhodamine B. $\phi(H_2)$ stands for effective quantum yield of dihydrogen (according to [36]).

photocatalytic system for the evolution of H_2 from water solutions of oxalic acid [36]. The new band is located in the visible region of the spectrum, and its position is clearly shifted to long-wave region in comparison to the absorption spectrum of homogeneous water solutions of Rhodamine B (see Fig. 8c). The last observation suggests that the sensitization occurs due to adsorbed but not water-dissolved dye molecules. Similar data were obtained also for sensitization of platinized powder of TiO_2 by organic dye Acridine Yellow [37].

Sensitization of TiO_2 with Pt-Alizarine complex.

One should mention that in the both above examples, the photocatalysts sensitized by the organic dyes, needed also a promotion to dihydrogen evolution that was made by deposition of small particles of a noble metal (e.g., Pt) on the surface of

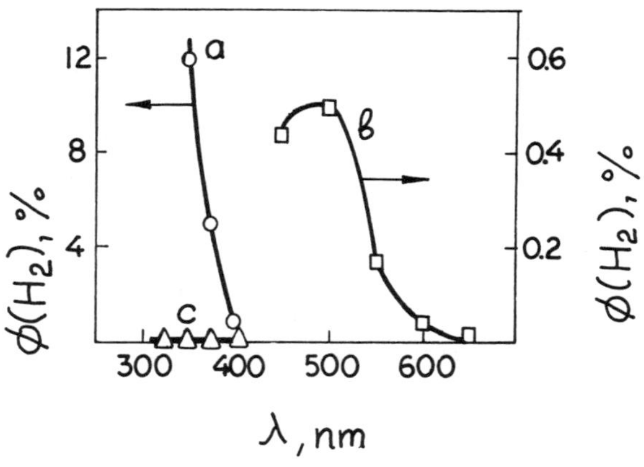

FIG. 9. Spectral dependencies of the effective quantum yield $\phi(H_2)$ for the photocatalytic evolution of dihydrogen from water solutions of disodium salt of ethylenediaminetetraacetic acid: (a) - for fundamental and (b) - for sensitized absorption bands of the TiO_2 with supported $Pt(Alizarine)_2$. (c) - The same dependence for non-sensitized and non-platinized TiO_2 in analogous conditions (according to [38]).

the semiconductor particles. Recently, we have found that the latter procedure can be avoided, if instead of organic dyes their complexes with noble metals are used as sensitizers [38]. Such complexes can be obtained, e.g., by using Alizarine dye as the ligand. The complexes of Alizarine with, say, platinum, Pt(Alizarine)$_2$, are strongly colored due to light absorption by the ligands. Being supported on a powder of TiO_2, such complex plays the roles of both a sufficiently efficient photosensitizer to visible light with wavelength of ca. 500 nm and catalyst promoting dihydrogen evolution (see Fig. 9).

Further works in this direction are needed to study the mechanisms of both photosensitization and catalytic promotion by dyes and their complexes with noble metals adsorbed on semiconductor particles.

2.2.2. Mechanistic studies of photocatalysis on dispersed CdS.

Advances in photocatalysis by dispersed semiconductors have been followed by development of new approaches to preparation of semiconductor particles, with heterojunctions or supported molecules of photosensitizers. For further progress in this field, mechanistic studies are needed of the processes occurring on the surface of such species as well as changes of their physical-chemical properties in the course of photocatalytic reactions. Unfortunately, the mechanistic data in this field are still quite scarce.

2.2.2.1. Photocleavage of H_2S on dispersed CdS.

The result of our mechanistic studies [40,41] of reaction (4) can be briefly summarized as follows:

Excitation of CdS particles caused the generation of the electron-hole pair:

$$CdS \xrightarrow{h\nu} e^- + h^+,$$

which can either recombine in the bulk of the semiconductor particle (with the rate constant ca. $(0.5\text{-}2)\cdot 10^6\ s^{-1}$ or reach its surface. For the particles with the sizes less than 200 Å, the probability of this latter process increases rapidly with the decrease of the particle size.

When reaching the surface the holes oxidize the surface sulfur vacancies:

$$h^+ + V_s \longrightarrow V_s^+,$$

providing surface oxidative states V_s^+. This process has the heterogeneous rate constant of (17±1) cm/s and can be accompanied by chemiluminescence. Particles V_s^+ can either oxidize adsorbed sulfide dianion S^{2-} to elementary sulfur:

$$2V_s^+ + S^{2-} \longrightarrow S^o$$

with the effective heterogeneous rate constant 0.02-0.2 cm/s, or recombine with electrons:

$$e^- + V_s^+ \longrightarrow V_s$$

The rate-determining step of the photocatalytic dihydrogen evolution appears to be the reduction of protons, which can be promoted, e.g., by metal particles attached to CdS. The promotion is more efficient for noble metals with small overpotentials for cathodic dihydrogen evolution. For the best metallic promotor, Pt, the heterogeneous rate constant of the reaction

$$e^- + H^+ \longrightarrow 1/2H_2$$

appeared to be equal to 10^{-3} cm/s.

2.2.2.2. The influence of the size of CdS particles on their photochemical properties.

Such influence was detected in our studies of photoinduced electron transfer from ultrafine (with radii R⩽50 Å) colloidal particles of CdS to reversible electron relay - methylviologen MV^{2+}, in the presence of L-cysteine as an irreversible electron donor [42]. It was found in the experiments on steady state photolysis, that for the colloidal particles of CdS with R=50 Å the kinetics of the accumulation of the reduced form of MV^{2+}

follows the equation:

$$\Delta[MV^{+}] \sim \ln(1 + \alpha t),$$

where t is the exposition time, α is the coefficient. In contrast to this, for larger particles with R > 100 Å, the formation of MV^{+} obeys the usual first order kinetics. The main reason of the unusual kinetic behavior of the ultrafine semiconductor particles seems to be a dramatical influence of the size of the CdS particles, on their reactivity. In the presence of a distribution over the sizes of the particles, the scatter in the reactivity for various particles results in the unusual logariphmic law for the accumulation of MV^{2+}.

Note, that the particle sizes of R < 50 Å, for which the unusual logariphmic kinetics is observed, indeed are in the region of sizes for which a strong influence of the particle size on its electronic properties (e.g., the width of the band-gap) is known [43]. The size of such colloidal particles is less than the characteristic length of diffusion of charge carriers, less than or comparable with the characteristic length of migration of "hot" charge carriers, and comparable with the size of exitons. It is known also that the solubility product for such ultrafine particles differs dramatically from that for bulk CdS (compare the value 10^{-5} M^2 for CdS colloids with R≅20 A°, and the value 10^{-27} M^2 for massive crystalline CdS [44]. It seems reasonable to assume, that the main parameter influenced by particles' size, is the activation energy for electron phototransfer. Indeed, the effective activation energy E_a for such redox process can be expressed as:

$$E_a = E_a^{PZC} + \beta\varphi_{DL} + (\beta + z)\varphi_{OHL},$$

where E_a^{PZC} is the effective activation energy in the point of zero charge of the semiconductor particle, φ_{DL} and φ_{OHL} are, respectively, the potential of the double layer on the surface of the particle and on the outer surface of Helmholz layer; β stands for electrochemical transfer coefficient and z=2 is the charge of

MV^{2+}. Note, that when deriving the above expression, φ_{DL} for the bulk solution was considered to be equal to zero. Thus, the size dependence of the activation energy E_a can be due to the size dependences of the work function (which is related to E_a^{PZC}), or φ_{DL} and φ_{OHL}. The value of the potential of the double layer is determined in a large extent by the heat of adsorption of potential-determining ions (in our case RS^- anions of cysteine).

To verify the above assumptions about the influence of R on E_a, we have performed direct electroforetic measurements of the ζ-potential of the double layer of CdS particles. These measurements have shown, that with diminishing the particle size ζ-potential of CdS colloids is shifted to cathodic region. E.g.,it was found, that at R>100 Å ζ=-20 mV (at ionic strength less than $1.5 \cdot 10^{-2}$ M), while at R≅80 Å and R≅50Å ζ=-35 mV and ζ=-55 mV,respectively. We have observed also the changes in the values of ζ-potentials for the same particles with R<100 Å in the course of the photochemical reaction. These changes are equal to ca. 10% and 30% of the ζ-potential value for CdS colloids with R≅80 Å and 50 Å, respectively, even at light intensity less than 1 mW/cm^2 (λ=365 nm), which corresponds to accumulation of not more than one electron per a colloidal particle.

One can assume that the observed photostimulated changes of ζ-potential are induced by either a change of the total charge of the particle due to different rates of the chemical reactions of photoinduced electrons and holes, or the changes in the coverage of the particles by potential-determining ions (note, that in our case RS^- anion is not only the potential-determining ion, but also the reagent for the hole). In the both cases the respective change of ζ-potential is expected to be determined by size-dependent rate constants of processes with the participation of electrons and holes on the surface of CdS particles.

The above results suggest that the observed kinetic non-equivalence of CdS particles of various sizes, can be determined by both the "biographical" difference in their physical-chemical properties, and photoinduced changes of the potential of the double layer.

To verify the existence of photoinduced non-equivalency of

various CdS particles in a colloidal solution, special flash-photolysis experiments were made with the particles having R > 100 Å, for which the "biographical" non-equivalency is not revealed in kinetic studies.

The flash-photolysis experiments were carried out as follows [42]. The colloidal solution of CdS was flash-photolyzed in the presence of O_2 as electron acceptor and L-cysteine as the electron donor. Under these conditions the colloidal particles were pumped with electrons, since the rate of reactions between the photogenerated holes and L-cysteine anions adsorbed on the surface of CdS, was much faster than the reaction between photogenerated electrons and O_2 molecules dissolved in water. As a result, a photobleaching of CdS is observed since excessive electrons occupy the lower energy levels in the conduction band.

We have studied the kinetics of the relaxation of the photobleached CdS to its initial state. According to our theoretical estimates, the change in optical density in the edge of the absorption band of a semiconductor particle, is proportional to $n^{2/3}$, where n is the local concentration of the excess electrons inside the particle. Having this in mind, we have analyzed the kinetic curves for the increase of the optical density in the process of its relaxation after the flash. The rate, v, of this process was found to follow the equation

$$v \sim ne^{\gamma n},$$

where γ is a coefficient. Note, that in our experiments with colloids having R > 100 Å, the local concentration of excess electrons just after the flash was equal ca. 10^{-7} M. This corresponds to ca. 20 excess electrons per a particle of CdS.

The factor γn in the above expression for v can be considered as a change of the activation energy E_a because of photoinduced change of the potential φ_{DL} of the double layer. Numeric estimates of these changes for ultrafine CdS particles using a simple electrostatic model, predict dramatic (about one Volt) changes in φ_{DL} upon accumulation of several excess electrons per a colloidal particle. Note, that in our experiments with colloids having R=50 Å, the number of excess electrons has reached ca. 100. In this case the change of φ_{DL} must be still much larger.

3. THERMOCATALYTIC SYSTEMS FOR SOLAR ENERGY CONVERSION.

In our lecture [1] at the I.P.S.-7 we have described thermocatalytic systems for solar energy conversion, storage and back release in the form of high-potential heat. These systems were based on cyclic accomplishment of the endothermic reactions:

$$CH_4 + H_2O \longrightarrow 3H_2 + CO \tag{5}$$

$$CH_4 + CO_2 \longrightarrow 2H_2 + 2CO \tag{6}$$

and subsequent back exothermic catalytic reactions of methanation.

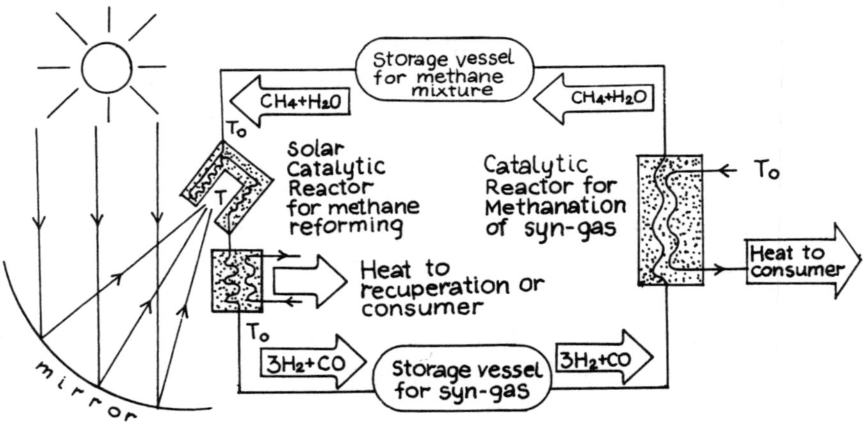

FIG. 10. A simplified principal scheme of a solar thermocatalytic power plant based on cyclic reactions "methane reforming - syn-gas methanation" (according to [1,2]).

Reactions (5) and (6) were accomplished under the action of heat collected from mirror concentrator of solar light. The syn-gas (e.i. the mixture of CO and H_2) produced in these reactions can be stored and then, when necessary, converted into

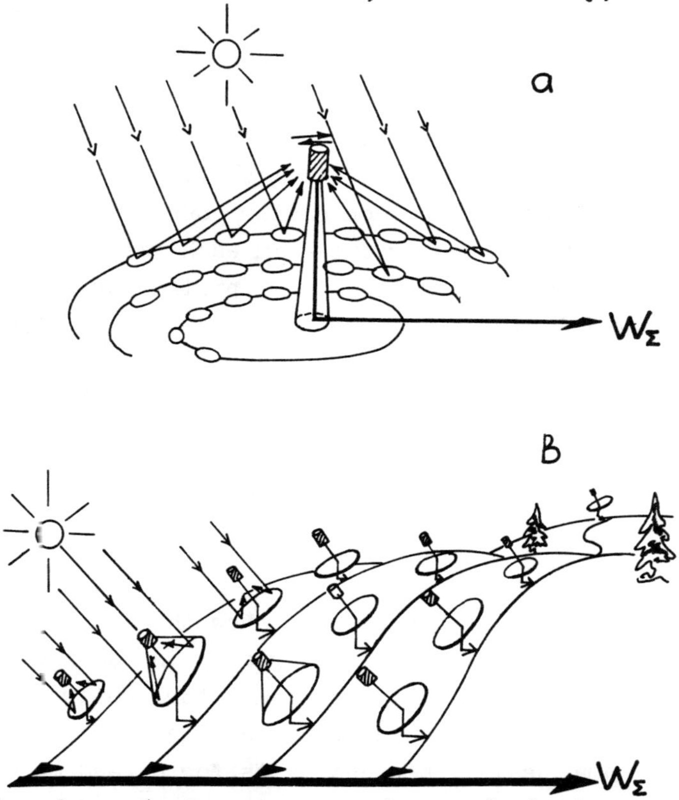

Fig. 11. Conceptual schemes of a single-tower (a) and a distributed modular (b) solar catalytic power-plant of the same total power W_Σ. In the case (a) the conversion of all the concentrated solar energy fluxes is done by a single catalytic reactor of a great power while in the case (b) - with the use of a field of standardized modular SCRs of a moderate power united by a fuel collector. The scheme (b) seems to be more simple for realization and reliable than scheme (a).

high-potential heat (with the temperature up to 950 K) in the methanation catalytic reactor. The principal scheme of the solar thermocatalytic power plant is presented in Fig. 10. The efficiency of solar-to-chemical energy conversion as high as 43% was experimentally achieved for reaction (5) in the solar catalytic reactor (SCR) with useful power of SCR ca. 2kW [45].

Fig. 11 presents two possible designs of a more powerful plant for solar-to-chemical energy conversion. The design, presented in Fig. 11a, implies the synchronized collection of solar light by a large number of mirrors which all focus this light on the solar catalytic reactor located on the top of the "solar" tower. The design presented in Fig. 11b implies conversion of solar-to-chemical energy in much smaller SCRs located in the focus of each mirror. Syn-gas produced in each reactor can then be collected into storing pipe-line. A preliminary assessment suggests that the scheme of Fig. 11b will be more reliable and efficient economically [46].

The work on design of the integrated scheme of Fig. 11b is underway.

4. CONCLUSION.

The data presented in this lecture show that certain progress has been made within the last two years in design of both photocatalytic and thermocatalytic systems for solar-to-chemical energy conversion. For molecular photochemical systems (which mimic plant photosynthesis), new catalysts for H_2 and O_2 evolution and new relays for electron transport were found. Mechanisms of electron transport across lipid membranes, which provides photo-induces charge separation in vesicle solutions, were elucidated. For semiconductor photochemical systems photocatalysts with heterojunctions and supported dyes were designed. They permit one to obtain dihydrogen via oxidation of H_2S and organic compounds in diluted solitions. These catalysts may occur useful in the future for processing with solar light of natural waters containing H_2S and waste waters containing H_2S and organic compounds. Such processing will provide detoxication of these waters with the simultaneous production of H_2 fuel. The advances made so far in thermocatalytic conversion of solar energy, suggest that this method may, perhaps, become important for the energetics of the XXI century.

Whether this actually happens or not, will depend primarily

upon two factors. The first factor is of economic character. At present all the existing methods of energy production via conversion of solar energy are far more expensive than the traditional ones, such as combustion of fossil fuels at electric plants, nuclear fission, etc. In this situation the shift towards a wide use of solar energy is expected to occur gradually, following the trends of solar energy becoming cheaper with every new scientific and technological breakthrough, and traditional energy sources becoming more expensive due to their gradual exhaustion.

The second factor is of ecological character. Much will depend on how dangerous will actually be global consequences of Earth pollution with manmade extra heat, chemicals, etc., associated with traditional types of energetics. Note, that nuclear fusion energetics, which sooner or later is anticipated to be developed, also is expected to pollute Earth with extra heat. If such pollution occurs intolerable, the development of solar energetics, which produces no extra heating of the Earth and in other respects also seems to be ecologically more pure than any other kind of energetics, may obtain high priorities.

REFERENCES.

1. V.N.Parmon and K.I.Zamaraev in: Photochemical energy conversion, J.R.Norris, Jr., and D.Meisel, eds. (Elsevier, New York 1989), pp.316-342.
2. V.N.Parmon and K.I.Zamaraev in: Photocatalysis. Fundamentals and Application, E.Pelizzetti and N.Serpone, eds. (John Wiley & Sons, New York 1989), pp.565-601.
3. M.Gratzel, ed., Energy Resources through Photochemistry and Catalysis (Academic Press, New York-London 1983).
4. K.I.Zamaraev and V.N.Parmon, eds.,Fotokatalitícheskoe Preobrazovanie Solnechnoi Energii (Photocatalytic Conversion of Solar Energy), Parts 1 and 2 (Nauka, Novosibirsk, 1985).
5. E.Pelizzetti and N.Serpone, eds., Photocatalysis. Fundamentals and Application, (John Wiley & Sons, New York, 1989).
6. I.Okura, Coord. Chem. Rev. 68, 53-99 (1985).

7. E.N.Savinov, E.R.Savinova and V.N.Parmon in: Fotokataliticheskoe Preobrazovanie Solnechnoi Energii, Part 2, K.I.Zamaraev, V.N.Parmon, eds., (Nauka, Novosibirsk 1985) pp. 107-151.
8. V.N.Parmon, S.V.Lymar and K.I.Zamaraev in: Hydrogen Energy Progress VI, vol. 1, T.N.Veziroglu, N.Getoff and P.Weinzierl, eds., (Pergamon Press, New York - Oxford 1986) pp. 351-364.
9. G.McLendon in: Energy Resources through Photochemistry and Catalysis, M.Gratzel, ed. (Academic Press, New York 1983) pp. 99-122.
10a E.R.Savinova, E.N.Savinov and V.N.Parmon, J. Molec. Catal., 48, 231-248 (1988).
10b E.R.Savinova, A.L.Chuvilin and V.N.Parmon, J.Molec. Catal., 48, 217-229 (1988).
11. G.L.Elizarova and V.N.Parmon in:Fotokataliticheskoe Preobrazovanie Solnechnoi Energii, Part 2,K.I.Zamaraev and V.N.Parmon, eds., (Nauka, Novosibirsk 1985) pp. 107-151.
12. T.S.Dzhabiev and A.E.Shilov, ibid., Part 1, pp. 169-192.
13. N.P.Luneva, V.Ya.Shafirovich and A.E.Shilov, J. Molec. Catal., 52, 49-62 (1989).
14. P.Comte, M.K.Nazeeruddin, F.P.Rotzinger, A.J.Frank and M.Gratzel, ibid., 52, 63-84 (1989).
15. O.V.Gerasimov, S.V.Lymar, I.M.Tsvetkov and V.N.Parmon, React. Kinet. Catal. Lett., 36, 145-149 (1988).
16. N.P.Luneva, V.Ya.Shafirovich and A.E.Shilov, Kinet. Katal., 30, 250 (1989).
17. G.L.Elizarova, O.P.Taran, L.G.Matvienko and V.N.Parmon Manganese Starch-Stabilized Catalysts for Oxidation of Water to Dioxygen. (will be published).
18. G.L.Elizarova, O.V.Gerasimov, L.G.Matvienko, N.V.Lozhkina and V.N.Parmon, Izvestia Sibirskoe Otdelenie Akademii Nauk SSSR, N3, 94-102 (1990).
19. Govindjee, ed., Photosynthesis, vol.1 (Academic Press, New York, 1982).
20. V.Ya.Shafirovich and A.E.Shilov in: Photochemical Energy Conversion, J.R.Norris, Jr., and D.Meisel, eds. (Elsevier, New York, 1989), pp. 173-183.

21. P.A.Christensen, A.Harriman, G.Porter and P.Neta, J. Chem. Soc. Faraday Trans. 2, 80, 1451 (1984).
22. A.Harriman, P.Neta and M.C.Richoux, J. Phys. Chem., 90, 3444 (1986).
23. J.Davila, A.Harriman, M.C.Richoux and L.R.Milgorm, J. Chem. Soc. Chem. Commun., 525 (1987).
24. G.S.Nagor, S.Mosseri, P.Neta and A.Harriman, J. Phys. Chem., 92, 4499 (1988).
25. O.V.Gerasimov, S.V.Lymar and V.N.Parmon. Photochem. Photobiol. (1990) (in press).
26. K.I.Zamaraev, S.V.Lymar, M.I.Khramov and V.N.Parmon, Pure Appl. Chem., N7, 1039-1946 (1988).
27. S.V.Lymar, V.N.Parmon and K.I.Zamaraev in: Photoinduced Electron Transfer in Chemistry and Biology, J.Mattey, ed. (Springer Verlag, Berlin, 1990).
28. E.V.Efimova, S.V.Lymar and V.N.Parmon, A Novel Membrane-Bound Mediator in the Reaction of Dihydrogen Evolution from Water Suspensions of Lipid Vesicles. (will be published).
29. N.P.Luneva, V.E.Maier, V.Ya.Shafirovich and A.E.Shilov, Kinet. Katal., 28, 505 (1987).
30. A.A.Krasnovsky in: Fotokataliticheskoe Preobrazovanie Solnechnoi Energii, Part 1, K.I.Zamaraev and V.N.Parmon, eds., (Nauka, Novosibirsk 1985) p94.
31. S.D.Varfolomeev, ibid., p. 119.
32. J.R.Norris, Jr., and D.Meisel, eds., Photochemical Energy Conversion (Elsevier, New York, 1989).
33. Yu.A.Gruzdkov, E.N.Savinov and V.N.Parmon in: Fotokataliticheskoe Preobrazovanie Solnechnoi Energii, V.N.Parmon, ed. (Nauka, Novosibirsk 1990).
34. Yu.A.Gruzdkov, E.N.Savinov and V.N.Parmon, Khimiya Vysokikh Energij, 20, 445-448 (1986).
35. E.N.Savinov, Yu.A.Gruzdkov and V.N.Parmon, Int. J. Hydrogen Energy, 14, 1-9 (1989).
36. V.S.Zakharenko and A.I.Yartsev, Izvestiya Akademii Nauk SSSR, Seria Khimicheskaya, 2524 2527 (1987).
37. V.S.Zakharenko and V.N.Parmon, React. Kinet. Catal. Lett., 34 (1987), pp 389-394.

Environment: Photochemical and Photocatalytic Processes. Degradation of Organic Compounds

R.W. MATTHEWS
CSIRO Division of Fuel Technology
PMB 7 Menai, N.S.W. 2234
Australia

ABSTRACT. The photocatalytic degradation of organic compounds is an important branch of the broader subject of photocatalysis. Natural photodegradation is a major purification process occurring in the aqueous environment. Photocatalytic degradation of organics, to some extent, mimics the indirect action of photons in the presence of natural sensitizers, but the sensitizer is a semiconductor photocatalyst. By this means oxidative degradation rates orders of magnitude greater than natural degradation rates can be achieved. In this report a brief historical survey of the development of the application of photocatalysis to water treatment is given, with particular attention to the rapidly expanding number of recent publications. Subjects considered are: the importance of total mineralization rates, the range of compounds photodegraded in the presence of semiconductors, factors affecting photocatalytic activity, doping and form of semiconductor, effect of promoters and inhibitors, kinetics of photocatalytic degradation and CO_2 formation, significance of kinetic parameters, multicomponent systems, analytical applications of photocatalysis and solar powered photocatalytic degradation rates.

1. Introduction

The increasing use of earth's natural resources in human activities has generated an increasing disposal problem of waste products. Many of these are organic chemicals. In some instances the problem occurs because the chemical becomes distributed far beyond the area for which it was intended. The application of insecticides, herbicides and pesticides to crops with subsequent run-off of residual chemical to catchment areas is a case in point. Other common examples are accidental leakages and spills and the continual discharge of waste byproducts in effluent streams from essential industries.

As community awareness of dangers to health and environment grows, stricter anti-pollution restrictions come into force with tighter controls on discharge limits. Manufacturers and industries are being compelled to implement better control of waste discharges [1] In many instances, additional treatment of wastes are necessary with substantial escalation in costs which must be passed on to the consumer. A healthy and clean environment is highly valued. The most economical way of achieving this without loss of production is a challenge to human ingenuity and responsibility.

Contamination of ground and surface water with organic pollutants is a pressing

E. Pelizzetti and M. Schiavello (eds.), Photochemical Conversion and Storage of Solar Energy, 427–449.

problem. For example, approximately half of the population in the U.S.A. depend on groundwater for drinking water supplies. Thousands of wells have been closed because of dangerous concentrations of chemicals [2]. More than 700 organic compounds have been identified in drinking water supplies in the US [3]. Chlorinated hydrocarbon contamination of drinking water is common in the US [4] and Europe [5]. Besides accidental pollution, contamination with trihalomethanes (THM) appears to be an inevitable consequence of chlorinating drinking water supplies; the higher the natural organic content of the water, the higher the THM content following chlorination. Very high THM concentrations have been reported for South Australian drinking waters [6], for example.

The atmosphere, hydrosphere and soil can all be contaminated with manmade organic chemicals. Light-energized purification procedures are an important means of dealing with organic contamination. In this report only photodegradation in the aqueous environment will be dealt with.

2. Photodegradation By Natural UV

Insofar as this is a solar energy conference particular attention will be paid to photodegradation by sunlight. It has long been recognized that air and sunlight have a purifying effect. The solar spectrum at sea level has an appreciable component of UV radiation extending down to 290 nm and overlapping the electronic absorption spectrum of the majority of organic compounds. The limit of 290 nm is due to the attenuation caused by ozone in the atmosphere [7]. The energy of natural UV light at sea level therefore extends to 98.6 kcal/mol which is sufficient to break many chemical bonds.

The near-UV region of the solar spectrum (290-400 nm) is particularly important in the photodegradation of organic chemicals. The long wavelength region (320-400 nm) is commonly called UV-A radiation and the more energetic (280-320 nm) region, UV-B. According to Getoff [8], the energy distributed in the 315-400 nm region of the solar spectrum at 30° latitude is 2.7% of the total of 925 Wm^{-2} irradiating the earth's surface for air mass ratio of 1. The solar irradiance varies with the season of the year and the latitude, particularly in the shorter wavelength region of the spectrum. A computer program in Fortran IV called SOLAR has been written by Zepp and Cline [7(d)] to calculate northern hemisphere solar irradiance values for wavelengths from 297.5 to 800 nm as a function of latitude and season of the year.

2.1. DIRECT ACTION

The continual exposure of natural water bodies to the UV radiation of the sun degrades organic compounds to simpler molecules and to inorganic products eventually. Equations describing the direct photoreaction of chemicals in the aqueous environment have been derived which allow the extrapolation of laboratory data to degradation rates under natural conditions [7(d)]. The reliability of the procedure has been confirmed by many workers, (see, for example, [9].)

In the natural aquatic environment very low concentrations of chemicals are frequently present and significant advances have been made in procedures for the measurement of reaction quantum yields for these conditions both using natural sunlight and monochromatic light [9(b),10].

2.2. INDIRECT ACTION

Not all organic chemicals absorb significantly at wavelengths above 290 nm and therefore the quantum yield for their photodegradation by direct action is essentially zero. However, it has been established that these chemicals still photodegrade as a result of indirect action, that is, natural sensitizers such as humic acids present in the water have broad absorption bands extending into the visible and these electronically excited molecules are capable of initiating degradation reactions in the non-absorbing molecules [11,12]. Some compounds which photodegrade by indirect action are: ethylene thiourea [13], 2,5-dimethylfuran [14,15], aldrin [16], methoxychlor [17], n-butoxyethyl ester of 2,4-dichlorophenoxyacetic acid [18], furfuryl alcohol [12], 2,4,5-trichlorophenoxyacetic acid [19] and 3,4-dichloroaniline [20].

The sensitizing action of the natural humic acids results from the formation of triplet excited state species following the absorption of sunlight. The triplet state may react with other species present in the water thereby initiating various possible photodegradation reactions. A sequence for indirect photosensitized reaction of an organic molecule, R, by a humic acid sensitizer, S, is shown as follows [21]:

$$S_o \xrightarrow{h\nu} S_1 \qquad (1)$$

$$S_1 \rightarrow So + \Delta \qquad (2)$$

$$S_1 \rightarrow T_1 \qquad (3)$$

$$T_1 \rightarrow S_o + \Delta \qquad (4)$$

$$T_1 + {}^3O_2 \rightarrow {}^1O_2 + S_o \qquad (5)$$

$${}^1O_2 + H_2O \rightarrow {}^3O_2 + H_2O^* \qquad (6)$$

$${}^1O_2 + R \rightarrow {}^3O_2 + R^* \qquad (7)$$

$${}^1O_2 + R \rightarrow RO_2 \qquad (8)$$

where S_o represents the sensitizer in the ground state, S_1 represents the first excited singlet state, T_1 represents the first excited triplet state, ${}^3O_2$ represents oxygen in the ground state, ${}^1O_2$ represents singlet oxygen and the asterisk denotes an unspecified excited state. Various reactive intermediates have been invoked in these indirect action photodegradation reactions. They include: triplet excited states of humic acid substances [11], singlet oxygen [11,12], alkylperoxy radicals [22,23], hydroxyl radicals [22(a), 23-25], perhydroxyl radicals and their anionic form, superoxide radical anions [26,27]. It is noteworthy that a pilot plant using solar powered disinfection of wastewater capable of processing 144,000 gal per day has been installed in Livingston Tenn. U.S.A. [28]. Methylene blue is added as the sensitizing agent and energy transfer to the organic contaminants occurs via dissolved oxygen.

The preceding is a brief introduction to one important aspect of the photodegradation of organic compounds. For more complete treatments of the subject the reader is referred to books and reviews [21, 29-32).

3. UV-Ozonation

Another important application of photodegradation is in the combination of ultraviolet light (<300 nm) with ozonation for the destruction of hazardous pollutants [33]. The photolysis of ozone in water forms hydrogen peroxide which reacts with ozone to form hydroxyl radicals [34,35]. Oxidation rates with UV/ozonation have been found to be 10 to 1000 times faster than with ozone alone [36]. This treatment procedure has recently been reviewed [35].

4. Photocatalysis

4.1. IN NATURE

In the natural environment the most significant photodegradation effects are caused by the direct absorption of sunlight by the chemical or indirectly by natural sensitizing agents dissolved in the water. A sensitization process that could be of importance is via suspended sediments. It is known that semiconductor powders such as TiO_2 can provide a striking enhancement of photodegradation rates and since these are common constituents of clays, sediments and soils, the rate of photodegradation of organic pollutants in water containing suspended sediments and clays was investigated [37(a)]. No increase in photodegradation rates was found. Iron oxide has also been investigated for possible photocatalytic activity in the environment since it is ubiquitous and known to have some photocatalytic activity. In the reactions studied, the activity of α-Fe_2O_3 was negligible [37(c)].

4.2. DEVELOPMENT OF RESEARCH ON PHOTOCATALYSIS

Photocatalytic degradation has been comprehensively reviewed previously by Pelizzetti et al. [38(a)] and its application to water purification by Ollis et al. [38(b)]. The present report will provide a brief historical review of the development of photocatalysis in water treatment with particular attention to recent developments. There has been a rapid growth of publications on this subject during the last few years especially because of the potential benefits of photocatalysis in enhancing photodegradation rates of organic contaminants in water.

An important contribution was made by Markham and Laidler in 1953 [39] using mainly zinc oxide as the semiconductor photocatalyst. Another major stimulus to interest in the photoactivity of semiconductors was provided in 1972 by Honda and Fujishima [40] who reported the solar powered splitting of water into hydrogen and oxygen using titanium dioxide-based photoelectrochemical cells. This particular application of semiconductors has been the focus of great attention and a major branch has grown from the pioneering experiments of Bard et al. [41]and others [42-45] on the photocatalytic reactions of semiconductor particles in aqueous suspensions. Whilst these experiments highlighted the possibility of useful photosynthesis and photooxidations based on heterogeneous photocatalysis, researchers in the surface coating industry had recognized deleterious photodegradation of paints as a result of photoactivity of TiO_2 based pigments [46]. There is therefore a dichotomy between the paint industry research designed to minimize photocatalytic activity and other research which seeks to select and, in some cases, maximize photocatalytic activity. This split, and the enhancement or suppression of photocatalytic activity, has been commented on in more detail by Heller et al. [47].

Enhanced photocatalytic activity is of great interest in the conversion of sunlight to stored chemical energy [48], in chemical conversions mediated by irradiated semiconductors [49] and in the degradation of organic contaminants in water [50]. The reported dechlorination and oxidizing ability of illuminated TiO_2 and ZnO led Ollis and coworkers to systematically study the degradation of halogenated hydrocarbon contaminants in water supplies by hetereogeneous photocatalysis resulting in a succession of publications in 1983 and 1984 [51].

In 1983 Hustert et al. [52] reported the degradation of 4-hydroxybenzoic acid, 2,6-dichlorothiamide, 2,4,5-trichlorophenol, atrazine, diethylphthalate, chlordane and a PCB in aqueous suspensions of TiO_2 illuminated with UV light and sunlight. No degradation products were identified but the significant destruction recorded suggested that the process could be a possible competitor with other destructive oxidation processes for intractable organic compounds in industrial wastewaters. Other workers had begun looking at the photocatalytic oxidation products from aromatics in illuminated TiO_2 suspensions [51(f), 53, 54, 55]. There was general agreement that hydroxylation of the aromatic ring is a primary reaction step and various phenolic compounds had been identified but disagreement on the extent of the oxidation. Kawaguchi studied phenol oxidation and suggested that phenol cannot be completely oxidized by TiO_2 photocatalysis [55]. Ollis et al. [51(f)] reported no evidence for ring opening with mono- or dichlorobenzenes. On the other hand, Barbeni et al. [54] reported quantitative formation of CO_2 and HCl in the TiO_2 promoted photodegradation of 4-chlorophenol.

The question of partial versus total oxidation is quite important in the field of water treatment since some partially oxidized compounds can be more toxic and odorous than the parent contaminant. Some attention was therefore directed to the formation of CO_2 from a range of compounds representing various families [56]. A list of the rates of formation of CO_2 from these compounds is shown in Table 1. Since there was little difference in the rates regardless of the class of compound it seemed clear that under suitable conditions with adequate oxygen, and an active grade of TiO_2, total oxidation to CO_2 would occur. This was subsequently confirmed by experiment with the determination of CO_2 in essentially stoichiometric equivalent amount to the 3,4-dichlorophenol [57] and monochlorobenzene [58] initially present. It has now been established that dozens of organic compounds are totally mineralized by photocatalytic oxidation [59] and this finding provides the basis of a new instrumental method for the determination of organic carbon in aqueous solution [60].

Pelizzetti et al. [57] studied the total degradation rates for a family of environmentally hazardous chlorinated aromatics using simulated sunlight illumination of TiO_2 suspensions. These results are shown in Table 2. The striking observation was made that half-lives of only minutes or hours were obtained in the presence of TiO_2 in contrast to the days and years obtained in its absence.

The photocatalytic degradation of many compounds in the presence of illuminated semiconductors has now been studied in detail. The additional compounds to those already referred to are listed in Table 3. Where comparisons with direct photodegradation rates were made the photocatalytic degradation rates were generally orders of magnitude greater.

TABLE 1: Rate of CO_2 formation from illuminated aerated suspensions of 0.2 g TiO_2 in 400 mL of 10^{-3} M solutions, 100 W Hg lamp.

Solute	pH[a]	R(CO_2) (cm^3h^{-1})	R(CO_2) (μmoles min^{-1})[c]
benzene	2.8	1.91±0.07	1.30±0.04
benzene	4.0[b]	4.75±0.07	3.24±0.05
phenol	2.8	2.25±0.07	1.53±0.04
phenol	4.0[b]	4.05±0.02	2.76±0.01
monochlorobenzene	2.8	3.11±0.12	2.12±0.08
monochlorobenzene	4.0[b]	5.24±0.11	3.57±0.07
nitrobenzene[e]	2.8	2.06±0.08	1.40±0.05
aniline	3.1	3.67±0.20	2.50±0.14
benzoic acid	3.1	4.56±0.12	3.11±0.08
catechol	2.9	3.21±0.04	2.19±0.02
resorcinol	2.9	3.81±0.05	2.60±0.03
hydroquinone	2.9	4.25±0.03	2.90±0.02
1,2 dichlorobenzene	3.0	4.48±0.10	3.06±0.06
2 chlorophenol	2.8	2.80±0.03	1.90±0.02
3 chlorophenol	2.8	3.17±0.03	2.16±0.02
4 chlorophenol	2.8	3.32±0.04	2.26±0.03
4 chlorophenol[d]	3.0	1.15±0.03	0.78±0.02
2,4 dichlorophenol	2.9	5.05±0.05	3.44±0.04
2,4,6 trichlorophenol	2.9	4.72±0.03	3.22±0.02
methyl viologen[e]	2.9	1.67±0.08	1.14±0.05
2 naphthol	2.8	3.90±0.07	2.66±0.04
chloroform	2.9	3.70±0.11	2.52±0.07
trichloroethylene	2.9	3.42±0.09	2.33±0.06
ethylene diamine	2.9	1.38±0.01	0.94±0.01
dichloroethane	2.9	1.66±0.03	1.13±0.02
dichloroethane	4.4[b]	3.16±0.03	2.15±0.02

a – pH adjusted with dilute H_2SO_4 unless otherwise indicated
b – no H_2SO_4 added
c – the number of micromoles per min. assuming that all the CO_2 is dissolved
d – La Porte Tiona drade TiO_2, 9 m^2g^{-1} surface area by BET method
e – non-linear response curving upwards

TABLE 2. Half-lives for the Total Photodegradation of Contaminants Assisted by TiO_2 on Exposure to Simulated Sunlight (·)

COMPOUND	CONCENTRATION (ppm)	pH	$t_{1/2}$(min)
4-chlorophenol (··)	6	3.0	14
3,4-dichlorophenol	18	3.0	45
2,4,5-trichlorophenol	20	3.0	55
pentachlorophenol (··)	12	3.0	20
sodium pentachlorophenate	12	10.5	15
chlorobenzene	45	2.5	90
1,2,4-trichlorobenzene	10	3.0	24
2,4,5-trichlorophenoxyacetic acid	32	3.0	40
4,4′-dichlorodiphenyltrichloroethane (···)	1	3.0	46
3,3′-dichlorobiphenyl (···)	1	3.0	10
2,7-dichlorodibenzo-*p*-dioxin (···)	0.2	3.0	46

(·) Concentration of catalyst, 2.0 g/l; aerated aqueous solutions; wavelength > 330 nm.
(··) Wavelength > 310 nm.
(···) Absorbed on TiO_2

TABLE 3. Compounds Photocatalytically Degraded in Aqueous Solution

Compounds	Reference
formic acid, formate	42,61-64
oxalic acid	65
glucose	66
acetic, propionic, butyric acids	67
pentane, heptane, isooctane, parafin, polyethylene, benzene, phenol, pyridine	68
methyl orange	69,70
dodecylbenzene sulfonate and other surfactants	71-73
phenol	74
fluorescein	75
atrazine	76
2,4,5-trichlorophenoxyacetic acid, 2,4,5-trichlorophenol	77
trimethyl phosphate and other organophosphorus cpds.	78
4-nitrophenyldiethylphosphate	79
polychlorinated dioxins and biphenyls	80
methylene blue	81
methane	82
2-ethoxy- and 2-butoxyethanol	83
halothane	84
chloral hydrate	85
DDT	86
bentazon	87
dodecane and dodecyl derivatives	88
nitrophenols	111

5. Semiconductor Photocatalysts

The photocatalysts commonly used are: TiO_2, ZnO, WO_3, CdS, ZnS, $SrTiO_3$, SnO_2 and Fe_2O_3 with TiO_2 being frequently reported as the most active in organic degradation experiments when comparisons are made. In water treatment processes involving semiconductors, non-toxicity and insolubility, both in the dark and under illumination, are important considerations. Titanium dioxide satisfies these requirements. Various combinations of semiconductors have been used (see, for example [89]) and also combinations of semiconductors and insulators [90-92] with some beneficial results being reported. It has been shown that for some organics the photodecomposition is significantly influenced by the acid-base properties of the semiconductor [91]. Semiconductors are sometimes combined with a small concentration of a metal such as Pt, Ru or Cu but platinization does not enhance photocatalytic activity in oxidative degradation reactions as a rule. Photocatalysts are generally used in particulate form in degradation studies but much work has been done with colloids, particularly with regard to the study of early processes [93, 82(c)]. Semiconductors may also be attached to or incorporated in immobile supports thereby enabling contaminated aqueous streams to be passed continuously over the catalyst and avoiding a filtration or settling step and resuspension. Examples of immobilized photocatalysts are: TiO_2 incorporated in Nafion and clay [94,95], TiO_2 on glass beads [96], on glass [97(a)(b)] and Teflon tubing [97(e)] and incorporated in silica gel [97(c)(d)]. Devices for continuous water purification based upon some of these have been suggested [97(a)(d)(f)].

5.1. FACTORS AFFECTING PHOTOCATALYTIC ACTIVITY

At present the quantum efficiency for the mineralization of organic compounds by photocatalytic oxidation is, for most compounds, on the order of a few per cent and less of the photoactivating light. In the case of TiO_2 the bandgap or photoactivating light is that of wavelength shorter than approximately 380 nm which comprises a few per cent of the solar spectrum. There is therefore much to be gained when using solar illumination in selecting a semiconductor whose photoaction spectrum has a significant overlap with the solar spectrum. Cadmium sulfide with a bandgap energy of 2.2 eV is well placed in this respect but photocorrosion and toxicity rule it out for water treatment. One approach has been to dope titanium dioxide with metal ions that shift the spectrum into the visible. Chromium doped titania has significant absorption in the visible but unfortunately decreases the photocatalytic activity [65, 98, 99].

The photosensitization of TiO_2 particles by zinc tetraphenylporphyrin bound to polyvinylpyridine has been reported [118]. The photodegradation of PCBs was more efficient with this composite semiconductor than with TiO_2 alone.

5.1.1. *Doping of Semiconductor.* The chief loss of quantum efficiency (desired by paint chemists and disdained by photochemists) is the recombination of photoelectrons and photoholes within or at the surface of the particle without chemical reaction. It is well estabished that doping the semiconductor with other metals or different valency states can markedly effect the recombination rate [82(b)]. Hydrogen reduced titania is frequently used where maximum photocatalytic activity is sought [47]. It should be recognized, however, that the promotion of photocatalytic activity for one particular type of reaction, eg, reduction, may not be beneficial in oxidation reactions. In water treatment, the solubility and toxicity of dopants are additional constraints.

5.1.2. *Form of Semiconductor.* Titanium dioxide exists in three crystalline forms: anatase, rutile, and brookite. Anatase and rutile are the most common forms. Both forms are photocatalytically active with anatase generally being recognized as the more active in photooxidation reactions. The bandgap energies are approximately 3.2 eV for anatase and 3.0 eV for rutile but the driving force for oxidative processes are similar. It has been suggested that the poorer photocatalytic activity of rutile is associated with its higher electron-hole recombination rate [100] and its lower oxygen photoadsorption capacity. In addition to the crystalline form of titania, Sclafani et al. [100] have shown that there are wide variations in activity within a single form. The heat treatment and surface area of the catalyst are important parameters. For titania prepared by a precipitation method the oxidative photocatalytic activity increased with heating time up to temperatures of 600°C. At higher temperatures decreases in activity were observed on prolonged heating. It is known that at high temperatures an irreversible dehydroxylation of titania surfaces occurs and that this inhibits oxygen adsorption at the surface [101]. Oxygen is essential for the oxidation [74(b)]. Therefore, the decrease in the rate may be associated with dehydroxylation of the titania surface.

5.1.3. *Effect of Ionic Species.* The adsorption of various anions and cations on the surface can influence the photocatalytic activity. For example, it is claimed that in the photocatalytic oxidation of aromatic compounds on TiO_2 the quantum efficiency in the presence of sulfuric acid and cupric ions was 16 times greater than in the additive-free system [44(a)]. The beneficial effects of Cu(II) and Fe(III) ions under these conditions were attributed to enhanced oxidation rates of the intermediate aromatic radical species by these ions. It is also known that electron acceptors such as Ag(I) ions can enhance oxidation rates of organics [67, 102]. The inhibiting effect of chloride ions has been noted [44(b), 51(c), 103] and the effect of protons [51(c)]. Recently, a systematic study of the effect of common inorganic anions on the photocatalytic oxidation rates over TiO_2 of several reactions was done showing inhibition by Cl^-, $SO_4^=$ and $H_2PO_4^-$ in all cases and no significant effect of NO_3^- and ClO_4^- [104].

5.1.4. *Effect of Hydrogen Peroxide.* The formation of H_2O_2 on illuminated TiO_2 surfaces has been proposed as an important source of hydroxyl radicals in the photocatalytic oxidation of aromatic compounds [44]. The reactions envisaged were the Haber-Weiss reaction:

$$H_2O_2 + O_2^{\cdot -} \rightarrow \cdot OH + OH^- + O_2 \qquad (9)$$

and reaction between H_2O_2 and the photoelectron:

$$H_2O_2 + e^-(TiO_2) \rightarrow \cdot OH + OH^- \qquad (10)$$

Remarkable enhancements to photocatalytic degradation rates of trichloro - and tetrachloroethylene by the addition of hydrogen peroxide have been reported by Tanaka et al. [106(a)]. Similar effects were observed with both TiO_2 and WO_3 semiconductors. The degradation rate for 3.5 x 10^{-4} M trichloroethylene in TiO_2 aqueous suspension was increased by 6 to 8 times in the presence of hydrogen peroxide in the concentration range 4 x 10^{-2} to 1.2 x 10^{-2} M. The photocatalytic degradation of chloral hydrate was also observed to be enhanced by hydrogen peroxide [106(b)]. In this case the maximum increase in degradation rate was 3.4 times.

A systematic study of the influence of hydrogen peroxide on the kinetics of phenol photodegradation in aqueous TiO_2 dispersion has recently been done by Augugliaro et al. [107]. Significant degradation of phenol was observed in TiO_2 suspensions containing H_2O_2 in the absence of oxygen and this rate was greater than the rate in the presence of oxygen and absence of H_2O_2. The maximum rate was observed when both H_2O_2 and oxygen were present and was about 5 times faster in the presence of 1.8×10^{-2} M H_2O_2 than in the absence of H_2O_2.

These experiments showing the enhancement of photocatalytic oxidation rates by H_2O_2 are of considerable significance because they suggest that the effective photoaction spectrum of H_2O_2 is shifted to longer wavelengths by the TiO_2. The extinction coefficient of H_2O_2 becomes negligibly small for wavelengths higher than 300 nm [108]. Consequently, there is little overlap with the solar spectrum. However, the shift to longer absorbed wavelengths in the presence of TiO_2 means that the oxidizing power of the H_2O_2 can be released by solar illumination.

6. Kinetics of Photocatalytic Degradation

6.1 SOLUTE DISAPPEARANCE

It has been repeatedly demonstrated that the photocatalytic degradation of various organic solutes obeys pseudo first-order kinetics.

$$-\frac{dS}{dt} = k[S] \tag{11}$$

where k is the pseudo first-order decay constant and $[S]$ is the concentration of solute. (Zero order kinetics have also been observed at high solute concentrations [114]). On integration, expression 11 yields

$$\text{In}[S_t] = \text{In}[S_o] - kt \tag{12}$$

where $[S_t]$, $[S_o]$ = the concentrations of solute at time t and 0 respectively.

It has also been found that the decay constant, k, decreases with increasing solute concentration. See, for example, the results in Table 4. The half-life for the reaction, $t_{0.5}$, shown in the last column is given by

$$t_{0.5} = \frac{0.693}{k} \tag{13}$$

Langmuir-Hinshelwood (L-H) kinetics have been useful in many studies of photocatalytic degradation rates in reducing the concentration dependence of the data to a minimum number of numerical parameters represented by a reactivity constant, k_1, and an adsorption equilibrium constant, K [51,59(a), 65].

The general expression describing the rate is

$$-\frac{dS}{dt} = k_1 . \frac{K[S]}{1 + K[S]} \tag{14}$$

which, on integration, gives

$$t = \frac{1}{k_1 K} \ln \frac{[S_o]}{[S_t]} + \frac{1}{k_1}([S_o] - [S_t]) \tag{15}$$

TABLE 4. k values from expression 11 for different solutes. Reference [97(b)].

solute	concn, μM	k', min^{-1}	$t_{0.5}$, min
salicyclic acid	1	0.1017 ± 0.0007	6.81
	2	0.1024 ± 0.0008	6.77
	4	0.0989 ± 0.0007	7.01
	10	0.0975 ± 0.0020	7.11
	20	0.0857 ± 0.0018	8.08
	30	0.0857 ± 0.0014	8.09
	40	0.0811 ± 0.0013	8.54
	50	0.0695 ± 0.0008	9.97
	100	0.0519 ± 0.0013	13.35
phenol	1	0.116 ± 0.001	5.97
	2	0.102 ± 0.002	6.79
	10	0.0966 ± 0.0019	7.17
	20	0.0742 ± 0.0017	9.34
	50	0.0562 ± 0.0002	12.3
	100	0.0408 ± 0.0005	17.0
2-chlorophenol	10	0.0843 ± 0.0013	8.22
4-chlorophenol	10	0.0793 ± 0.0017	8.73
benzoic acid	10	0.1002 ± 0.0016	6.92
	20	0.0934 ± 0.0007	7.42
	50	0.0704 ± 0.0016	9.84
2-naphthol	1	0.0930 ± 0.0007	7.45
	2	0.0848 ± 0.0010	8.17
	5	0.0889 ± 0.0008	7.80
	10	0.0812 ± 0.0005	8.53
	20	0.0671 ± 0.0010	10.3
	50	0.0617 ± 0.0015	11.2
	100	0.040 ± 0.0009	17.1
naphthalene	1	0.174 ± 0.0015	3.98
	2	0.176 ± 0.0007	3.94
	5	0.165 ± 0.0008	4.20
	10	0.160 ± 0.0005	4.33
	20	0.154 ± 0.0007	4.50
	50	0.126 ± 0.0012	5.50
fluorescein	1	0.131 ± 0.0013	5.29
	2	0.125 ± 0.0009	5.54
	5	0.119 ± 0.0015	5.82
	10	0.108 ± 0.0018	6.41
	20	0.091 ± 0.0020	7.61

It follows that $$t_{0.5} = \frac{0.693}{k_1 K} + \frac{0.5[S_o]}{k_1} \tag{16}$$

and therefore a plot of $t_{0.5}$ versus $[S_o]$ should yield a straight line whose slope is $0.5/k_1$ and whose intercept is $0.693/(k_1 K)$. Treatment of the data in Table 4 in this way yielded the appropriate k_1 and K parameters for the solutes. Substitution of these parameters in expression 15 gave curves that were in good agreement with the experimental data. See, for example, Fig. 1. The curves were approximately linear on a log[S_t] versus time plot with slopes decreasing with increasing [S_o] The approximately first-order plots with rate constants decreasing with increasing solute concentration and tending to zero order are therefore consistent with L-H kinetics.

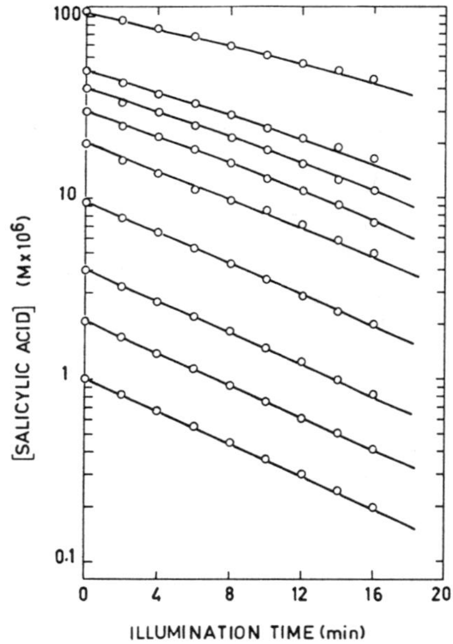

Figure 1. Effect of initial concentration on the photocatalytic destruction of salicylic acid. Circles experimental points, lines calculated using expression 15. Reference [97(b)].

6.2. CARBON DIOXIDE FORMATION

The rate of CO_2 formation with increase in solute concentration has the typical shape of a Langmuir adsorption isotherm (Fig. 2) irrespective of whether the semiconductor is used in free suspension or immobilized [109]. When the photocatalyst is immobilized mass transfer effects cannot be ignored [110] and flow rate becomes important.

6.3. SIGNIFICANCE OF L-H PARAMETERS

The two parameters, k_1 and K in the L-H expression are usually interpreted as respectively reflecting the limiting rate at high concentrations of solute when essentially all available sites capable of adsorbing the solute are filled, and the adsorption equilibrium constant for the solute. The latter becomes progressively important as the solute concentration is decreased. Thus a solute recognized as difficult to oxidize at high concentrations may be oxidized comparatively rapidly at low concentrations because it appears to be strongly adsorbed. However, the physical significance of these parameters in photocatalytic degradation has not been definitively established. They are empirical constants which describe the rate of degradation or mineralization for particular experimental conditions covering a range of solute

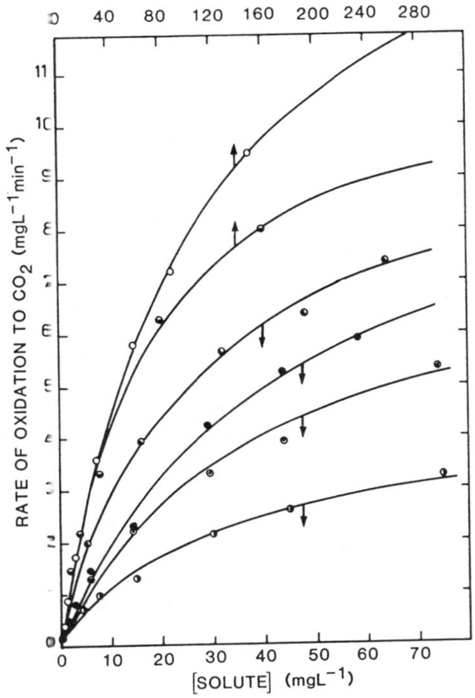

FIG. 2. Rates of oxidation to CO_2 versus initial solute concentrations. (◑) 2-Propanol; (⊙) acetic acid; (●) phenol; (◐) 4-chlorophenol; (◒) methanol; (○) chloroform.

concentrations. Recently, a detailed study has been published [112(a)] seeking to obtain a better insight into the fundamental meaning of these parameters. Four possible mechanisms were evaluated, all based on OH radical attack of the organic solute. All of the kinetic expressions were of similar form to the L-H expression but the derived kinetic parameters represented fundamentally different reactions and properties to the traditional L-H constants.

6.4. MULTICOMPONENT SYSTEMS

Although one particular contaminant may be dominant in a waste stream it is unlikely to be the only component. It is therefore important to establish that the kinetic expressions from single organic component systems can be translated to the multicomponent systems of the real world. Several studies are relevant to this point [112(b), 113, 114]. It has been shown that organic solutes and their intermediate degradation products compete for the primary oxidizing radical species of the illuminated photocatalyst. It has also been shown that the parameters from single component studies substituted in the appropriate L-H expressions and numerically integrated give good agreement with observed product yields in some bicomponent mixtures [112(b)]. Significant disagreements were observed in other instances, however, indicating possible deficiencies in the model.

7. Analytical Applications

Another application of photocatalysis is in the development of instruments for chemical analysis, some of which have considerable potential for use in safeguarding the environment [116(a)]. A flow-through photoelectrochemical detector for attachment to HPLC has been described [115]. Photocatalytic oxidation using illuminated TiO_2 suspensions or TiO_2 immobilized on glass tubing is the basis for a new instrumental method for the determination of total dissolved organic carbon in water [116(b)]. Since light of wavelength up to 380 nm is effective in photoactivating the catalyst the apparatus may be constructed using glass instead of quartz with considerable savings in material costs and ease of fabrication. Using high power compact lamps, total oxidation to CO_2 of 10 mL samples containing 5 ppm of carbon occurs within approximately 2 min for some compounds [116(c)]. The low power version of the instrument is very useful for the measurement of oxidation rates to CO_2 for various solutes [109(a)] and also for determining the relative activities of various photocatalysts [104].

Other possible applications are in the determination of organic heteroatoms such as N, P, S and halogens, common in fungicides, pesticides and herbicides. Very low concentrations of these can be determined by techniques such as ion chromatography following oxidation over immobilized titania photocatalyst. Phosphorus, sulfur and halogens reportedly form single anions following oxidation but nitrogen compounds give both ammonium and nitrate ions [116(d)].

8. Solar Degradation Rates

It is customary to report photodegradation rates for organics in the environment in terms of half-lives. Frequently, a number of degradation products are formed, some of which may be stable compounds that are as undesirable environmentally as the parent compound. In water treatment, complete mineralization of organic compounds to carbon dioxide, water and other inorganic products is desired and therefore a key parameter in a degradation process is the rate of conversion to carbon dioxide. The rates of oxidation to CO_2 of a number of solutes have recently been determined as a function of solute concentration [109]. Both immobilized TiO_2 and TiO_2 in free suspension were used as catalysts. The illumination sources were a 20 W blacklight blue fluorescent tube and an 0.3 m^2 parabolic trough. The rates of oxidation found with the parabolic trough were 77% of those found with the 20 W lamp. The solar experiments were done at the end of summer. The oxidation rates found for 20 W lamp illumination and 40 mL of 0.1% TiO_2 suspension are summarised in terms of L-H kinetic parameters in Table 5.

TABLE 5. Summary of L-H parameters and oxidation rates for various solutes in 0.1% TiO_2 suspension. Reference [109(b)]

Solute	K (mg^{-1} L)	k_1 (mg L^{-1} min^{-1})	R(S ->CO_2) (mg L^{-1} min^{-1})			$t_{0.5}$ (min)
			1[a]	10[a]	50[a]	[b]
Benzoic acid	0.082±0.005	8.2±0.2	0.62	3.69	6.58	40
Salicylic acid	0.074±0.006	8.6±0.4	0.59	3.67	6.79	39
Phenol	0.072±0.017	7.2±0.9	0.48	3.01	5.63	47
2-Chlorophenol	0.050±0.008	10.1±0.8	0.48	3.37	7.21	38
3-Chlorophenol	0.027±0.005	12.2±1.2	0.32	2.59	7.01	41
4-Chlorophenol	0.045±0.010	10.1±1.0	0.43	3.12	6.96	39
Nitrobenzene	0.060±0.008	8.1±0.6	0.45	3.02	6.05	44
Methanol	0.048±0.004	9.1±0.3	0.42	2.96	6.44	42
Ethanol	0.038±0.004	9.0±0.5	0.33	2.47	5.87	47
Acetic acid	0.028±0.001	14.0±0.2	0.38	3.06	8.17	35
Formic acid	0.026±0.003	29.4±1.5	0.74	6.07	16.6	17

[a] Initial concentration of solute, S, in mg L^{-1}.

[b] Time for mineralization of 50% of solute in one L of 0.1% suspension initial solute concentration 50 mg L^{-1}, one m^2 solar illumination, see text.

The rate of oxidation of each solute, R in mg L^{-1} min^{-1} at any particular solute concentration and 20 W lamp illumination is given by substituting the appropriate k_1 and K values in the expression:

$$R = k_1 \frac{K[S_o]}{1 + K[S_o]} \tag{17}$$

Apart from formic acid, there is surprisingly little difference in the rates for the rest of the solutes and a general approximation for these is given by the expression:

$$R = 9.66x \frac{0.0524[S_o]}{1 + 0.0524[S_o]} \tag{18}$$

Assuming the lamp manufacturer's specification of 4 watts of radiative power centred at 365 nm and assuming that all this radiation was absorbed by the 40 mL of suspension it can be shown that if every photon resulted in the oxidation of one molecule of solute to carbon dioxide the rate of oxidation would be 732 μmoles/min. The average rate of oxidation for 50 mg/L solute is according to expression 18, 7.0 mg/L/min and hence 0.28 mg/min for 40 mL of suspension. For a mean solute molecular weight of 100 this would be 2.8 μmoles/min. The quantum efficiency for the conversion of solute to CO_2 is thus ~2.8 x 100/732 = 0.38%. Recognizing that not all the suspension was exposed to the light at any point in time and that not all the radiation could be absorbed because of scattering this estimate is a lower limit.

As far as solar illumination is concerned, the parameters, k_1, in Table 5 have to be multiplied by 0.77 to give the corresponding solar oxidation rates for the 0.3 m^2 parabolic trough used or 0.77 x 3.33 for the estimated rates with a one m^2 parabolic trough. The $t_{0.5}$ values for one litre of suspension per square metre of collector, obtained by substitution in expression 16, are given in column 7 of Table 5.

These values were all obtained using a 7 m long 6 mm dia borosilicate glass tube wrapped in a spiral as the reaction vessel. It remains to be shown what CO_2 oxidation rates can be obtained in lagoon simulating vessels using unfocussed sunlight and photocatalysts in suspension or immobilized on suitable supports.

9. Pilot Plant Studies

The focussed sunlight reactor concept has been taken to the pilot plant stage by workers at Sandia National Laboratories with impressive and surprising results [117]. The object of the research is to demonstrate the economic competitiveness of the solar process for the detoxification of water. Parabolic troughs (219 m long) and heliostats were used to concentrate the sunlight on straight glass tubes and falling film reactors respectively. In the parabolic trough system, flow rates of 100 litres/min were used with a residence time of about 2.5 min for the 0.1% TiO_2 suspensions. In the presence of 100 ppm hydrogen peroxide, the concentration of trichloroethylene was decreased from 1 ppm to less than 0.05 ppm in less than 5 min.

9.1. EFFECT OF LIGHT INTENSITY

By comparing rates of destruction of model contaminants, salicylic acid and trichloroethylene, in the presence of low concentrations of hydrogen peroxide under different solar intensities, it was shown that the degradation rates increased linearly with light intensity. This is a surprising and pleasing result because a dependence of the rate on the square root of intensity has been previously reported for additive-free TiO_2 suspensions [46,74(b)]. Apparently the hydrogen peroxide, which increases the degradation rates by about ten times, inhibits the recombination reactions responsible for the square root dependence.

Irrespective of whether natural or concentrated sunlight or electrically generated UV is used to photodegrade organic pollutants in water, in most cases the addition of quite low concentrations of TiO_2 (~0.1%) enhances the degradation rates by orders of magnitude. Any water treatment process that incorporates the natural UV from sunlight as a final polishing step will be made much more efficient by the use of TiO_2 as a sensitizer.

The use of sunlight to convert potentially harmful, wayward organic chemicals to harmless mineral products at accelerated rates by the use of a comparatively cheap and non-toxic photocatalyst is an attractive alternative in processes aimed at the restoration and more responsible management of the environment. Much information has been gathered on the process, particularly during recent years but many questions have not yet been definitively answered. Further work should provide a better understanding of the fundamental processes occurring, additional significant enhancements to the efficiency of the process, and simple and economic photoreactors that are specifically designed to maximize the photoaction characteristics of these photocatalysts.

Acknowledgements

Thanks are due Drs to Augugliaro, Schiavello, Palmisano, Serpone, Pelizzetti, Pacheco and Tyner for preprints of their contributions prior to publication.

References

1. Chem. and Eng. News (1987) 'EPA water standards: major impact on chemical plants', anon. Oct 12, p4.
2. Sun M. (1986), Science 232, 1490-1493.
3. Stachka J.H. and Pontius F.W. (1984), J. Am. Water Works Assoc. 76, 73-85.
4. (a) Symons J.M., Bellor T.A., Carswell J.K., DeMarco J., Kropp K.L., Robeck G.G., Seegar D.R., Slocum C.J., Smith B.L. and Stevens A.A. (1975), J. Amer. Water Works Assoc. 67, 634-647.
   (b) Foley P.D. and Missingham T.A. (1976), J. Amer. Water Works Assoc. 68, 105-111.
5. Rock J.J. (1976), J. Amer. Water Works Assoc. 68, 168-172.
6. Nicholson B.C. (1984) 'Australian water quality criteria for organic compounds', Australian Water Resources Council, Technical Paper No. 82, Australian Government Publishing Service, Canberra p 201.
7. (a) Demerjian K.L., Schere K.L. and Peterson J.T. (1980) 'Advances in Environmental Science and Technology,' ed. by Pitts J.N., Metcalf R.L. and Grosjean D., Volume 10, John Wiley and Sons, p397.
   (b) Leighton P.A. (1961) Photochemistry of Air Pollution, Academic Press, New York, Chapter II, pp 6-41.
   (c) Peterson J.T. (1976) Calculated Actinic Fluxes (290-700 nm) for Air Pollution Photochemistry Applications; U.S. Environmental Protection Agency, U.S. Government Printing Office: Washington DC; EPA 600/4-76-0025.
   (d) Zepp R.G. and Cline D.M. (1977). Environ. Sci. Technol. 11, 359-366.
8. Getoff N. (1984), Int. J. Hydrogen Energy 9, 997-1004.
9. (a) Mill T., Mabey W.R., Lau B.Y. and Baraze A. (1981), Chemosphere 10, 1281-1290.
   (b) Smith J.H., Mabey W.R., Bohonos N., Holt B.R., Lee S.S., Chou T.-W., Bomberger D.C. and Mill T. U.S. Environmental Protection Agency. U.S. Government Printing Office: Washington DC, 1977, EPA 600/7-77-113; Part II. Laboratory Studies, ibid., 1978, EPA 600/7-78-074.
   (c) Wolfe N.L., Zepp R.G., Baughman G.L., Fincher R.C. and Gordon J.A. U.S. Environmental Protection Agency. U.S. Government Printing Office: Washington DC, 1976; EPA 600/3-76-067.
10. (a) Dulin D. and Mill T. (1982), Environ. Sci. Technol. 16, 815-820.
   (b) Mill T., Davenport J.E., Dulin D.E., Maybe W.R. and Bawol R. (1981), U.S. Environmental Protection Agency, U.S. Government Printing Office: Washington DC, EPA 560/5-81-003.
   (c) Zepp R.G. (1978), Environ. Sci. Technol. 12, 327-329.

(d) Zepp R.G. (1981) in the Handbook of Environmental Photochemistry, ed. by O. Hutzinger, Springer-Verlag, New York, vol 2, part B, p 19.

11. (a) Zepp R.G., Baughman G.L. and Schlotzhauer P.F. (1981), Chemosphere 10, 109-117; 119-126.

(b) Zepp R.G., Schlotzhauer P.F. and Merritt Sink R. (1985), Environ. Sci. Technol. 19, 74-81.

12. (a) Haag W.R., Hoigne J. Gassman E. and Braun A.M. (1984), Chemosphere 13, 631-640; 641-650.

(b) Haag W.R. and Hoigné J. (1986), Environ. Sci. Technol. 20, 341-348.

13. Ross R.D. and Crosby D.G. (1973), J. Agric. Food Chem. 21, 335-337.

14. Zepp R.G., Wolfe N.L., Baughman G.L. and Hollis R.C. (1977), Nature 267, 421-423.

15. Wolff C.J.M., Halmans M.T.H. and Van der Heijde H.B. (1981), Chemosphere 10, 59-62.

16. Ross R.D. and Crosby D.G. (1975), Chemosphere 5, 277-282.

17. Zepp R.G., Wolfe N.L., Gordon J.A. and Fincher R.C. (1976), J. Agric. Food Chem. 24, 727-733.

18. Zepp R.G., Wolfe N.L., Gordon J.A. and Baughman G.L. (1975), Environ. Sci. Technol. 9, 1144-1150.

19. Skurlatov Y.I., Zepp R.G and Baughman G.L. (1983), J. Agric. Food Chem. 31, 1065-1071.

20. Miller G.C., Zisook R. and Zepp R.G. (1980), J. Agric. Food Chem 28, 1053-1056.

21. Leifer A. (1988) The Kinetics of Environmental Aquatic Photochemistry Theory and Practice, ACS Professional Reference Book, American Chemical Society, p 266.

22. (a) Mill T., Hendry D.G. and Richardson H. (1980), Science 207, 886-887.

(b) Mill T., Maybe W.R., Bomberger D.C., Chou T.-W., Hendry D.G. and Smith J.H. (1982), U.S. Environmental Protection Agency. U.S. Government Printing Office: Washington DC, EPA 600/3-82-002.

23. Draper W.M. and Crosby D.G. (1981), J. Agric. Food Chem. 29, 699-702.

24. Haag W.R. and Hoigné J. (1985), Chemosphere 14, 1659-1671.

25. Cooper W.J. and Zika R.G. (1983), Science 220, 711-712

26. Draper W.M. and Crosby D.G. (1984), J. Agric. Food Chem. 32, 231-237.

27. Draper W.M. and Crosby D.G. (1983), Arch. Environ. Contam. Toxicol. 12, 121-126.

28. Chem. and Eng. News (1985) 'Solar wastewater treatment plant under test', Oct 14, anon. p 41.

29. Turro N.J. (1978) Modern Molecular Photochemistry, Benjamin-Cummings, Menlo Park, Calif.

30. Crosby D.G. et al. (1972) in Environmental Toxicology of Pesticides, ed. by F. Matsumura, Academic, New York, pp 423-33.

31. Sundstrom G. and Ruzo L.O. (1978) in Aquatic Pollutants: Transformations and Biological Effects, ed. by O. Hutzinger, I.H. van Lelyveld and B.C.J. Zoeteman, Pergamon, Oxford, pp 237-63.

32. Hwang H.-M., Hodson R.E. and Lee R.F. (1987) in Photochemistry of Environmental Aquatic Systems, ed. by R.G. Zika and W.J. Cooper, ACS Symposium Series 327, American Chemical Society: Washington DC pp 27-43.

33. Khan S.R., Huang C.R. and Bozzelli J.W. (1985), Environ. Progress 4, 229-237.

34. Staehelin J. and Hoigné J. (1985), Environ. Sci. Technol. 19, 1206-13.

35. Peyton G.R. and Glaze W.H., in Photochemistry of Environmental Aquatic Systems, ed. by R.G. Zika and W.J. Cooper ACS Symposium Series 327, American Chemical Society, Washington DC (1987) pp 76-88.

36. (a) Prengle H.W. and Mauk C.E. 'Ozone/UV Oxidation of Pesticides in Aqueous Solution'. Ozone/Chlorine Dioxide Oxidation Products of Organic Materials, International Ozone Association (1978).

(b) Prengle H.W., Hewes C.G. and Mauk C.E. "Oxidation of Refractory Materials by Ozone with Ultraviolet Radiation". Proc. Second International Symposium on Ozone Technology, International Ozone Institute (1975).

(c) Prengle H.W. and Mauk C.E. Water-1977 AIChE Symp. Ser. 74, 288 (1978).

37. (a) Oliver B.G., Cosgrove E.J. and Carey J.H. (1979), Environ. Sci. Tech. 13, 1075-1077.

(b) Carey J.H., Lawrence J. and Tosine H.M. (1976), Bull. Environ. Contam. Toxicol. 16, 697-701.

(c) Kormann C., Bahnemann D.W. and Hoffmann M.R. (1989), J. Photochem. Photobiol. A Chemistry 48, 161-169.

38. (a) Pelizzetti E., Minero C., Pramauro E., Serpone N. and Borgarello E. Photocatalysis and Environment Trends and Applications. ed. by M. Schiavello, NATO ASI Series, Klewer Academic Publishers, Dordrecht/Boston/London (1988) pp 469-497.

(b) Ollis D.F., Pelizzetti E. and Serpone N. 'Photocatalysis Fundamentals and Applications', ed. by Serpone N. and Pelizzetti E., John Wiley and Sons, New York (1989) pp 603-634.

39. Markham M.C. and Laidler K.J. (1953), J. Phys. Chem. 57, 363-369.

40. Fujishima A. and Honda K. (1972), Nature 238, 37-38.

41. (a) Kraeutler B. and Bard A.J. (1978), J. Amer. Chem. Soc. 100, 5985-5992.

(b) Izumi I., Dunn W.W., Wilbourn K.O., Fan F.-R.F. and Bard A.J. (1988), J. Phys. Chem. 84, 3207-3210.

(c) Izumi I., Fan F.-R.F. and Bard A.J. (1981), J. Phys. Chem. 85, 218-223.

42. Bideau M., Claudel B. and Otterbein M. (1980), J. Photochem. 14, 291-302.

43. (a) Kawai T. and Sakata T. (1980), Nature 286, 474-476.

(b) Kawai T. and Sakata T. (1980), J.C.S. Chem. Commun. 694-695.

44. (a) Fujihira M., Satoh Y. and Osa T. (1981), Nature 293, 206-208.

(b) Fujihira M., Satoh Y. and Osa T. (1982)', Bull. Chem. Soc. Jpn. 55, 666-671.

45. Pichat P., Herrmann J.-M. Disdier J., Courbon H. and Mozzanega M.-N. (1981), Nouv. J. Chim. 5, 627-636.

46. Egerton T.A. and King C.J. (1979), J. Oil Col. Chem. Assoc. 62, 386-391.

47. Heller A., Degani Y., Johnson, Jr. D.W. and Gallagher P.K. (1987), J. Phys. Chem. 91, 5987-5991.

48. Grätzel M. (1988), In Photoinduced Electron Transfer. Part D. ed by Fox M.A. and Chanon M., Elsevier, Amsterdam Oxford New York Tokyo, pp 394-440.

49. Fox M.A. (1983), Acc. Chem. Res. 16, 314-321.

50. Ollis D.F. (1985), Environ. Sci. Technol. 19, 480-484.

51. (a) Pruden A.L. and Ollis D.F. (1983), Environ. Sci. Technol 17, 628-631.
(b) Pruden A.L. and Ollis D.F. (1983), J. Catal. 82, 404-417.
(c) Hsiao C.-Y., Lee C.-L. and Ollis D.F. (1983), J. Catal. 82, 418-423.
(d) Nguyen T. and Ollis D.F. (1984), J. Phys. Chem. 88, 3386-3388.
(e) Ahmed S. and Ollis D.F. (1984), Solar Energy 32, 597-601.
(f) Ollis D.F., Hsiao C.-Y., Budiman L. and Lee C.-L. (1984), J. Catal. 88, 89-96.

52. Hustert K., Kotzias D. and Korte F. (1983), Chemosphere 12, 55-58.

53. Matthews R.W. (1984), J. Chem. Soc. Faraday Trans I 80, 457-471.

54. Barbeni M., Pramauro E., Pelizzetti E., Borgarello E., Grätzel M. and Serpone N. (1984), Nouv. J. Chim. 8, 547-550.

55. Kawaguchi H. (1984), Environ. Technol. Lett. 5, 471-474.

56. (a) Matthews R.W. (1985), Sunworld 9, 3-5.
(b) Matthews R.W. (1986), Wat. Res. 20, 569-578.

57. Pelizzetti E., Barbeni M., Pramauro E., Serpone N., Borgarello E., Jamieson M.A. and Hidaka H. (1985), La Chim. & L'Industria 67, 623-625.

58. Matthews R.W. (1986), J. Catal. 97, 565-568.

59. (a) Matthews R.W. (1987), Aust. J. Chem. 40, 667-75.
(b) Matthews R.W. (1988) unpublished results.

60. Matthews R.W., Abdullah M. and Low G.K.-C. (1990), Anal. Chim. Acta (in press).

61. Aurian-Blajeni B., Halmann M. and Manassen J. (1982), Photochem. Photobiol. 35, 157-162.

62. Henglein A, Gutierrez M. and Fischer Ch.-H. (1984), Ber. Bunsenges. Phys. Chem. 88, 170-175.

63. Matthews R.W. (1986), J. Catal. 100, 275-278.

64. Yoneyama H., Matsumoto N. and Tamura H. (1986), Bull. Chem. Soc. Jpn. 59, 3302-3304.

65. Herrmann J.-M., Mozzanega M.-N. and Pichat P. (1983), J. Photochem. 22, 333-343.

66. St John M.R., Furgala A.J. and Sammells A.F. (1983), J. Phys. Chem. 87, 801-805.

67. Sakata T., Kawai T. and Hashimoto K. (1984), J. Phys. Chem. 88, 2344-2350.

68. Hashimoto K., Kawai T. and Sakata T. (1984), J. Phys. Chem. 88, 4083-4088.

69. Brown G.T. and Darwent J.R. (1984), J. Phys. Chem. 88, 4955-4959.

70. Darwent J.R. and Lepre A. (1986), J. Chem. Soc. Faraday Trans. 2 82, 1457-1468.

71. Hidaka H., Kubota H., Grätzel M., Serpone N. and Pelizzetti E. (1985), Nouv. J. Chim. 9, 67-69.

72. (a) Hidaka H., Yamada S., Suenaga S., Kubota H., Serpone N., Pelizzetti E. and Grätzel M. (1989), J. Photochem. Photobiol. A: Chemistry 47, 103-112.
(b) Hidaka H., Yamada S., Suenaga S., Zhao J., Serpone N. and Pelizzetti E. (1990), J. Molec. Catal. 59, 279-290.

73. Pelizzetti E., Minero C., Maurino V., Sclafani A., Hidaka H. and Serpone N. (1989), Environ. Sci. Techol. 23, 1380-1385.

74. (a) Okamoto K., Yamamoto Y., Tanaka H., Tanaka M. and Itaya A. (1985), Bull. Chem. Soc Jpn. 58, 2015-2022.
(b) Okamoto K., Yamamoto Y., Tanaka H. and Itaya A. (1985), Bull. Chem. Soc Jpn 58, 2023-2028.

75. Matthews R.W. (1986), Appl. Radiat. Isot. 37, 1247-1248.

76. Pelizzetti E., Minero C., Pramauro E., Barbeni M., Maurino V. and Tosato M. (1987), La Chim. L'Ind. 69, 88-89.

77. Barbeni M., Morello M., Pramauro E., Pelizzetti E., Vincenti M., Borgarello E. and Serpone N. (1987), Chemosphere 16, 1165-1179.

78. Harada K., Hisanaga T. and Tanaka K. (1987), New J. Chem. 11, 597-600.

79. Grätzel C.K., Girousek M. and Grätzel M. (1987), J. Molec. Catal. 39, 347-353.

80. Pelizzetti E., Borgarello M., Minero C., Pramauro E., Borgarello E. and Serpone N. (1988), Chemosphere 17, 499-510.

81. Matthews R.W. (1989), J. Chem. Soc. Faraday 1 85, 1291-1302.

82. (a) Grätzel M., Thampi K.R. and Kiwi J. (1989), J. Phys. Chem. 93, 4128-4132.
(b) Grätzel M. and Howe R.F. (1990), J. Phys. Chem. 94, 2566-2572.
(c) Rothenberger G., Moser J., Grätzel M., Serpone N. and Sharma D.K. (1985), J. Amer. Chem. Soc. 107, 8054-8059.

83. Minero C., Maurino V., Campanella L., Morgia C. and Pelizzetti E. (1989), Environ. Tech. Lett. 10, 301-310.

84. Bahnemann D.W., Mönig J. and Chapman R. (1987), J. Phys. Chem. 91, 3782-3788.

85. Tanaka K., Hisanaga T. and Harada K. (1989), J. Photochem. Photobiol. A: Chemistry 48, 155-159.

86. Borello R., Minero C., Pramauro E., Serpone N. and Hidaka H. (1989), Environ. Toxicol. Chem. 8, 997-1002.

87. Pelizzetti E., Maurino V., Minero C., Zerbinati O. and Borgarello E. (1990) 'Photocatalytic degradation of bentazon by TiO_2 particles' Chemosphere (in press).

88. Pelizzetti E., Minero C., Maurino V., Hidaka H., Serpone N. and Terzian R. (1990), Ann. Chim. 80, 81-87.

89. Serpone N. (1989), In 'Photochemical Energy Conversion', ed. by Norris Jr. J.R. and Meisel D. Elsevier Science Publishing Co. Inc. pp 297-315.

90. Palmisano L., Schiavello M., Sclafani A., Coluccia S. and Marchese L. (1988), New J. Chem. 12, 847-852.

91. Palmisano L., Augugliaro V., Schiavello M. and Sclafani A. (1989), J. Molec. Catal. 56, 284-295.

92. Anpo M., Kawamura T., Kodama S., Maruya K. and Onishi T. (1988) 'Photocatalysis on Ti-Al binary metal oxides: enhancement of the photocatalytic activity of TiO_2 species', J. Phys. Chem. 92, 438-440.

93. Henglein A. (1982), Ber. Bunsenges. Phys. Chem. 86, 241-246.

94. Fan F.-R.F., Liu H.-Y. and Bard A.J. (1985), J. Phys. Chem. 89, 4418-4420.

95. Yoneyama H., Haga S. and Yamanaka S. (1989), J. Phys. Chem. 93, 4833-4837.

96. Serpone N., Borgarello E., Harris R., Cahill P. and Borgarello M. (1986), Solar Energy Mat. 14, 121-127.

97. (a) Matthews R.W. (1987), Solar Energy 38, 405-413.

(b) Matthews R.W. (1987), J. Phys. Chem. 91, 3328-3333.

(c) Matthews R.W. (1988), Chem. Ind. 4 January, 28-30.

(d) Matthews R.W (1988), J. Catal. 113, 549-555.

(e) Matthews R.W. and Low G.K.-C (1990) 'Flow injection determination of organic contaminants in water using an ultraviolet-mediated titanium dioxide film reactor', Anal. Chim. Acta (in press).

(f) Matthews R.W. (1987), Solar 87 Creating the future. Vol 1 ed. by Lee T., Proceedings of the Annual Conference of the Australian and New Zealand Solar Energy Society, 26-28 November 1987, Australian National University, Canberra, published by Australian and New Zealand Solar Energy Society 1987, pp 277-286.

98. Palmisano L., Augugliaro V., Sclafani A. and Schiavello M. (1988), J. Phys. Chem. 92, 6710-6713.

99. Ashokkumar M. and Maruthamuthu P. (1989), J. Photochem. Photobiol. A Chemistry 49, 249-258.

100. Sclafani A., Palmisano L. and Schiavello M. (1990), J. Phys. Chem. 94, 829-832.

101. Boonstra A.H. and Mutsaers C.A.H.A. (1975), J. Phys. Chem. 79, 1694-1698.

102. Niskimoto S.-I., Ohtani B., Kajiuara H. and Kagiya T. (1985), J. Chem. Soc. Faraday Trans I 81, 61-68.

103. Augugliaro V., Palmisano L., Sclafani A., Minero C. and Pelizzetti E. (1988), Toxicol. Environ. Chem. 16, 89-109.

104. Matthews R.W., Abdullah M. and Low G.K.-C. (1990) 'Effects of common inorganic anions on rates of photocatalytic oxidation of organic carbon over illuminated titanium dioxide', J. Phys. Chem (in press).

105. Sclafani A., Palmisano L. and Davi E. (1990) 'Photocatalytic degradation of phenol by TiO_2 aqueous dispersions: rutile and anatase activity', New J. Chem. 14 (in press).

106. (a) Tanaka K., Hisanaga T. and Harada K. (1989), New J. Chem. 13, 5-7.

(b) Tanaka K., Hisanaga T. and Harada K. (1989), J. Photochem. Photobiol. A: Chemistry 48, 155-159.

107. Augugliaro V., Davi E., Palmisano L., Schiavello M. and Sclafani A. (1990) 'Influence of hydrogen peroxide on the kinetics of phenol photodegradation in aqueous titanium dioxide disperson', J. Appl. Catal. (in press).

108. Ogata Y., Tomizawa K. and Yamashita Y. (1980), J. Chem. Soc. Perkin II, 616-619.

109. (a) Matthews R.W. (1988), J. Catal. 111, 264-272.

(b) Matthews R.W. (1990) 'Purification of water with near-UV illuminated suspensions of TiO_2', Water Res. in press.

110. Turchi C.S. and Ollis D.F. (1988), J. Phys. Chem. 92, 6852-6853.

111. Augugliaro V. and Palmisano L. (1988), 2nd International Conference on Environment Protection, S. Angelo, Ischia (NA) Italy, 5-7 Oct 1988, Proceedings vol I pp 2.A-59 to 2.A-66.

112. (a) Turchi C.S. and Ollis D.F. (1990), J. Catal. 122, 178-192.

(b) Turchi C.S. and Ollis D.F. (1989), J. Catal 119, 483-496.

113. Matthews R.W. (1983), J. Chem. Soc. Chem. Commun. 177-179.

114. Al-Ekabi H., Serpone N., Pelizzetti E., Minero C., Fox M.A. and Draper R.B (1989), Langmuir 5, 250-255.

115. Fox M.A. and Tien T.-P (1988), Anal. Chem. 60, 2278-2282.

116. (a) Matthews R.W., Abdullah M. and Low G.K.-C. (1990), Chemistry in Australia, Royal Australian Chemical Institute, April pp 85-89.

(b) Matthews R.W., Abdullah M. and Low G.K.-C. (1990) 'Photocatalysis: a new method of oxidation for total organic carbon analysis, Anal. Chim Acta (in press).

(c) Matthews R.W. and S. McEvoy (1990), unpublished results.

(d) Low G.K.-C., McEvoy S.R. and Matthews R.W. (1990) 'Formation of ammonium and nitrate ions from photocatalytic oxidation of ring nitrogenous compounds over titanium dioxide', Chemosphere (in press).

117. Pacheco J.E. and Tyner C.E. (1990) 'Enhancement of processes for solar photocatalytic detoxification of water' 1990 ASME Solar Energy Division, International Solar Energy Conference, April 1-4, Miami Florida.

118. Menassa P.E., Mak M.K.S. and Langford C.H. (1988), Environ. Tech. Lett. 9, 825-832.

HETEROGENEOUS PHOTOCATALYSIS: PHOTOCHEMICAL CONVERSION OF INORGANIC SUBSTANCES IN THE ENVIRONMENT: Hydrogen Sulfide, Cyanides, and Metals.

N. SERPONE,[a*] D. LAWLESS,[a] R. TERZIAN,[a]
C. MINERO[b] and E. PELIZZETTI[b]
[a] *Department of Chemistry, Concordia University, Montréal, Québec,Canada H3G lM8.*
[b] *Dipartimento di Chimica Analitica, Università di Torino, 10125 Torino, Italia.*

ABSTRACT Industrial activities release many toxic agents into the environment. Society faces two major hurdles: (i) the removal of undesirable substances from chemical and nuclear process effluents and (ii) the need to obtain and/or recover some substances from progressively leaner initial sources. **Hydrogen sulfide** is a major constituent of natural gas and is the principal product in the hydrodesulfurization of petroleum and coal. Although it is a toxic pollutant, it is an important energy vector. **Cyanide** is a very toxic substance frequently found above safe threshold levels in aquatic ecosystems. Unlike H_2S and CN^-, **metals** are not biodegradable. They possess infinite lifetimes and cannot be detoxified by chemical means. In recent years, alternative biological and photochemical methods have surfaced as potential new technologies for treating water effluents to transform H_2S and CN^- into less harmful substances, and to recover *Pt, Pd, Au, Ag, Rh, Cr, Cu, Pb,* and *Hg*. This paper focusses on photocatalysis, which uses semiconductor particulates as light harvesters and photocatalysts to transform/recover undesirable inorganic substances from the environment. Actual industrial effluents can be detoxified by this heterogeneous photocatalytic technique. A few examples are given.

1. Introduction

Photochemical reactions in the aquatic ecosystem are the object of increased attention. In today's highly industrialized society, the environment is charged with a multitude of toxic and potentially toxic agents. Pollution abatement and cleanup require major efforts. Engineering processes such as (i) trapping methods, (ii) chemical changes of wastes, and (iii) process changes have been developed. Laws have been enacted (and implemented?) to curb and handle domestic and industrial wastes. The generation of wastes remains high. It is presumptuous to believe that pollutants can be diluted to

E. Pelizzetti and M. Schiavello (eds.), Photochemical Conversion and Storage of Solar Energy, 451–475.

below threshold levels to disappear by assimilation by the natural ecosystem.

The handling and disposal of hazardous chemicals from the moment of their manufacture to their ultimate disposal is a major environmental issue. The more alarming aspect is the long-term unknown consequences of unregulated or unsatisfactory disposal methods to future health and ecological problems. By definition, every industrial process and use of an industrial product creates environmentally unacceptable hazardous wastes: chemical solvents in manufacturing processes, textile mill effluents, acids from glass production. Hazardous and other wastes are also produced as by-products of manufacturing: (a) every ton of manufactured raw steel produces 14 Kg of air pollutants, 1 Kg of water pollutants and 9 Kg of solids; (b) every ton of aluminum yields 320 Kg of air pollutants, 20 Kg of water pollutants and 40 Kg of solids [1]. Considering the world's production of just these two products, the numbers are staggering. More than one ton of toxic materials are produced for every person in industrialized countries [2]. Of the billions of tonnes of regulated hazardous wastes produced each year, less than 0.5% is recycled and about 1% is incinerated [3]. The rest is dumped or discharged directly into the ecosystem. Wastewaters contain chemical solvents, corrosive acids, metals, pesticides, and other toxic substances. *"Nothing short of a second chemical revolution will gain control over toxic chemicals"* [4].

In any detoxification process designed to remove toxic constituents from wastewaters, other raw materials, new processes, and new technologies must be considered to minimize the amounts of pollutants in the waste effluents. Presently, the various means of treating industrial waste waters are a combination of those summarized in Table 1, particularly when the wastes contain a mixture of pollutants.

In this paper we consider a relatively novel technology, photocatalysis, which has been under development during the last decade to seek ways to safely dispose many of the toxic agents found in the aquatic environment. Specifically, we delve into recent research carried out to photoconvert H_2S and CN^- to less toxic materials, and to recover various strategic and/or toxic metals from waste effluents.

2. The Photocatalytic Process

Absorption of light of energy $\geq E_{bg}$, the forbidden gap of a semiconductor material (e.g., TiO_2, CdS, WO_3, ZnO) which serves as the photocatalyst in these processes, generates conduction band electrons (e^-) and valence band holes (h^+). Semiconductor particulates are excellent light harvesting units offering distinct advantages in a heterogeneous photocatalyzed process: (i) high absorption cross sections of the light photons; (ii) fast carrier diffusion to the interface; and (iii) suitable redox levels of the valence and conduction band edges that can yield high efficiencies in converting light energy to useful redox events (reduction of an oxidizing substrate and oxidation of a reducing substrate) at the solid/solution interface. The charge carriers are trapped at the surface {e^- as Ti(III) centres and h^+ as surface bound OH radicals [5]. Competing with these events is the radiative and nonradiative recombination of e^- and h^+. The

TABLE 1.- Principal Physicochemical and chemical processes for the purification of waste waters[a]

Physicochemical process mainly for		Chemical process
Dispersed substances	substances in true solution	
flocculation	stripping	ion exchange
ultrafiltration	evaporation	precipitation
electrophoresis	reverse osmosis	hydrolysis
	electrodyalysis	neutralization
	adsorption	oxidation
extraction	extraction	wastewater incineration
		electrolytic reduction

[a] From ref.[6].

surfaces of the particulates can be engineered to improve both the light harvesting and the redox characteristics. Moreover, the photoredox events occurring at the interface between the excited semiconductor photocatalyst and the redox species in solution are unidirectional and generally nonreversible [7].

The above principles (Figure 1) have found extensive application in the field of light-induced generation of fuels [8], in organic phototransformations [9], and only recently in the area of environmental chemistry and pollution abatement. Except for a few reports concerning some pollutants, photoeffects at semiconductor surfaces have been largely ignored by environmental chemists [10].

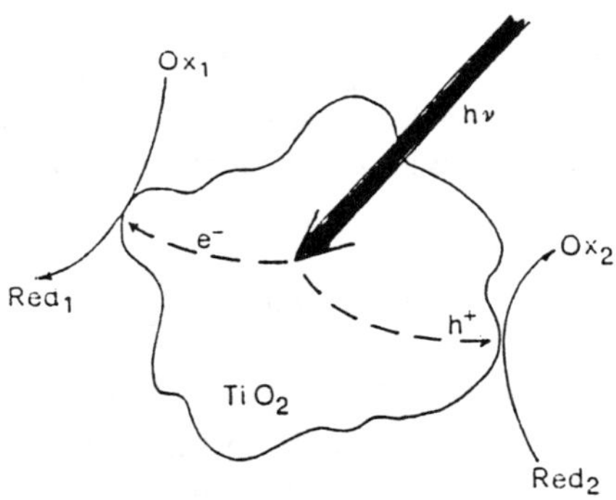

Figure 1.- Schematic representation of the formation of an electron/hole pair and electron transfer reaction at the semiconductor particle.

3. Photocatalyzed Conversion of Hydrogen Sulfide

Hydrogen sulfide is a group VI hydride and like water is a potential source of hydrogen, an excellent energy vector. Its is a major constituent of natural gas [11-14] and is the principal product of the hydrodesulfurization of petroleum and coal. Industrially, this sulfide is a major precursor to sulfuric acid and to a variety of other inorganic and organic compounds such as thiophene and thiols. It is also used extensively with other organic reagents in the development of high pressure lubricants and cutting oils; substantial quantities of the sodium salt are used in the leather industry [15]. Over 10^6 metric tonnes of H_2S are produced each year from all sources [12-14]. Unlike water, oxidation of H_2S is thermodynamically facile, requiring 4 redox equivalents (eqn 1).

$$H_2S \xrightarrow[\text{semiconductor photocatalyst}]{\text{light}} H_2 + S \qquad (1)$$

At high concentrations, H_2S is poisonous, almost as toxic as HCN; its action may be just as rapid [15]. The characteristic olfactory response to this sulfide is an important feature of its toxicology: the typical odor of H_2S is detected at concentr-ations as low as 0.035 ug/L in air. At $[H_2S] > 280$ ug/L, hydrogen sulfide is definitely toxic. The impact of H_2S on the environment must also be assessed for its role in SO_2 emissions and thus for its role on acid rain [16,17]. One process that transforms H_2S to a less harmful substance is the Claus process [18]. The sulfide is converted to water and sulfur. The first oxidation step yields SO_2 which is then reacted with fresh H_2S to form sulfur.

Reaction 1 has been examined in some detail using CdS as the photocatalyst [19-21]. Although CdS is not a particularly good electrode for water reduction, photo-evolution of H_2 does proceed at reasonable rates without the intervention of a redox catalyst (Pt or RuO_2). The complete cleavage of H_2S at pH 14 in the presence of SO_2 yields dihydrogen and thiosulfate, $S_2O_3^{2-}$, in a 1:1 stoichiometric ratio (reaction 2) [420,21]. Coupling the process in reaction 2 with one capable of recycling the $S_2O_3^{2-}$

$$S^{2-} + SO_3^{2-} + 2H_2O \xrightarrow{2h\nu} S_2O_3^{2-} + 2OH^- + H_2 \qquad (2)$$

back to S^{2-} and SO_3^{2-} can sustain dihydrogen evolution in a cyclic system (reaction 3) [22]. No accumulation of S and $S_2O_3^{2-}$ would take place in such a photosystem.

Some time ago we introduced a new strategy in photocatalysis [23]:

$$3H_2O + S^{2-} \xrightarrow{h\nu} SO_3^{2-} + 3H_2 \qquad (3)$$

the interparticle electron transfer pathway (IPET) between two semiconductors to accomplish vectorial displacement of charges. We first applied this novel strategy to the visible light-induced conversion of H_2S [23-26] and later the photo-oxidation of alcohols [27]. IPET is depicted in Figure 2 and summarized by reactions 4-8 for H_2S. Loading the TiO_2 semiconductor particles with a suitable redox catalyst (e.g., RuO_2)

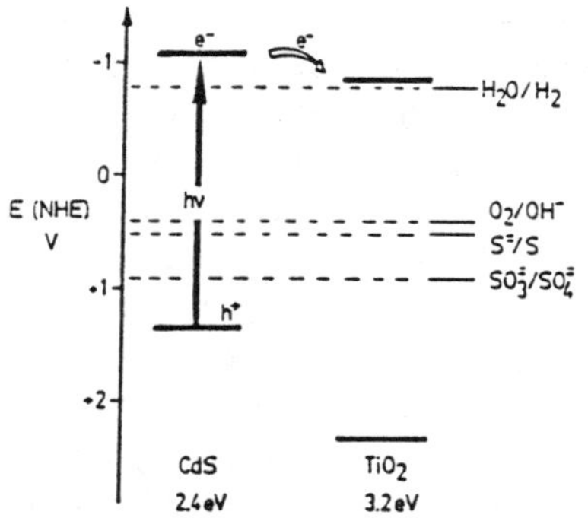

Figure 2.- Flatband potentials of CdS and TiO_2 at pH 14 and 0.1 M Na_2S in aqueous media. Positions of other redox couples are also indicated.

$$CdS \xrightarrow{h\nu} e^- (CdS) + h^+ (CdS) \qquad (4)$$

$$2h^+ (CdS) + S^{2-} \longrightarrow S \qquad (5)$$

$$2e^- (CdS) + 2H_2O \longrightarrow H_2 + 2OH^- \qquad (6)$$

$$e^- (CdS) + TiO_2 \longrightarrow e^- (TiO_2) + CdS \qquad (7)$$

$$2e^- (TiO_2) + 2H_2O \longrightarrow H_2 + 2OH^- \qquad (8)$$

leads to more efficient conversion of the H_2S to less harmful but useful products. The energy conversion efficiency is 1%; in the presence of SO_3^{2-} ions the efficiency is about 2%. Through the combination of the two semiconductors, the extent of e^-/h^+ recombination is diminished. Charge separation is achieved through the selective transfer of e^- from CdS to TiO_2; the transfer of h^+ is thermodynamically inhibited [23]. An analogous process was noted in the photo-oxidation of alcohols; the coupled semiconductor materials are CdS and platinized TiO_2. The IPET strategy in photocatalysis has since been confirmed photoelectrochemically [28-30], by photoconductance [31] and by spectrofluorimetric techniques [32].

Evolution of H_2 occurs with a quantum efficiency of ~23% and an energy conversion efficiency close to 3% for a Rh-loaded CdS catalyst in alkaline media of

H_2S and SO_2 has been achieved [26]. The high quantum yield and the facile preparation of the CdS/Rh catalyst make it practical in photogenerating H_2 and disposing of two hazardous pollutants H_2S and SO_2.

4. Photocatalyzed Conversion of Cyanide

Cyanide is among the most hazardous of water pollutants. It is a common metabolite formed from a variety of precursors widely distributed in nature [33]. With increasing use of cyanide in various industrial processes, the amount of cyanide dispersed in the environment is increasing. The aqueous streams from the electroplating industry initially contain up to 20% cyanide and while efforts have been made to control this cyanide waste, an appreciable amount is still being discharged into the environment [34]. The largest future potential source of cyanide contamination of the water supplies is the coal gasification process for making syngas.

Treatment methods attempt to destroy CN^- or transform it to less toxic forms (eg., OCN^- or SCN^-). Such methods include oxidation of CN^- by alkaline chlorination, ozonation, and permanganate, and electrolytic decomposition [35]; other techniques include exposure to ionizing radiation, acidification, reaction with an aldehyde, ion exchange, evaporation, lagooning, and biological degradation [34].

Rivers polluted by peak discharges from galvanic industries show CN^- concentration levels to 200 ug/l [36]. In mine tailing ponds, levels as high as 10 to 30 mg/l have been uncovered [37]. By comparison, the water quality objective is $\leq$ 10 ug/l of total cyanide [38]. Cyanide has a low persistence in air and water under normal conditions, however the persistence of cyanide contamination in underground water is longer [34]. Clearly additional ways are needed to convert CN^- to the less toxic OCN^- and SCN^-. Both are formed *in vivo* in the course of cyanide detoxification, with the body handling SCN^- as if it were a halide ion; its toxicity is some two orders of magnitude lower than that of CN^- [39].

4.1 PHOTOCATALYZED TRANSFORMATION OF CN^- TO SCN^-

The quantum efficiency of converting CN^- to OCN^- over TiO_2 anatase dispersions is ~ 6% [40]. Recent work [41] with Rh-loaded CdS dispersions has shown relatively high quantum efficiencies (> 25%), the use of a waste by-product (H_2S) as the oxidizing agent, irradiation with visible light (possibility of using natural sunlight), and the relative long-term stability of the CdS/Rh photocatalyst to cyanide attack and to photocorrosion.

The stoichiometry of the transformation reaction is:

$$CN^- + S^{2-} + 2H_2O \xrightarrow[> 405\ nm]{CdS/0.2\ wt\%\ Rh} SCN^- + H_2 + 2OH^- \qquad (9)$$

After light absorption by the semiconductor photocatalyst to produce electron/hole pairs, h^+ oxidize S^{2-} to S which dissolves in alkaline media as polysulfides, S_n^{2-}, containing the sulfane sulfur. This sulfur reacts with CN^- to give SCN^-. No transformation occurs between S^{2-} and CN^- in the absence of the photocatalyst CdS/Rh and/or light. This shows that h^+ oxidize sulfide and mediate the CN^- to SCN^- conversion. Conduction band electrons reduce water to H_2 and are involved in e^-/h^+ recombination.

Disposal of CN^- by photochemical processes with the participation of Rh-loaded CdS photocatalyst opens up an additional practical alternative to the methods above. This method employs relatively inexpensive materials and abundant sunlight [42]. The conversion is efficient (> 25%) even in the presence of air and at very high $[CN^-]$. Hydrogen produced concomitantly with SCN^- is a useful by-product. Where both CN^- and S^{2-} are available, the process converts two toxic materials, CN^- and H_2S, to less harmful products [33,39]. In a practical process it will be important to constantly verify that no Cd^{2+} ions are discharged in the wastes, because of the toxicity of Cd^{2+} [43].

4.2 LIGHT-INDUCED PEROXIDE OXIDATION OF CN^- AND PHOTODEGRADATION OF CYANOMETALLATES

The detoxification of CN^- is well documented in the literature (see also ref.[44]). Precious metals and cyanide are present in the waste effluents from various industrial processes (electroplating, mining, jewellery). For instance, gold is extracted from ores by the cyanidation process [45].

We have been concerned with metal toxicity in industrial waste effluents and have carried out a systematic study into their removal from solutions and wastewaters (details below). We showed earlier [46] that gold can be recovered efficiently from aqueous chloride media via a light-induced process mediated by TiO_2 dispersions. Since some of the industrial CN^- wastes contain hundreds of ppm of gold and other metals [47], prior disposal of CN^- is imperative before recovery of the metals can be undertaken. Detoxification of actual industrial wastes has been carried out by treating the wastes with H_2O_2 in the dark and under illumination.

4.2.1 *Effect of Light on the Oxidation of CN^- by H_2O_2* - Figure 3 shows the results from a typical experiment in which 184 ppm (7.08 mM) CN^- was oxidized by H_2O_2 (200 mM, 0.6 vol %) in the dark and under UV light (25°C). Oxidation is rapid under illumination. No CN^- was detected after 30 min of UV irradiation; 24% of CN^- decomposed in the dark. Simulated sunlight had no effect on the (dark) process.

4.3 PHOTOCATALYZED DEGRADATION OF A Au-CN^- COMPLEX

Illumination of a TiO_2 suspension containing Au(III) and CN^- ions leads to oxidation of CN^- to OCN^- as for free CN^- solutions [40]; the gold/cyanide complex is not decomposed in excess cyanide. Figure 4 shows the results when a solution 18.3 mM in CN^- and 0.93 mM in Au(III) is illuminated (simulated sunlight) in the presence of

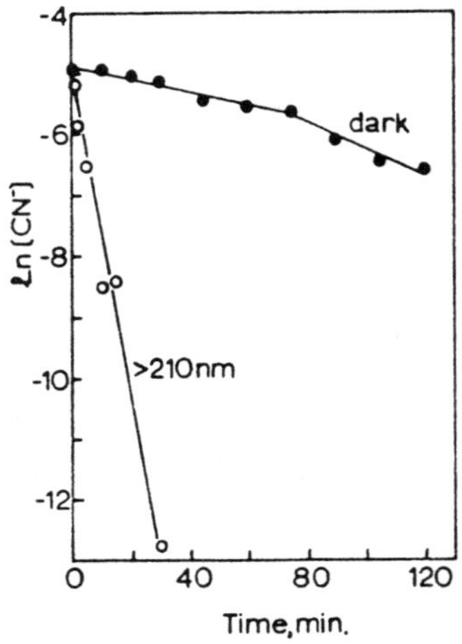

Figure 3.- Ln[CN^-] vs time for the degradation of CN^- via oxidation by H_2O_2 in the dark and UV light (>210 nm); [H_2O_2] = 200 mM; 25°C; initial [CN], 7.08 mM (dark process) or 7.94 mM (UV light). Reproduced from ref.[47].

2g/l TiO_2 for up to 15 hr. Results from the H_2O_2 (200 mM) oxidation of CN^- (19.6 mM, [Au] = 0.93 mM), under the same experimental conditions in the absence of TiO_2 are included for comparison. Peroxide oxidation of CN^- occurs faster than oxidation by TiO_2. On complete removal of unreacted peroxide, addition of TiO_2 followed by UV irradiation leads to degradation of the gold/cyanide complex on the surface of the TiO_2 particulates [46]. Removal of unreacted peroxide is important because of its action on TiO_2 [48]. The photodegradation of the gold/cyanide complex is most efficient under acidic conditions (Figure 5).

In CN^- media, the reduction of Au(III) is slower than in Cl^- media.In a typical experiment, 155 ppm of Au(III) were brought into contact with TiO_2 in the presence of CN^- with [CN^-] > 10 [Au(III)] at pH 10.3. After 15 hrs irradiation, 20% of the gold deposited on TiO_2; with CH_3OH present, 80% of the gold deposited after 5.5 hrs. The slow photoreduction of gold in CN^- media is understandable inasmuch as the reduction potential of $Au(CN)_4^-$ is -0.4 V (vs. NHE). Addition of methanol suppresses e^-/h^+ recombination and formation of Au^o becomes efficient.

As a practical test, actual industrial samples (from jewellers wastes) were treated with the photocatalytic method to dispose of CN^- and to recover the strategic materials. The results are summarized in Table 2.

Two different samples were analyzed and treated for disposal of free and complexed CN^- (with gold and other metals). Sample A contained 463 ppm CN^- (pH 10.2), ~ 100 ppm gold, and varying amounts of other metals such as Zn, Ag, and Cu

(Table 2). Sample B contained 2600 ppm of cyanide (pH 11.6), 100 ppm of gold and Ag, Cu, and Zn. The two samples A and B (25 ml) were treated with 1 ml and 5 ml, respectively, of a 30 vol.% H_2O_2 solution in a large volume reactor illuminated with

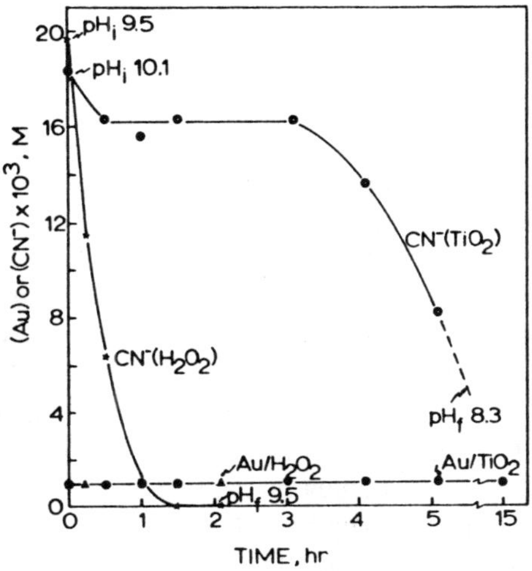

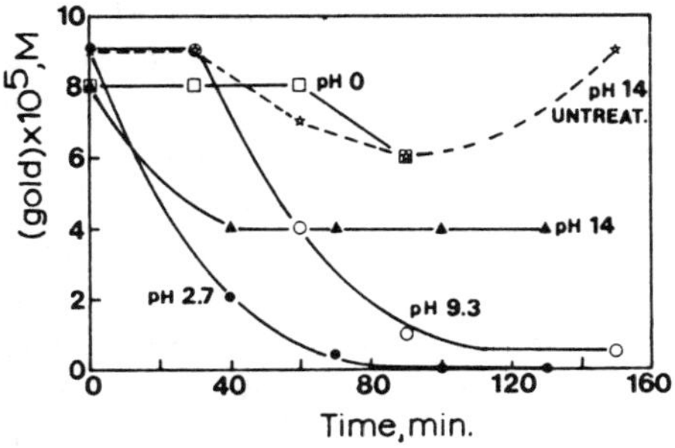

Figure 4.- Changes in [Au] and CN^-] as a function of time that compare treatments of Au-CN^- solutions by H_2O_2 and TiO_2 under AM1 simulated sun-light (λ> 310 nm): initial [CN^-], 18.3 mM; [Au] = 0.931 mM; TiO_2 = 2 g/l; [H_2O_2] = 200 mM; 35°C. Reproduced from ref.[47].

Figure 5.- Reduction of Au(III) in the presence of TiO_2, initial pH 0, 2.7, 9.3, and 14 after treatment of the CN^- solution with H_2O_2; for the untreated solution, initial pH 14; 25 ml solution; [Au(III)] = 80-90 uM; TiO_2 = 2 g/l; [CH_3OH] = 10 vol %; 35°C. Reproduced from ref.[47].

TABLE 2.- Selective separation of metals from jewellers wastes after disposal of the free CN^- (463 ppm) with H_2O_2 {UV irradiation, 1 hr}

	Initial Conc'n (ppm)	Conc'n after irrad'n (ppm)	Quantity recovered (ppm)
Au	97	0	78
Cu	95	56	26
Ag	3.4	1.4	0.7
Zn	>120	>120	0

Conditions: 50 mg TiO_2, 25 ml solution, pH 0.8, 17 vol.% methanol.

simulated sunlight (at this concentration, H_2O_2 absorbs UV light from the solar lamp). Total degradation of the uncomplexed CN^- occurred. Subsequent addition of TiO_2 and methanol followed by irradiation led to degradation of the CN^- complexes, as noted by the total reduction of the 100 ppm of gold present in the two samples {purple of Cassius, Au° deposited on a metal oxide surface}.

5. Photocatalyzed Reductions and the Selective Recovery of Metals

An important dimension in metals disposal and recovery is the potential for resource conservation. The demand for certain metals could exceed the known reserves of minerals and ores by the year 2000 [49]. The metals Al, Cd, Cr, Co, Cu, Fe, Pb, Hg, Ni, and Zn are on the list of strategic minerals. Thus waste metals recovery for industrial re-use could resolve two problems: metals pollution and resource conservation. In developing the technology for metals recovery, the metals toxicity from landfill sites and the strategic importance of the metal(s) will need to be considered [49].

The photochemical deposition of metals on semiconductor powders, particularly in regard to the preparation of photocatalysts, has been exploited by many [50]. Several reports have dealt with the photoreduction of metals on semiconductor particles as a route to suppress e^-/h^+ recombination by scavenging e^- and allowing the h^+ to undergo the appropriate anodic reaction(s). This technique has proven useful in detailed studies of processes occurring at the anodic side of the excited semiconductor [51].

A potential and attractive practical application of the light-driven reduction process is in the recovery of metals of strategic and economic importance (Pt, Pd, Cr, Au, Rh, Ag, and others) from industrial waste effluents. The method affords selective recovery of one specific metal over several others in solution by appropriately chosen experimental conditions. The metal is reduced on the particle surface and is subsequently extracted from the slurry by mechanical and/or chemical means (*aqua regia*).

We discuss below our recent work [52-54] and note that of others [55-59] on this subject; also included is our work on the removal of such toxic metals as Hg(II) and CH_3Hg(II) [60], and Pb(II) [61] from industrial wastes. Tanaka and co-workers [57] have examined TiO_2 powders in light-driven processes to eliminate Pb(II), Mn(II), Hg(II) Tl(I), and Co(II) from solutions.

5.1 GOLD(III) IN CHLORIDE MEDIA

Two semiconductor materials, TiO_2 and WO_3, have been examined. Quantitative reduction and deposition of gold occurs in ca. 20 min of UV irradiation for a 100 mg TiO_2 suspension containing 84 ppm of $AuCl_3$ (pH 3.1); on WO_3 (100 mg, pH 3.1), total reduction and deposition of Au(III) necessitates ca. 25 min with visible illumination ($\geq$ 405 nm). We investigated this TiO_2 semiconductor material in more detail. Oxygen has little or no effect on the reduction process at pH 3.1. Adsorption of Au(III) on the TiO_2 particle surface is pH sensitive; at pH 3.15, 57% of $AuCl_3OH^-$

is adsorbed (dark); 14% of $AuCl_4^-$ is adsorbed at pH 0.

Figure 6 depicts the interactions of gold(III) with the TiO_2 particles in the dark (adsorption process, no reduction) and under UV illumination where Au(III) is photoreduced to Au° on the metal oxide surface [62]; pH 1.5. Gold(III) is not photoreduced at pH 0 in the presence of TiO_2, even in the absence of O_2 which

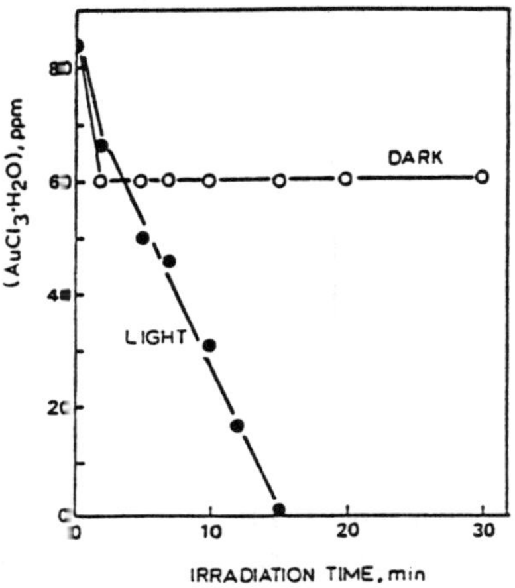

Figure 6.- Changes in the concentration of gold(III) in solution as a function of irradiation time: TiO_2 100 mg/50 ml solution; UV irradiation; pH 1.5; 900 Watt Hg/Xe lamp From ref.[52].

competes with metal ions or the e^-. However, addition of 4 vol.% methanol, a hole scavenger, favors reduction. The reduction quantum efficiency is ~ 3% at pH 14 and ~ 15% at pH 5-6. At this pH, photoreduction of gold is optimal.

The light-driven reduction process is suitable to recover the metals from a gold alloy under conditions that may find practical applications. The results are collected in Table 3.

TABLE 3.- Selective photoreduction and recovery of gold from an alloy in 1 M HCl (50 ml solution containing 4 vol.% methanol; pH 0) under UV irradiation (1 hr). From Ref.[52].

	Initial Conc'n (ppm)	Conc'n after irrad'n (ppm)	Conc'n after recovery (ppm)
Cu	100	96	6
Au	260	0	242
Ni	11	10	1
Zn	13	12	61

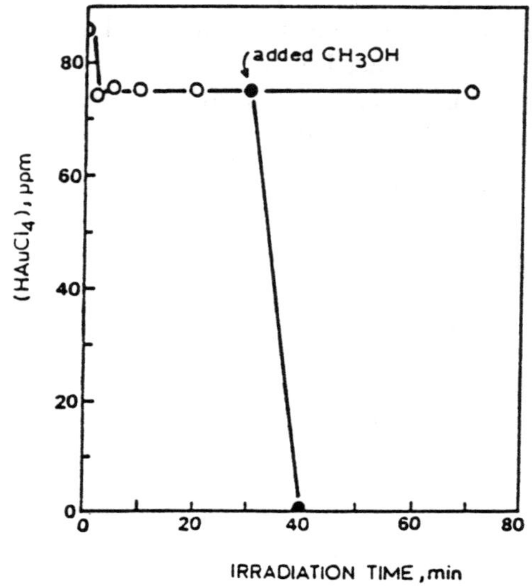

Figure 7.- Variations in [$HAuCl_4$] with illumination time; pH 0; TiO_2 100 mg/50 ml solution; 900 Watt Hg/Xe lamp; UV light; 4 vol.% CH_3OH added after 30 min. From ref. [52].

The relevant redox reactions in the photocatalyzed deposition of gold are:

$$TiO_2 \xrightarrow{h\nu} e^- + h^+ \qquad (10)$$

$$Au(III) + 3\,e^- \longrightarrow Au^o \qquad (11)$$

$$2\,H_2O + 4\,h^+ \longrightarrow 4\,H^+ + O_2 \qquad (12)$$

and the net reaction is:

$$4\,Au(III) + 6\,H_2O \xrightarrow[TiO_2]{h\nu} 4\,Au^o + 12\,H^+ + 3\,O_2 \qquad (13)$$

5.2 CHROMIUM(VI)

The photochemical reduction of chromium(VI), a common component of industrial wastewaters, has been examined in dispersions of ZnO, CdS, ZnS, and WO_3 [63,64].

Irradiation of air-equilibrated aqueous Cr(VI) solutions containing CdS, ZnS, or WO_3 powders leads to a decrease in [Cr(VI)]; the efficiency follows the order CdS > ZnS > WO_3 [63]. The low efficiency of WO_3 is due to the low reducing power of the conduction band electrons. Since O_2 can also scavenge e^-, reduction of Cr(VI) is more efficient in de-aerated media [63]. Also, the efficiency of Cr(VI) removal from solution increases with a decrease in the initial [Cr(VI)] because of an inner filter effect by Cr(VI) (absorption at 350 nm). In ZnO suspensions, the quantity of Cr(VI) eliminated from solution increases with irradiation time and decreases with decreasing initial [Cr(VI)] for a given irradiation period [64]. The process follows apparent first-order kinetics, 3.64 x $10^{-2}min^{-1}$, analogous to observations on TiO_2 [65].

The photoreduction of Cr(VI) with irradiated ZnO consumes O_2 and generates H_2O_2; the apparent rate of this process is smaller than in O_2-free media, probably a result of a homogeneous reduction of Cr(VI) by H_2O_2 [64]. Chromium-doped and H_2-reduced ZnO are somewhat more efficient. Photoreduction also occurs in natural sunlight [64].

5.3 COPPER(II)

The photoreduction and deposition of copper occurs on irradiated TiO_2 and WO_3 semiconductor powders [66] and in illuminated ZnO suspensions [55,67] containing acetate buffer (pH 5); this prevents the formation/precipitation of $Cu(OH)_2$ and diminishes the extent of photocorrosion of the ZnO photocatalyst. The rate of deposition of Cu decreases with increasing irradiation time; this has been attributed to (i) copper deposits on ZnO inhibiting light absorption by the photocatalyst and (ii) the ability of metal deposits to enhance surface e^-/h^+ recombination [55]. A large drop in the rate of Cu deposition occurs if the ZnO is pre-loaded with copper. At constant initial conditions, copper deposition is light intensity dependent and is more efficient in the absence of O_2. Sunlight experiments have also been carried out. Reduction of Cu^{2+} ions in a 10 mM $CuSO_4$ solution in 0.1M acetic acid/acetate buffer is complete in 7 hrs in the presence of sufficient ZnO}; the process is more efficient at the low copper concentrations [55,67].

In spite of its efficiency in reducing various metals such as Cr(VI) and Cu(II), a major problem with ZnO is photocorrosion. This arises from the reaction of photogenerated h^+ with ZnO to produce Zn^{2+} in solution, a thermodynamically favored process [55,68]. Addition of suitable redox substrates such as acetate reduces photocorrosion by competition of acetate for the valence band holes [67,69].

5.4 SILVER(I)

The photoreduction of silver on TiO_2 [70-75] and ZnO [70,76] has been examine extensively. Thie process is particularly relevant because of the potential recovery of silver from wastewaters.

Figure 8 [76] shows the increase in the amount of silver reduced in a ZnO suspension containing $AgClO_4$ with irradiation time at 365 nm for different initial $[Ag^+]$. The quantum yields for the photoreduction of Ag^+ are 0.39, 0.30, and 0.2]4 for $[AgClO_4]$ 1.0 mM, 0.1 mM, and 0.05 mM, respectively, (pH's 6.8, 6.6, and 6.5) [76].

The nature of the counterion {NO_3^-} has no effect on the quantum yields [76].

The use of TiO_2 is preferred over ZnO for silver reduction because of the dissolution of the latter [74]. Figure 9 [74] shows the increase in the quantity of silver reduced in TiO_2 suspensions containing $AgClO_4$ versus irradiation time at 365 nm for different initial [Ag^+]. The quantum yields of photoreduction of Ag^+ range between 0.084 and 0.191 for [$AgClO_4$] between 0.1 mM and 10 mM; pH 4.5 [74]. While the e^- reduce Ag^+, the h^+ oxidize H_2O to give O_2; the O_2/Ag ratio was close to expected, 1/4 (reaction 14) [74].

$$4Ag^+ + 2H_2O \text{ --------> } O_2 + 4H^+ + 4Ag^o \qquad (14)$$

The photodeposition of Ag and formation of O_2 by irradiated TiO_2 in aqueous solutions of various silver salts was further investigated to determine the source of O_2 from the oxidation path, the activity of the photocatalyst used (anatase vs rutile TiO_2), and the effect of surface properties on the processes [73]. The apparent reaction rate of Ag deposited depends on the nature of the counterion: $ClO_4^- < NO_3^- \sim SO_4^{2-} < F^-$ ($\sim PO_4^{3-}$). The photocatalytic activity of rutile is less sensitive to the nature of the anion. Except for AgF and Ag_3PO_4, rutile is nearly two-fold more active than anatase for both oxygen formation and silver deposition.

Isotopic studies have confirmed that during silver reduction O_2 origin-ates from oxidation of water in TiO_2 dispersions [73]. Neither O_2 production nor Ag^+ reduction takes place at pH < 2; on addition of base, TiO_2 recovers its activity. Addition of 2-propanol to the acidic TiO_2 suspension (pH < 2) restores the activity of the photocatalyst for Ag^+ reduction but **not** for O_2 production [73]. In the presence of 2-propanol, a known hole scavenger, h^+ oxidize the alcohol to acetone rather than yield O_2 from water oxidation.

The photocatalytic activity of TiO_2 for Ag^+ reduction is pH dependent. The initial rate of silver deposition decreases with decreasing pH for both anatase and rutile TiO_2 and becomes almost negligible at pH 2 [71]. The extent of Ag^+ ion ads-orption in the dark also decreases at the lower pH's, no doubt the result of the positively charged photocatalyst surface in acid media. Oosawa and Gratzel [77] have noted that the photocatalytic activity of TiO_2 suspended in an aqueous $AgNO_3$ solut-ion is proportional to the reciprocal of the density of surface hydroxyl groups.

The initial rate of silver deposition on powdered TiO_2 varies with the initial [Ag^+] according to a Langmuir-Hinshelwood mechanism [75]. The rate of Ag deposit-ion is not temperature dependent around 300 K, but is dependent on the radiant flux of photons absorbable by TiO_2 (300 nm - 400 nm) at least up to a value of 7.6×10^{15} photons cm^{-2} s^{-1} (apparent initial quantum yield ~ 0.16 for complete coverage of the surface) [75].

Removal of Ag from a solution of Cu^{2+} and Ag^+ is selective in TiO_2 suspens-ions; copper is not photoreduced under the conditions used [75].

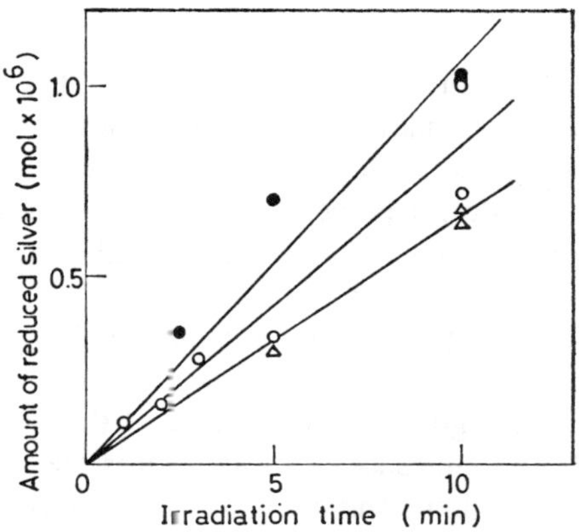

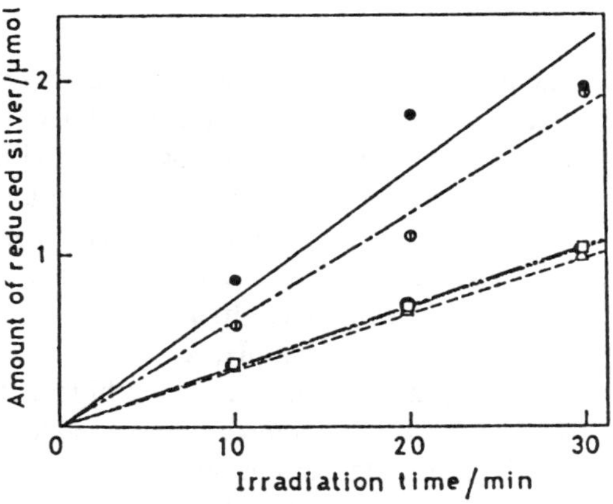

Figure 8.- Growth of amount of reduced silver in ZnO suspensions containing $AgClO_4$ with irradiation time (min): ●, $[Ag^+]$ = 1.0 mM; O, 0.1 mM; and ▲, 0.05 mM. From ref.[76].

Figure 9.- Growth of the amount of reduced silver in TiO_2 suspensions of $AgClO_4$ and 0.5M KNO_3 with irradiation time: ●, $[AgClO_4]$ = 10 mM; ⦶, 2 mM; O, 1 mM; □, 0.5 mM; and △, 0.1 mM. From ref.[74].

5.5 RHODIUM(III)

Complete reduction of Rh(III) occurs in O_2-free (chloride) solutions within 45 min of irradiation in the presence of 20 vol.% methanol [53]. In acid media, reduction of O_2 by TiO_2 e^+ is thermodynamically favored and interferes with the reduction of the metal ions, in contrast to the photoreduction of gold(III) where O_2 is inconsequential [52]. At pH 0 no light-induced reduction of Rh(III) occurs without methanol. In the dark at pH 6.74, nearly 90% of Rh(III) is adsorbed on the semiconductor. Under AM1 simulated sunlight, the amount of Rh(III) adsorbed/reduced is ~ 55%. Though the photoreduction of rhodium(III) is thermodynamically favored the process is slower than the photoreduction of gold(III).

5.6 PALLADIUM(II)

Palladium(II) is photochemically reduced on TiO_2 in acid media (chloride, pH 3.16) [53]. About 68 ppm of Pd(II) (air-equilibrated, 10 vol.% methanol) is completely reduced in < 45 min under AM1 simulated sunlight illumination. O_2 competes with Pd(II) for the e^- but its interference is less marked than for the Rh(III) case. At pH 0 no reduction occurs, even in the presence of methanol and O_2, at least for < 45 min of irradiation. At pH > 0, reduction takes place, optimally in the pH range 3-5.

5.7 PLATINUM(IV)

Kinetic and mechanistic studies have been done [78] for the deposition of Pt on UV-irradiated TiO_2 suspended in aqueous $PtCl_6^{2-}$ solutions under different conditions of light intensities, salt concentrations, and pH among others. Deposition occurs in two stages: a rapid initial process followed by a second slower process in which the rate is less sensitive to process parameters. The main conclusions [78] are (a) the initial rate of Pt deposition varies as the square root of the light intensity, (b) the process quantum yield varies only slightly with initial [$PtCl_6^{2-}$] (for 10-fold increase) and decreases with increasing pH, (c) the ionic strength affects only the slower second stage with greater efficiencies at the higher ionic strengths, (d) both stages of the deposition process are temperature dependent, (e) the presence of O_2 decreases the rate of the initial deposition but has no effect on the slower process, and (f) increasing photocatalyst (TiO_2) loading increases the process quantum yield. From a practical standpoint, the otpimal conditions for the photoreduction and deposition of Pt on irradiated TiO_2 are low pH, low light fluxes, high ionic strength, low oxygen content, and higher than ambient temperatures.

The photocatalyzed reduction and deposition of Pt on TiO_2 using different Pt(IV) complexes {H_2PtCl_6, Na_2PtCl_6, $H_2Pt(OH)_6$, and $Pt(NH_3)_2(NO_2)_2$} have also been examined; for the latter Pt(II) complex the rate of deposition is smaller than for the Pt(IV) species [79].

CdS is also an excellent photocatalyst for the reduction and deposition of Pt [58]. Unfortunately this photocatalyst photocorrodes, particularly during the first 30 min of irradiation; the ratio of Cd^{2+} formed to Pt(IV) reduced is ~ 2, as expected if the photogenerated h^+ reacted with CdS. At longer irradiation time, CdS dissolution is slower owing to Pt deposits on the CdS particle surface which suppress further corrosion.

5.8 LEAD(II)

The photoreduction and deposition of Pb(II), Hg(II), Mn(II), Tl(I), and Co(II) onto irradiated platinized TiO_2 in solutions of their respective salts has been reported [57]. On naked rutile TiO_2, the process efficiency is significantly smaller for lead. No elimination of Pb^{2+} from a 100 ppm solution occurs on irradiated naked TiO_2 anatase (2 g/L) when O_2 is present [61]. Elimination of lead from solution with irradiated Pt-loaded TiO_2 (anatase) takes place only in the presence of O_2. It appears that Pb^{2+} is both reduced to Pb^{o} and oxidized to PbO_2 [57,61]. Figure 10 summarizes the process changes under a variety of conditions.

5.9 MERCURY(II) AND METHYLMERCURY(II)

5.9.1 *Mercury(II) Salts* - Mercury and its salts have attracted much attention as environmental contaminants. Their removal from aqueous media is efficient in air-equilibrated suspensions of TiO_2 on irradiation by AM1 simulated sunlight, on alteration of the pH, and on addition of methanol to the slurry.

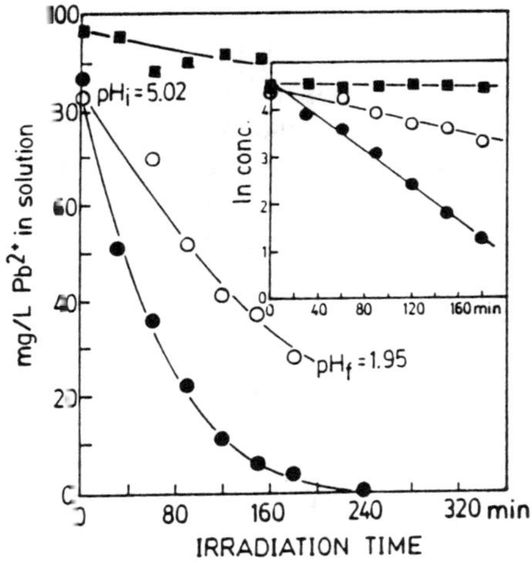

Figure 10.- Plots of mg/L of Pb^{2+} remaining in solution against irradiation time (min) for aqueous TiO_2/5 wt.% Pt dispersions; initial pH 1.4 (unless noted otherwise). Concentration of photocatalyst, 2 g/L. From ref.[61].

The chemistry of Hg(II) is complex and is very sensitive to the nature of the medium. In aqueous media and at natural pH, $HgCl_2$ is 99% undissociated; only 1% of $HgCl_2$ exists as Hg^{2+} ($K_{diss} = 6.0 \times 10^{-14}$). In 1 M Cl^- media, $HgCl_4^{2-}$ predominates; however in 0.1 M Cl^-, the species present are $HgCl_2$, $HgCl_3^-$, and $HgCl_4^{2-}$. Some of these species hydrolyze in water (reaction 15); $[Hg^{2+}] = [Hg(OH)_2]$ at pH 3.0.

The photoreduction of $HgCl_2$ in air-equilibrated TiO_2 aqueous suspensions (initial pH 4.65) occurs efficiently in < 20 min of irradiation (Figure 11). Photoreduction of 0.5 mM of Hg(II) yields 1 mM H^+ as per the stoichiometry of reaction 15; the final pH after total disposal of Hg was 3.0. At

$$Hg(II) + H_2O \xrightarrow[TiO_2]{h\nu} 2\,H^+ + 1/2\,O_2 + Hg(0) \qquad (15)$$

pH 0, no photoreduction takes place in aqueous media and no adsorption of $HgCl_4^{2-}$ is evident from the data of Figure 12. Addition of 20 vol.% methanol leads to total reduction in ca. 3 hrs. At more alkaline pH's, photoreduction occurs but in a very complex way, the result of the complex and varied nature of the chemistry of the mercury species present.

The elimination of Hg(II) from aqueous solutions by photocatalytic action of

UV-irradiated WO_3 [80], ZnO [81], and TiO_2 [57,80] has been reported by others under various conditions. Unlike the process noted for Cu, the photoreduction of Hg(II) over ZnO occurs via complex kinetics: reaction order, 7/5; apparent rate constant, 2.1×10^{-4} $mol^{-2/5}$ $L^{7/5}$ s^{-1} [81]. The process is nearly 4-fold more efficient in O_2-free suspensions and depends on the light intensity and on the medium pH [81]. Use of sunlight also leads to efficient elimination of Hg {92% reduction after 4 hrs for $[Hg^{2+}] = 0.1$ mM}. WO_3 is a poor electrode for the photoreduction of mercury [80]. The ions SO_4^{2-} and NO_3^- ions have no effect on the rate of photoreduction of Hg(II) in TiO_2 aqueous dispersions; However, Cl^- decreases the efficiency of the process because of formation of stable Hg(II)-Cl^- complexes [80]. Addition of ethanol and the absence of O_2 did have a beneficial effect on the reduction rate.

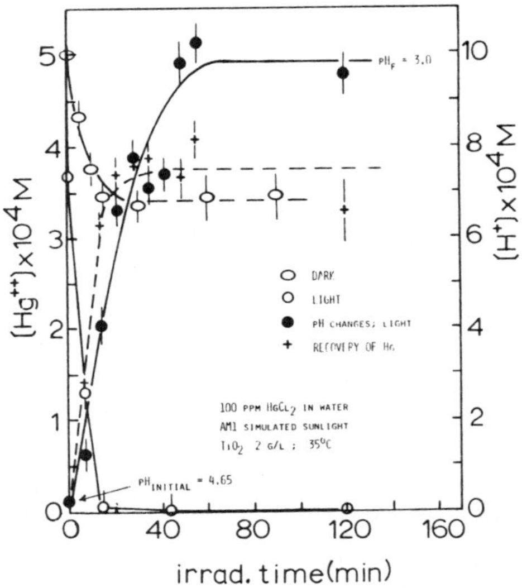

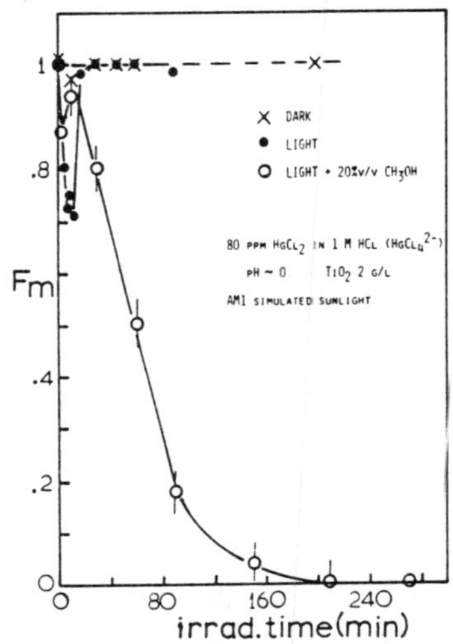

Figure 11.- Photochemical reduction of mercury(II) (100 ppm) on TiO_2 (2 g/l) as a function of irradiation time;temperature 35°C. From ref.[60].
Figure 12.- Photochemical reduction of Hg(II) (80 ppm) on TiO_2 (2 g/l) described by the fraction of metal remaining in solution as a function of irradiation time; AM1 simulated sunlight; slurry contains 20 vol.% methanol. From ref.[60].

5.9.2 *Methylmercury(II)* - The aqueous chemistry of CH_3Hg(II) chloride is mostly dominated by pH-dependent reactions yielding (pH 5.7) the species CH_3HgOH, $(CH_3Hg)_2OH^+$, CH_3Hg^+, and CH_3HgCl. Methylmercury(II) is **not** photoreduced at pH 5.7 in a TiO_2 aqueous suspension. In methanolic aqueous media (Figure 13), photoreduction occurs readily and complete elimination of the CH_3Hg(II) occurs after 30 min of AM1 solar illumination. The stoichiometry of the reduction is indicated in reaction 16

$$CH_3HgCl + H_2O \xrightarrow[TiO_2]{h\nu} 1/2\ O_2 + H^+ + Cl^- + Hg + CH_4 \qquad (16)$$

5.10 SELECTIVE PHOTOREDUCTION AND DEPOSITION OF METALS ON TiO_2

The selectivity of the photochemical reduction process in a TiO_2 dispersion was examined to selectively separate and recover the metals Au, Pt, and Rh from a mixture that might mimic industrial effluents. The important parameters in the reduction step are (i) pH, (ii) O_2 presence, (iii) an electron donor (e.g.,methanol) and (iv) the quantity of TiO_2 used [53].

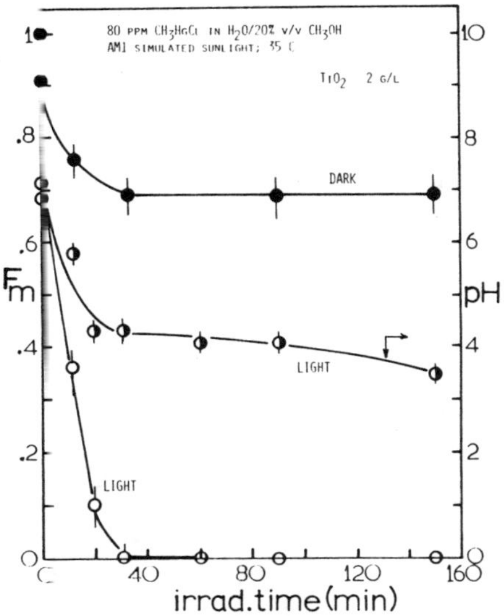

Figure 13.- Photoreduction of 80 ppm $CH_3Hg(II)$ in water/20 vol.% methanol on TiO_2 (2 g/l) powder; 35°C; AM1 simulated sunlight. From ref.[60].

Figure 14 summarizes the results of a typical experiment in which an air-equilibrated mixture consisting of 30 ppm Au(III), 13 ppm Rh(III), and 45 ppm Pt(IV) was irradiated in the presence of 2 g/l TiO_2 at pH 7.39. Pt(IV) was reduced in < 10 min of illumination, Au(III) was reduced within 30 min; in this time frame, Rh(III) remained in solution. The presence of methanol leads to rapid reduction of Au(III) and Pt(IV) in < 5 min; however, Rh(III) is also photoreduced with 32% remaining in solution. The presence of O_2 is crucial in preventing the photoreduction of rhodium within the time and conditions of the experiments of Figure 14.

Selective reduction and recovery can also be obtained in solutions at pH 0. Gold(III) is easily photoreduced in air-equilibrated TiO_2 dispersions (0.2 g/l) containing 10 vol.% methanol, 58 ppm Au(III), 49.4 ppm Pt(IV) and 22 ppm Rh(III). The results are summarized in Figure 15. Gold(III) is reduced in < 30 min of illumination; Pt(IV) and Rh(III) remain in solution. In the absence of methanol, 90 min of irradiation are required to reduce 80% of the initial quantity of gold. This finding is obtained only for the mixture since Au(III) is not photoreduced on TiO_2 at pH 0 without methanol. The rate of reduction of gold in the three-metal mixture is nearly independent of the quantity of TiO_2 employed; limiting rate 2.7×10^{-7} M s^{-1}.

Once Au is reduced and separated from Pt and Rh, the latter two metals can also be reduced and separated. Pt(IV) is selectively reduced, leading to a facile separation of the two metals.

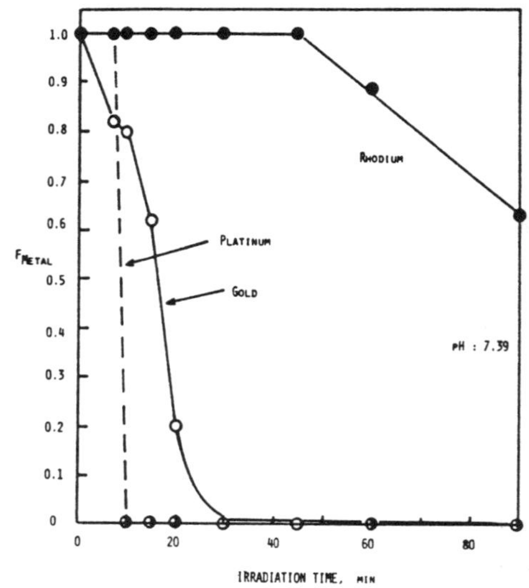

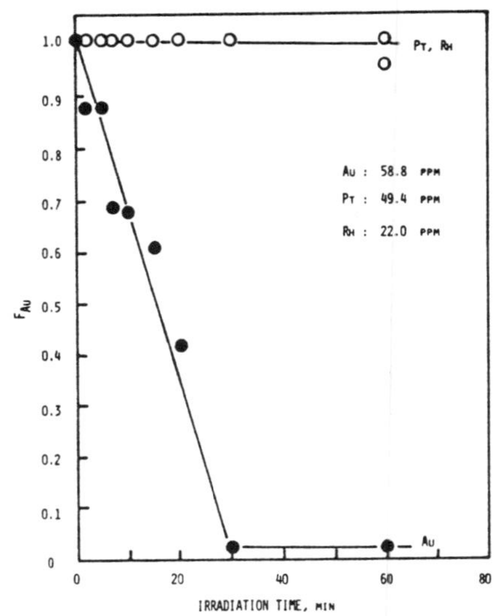

Figure 14.- Photochemical separation of Rh(III) from Au(III) and Pt(IV) on TiO_2 as a function of AM1 simulated sunlight irradiation (air-equilibrated solutions) From ref. [53].

Figure 15.- Photoreduction and recovery of Au(III) (58.8 ppm) from Pt(IV) (49.4 ppm) and Rh(III) (22 ppm) on 2 g/l TiO_2. Irradiation with AM1 simulated sunlight; pH = 0; air-equilibrated solutions with 10 vol.% methanol. From ref. [53].

6. Concluding Remarks

It is clear that photocatalysis mediated by semiconductor materials under illumination by either UV light or by AM1 simulated sunlight (depending on the bandgap of the material) represents an attractive approach to dispose of toxic pollutants from the ecosystem and to retrieve precious and strategic metals from dilute solutions and/or industrial effluents. This developing technology could prove useful in eliminating undesirable toxic metals from waste effluents as a stand-alone technology or in combination with established technologies. An important point worth making is that photocatalysys and pollution abatement are best done at the source.

7. Acknowledgements

Our work is supported by the Natural Sciences and Engineering Research Council of Canada (to NS) and by the Consiglio Nazionale delle Ricerche Rome and by the Regione Piemonte (to EP). We are particularly grateful to the North Atlantic Treaty Organization for an exchange grant (NATO Grant No. CRG-890746).

8. References

[1] A.H. Purcell, *"The Waste Watchers: A Citizen's Handbook for Conserving Energy and Resources"*, Garden City, N.Y., Anchor Books, 1980.

[2] D. Wann, *"Toxic Waste - America is Mobilizing to Meet the Challenge"*, Rocky Mountain News, Denver, CO, October 10, 1987.

[3] J.L. Graham and B. Dellinger, *"Solar Thermal/Photolytic Destruction of Hazardous Organic Wastes"*, *Energy*, vol.12, No.3/4, March-April 1987, pp. 303-310.

[4] R.E. Train, *"Big Questions Facing the Cleanup"*, *EPA Journal*, vol.13, No. 1, January-February 1987, pp.8-10.

[5] C. Turchi and D.F. Ollis, *J.Catal.*, in press (1990).

[6] D. Martinetz, in *"Chemical Waste - Handling and Treatment"*, J. Bromley, J.T. Fraquhar, P.T. Gidley, S. James, D. Martinetz, A. Robin, N.B. Schomaker, R.T. Stephens, and D.B. Walters, Eds., Springer-Verlag, Berlin, 1986, p.113, and references therein.

[7] M. Gratzel, in *"Modern Aspects of Electrochemistry"*, R.E. White, J. O'M. Bockris, and B.E. Conway, Eds., Plenum Press, New York, 1983, p. 83.

[8] K. Kalyanasundaram, M. Gratzel, and E. Pelizzetti, *Coord.Chem.Rev.*, **69**, 57 (1986)

[9] M.A. Fox, *Acc.Chem.Res.*, **16**, 314 (1983).

[10] B.G. Oliver and J.H. Carey, in *"Homogeneous and Heterogeneous Photocatalysis"*, E. Pelizzetti and N. Serpone, Eds., Reidel Publ.Co., Dordrecht, The Netherlands, 1986, p. 629.

[11] J.F. Reber, in *"Photoelectrochemistry, Photocatalysis and Photoreactors"*, M. Schiavello, Ed., NATO ASI Series C146, Reidel Publ.Co., Dordrecht, The Netherlands, 1985, p. 321.

[12] T.N. Veziroglu, W.D. VanVorst, and J.H. Kelly, *"Hydrogen Energy Progress IV"*, Pergamon Press, New York, 1982.

[13] T.N. Veziroglu and A. Taylor, Eds., *"Hydrogen Energy Progress V"*, Pergamon Press, New York, 1984.

[14] E. Pelizzetti, *Energia e Materie Prime*, **17**, 31 (1981).

[15] E. Borgarello, R. Pappa, N. Serpone, and E. Pelizzetti, in *"Photocatalysis and Environment - Trends and Applications"*, M. Schiavello, Ed., NATO ASI Series C237, Kluwer Academic Publ., Dordrecht, The Netherlands, 1988, p. 567.

[16] *Canadian Chemical News*, January 1986.

[17] *Chem.& Eng.News*, August 12, 1985, p.4.

[18] A. Kohl and F. Riensfeld, *"Gas Purification"*, Gulf Publ.Co., Houston, Texas, 1979.

[19] E. Borgarello and M. Gratzel, in ref.12, p. 739.

[20] E. Borgarello, W. Erbs, M. Gratzel, and E. Pelizzetti, *Nouv.J.Chim.*, **7**, 195 (1983).

[21] D.H.M.W. Thewissen, K. Timmer, M. Eemwhorst-Reinten, A.H.A. Tinnemans, and A. Mackor, *Nouv.J.Chim.*, **7**, 191 (1983).

[22] E. Borgarello, J. deSilvestro, M. Gratzel and E. Pelizzetti, *Helv.Chim.Acta*, **66**, 1827 (1983)

[23] N. Serpone, E. Borgarello, and M. Gratzel, *J.Chem.Soc.Chem.Commun.*, 342 (1984).

[24] M. Barbeni, E. Pelizzetti,E. Borgarello, N. Serpone, M. Gratzel, L. Balducci, and M. Visca, *Int.J.Hydrogen Energy*, **10**, 249 (1985).

[25] E. Borgarello, N. Serpone, M. Gratzel, and E. Pelizzetti, *Inorg.Chim.Acta*, **112**, 197 (1986).

[26] E. Borgarello, N. Serpone, E. Pelizzetti, and M. Barbeni, *J.Photochem.*, **33**, 35 (1986).

[27] N. Serpone, E. Borgarello, E. Pelizzetti, and M. Barbeni, *Chim.Ind.(Milano)*, **67**, 318 (1985).

[28] H. Gerischer and M. Lubke, *J.Electroanal.Chem.*, **204**, 225 (1986).

[29] N. Serpone, P. Pichat, J.-M. Herrmann, and E. Pelizzetti, in *"Supramolecular Photochemistry"*, V. Balzani, Ed., NATO ASI Series C214, Reidel Publ.Co., Dordrecht, The Netherlands, 1987, p. 415.

[30] N. Serpone, E. Borgarello, and E. Pelizzetti, *J. Electrochem.Soc.*, **135**, 2760 (1988).

[31] P. Pichat, E. Borgarello, J. Disdier, J.-M. Herrmann, E. Pelizzetti, and N. Serpone, *J.Chem.Soc.Faraday Trans.1*, **84**, 261 (1988).

[32] L. Spanhel, H. Weller, and A. Henglein, *J.Am.Chem.Soc.*, **109**, 6632 (1987).

[33] L.P. Solomonson, in *"Cyanide in Biology"*, B. Vennesland, E.E. Conn, C.J. Knowles, J. Westley, and F. Wissing, Eds., Academic Press, New York, 1981, p. 11; and references therein.

[34] J.L. Way, in ref. 33, p. 29; and references therein.

[35] Q.D. Mehrkam, *Met.Prog.*, **108**, 103 (1975).

[36] H. Krutz, in ref. 33, p. 479; and references therein.

[37] Y. Berube and R. Gilbert, *"Investigation of Water Quality for two Mines in the Yellowknife Area"*, Department of Indian Affairs and Northern Development, Government of Canada, Ottawa, 1971.

[38] G. Leduc, in ref. 33, p. 487; and references therein.

[39] *"Registry of Toxic Effects of Chemical Substances"*, H.E. Christensen and E.J. Fairchild, Eds., U.S. Department of Health, Education and Welfare, Rockville, MD., 20852, 1976; HEW Publ.No. (NIOSH) 76-191.

[40] S.N. Frank and A.J. Bard, *J.Am.Chem.Soc.*, **99**, 303 (1977); *J.Phys.Chem.*, **81**, 1484 (1977).

[41] E. Borgarello, R. Terzian, N. Serpone, E. Pelizzetti, and M. Barbeni, *Inorg. Chem.*, **25**, 2135 (1986).

[42] M. Schiavello, Ed., *"Photoelectrochemistry, Photocatalysis, and Photoreactors"*, Reidel Publ.Co., Dordrecht, The Netherlands, 1985.

[43] L. Friberg, M. Piscator, G.F. Nordberg, and T. Kjellstrom, *"Cadmium in the Environment"*, CRC Press, Cleveland, Ohio, 1974.

[44] O. Masson, *Proc.Chem.Soc.*, **23**, 117 (1907); *J.Chem.Soc.*, **91**, 1449 (1908); W. Henry and R. Wurinser, *C.R.Acad. Sci.*, **284**, 157 (1913); J.H. Matthews and H.A. Curtis, *J.Phys.Chem.*, **18**, 166, 521 (1914); S.G. Siderova, *Farmakol.Toksikol.*, **7**, 39 (1944); *Chem.Abstr.*, **39**, 3843 (1945); J.T. Byrne, W.S. Turnley, and A.K. Williams, *J.Electrochem.Soc.*, **105**, 607 (1958); J. Broucek, *Koroze Ochr.Mater.*, 95 (1962); R.C. Sheridan and E.H. Brown, *J.Org.Chem.*, **30**, 668 (1965); H. Claude and J.C. Boeglin, *Trib.CEBEDEAU*, **331**, 282 (1971); A. Nergarazian and M.C. Merkaham, *J.Phys. Chem.* **67**, 2679 (1963); S.N. Frank and A.J. Bard, *J.Am.Chem.Soc.*, **99**, 4667 (1977); **99**, 303 (1977); *J.Phys.Chem.*, **81**, 1484 (1977); T.L. Rose and C. Nanjundiah, *J.Phys.Chem.*, **89**, 3766 (1985);K. Katsuyuki, Y. Hiroshi, and T. Ideo, *J.Phys.Chem.*, **84**, 1705 (1980); A.T. Kuhn, *J.Appl.Chem.Biotechnol.*, **29**, 21 (1971); A.T. Kuhn in J.O'Brockis and B.E. Conway, Eds., *"Modern Aspects of Electrochemistry"*, vol.8, Plenum Press, New York, 1972, p. 273; W.E. Beckenn, *Electroplat.Met.Finishing*, **25**, 20 (1972); R.H. Cherry, *Finishers Managements*, July 1971, p. 39; B. Drabeik and V. Komendova, *Fortsch. Wasserchem.*, **3**, 103 (1965).

[45] N. Hedley and H. Tabachnick, *"Chemistry of Cyanadation, Mineral Dressing Notes"*, No. 23, American Cyanamid Co. December, 1968.

[46] E. Borgarello, R. Harris, and N. Serpone, *Nouv.J.Chim.*, **9**, 743 (1985).

[47] N. Serpone, E. Borgarello, M. Barbeni, E. Pelizzetti, P. Pichat, J.-M. Herrmann, and M.A. Fox, *J.Photochem.*, **36**, 373 (1987).

[48] A.I. Vogel, *"Macro and Semimicro Qualitative Analysis"*, Longmans, London, 4th edn., 1954.

[49] J.W. Patterson, Director of the Industrial Waste Elimination Research Centre, Illinois Institute of Technology, as quoted by J. Haggin in *Chem.& Eng.News,* September 8, 1986, p.37.

[50] B. Kreautler and A.J. Bard, *J.Am.Chem.Soc.*, **100**, 4317 (1978); E. Borgarello, J. Kiwi, E. Pelizzetti, M. Visca, and M. Gratzel, *J.Am.Chem.Soc.*, **103**, 6324 (1981); E. Borgarello and E. Pelizzetti, *Chim.Ind.(Milano)*, **65**, 474 (1983); J.S. Curran, J. Domenech, N. Jaffrezic-Renault, and R. Philippe, *J.Phys.Chem.*, **89**, 957 (1985); T.M. Tricot and J.H. Fendler, *J.Am.Chem.Soc.*, **106**, 2475 (1984); F. Mollers, H.J. Tolle, and R. Memming, *J.Electrochem.Soc.*, **121**, 1160 (1974); H. Hada, H. Tanemura, and Y. Yonezawa, *Bull.Chem.Soc.Jpn.*, **51**, 3154 (1978); H. Hada, Y. Yonezawa, and M. Saikawa, *Bull.Chem.Soc.Jpn.*, **55**, 2010 (1982).

[51] J.R. Darwent and A. Mills, *J.Chem.Soc.Faraday Trans.II*, **78**, 359 (1978); E. Borgarello and E. Pelizzetti, *Inorg.Chim.Acta*, **91**, 295 (1984); N. Seiichi, O. Bunsho, K. Hiroshi, and K. Tsutumu, *J.Chem.Soc.Faraday Trans.I*, **79**, 2685 (1983); Y. Oosawa and M. Gratzel, *J.Chem.Soc.Chem.Commun.*, 1624 (1984); N.M. Dimitrijevic, Li Shuben, and M. Gratzel, *J.Am.Chem.Soc.*, **106**, 6565 (1984); H. Reiche, W.W. Dunn, and A.J. Bard, *J.Phys.Chem.*, **83**, 2248 (1979); W. Erbs J. DeSilvestro, E. Borgarello, and M. Gratzel, *J.Phys.Chem.*, **88**, 4001 (1984).

[52] E. Borgarello, R. Harris, and N. Serpone, *Nouv.J.Chim.*, **9**, 743 (1985).

[53] E. Borgarello, N. Serpone, G. Emo, R. Harris, E. Pelizzetti, and C. Minero, *Inorg.Chem.*, **25**, 4499 (1986).

[54] N. Serpone, E. Borgarello, M. Barbeni, E. Pelizzetti, P. Pichat, J.-M. Herrmann, and M.A. Fox, *J.Photochem.*, **36**, 373 (1987).

[55] J. Domenech and A. Prieto, *Electrochim.Acta*, **31**, 1317 (1986).

[56] J. Domenech, M. Andres, and J. Munoz, *Tecnologia del Agua,* **29**, 35 (1986).

[57] K. Tanaka, K. Harada, and S. Murata, *Sol.Energy*, **36**, 159 (1986).

[58] J. Domenech, J. Curran, N. Jaffrezic-Renault, and R. Philippe, *J.Chem. Research (S)*, 226 (1986).

[59] J.S. Curran, J. Domenech, N. Jaffrezic-Renault, and R. Philippe, *J.Phys.Chem.,* **89**, 957 (1985).

[60] N. Serpone, Y.K. Ah-You, T.P. Tran, R. Harris, E. Pelizzetti, and H. Hidaka, *Sol.Energy,* **39**, 491 (1987).

[61] D. Lawless, A. Res, R. Harris, N. Serpone, C. Minero, and E. Pelizzetti, *Chim. Ind.(Milano)*, in press (1990).

[62] The colour of the powder during irradiation of Au(III) solutions changes from white (t = 0) to purple indicating the formation of Au(0) colloidal deposits on TiO_2 [see J.W. Mellor, *"A Comprehensive Treatise on Inorganic and Theoretical Chemistry"*, vol.III, Longmans, Green and Co., New York, 1923, p.491.

[63] J. Domenech and J. Munoz, *J.Chem.Res.,(S)*, 106 (1987).

[64] J. Domenech and J. Munoz, *Electrochim.Acta*, **32**, 1383 (1987).

[65] M. Miyake, H. Yoneyama, and H. Tamura, *Bull.Chem.Soc.Jpn.*, **50**, 1492 (1977).

[66] H. Reiche, W.W. Dunn, and A.J. Bard, *J.Phys.Chem.*, **83**, 2248 (1979).

[67] J. Domenech, A. Pietro, and C. Franco, *Affinidad*, **XLIV**, 330 (1987).
[68] A.J. Bard and M.S. Wrighton, *J.Electrochem.Soc.*, **124**, 1706 (1977).
[69] B. Kraeutler and A.J. Bard, *J.Am.Chem.Soc.*, **100**, 5985 (1978).
[70] P.D. Fleischauer, H.K. Alan Kan, and J.R. Shepherd, *J.Am.Chem.Soc.*, **94**, 283 (1972).
[71] B. Ohtani, Y. Okugawa, S. Nishimoto, and T. Kagiya, *J.Phys.Chem.*, **91**, 3550 (1987).
[72] B. Ohtani, M. Kakimoto, H. Miyadzu, S. Nishimoto, and T. Kagiya, *J.Phys. Chem.*, **92**, 5773 (1988).
[73] S. Nishimoto, B. Ohtani, H. Kajiwara, and T. Kagiya, *J.Chem.Soc.Faraday Trans.1*, **79**, 2685 (1983) and references therein.
[74] H. Hada, Y. Yonezawa, and M. Saikawa, *Bull.Chem.Soc.Jpn.*, **55**, 2010 (1982) and references therein.
[75] J.-M Herrmann, J. Disdier, and P. Pichat, *J.Catal.*, **113**, 72 (1988).
[76] H. Hada, H. Tanemura, and Y. Yonezawa, Bull.Chem.Soc.Jpn., **51**, 3154 (1978).
[77] Y. Oosawa and M. Gratzel, *J.Chem.Soc.Chem.Commun.*, 1629 (1984).
[78] J.S. Curran, J. Domenech, N. Jaffrezic-Renault, and R. Philippe, *J.Phys.Chem.*, **89**, 957 (1985).
[79] J.-M. Herrmann, J. Disdier, and P. Pichat, *J.Phys.Chem.*, **90**, 6028 (1986).
[80] J. Domenech and M. Andres, *Gazz.Chim.Ital.*, **117**, 495 (1987).
[81] J. Domenech and M. Andres, *New J.Chem.*, **11**, 443 (1987), and references therein.

DIRECT AND INDIRECT PHOTOTRANSFORMATIONS OF CHLOROPHENOLS AND CHLOROBENZENES IN LABORATORY CONDITIONS RELEVANT TO THE AQUATIC ENVIRONMENT

J. LEMAIRE, P. BOULE, T. SEHILI, C. RICHARD
Laboratory of Photochemistry (URA CNRS 433)
University Blaise Pascal (Clermont-Ferrand II)
63177 Aubière Cedex
France

ABSTRACT. Abiotic degradation through direct and indirect phototransformation must be assessed to predict the fate of xenobiotics when released into the environment. Experimental procedures must supply data which are intrinsic characteristics of the chemicals examined. A procedure for hydrosoluble compounds which absorbed daylight is essentially based on quantum yield determination. Identification of photoproducts is easier in photoreactions with ionic intermediates than with radicalar photoreactions. A procedure for hydrosoluble compounds, which do not absorb daylight, is proposed which is based on the selection of precise sources for oxygenated reactive species. Examples of indirect phototransformations are given. ZnO-photocatalyzed conversion of dichlorobenzenes and chlorophenols and NO_3^-- and NO_2^--photoinduced conversion of phenol, nitrophenols and dichlorophenols are presented.

International regulations, like for example the European Communities 6th Amendment to the 1967 Directive on the Classification, Labelling and Packaging of Dangerous Substances, require a better prediction of the fate of xenobiotic chemicals released, accidentally or on purpose, into the three compartments of the environment. Such a prediction can only be based on well-defined experiments in laboratory conditions. Actual evolutions in natural conditions afford localized data which are very difficult to adapt to another natural situation.

In the aquatic compartment, non-volatile and hydrosoluble xenobiotic chemicals which are not biodegradable can be removed either through hydrolysis or through direct or indirect phototransformation. Where hydrolysis prevails, the lifetime of the compounds in the aqueous phase can be derived from classical kinetic determinations at concentrations for which no interactions between solute molecules are observed. A relevant hydrolysis rate must be checked to be concentration-independent, it

E. Pelizzetti and M. Schiavello (eds.), Photochemical Conversion and Storage of Solar Energy, 477–495.

is not really necessary to determine the hydrolysis rate at the very low concentrations of environmental conditions.

For hydrosoluble and non volatile chemicals compounds , not hydrolysed and not-biodegraded, phototransformation can be considered as a potential route for abiotic degradation. A correct evaluation of phototransformation as natural depollution must included data on photodegradability and, when possible, identification of the main primary photoproducts ; the fate of these primary photoproducts has to be determined to evaluate the real benefit of the photoconversion. Considering the actual difficulties of the assessment of mass balance in dilute conditions, international regulations are more demanding on photodegradability than on the nature of photoproducts. Nevertheless cases in which photoproducts are more toxic and more photostable than the parent substance can be observed.

At the present time, the organic compounds whose fate is controlled by phototransformation should be classified in 3 groups, considering the various experimental procedures which are very inequally difficult and therefore not necessary generally accepted. The 3 followings groups should be considered :

<u>Class A</u> : hydrosoluble compounds absorbing daylight
<u>Class B</u> : hydrosoluble compounds, which do not absorb daylight
<u>Class C</u> : insoluble compounds (or compounds with low solubility in water).

Experimental procedures for the characterization of class A compounds are now widely accepted [1]. Experimental procedures for class B compounds are still the subject of controversies and discussions. Experimental procedure for class C compounds is in a very preliminary status.

DIRECT PHOTOTRANSFORMATION OF CLASS A COMPOUNDS

For class A compounds, the lifetime in environmental conditions can be computed from :

- a- the direct phototransformation quantum yield ;
- b- the absorption spectrum of the xenobiotic compound under investigation ;
- c- the spectral distribution of daylight and the absorption spectra of the aquatic phases.

The conversion quantum yield in the aqueous phase is measured in laboratory conditions relevant to natural exposure and its variations with the excitation wavelength (in monochromatic light), initial concentration of the solute, and light intensity must be determined in presence of dissolved oxygen. To be relevant, the quantum yield $\Phi(\lambda)$ must be checked to be concentration- and light intensity-independent. The phototransformation quantum yield $\Phi(\lambda)$ is an intrinsic evaluation of the photodegradability of the chemical compound which is independent on experimental conditions and which can be used as a common base for inter-laboratories comparison.

The lifetime is computed as :

$$\tau = \frac{1}{2303 \int_{\lambda_1}^{\lambda_2} \Phi(\lambda) . I_0 (\lambda) . \varepsilon(\lambda) \, . \, d\lambda} \qquad (\tau_{1/2} = \tau \ln 2)$$

where - λ_1 and λ_2 are the limits of the overlap between the absorption spectrum of the compound and the daylight spectrum ;

- I_0 (λ) is the incident daylight in the different layers of the aquatic compartment (considering the opacity of natural media) ;
- $\varepsilon(\lambda)$ is the molar extinction coefficient.

The kinetic analysis should be completed for class A compounds by the identification of the major primary photoproducts. It is known from experience that phototransformations in aqueous media imply complicated stoichiometries. Many primary photoproducts can be formed and secondary photoreactions can occur even at low conversion degree. These general features are specially true in phototransformations occuring through radical processes. In such mechanisms, loss of specificity is generally observed and, in most cases, a material balance is very difficult to establish. Moreover, the material balance is quite impossible to ascertain with photoreactions characterized by a quantum yield lower than 10^{-2}.

A new trend is now appearing in environmental aqueous photochemistry. Water as a polar solvent, can favor photochemical processes which imply polarized (zwiterrionic) or ionic excited states. In such processes, a better specificity is expected. Water is indeed a very unusual solvent in organic photochemistry, and very few examples of specific photoreactivities in water have been reported before 1980 [2,3]. Such isolated examples underlined, nevertheless, the special role of the aquatic media. More recently, other examples of specific photoreactions in the aqueous phase have been reported.

- ring photocontractions of 2-halogenophenolates have been observed with 2-chlorophenolate (pH>9), 2-bromo- and 2-fluorophenolate, 2,4-dichlorophenolate, 2-chloro-6-methylphenolate, 2,4-dichloro 6-methyl phenolate. The conversion quantum yields are generally high (0.40 for 2-bromophenolate) and the reaction is very specific [4,5,6].

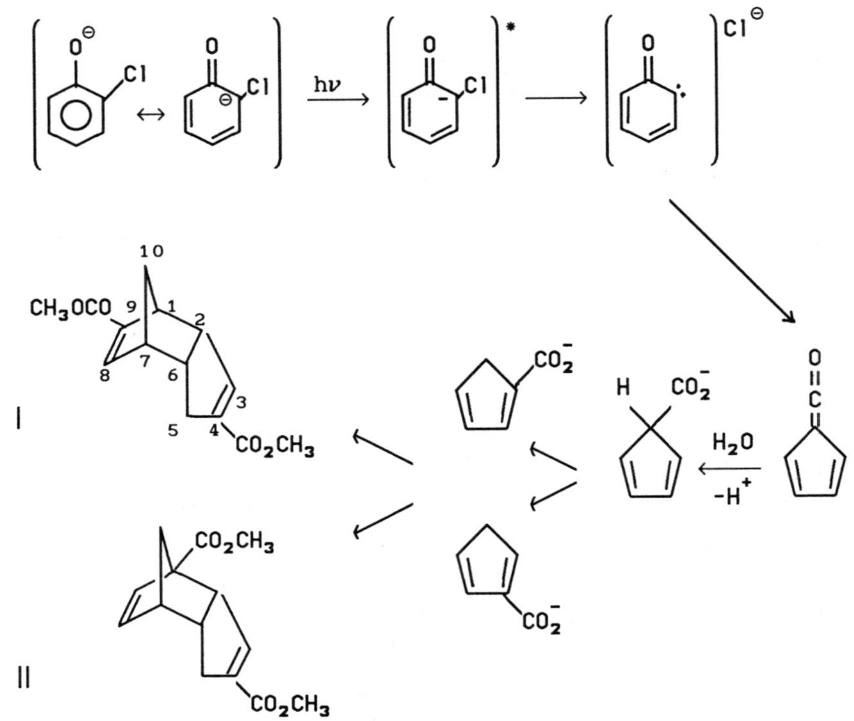

The ring contraction can be observed with the molecular forms, the corresponding yields are lower than with phenolates (0.01 vs 0.30 for 2-chlorophenol) and the reaction is not specific : the formation of catechol is also observed.

• photohydrolysis of 3-chloro, 3,5- and 3,4-dichlorophenol, monochlorobenzene, di- and trichlorobenzenes, which have been observed to occur at any pH (1-10), is a specific substitution of chlorine by an OH group [7-11]. These photoreactions could not be identified to photosubstitutions which would be occuring in basic media. The proposed mechanism was based on heterolytic scission of the polarized CCl bond in the excited state and fast hydrolysis of the cation formed (concerted or not).

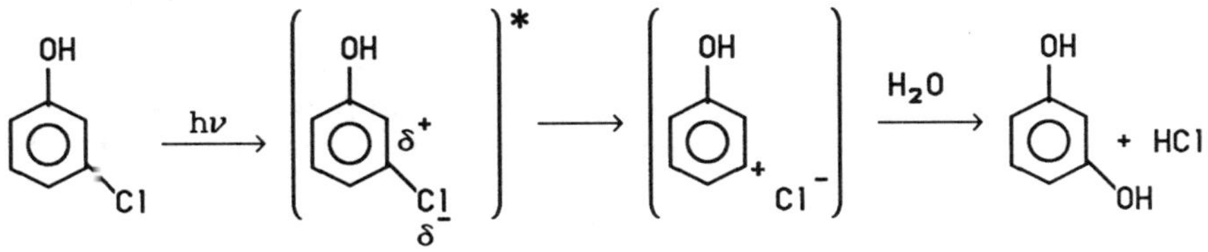

It should be noted that the photohydrolysis quantum yield decreased as the number of chlorine atoms on the ring increases and when substituents are located on conjugated sites. This can be rationalized considering the electro-attracting influence of chlorine atoms.

• <u>photocyclization</u> can be specific too in the aqueous phase. Photocyclization of diphenylamine derivatives into carbazole derivatives has been observed several years ago [2]. Very recently photocyclisation of *o*-phenylbenzoquinone into 2-hydroxydibenzofuran has been reported [12].

INDIRECT PHOTOTRANSFORMATIONS OF CLASS B COMPOUNDS

In natural media, indirect phototransformation of a well-defined xenobiotic compound implies that daylight is absorbed by an external chromophore which can be ill-defined. This is a main reason of the difficulty to assess the importance of indirect transformation in laboratory conditions. Indeed from excitation of external chromophores in the aquatic compounds *i.e.* excitation of molecular organic or inorganic compounds (other micropollutants, absorbing anions, coordination complexes ...), of natural macromolecules or solid crystals (solid particules in suspension), photosensitization, photoinduction or photooxidation can account for the conversion of the compound under scope. Real photosensitizations involving energy transfer are not very usual in environmental conditions, since donor molecules must present a large number of prerequisites (specially high energy and long-lived triplet levels). Real photoinductions which account for the faster mutual conversion of two micropollutants are on opposite very usual. However, it is not an easy recommandation to examine a large number of pollutants couples to determine the photodegradability of one single compound.

The only indirect phototransformation, which is at the present time worthy to handle in laboratory conditions, concern the oxidation of the compounds by reactive oxygenated species like •OH, $O_2^*(^1\Delta_g)$, $O_2^{\bullet -}$ and $HO_2^{\bullet}$ and H_2O_2 which are formed photochemically on external chromophores. Two different approaches are proposed to examine the indirect photooxidation of xenobiotics and these approaches differ fundamentally.

In the first approach, rather proposed by the specialists interested in the practical aspects of environmental degradations, it is intended to characterize the reactive oxygenated species present in the natural media by specific reagents and to measure their actual concentrations The fate of a precise compound is therefore predicted from these concentrations and from the rate constants measured in laboratory conditions using specific sources for the corresponding reactive species. Indeed, if specific reagents and specific sources can be found for very reactive species like •OH, they cannot be easily defined for less reactive species like O_2^*, $O_2^*(^1\Delta_g)$ and $O_2^{\bullet -}$.

In the second approach, modelization in laboratory conditions is proposed using identified and relevant sources of reactive oxygenated species. On exposure to daylight or to artificial light, absorption by these sources can be controlled and quantified and the indirect phototransformation quantum yields precisely measured. The following relevant sources can be recommanded :

- photoactive oxides like TiO_2, ZnO, ZnS ;
- absorbing anions like NO_3^- and NO_2^-.

Reference experiments can be based too on the use of metallic cations like Fe(III), Cu(II), which are complexed by non-absorbing ligands and which, under UV or visible exposure, are involved in the coordinated state in photo-oxydoreduction processes accounting for the conversion of the ligand.

EXAMPLES OF INDIRECT PHOTOTRANSFORMATION THROUGH ZnO-PHOTOCATALYZED CONVERSION

The oxidative transformation of **dichlorobenzenes** photocatalyzed by ZnO in aqueous solution has been studied using high performance liquid chromatography (HPLC) and gas chromatography-mass spectrometry (GC-MS) coupling [13]. The ZnO suspensions were irradiated at 365 nm (half-width 50 nm), at 2 g/l (which was the optimal concentration), at 15°C or 20°C.

The conversion quantum yields at initial time were found to be around 0.05 (0.046±0.008 for 1,2- and 1,4-dichlorobenzene, 0.038±0.008 for 1,3-dichlorobenzene) and kinetics show that reaction occurs mainly on the ZnO surface. It was indeed hardly possible to distinguish between primary and secondary photoproducts. Indeed, adsorption and desorption kinetics are slow compared with the photocatalytic oxidation and the kinetics of the formation of most photoproducts are explained by successive reactions of hydroxyl radicals with the same molecule being progressively transformed in the adsorbed phase. Competition between the starting material and the photoproducts for the •OH radical scavenging would account for the self-inhibition on the dichlorobenzene conversion.

Through addition of ethanol (10^{-1} mol.l^{-1}), it was possible to inhibed totally the conversion of any dichlorobenzene.

A complex mixture of photoproducts was obtained when a solution of 1,3-dichlorobenzene (3×10^{-4}mol.l^{-1}) was irradiated in the presence of ZnO (2 g.l^{-1}). The following products were identified by HPLC analysis using the following reference compounds : 2,4- and 2,6-dichlorophenol, chlorohydroquinone, 4-chlorocatechol and chlorobenzoquinone. The formation of 2,6-dichlorohydroquinone was established by GS-MS (m/z=178-180-182 in good agreement with the ratio ^{37}Cl/^{35}Cl ; main fragments, 142, 114, 86 and 79) and by 300 MHz NMR. The formations of tetrachlorobiphenyl, trichlorohydroxybiphenyl, dichlorodihydroxybiphenyl and dichlorotrihydroxybiphenyl were identified on GC-MS chromatograms. The mass balance obtained from a solution irradiated for 10 min, up to a conversion equal to 0.23 and filtered to eliminate ZnO, is given below in terms of the percentage of 1,3-dichlorobenzene transformed.

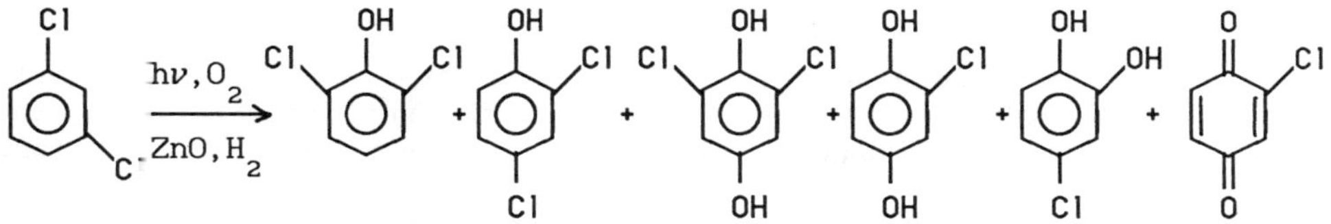

Conversion Extent = 0.23   0.9%   10.5%   17%   18.5%   1.5%   5%

The total mass balance, considering the products present in the aqueous phase and the absorbed photoproducts accounts for more than 57% of the converted 1,3-dichlorobenzene. To appreciate the influence of adsorption/desorption kinetics, a mass balance was calculated using a solution of the same concentration (3×10^{-4} mol.l^{-1}), and then irradiated ten times for 1 min and stirred for 10 min between each irradiation. The concentrations of photoproducts in the filtered solution, given below in per cent of transformed dichlorobenzene, were somewhat higher than in the first experiment.

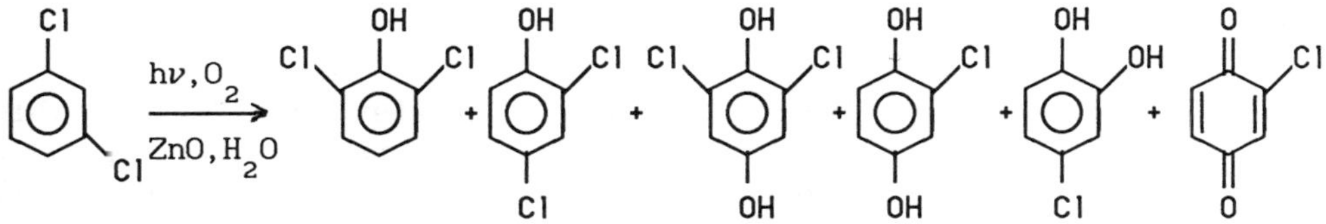

2%   11%   21%   20%   2%   6%

A solution of 1,2-dichlorobenzene (4.6×10^{-4} mol.l^{-1}) and a solution of 1,4-dichlorobenzene (1.5×10^{-4} mol.l^{-1}) were continuously irradiated for 10 min in the presence of ZnO (2 g.l^{-1}) and the products analysed in the solution and on ZnO after washing with methanol. The following products were obtained and titrated :

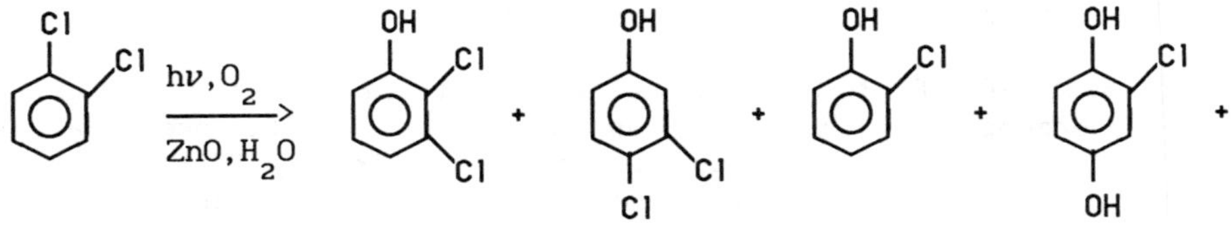

Conversion	in sol:	3.6%	3.8%	1.3%	5.9%
Extent=0.24	on ZnO:			1.3%	3.2%

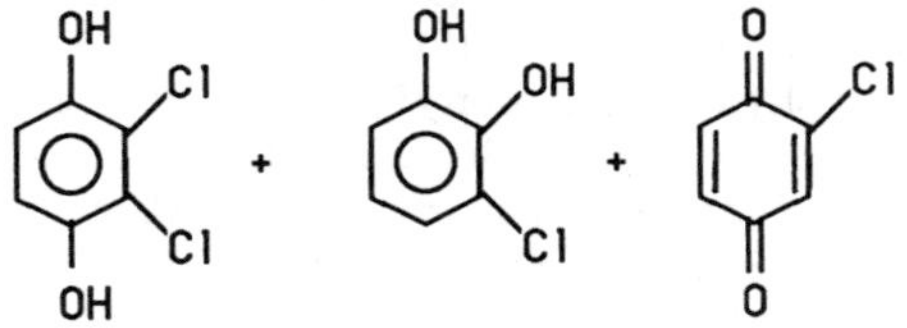

in sol:	32%	0.8%	2.7%
on ZnO:	2.7%		0.8%

2,3-dichlorohydroquinone was isolated with preparative HPLC and identified by MS and NMR spectra. An authentic sample was prepared using the method of Peratoner and Genco [14] to obtain a quantitative titration. The total mass balance, including dissolved and adsorbed photoproducts, accounts for more than 60% of the 1,2-dichlorobenzene transformed.

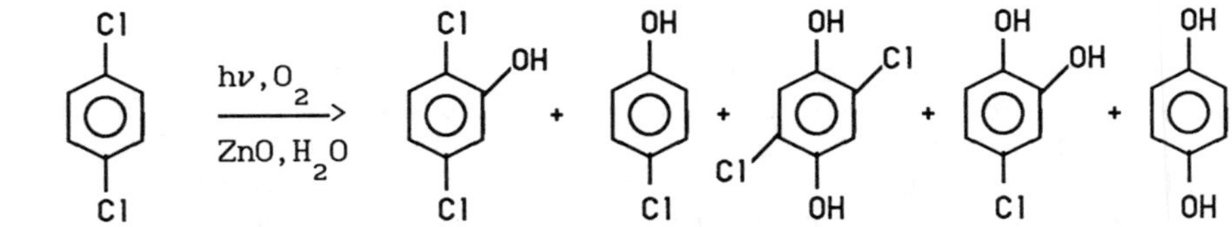

Conversion	in sol:	10%	1.2%	5%	2.5%	31%
Extent=0.24	on ZnO:			7.2%	3.8%	

The total mass balance accounts for more than 60% of 1,4-dichlorobenzene transformed. In contrast with 1,3-dichlorobenzene, substitution of a chlorine atom can occur with 1,2- and 1,4-dichlorobenzene.

When excited with radiation of wavelengths shorter than 388 nm (this wavelength corresponds to the energy of the forbidden gap, *i.e.* 3.2 eV), ZnO can have oxidative photocatalytical properties. It is generally agreed that the electron of the conduction band is scavenged at least partially by oxygen and leads to the formation of H_2O_2 [15,16].

$$ZnO + h\nu \longrightarrow h^+ + e^-$$

$$e^- + O_2 \longrightarrow O_2^{\bullet -}$$

$$O_2^{\bullet -} + H^+ \rightleftharpoons HO_2^{\bullet} \quad (pKa = 4.75)$$

$$HO_2^{\bullet} + O_2^{\bullet -} \longrightarrow HO_2^- + O_2$$

$$HO_2^- + H^+ \rightleftharpoons H_2O_2 \quad (pKa = 11.6)$$

Kormann *et al.* [17] have evaluated the quantum yield of H_2O_2 formation to be 0.15 with oxygen-saturated ZnO colloïds irradiated at 330 nm. The positive holes h^+ can react with water and be converted into hydroxyl radicals. It is also possible that a "surface hydroxyl group" or lattice oxygen O^{2-} is oxidized by the hole, accounting for •OH production [18]. Hydroxyl radicals are strongly oxidizing species and react with most of the aromatic compounds through a diffusion-controlled process [19,20]. Other authors have suggested that positive holes can be directly involved in a charge transfer process with the adsorbed substrate [21].

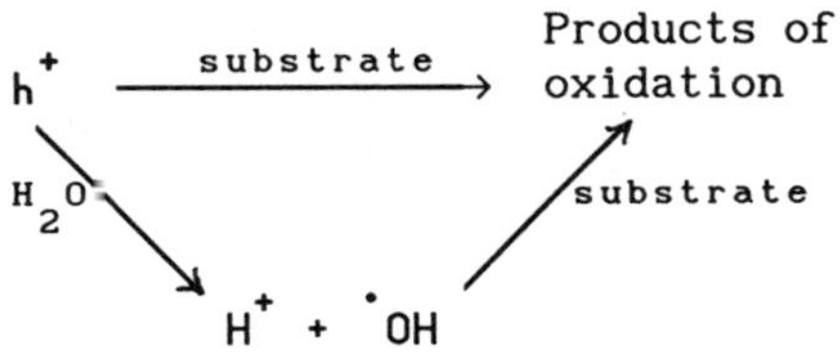

In the above mechanism, two photons are required for the formation of one molecule of H_2O_2, and the two corresponding positive holes lead to the formation of two radicals. The initial conversion rate of dichlorobenzene is not significantly different from the formation rate of H_2O_2. The initial transformation of one molecule of dichlorobenzene involves two hydroxyl radicals.

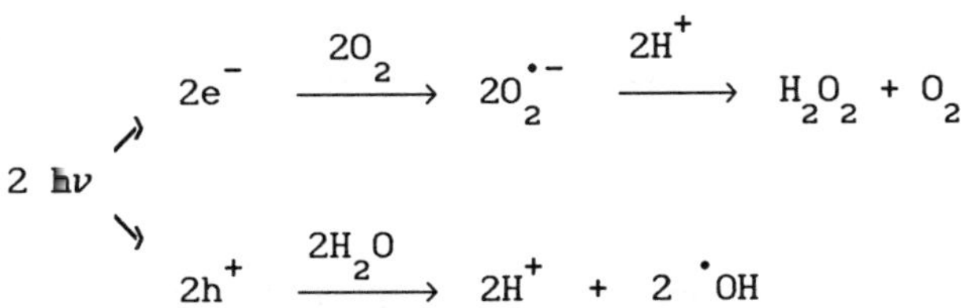

The primary products formed from 1,3-dichlorobenzene are the result of hydroxylation in either the *ortho* or the *para* position with respect to the chlorine atoms. This orientation corresponds to the electrophilic behaviour of hydroxyl radical.

The formation of an intermediate adduct •OH-dichlorobenzene can be suggested according to the results of Ebert and coworkers on the radiolysis of phenol [22] and 4-nitrophenol [23], and of Getoff and Solar on chlorophenols [24]. This adduct R• can react with •OH or disproportionate. However, using the value given bo Cercek and Ebert for the rate constant of disproportionation of the •OH-nitrophenol adduct (10^9 $mol^{-1}.l.s^{-1}$) [23], it is probable that disproportionation of R• is much faster than its oxidation by •OH.

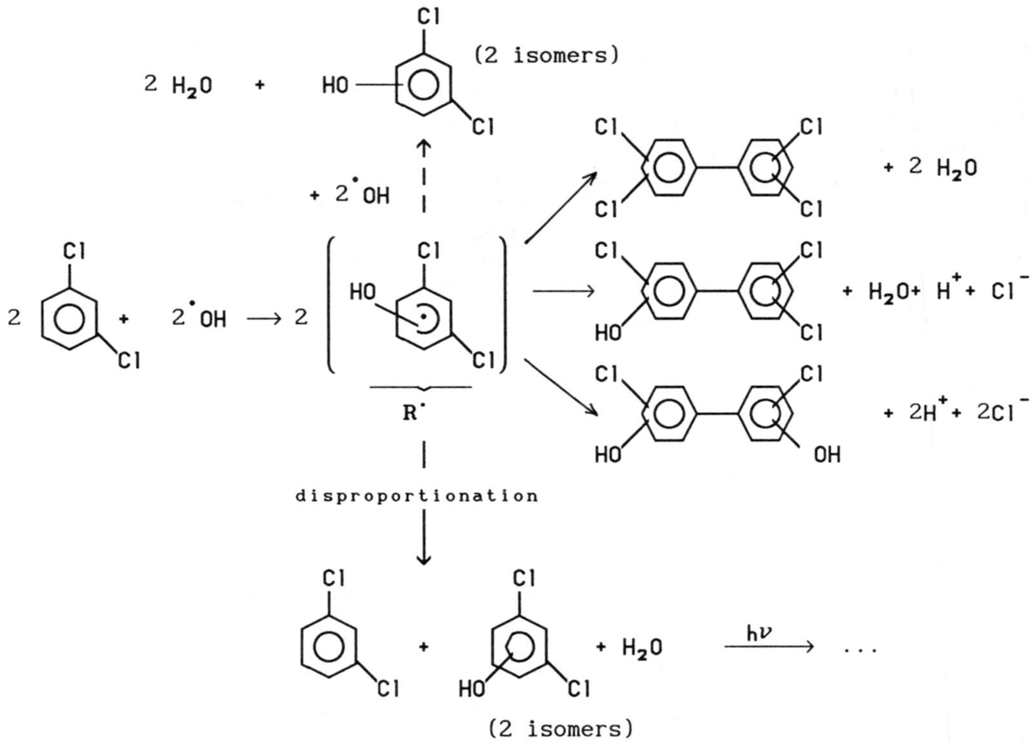

With 1,4- and 1,2-dichlorobenzene similar hydroxylations lead respectively to the formation of 2,5-dichlorophenol and a mixture of 2,3- and 3,4-dichlorophenol. However, substitution of a chlorine atom by •OH can also occur.

The oxidative transformation of **2-, 3- and 4-chlorophenol** photocatalyzed by ZnO in aqueous solution has been studied too using the same experimental techniques and conditions (λ_{exc} = 365 nm, 2 $g.l^{-1}$, 15-20°C).

The initial quantum yields of indirect phototransformations in the presence of ZnO were found to be 0.04 with the three chlorophenols. The quantum yields increase with increasing pH ($\Phi \cong 0.09$ at pH=12). The initial rate of H_2O_2 formation was not significantly different from the initial rate of chlorophenol conversion. The H_2O_2 formation was a linear function of irradiation time whereas the transformations rates decrease with increasing extent of the transformation. This phenomenon is again attributed to the presence of adsorbed photoproducts on ZnO.

When a solution of 2-chlorophenol (5.5×10^{-4} $mol.l^{-1}$) was irradiated for 10 min in the presence of ZnO, four photoproducts, namely chlorohydroquinone, chlorobenzoquinone, catechol and 3-chlorocatechol, were identified by HPLC in the filtered solution using reference compounds. Complementary analysis using GC-MS coupling after dissolution of ZnO with H_2SO_4 and ether extraction confirmed the identification. It may be noted that 3-chlorocatechol can also be formed from 3-chlorophenol and 1,2-dichlorobenzene, but not from 4-chlorophenol. The mass balance given in the following scheme in per cent of transformed 2-chlorophenol was obtained from the filtered solution.

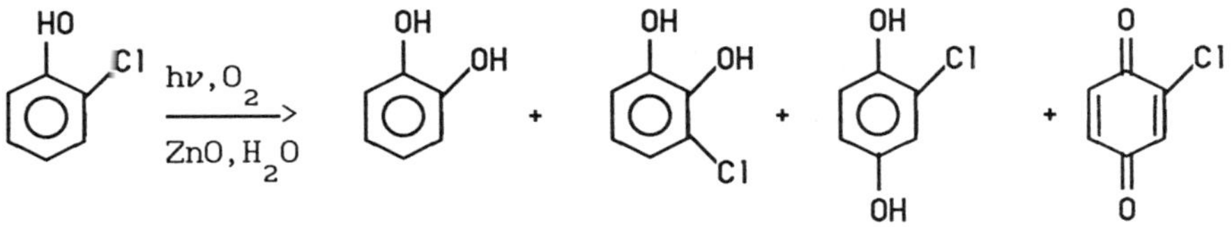

transf. extent 14.5% 1% 6% 49% 6.5%

The products absorbed on ZnO were extracted with methanol. Chlorohydroquinone and catechol accounted for 2.5% and 1.6% of the transformed 2-chlorophenol respectively. Thus, the total mass balance accounts for 66.6% of the transformation. In contrast with the case of dichlorobenzenes, no biphenyl derivative was detected by GC-MS after dissolution of ZnO and ether extraction. It appears that the concentrations of the four photoproducts in solution reach a maximum value after an irradiation time shorter than 30 min.

To evaluate the role of hydroxyl radical in the oxidation of chlorophenols, solutions were irradiated in the presence of ZnO with ethanol. It appears that the photocatalyzed transformations of 2- and 3-chlorophenols were about 70% inhibited by ethanol 0.08% v/v (*i.e.* 1.4×10^{-2} $mol.l^{-1}$) and the transformation of 4-chlorophenol was about 60% inhibited. The inhibition ratio unchanged for ethanol concentrations as high as 2% v/v. It was noticed that the formation of chlorohydroquinone and chlorobenzoquinone was inhibited less than that of catechol and 3-chlorocatechol. In contrast, with dichlorobenzenes the transformation is quantitatively inhibited by ethanol about 1% v/v. It may be concluded therefore that two different pathways were involved in the photocatalysed transformation of chlorophenols.

The photoproducts formed from a solution of 4-chlorophenol and from a solution of 3-chlorophenol (4.1×10^{-4} mol.l^{-1}) irradiated for 10 min in the presence of ZnO (2 g.l^{-1}) were analysed in the same manner using HPLC and GC-MS. The mass balances obtained by HPLC from the filtered solutions and after the washing of ZnO with methanol are given below. In both cases no biphenyl derivative was detected by GC-MS.

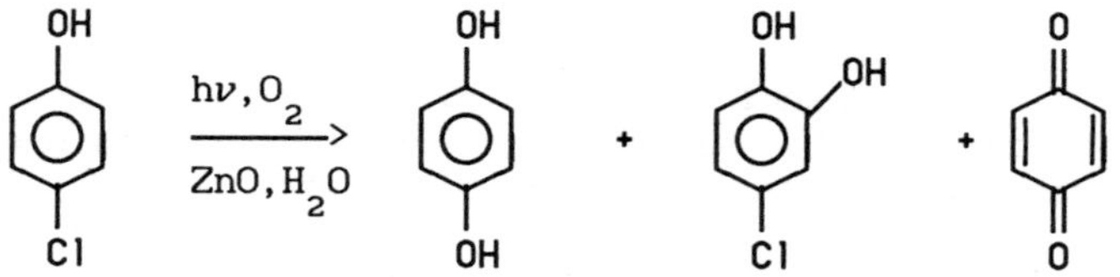

transf. extent 30% — in sol: 9%, 16%, 1.4%; on ZnO: 5.3% } total 31.7%

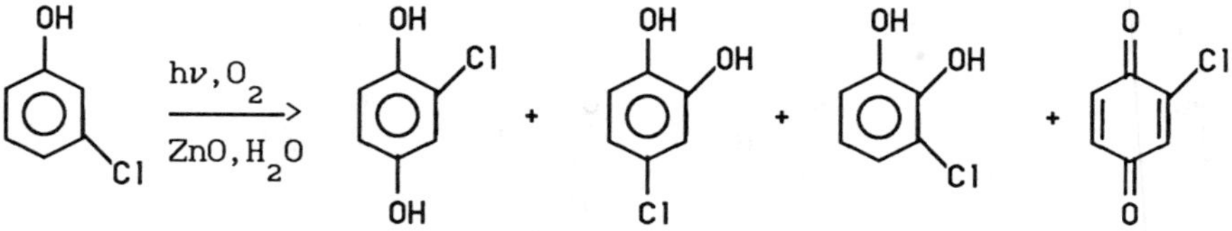

transf. extent 30% — in sol: 32%, 10%, 9%, 8%; on ZnO: 1.8%, 0.8%, 3.3%

The total mass balance accounts for 64.9% of the transformed 3-chlorophenol. In contrast with 2- and 4-chlorophenol no substitution of the chlorine atom was observed with 3-chlorophenol.

The main pathway (about 65%), inhibited by ethanol, is similar to the mechanism of photocatalysed transformation of dichlorobenzenes. Oxidation by hydroxyl radicals is suggested. The formation of one H_2O_2 molecule results from the disproportionation of two superoxide ions after protonation of one of them.

The transformation of one molecule of chlorophenol needs two hydroxyl radicals. The hydroxylation occurs in the *ortho* or *para* position with respect to the phenol function, since this site corresponds to the highest electron density on the molecule. The formation of an intermediate adduct between the •OH radical and 2-chlorophenol was observed by Getoff and Solar [24] and it may be assumed that the same reaction occurs with 3- and 4-chlorophenol. This adduct can disproportionate or can be oxidized into chlorodiphenol.

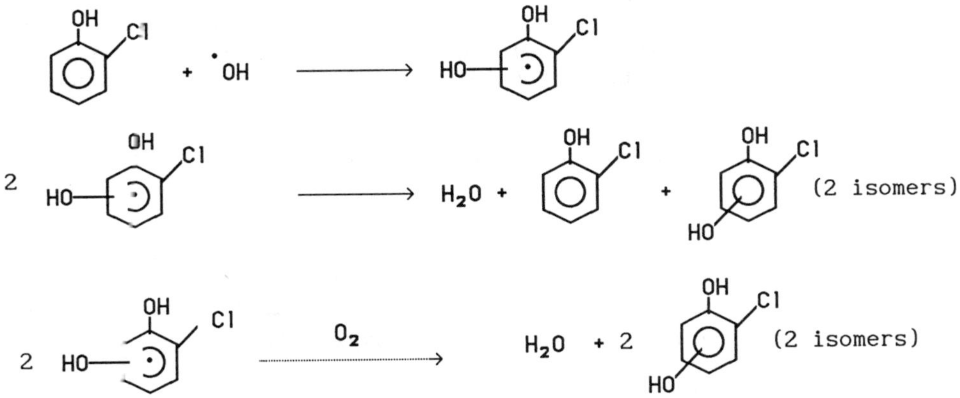

If the rate constant of disproportionation is similar to the rate constant calculated by Cercek and Ebert [23] for 4-nitrophenol (10^9 $mol^{-1}.l.s^{-1}$), the disproportionation is the main pathway. With 2- and 4-chlorophenol, where one of the reactive sites is chlorinated, a dechlorination can occur :

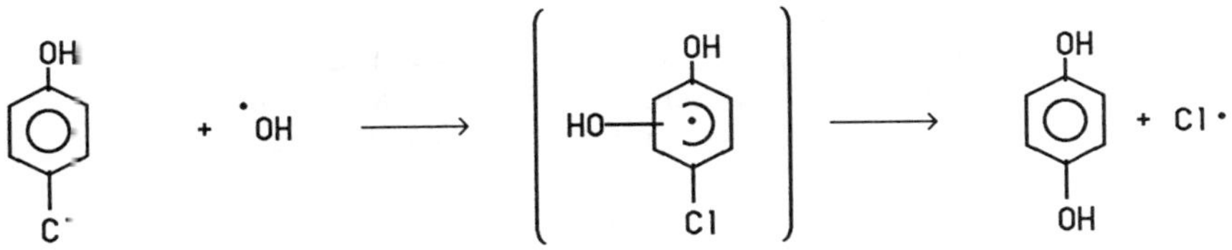

The chlorine radical is probably converted into a chloride ion on the ZnO surface.

Hydroxylation is the main primary reaction obtained when a suspension of ZnO is excited in the presence of chlorophenols, as observed with the dichlorobenzenes. This electrophilic reaction occurs preferentially in the *ortho* and *para* positions with respect to the phenol function. Substitution of a chlorine atom can occur when the *ortho* or *para* sites are chlorinated. It is a minor pathway with 2-chlorophenol. With each chlorophenol the hydroxylation parallels the formation of H_2O_2.

The photocatalysed oxidation results in reactions clearly different from the direct phototransformations. It was previously pointed out that the behaviour of chlorophenols under direct excitation depends on the position of the chlorine atom in the ring. In the photocatalytic transformation on ZnO, the oxidation occurs mainly in the adsorbed phase and the kinetics of adsorption and desorption play an important role on the mass balance. Primary photoproducts can be reoxidized several times before desorbing and this phenomenon lowers the transformation rate or the starting compound. From the partial inhibition by ethanol, it may be suggested that hydroxylation by •OH radicals is the main but not the only process involved in the photocatalysed oxidation of chlorophenols.

EXAMPLES OF INDIRECT PHOTOTRANSFORMATION THROUGH NO_3^- AND NO_2^- PHOTO-INDUCED CONVERSION

Nitrate ions and nitrite ions absorb daylight. The maxima of their UV-absorption bands are respectively located at 302 and 354 nm with molar extinction coefficients equal to 7 and 22 $mol^{-1}.l.cm^{-1}$. Under irradiation, they dissociate with formation of hydroxyl radicals :

$$NO_2^{-*} \longrightarrow \cdot NO + O^{\bullet -} \quad (I)$$

$$NO_3^{-*} \longrightarrow \cdot NO_2 + O^{\bullet -} \quad (II)$$

$$O^{\bullet -} + H^+ \rightleftharpoons \cdot OH \quad (pK = 12)$$

The quantum yield of the reaction I is wavelength dependent. It is lower at 365 or 370 nm (≃0.015) than at 300-320 nm (0.07-0.08) or at 254 nm (0.11) [25,26]. This phenomenon does not appear in the excitation of nitrous acid at pH≤3.4. The quantum yield was evaluated to be about 0.45 in the range 254-365 nm. Wavelength effect was also observed with nitrate ions which are involved in a competitive dissociation :

$$NO_3^{-*} \longrightarrow NO_2^- + O \quad (III)$$

Warneck and Wurzinger [27] estimated the quantum yields of processes II and III at 310 nm to be respectively 9.2×10^{-3} and 1.1×10^{-3}. The formation of •OH is about 8 times higher than the formation of O . Atomic oxygen is less reactive than hydroxyl radical, both in gas phase [28] and in aqueous solution [29] and it is generally admitted than hydroxyl radicals are involved in the induced oxidations. Oxidations are not the only reactions observed : when nitrate ions were excited in the presence of **phenol**, many photoproducts were identified [30] :

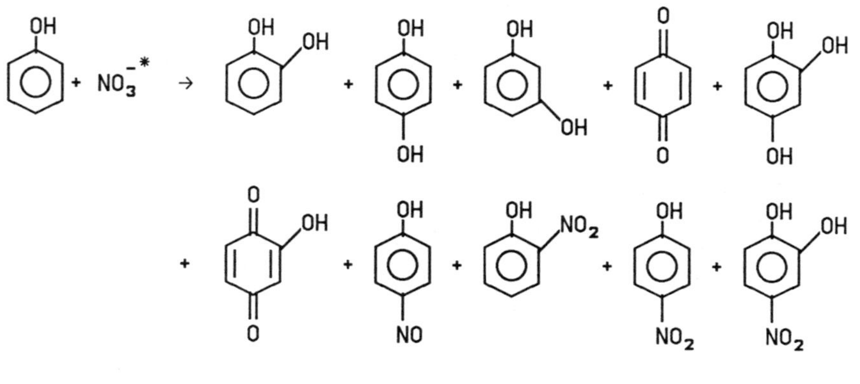

These photoproducts can be separated into oxidation, nitrosation and nitration products. It was observed that the ratio (nitrosation + nitration) / (oxidation) increases with decreasing pH. Nitrocatechol was observed to be secondary product formed from catechol. Benzoquinone, 1,2,4-trihydroxybenzene and hydroxybenzoquinone were formed mainly from hydroquinone which did not accumulate in the solution. This phenomenon was attributed to direct excitation of the hydroquinone formed. Indeed hydroquinone absorbs in the same range than nitrate ions, and its quantum yield of oxidation is drastically enhanced by the presence of nitrate ions.

The behaviours of other phenolic compounds were studied in order to determine the influence of substituents on the oxidation of the reaction. **Hydroxybiphenyls** behaves almost like phenols, but no formation of nitroso derivative was observed and phenylbenzoquinone photocyclized into 2-hydroxydibenzofuran :

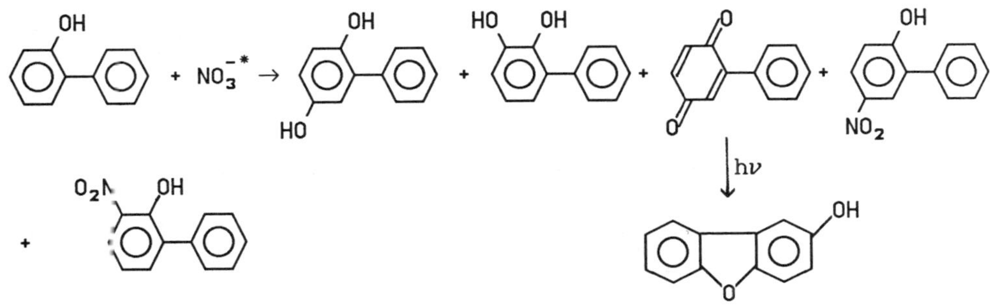

With **chlorophenols**, the main products formed from the excitation of nitrate ions are chlorodihydroxybenzenes. No formation of nitro or nitroso derivative was observed :

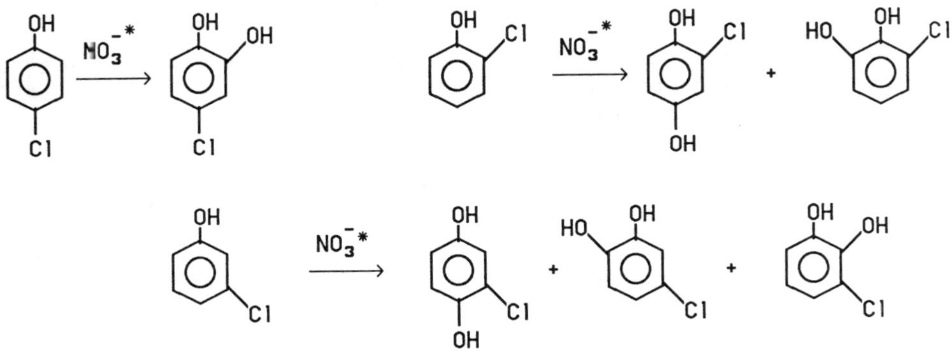

The behaviour of **nitrophenols** was studied, in the presence of nitrate ions excited at 254 or at 365 nm. The only products resulted from a hydroxylation in *ortho* or *para* positions with respect to the phenol function. Similarly to chlorophenols, 4-nitrocatechol was formed from 4-nitrophenol, nitrohydroquinone and 3-nitrocatechol from 2-nitrophenol, nitrohydroquinone, 3-nitro- and 4-nitrocatechol from 3-nitrophenol.

With **nitrite ions**, it was observed that the quantum yield of the reaction is higher in acidic solution than in neutral medium ; it increased with increasing nitrophenol concentration or decreasing nitrite concentration. The influence of the concentrations can be explained by the following competition :

$$\begin{cases} \cdot OH + NO_2^- \longrightarrow \cdot NO_2 + OH^- \text{(IV)} & k=10^{10}\ mol^{-1}.l.s^{-1} \\ \cdot OH + \text{nitrophenol} \longrightarrow \text{complex} & k\simeq 5\times 10^{9}\ mol^{-1}.l.s^{-1} \end{cases} \quad [20]$$

Then the induced oxidation of nitrophenol is self-inhibited by nitrite ions and the highest yield is obtained for the lowest ratio [nitrite]/[nitrophenol].

A more detailed study was carried out with **resorcinol** in order to determine the mechanisms involved in the reaction. When nitrate ions (5.10^{-2} M) were excited in the presence of resorcinol (5.10^{-4}M) in unbuffered distilled water, the formations of products of oxidation, 4-nitrosoresorcinol and 4-nitroresorcinol, were observed.
2,4-dinitrosoresorcinol was formed in a further stage of the reaction.

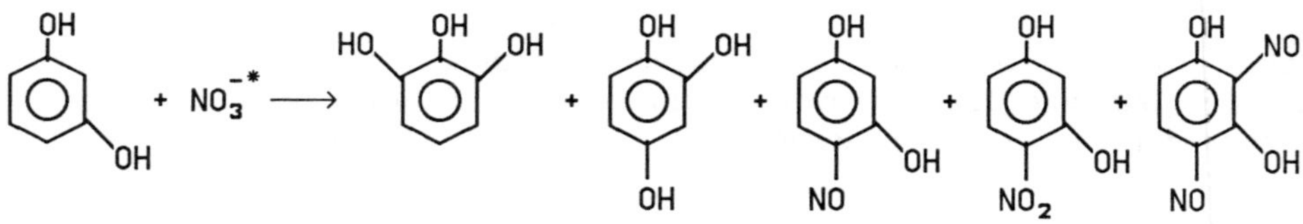

A solution was irradiated, at the same pH, in the presence of formate ions (5.10^{-2}M) to eliminate hydroxyl radicals. The oxidations were completely inhibited, but the nitrosation was only slightly affected. It can be concluded that hydroxyl radicals are not involved in the nitrosation. On the other hand, it was established that nitrogen dioxide or a mixture NO_2/NO induce the formation of 4-nitrosoresorcinol at pH=7-7.5, *i.e.* in a range where HNO_2 is not formed. This reaction does not occur with pure NO dissolved in the solution. Then the following mechanism can be suggested :

$$NO_3^{-*} \longrightarrow \cdot NO_2 + O^{\bullet -}$$

$$2\ \cdot NO_2 \rightleftharpoons N_2O_4 \rightleftharpoons (NO^+ \ldots NO_3^-)$$

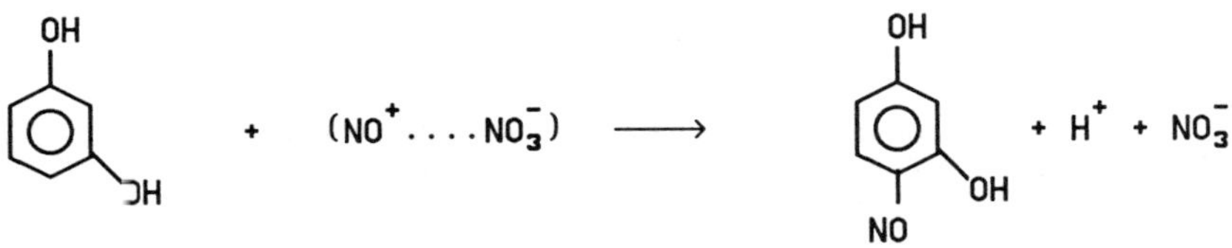

$$N_2O_4 \rightleftharpoons (NO_2^+ \ldots NO_2^-) \quad \text{minor pathway}$$

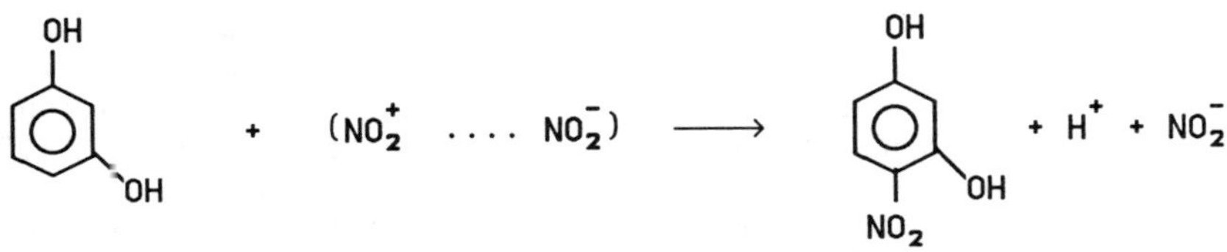

The enhancing effect of decreasing pH on nitration and nitrosation may be attributed to the influence of pH on the competitive disproportionation of nitrogen dioxide in water.

$$2\ \cdot NO_2 + 2\ OH^- \longrightarrow NO_2^- + NO_3^- + H_2O$$

When nitrite ions were excited in the presence of resorcinol, 4-nitrosoresorcinol was the only photoproduct detected, and the reaction was slightly inhibited by formate ions. The absence of trihydroxybenzenes can be explained by the reaction of hydroxyl radicals with nitrite ions

CONCLUSIONS

The relative importance of direct and indirect phototransformations of xenobiotics should be assessed in relevant and well controlled laboratory conditions. Experimental procedures should supply data which characterize intrinsically chemical compounds. Direct determinations of the fate of chemicals in natural media which are helpful for control, are too localized to allow the prediction of the fate in other environmental conditions.

For class A compounds, the quantum yield of the phototransformation is one of these reference data which can be measured in any laboratory. For class B compounds, the quantum yield of indirect photottrasnformation is depending on the reference sources for oxygenated reaction species.

These sources should be defined and their relevancy controlled in natural media. Evaluation of the fate of chemicals with low solubility is in a very preliminary state and procedure cannot be yet recommanded.

REFERENCES

1- European Chemical Industry Ecology and Toxicology Center (ECETOC) (1984)'The phototransformation of chemicals in water: results of a ring-test.' 12th Report Brussels

2- Joussot-Dubien, J. and Houdard J. (1967),'Reversible photolysis of pyridine in aqueous solution.' Tetrahedron Lett. 44, 4389-4391

3- Lopez, D., Boule, P. and Lemaire, J. (1980),'Photochimie et Environnement.I-Comportement photochimique de la diphénylamine et d'anti-oxydants dérivés en solution aqueuse.' Nouveau J. Chimie 4, 615-620

4- Guyon, C., Boule, P. and Lemaire, J. (1982),'Photochimie et Environnement. III-Formation d'acides cyclopentadiéniques par irradiation du chloro-2 phénol en solution aqueuse.' Tetrahedron Lett. 23, 1581-1584

5- Guyon, C., Boule, P. and Lemaire, J. (1984),'Photochimie et Environnement. IX-Photocontraction du cycle aromatique par irradiation des halogéno-2 phénols en solution aqueuse ou alcoolique.' Nouveau J. Chimie 8, 685-692

6- Boule, P., Guyon, C. and Lemaire, J. (1984),'Photochimie et Environnement. VIII-Comportement photochimique des dichlorophénols en solution aqueuse diluée.' Chemosphere 13, 603-612

7- Tissot, A., Boule, P. and Lemaire, J. (1983),'Photochimie et Environnement. V-Photohydrolyse du monochlorobenzène en solution aqueuse diluée.' Chemosphere 12, 859-872
Tissot A. (1984), Docteur-Ingénieur Thesis,'Phototransformation directe, sensibilisée ou induite de chlorobenzènes en solution aqueuse diluée.' University of Clermont II, Clermont-Ferrand, France

8- Tissot, A., Boule, P. and Lemaire, J. (1984),'Photochemistry and Environment. VII-The photohydrolysis of chlorobenzene:studies of the excited state involved.' Chemosphere 13, 381-389

9- Boule, P., Tissot, A. and Lemaire J. (1985),'Photochimie et Environnement. XI-Règles d'orientation de la photohydrolyse de dérivés aromatiques chlorés.' Chemosphere 14, 1789-1793

10- Boule, P., Guyon, C. and Lemaire, J. (1982), 'Photochemistry and Environment. IV-Photochemical behaviour of monochlorophenols in dilute aqueous solution.'Chemosphere 11, 1179-1186

11- Boule, P., Guyon, C. and Lemaire, J. (1984),'Photochemistry and Environment. VI-Direct phototransformation of chlorophenols and interactions with phenol on UV exposure in aqueous solution.' J. Toxic. and Environ. Chem. 7, 97-110

12- Sarakha, M., Rossi, A., Guyot, G. and Boule, P. (1988),'Photocyclisation de la phénylbenzoquinone.' C.R. Ac. Sci. 307 Serie II, 739-744

13- Sehili, T., Boule, P. and Lemaire J. (1989),'Photocatalysed transformation of chloroaromatic derivatives on zinc oxide II- dichlorobenzenes.' J. Photochem. Photobiol. 50, 103-116

14- Peratoner, W. and Genco (1894), Gazz. Chim. Ital. 24 II, 377, 388, 394, Beilsteins Handbuch Org. Chem. 6, II, 845

15- Munuera, G., Rives-Arnau, V. and Saucedo, A.' (1979),'Photo-adsorption and photo-desorption of oxygen on highly hydroxylated TiO_2 surfaces. Part 1-Role of hydroxyl groups in photo-adsorption.' J. Phys. Chem. 75, 736-747

16- Fujihira, M., Satoh, Y. and Osa, T.,(1982) 'Heterogeneous photcatalytic reactions on semiconductor materials. III.Effect of pH and Cu^{2+} ions on the photo-Fenton type reaction.'J. Chem. Soc. Japan (1982), 55, 666-671

17- Kormann, C., Bahnemann, D.W. and Hoffmann, M. R. (1988),'Photocatalytic production of H_2O_2 and organic peroxides in aqueous suspensions of TiO_2, ZnO and desert sand.' Environ. Sci. Technol. 22, 798-806

18- Harbour, J.R., Hair, M.L. (1979),'Radical intermediates in the photosynthetic generation of H_2O_2 with aqueous ZnO dispersions.'J. Phys. Chem. 83, 652-656

19- Klöpffer, W., Kaufmann, G. and Frank, R. (1985),'Phototransformation of air pollutants: rapid test for the determination of $k_{OH\bullet}$.' Z. Naturforsch., Teil A 40, 686-692

20- Buxton, G.V., Greenstock, C.L., Helman, W.P. and Ross A.B. (1988),'Critical review of rate constants for reactions of hydrated electrons, hydrogen atoms and hydroxyl radicals ($\bullet OH/\ O^{\bullet-}$) in aqueous solution.' J. Phys. Chem.Ref. Data, 17, 513-886

21- Hashimoto, K., Kawai, T. and Sakata, T. (1984),'Photocatalytic reactions of hydrocarbons and fossil fuels with water, hydrogen production and oxidation.' J. Phys. Chem. 88, 4083-4088

22- Land, E.J. and Ebert, M. (1967),'Pulse radiolysis studies of aqueous phenol.' Trans.Faraday Soc. 63, 1181-1190

23- Cercek, B. and Ebert, M. (1968),'Radiolytic transients from p-nitrophenol and their inter- and intramolecular reactions.' Adv. Chem. Ser. 81, 210-221

24- Getoff, N. and Solar, S. (1986),'Radiolysis and pulse radiolysis of chlorinated phenols in aqueous solutions.' Radiat. Phys. Chem. 28, 443-450

25- Zafiriou, O.C., Bonneau, R. (1987),'Wavelength-dependent quantum yield of OH radical formation from photolysis of nitrite ion in water.'Photochem. Photobiol. 45, 723-727

26- Alif, A. (1990), Thesis,'Phototransformations directe et indirecte des nitrophénols en solution aqueuse.' Clermont-Ferrand, France

27- Warneck, P., Wurzinger, C. (1988),'Product quantum yields for the 305nm photodecomposition of NO_3^- in aqueous solution.' J. Phys. Chem. 92, 6278-6283

28- Huie, R.E. and Herron, J.T. (1975),'Reactions of atomic oxygen with organic compounds.' Prog. Reac. Kinet. 8, 3-80

29- Vysotskaya, N.A. (1973), Russian Chem. Rev. 42, 851

30- Niessen, R., Lenoir, D. and Boule, P. (1988),'Phototransformation of phenol induced by excitation of nitrate ions.' Chemosphere 17, 1977-1984

PHOTOCHEMICAL CONVERSION OF SOLAR ENERGY IN THE ENVIRONMENT

Richard G. Zepp
Environmental Research Laboratory
U.S. Environmental Protection Agency
Athens, Georgia 30613

ABSTRACT. Past research on photochemistry in the environment has focused on vapor-phase reactions in the atmosphere. Recently, however, environmentally significant photoreactions have been discovered in natural waters (i.e. the sea, lakes, rivers), on soil surfaces, and in atmospheric condensed phases. These new investigations have been stimulated in part by interest in developing a scientific understanding of the role of photochemical processes in the biogeochemical cycles of various elements. In addition, other studies have explored the role of natural photochemical processes in cleansing the environment of various waste materials or, in some cases, in converting the wastes to more toxic substances. In this paper, current research results on the photochemical conversion of solar energy in aquatic environments and on soil and metal oxide surfaces are presented. Rate equations and products for selected homogeneous and heterogeneous photoreactions that occur in these systems are described. Data are presented for direct and sensitized photoreactions and for sunlight-initiated free radical reactions.

1. INTRODUCTION

1.1. Historical Perspective

The origins of the field of photochemistry date back to a period in which sunlight was the only available light source for photochemical studies. For example, pioneering research was conducted at the turn of this century in a photochemical laboratory located on the rooftops of the University of Bologna (Heindel and Pfau, 1965; Balzani and Moggi 1990). In sunlight experiments at the Bologna laboratory, the great Italian scientist, Giacomo Ciamician, conducted photochemical studies that laid the groundwork for organic photochemistry. Interestingly, Ciamician recognized the potential of solar energy conversion (Ciamician, 1912) at about the same time that the Swedish chemist, Svante Arrhenius (1903), was warning that the widespread use of fossil fuels would cause significant global warming through what today is widely referred to as the "greenhouse effect." Both of these topics continue to be of great interest as we approach the end of this century. With the advent of artificial light sources, the great bulk of photochemical studies during the middle part of the 1900s no longer used sunlight due to its variable intensity and polychromatic nature.

Interest in the photochemical conversion of solar energy in the environment has been rekindled since the 1960s through the emergence of various environmental problems. First, it was recognized that photoreactions play an important role in smog formation (Leighton, 1961), stratospheric ozone depletion (Rowland and Molina, 1975), acid rain (Calvert et al., 1985), and the cleansing of the aquatic and

E. Pelizzetti and M. Schiavello (eds.), Photochemical Conversion and Storage of Solar Energy, 497–515.

terrestrial environment of the waste materials derived from human activities (Zafiriou et al., 1984). Second, recent studies indicate that photoreactions in the sea have significant effects on the cycling of carbon, oxygen, sulfur, and various trace metals that are biologically important (Zafiriou et al., 1984). Finally, as foretold by Ciamician, concerns over global climate change and depletion of fossil sources of energy have stimulated interest in developing photochemical systems that can be used to harvest solar energy.

1.2. Rates and Types of Environmental Photoreactions

The photochemical conversion of solar energy in the environment can be quantified through the use of rate equations. The rate (conversion per unit time) of any type of photoreaction depends upon the rate of light absorption by the photoactive species, I_a, and the quantum efficiency of the reaction, ϕ_r (eq. 1).

$$\text{Rate} = I_a \, \phi_r \tag{1}$$

The rate of light absorption is affected, in turn, by the spectral overlap between sunlight and the spectrum of the chemical that initiates the photoreaction. The quantum yield for reaction is an important parameter that describes the fraction of absorbed radiation that results in reaction. When the substrate is very dilute, the rate is usually directly proportional to substrate concentration, that is, a pseudo first order rate expression applies (Zepp and Cline, 1977; Zepp, 1982) .

The light absorption rate depends both on the solar spectral irradiance as it is transmitted into the environmental medium (e.g. water, soil) and on the geometry of the light field (Zepp and Cline, 1977; Miller and Zepp, 1979). Natural substances in the photic zone influence the irradiance, and thus photolysis rates, through both light absorption and light scattering (Miller and Zepp, 1979; Baker and Smith, 1982). The variation of the solar spectral irradiance and photolysis rates with time and location can be described using computer programs such as GCSOLAR (Zepp and Cline, 1977; Zepp et al., 1987b). (GCSOLAR is available on request.)

Environmental photochemical processes can be subdivided into two main categories--direct and indirect. Direct photolysis processes involve the direct absorption of light by the substrate.

$$\text{P} + \text{light} \longrightarrow \text{P}^* \longrightarrow \text{products} \tag{2}$$

In eq. 2, P is some photoreactive substance and P* is the excited state that reacts to form a product. On the other hand, photoreactions also can involve indirect photoprocesses that are initiated through light absorption by another substance that is in the system with the substrate. Chromophores that initiate indirect photoreactions are symbolized here by "S." Such substances are sometimes referred to as photosensitizers. Indirect photoreactions occur through the intermediacy of various short-lived reactive transients, such as excited states, S*, or species produced therefrom, I, that react with the substrate (eqns. 3-5).

$$S + \text{light} \longrightarrow S^* \qquad (3)$$

$$S^* + P \longrightarrow \text{products} \qquad (4)$$

and/or

$$S^* \xrightarrow{(O_2)} I \xrightarrow{+P} \text{products} \qquad (5)$$

Oxygen often is involved in the formation of the reactive transients. These reactions involve electronic energy transfer, electron transfer, and free radical reactions that are mediated by these transients (Zafiriou et al., 1984; Hoigné et al., 1989; Cooper et al., 1989; Zepp, 1988). Recent studies have shown that indirect photoreactions are an important pathway for the degradation of xenobiotics and for the geochemical cycling of elements in aquatic environments. Components of dissolved organic matter are the most important chromophores involved in indirect photoreactions in the sea and freshwaters, although certain inorganic substances such as nitrate, nitrite, and trace metals also are involved in these reactions.

In the next three sections, photoreactions of dissolved organic and inorganic substances in natural waters are discussed. Then, recent research on heterogeneous reactions is considered, including reactions on soil and vegetative surfaces, metal oxide redox reactions, and reactions in atmospheric condensed phases.

2. PHOTOREACTIONS OF NATURAL ORGANIC MATTER

2.1. Overview of Investigative Procedures

Research conducted during the past few years indicates that a significant portion of the photochemistry that occurs in aquatic environments and on soil surfaces is associated with non-living organic matter. This organic matter is a mixture of substances derived mainly from the decay of terrestrial vegetation or of marine algae in the open sea. A large portion of these organic compounds are humic substances (Thurman, 1985). The term "humic substances" is used in this paper to refer to the colored organic matter that has been isolated from natural waters or from soils. The term "dissolved organic matter" (DOM) is used to denote the organic matter in natural waters when it has not been isolated. This section describes spectroscopic, laser flash photolytic, and continuous irradiation methods that have been used to investigate the excited state processes that occur when natural organic matter absorbs light.

2.2. Spectroscopic Properties

The first law of photochemistry, originally formulated by Grotthus and Draper, states that light must be absorbed in order for photochemical reactions to occur. With this principle in mind, most studies of natural water photochemistry have strongly emphasized the colored organic matter that strongly absorbs sunlight. Approximately half of the organic matter and nearly all of the colored organic matter in freshwaters are humic substances (Thurman, 1985). The colored organic matter in sea water is often referred to as "Gelbstoffe" by marine scientists. The electronic absorption spectra of humic substances and DOM typically are structureless, as illustrated in Figure 1.

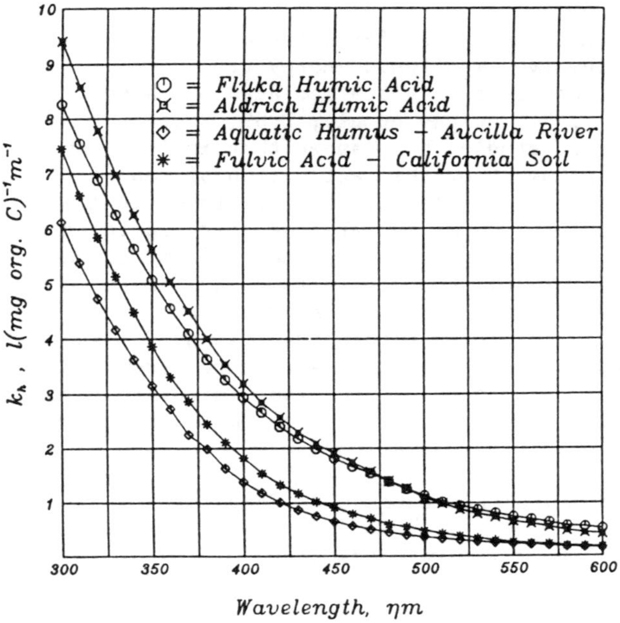

Figure 1. Electronic absorption spectra of humic substances (Zepp and Schlotzhauer, 1981).

The absorption coefficient, $k_h(\lambda)$, of humus or DOM solutions at wavelength λ can be represented by an equation of the following form.

$$k_h(\lambda) = A\ e^{B(1-\lambda)}\ 1\ (\text{mg org. C})^{-1}\ m^{-1} \qquad (6)$$

where A is defined as the absorption coefficient (m^{-1}) of the solution of humic substance or DOM at 450 nm and B is the slope of a natural log plot of $k_h(\lambda)$ versus wavelength (nm). The spectra of humic substances and of colored natural water samples are remarkably similar considering the diversity of vegetative and algal precursors from which they are derived. Fulvic acids isolated from soils and the DOM in wetlands, lakes, rivers, and coastal areas have similar values of B, typically about 0.014. Carder et al. (1989) have related spectral properties of humic substances to their molecular weights. Generally, as average molecular weight of the humic substance increases, absorptivity in the visible region also increases and the change in absorptivity with wavelength decreases. Fluorescence spectra also are similar, as are fluorescence quantum yields, which typically are low (on the order of 0.1 -1.0%) (Zepp and Schlotzhauer, 1981).

2.3. Direct Photoreactions

That the organic matter from soils and in water bodies is photoreactive has been established by several recent studies (Kouassi, 1986; Kouassi et al., 1986; Frimmel and Bauer, 1987). Evidence for such photoreactivity derives from observed changes in the electronic absorption spectrum of the organic matter as well as in the fluorescence intensity and spectrum. Fading is generally retarded in the absence of dioxygen (Gjessing, 1976; Frimmel and Bauer, 1987). Both our observations and those of Gjessing (1976) indicate that the sunlight-induced decrease in absorbance is not accompanied by a corresponding change in dissolved organic carbon content (DOC). Other studies showed that sunlight fades the fluorescence intensity of humic substances more rapidly than the absorbance (Kouassi, 1986; Kouassi et al., 1986). Results of Ertel (1990) indicate that photooxidation of the lignin component of DOM helps account for these spectral changes. The photoproducts of the biologically refractory DOM in sea water include low molecular weight compounds that are both biologically labile, e.g. formaldehyde, acetaldehyde, and the alpha-keto acid glyoxylate (Kieber et al., 1989) as well as volatile, such as carbon monoxide (Conrad et al., 1982; Gammon and Kelly, 1988) and carbonyl sulfide (Ferek and Andreae, 1984; Zepp and Andreae, 1989) (Figure 2).

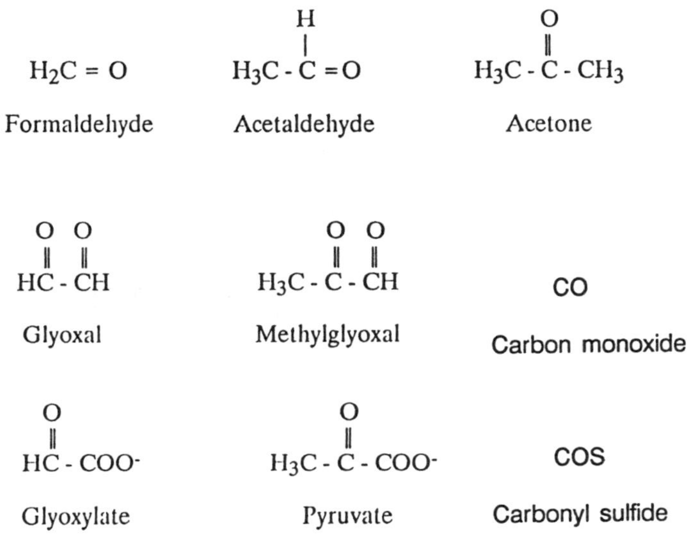

Figure 2. Photoproducts of marine dissolved organic matter (Kieber et al., 1989; Zepp and Andreae, 1989; Valentine and Zepp, 1990).

Photoproduction rates of the carbonyl compounds and carbonyl sulfide (COS) are greatest for inland and coastal waters and least for open ocean waters. Mopper and Kieber (1990) have found that the rate of production of the carbonyl compounds was linearly related ($r^2 > 0.98$) to initial absorbance at 300 nm and initial fluorescence (excitation 360 nm; emission 460 nm). Solar ultraviolet radiation is most effective at inducing the formation of these photoproducts (Zepp and Andreae, 1989; Kieber et al., 1989; Valentine and Zepp, 1990), as illustrated by the action spectrum for the COS photoreaction in a water sample from the North Sea (Figure 3). The formation of COS involves the photosensitized oxidation of organosulfur compounds in seawater.

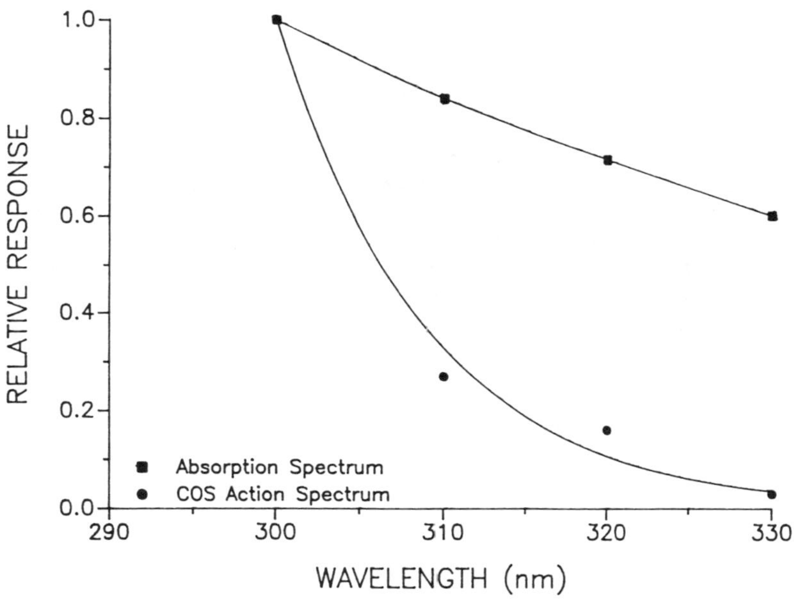

Figure 3. Action spectrum for the photochemical formation of carbonyl sulfide in coastal North Sea water (Zepp and Andreae, 1989).

2.4. Primary Photochemical Processes

Various studies described in the following section have identified and quantified the primary photochemical processes of natural organic matter. An abbreviated scheme that illustrates the typical partitioning of the photons absorbed by DOM in the ultraviolet region is shown in Figure 4.

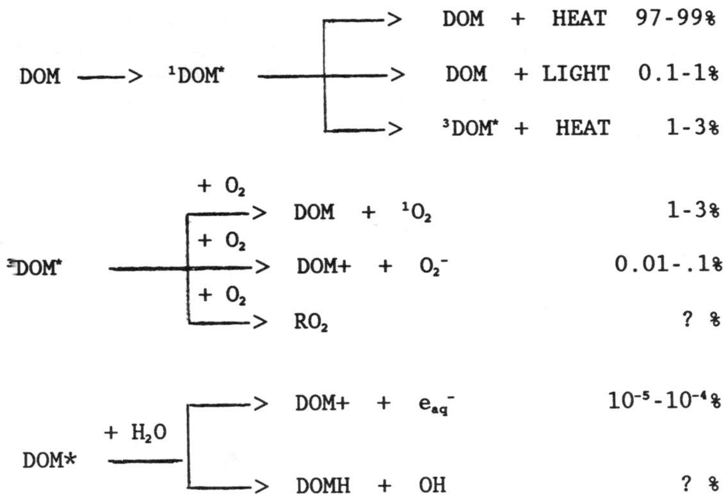

Figure 4. Partitioning of ultraviolet energy absorbed by dissolved organic matter, expressed as % of initially-formed ^{1}DOM*.

In Figure 4, the symbols ^{1}DOM* and ^{3}DOM* represent singlet and triplet excited states, respectively, $^1O_2$ represents singlet molecular oxygen, DOM+ represents the radical cation(s) formed on photoionization of DOM, O_2^- denotes superoxide ion, RO_2 symbolizes organoperoxyl radicals, e_{aq}^- represents the solvated electron, and OH is the hydroxyl radical. These reactive transients mediate indirect photoreactions that are initiated through absorption of solar radiation by DOM (see earlier discussion in Section 1.2). Figure 4 is based on a number of recent studies (e.g. Faust and Hoigné, 1987; Fischer et al., 1987; Haag et al., 1984; Mill et al., 1980; Petasne and Zika, 1987; Zepp et al., 1977, 1985, 1987a, 1987b, 1988). These transients, no doubt, are involved in the DOM photoreactions that were discussed in Section 2.3. The following section describes experimental approaches that have been used in these investigations.

2.5. Experimental Approaches

Two general approaches have been used to examine the transients that are produced as a result of irradiation of natural organic matter, laser flash photolysis, and continuous irradiations. With the important exception of hydrogen peroxide (Cooper et al., 1989; Petasne and Zika, 1987), the transients involved in these photoreactions are usually very short-lived and too dilute to be directly observed on continuous irradiation by commonly used light sources.

2.5.1 Laser flash photolysis. In some cases, it is possible to directly observe the transients through laser flash photolysis (Fischer et al., 1987; Frimmel et al., 1987; Kanner et al., 1989; Power et al., 1987; Zafiriou and Bonneau, 1987; Zepp et al., 1987a).

Triplet states, singlet oxygen, and solvated electrons have been studied using this technique. For example, Figure 5 shows the absorption spectrum of a short-lived transient observed by Zepp and co-workers (1987a) in the laser flash photolysis of a solution of

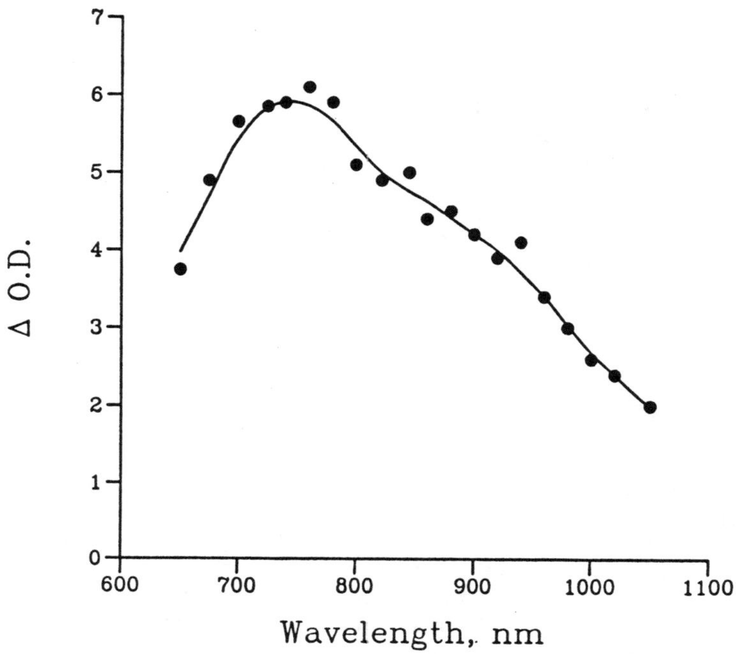

Figure 5. Absorption spectrum of 1.5 microsecond transient observed in laser flash photolysis of Suwannee R. DOM (Zepp et al., 1987).

humic substances obtained from the Suwannee River in Florida. This absorption spectrum, coupled with mechanistic tests involving various quenchers such as nitrous oxide, indicated that the transient is the solvated electron, a powerful reductant that reacts rapidly with electronegative xenobiotics such as chlorinated organics as well as with inorganics such as dioxygen (Zepp et al., 1987a).

Presently available laser photolysis equipment permits studies within time domains in the picosecond range. For thorough discussions of this research area, papers by Frimmel and co-workers (1987), Power and co-workers (1986; 1987), Fischer and co-workers (1987), and Milne and co-workers (1987) are recommended.

2.5.2. Continuous irradiations. In most cases, the identity and concentration of transients produced on irradiation of natural waters have been inferred indirectly through steady-state analysis of kinetic results obtained in continuous irradiations with monochromatic light or with sunlight (Zepp, 1988; Hoigné et al., 1989; Cooper et al., 1989). This approach involves the addition of a readily detected "probe" substance to the system that reacts with the transient in a well defined way. Lead references to the application of steady-state irradiations in studies of photoreactions involving humic substances appear in recent journal articles by Zepp and co-workers (1985), Haag and Hoigné (1986), and Frimmel and co-workers (1987). For example, nitroxides have been used to scavenge photochemically produced free radicals (Blough, 1988), halocarbons have been used to trap solvated electrons (Zepp et al., 1987a), and aromatic amines have been used to investigate carbonate radicals (Larson and Zepp, 1988).

In the case of poorly defined chromophores such as those that comprise the DOM in marine and freshwater environments, wavelength studies have primarily involved obtaining action spectra and/or quantum yields for transient production. For example, it has been shown that the formation of singlet oxygen and triplet states and of hydrogen peroxide (and thus of superoxide ions) is mainly induced by solar ultraviolet radiation, especially by the ultraviolet-B portion (280--320 nm) (Zepp et al., 1985; Cooper et al., 1989).

3. PHOTOREACTIONS OF NATURAL INORGANIC SUBSTANCES

Inorganic substances photoreact to produce free radicals in aquatic environments. Both nitrate and nitrite photoreact in water to produce hydroxyl radicals (Figure 6). (Zepp et al. 1987b; Warneck and Wurzinger, 1988; Zafiriou and Bonneau, 1987). Zepp and co-workers (1987) have presented evidence that the photolysis of nitrate is the dominant source of hydroxyl radicals in some freshwaters.

Hydrogen peroxide is a ubiquitous substance in the sea and freshwaters that forms through the dismutation of photochemically produced superoxide ions (Petasne and Zika, 1987; Cooper et al., 1989). Both homogeneous (Cooper et al., 1989) and heterogeneous (Kormann et al., 1988) photochemical reactions result in the environmental formation of H_2O_2. H_2O_2 is sufficiently long-lived that it can be transported out of the photic zone to dark regions of water bodies. Thus, the formation of hydrogen peroxide provides an important natural mechanism for the storage of solar energy in aquatic environments. This stored energy is released via various redox reactions involving hydrogen peroxide. In addition to its participation in heterogeneous reactions that are discussed in Section 5, hydrogen peroxide is the the most important oxidant of reduced trace metals (iron and copper) in sea water (Moffett and Zika, 1987). Recent studies by Zepp et al.(1990) indicate the reaction of Fe(II) with hydrogen peroxide produces hydroxyl radicals with high efficiency at pH values commonly found in natural waters (pH 4 to 8). In view of the widespread occurrence of Fe(II) and hydrogen peroxide in aquatic environments, it is likely that their reaction is another important source of hydroxyl radicals.

Hydroxyl radicals react predominantly with DOM, carbonate (Larson and Zepp, 1988) and/or bromide (in sea water) to form other more selective radicals. Sources and sinks of OH radicals in the sea or freshwaters are summarized in Figure 6.

4. REACTIVE PHOTOCHEMICAL TRANSIENTS IN NATURAL WATERS

Table 1 summarizes kinetic data on the formation, decay, and steady-state concentrations of transients produced in natural waters on full exposure to sunlight. The table provides the chromophore believed to be the major source(s) of the transient and the range of transient concentrations observed at midday during summer with full exposure to sunlight. The table also includes the steady-state concentrations

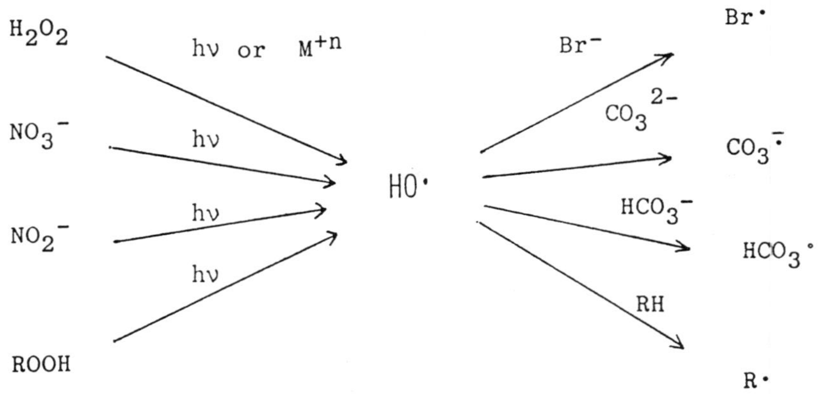

Figure 6. Sources and sinks of hydroxyl radicals in natural waters. "R" represents any organic moiety.

that pertain to a small lake in Switzerland, the Greifensee, and computed half-lives of selected substrates, that are based on these Greifensee concentrations (see Hoigné et al., 1989 for more details).

It should be noted that the reactivities and selectivities of the transients vary considerably. The hydroxyl radical, OH, is the most reactive transient, and it is capable of oxidizing most organic compounds. Solvated electrons are highly reactive but their reactions are limited to good electron acceptors such as halocarbons. Singlet oxygen, carbonate radicals, peroxyl radicals, Br_2^- (from Br^-), and superoxide radicals are much more selective, electrophilic transients. Rate constants for reactions of these transients are available in the radiation chemistry literature. Tabulations of these data are cited in the references at the bottom of Table 1.

5. HETEROGENEOUS PHOTOREACTIONS IN THE ENVIRONMENT

Research on heterogeneous photoreactions in the aquatic and terrestrial environment has received much recent attention (Waite, 1986; Zepp and Wolfe, 1987; Miller et al., 1989). In this section, the focus is on photoreactions on soil and vegetation surfaces and photoreactions of metal oxides.

Table 1. Photochemical Reactive Transients in Natural Waters

Transient	Source	Near-surface Range of $[X]_{ss}$, M	Greifensee $[X]_{ss}$, M	Substrate	Computed Mid-range $t_{1/2}$, h.
$^1O_2$	DOM	10^{-14}-10^{-12}	8×10^{-14}	Dimethyl-furan	3.5
$^3S^*$	DOM	$0.2[^1O_2]$	2×10^{-14}	1,3-Penta-diene	14
OH	NO_3^- NO_2^- M^{n+}/H_2O_2	10^{-17}-10^{-15}	2×10^{-16}	Chloro-benzene	170
RO_2	DOM	??	10^{-11}	Trimethyl-phenol	4.3
e_{aq}^-	DOM	??	2×10^{-17}	Carbon tetra-chloride	350
CO_3^-	same as OH	??	3×10^{-14}	Aniline	10
O_2^-	DOM	10^{-8} (sea water)	??	Ferric EDTA	2

Sources: Faust and Hoigné, 1987; Haag and Hoigné, 1986; Hoigné and co-workers, 1989; Larson and Zepp, 1988; Petasne and Zika, 1987; Zafiriou and co-workers, 1984; Zepp and co-workers, 1987a; Zepp and co-workers, 1987b; Zepp and co-workers, 1985; Waite et al., 1988a.

5.1 Soil and Vegetation Surfaces

5.1.1. Reactions on soil and plant surfaces. A variety of chemical and physical factors affect photoreactions on soil surfaces. These include light screening, changes in rates of direct photoreactions due to spectral and quantum yield shifts, indirect photoreaction involving transients such as singlet oxygen and free radicals, semiconductor mechanisms, and transport to the irradiated surface.

Attenuation of light by soils has been investigated by irradiating immobile photoreactive organic compounds that were incorporated uniformly into soils (Miller et al., 1989). These studies indicate that the photic zone in soils is limited to depths of between 0.1 mm and 0.7 mm. Light attenuation in soils is either due to bulk attenuation or to the inner filter effect discussed by Yokley and coworkers (1986).

Direct photoreactions of chlorinated aromatics can be strongly modified on soil surfaces in comparison to photoreaction in solution. An example of such a difference is provided by the photoreaction of octachlorodibenzo-p-dioxin (OCDD) (Miller et al., 1989). Irradiation of OCDD on thin layers of soil for up to 15 days resulted in production of lower chlorinated dioxin congeners. Isomer-specific analysis of the irradiated soils indicated that preferential reduction occurred at the 1-,4-,6-, and 9- chlorines to give the more toxic 2,3,7,8 substituted congeners. For example, of the 22 possible tetrachlorinated dioxins, 2,3,7,8-tetrachlorodibenzo-p-dioxin (2,3,7,8-TCDD) accounted for approximately one-half of these dioxins recovered from the irradiated soils. In solution, the photolysis rate was substantially more rapid (half-life of 1 hour in midday sunlight) and 2,3,7,-8-TCDD was not observed. No preferential photoreduction at the 1,4,6,9 positions was observed. Changes in photoproduct distribution are attributable to effects of the soil surface on the excited state chemistry of both OCDD and its photoproducts.

Quantum yields for direct photoreactions on soil surfaces are reduced by excited state quenching, probably via energy transfer from excited states to soil components. For example, in the case of polycyclic aromatic hydrocarbons sorbed on humic substances, fluorescence is strongly quenched (Gauthier et al., 1986). On the other hand, indirect photochemical processes involving singlet oxygen on soil surfaces are known to enhance the photoreaction rates of certain substrates on soil surfaces. Photosensitized oxygenations on soil surfaces have been reported by Gohre et al. (1986) (Figure 7).

Few studies have been conducted of photoreactions on vegetation surfaces. Past efforts in this area have focused on the possible effects of such photoreactions on exposure of workers to toxic photoproducts on reentry into fields sprayed with various agrochemicals (Gunther et al., 1977). Recently, Berrenbaum and Larson (1988) have demonstrated that singlet oxygen is photosensitized on the surfaces of certain plants. This photoprocess appears to occur most efficiently with plants known to be phototoxic to insect predators, suggesting that singlet oxygen photoproduction may be a defense mechanism for these plants.

5.1.2. Peroxidase catalyzed reactions. As discussed in Section 3, hydrogen peroxide is formed through the photooxidation of organic matter in natural waters. Recent studies have indicated that hydrogen peroxide may activate peroxidases that mediate biological oxidation of organic substrates such as phenols or the biological formation of halocarbons in the sea. Cooper and Zepp (1990) have provided evidence that hydrogen peroxide activates soil-associated peroxidases that can efficiently oxidize anilines.

FeHeme and vanadium haloperoxidases also can be activated by exogenous peroxide to produce volatile halocarbons such as methyl chloride and bromoform (Dawson, 1988; Wever, 1988). The amounts of halocarbons produced in the sea are comparable to those produced commercially (Gschwend et al., 1985).

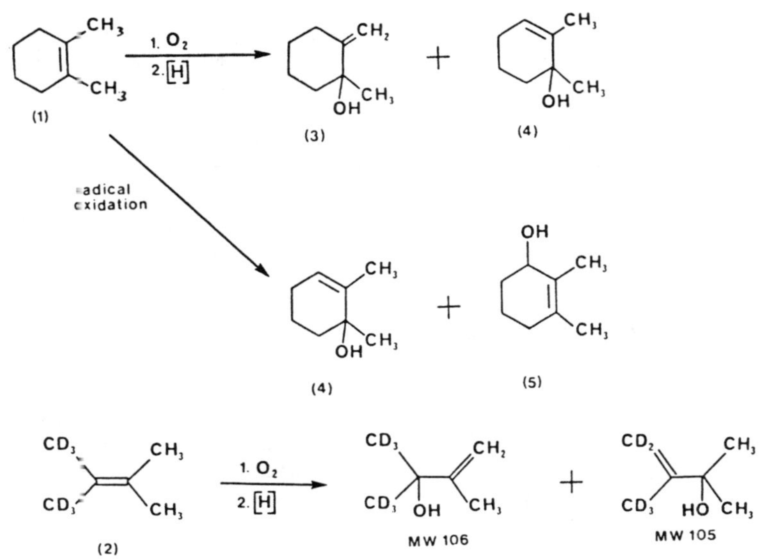

Figure 7. Photooxidations on soil surfaces (Gohre et al., 1986).

5.2 Metal Oxide Photoreactions

Much attention recently has been devoted to the study of photodissolution of iron and manganese oxides in aquatic environments. Both iron and manganese are essential trace elements and these photoreactions have important effects on their bioavailability. Because both iron and manganese are limiting nutrients in parts of the sea (Brand et al., 1986; Martin and Fitzwater, 1988), e.g., the northeast Pacific, these photoreactions may play a very important role in the global carbon cycle.

Field studies have provided ample evidence that the photoreduction of iron or manganese oxides occurs in oxygenated natural waters (e.g., Tipping, 1986; Behra and Sigg, 1990; McKnight et al., 1988; Sunda et al., 1983; Hong and Kester, 1986; Sulzberger et al., 1989). Natural organic matter (Tipping, 1986; Waite, 1986; Waite et al., 1988b; Sulzberger et al., 1989; Bertino and Zepp, 1990) or inorganic reductants such as S(IV) (Faust et al., 1989) or hydrogen peroxide (in the case of manganese oxides) play an important role in the photodissolution of these metal oxides.

The mechanisms of metal oxide photoreduction have received much attention. These reactions involve two general mechanisms: direct photolysis of inner-sphere surface complexes of the metal with various ligands (Faust et al., 1989; Waite and Morel, 1984; Waite, 1986) or indirect photoreaction that is initiated through light absorption by photoreactive components of the oxide surface. The latter mechanism includes the reactions mediated by excited states of the reductant sorbed on the oxide surface, e.g., the photoreduction of manganese oxides by humic substances and quinones (Waite et al., 1988b; Bertino and Zepp, 1990), or possibly a semiconductor-type mechanism, such as the photoreduction of certain iron oxides (Leland and Bard, 1987). The photoreduction of iron oxides by electron donors such as organic acids apparently can occur by either the direct mechanism or by the semiconductor pathway (Faust et al., 1989; Leland and Bard, 1987); it is likely that the pathway depends on the nature of the oxide. The reduction of metal oxides is accompanied, of course, by the oxidation of the electron donor and such oxidations may be environmentally significant.

5.3 Atmospheric Condensed Phases

At most only about 1 part in 10^7 of the volume of the atmosphere is available for liquid phase reactions. Therefore, in order to successfully compete with gas phase reactions, atmospheric reactions in condensed phases must be very efficient. Reactants must have a high tendency to partition into water from air. Changes in chemical speciation through hydrolysis or other reactions enhance partitioning into clouds, fog, and rain. Important tropospheric oxidants, such as HO_x ($HO_x = OH + HO_2$), ozone, and hydrogen peroxide, partition efficiently into cloud water. So do sulfur dioxide and formaldehyde, due in part to their hydrolysis to highly water soluble species on entry into the aqueous phase.

Reactions in cloud water involving oxidation of sulfur dioxide to sulfuric acid have received much attention due to concerns about the acidification of precipitation (Calvert et al.,1985). Crutzen and Lelieveld (1990) have recently demonstrated through a modeling study that chemical reactions within cloud droplets can have much broader impacts on tropospheric chemistry than acidification. For example, clouds act as an important sink for ozone, due primarily to efficient reactions between superoxide ions and ozone within the droplets.

While the research of Crutzen and Lelieveld demonstrated the sensitivity of condensed phase processes to oxidants derived from the vapor phase, other studies have demonstrated that photochemical reactions within the cloud droplets also can play an important role in tropospheric chemistry. For example, Graedel and Wechsler (1981) and Chameides and Davis (1982) have both pointed out the possible importance of the photoreduction of trace metals to chemical reactions that occur in cloud water. Faust et al. (1989) have shown that the autoxidation of S(IV) in cloud water is accelerated through photoreactions involving iron oxides. Faust and Hoigné (1990) have shown that Fe(III) is efficiently photoreduced by solar ultraviolet-B radiation in acidic water, with concurrent formation of hydroxyl radicals. This reaction, along with the Haber-Weiss reaction of Fe(II) with hydrogen peroxide, provides a significant source of OH for oxidations in atmospheric condensed phases (Zepp et al., 1990)

In addition, studies of Faust and Allen (1990) have provided initial evidence for the formation of singlet molecular oxygen and alkylperoxy radicals in irradiated cloud water samples. These oxidants are derived from photoreactions of complex, hydrophilic organic chromophores that accumulate in the cloud water. As discussed in the Section 4, these oxidants can play an important role in the oxidation of certain organic substrates.

6. REFERENCES

Arrhenius, S.A. (1903) Kosmische Physik, 2.

Balzani, V. and Moggi, L. (1990) 'Photochemistry of coordination compounds: A glance at past, present, and future', Coord. Chem. Rev. 97, 313-325.

Baker, K.S. and Smith, R.C. (1982) 'Spectral irradiance penetration in natural waters', in J. Calkins (ed.), The Role of Solar Ultraviolet Radiation in Marine Ecosystems, Plenum Press, New York, pp. 79-91.

Behra, P. and Sigg, L. (1990) 'Evidence for redox cycling of iron in atmospheric water droplets', Nature, 344, 419-421.

Berrenbaum, M.R. and Larson, R.A. (1988) 'Flux of singlet oxygen from leaves of phototoxic plants', Experentia 44, 1030-1031.

Bertino, D.J. and Zepp, R.G. (1990) 'Effect of solar radiation on manganese oxide reactions with organic compounds' submitted for publication.

Blough, N.V. (1988) 'Electron paramagnetic resonance measurements of photochemical radical production in humic substances. 1. Effects of O_2 and charge on radical scavenging by nitroxides', Environ. Sci. Technol., 22,77-82.

Brand, L.E.; Sunda, W.G. and Guillard, R.R.L. (1986) 'Reduction of marine phytoplankton reproduction rates by copper and cadmium', J. Exp. Mar. Biol. Ecol., 96, 225-250.

Carder, K.L., Steward, R.G.; Harvey, G.R. and Ortner, P.B. (1989) 'Marine humic and fulvic acids: Their effects on remote sensing of ocean chlorophyll', Limnol. Oceanogr., 34, 68-81.

Calvert, J.G.; Lazrus, A.; Kok, G.L.; Heikes, B.G.; Walega, J.G.; Lind, J. and Cantrell, C.A. (1985) 'Chemical mechanisms of acid generation in the troposphere', Nature, 317, 27-35.

Chameides, W.L. and Davis, D.D. (1982) 'The free radical chemistry of cloud droplets and its impact upon the composition of rain', J. Geophys. Res., 87, 4863-4877.

Ciamician, G. (1912) 'The photochemistry of the future', Science 36, 385-394.

Conrad, R.; Seiler, W.; Bunse, G. and Giehl, H. (1982) 'Carbon monoxide in sea water (Atlantic Ocean)', J. Geophys. Res., 87, 8839-8852.

Cooper, W.J. and Zepp, R.G. (1990) 'Hydrogen peroxide decay in waters with suspended soils: Evidence for biologically-mediated processes', Can. J. Fish. Aq. Sci., 47, 888-893.

Cooper, W.J.; Zika, R.G.; Petasne, R.G. and Fischer, A.M.. (1989) 'Sunlight induced photochemistry of humic substances in natural waters: major reactive species', in P. MacCarthy and I.H. Suffet (eds.), Influence of Aquatic Humic Substances on Fate and Treatment of Pollutants, ACS Symposium Series 219, Washington, D.C., pp. 333-362.

Crutzen, P.J. and Lelieveld, J. (1990) 'Influences of cloud photochemical processes on tropospheric ozone', Woods Hole Oceanographic Institution Report No. WHOI-90-09, January 1990, pp. 39-41; Nature, in press.

Dawson, J.H. (1988) 'Probing structure-function relations in Heme-containing oxygenases and peroxidases', Science, 240, 433-439.

Donard, O.F.X.; Belin, C. and Ewald, M. (1987) 'Corrected fluorescence excitation spectra of fulvic acids. Comparison with the UV/visible absorption spectra', Sci. Tot. Environ., 62,157-161.

Ertel, J.R. (1990) 'Photochemistry of dissolved organic matter: An organic geochemical perspective', Woods Hole Oceanographic Institution Report No. WHOI-90-09, January 1990, pp.77-78.

Faust, B.C. and Allen, J.M. (1990) "Aqueous-phase photochemical sources of oxidants in clouds' Woods Hole Oceanographic Institution Report No. WHOI-90-09, pp. 48-50.

Faust, B.C. and Hoigné, J. (1990) 'Photolysis of Fe(III)-hydroxy complexes as sources of OH radicals in clouds, fog and rain', Atmos. Environ., 24A, 79-89.

Faust, B.C.; Hoffmann, M.R. and Bahnemann, D.W. (1989) 'Photocatalytic oxidation of sulfur dioxide in aqueous suspensions of α - Fe_2O_3', J. Phys. Chem., 93, 6371-6381.

Faust, B.C. and Hoigné, J. (1987), 'Sensitized photooxidation of phenols by fulvic acid and in natural waters, Environ. Sci. Technol., 21, 957-964.

Ferek, R.J. and Andreae, M.O. (1984) 'Photochemical production of carbonyl sulfide in marine surface waters', Nature, 307,148-150.

Fischer, A.M.; Winterle, J.S. and Mill, T. (1987) 'Primary photochemical processes in photolysis mediated by humic substances', in R.G. Zika and W.J. Cooper (eds.), Photochemistry of Environmental Aquatic Systems, ACS Symposium Series 327, pp. 141-156.

Frimmel, F. and Bauer, H. (1987) 'Influence of photochemical reactions on the optical properties of aquatic humic substances gained from fall leaves', Sci. Tot. Environ., 62, 139-148.

Frimmel. F.H.; Bauer, H.; Putzien, J.; Murasecco, P. and Braun, A.M. (1987) 'Laser flash photolysis of dissolved aquatic humic material and the sensitized production of singlet oxygen', Environ. Sci. Technol., 541-545.

Gammon, R.H. and Kelley, K.C. (1988) 'Reassessment of the flux of CO from the global ocean to the atmosphere', Trans. Am. Geophys. Union, 69, 1072.

Gauthier, T.D.; Shane, E.C.; Guerin, W.F.; Seitz, W.R.; and Grant, C.L. (1986) 'Fluorescence quenching method for determining equilibrium constants for polycyclic aromatic hydrocarbons binding to dissolved humic materials', Environ. Sci. Technol., 20, 1162-1166.

Gjessing, E.T. (1976) 'Physical and Chemical Characteristics of Aquatic Humus', Ann Arbor: Ann Arbor Sci. Publisher Inc.

Gohre, K.; Scholl, R. and Miller, G.C. (1986) 'Singlet oxygen reactions on irradiated soil surfaces', Environ. Sci. Technol., 20, 934-938.

Graedel, T.E. and Weschler, C.J. (1981) 'Chemistry within aqueous atmospheric aerosols and raindrops', Rev. Geophys. Space Phys., 19, 505-538.

Gschwend, P. M.; MacFarlane, J. K. and Newman, K. A. (1985) 'Volatile halogenated organic compounds released to seawater from temperate marine macroalgae', Science, 227, 1033-1035.
Gunther, F.A.; Iwata, Y.; Carman, G.E. and Smith, C.A. (1977) 'The citrus reentry problem: Research on its causes and effects, and approaches to its minimization', Residue Reviews, 67, 1-139.
Haag, W.R. and Hoigné, J. (1986) 'Singlet oxygen in surface waters - Part III. Photochemical formation and steady-state concentrations in various types of waters', Environ. Sci. Technol., 20, 341-348.
Haag, W.R.; Hoigné, J.; Gassmann, E. and Braun, A.M. (1984) 'Singlet oxygen in surface waters. Part II: Quantum yields of its production', Chemosphere, 13, 641-650.
Heindel, N.D. and Pfau, M. (1965) 'A profitable partnership: Giacomo Ciamician and Paul Silber', J. Chem. Ed. 42, 383-385.
Hoigné, J.; Faust, B.C.; Haag, W.R. and Zepp, R.G. (1989) 'Aquatic humic substances as sources and sinks of photochemically produced transient reactants', in P. MacCarthy and I.H. Suffet (eds.), Influence of Aquatic Humic Substances on Fate and Treatment of Pollutants, ACS Symposium Series 219, Washington, D.C. pp. 363-384.
Hong, H. and Kester, D.R. (1986) 'Redox state of iron in the offshore waters of Peru', Limnol. Oceanogr., 31(3), 512-524.
Kanner, R.; Zepp, R.G. and Foote, C.S. (1989) 'Laser flash photolytic studies of singlet molecular oxygen production from humic substances', in preparation.
Kieber, D.J.; McDaniel, J.A. and Mopper, K. (1989) 'Photochemical source of biological substrates in seawater: Implications for carbon cycling', Nature, 341, 637-639.
Kormann, C.; Bahnemann, D.W. and Hoffmann, M.R. (1988) 'Photocatalytic production of H_2O_2 and organic peroxides in aqueous suspensions of TiO_2, ZnO, and desert sand', Environ. Sci. Technol., 22, 798-806.
Kouassi, A.M. (1986) 'Light induced alteration of the photophysical properties of dissolved organic matter in seawater', M.S. Thesis, Univ. of Miami, Coral Gables, FL.
Kouassi, M.; Zika, R.G.; Plane, J.M.C. and Gidel, L. (1986) 'Photochemical modelling of marine humus fluorescence in the ocean', Trans. Am. Geophys. U., 66, 1266.
Larson, R.A. and Zepp, R.G. (1988) 'Reactivity of the carbonate radical with aniline derivatives', Environ. Toxicol. Chem., 7, 265-274.
Leighton, P.A. (1961) 'Photochemistry of Air Pollution', Academic Press, New York.
Leland, J.E. and Bard, A.J. (1987) 'Photochemistry of colloidal semiconducting iron oxide polymorphs', J. Am. Chem. Soc., 91, 50-76.
Martin, J.H. and Fitzwater, S.E. (1988) 'Iron deficiency limits phytoplankton growth in the northeast Pacific subarctic', Nature, 331, 341-343.
McKnight, D.M.; Kimball, B.A. and Bencala, K.E. (1988) 'Iron photoreduction and oxidation in an acidic mountain stream', Science, 240, 637.
Mill, T.; Hendry, D.G. and Richardson, H. (1980) 'Free radical oxidants in natural waters', Science, 207, 886-887.

Miller, G.C.; Hebert, V.R. and Miller, W.W. (1989) 'Effect of sunlight on organic contaminants at the atmosphere-soil interface', in Reactions and Movement of Organic Chemicals in Soils, Soil Science Society of America Special Publication No. 622, pp. 99-110.

Miller, G. C. and Zepp, R. G. (1979), Effects of suspended sediments on photolysis rates of dissolved pollutants. Water Res. 13: 453-459.

Milne, P.J., Odum, D.S.; and Zika, R.G. (1987) 'Time-resolved fluorescence measurements on dissolved marine organic matter', in R.G. Zika and W.J. Cooper (eds.), Photochemistry of Environmental Aquatic Systems, ACS Symnposium Series 327, Washington, D.C., pp. 132-140.

Moffett, J.W. and Zika, R.G. (1987) 'Reaction kinetics of hydrogen peroxide with copper and iron in seawater', Environ. Sci. Technol., 21(8), 804-810.

Mopper, K. and Kieber, R.J. (1990) 'Abiotic formation of formaldehyde, acetaldehyde, and glyoxalate from UV-B induced photodegradation of humic substances in natural waters', Woods Hole Oceanographic Report No. WHOI-90-09, pp. 169-175.

Petasne, R.G. and Zika, R.G. (1987) 'Fate of superoxide in coastal sea water', Nature, 325, 516-518.

Power, J.F.; Sharma, D.K.; Langford, C.H.; Bonneau, R. and Joussot-Dubien, J. (1987) 'Laser flash photolytic studies of a well-characterized soil humic substance', in R.G. Zika and W.J. Cooper (eds.), Photochemistry of Environmental Aquatic Systems, ACS Symnposium Series 327, Washington, D.C., pp. 157-173.

Power, J.F.; LeSage, R.; Sharma, D.K. and Langford, C.H. (1986) 'Fluorescence lifetimes of the well-characterized humic substance, Armadale fulvic acid', Environ. Tech. Letters, 7, 425-430.

Rowland, F.S. and Molina, M.J. (1975) 'Chlorofluoromethanes in the environment', Rev. Geophys. Space Phys., 13, 1-36.

Sulzberger, B.; Suter, D.; Siffert, C.; Banwart, S.; and Stumm, W. (1989), Mar. Chem. 28, 127-144.

Sunda, W.G.; Huntsman, S.A. and Harvey, G.R. (1983) 'Photoreduction of manganese oxides and the supply of manganese to marine plants', Nature, 301, 234-236.

Thurman, E.M. (1985) 'Organic Geochemistry of Natural Waters', Boston: Martinus Nijhoff/DR W. Junk Publishers.

Tipping, E. (1986) 'Some aspects of the interactions between particulate oxides and aquatic humic substances', Mar. Chem. 18, 161-169.

Valentine, R. and Zepp, R.G. (1990) 'Photoproduction of carbon monoxide in wetland waters', in preparation.

Waite, T.D.; Sawyer, D.T. and Zafiriou, O.C. (1988a) 'Panel 1: Oceanic reactive chemical transients', Appl. Geochem., 3, 9-17.

Waite, T.D.; Wrigley, I.C. and Szymczak, R. (1988b) 'Photo-assisted dissolution of a colloidal manganese oxide in the presence of fulvic acid', Environ. Sci. Technol., 22, 778.

Waite, T.D. (1986) 'Photoredox chemistry of colloidal metal oxides', in J.A. Davis and K.F. Hayes (eds.), Geochemical Processes at Mineral Surfaces, ACS Symposium Series 323, American Chemical Society, Washington, D.C., pp. 426-445.

Waite, T.D. and Morel, F.M.M. (1984) 'Photoreductive dissolution of colloidal iron oxides in natural waters', Envir. Sci. Technol., 18, 860.

Warneck, P. and Wurzinger, C. (1988) 'Product quantum yield for the 305-nm photodecomposition of NO_3 in aqueous solution', J. Phys. Chem., 92, 6278-6283.

Wever, R. (1988) 'Ozone destruction by algae in the Arctic atmosphere', Nature, 335, 501.

Yokley, R.; Garrison, A.; Wehry, E. and Mamantov, G. (1986) 'Photochemical transformation of pyrene and benzo(a)pyrene vapor-deposited on eight coal stack ashes', Environ. Sci. Technol., 20, 86.

Zafiriou, O.C. and Bonneau, R. (1987) 'Wavelength-dependent quantum yield of OH radical formation from photolysis of nitrite ion in water', Photochem. Photobiol., 45, 723-727.

Zafiriou, O.C.; Joussot-Dubien, J.; Zepp, R.G. and Zika, R.G. (1984) 'Photochemistry of natural waters', Environ. Sci. Technol., 18, 358A-371A.

Zepp, R.G. (1982) 'Experimental approaches to environmental photochemistry', in O. Hutzinger (ed.), The Handbook of Environmental Chemistry, Volume 2/Part B, Springer-Verlag, Berlin, pp. 19-41.

Zepp, R.G and Andreae, M.O. (1989) 'Factors affecting the photochemical formation of carbonyl sulfide in sea water', Trans. Am. Geophys. Un. 70, 1023.

Zepp, R.G (1988) 'Environmental photoprocesses involving natural organic matter', in F.H. Frimmel and R.F. Christman (eds.), Humic Substances and Their Role in the Environment, Wiley & Sons Ltd., London, pp. 193-213.

Zepp, R.G.; Braun, A.M.; Hoigné, J. and Leenheer, J. (1987a) 'Photoproduction of hydrated electrons from natural organic solutes in aquatic environments', Environ. Sci. Technol., 21, 485-490.

Zepp, R. G., and Cline, D. M., 1977. Rates of direct photolysis in aquatic environment. Environ. Sci. Technol., 11, 359-366.

Zepp, R.G.; Faust, B.C. and Hoigné, J., 1990.'The Photo-Fenton reac tion in aqueous solution (pH 4-8)', in preparation.

Zepp, R.G.; Hoigné, J. and Bader, H. (1987b) 'Nitrate-induced photooxidation of trace organic chemicals in water', Environ. Sci. Technol., 21, 443-450.

Zepp, R.G.; Schlotzhauer, P.F. and Sink, R.M. (1985) 'Photosensitized transformations involving electronic energy transfer in natural waters: Role of humic substances', Environ. Sci. Technol., 19, 48-55.

Zepp, R.G. and Schlotzhauer, P.F. (1981) 'Comparison of photochemical behavior of various humic substances in water: III. Spectroscopic properties of humic substances', Chemosphere, 10, 479-486.

Zepp, R.G. and Wolfe, N.L. (1987) 'Abiotic transformation of organic chemicals at the particle-water interface', in W. Stumm, Ed., Aquatic Surface Chemistry, Wiley, New York, pp.423-451.

Zepp, R.G.; Wolfe, N.L.; Baughman, G.L. and Hollis, R.C. (1977) 'Singlet oxygen in natural waters', Nature, 267, 421-423.

Utilization of Carbon Dioxide : A Strategy for the Control of its Level in the Atmosphere. The Role of the Photoconversion Technology.

Michele Aresta, Eugenio Quaranta and Immacolata Tommasi,
Department of Chemistry, Campus Universitario,
University of Bari, 70126 Bari, Italy.

Abstract.
Carbon dioxide is one of the gases responsible for the enhancement of the Greenhouse Effect (GHE): its emission into the atmosphere is estimated today to range around 5 Gt per year as carbon. The forecast is that the amount of carbon dioxide discarged into the atmosphere by the 2010 will range around 6.8 Gt per year (as carbon). Therefore, a severe cut of this amount has been recently agreed upon.
Several strategies have been proposed for putting this plan into action.
A very stimulating one, from the intellectual point of view, is the "recovery and utilization of carbon dioxide".
It appears of interest to develop new carbon dioxide based synthetic procedures of fine and bulk chemicals if carbon dioxide is used as the starting material in place of those toxic or more expensive species that are presently used; if more direct synthetic processes can be operated; if less drastic operative conditions can be adopted.
An attractive utilization of carbon dioxide is as "**hydrogen carrier**" or as a carbon source for the synthesis of **fuels**.
The conversion of carbon dioxide into **fuels** poses the problem of the **source of energy**.
The utilization of **solar energy** is a must in this case. It might be used for driving thermal, electrochemical or photoelectrochemical reactions.
All reactions aimed to use carbon dioxide need a catalyst for the activation and fixation of the cumulene: **transition metal systems** are good candidates for this role. To date they have been used as **chemical, electrochemical, photochemical, photoelectrochemical** catalysts in a number of reactions.
The extension of the use of carbon dioxide from the actual level to that considered as appropriate for the utilization of carbon dioxide to become an operative strategy for the control of its level into the atmosphere requires a medium- to long-term programme and a great investment, both in terms of funds and of intellectual efforts, for developing new **catalysts** having a higher turnover number and higher selectivity, **new materials** and **new technologies**.

E. Pelizzetti and M. Schiavello (eds.), Photochemical Conversion and Storage of Solar Energy, 517–550.

1. *Introduction*

The emission into the atmosphere of those gases enhancing the "Greenhouse Effect" (GHE) can produce a change in the average seasonal temperature of our planet.[1]
Carbon dioxide generated in the following human activities:
- Electric power generation (36%)
- Transportation (30 %)
- Industry (24%)
- Residences (10 %).

is one of those gases and is estimated to be responsible of the highest rela-tive contribution to the global warming (Table 1).

TABLE 1
Percentage Contribution to the Global Warming.

Gas	Percent	Main Source	Time of residence in the atmosphere (years)
Carbon dioxide	50	Coal (10) Gas (6) Oil (14) Cement and other (5) Land use and Deforestation (15)	7
Methane	18	Agriculture, livestock, biomass, natural gas	10
Dinitrogen oxide, N_2O	10	Biomass, transportation, fuel burning, fertilizers, natural emission.	180
Chlorofluorocarbons	13	Sprays, air conditioners, refrigerators, electronic cleaners, plastics, insulators.	Several centuries?
Ozone	15	Transportation means.	0.3
Others	5		

Table 2 gives the estimated preindustrial era and present concentration of GHE gases in the atmosphere.

TABLE 2
Variation Trend of the Concentration of GHE Gases since the Beginning of the Industrial Era.

Gas	Concentration/ppm 1800	1987
Carbon dioxide	275	358
Methane	0.7	1.8
Dinitrogen oxide	0.29	0.32
Chloro-fluoro-carbons	0.0	0.0006

2. *How can the Concentration of CO_2 in the Atmosphere be Limited ?*

Scientists are facing the problem of how to contain the increase in concentration of these gases in order to avoid serious negative consequences for life on our Planet.[2-3] The problem of limiting carbon dioxide emission is quite serious as this species is generated in most metabolic and combustion processes, all essential to man's energy needs and strictly related to the living standards.

As reported in the introduction, the total amount of carbon dioxide emitted comes from several sources. The figures comment on the real dimension of the problem of controlling the carbon dioxide emission.

A cut down of 20 % in the emission of carbon dioxide by the year 2010, as agreed at the 1987 Montreal Conference, is equivalent to a diminution of 40 % on the 6.8 Gt (as carbon) emission forecast for that year. This poses the serious problem of finding valid strategies to be operative in the short- and medium-term.

An interesting comprehensive analysis of the technologies we have at hand has been made at the IEA-OECD Expert Seminar (Paris,12-14April 1989).[4]

They can be categorized as follows.

2.1 Increase in the efficiency of those energy producing processes that use carbon based fuels.

The amelioration of the combustion process and recovery of heat from flue gases produced in power plants are examples.

2.2 The utilization of the energy with greater efficiency and reduction of any excess and avoidance of any misuse.

For example we might :
increase the end-use efficiency for oil, coal, gas;
improve the efficiency in electricity generating plants;
improve some industrial processes and essentially steel production;
substitute industrial- and space-heating fuel oil by gas;
improve the efficiency of energy transfer;
ameliorate building insulation;
use electricity only if strictly necessary and avoid its use for heating;
limit the speed of cars.

2.3 The possible use of alternative energy sources to complement the burning of carbon based fuels.

The utilization of other energy sources (geothermal, wind- and water- power, solar) are of great importance in order to reduce the emission of carbon dioxide and to avoid any environmental pollution.

2.4 The production and use of energy from biomass and reforestation.

The role of plants in fixing carbon dioxide should not be underestimated as the cost of carbon dioxide fixing through plants is one of the lowest considering the technologies currently available. Biomasses might be used for energy production. This strategy is a priority choice.

2.5 Recovery and disposal or utilization of carbon dioxide.

This technology appears to be quite expensive today.
It has been estimated that the cost of carbon dioxide removal ranges around 13 USA $ per t of carbon dioxide. This cost must be taken into account when we decide about which strategy for reducing the carbon dioxide concentration in the atmosphere should have priority today.
Interestingly, if carbon dioxide is utilized as a raw material in the chemical industry, then the whole economical aspect of its recovery must be reconsidered.

3. Recovery of Carbon Dioxide : is this a Priority Choice Today ?

The recovery of carbon dioxide can be accomplished in two different ways :

- from concentrated sources such as flue gases from power plants (ca. 20% of carbon dioxide) and from industries (up to 99 % pure carbon dioxide).
- from diluted sources such as the atmosphere where carbon dioxide is present at a low concentration (0.03 % in volume).

This aspect will not furtherly be considered as we can assume that plants and algae are by far the best candidates for utilization of CO_2 from the atmosphere.

As we have reported in §1, 60% of the total amount of emitted carbon dioxide comes from power plants (36% of the total) together with industrial processes (fermentation and other industrial activities, 24%) alone.
The cost of the recovery of carbon dioxide from the flue gases of a power station has been evaluated considering both the use of:

a) air [5] ;
and of
b) pure dioxygen
for burning carbon fuels.

(**a**) If air is used for burning carbon fuels, the recovery of carbon dioxide will require a technology for the separation of emitted CO_2 from dinitrogen, nitrogen- and sulphur-oxides (NOx and SOx).
This can be achieved using either :
(i) condensed phases as a trap for carbon dioxide (Table 3)
or
(ii) selective membranes.

The first technology is by far the most used nowadays, selective membranes are very promising and may be the technology of the future, once the technological problems related to the dinitrogen separation have been solved.
Using MEA ($HOCH_2CH_2NH_2$) the actual cost of the recovery ranges around 9.4 USA $ per ton of carbon dioxide. The utilization of selective membranes can reduce the cost of carbon dioxide recovery, because of the lower cost of both recovery plants and chemicals.

TABLE 3
Processes for Carbon Dioxide Recovery.
Trapping by Liquid Phases.
- Basic water solutions.
- Amines (monoethanolamine, MEA, is widely used).
- Alcohols and glycols.
- Other solvents (Organic carbonates, ethers).

Trapping by Solid Phases.
- Basic oxides.
- Molecular sieves.

(b) When dioxygen is used for the combustion of carbon based fuels, we must consider the cost of air distillation. If we must keep in use the same combustion plants we use today, dioxygen should be mixed with emitted carbon dioxide prior to being used in the combustion. The flue gases obtained in this way consist essentially of carbon dioxide that can be compressed directly into pipelines, with or without SOx and NOx impurities elimination.
If the end fate of the carbon dioxide thus recovered is its disposal in natural fields or in the oceans, to the recovery costs we must add the cost of the pipeline, of carbon dioxide pressurizing and pumping.
A reasonable forecast of the cost is 3.8 USA $ per ton of carbon dioxide when a 600 MW sized power plant is considered (such a plant would dispose 450 t of carbon dioxide per hour).[6]
The total cost for recovery and disposal ranges, thus, around 13.2 USA $ per ton of carbon dioxide that corresponds to an estimated increase of the price of electricity of ca. 0.01 USA $ per kWh.

4. The Utilization of Carbon Dioxide.

In these days only a few industrial processes based on carbon dioxide are operated and these can be divided into two main categories:
i) Technological use of carbon dioxide (Table 4):
ii) Conversion of carbon dioxide (fixation of the entire molecule or reduction of carbon dioxide to other C1 molecules) (Table 5).

TABLE 4
Technological Applications of Carbon Dioxide (ca. 8 Mt per year).
- Waste water treatment.
- Addition to beverages and drinks.
- Food packaging and freezing.
- Refrigerators.
- Extraction of active components from natural products.
- Extraction of oil.
- Welding, moulding.
- Fire-estinguishers.
- Sprays.

The total amount used ranges around 100 Mt per year.
The sources of the carbon dioxide so used are reported in Table 6.

Surprisingly, a considerable amount of carbon dioxide is extracted from natural wells, which can give quite pure (> 99%) carbon dioxide.
This can be in some cases the main source, depending on its availability in the country concerned.

TABLE 5
Uses of Carbon Dioxide in Synthetic Chemistry (ca. 90 Mt per year).
- Synthesis of Urea (Fertilizers, animal feeding, additive, resins).
- Organic carbonates (Polycarbonates, other chemicals).
- Inorganic carbonates (First and second group elements, other salts used as pigments or in the glass industry, etc.).
- Pharmaceuticals (Aspirin, others).
- Additive in the synthesis of methanol.

A considerable amount of carbon dioxide is recovered from fermentation and other industrial processes that can give very pure CO_2. Its cost is roughly comparable with that of the extracted carbon dioxide.

TABLE 6
Industrial Sources of Carbon Dioxide.
- Natural wells (high purity, > 99%).
- Industrial processes (high purity, > 99%).
- Fermentation processes (high purity, > 99%)
- Cement manufacture
- Fuel (coal, natural gas, oil, wood) burning.

It must be pointed that while extraction costs are for an operating technology, the costs of recovery from flue gases are evaluated for an untried technology and this means that they might be higher or lower. Anyway, it is clear that the recovery of carbon dioxide from flue gases might enter into practice if we solve the problem of the final fate of CO_2.
If the utilization of carbon dioxide has to be considered seriously, then we must analyze a few key points:
1) The lifetime of the products obtained from carbon dioxide.
2) The amount of carbon dioxide we must use for the utilization to be an effective tool for the carbon dioxide level control.
3) The source of the energy required in the conversion processes.
We shall analyze these points in order to have a clear picture of the potential application of the recovery and utilization of carbon dioxide[7] in the short- and medium-term.

4.1 The lifetime of the products obtained from carbon dioxide.

As we have already specified, the final aim of the recovery of carbon dioxide can be either its disposal in natural fields or its utilization.
The former can be considered as a practice that eliminates carbon dioxide from the environment forever.
Conversely, if we fix carbon dioxide into chemicals when we use that species we convert it back into carbon dioxide. For this reason the "lifetime" of chemicals and goods obtainable from carbon dioxide (Table 7) is an essential parameter.
"Lifetime" is here considered as the average time required for carbon dioxide release from a species when it comes into its typical utilization.

TABLE 7
Lifetime of Species Obtained from Carbon Dioxide.

Classification	Lifetime	Example
Imperishable	>> E2 Centuries	Calcium carbonate
Long life	Decades - century	Polymers.
Short life	Months - year	Agrochemicals (urea, herbicides etc)
Very short life	< Hours	Pharmaceuticals, fuels, etc.

Chemicals of the type of calcium carbonate are real "carbon dioxide fixing agents" as they can last for an illimitated time, at least if standard ambient conditions (298 K and 0.103 MPa) are considered. Conversely, to fix carbon dioxide into chemicals that have a lifetime of the order of "years" cannot make a substantial contribution to the problem of the control of carbon dioxide concentration in the atmosphere, even if considerable amounts of carbon dioxide are used. This is the case of urea and of similar products (carbamates used as agrochemicals: pesticides, herbicides, etc). Nevertheless, it would be of great economic interest to develop new carbon dioxide based synthetic procedures for obtaining fine chemicals if:

- carbon dioxide is used as the starting material in place of those toxic or more expensive species that are presently used;
- more direct synthetic processes are operated;
- less drastic operative conditions can be adopted.

In any case this would allow energy to be saved and would ensure safer working conditions in the industry and laboratories.
The utilization of carbon dioxide in the synthesis of polymers such as polyurethanes and polycarbonates is of interest for the lifetime

of these macromolecules and for their properties. Indeed, polyurethanes might contribute in two concurrent ways to the limitation of carbon dioxide emission in the atmosphere: in a direct way as carbon dioxide is fixed in the molecule; and in an indirect way as they might be used as insulators in buildings.

4.2 Utilization of CO_2 in the synthesis of fuels.

An alternative utilization of carbon dioxide is as a "hydrogen carrier" or as the starting material for the synthesis of fuels. In the first case carbon dioxide can be converted into formic acid, HCOOH, which is easily converted back into hydrogen and carbon dioxide. This peculiar utilization might be of interest also in the case of a "hydrogen based economy" as to store and transport hydrogen as formic acid would reduce both the cost and the risks.
The conversion of carbon dioxide into fuels poses the problem of the source of energy to be used for the conversion and of the amount of carbon dioxide.
In this case, the point of great interest is the rate of production of such fuels rather than the "lifetime of the product".
In fact, as far as fuels are considered we must think of the utilization of carbon dioxide in terms of a "man made cycle" (Fig. 1) that cooperates with the "natural photosynthetic cycle".

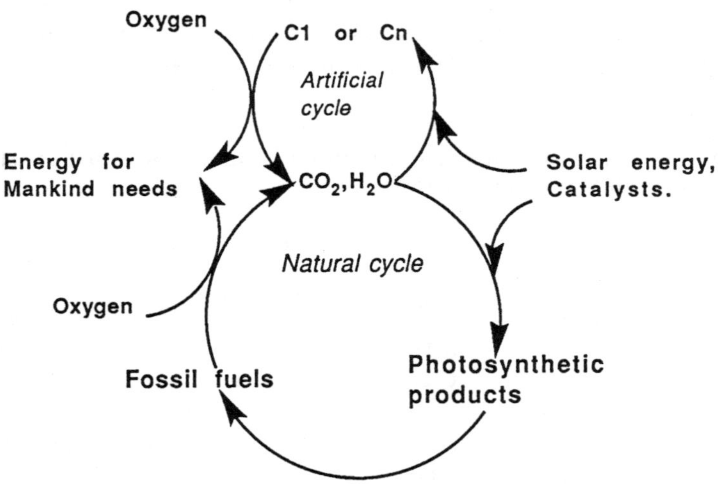

Fig.1 "Man made carbon dioxide cycle" cooperating with the "natural" one.

To make this process operative means to have the possibility of converting several Mt of carbon dioxide per year and this would give a net contribute to limiting the CO_2 emission in the atmosphere.
Fundamental parameters in the artificial cycle are, thus: the rate of conversion of carbon dioxide and the source of the energy required in the reduction process.

4.3 Energetics of the processes based on carbon dioxide and the source of energy.

From a formal point of view all the processes that utilize carbon dioxide have to be considered as reduction reactions and thus require energy. Indeed, all those processes that use the entire CO_2 molecule (the carboxylation reactions or, in general, the reactions affording the formation of C-C, C-N, C-O bonds) present (if any) a low energy demand (of the order of a few kJ mol^{-1}), while those processes in which a deoxygenation reaction takes place require a higher amount of energy (often greater than 100 kJ mol^{-1}). As an example, the $\Delta G°$ (at 298 K) for the conversion of CO_2 into CO and oxygen atom is 257.15 kJ mol^{-1}.
Nevertheless, CO_2 cannot be considered an unreactive species and, in fact, it undergoes several reactions under very mild conditions provided that an electron rich species is available as the energy source. Table 8 gives values for the energy required in a few reduction processes in the presence or absence of hydrogen.

TABLE 8
Thermodynamic data for carbon dioxide conversion to other C1 molecules. (All species in the gas phase).

						ΔG/kJ mol^{-1}
CO_2			=	CO	+ 1/2 O_2	257.15
CO_2	+	H_2	=	HCOOH		58.66
CO_2	+	$2H_2$	=	CH_2O	+ H_2O	55.73
CO_2	+	H_2	=	CO	+ H_2O	28.49
CO_2	+	$3H_2$	=	CH_3OH	+ H_2O	3.9
CO_2	+	$4H_2$	=	CH_4	+ $2H_2O$	-113.9

It is worth noting that there is a strong entropic effect in these reactions as reported in Table 9.
Among C1 species formic acid and methanol are, in our opinion, the most interesting ones as they can be considered as energy vectors and find a wide utilization as starting materials in the chemical in-

TABLE 9
Carbon Dioxide Reduction to Methanol: Role of the Entropic Contribution. (Evaluated at 298 K)

Reaction	$\Delta H°$	$\Delta G°$ (kJ mol^{-1})
$CO_2(g) + 3H_2(g) \rightarrow CH_3OH(g) + H_2O(g)$	-49.50	+3.89
$CO_2(g) + 3H_2(g) \rightarrow CH_3OH(l) + H_2O(l)$	-130.8	-9.04

dustry.
The synthesis of methane from carbon dioxide is an exoergonic process that needs hydrogen, but CH_4 does not seem to be the most interesting species to be obtained from CO_2 as its chemistry is not very rich.
The carbon dioxide reduction to other C1 molecules requires, thus, energy and/or hydrogen. It seems obvious to think to the possibility of coupling this reaction with the water splitting reaction that generates hydrogen and oxygen from water. The overall reaction, after the utilization of dihydrogen in the synthesis of the hydrogenated C1 species, would be mimic of the photosynthetic reduction of carbon dioxide.

$$CO_2 + H_2O \xrightarrow{\text{Solar energy}} 1/n\,(CH_2O)_n + O_2$$

The utilization of water as the source of hydrogen and electrons is of fundamental importance for the carbon dioxide conversion.
Other bulk chemicals can be also synthetized from CO_2. Table 10 comparates ΔH and ΔG values for some selected processes based on CO and CO_2.
Reactions based on CO_2 are, thus, thermodinamically possible, and require a serious investigation. However, it would be of great interest to substitute carbon dioxide to CO in the synthesis of methanol and in Fischer-Tropsch processes (one extra mole of hydrogen is demanded) or convert carbon dioxide into CO (may we use carbon dioxide as an oxidant?).
We shall deal in the next paragraphs with the utilization of photochemical systems able to convert carbon dioxide and we'll try to point out which are the actual limitations and the work to do for a real utilization of carbon dioxide.

TABLE 10
CO_2 versus CO Utilization in the Synthesis of Bulk Chemicals. (At 298 K)

	Reactants		Products	ΔH°	ΔG° kJ mol^{-1}
1)	$CO_2(g)$ + $3H_2(g)$	--->	$CH_3OH(l)$ + $H_2O(l)$	-130.9	-9.04
1')	$CO(g)$ + $2H_2(g)$	--->	$CH_3OH(l)$	-128.0	-28.9
2)	$CO_2(g)$ + $CH_3OH(l)$ + $H_2(g)$	--->	$CH_3COOH(l)$ + $H_2O(l)$	-140.0	-69.28
2')	$CO(g)$ + $CH_3OH(l)$	--->	$CH_3COOH(l)$	-137.9	-88.9
2")	$CO_2(g)$ + CH_4	--->	$CH_3COOH(l)$	-18.61	+52.7
3)	$CO_2(g)$ + $C_2H_2(g)$ + H_2	--->	$CH_2{=}CHCOOH(l)$	-217.8 (-169.45 g)	-88.87 (-100.88 g)
3')	$CO(g)$ + $C_2H_2(g)$ + $H_2O(l)$	--->	$CH_2{=}CHCOOH(l)$	-214.9 (-210.6 g)	-108.7 (-129.54 g)
4)	$CO(g)$ + $C_2H_4(g)$	--->	$CH_2{=}CHCOOH(l)$	-43.3 (+5.02 g)	+51.80 (+40.21 g)
4')	$CO(g)$ + $C_2H_4(g)$ +1/2 O_2	--->	$CH_2{=}CHCOOH(l)$	-326.3 (-277.9 g)	-205.0 (-216.90 g)

5. Photoreduction of Carbon Dioxide.

The direct photolysis of CO_2 to CO and O requires high energy radiations (< 167 nm). The utilization of solar energy for the reduction of carbon dioxide is of great interest for the potential application of this technology to the synthesis of other C1 molecules richer in energy. The storage of solar energy in the form of liquid or gaseous fuels would allow to reduce the extraction of fossil fuels. If the artificial photosynthetic cycle shown in Fig. 1 can be operated efficiently then the recovery and utilization of carbon dioxide would be a meaningful and feasible process and this would make possible the control of accumulation of CO_2 in the atmosphere. Key parameters in the photoreduction process are the efficency of light energy conversion and the selectivity.
Moreover, the study of the non-biological photofixation of CO_2 can furnish anologies between photosynthetic CO_2 reduction in nature and in vitro.
Interestingly, the reduction can be accomplished using not only soluble metal complexes but also heterogeneous catalysts (such as semiconductor powders or crystals) from which we can get

information on the carbon dioxide reduction, and dioxygen formation, that most probably were promoted by the oxides present in minerals on the earth surface in the prebiotic era.

5.1 Heterogeneous catalysts for CO_2 reduction.

The utilization of semiconductors powders and solar light corresponding to the bandgap has been explored by several research groups for CO_2 reduction in water.

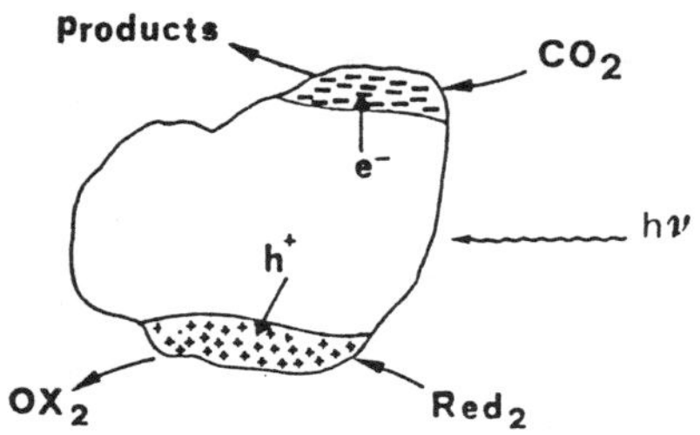

Fig.2 Semiconductor at work in CO_2 photoreduction.

The mode of action of a semiconductor particle can be depicted as in Fig 2. The absorption of a photon generates an electron and a hole that, if reach the surface, can give a reaction with water or organic species. The electrons generated in this way will be eventually consumed in a reduction process involving carbon dioxide, its reduction products or water.

The semiconductor particle behaves, thus, as a short-circuited photochemical cell.

Several oxides characterized by different energy levels[8,9] have been used so far (Fig. 3) and their energy-gaps must be compared with CO_2 reduction potentials (Fig. 4)

Several systems have been used to date, sometimes with interesting results.

The knowledge of the reduction mechanism requires further investigation: if a one- or multi-electron process is operating is not clearly demostrated, but it has been ascertained that the one electron transfer to CO_2 for generating the radical anion $CO_2^{\cdot -}$ is the process demanding the higher energy.*

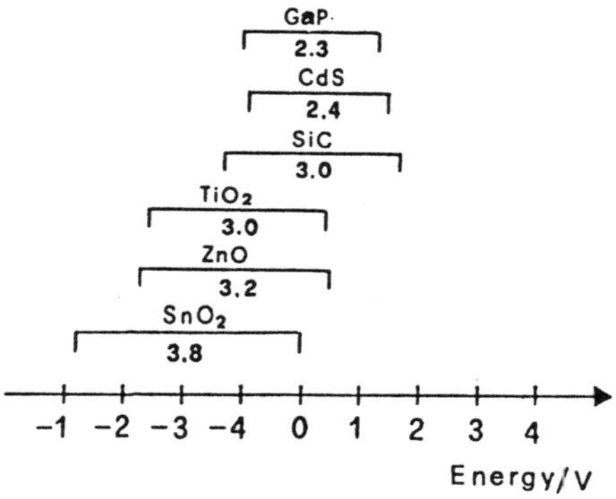

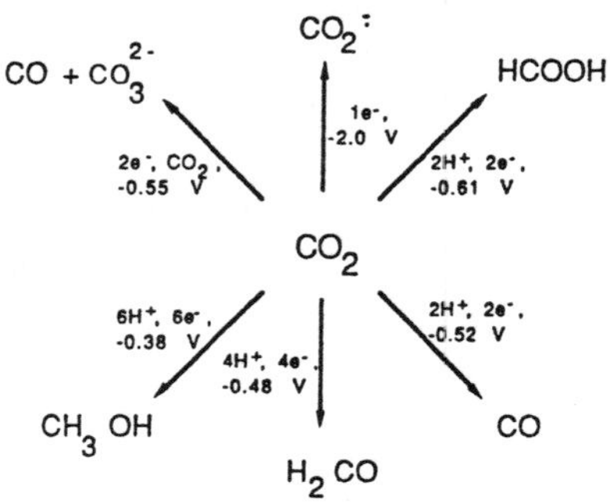

Fig.3 Energy-gap for some semiconductors. Fig.4 Mono- and poly- electronic reduction reactions of carbon dioxide.

A great emphasis has been put in the study of rutile TiO_2 which has a conduction band located approxymately at the level of the H^+/H_2 couple.
This makes rather difficult the CO_2 reduction by TiO_2 in the presence of water.

Titanates,[9] such as $SrTiO_3$ that shows a conduction band at a potential ca. 0.2 V more negative than TiO_2, have encountered a great deal of interest. They appear to afford the best chemical conversion of the absorbed photons and this is explained also in terms of a strong absorption of carbon dioxide.[11]
The addition of ruthenium or platinum family (Rh, Ir, Pt) metal oxides occur to influence the selectivity.[12]
Formic acid seems to be the predominant reduction product and IrO_x apppears to be the most effective catalyst for HCOOH generation. Methanol is the most abundant C1 species when RuO_x is used as co-catalyst. The addition of the above mentioned oxides also causes the light adsorption into the visible region, which has been found to be characteristic of titanates added with $LaCrO_3$, too.[13] The

* There is some controversy in the literature [10] about the energy of the CO_2/CO_2^- couple. A value of -2.8 V has been found for strictly anhydrous organic solvents, compared to - 2.0 V found in wet solvents.

visible absorption can be assigned to a metal-to-metal charge transfer.

TABLE 11

CO_2 Reduction in Aqueous Medium**

	% mol		Products/μmol h^{-1}					Efficiency
			HCOOH	HCHO	MeOH	MeCHO	EtOH	η
TiO_2	6.83	10.81						
RhO_x	4.55		0.76	0.015	0.029	0.002	0.003	0.0015
IrO_x	4.55		1.1	0.020	0.011	0.007	0.034	0.022
PtO_x	4.55		0.7	0.014	0.016	0.003	0.0033	0.013
RuO_x	4.55		0.56	0.01	0.04	0.001	0.002	0.012
RuO_x		0.57	1.25	0.04	0.10	0.08	0.00	0.032
RuO_x		1.14	0.82	0.01	0.083	0.04	0.00	0.020
RuO_x		2.28	0.94	0.02	0.05	0.006	0.006	0.019
RuO_x		4.55	0.56	0.01	0.04	0.001	0.002	0.012

** $SrTiO_3$ /$MOx\cdot TiO_2$. O.8-1.0 g in 160 mL. CO_2 flow rate 100 mL min^{-1}. 75 W Hg lamp (HP) at 333 K.

The amount of the added species causes a shift of the light efficiency and selectivity (Table 11).[12] In the case of RuO_2 as additive the decrease of the light efficiency with the amount of RuO_2 can be interpreted on the basis of the oxidant capacity of RuO_2: in fact, this species might be involved into the back-oxidation of the product to CO_2.[14]

At the Weizmann Institute experiments using solar light as the energy source for CO_2 reduction have been carried out.[12] The solar energy conversion efficiency ranges from 0.003 to 0.016 % using $SrTiO_3$ added with $Rh_2O_3 \cdot TiO_2$ or $LaCrO_3$ (Table 12).

These results show that from the termodynamic point of wiew the CO_2 reduction by water using semiconductor oxides is possible. Still several points require further investigation: kinetic factors are of fundamental importance in order to ameliorate the yield. One of the major points, CO_2 reduction prior to the H_2 evolution, has been investigated in details so far and the results of studies by Halmann and Mackor show that the pH of the solution can play an important role. The lack of selectivity represents another problem with these systems. New catalysts must be tested.

TABLE 12
Carbon Dioxide Reduction in Water under Sunlight Irradiation.

Catalyst	Additive (% mol)	Products HCOOH	HCHO	MeOH	MeCHO	EtOH	η %
$SrTiO_3$		6.4	0.7	0.4	0.26		0.003
$SrTiO_3$	Rh_2O_3(4.55)-TiO_2(6.83)	6.3	0.5	0.16	0.11	0.1	0.004
$SrTiO_3$	$LaCrO_3$ (0.5)	18.0	0.33	0.15	0.32		0.016

A quite unique result has been reported by Rophael and Malati[15] who have irradiated aqueous carbonate solutions with u.v.-visible (254 nm) radiations in the presence of Pt-doped titanate powders. The main reduction product was carbon, accompanied by formaldehyde. In the most effective run, 11 h irradiation with a MP Hg-lamp (125 W), generated 2 mmol of C and 2.4 μmol of CH_2O. This reaction also requires further investigation.
Most recently other semiconductors have been used, namely SiC and ZnSe[16] either pure or coated with Pd, Pt, Rh for the synthesis of C2 molecules, such as ethanol, from CO_2 and water. The authors have reported that the main products were C_2H_5OH and CH_3OH, while C1 species such as formic acid, formaldehyde and methane, were not detected.
The best yields in ethanol range around 18 μmol with an irradiation time of (500W Xe lamp) using a 1000 Mesh SiC. This photocatalyst also yielded 12 μmol of methanol as the coproduct.
The utilization of metal colloids (Pt, Ru or Os) catalysts[17] in the presence of Ru(II) complexes, stabilized by the dipyridine ligand, as photosensitizer, triethanolamine as electron donor, and relays of the methylviologen type, has brought to CO_2 reduction to methane. Ethylene, ethane and dihydrogen are formed as coproducts according to the system used.
The authors have demonstrated that the H_2 evolution process can be controlled and inhibited by addition of nitrogen ligands like dipyridine.
The use of these systems is of interest as they are reminiscent from one hand of heterogeneous catalysts (Ru, Os) used in the thermal methanation of CO_2 with dihydrogen[18-19] and, on the other hand, of the utilization of metal complexes with macrocyclic ligands used as catalysts in the electro- and photo-electrochemical reduction of CO_2.

A dependance on the pH of the aqueous medium has been demonstrated, and the most favourable conditions for CO_2 reduction (over proton reduction) have been demonstrated to be the basic ones.[20] In these conditions carbon dioxide exists in solution essentially as the hydrogen carbonate species and the question may arise wether HCO_3^- more than CO_2 is the reactive species. This question is of interest for its relevance to natural utilization of carbon dioxide. In Nature either HCO_3^- or CO_2 can be selectively used. Nature has developed very selective systems for the activation of HCO_3^- which has a lower electrophilicity than CO_2.[21] Moreover, the hydrogencarbonate anion has been shown to be reduced electrochemically to formate at a palladium electrode.[21-22]

In the experiments with Ru and Os colloids, CO_2 is assumed to be reduced to methane and to other C2 products (ethylene and ethane) according to the electron relay systems used (Table 13). All components are necessary for methane formation and CO_2 reduction is proposed to proceed through hydrogenation. When Ru^{2+}-bipyridine complexes are used as photosensitizer a kinetic barrier towards H_2 evolution is proposed to rise.[23]

TABLE 13***
Methane, Ethylene and Dihydrogen Formation from CO_2 Catalyzed by Ru or Os Colloids.

		ΦRu/ ΦOs			ΦRu · 10^3		
Relay	E°	H_2	CH_4	C_2H_4	H_2	CH_4	C_2H_4
TQ^{2+}	-0.794	1.37	2.71	1.83	2.6	0.57	0.019
DQ^{2+}	-0.723	0.57	4.51	2.52	1.7	0.23	0.0073
MQ^{2+}	-0.652	0.20	2.30	1.61	1.8	0.14	0.0108
$MPVS^0$	-0.551	0.44	1.67	1.2	0.28	0.020	0.0018

*** $[Ru(bipy)_3]^{2+}$ as photosensitizer; TEOA as electron source; HCO_3^-/H_2O solution medium.

As a result, methane, ethylene and ethane are formed. The formation of the C2 hydrocarbons has been proposed taking into account the formation of carbene- or methyl-species: Ru=CH_2, Ru-CH_3, Os=CH_2. Very recently RuO_2 dispersed on a TiO_2 support has been used by Grätzel and coworkers for CO_2 to CH_4 reduction.[24] High yields were obtained especially when transition metal complexes were used as photosensitizers.[25]

These systems appear of greater interest than semiconductors used in the early 80's and discussed above.

The reduction of CO_2 to afford C2 acids more than C1 species has been reported quite recently as promoted by colloidal CdS.[26] The formation of glyoxylic acid, of acetic acid and, sometimes, of acetaldehyde was detected at pH 6.
Formaldehyde, formate and methanol were formed besides C2 species and resulted to be the major products when CO_2 was photolyzed in water: glyoxylate was not detected in these conditions. The addition of tetramethylammonium chloride caused the appearance of glyoxylates, the increase of acetate yield and the production of large quantity of formate. Methanol was not analyzed in these conditions.
Oxalate, invoked as a likely intermediate in the synthesis of glyoxylate, was not found in the reaction mixture. Conversely, photolysis of oxalate with CdS produced glyoxylate, tartrate, formate (as the major products) and CO_2. The low photon-electron quantum efficiency (ca.$10^{-4}\div10^{-3}$) was explained in terms of the possible reoxidation of the products, of the conversion of CdS into CdO and involvement of two electrons-two holes mechanism.[27] The low stability of the catalyst and low quantum yields are major problems with metal oxides so that in these days there is not too much interest in their utilization in CO_2 reduction. Obviously, the synthesis of new semiconductors with higher efficiency is one of the research objectives in this area. Coupling these oxides with transition metal complexes can also reveal of practical utility as semiconductors can be used as light harvesting and charge transfer systems and transition metal complexes as catalysts. These combined systems might assure higher selectivity.

5.2 Carbon dioxide reduction in homogeneous conditions.

The phoreduction of CO_2 using homogeneous catalysts can be achieved in several ways. Aqueous solutions of metal ions[28] or of organic dyes[29] have been used in the earlier experiments. Most recently transition metal complexes[30] have been employed and organic solvents or water were used as electron sources. The monoelectronic transfer process to CO_2 being the one requiring the higher energy (ca. 4.5 e V), not available in the solar spectrum, multielectron reduction of CO_2 to CO or formate.[31] In principle, ca. 40 % of solar energy could be stored in chemicals generated in this way.[32] As H_2O and CO_2 do not absorb in the visible region, it is necessary to use a photosensitizer and a catalyst for such a reduction reaction. The catalyst must be able to transfer more than one electron and, usually, Co(I), Rh(I), Re(0) and Ru(I) complexes are used that are oxidized to Co(III), Rh(III), Re(II) and Ru(III). There are a few musts about these systems and they deal with:

i) the utilization of water as electron source;
ii) the use of selective catalysts that show also a high quantum yield;
iii) the search of systems able to forward the CO_2 reduction reaction and to inhibit the back reaction, i.e. the products oxydation.
iv) the use of compatible photosensitizers, relay systems and catalysts.

According to the data available in the literature, quite interesting appear the selective reduction of CO_2 to CO or formate using water as electron source.

The first photochemical system for carbon dioxide reduction in homogeneous conditions[33] consisted of $[Ru(bipy)_3]^{2+}$ (bipy=2,2 bipyridine) and $[Co(bipy)_3]^{2+}$ or Co^{2+} species in organic solvent and tertiary amines.

The products of the reaction were CO and H_2 and the maximum quantum yield and the CO/H_2 ratio obtained depended on:

- the amount of free ligand (2,2'-bipyridine) present in the medium (the best yield in CO was obtained in the absence of free ligand).
- the nature of the ligand (with 2,9 dimethyl-1,10 phenantrolyne maximum quantum yield of 1.2 % for CO and 6.5% for H_2 were obtained but when dimethyglyoxime was used only H_2 production was registred with 16% of light energy conversion into H_2).
- the nature of the tertiary amine (ethylamine or hydroxylated amines).

A system affording a good CO/H_2 ratio has been studied by Spreer[34] and consists of $[Ru(bipy)_3]^{2+}$ and $[Ni(cyclam)]^{2+}$ (cyclam=1,4,8,11-tetraazacy-clotetradecane) complexes in aqueous medium and ascorbate buffer as electron donors.

The dependence of CO/H_2 ratio on the pH of the medium was studied in the interval of pH 3 ÷ 6 and the obtained plot is shown in Fig. 5

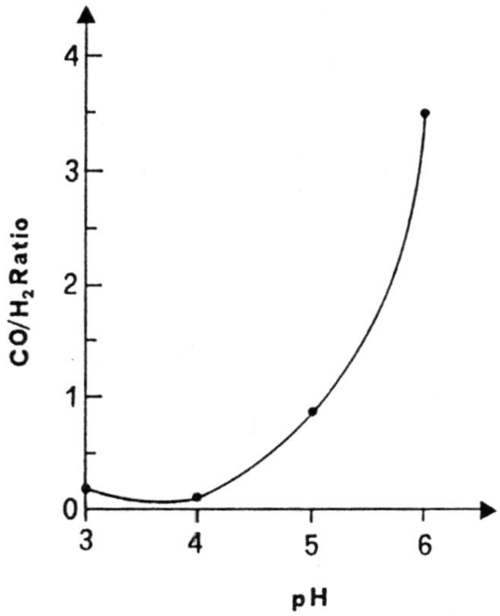

Fig. 5 pH dependence of the CO/H_2 ratio in CO_2 reduction.

Although the best CO/H_2 ratio was observed at pH=6, the maximum amount of CO produced was registred at pH=5 .

A system involving only $[Re(bipy)(CO)_3Cl]$[33] acting both as photosensitizer and electron mediator was found able to reduce carbon dioxide to CO in a DMF/TEAO mixture (DMF= dimethylformamide; TEOA= triethanolamine) with 14% quantum yield in presence of an excess of chloride anion.

The overall process has been monitored by ^{13}C n.m.r. spectroscopy that permitted:

- to observe the conversion of $^{13}CO_2$ (disapparence of the signal at 125 ppm) into ^{13}CO (signal at 184 ppm);
- to monitor the appearence of a new metal-carbonyl species in solution (signals at 198 and 194 ppm);
- to detect the intense signal (167 ppm) which in absence of proton decoupling splits in two signals (J_{CH} = 197 Hz), indicating the formation of a formate species in solution.

The metal complexes formed in the reaction were isolated and characterized. Major products were fac-$[Re(bipy)(^{13}CO)_3Cl]$ and $Re(bipy)(^{13}CO)_3(OO^{13}CH)$ that was isolated and characterized by X-ray spectroscopy. According to the authors, the metal-formate complex is not an intermediate in the formation of ^{13}CO as its formation occurs simultaneously to the ^{13}CO complex formation. Quantum yields obtained from the discussed systems are summarized in Table 14.

TABLE 14
Maximum Quantum Yields for CO_2 Photochemical Reduction to CO .

Photosensitizer	Catalyst	Electron Donor	Products CO	H2
$[Ru(bipy)_3]^{2+}$	$Co^{(II)}$*	Tertiary amines	1.2 %	6.5 %
$[Re(bipy)(CO)_3Cl]$	$[Re(bipy)(CO)_3Cl]$	TEOA	14 %	

* The reported yield has been obtained using 2,9 dimethyl-1,10 phenantrolyne as ligand for the $Co^{(II)}$ metal.

The high quantum yield obtained with the Re complex seems to be very promising and this system will certainly become of practical application supposed that water can be used in place of TEOA.
Because of the evidences for metal-formate species formation, interest increased in the search of photochemical systems for selective carbon dioxide reduction to formate.
In particular Lehn and Ziessel[35] have studied the catalytic formate generation by carbon dioxide reduction in two different systems:

1) using the species $[Ru(bipy)_3]^{2+}$ either as photosensitizer or as precursor of carbon dioxide activating complex in an aquo-organic mixture DMF/d7-DMF/TEOA/H_2O (3:1:1:1). (DMF= dimethylformamide; d7-DMF= perdeuteriated dimethylformamide; TEOA= triethanolamine);

2) using $[Ru(bipy)_3]^{2+}$ as photosensitizer and $[Ru(bipy)_2(CO)(Cl)]^+$ or $[Ru(bipy)_2(CO)H]^+$ as complexes for the activation of carbon dioxide in dry DMF/TEOA solution.

Studying the first system by ^{1}H-n.m.r. spectroscopy, they have observed that irradiation by visible light of $Ru(bipy)_3]^{2+}$ alone (that shows a triplet at 8.22 ppm assigned to the aromatic protons of the bipyridine), did not afford formate while in the presence of water (15%) a new Ru-bipyridine complex (showing a doublet at

8.36 ppm, assigned to the aromatic protons of the bipyridine) and formate (doublet at 8.56 ppm) were formed.
This observation together with the known photolabilisation of Ru(II)-trischelate complexes in water allowed to conclude that $[Ru(bipy)_3]^{2+}$ plays the role of photosensitizer and that the electron mediator and carbon dioxide activating species are generated by its photolabilization.
In these conditions, in spite only 0.001% conversion of the $[Ru(bipy)_3]^{2+}$ complex into its active catalytic form was observed, formate was produced with a maximun quantum yield of 4.9%. Addition of water, up to 15% in volume as specified above, increases the efficiency in formate production, while further dilution causes the formate generation to decrease because of the competitive formation of CO and H_2.
In order to improve both yield and selectivity in HCOOH production, $Ru(bipy)_3]^{2+}$ complex was used together with Ru(II) bis- or mono-bipy complexes in absence of water. In this case, in $^{13}CO_2$ saturated solutions, ^{13}C n.m.r. spectroscopy allowed to study the $^{13}CO_2$ disapparence (signal at 125 ppm) and the simultaneous selective formate formation (167 ppm; J_{CH}=187 Hz) with a maximum quantum yield of 15%.
The mechanism for carbon dioxide reduction to formate is unknown.
N.m.r. and i.r. spectroscopy have shown that by irradiation of solutions containing $[Ru(bipy)_3]^{2+}$ and $[Ru(bipy)_2(CO)H]^+$, the characteristic signal for hydride species at -11.3 ppm and the ν(CO) vibration at 1930 cm^{-1} disappeared after three hours and a new ν(CO) vibration was observed at 1980 cm^{-1}. These results seem to suggest that hydride species are not involved in carbon dioxide reduction, but carbonylic species can be intermediate in the process.

This observation has been confirmed by Tanaka's[36] studies.
A photochemical system has been studied with $[Ru(bipy)_3]^{2+}$ and $[Ru(bipy)_2(CO)_2]^{2+}$ dissolved in TEOA)/DMF (1:4 v/v) mixture saturated with CO_2. This system catalyzes a selective CO_2 reduction to formate with a maximum quantum yield of 14 %. The products of the reaction have been detected by h.p.l.c. technique.

In order to mimic biological systems also BNAH (1-benzyl-1,4-dihydronicotinamide) was used as electron source in aqueous medium.[36]
In fact, BNAH can be considered as a model for NADPH (nicotinamide-adenin-dinucleotidephosphate), the most important electron and proton source in Nature. In this case photoirradiation at λ above 400 nm of a CO_2 saturated H_2O/DMF solution (1:9 v/v) containing $[Ru(bipy)_3]^{2+}$ and $[Ru(bipy)_2(CO)_2]^{2+}$ gave HCOOH (maximum quantum yield of 2.7%) and CO (maximum quantum yield of 14.8%).

5.2.1 Mechanistic aspects in carbon dioxide photoreduction.

Although very few data are available to build up a mechanism for photo-catalytic carbon dioxide reduction, several models have been proposed. The overall process consists of two cycles: a photochemical one and a thermal one.
In the photochemical cycle the photosensitizer $[Ru(bipy)_3]^{2+}$ or $[Re(bipy)(CO)_3Cl]$ is excited by light energy and is reductively quenched by the electron donors.

$$[Ru(bipy)_3]^{2+} \dashrightarrow [Ru(bipy)_3]^{2+*} \quad (1)$$

$$[Ru(bipy)_3]^{2+*} + E.D. \dashrightarrow [Ru(bipy)_3]^{+} + E.D^{*} \quad (2)$$

E.D. = electron donor.

TEOA, BNAH or ascorbic acid have been used as electron donors in some systems, and $[Ru(bipy)_3]^{+}$ species is generated in the medium.
The $[Ru(bipy)_3]^{+}$ generation in presence of tertiary amines has been demonstrated in flash photolysis experiments.[37]
Since the redox potential of the couple $[Ru(bipy)_3]^{2+/+}$ is -1.23 V vs. S.H.E.[35] this complex, in its reduced form, is able to reduce species as:

- $[Co(bipy)_3]^{2+}$ [20-30a,b]
- $[Ni(ciclam)]^{2+}$ (E° $Ni^{2+/+}$= -1.1 V vs. S.H.E.)[34]
- cis-$[Ru(bipy)_2(CO)Cl]^{+}$ (E° $Ru^{2+/+}$ = -1.19 V vs S.H.E.)[38-39]
- $[Ru(bipy)_2(CO)_2]^{2+}$ (E° $Ru^{2+/+}$= - 0.88 V vs. S.H.E.)[36]

Because monoelectronic carbon dioxide reduction is a process requiring at least -2 V, in these conditions a trasfer of two electrons has to be postulated . One can suppose that the following reaction occurs:

$$2\,[Ru(bipy)_3]^+ + M^{(II)}Ln \text{ -----> } 2\,[Ru(bipy)_3]^{2+} + M^{(0)}Ln \quad (3)$$

$M^{(II)} = Co^{(II)}, Ni^{(II)}, Ru^{(II)}$
$M^{(0)} = Co^{(0)}, Ni^{(0)}, Ru^{(0)}$
L= bipy, phen, cyclam, CO, Cl.

This process can be schematically represented as shown in Fig. 6.

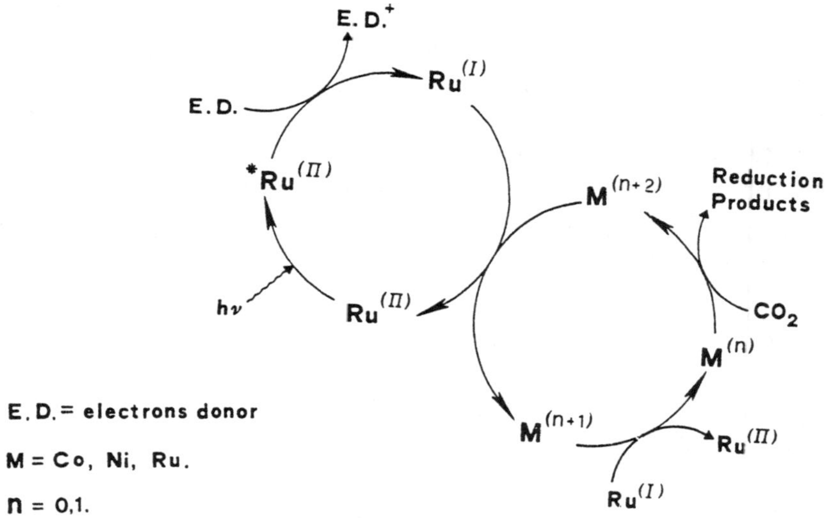

Fig.6 Schematic representation of the catalytic reduction of CO_2 by homogeneous catalysts.

Also unknown are both the coordination mode of carbon dioxide to the metal center and the mechanism for CO or HCOOH generation. For carbon dioxide reduction to formate Lehn et coll.[35] have proposed the following mechanism:

$$[Ru(bipy)_2(CO)]^0 + CO_2 \text{ -------> } [Ru(bipy)_2(CO)CO_2]^{2+} \quad (4)$$

$$[Ru(bipy)_2(CO)CO_2]^0 + H^+ \text{ -------> } [Ru(bipy)_2(CO)(COOH)]^+ \quad (5)$$

$$[Ru(bipy)_2(CO)CO_2]^0 + H^+ \text{ --------> } [Ru(bipy)_2(CO)HCOO]^+ \quad (6)$$

It is worth to note that complex $[Ru(bipy)_2(CO)COOH]^+$ prepared by an independent route does not rearrange to a formate complex. If a comparison with dark processes is meaningful, it might be considered that protonation of carbon dioxide bonded to Ni(0)[40a] in the η^2-C,O mode of bonding, affords with high selectivity (100 %) CO. Reaction with dihydrogen of the Ni-complex affords formate.[40b]
The mechanism of formate formation in light driven reactions requires further investigation.

5.3 Other systems able to reduce CO_2 in homogeneous conditions.

The photochemical conversion of carbon dioxide has been also achieved using tungsten carbonyl complexes.[41] The reactive species is a 19-e^- metal complex photogenerated by irradiation with visible light (λ >420 nm) of the dimer $W_2(CO)_{10}^{2-}$ that shows adsorption at 347 nm (ε = 740) and 390 nm (sh, ε = 4840) assigned to the $\sigma \rightarrow \sigma^*$ and $\pi \rightarrow \sigma^*$ transitions, respectively.[42]
The irradiations were therefore into the low energy tail of the $d\pi \rightarrow \sigma^*$ transition.
The products of CO_2 conversion were $NaHCO_2$ (20%), Na_2CO_3 (39%) and $NaHCO_3$ (18%).

The proposed mechanism considers that the 19-e^- species $[W(CO)_5PR_3]^-$ can be generated by irradiation of $[W_2(CO)_{10}]^{2-}$ in the presence of PR_3. This 19-e^- species would be an agent capable of reducing CO_2: the reduction potential of $[W(CO)_5PR_3]^-$ is estimated to range around -2.6 V vs. SCE for analogy with the reduction potential of $[W(CO)_6]$.[43]
This photochemical conversion is reminiscent of the so called "disproportionation" of carbon dioxide

$$2CO_2 + 2e^- \longrightarrow CO + CO_3^{2-} \qquad (6)$$

promoted by several metal systems.[44]
The authors have proposed that residual water in the system is the source of hydrogen for the formate, but they exclude that a hydrido species of the type $HW(CO)_5^-$ would generate formate by reaction with CO_2.
Quantum yields are of the order of 0.15 at λ = 436 nm

In this case phosphine oxide, that can be formed by reaction of free phosphine with coordinated CO_2,[45] is not observed.

The photoinducted disproportionation of CO_2 (Eq. 6) in the presence of transition metal systems has been documented quite recently[44] for $Cp_2Mo(CO_2)$ irradiated at 330 nm. In this case formate was not detected and only CO and $CO_3^{=}$ were formed.

Another attempt to use light for CO_2 conversion in the presence of transition metal complexes has been made with the aim of using excited states of the metal systems.

In particular, the dimer $Ni_2(\mu\text{-CNMe})(CNMe)_2(dppm)_2$ [dppm=bis-(diphenylphosphino)methane] reacts under irradiation (λ 355 nm) with CO_2 to afford $\{Ni_2[\mu\text{-CN(Me)(COO)}](CNMe)_2(dppm)_2\}$.

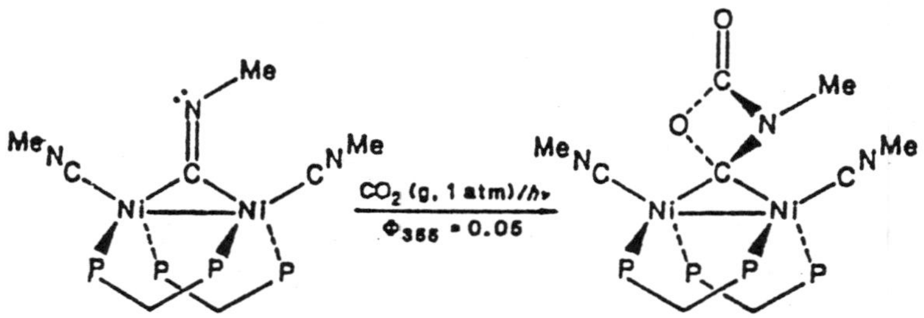

Fig.7 Photochemical coupling of CO_2 with RNC.

According to the authors the same product can be obtained by the thermal reaction of the nickel dimer with liquid CO_2.

The active form in the photochemical process has been proposed to be a complex with high charge separation (Fig 8) able to react with CO_2.

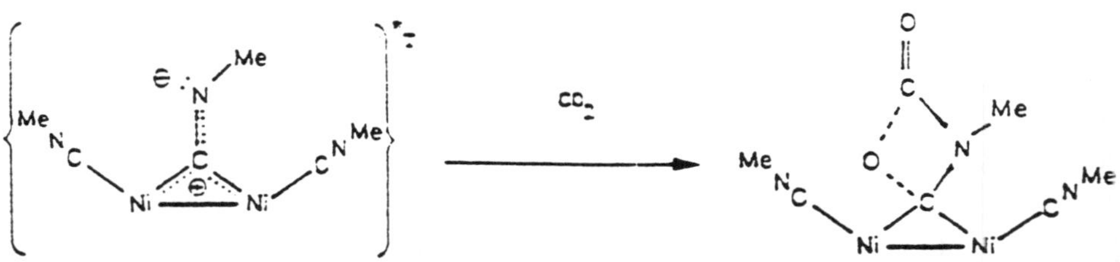

Fig. 8 Charge separation.

5.4 Photochemical production of fine- and speciality chemicals from CO_2 Catalyzed by transition metal systems.

An area of increasing interest is the photochemical synthesis of fine- or speciality-chemicals from CO_2 in the presence of transition metal systems acting as catalysts. Only a few example are available in the literature and the full potentiality of this technique has not been explored.
Transition metal complexes excited states can be dissociative with respect to metal-ligand or metal-metal bond. This can generate unsaturated species prompt to react with organic substrates with subsequent C-H or C-C activation.[46] Alternatively, the excited state can interact with a substrate affording reactions that are or are not thermally allowed. An example of the interaction of M->LCT with CO_2 has been discussed earlier.
Reactions of great interest for their practical application are those depicted below:

$$\text{C-H} + CO_2 \dashrightarrow \text{C-COOH} \qquad (7)$$

$$\text{C-C} + CO_2 \dashrightarrow \text{C-C}(=\text{O})\text{-O-C} \qquad (8)$$

These reactions do not occur easily by the thermal way. The C-H[47] and C-C bond[48] cleavage catalyzed by transition metal systems are documented in the literature. Coupling the action of radiations with the catalytic action of TM systems can bring to new synthetic procedures.
The irradiation of $RhClP_3$ (P=PMe_2Ph) solutions in toluene at 250 K under CO_2 (20 atm) and cyclopropane (3 atm) pressure in an autoclave (Fig. 9) equipped for irradiation experiments under controlled temperature[49] afforded the synthesis of butyrrolactone (Scheme 1).
This product can be considered as derived from a formal CO_2 insertion into a C-C bond.
Interestingly, the thermal way does not afford butyrrolactone, the major CO_2 conversion product being the reduction to CO by the phosphane ligand.[45]

$$CO_2 + PMe_2Ph \dashrightarrow (CO)_{\text{bound or free}} + Me_2PhP{=}O \qquad (10)$$

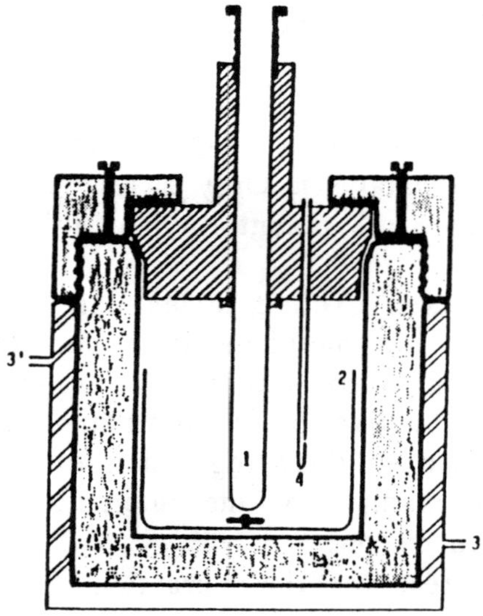

Fig. 9 Schematic diagramme of the irradiation reactor under pressure. (1) Quartz tube for UV lamp; (2) glass vessel; (3,3') thermostat inlet/outlet; (4) thermometer well.

The resulting carbonyl complex does not catalyze the lactone formation.

That coordinatively unsaturated Rh(I) complexes are able to activate C-H bonds is well documented in the literature.[50]

Similarly it is known that allyl Ni complexes can promote the allyl-CO_2 coupling with subsequent lactone elimination.[51] σ-π-Allyl equilibrium has been proposed as operative in several organometallic systems derived from butadiene and able to fix carbon dioxide.[52]

Another example of photoassisted fixation of CO_2 is reported in Scheme 2.

The isomerization allyl --> propenyl aniline is also observed in a thermal reaction,[50] but the formation of the oxazolidine-2-one is a light driven reaction that most probably goes through a N-H activation.

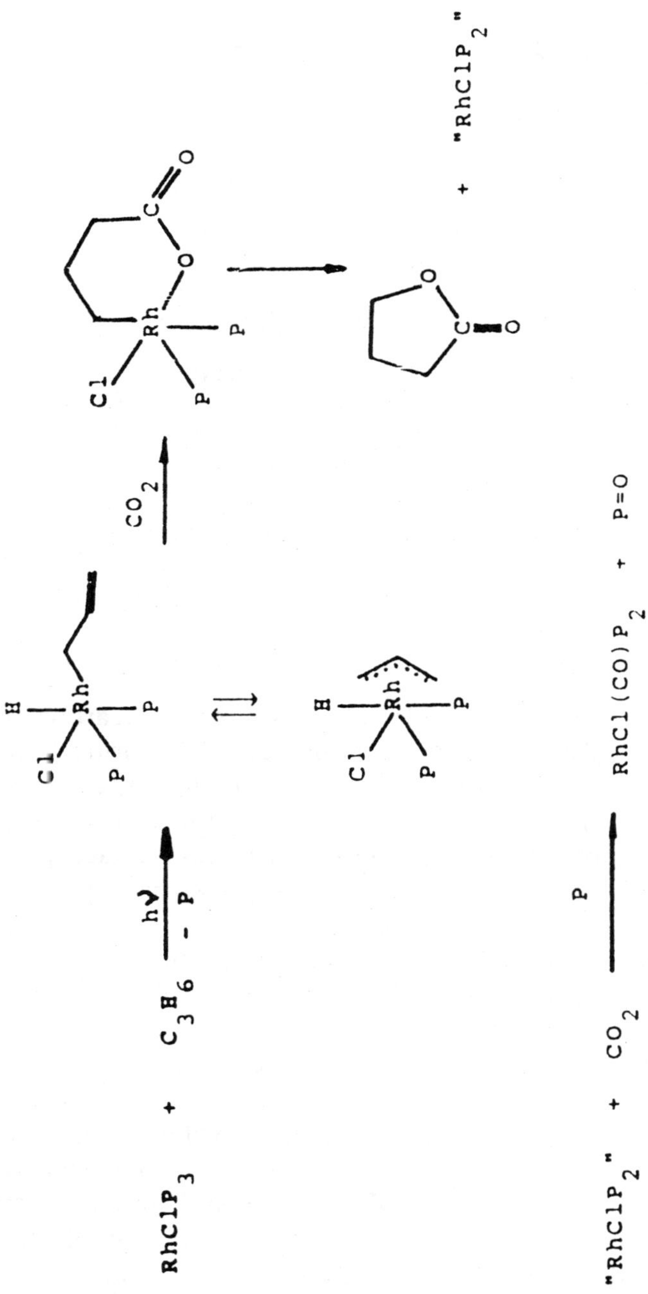
RhClP3
+
C3H6
hν
- P
Cl
H
Rh
P
CO2
O
C
"RhClP2"
+
"RhClP2"
+
CO2
P
RhCl(CO)P2
+
P=O

Scheme 1

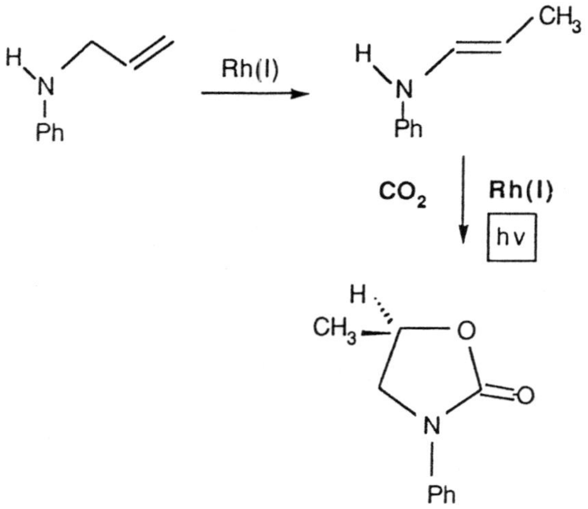

Scheme 2

Conclusions.

The utilization of carbon dioxide as a source of carbon for the synthesis of chemicals and fuels is a process of enormous potential practical application. It requires energy and the use of sunlight seems an attractive hypothesis. Besides the utilization of biological systems such as photosynthetic bacteria, enzymes, algae, plants etc. (not discussed in this paper), artificial ones can play a very important role.

Harvesting solar energy remains the crucial point. This energy might be used to drive chemical (as heat), photochemical and photoelectrochemical reactions.

Both heterogeneous and homogeneous catalytic systems can be operated for CO_2 utilization.

The entire molecule can be used in carboxylation reactions or, alternatively, its reduced forms can be used in the synthesis of chemicals. HCOOH, H_2CO, CH_3OH can act as hydrogen carrier and energy vectors. The utilization of water as a source of electrons and protons in the synthesis of these chemicals is a key point.

In particular the synthesis of methanol is of great interest as this molecule is widely used in the chemical industry as a starting material. The possibility of synthetizing C2 molecules (oxalates,

glyoxylic acid, acetaldehyde, acetic acid, glycols etc.) directly from CO_2 and water also requires attention for the importance those species have in the chemical industry.
The reaction mechanisms must be clarified both in homogeneous and heterogeneous processes. This would allow to design new catalysts for achieving high conversion rates and good selectivity. Comparisons with reaction mechanisms ascertained for dark carbon dioxide reduction reactions can be helpful for understanding the role of the catalysts in the reduction process and for designing new more selective and efficient ones. It may turn out that coupling efficient artificial antennae and fast charge transfer systems to selective transition metal complexes acting as catalysts can afford very good carbon dioxide reduction systems, may be more active than natural ones. This will require the cooperation of Scientists working in different areas and long-term international programmes.

References.

1)a) M. Aresta, in "Carbon dioxide as a source of carbon", M.Aresta and G. Forti Eds., Reidel Publ., p. 1-22, 1987.

b) I.M.Smith, IEACR/O7, London, 1988, IEA Coal Research.

c) M.Shepard, EPRI Journal, USA 1986.

d) J. Mathews, Science and Technology, 3, 57,1987 .

2) "The changing atmosphere: implications for global security". in "The changing atmosphere conference", Toronto, Ontario, Canada, 27-30 June, 1988. Downsview, Ontario, Canada, Environment Canada.

3)a) I.M. Torrens, Conference of global climate linkages; acid rain, air quality and stratospheric ozone, Washington DC, USA, 15-16 November, 1988.

b) M.Steinberg and H.C. Cheng, Proceedings APCA Annual Meeting, Dallas, TX, USA, 1988.

4) IEA-OECD Expert Seminar on *"Energy technologies for reducing emissions of greenhouse gases"*, Paris, 12-14 April, 1989.

5) M.Steinberg and C.H.Cheng, *"A system study for the removal, recovery and disposal of carbon dioxide from fossil fuel power plants in the United States"*, 1984.

6) C. Blok, C. Hendriks, W.C.Turkenburg, IEA-OECD Expert Seminar, Paris, 12-14 April, 1989.
7) M.Aresta, IEA-OECD Expert Seminar, Paris, 12-14 April, 1989.
8) T.Inoue, A.Fujishima, S.Konishi and K.Honda, Nature, **277**,637,1974.
9) M.Ulmann, B. Aurian-Blajeni and M.Halmann, Chem. Tech.,235,1984.
10) G.Silvestri, in "Carbon dioxide as a source of carbon: biochemical and chemical uses", M.Aresta and G.Forti Eds., Reidel Publ., p. 339, 1987.
11) A.Monnier, J.Augustynski and C.Shalder, J.Electroanal.Chem., **112**, 383, 1980.
12) M.Ulman, A.H.A.Tinnemens, A.Mackor, B.Aurian-Blajeni and M.Halmann, Int. J. Solar Energy, **1**, 213, 1982.
13) A.Mackor, and G.Blasse, Chem. Phys. Letters, **77**, 6, 1981.
14) A.H.A.Tinnemans, T.P.M.Koster, D.H.M.W.Thewissen and A.Mackor, Nouv. J. Chim., **6**, 373, 1982.
15) M.W.Rophael and M.A.Malati, J. Chem. Soc. Chem. Comm., 1418,1987.
16) S.Yamamura, H.Kojima, J.Iyoda and W.Kawai, J. Electroanal. Chem., **225**, 287, 1987.
17) I.Willner, R. Maidan, D.Mandler, H.Dürr, G. Dörr and K.Zengerle, J. Am. Chem. Soc., **109**, 6080, 1987.
18) F.Solymosi, A.Erdonelyi and M.Kocsis, J.Chem. Soc., Faraday Trans. I, **77**, 1003, 1981 and references therein.
19) P.Moggi, G.Albanesi, G.Predieri and E. Sappa, J. Organometall. Chem., **252**, C89, 1983.
20) F.R.Keene, C.Creutz and N.Sutin, Coord. Chem. Rev., **64**, 247, 1985.
21) M.Aresta, A.Ciccarese and E.Quaranta, Gazzetta Chim. Ital., in press.
22) M.Ulmann and J.Augustynski, Communication C19 in "Enzymatic and model carboxylation and reduction reactions for carbon dioxide utilization", NATO-ASI Summer School, Riva dei Tessali (Italy), 1989.
23) H.Dürr, G.Dörr, K.Zengerle, E.Meyer, J.-M.Curchod and A.-M.Braun, Nouv. J. Chim., **9**, 717, 1985.
24) R.K.Thampi, J.Kiwi and M.Grätzel, Nature, **327**, 506, 1987.
25) N.Vlachopoulos, P.Liska, J.Augustynski and M.Grätzel, J. Am. Chem. Soc., **110**, 1216, 1988.

26) B R.Eggins, J.T.S.Irvin, E.P.Murphy and J.Grimshaw, J.Chem. Soc. Chem. Comm., 1123, 1988.
27) A Henglein, Ber. Bunsenges. Phys. Chem., **86**, 301, 1982, ; Pure Appl.Chem., **56**, 1215, 1984.
28) N.Getoff, Z. Naturforcsh., **B 17**, 87, 1962; **B 18**, 169, 1963.
29) S.Tazuka and N.Kitamura, Nature, **275**, 301, 1978.
30) a) J.M.Lehn and R.Ziessel, Proc. Natl. Acad. Sci. USA, **79**, 701, 1982.
b) R.Ziessel, J.Hawecker and J.M.Lehn, Helv. Chim. Acta, **69**, 1065, 1986.
c) J.Hawecker, J.M.Lehn and R.Ziessel, J. Chem. Soc. Chem. Comm., 536, 1983; ibid., 56, 1985.
31) R.Ziessel, in "Enzymatic and model carboxylation and reduction reactions for carbon dioxide utilization", M.Aresta and J.V.Schloss Eds., Kluwer Publ., p. 79, 1990.
32) V.Balzani, L.Moggi, M.F.Manfrin, F.Bolletta and M.Gleria, Science, **189**, 852, 1975.
33) R.Ziessel, in "Carbon dioxide as a source of carbon: biochemical and chemical uses", M.Aresta and G.Forti Eds., Reidel Publ., p.113, 1987.
34) J.L.Grant, K.Goswami, L.O.Spreer, J.W.Otvos and M.Calvin, J. Chem Soc., Dalton Trans., 2105, 1987.
35) J.M Lehn and R.Ziessel, J. Organometall. Chem., **382**, 157, 1990.
36) H.Ishida, T.Terada, K.Tanaka and T.Tanaka, Inorg. Chem., **29**, 905, 1990.
37) C.P.Anderson, D.J.Salmon, T.J.Meyer and R.G.Young, J. Am. Chem. Soc., **99**, 1980, 1977.
38) J.M.Kelly, C.M.O'Connell and J.G.Vos, J. Chem. Soc. Dalton Trans., **253**, 1, 1986.
39) J.M.Kelly and J.G.Vos, J. Chem. Soc. Dalton Trans., 1045,1986.
40) a) M.Aresta, E.Quaranta and I.Tommasi, JCS Chem. Comm., 450, 1988.
b) M.Aresta, E.Quaranta and I.Tommasi, manuscript in preparation.
41) N.D.Silavwe, A.S.Goldman, R.Ritter and D.R.Tyler, Inorg. Chem., **28**, 1231, 1989.
42) D.G.Harris and H.B.Gray, J. Am. Chem. Soc., **97**, 3073, 1975.
43) C.J.Pickett and D.J.Plechter, Chem. Soc. Dalton Trans., 879, 1975.

44) K.A.Belmon, R.A.Vanderpool, J.-C. Tsai, M.A.Kahn and K.M.Nicholas, J. Am. Chem. Soc., **110**, 2004, 1988.
45) M.Aresta and C.F.Nobile, Inorg. Chim. Acta, **24**, L49, 1977.
46) R.H.Crabtree, Chem. Rev., **85**, 245, 1985.
47) D.M.Haddleton and R.N.Perutz, J.Chem. Soc. Chem. Comm., 1734, 1986.
48) W.E.Farneth and M.W.Thomsen, J. Am. Chem. Soc., **105**, 1843, 1983.
49) M.Aresta, A.Ciccarese and E. Quaranta, C1 Mol. Chemistry, **1**, 267, 1985.
50) M.Aresta, E.Quaranta, S.Treglia and J.A.Ibers, Organometallics, **7**, 577, 1987.
51) T.Tsuda, Y.Chujo and T.Saegusa, Synth. Commun., **9**, 427, 1979.
52) a) A. Behr, in "Carbon dioxide as a source of carbon: biochemical and chemical uses", M.Aresta and G.Forti Eds., Reidel Publ., p.295, 1987.
b) A.Behr "Carbon dioxide activation by metal complexes" Werlag Chemie, Weinheum, 1988.

PROGRESS IN THE APPLICATIONS OF PHOTOCHEMICAL CONVERSION AND STORAGE

A.M. BRAUN
Institut de chimie physique
Ecole Polytechnique Fédérale de Lausanne
1015 Lausanne,
Switzerland

ABSTRACT. Recent research results in solar photochemistry, in particular in the domains of photochemical conversion of solar energy and of photocatalyzed oxydative degradation are reviewed with respect to their up-scaling potential.

1. Introduction

Status reports on technical developments and applications in a given field can only reflect information already released to the public domain. They are thus necessarily incomplete as far as results and techniques and their potential impact on further experimental research are concerned. On the other hand, this report is not meant to furnish an up-dated literature survey covering all the work published in the domain of solar photochemistry since IPS 7 two years ago. In fact, for an evaluation of the status of research and development in solar photochemistry, in general, and of photochemical conversion of solar energy, in particular, actual and potential efficiency levels of competing, non-photochemical techniques are used as a reference, and technically promising developments will be discussed in placing their up-scaling potential on the same scale.

In the last few years, the domain of photochemical conversion of solar energy expanded definitely from its puristic restriction of dealing exclusively with techniques and materials producing chemicals containing high energy potentials at minimum molecular weight under the impact of solar light. In fact, a large section of IPS 8 is dedicated to photocatalytic oxidation reactions to be exploited on an industrial level, hence, reflecting, at least for their final stage of development, only an interest in energy economy. Based on the level of photochemical engineering already achieved and due to the experience in the domain of industrial photochemistry [1], the transfer of photochemical techniques from applications using artificial light sources to developments exploiting solar illumination may be achieved without major difficulties. Such success in energy economy should, however, not deviate the still gigantic research effort needed in order to reach a technical solution in the domain of solar energy conversion and storage, on which may depend further improvements of the living conditions on this planet to a large extent.

E. Pelizzetti and M. Schiavello (eds.), Photochemical Conversion and Storage of Solar Energy, 551–560.

2. Energy conversion for storage

2.1. PHOTOVOLTAIC SYSTEMS

Based on successful pilot experiments with large surface silicon captors and on parallel improvements in the construction of water electrolysis cells, small to medium size hydrogen production plants (~ 10^5 $m^3 \cdot y^{-1}$), combining these two techniques, are planned or under construction in some of the highly industrialized countries [2]. At present, storage is avoided in all known projects, hydrogen being introduced into existing gas supply systems which already satisfy higher security levels [3]. These projects represent an already strong effort to reduce the consumption of fossil fuels or its derivatives and to diminish the consequential ecologic and climatic problems. Despite of these early large investments in pioneering production plants, the combination of electricity producing captors and water electrolysis is the topic of considerable efforts in both research and development, and new materials and techniques will lead in a near future to efficiencies of hydrogen production of ~ 30%, related to the absorbed radiative energy [4].

Besides improvements in solid state photovoltaic captors, regenerative photoelectrochemical cells occupy the center of interest of a number of research groups [5]. The development of such cells may be seen as a logic continuation of the pioneering experiments on light induced charge separation in microheterogeneous, heterogeneous and supramolecular systems [6-8].

In fact, successes in supramolecular photochemistry reflect the enormous research effort undertaken towards a detailed description and the modelling of the complex mechanisms of photosynthesis [9].

To be designed for the task to produce large quantities of a chemical energy vector on a limited surface, such regenerative photochemical conversion systems must necessarily show at least comparable efficiency and stability characteristics when compared to solid state photovoltaic devices already existing or under development. Latest results of experiments using functionalized semiconductor surfaces show a net breakthrough as far as efficiency is concerned. In fact, overall light-to-current conversion efficiencies of > 70% have been found [5]. As the spectral response of these devices is changing with the "sensitizer" attached to the semiconductor surface, and polychromatic excitation is used in pragmatically designed laboratory experiments, information on quantum yields of charge injection into the electrochemical circuit is, however, not available.

Light stability of photochemical conversion systems involving complex organic molecules is of major concern and has to be tested under conditions of actual utilisation, before any forecast on their technical potential can be given.

The efficiency of electricity producing molecular devices is by far superior to that of earlier systems where charge separation was primarily controled by cage escape efficiencies or was enhanced by exploiting differences of potenitials across charged interfaces [10]. On the other hand, photochemical conversion systems consisting of larger than colloidal surfaces (large particles, (functionalized) membranes and electrodes) were known to exhibit unsatisfactory absorption characteristics compared to homogeneous or microheterogeneous systems, their surfaces reflecting a large part of the incident light. This disadvantage has apparently been overcome as new techniques allow the optimization of the morphology of the semiconductor surfaces towards a minimum reflectance of the incident irradiance [5].

Further problems to be solved in technical development will include the optimal chemical redox system, as light induced radical generation has to be avoided for the sake of light stability. Early solutions to this problem might involve compartmentalization of the energy conversion part, although this would probably abort final development for practical and economical reasons.

2.2. THERMOELECTRIC SYSTEMS

Despite the uneconomical on-off mode of operation and the many consequential technical problems, thermoelectrical pilot plants reach an over-all efficiency of 17% and represent, at present, a realistic alternative for relative large scale production of electricity. Latest pilot plants are designed for the yearly production of up to 50 MW electrical power [3].

2.3. HYDROGEN OR HYDROCARBON ECONOMY

On paper, the combustion of hydrogen is without ecological impact. However, due to the high temperature of combustion, nitrogen oxides will also be produced during oxidation under normal atmospheric conditions. In addition, no precise model calculations have been made so far which could describe potential climatic changes due to the highly delocalized and dense water vapor production at times of a generalized hydrogen economy.

On shorter and more pragmatic terms, huge investments for a safe distribution system of hydrogen are certainly a major argument to direct research policy and projects towards methane as an energy vector. Hydrogen, produced by water electrolysis, would then be used for the reduction of carbon dioxide (Sabatier reaction (1)), although a net loss of the efficiency of the total energy conversion system would have to be taken into account.

$$CO_2 + 4\ H_2 \longrightarrow CH_4 + 2\ H_2O \qquad (1)$$

A recent literature survey has revealed that up to the present no efficient methanation process is yet known [11a,b]; methane (and methanol) production being industrially produced from carbon monoxide, practically all further development efforts concern the reduction of CO. This situation leaves a large field of research open to scientists specialized on mechanisms of catalytic reactions and on the formulation and testing of new catalysts.

However, electricity produced by photovoltaic devices could be used for the electrochemical CO_2 reduction which has been found to produce CH_4 and C_2H_4 with efficiencies of already up to 50 and 10%, respectively [11c].

2.4. PHOTOCHEMICAL CONVERSION SYSTEMS

In contrast to the photovoltaic system, photochemical conversion systems are capable to produce directly a compound of defined composition and structure to be used as an energy vector. Historically, this domain of research developed from interdisciplinary experience in such fields as transition metal complex chemistry, surfactant chemistry , electrochemistry, colloid chemistry ... It is interesting to note that the initially explored reaction cycle, consisting of $Ru(bipy)_3{}^{2+}$, methylviologen (MV^{2+}) and colloidal platinum and using ethylenediamine tetraacetic acid (EDTA) as a reductant ((2) to (5)), is still considered a standard against which all new "sensitizer", relay molecules and catalyst systems are compared [12].

$$Ru(bipy)_3^{2+} \xrightarrow{h\nu} (Ru(bipy)_3^{2+})^* \quad (2)$$

$$(Ru(bipy)_3^{2+})^* + MV^{2+} \longrightarrow Ru(bipy)_3^{3+} + MV^{+\cdot} \quad (3)$$

$$MV^{+\cdot} + H_2O \xrightarrow[Pt]{} MV^{2+} + 1/2\ H_2 + HO^- \quad (4)$$

$$Ru(bipy)_3^{3+} + EDTA \longrightarrow Ru(bipy)_3^{2+} + EDTA_{ox} \quad (5)$$

$$Ru(bipy)_3^{3+} + MV^{+\cdot} \longrightarrow Ru(bipy)_3^{2+} + MV^{2+} \quad (6)$$

As the efficiency of electron transfer from the excited "sensitizer" to the relay molecule (3) is limited by the corresponding back reaction (6), it was thought that "sensitizers" working in accord with the reductive mechanism ((7) to (9) and (4)) would lead to higher quantum yields of generation of reduced relays; in fact, this can be shown when a dissolved reductant is used [13].

$$Ru(bipz)_3^{2+} \xrightarrow{h\nu} (Ru(bipz)_3^{2+})^* \quad (7)$$

$$(Ru(bipz)_3^{2+})^* + EDTA \longrightarrow Ru(bipz)_3^{+} + EDTA_{ox} \quad (8)$$

$$Ru(bipz)_3^{+} + MV^{2+} \longrightarrow Ru(bipz)_3^{2+} + MV^{+\cdot} \quad (9)$$

However, no such system is ready yet for technical development, where the reductant would be substituted by an electrode injecting electrons into the electronically excited "sensitizer" linked to its surface: one of the concepts for the hydrogen producing part of the combined cycles of photocatalyzed water photolysis.

Experiments with the above mentioned sacrificial system of hydrogen production (oxidative mechanism, ((2) to (5)) using an advanced solar photochemical pilot reactor (> 30 l) showed important obstacles still to be overcome and which cannot be seen or easily investigated in laboratory scale experiments [2,3,14]:

(a) colloidal catalyst systems are easily precipitated, when important temperature variations are applied. In fact, in planar solar captors, temperature differences between day and night time operation reach 40 to 50°C, without any amplification of the incident irradiance and despite the use of a heat exchanger (tap water) connected to the reservoir installation during the period of solar illumination. If this concept of hydrogen production should be continued, different solutions for catalyst dispersion must be adopted, e.g. supported catalysts using fluidized bed type solar reactors.

b) Several research groups have studied the hydrogenation of organic molecules contained in the aqueous hydrogen producing system. In fact, platinum catalyzed hydrogenation of methylviologen is the most important secondary reaction in up-scaled experiments, reducing electron transfer efficiency from the electronically excited "sensitizer" (oxidative mechanism), and, consequently, the light stability of the "sensitizer". Considerable efforts are presently undertaken in a number of research groups focussed on complex preparative projects in order to develop supramolecular structures, such as (cryptand)-type "sensitizers" with improved light stability due to strongly limited photoannation characteristics [7,15,16]. Such molecular devices could also consist of a direct "sensitizer"-catalyst contact which could improve overall efficiency of hydrogen production due to an intramolecular, hence, not diffusion dependent electron transfer and excluding a competing redox reactions of the relay molecule.

From the technical point of view, photochemical conversion systems will use different captor systems from those likely to be designed for photovoltaic applications or for photocatalyzed oxidative degradation processes. In fact, the incident photonic irradiance of solar irradiation on a given reactor surface is important enough in order to be compared to that of artificial light sources of medium electrical power. Consequently, similar problems of secondary reactions in reactor (captor) sections of high absorbed photon fluxes are to be overcome [1]. As one final result from up-scaled experiments, we deduce that attempts to enhance solar irradiation of photochemical reaction systems containing organic material will lead to a sharp increase of secondary products and hence to a abortive diminution of the number of reaction cycles. In fact, nature is teaching us the efficient use of extended antenna systems for the optimization of photosynthetic processes in complex organic and metal organic reaction systems [7].

Moreover photochemical conversion systems based on the catalyzed photolysis of water consist finally of both a hydrogen and an oxygen producing cycle. For the latter, improvements of its efficiency are still far from a result equivalent to that of the photocatalyzed hydrogen production and, hence, coupling of the two cycles has not yet been achieved in a self-consistent reaction system. In actual concepts for the combination of the two cycles, "sensitizer" functionalized membranes or electrodes are preferred, as they allow for an enhanced flow of electrons across a barrier configuration which provides physical separation of oxidation and reduction products. Four years ago, a water splitting device has been presented to this conference producing stoichiometric quantities of hydrogen and oxygen in separated flasks [17]. As no further development of this promising system has been reported, we must deduce from the information available, that, in general, the technical realization of photochemical conversion systems is far less advanced than that of photovoltaic systems mentioned above.

Besides hydrogen and oxygen production from photocatalyzed water photolysis, artificial photosynthesis, where, e.g., CO_2 would be directly reduced, is an important topic for a number of research groups. Although, no CH_4 could be isolated yet, CH_3OH with efficiencies of up to 80% has been produced on a laboratory scale [11c]. Photosynthetic N_2 fixation has qualitatively been demonstrated using a p-GaP photocatalyst and Al as a sacrificial electron donor [11c]; earlier reports have shown that Al could be produced from Al_2O_3 and C in a high-temperature solar oven [18].

Several research groups have proposed to undertake photochemical reactions on a preparative scale using solar irradiation. These projects are basically a "replay" of what has been advocated many years ago by eminent photochemists, such as Ciamician and Schönberg. The incident solar photonic irradiance in the spectral domain of 320 to 500 nm is ~ 10 $einstein \cdot m^{-2} \cdot d^{-1}$, hence, assuming a photochemical reaction with $\Phi = 1$, 2 kg of a product with a molecular weight of 200 could be prepared. In the present context, leaving the well known state of empirical product isolation and analysis, such projects might find some applications in the domains of photohalogenation, photooxychloration and sensitized oxidation [1], but, for the sake of a progress in photophysics and mechanistic photochemistry, should include investigations of activation spectra for primary and secondary product formation, quantum yield determinations under polychromatic excitation, as well as solar photochemical reactor design.

3. Photocatalyzed oxidative degradation

The TiO_2 photocatalyzed oxidative water splitting is known now for more than half a century, but the idea of its potential exploitation has not been taken up until the last five years. Despite the fact that there are several mechanisms involved in oxidative degradation of organic matter by this method [19], the numerous results published, in particular those with respect to the degradation of perfluorinated and perchlorinated compounds, lead us to include this technique into the class of oxidations by hydroxyl radicals ((10) to (13)). The reaction is yet another method of producing hydroxyl radicals which are used for the oxidative degradation (partial or total mineralization) of organic molecules contained in the aqueous systems.

$$(TiO_2) \xrightarrow{h\nu} h^+ + e^- \tag{10}$$

$$h^+ + H_2O \longrightarrow H^+ + HO^\cdot \tag{11}$$

$$HO^\cdot + RH \longrightarrow H_2O + R^\cdot \tag{12}$$

$$R^\cdot + O_2 \longrightarrow ROO^\cdot \longrightarrow \longrightarrow \longrightarrow \ldots \tag{13}$$

$$e^- + O_2 \longrightarrow O_2^- \tag{14}$$

Table 1 summarizes different techniques used or under development in order to cope with cases of severe pollution of the phreatic drinking water reserves, of surface waters used for the same purpose, as well as for a non-biological treatment of residual waters from different industries.

It becomes clear that the TiO_2 catalyzed degradation method has many technically well developed and attractive competitors. However, some of their draw-backs should be mentioned:

a) H_2O_2 is produced by a small number of manufacturers, and its price is one of the major constituents of an economic feasibility calculation.

The H_2O_2 photolysis is practically limited to the UV-C domain [20], and its efficiency might be strongly diminished by inner filter effects due to light absorption by dissolved or dispersed pollutants, as well as naturally contained humic materials [21].

Numerous catalysts have been developed for H_2O_2 decomposition, however, their use will be limited in applications where drinking water is involved. Means to maintain a critical Fe^{2+} concentration in a generally oxidative reaction environment are presently developed.

b) Photochemical and thermal ozone decompostions suffer from the economical draw-back of very high investment and maintenance (security) costs [22].

As the solubility of O_3 in water is rather limited, the absorption cross section of the dissolved or finally dispersed reactant is of primary concern for reactor design. For the same reason of limited solubility, care must be taken for efficient mixing, relative long contact or irradiation times and for the recycling of the unreacted oxidant. All these problems are still topics of new inventions in reactor design and construction.

Alkali addition for thermal O_3 decomposition would require pH adjustments after treatment which could have negative efects on the salinity in drinking water applications.

TABLE 1. - Present level of technical development of oxidative degradation techniques based on hydroxyl radical generation

Technique	Effidiency	Spectral domain	Level of development
H_2O_2 / hν	h	UV-C a)	Production a)
H_2O_2 / catalyst	h a)	---	Production a)
O_3 / hν	m b)	UV-B	Pilot b)
O_3 / pH	m b)	---	Production b)
H_2O_2 / O_3	h c)	---	Pilot b)
TiO_2 / hν	s d)	UV-A	Pilot d)
(Electrochemistry)	s f)	---	Pilot e)

Efficiency related to short residence times: h: high, m: medium, s: small;
a,b,c,d, e, f) see comments in text.

c) Although, suffering from the different economical draw-backs already mentioned, the thermal generation of HO' radicals in the combined application of H_2O_2 and O_3 is often forgotten, but has found the interest of the water treatment industry.

d) TiO_2 has the advantage of an absorption spectrum reaching through the ultra-violet to the limit of the visible spectral domain and, hence, may be used for solar photocatalyzed decontamination processes, whereas competing photochemical processes, such as H_2O_2 or O_3 photolysis, as well as the VUV photolysis of water, remain limited to the use of artificial light sources. Under conditions of solar light excitation, the absorption efficiency of TiO_2 may be further increased when inorganic "sensitizer" functionalized semiconductor surfaces are used. Furthermore, the concept of TiO_2 photocatalysis includes the recyclization of the catalyst, hence, no chemical product is consumed in a first approach towards the technical development of such a process.

All the above mentioned oxidative degradation processes may lead to complete mineralization of the organic material contained in aqueous reaction systems.

e) This does not seem the case for electrochemical processes [23] which actually do not belong to the class of oxidation reactions by hydroxyl radicals. In fact, only organic compounds will be oxidized by this technique which show smaller oxidation potentials than water. The incomplete mineralization might be due to the decreasing rate of oxidation of a diminishing concentration of organic material encountered at the electrodes.

The same might be true for all reactor designs, where the photocatalyst (TiO_2) is fixed on a practically immobile support [24]. Nevertheless, reactor designs based on an immobilized photocatalyst seem attractive as complete mineralization is not a necessary condition, neither for drinking water purification nor for abiotic waste water treatment. Their efficiencies seem, however, generally lower than those of reactors where dispersed TiO_2 particles are used. A disadvantage of the latter technique might consist in the application of rather sophisticated methods of separation and recyclization of the photocatalyst, as technical quantities of the material are only available as a very fine powder. Preparation of larger TiO_2 particles of high mechanical stability has been anounced [25], but no large scale pilot tests have yet been reported. Larger particles consisting of TiO_2 coated inert material would probably be of limited use due to abrasion of the photocatalyst in a highly turbulent reaction system. It seems, however, that fixed bed reactor designs might lead to efficiencies equivalent to those found by using immobilized TiO_2 coated surfaces yet using smaller surfaces of irradiation. Much work is still required in order to quantify those early qualitative results in using generally applicable methods of determining energetic or photonic irradiance [26].

A main concern for the technical development of any light induced oxidative degradation technique, and of the TiO_2 photocatalyzed oxidative degradation in particular, is the limited number of photons available (limited irradiance) and, hence, the overall rate of mineralization when compared with competitive thermal techniques as those mentioned above (footnote f), Table 1). In fact, our pilot scale investigations have shown so far rates of degradation of $\leq$ 100 ppm $TOC \cdot h^{-1} \cdot kW^{-1} \cdot l^{-1}$, at TOC > 100 ppm, whereas for some of the above mentioned photochemical methods, rates of mineralization higher by at least a factor of 10 have been measured, and thermal degradation processes are even faster.

It should be mentioned at this point, that investigations on photochemical or thermal degradation techniques, where only the diminution of the substrate concentration is analyzed, but where no results on the further oxidation of the subsequent products or on the diminution of TOC are available, are in many cases useless for further development. Intermediate products in the oxidative degradation might prove to be either bactericidal or non-biodegradabale, and their production or incomplete removal in waste water by any of these abiotic degradation methods prior to its introduction into conventional treatment stations could cause severe operational problems. Moreover, limits on flux and TOC for the introduction of residual waters into conventional treatment stations or into effluents are imposed in most cases.

We deduce from the collected information that there are potential domains of application of TiO_2 photocatalyzed degradation processes, e.g. for the detoxification and TOC diminution in residual waters of industrial activities with a lower operational TOC limit of 100 ppm and for practically complete mineralization (operational limits: 0.01 ppm < TOC < 100 ppm) for waters mainly used in pharmaceutical and microelectronic research and manufacturing and for drinking water decontamination, where in both cases relative low volumes per unit of time are consumed. In most of these applications, the TiO_2 photocatalyzed degradation techniques would be used in combination with other methods of water treatment. Also, short term drinking water sterilization might be possible by TiO_2 photocatalyzed oxidative water splitting, a process which probably could compete with the energy consuming ultra-violett irradiation used in many places. However, no information as to this application has been available.

Recent investigations have shown that there is only a small temperature effect connected with photocatalyzed TOC diminution. For this reason, the use of solar photochemical reactors in this application would not yield a great advantage over the presently used designs based on artificial light sources. Solar captors enhancing the incident irradiance could, however, be an economic factor when operational costs have to be compared with those of artificial light sources of high electrical power.

Consistent with all the other hydroxyl radical generating processes, TiO_2 photocatalyzed degradation is practically inefficient for the mineralization of perfluorinated and perchlorinated compounds as well as of many perhalogenated (F,Cl) carboxylic acids. Their destruction can be achieved by photochemical methods which, however, are not within the scope of this report.

4. References and suggested reading

[1] (a) Braun, A.M., Maurette, M.-T. and Oliveros, E. (1986) Technologie photochimique, Presses Polytechniques Romandes, Lausanne.
(b) Braun, A.M., Maurette, M.-T. and Oliveros, E., (Ollis, D.F. and Serpone, N., transl.) (1991) Photochemical Technology, Wiley, Chichester

[2] Braun, A.M. (1989) 'Energie und Umwandlungsprozesse. Gibt es Alternativen zur Konzentration?', Civitas (4,5) 113-119.

[3] Weber, R. (1988) Der sauberste Brennstoff. Der Weg zur Wasserstoff-Wirtschaft, Olynthus, Oberbözberg.

[4] Deb, S K.(1991) 'Curreent status of solid state photovoltaic research at SERI', in Pelizzetti, E. and Schiavello, M. (eds.) Photochemical conversion and storage of solar energy, Kluwer Academic Publishers, Dordrecht.

[5] e.g. Grätzel, M. (1990) 'Artificial Photosynthesis', in Scott, A. (ed.) Frontiers of Science, Cambridge University Press, Cambridge, chap. 7.

[6] Kalyanasundaram, K., Grätzel, M. and Pelizzetti, E. (1986) 'Interfacial electron transfer in colloidal metal and semiconductor dispersions and photodecomposition of water', Coord. Chem.Revs 69, 57-125.

[7] Denti, G., Campagna, S., Sabatino, L., Serroni, S., Ciano, M. and Balzani, V. (1991) 'Towards an artificial photosynthesis. Di-, tri and hepta-nuclear luminescent and redox reactive metal complexes', in Pelizzetti, E. and Schiavello, M. (eds.) Photochemical conversion and storage of solar energy, Kluwer Academic Publishers, Dordrecht.

[8] Gust, D., Moore, T.A., Moore A.L., Lee, S.-J., Bittersmann, E., Luttruli, D.K., DeGraziano, J.M., Ma, X.C. and Gao, F (1990) 'Multistep photoinitiated electron transfer in molecular pentads, tetrads and triads', I.P.S.-8, Book of Abstracts, Palermo, p. 105.

[9] Hall, D.O., Rao, K.K. and Park, I.H. (1991) 'Immobilized photosynthetic systems for the production of fuels and chemicals', in Pelizzetti, E. and Schiavello, M. (eds.) Photochemical conversion and storage of solar energy, Kluwer Academic Publishers, Dordrecht.

[10] (a) Turro, N.J., Grätzel, M and Braun, A.M. (1980) 'Photophysikalische und photochemische Prozesse in micellaren Systemen', Angew. Chem. 92, 712-734.
(b) Turro, N.J., Grätzel, M and Braun, A.M. (1980) 'Photophysical and photochemical processes in micellar systems', Angw. Chem.Intl.Ed. 19, 675-696.

[11] (a) Braun, A.M. and Skolnick, D.G. (1988) 'Organic chemistry and the potentials of the bulk-reduction of CO_2', Chimia 42, 104-105.
(b) Haruki, E. (1982) 'Organic synthesis with carbon dioxide' in Inoue, S. and Yamazaki, N (eds.) Organic and bio-organic chemistry of carbon dioxide, Halsted Press, New York, pp. 5-78.
(c) Tributsch, H. (1989) 'Photoelectrocatalysis', in Serpone, N. and Pelizzetti, E. (eds.) Photocatalysis, fundamentals and applications, Wiley, pp. 339-383.

[12] (a) Juris, A., Balzani, V., Barigeletti, F., Campagna, S., Belser, P. und Von Zelewsky, A. (1988) 'Ru(II) polypyridine complexes: photophysics, photochemistry, electrochemistry and chemiluminescence', Coord.Chem.Revs. 84, 85-277.
(b) Orellana, G. and Braun, A.M. (1989) 'Quantum yields of ^{3}MLCT state formation and triplet-triplet absorption spectra of ruthenium(II) tris-chelate complexes containing five- and six-membered heterocyclic moieties', J.Photochem.Photobiol. A 48, 277-289.
(c) Dürr, H., Thiery, U., Infelta, P.P. and Braun, A.M. (1989) 'A sensitizer-relay assembly with improved properties in electron transfer processes', New.J.Chem. 13, 575-577.

[13] Dürr, H., Dörr, G., Zengerle, K., Mayer, E., Curchod, J.-M. and Braun, A.M. (1985) 'Possibilités et limites des diazacomplexes de Ru^{2+} en tant que sensibilisateurs.induisant la photolyse de l'eau', Nouv.J.Chim. 9, 717-720.

[14] Curchod, J.-M. and Braun, A.M. (1986) 'Pilot level solar photochemical reactor', I.P.S.-6, Book of Abstracts, Paris, G9.

[15] Lehn, J.-M. (1990) 'From molecular ecognition towards information and signal processing at the molecular level' in Schneider, H.-J. and Dürr, H. (eds.) Supramolecular Organic Chemistry and Photochemistry, VCH, Weinheim, in print.

[16] Beuerlein, A. (1990) 'Bipyridazine in Photoelektronentransfer-Reaktionen. Synthese und photophysikalische Studien', PhD thesis, Universität des Saarlandes, Saarbrücken.

[17] Smotkin, E., Bard, A.J., Campion, A, Fox, M.A., Mallouk, -t., Webber, S.E. and White, J.M. (1986) 'Bipolar TiO_2/Pt semiconductor photoelectrodes and multielectrode arrays for unassisted photolytic water splitting', J.Phys.Chem. 90, 4604-4607.

[18] Eicher, H.-P. und Günthrodt, H.-J. (1986) 'Mechanism of reactions and side-reactions of the carbo-thermic reduction of α-alumina and silica with solar energy', in Becker, M. (ed.) solar thermal central receiver systems, Springer, Berlin, vol. 1, pp. 737-744.

[19] (a) Fox, M.A. (1991) 'Control of photocatalytic oxidative selectivity on irradiated TiO_2 powders: a diffuse reflectance kinetic study', in Pelizzetti, E. and Schiavello, M. (eds.) Photochemical conversion and storage of solar energy, Kluwer Academic Publishers, Dordrecht.
(b) Pichat, P. (1991) 'Catalysts, adsorption and new catalysts', in Pelizzetti, E. and Schiavello, M. (eds.) Photochemical conversion and storage of solar energy, Kluwer Academic Publishers, Dordrecht.

[20] Doré, M. (1989) Chimie des oxydants et traitement des eaux, Lavoisier, Paris

[21] Frimmel, F.H. (1990) 'Characterization of organic acids in freshwaters: a current status and limitations', in Perdue, E.M. and Gjessing, E.T. (eds.) Organic acids in aquatic ecosystems, Wiley, Chichester, pp. 5-24.

[22] Ollis, D.F. (1987) 'Process economics for water purification: a comparative assessment', in Schiavello, M. (ed.) Photocatalysis and environment, trends and applications, Kluwer Academic Publishers, Dordrecht, 663-677.

[23] Comninellis, C. and Plattner, E. (1988) 'Electrochemical wste water treatment', Chimia 42, 250-252.

[24] e.g. (a) Al-Ekabi, H., Safarzadeh-Amiri, A., Story J. and Sifton, W. (1990) 'Advanced technology for destruction of organic pollutants by photocatalysis', Nulite, London, Ontario N6E 2S8, CANADA.
(b) Braun, A.M., Grätzel, Greulich, N., M., Jakob, L. and Kottelat, J.-P. (1991) 'Photocatalyzed oxidative degradation of organic materials in aqueous systems using TiO_2 immobilized on a glass supports' Ecole Polytechnique Fédérale de Lausanne and Schott Glaswerke, Mainz.

[25] Pelizzetti, E., Minero, C., Carlin, V., Bogarello, E., Miano, F. and Serpone, N. (1990) 'Photocatalytic activity of differently prepared titania particles towards pollutants degradation' I.P.S.-8, Book of Abstracts, Palermo, p. 261.

[26] e.g. Rizzuti, L. and Brucato, A. (1991) 'Recent developments in heterogeneous photoreactor modelling' in Pelizzetti, E. and Schiavello, M. (eds.) Photochemical conversion and storage of solar energy, Kluwer Academic Publishers, Dordrecht.

RECENT DEVELOPMENTS IN HETEROGENEOUS PHOTOREACTOR MODELLING

L. RIZZUTI, A. BRUCATO
Dipartimento di Ingegneria Chimica dei Processi e dei Materiali,
Università di Palermo, Viale delle Scienze
90128 Palermo
Italy

ABSTRACT. A selection of recently proposed heterogeneous photoreactors is presented. In some cases efforts have been made for the formulation of models able to describe the photoreactor behaviour with regard to the absorption of radiation. Examples of proposals of both novel photoreactor and models are critically reviewed.

Due to the complexity of a formulation of a complete heterogeneous photoreactor model, only partial models have been proposed in the pertinent literature so far, although the importance of such a complete model has been pointed out by several investigators. This feature limits the number of the cases presented here to those having possible developments towards either a complete model formulation or for the possible application of the investigated photoprocess to real cases.

1. Introduction

Photocatalytic processes have recently received a great research interest for several reasons, among which the possible utilization of solar radiation is a promising way for energy saving. Moreover, photocatalysis has been found to be an efficient route for pollutants degradation; for this reason also the research interest towards this very important task has received a great concern.

In order to engineer the results of these researches for an industrial utilization, systematic studies of heterogeneous photoreactors have to be performed. This research area is lacking of specific studies, in particular of those concerning the modelling of heterogeneous photoreactors. This scarcity is most likely due to the fact that heterogeneous photocatalysis studies are still at the stage of chemical process understanding, so much more attention has been devoted to a deep investigation of the basic principles of the process itself, rather than to develop a more practical aspect. Probably due to this reason, a variety of heterogeneous photoreactors is described in the current literature, because many research groups utilize reactors designed for specific investigation purposes.

It was, then, impossible to rationalize, in terms of photoreactor type (or any other classification) the presentation of this report. For this reason only a selected number of proposed models and of reactors having particular features will be quoted and discussed, with the only time limitation to the last decade.

E. Pelizzetti and M. Schiavello (eds.), Photochemical Conversion and Storage of Solar Energy, 561–574.

2. Heterogeneous Photoreactor Modelling

The complexity of modelling photoassisted chemical reactors has been pointed out in two excellent review papers by Alfano et al [1, 2]. This complexity arises from the need of taking into account, for a complete definition of the whole model, at least the following four models: a) a light source model, b) a kinetic model, c) a fluid-dynamic model, and d) a light absorption model. By gathering these four "sub-models", taking into account all their connections, a complete model of a photoreactor can be obtained. Heterogeneity, of course, implies that b), c), and d) "sub-models" are very complex. In particular, light absorption is greatly affected by scattering phenomena which are also linked to the complex fluid-dynamics exhibited by these multi-phase systems This increases the intricacy of complete photoreactor models, whose formulation has not been attempted so far.

It is worth noting, however, that the set-up and development of such models is essential for a rational design of any type of industrial application of a photocatalytic process. In fact, as the research approach of a novel photoprocess often comprises the proposal of a new type of photoreactor, scale-up problems arise, that can be solved by utilizing a proper model.

As the aim of this paper is a status report on recent developments of heterogeneous photoreactor modelling, no general discussion nor complete mathematical treatment of the complete model will be reported. Only comments and descriptions of selected models will be made together with mathematical formulations, if any. The interested reader can find general comments and formulations of photoreactor models in recent publications [1-3] and in the paper by Cassano [4] in this book, where the extreme importance of the local volumetric rate of energy absorption (LVREA) together with other model parameters is pointed out for the case of homogeneous photoreactors. As a matter of fact, some of the concepts developed for the case of homogeneous photoreactors can be suitably modified for application to heterogeneous photoreactors.

3. Heterogeneous Photoreactors

3.1 ANNULAR MECHANICALLY STIRRED GAS-SOLID PHOTOREACTOR

A mechanically stirred moving bed photoreactor has been designed by Daroux et al. [5], with the aim of a most effective use of radiation. The reactor is of annular shape, has a volume of 3.5 liters, is equipped with a light filter, and contains an agitator specifically designed to assure a uniform concentration of gas, and in particular to maintain the catalyst particles in continuous movement. This agitator is cross-shaped, containing slits for gas feeding. An interesting and useful catalyst filtering device uses ultrasonic vibration to keep the catalyst particles inside the photoreactor. It has been described in detail by Parent [6].

The light intensity can be varied easily by using different concentrations of aqueous solutions of a salt chosen for the specific wavelength utilized.

From the point of view of mixing, the photoreactor was experimentally tested, and exhibited a perfect mixing of the gaseous phase.

Admittedly this reactor has a great versatility also for the several advantages showed:

i) the volume and the mixing device allow relatively large quantities of catalyst (up to 5 g) to be used, which are completely irradiated;
ii) the perfect mixing, that is very useful for kinetic studies;

iii) the possibility of varying both wavelength and light intensity of the irradiation, permitting a complete study of their effect on the rates of reaction.

The photocatalytic oxidation of ethane has been studied in this photoreactor [7]. A mercury lamp has been used as radiation source, and the intensity of radiation has been varied by means of a aqueous solutions of potassium chromate. Degussa Titanium Dioxide P25 has been used as photocatalyst. The studied parameters comprise the partial pressure of reactants, the UV light intensity, the amount of catalyst and the residence time. Their effect on the rate of ethane oxidation has been evaluated.

The kinetic model obtained by the study describes that ethane physisorbed on the surface of TiO_2 according to the Langmuir isotherm reacts with the absorbed oxygen. The reaction rate has the following form:

$$r=\frac{k_r K_a k_o P_o P_c}{\left[k_r K_a P_c + k_o P_o\left(1+K_a P_c\right)\right]}$$

Where k_r is the rate constant for the surface reduction, K_a the equilibrium constant for ethane physisorption, k_o the rate constant for the surface oxidation, P_o and P_c the partial pressures of oxygen and ethane respectively. All the kinetic and equilibrium constants of this equation have been determined, and are linear functions of the UV intensity.

The results on the effect of the amount of catalyst revealed the existence of a critical mass (1.5 g) that, apparently, is sufficient to detain all the radiation entering the photoreactor. It means that the mass of catalyst in excess of 1.5 g, independently of the photon flux and of the residence time, is not involved in the reaction for the very short life-time of TiO_2 photoactivation. This critical mass of catalyst has been compared with previous findings and a general expression for its evaluation has been suggested for any type of photocatalytic reactor. This expression has the following form:

$$m_{crit}= 10^{5} d_p \rho S_b$$

where m_{crit} is the critical mass of catalyst, d_p the diameter of particles, ρ the true density of TiO_2, and S_b the surface of the irradiated catalytic bed. By using the previous equation the critical mass of catalyst can be evaluated from the reactor geometry and from the properties of the catalyst particles.

In conclusion, it is claimed that this new type of stirred reactor allows a very good catalyst irradiation and a simple interpretation of kinetic data.

3.2 LIQUID-SOLID PHOTOREACTOR WITH SUPPORTED CATALYST

In a liquid reacting medium with suspended catalyst particles the problems caused by the entrainment of catalyst particles by the exit stream are worsened. In order to avoid these problems, the catalyst can be supported. This technique has been widely used by Matthews [8, 9], Borgarello et al. [10] and Serpone et al. [11]. More recently Al-Ekabi and Serpone [12] proposed two photoreactors with supported TiO_2. The first was a Pyrex glass tube wound to form a coil with the catalyst supported over the internal surface. Light irradiation was provided by either a Solarbox lamp (solar simulator) or six fluorescent tubes. The second photoreactor was assembled by filling with glass beads, coated with TiO_2, the spacing between two coaxial Pyrex glass tubes. In this case the irradiation was supplied by six fluorescent tubes. The reactors operated both in continuous recirculation and single-pass mode. The reactions performed in aerated media were the photodegradations of phenol, 4-chlorophenol, 2,4-dichlorophenol and 2,4,5-trichlorophenol. The experimental results have been interpreted in terms of modified (for liquid-solid reactions) Langmuir-Hinshelwood kinetics, and the values of the first-order rate constants have been reported.

Only a kinetic study has been performed, although the authors conclude that a future challenge will be the improvement of the photoreactors to obtain more efficient degradation in shorter times with simulated sunlight. It is worth noting, however, the importance of reliable values of kinetic constants especially for decontamination processes, as in this case. In fact the kinetic model can be considered the first step for a complete modelling of the process as a whole.

3.3 THREE-PHASE ANNULAR PHOTOREACTOR

Operating in a three-phase system, continuous with respect to the gas phase, a more complete kinetic study can be performed. In fact effects such as those of partial pressure of products, of gas velocity through the reactor on mass transfer, and of stability of the catalyst can be determined. Moreover a radiation model can be easily established if an annular photoreactor is used. Thus a three-phase annular photoreactor can be considered very convenient for the formulation of a mathematical model of a photoprocess. Furthermore the mixing of the reacting medium can be achieved by bubbling the gas in the reacting suspension, thus avoiding all the difficulties arising from a mechanical agitation. A photoreactor of this type (annular, solid-liquid-gas phase and continuous) has been designed and proposed by Simarro et al. [13]. It comprises a light filtering solution jacket, and particular care has been had in the design of the gas bubblers, as the gas bubbling must assure a perfect mixing and a good contact of the three reacting phases.

This photoreactor has been previously used by Simarro et al. [14] to study water photolysis in liquid phase, using neutral aqueous suspensions of $RuO_x/Pt/TiO_2$ particles and UV light.

In a previous paper Esplugas et al. [15] applied the spherical light model to the lamp of an annular photoreactor in order to evaluate the radiation intensity profiles at the inner wall, and the flow rate of radiation entering the reactor. Their theoretical treatment led to a number of parametric curves showing the importance of lamp centring for good reactor performance.

A more complete model has been presented by Esplugas et al. [16], who treated their heterogeneous system as being pseudo-homogeneous. The photoreactor model has been developed by utilizing mass balances and radiant energy balance, for whose formulation the linear spherical model has been applied to the lamp. The photon flux entering the reactor has ben analytically calculated. When this flux coincides with the photon absorption rate by the reacting medium, the mass balance equations can be analytically solved and the model is able to completely describe, under the pseudo-homogeneity hypothesis, the photoreactor behaviour as a function of geometrical and radiation parameters. The experimental results of Simarro et al. [14] have been interpreted by means of this approach. Curves depicting hydrogen generation rate and radiation absorption rate have been obtained. These curves show that there is a maximum in the hydrogen production rate per unit mass of catalyst for a catalyst concentration of about 0.5 g/l, and that the rate limiting factor of the process is the amount of catalyst in the concentration range 0.0-0.5 g/l, while for values close to 2 g/l the limiting factor is uniquely light. Admittedly, the decrease of hydrogen photogeneration rate for low values of catalyst concentration can be related to some deactivation processes, such as electron-hole recombination, changes in the oxidation states of the catalyst, oxygen photoabsorption, etc., whose importance increases with the amount of radiation per unit mass. On the other hand, the decrease of hydrogen photogeneration rate per unit mass of catalyst for higher values of catalyst concentration has been explained by observing that in this case a lower number of photons reaches each catalyst particle, so then fewer hydrogen molecules are photogenerated per each particle. These studies clearly demonstrate the importance of a correct photoreactor modelling, whose theoretical predictions can be utilized for a proper design of the photoprocess as a whole, as for instance for photoreactor sizing.

In a more recent paper Escudero et al. [17] reported another experimental investigation on the same system utilized by Simarro et al. [14] (water photolysis in the same reactor type over several $Pt(RuO_2)$-TiO_2 catalysts). This last study comprises a more detailed investigation on the influence of the various catalysts utilized. Moreover, three steps of the photocatalytic process have been identified and critically discussed: the luminic step, the chemical step, and the mass-transfer step. The first consists of photons collecting and absorption with generation of electron/hole pairs in the semiconductor. The second corresponds to the electron transfer towards catalytic centres and consequent reduction of protons to hydrogen; in the same time the hole transfer occurs, causing hydroxyl ion oxidation. Finally, the third step is the diffusion of the produced hydrogen from the semiconductor surface to the bulk of the liquid, and successive transport to the gas stream bubbling through the suspension. A complete mathematical treatment is reported only for the last, leading to an equation expressing the hydrogen production rate, that matches well the experimental data.

3.4 COMPARISON BETWEEN TWO LAMP MODELS

An annular UV disinfection reactor has been used by Suidan and Severin [18] to compare two lamp models. The process studied was the inactivation of f2 virus. Although this system is heterogeneous for the presence of the f2 viruses suspended in the liquid he authors model the system as being pseudo-homogeneous and utilize the experimental results of virus inactivation for comparison with model predictions. The results are interesting as they compare two different approaches to lamp modelling with experimental data.

The finite-length lamp model was compared to the infinite-length lamp model (parallel radiation model) for annular reactors. In the first model the lamp is approximated by a series of point sources located along a line segment. Light is emitted spherically from all points on the lamp axis. In the second model all light is presumed to be emitted radially from the line source, so that a monodimensional problem results .

The results have been reported in terms of the ratio of the average point intensity in the reactor to the surface flux at the tube. This ratio is expressed by the following equation for the first model and for non-UV-absorbing materials in the water (finite-length lamp):

$$\frac{\bar{I}}{I_0}=\frac{2\lambda}{\beta\left(\rho^2-1\right)}\left[\frac{\rho}{\lambda}\tan^{-1}\left(\frac{\lambda}{\rho}\right)-\frac{1}{\lambda}\tan^{-1}(\lambda)+\frac{1}{4}\ln\frac{\lambda^2+\rho^2}{\lambda^2+1}-\frac{1}{4}\left(\frac{\rho}{\lambda}\right)^2\ln\frac{\lambda^2+\rho^2}{\rho^2}+\frac{1}{4\lambda^2}\ln\left(\lambda^2+1\right)\right]$$

while for the infinite-length lamp model it takes the form:

$$\frac{\bar{I}}{I_0}=\frac{0.8685}{\alpha\left(\rho^2-1\right)}\left[1-e^{-2.303\,\alpha(\rho-1)}\right]$$

In these two equations λ is the ratio of lamp length to radius of tube, β a transmission factor relating the ratio of energy transmitted to the reactor to the total energy emitted by the finite line source, ρ the ratio of reactor outer radius to reactor inner radius, and α the absorbance of water.

The results of the comparison between theoretical curves obtained by the mathematical development of the two lamp models show that, for the situation studied, both models predict almost the same average light intensity. Thus there is no practical reason of using the much more complex finite-length lamp model instead of the simpler infinite-length one. This interesting finding, however, might not be valid for configurations different from that studied in the paper. A comparison between experimental data and the simpler model prediction has been finally performed, and a good agreement has been found.

3.5 RANDOM-WALK MODEL FOR LIGHT SCATTERING SIMULATION

In a recent paper Yokota et al. [19] proposed a random-walk model for Monte Carlo simulation of light scattering in a solid-liquid photoreactor. The model is based on three assumptions. The first is that the photoreactor is supposed to be divided into small cubes of the same volume. Each cube contains only one particle in order that the probability of a light ray to impinge a particle is equivalent in all the reactor volume. The second assumes that light rays can proceed only along the six Cartesian coordinates directions. The third considers non-isotropic scattering by attributing a fixed probability to one direction and equal probabilities for the other five. On these grounds the program for simulating most photoreactors is easy to write.

In order to simplify the calculations and reduce the number of parameters the following constraints have been imposed also, so that when one of the three probabilities is assigned, the others are determined consequently:

$$p_f + 4 p_s + p_b = 1$$

$$p_s = p_b \quad \text{or} \quad p_f = p_s$$

where p_f, p_s and p_b are the probabilities of forward, side and backward scattering, respectively. For the particles for which mirror reflection does not occur, i. e. for most particle types, the relation $p_f = p_s$ should be used.

The procedure to obtain trajectories of scattering rays is described in detail in the paper. This method allows the calculation of the intensity taken by a ray at each position after proceeding along a distance equal to the side length of a cube. Upon entering a cube, a ray is assumed to either hit or not the particle, depending on the comparison between randomly generated numbers and the size ratio of particle to cube.

For non hitting rays the intensity is:

$$I_i = I_{i-1} e^{-\mu_c l_m}$$

while for hitting rays is:

$$I_i = I_{i-1} \tau e^{-\mu_c (l_m - 2/3\, d_v)}$$

where I is the incident light intensity, τ represents the light absorption property of particles (either a reflectivity or transparency), μ_c the absorption coefficient of liquid phase in the suspension, l_m the side length of cube, and d_v the particle equivalent diameter. The procedure is iterative, and eventually allows the determination of the light intensity distribution in the photoreactor.

In order to have a closed system the two parameters p_f and τ must be experimentally determined. The experiments have been performed, for both side-scattering light and rear-scattering light measurements, in a double cell that is described in detail in the paper. Three types of particles have been used: one cut of transparent plastic beads, two cuts of opaque (white) porcelain particles, and two cuts of translucent (brownish) Toyura sands. Experimental measurements have been performed by chemical actinometry. From the experimental data the two model parameters have been determined by means of an estimation procedure. The results show that the value of τ increases with the decrease of particle size, and is smaller for the coloured sands. Reversely, the values of p_f decrease with the decrease of particle size and are smaller for the opaque porcelain particles, taking values close to 1/6 of that for isotropic scattering.

Experimental data have been compared with simulation results as normalized light intensity escaping from both rear and side reactor walls. The agreement between the experiments and the model results is satisfactory. The results show how the amount of scattering is related to the optical properties of the particles. In particular, for transparent plastic particles, an increase of suspended particle fraction corresponds to a small decrease of rear light intensity and to a moderate increase of the side one. For opaque porcelain particles and Toyura sands rear intensity diminishes rapidly with the increase of particle fraction. Side intensity shows a peak for a particular value of particle fraction for porcelain, while is very small for sand.

With the aim of verifying the validity of the model two additional series of experiments have been performed. In the first the continuous phase was a nickel sulphate solution to study the effect of light attenuation in the liquid phase. The second was performed in a cell having a volume about four times larger for investigating scale-up effects. Both series corfirmed the model validity.

Correlations between model parameters and optical properties of particles have been studied in a successive paper by Yokota et al. [20]. Relative refractive index and diffuse reflectance are the optical properties utilized for correlating the model parameters, as they are measurable properties. The relative refractive index, N, has been modified by varying the liquid phase where the experiment was carried out. The following relationships satisfactorily correlate the experimental data:

$$p_f = 1 - 7.8\,(1 - N)^{1.7} \qquad \text{for } N < 1$$

$$p_f = 1 - 0.5\,(1 - N)^{1.7} \qquad \text{for } N > 1$$

The diffuse reflectance has been varied by colouring the original particles with several dyes. The equation

$$(1 - R) = (1 - \tau)^m + \tau\,(1 - R_0)$$

has been derived by a photon balance and simple absorption and transmission models. In the previous equation the fitted parameter m is a constant for particles without dye coating, R is the reflectance averaged over a wavelength range, τ a light intensity decrease parameter, and R_0 the reflectance of supporting particles.

In a third paper [21] the light absorption rate has been estimated by means of this random-walk model for quantitatively evaluating the effect of sensitizer concentration on heterogeneous reaction rate of oxygenation. The reaction studied was the photo-oxygenation of 2,3-dimethyl-2-butene over dye-supported solid particles as photosensitizer. The photo-oxygenation rate has been found to be proportional to a power of the light absorption rate a little less than unity. Moreover higher-level dye supportation lowers the photo-oxygenation rate.

3.6 FLAT FLUIDISED BED GAS-SOLID PHOTOREACTOR

When a photocatalytic reaction is carried out in a gas-solid reactor, in addition to the requirement of good contact between gaseous reactants and solid catalyst, the achievement of an efficient exposure of the catalyst to irradiation is also necessary. A reactor of the packed bed type, often used for performing gas-solid reactions, is not suitable because impinging photons can not penetrate into the interior of the bed. Gas-solid fluidisation, besides assuring a very good contact between the reacting phases, thanks to the vigorous mixing of the solid phase allows the irradiation of all catalyst particles. Moreover, due to the increase of bed voidage fluidisation enhances light penetration. Another important consideration is the reactor configuration. It is well known that a parallel plate type of geometry is very efficient when an external parallel irradiation is used. This is the case of photoreactors which are utilised for photoprocesses promoted by solar irradiation. Therefore, a vertically mounted fluidised bed photoreactor, with transparent front walls separated by a small gap should be a very effective type of photoreactor. This parallel-wall configuration is known in fluidisation field as the so-called "two-dimensional fluidised bed".

On the grounds of the previous considerations a photoreactor of this type has been designed and used by Yue et al. [22] to study the photocleavage of water and the photofixation of nitrogen. The photoreactor was found to be very efficient. The need of fundamental research on the interaction between all the involved optical phenomena and fluidisation was also pointed out for a sound basis for photoreactor modelling.

When a flat bed is irradiated the light can be divided into many portions. Part will be reflected by the front wall and is lost. The rest enters the bed and can either be absorbed or reflected by the particle or transmitted through the back wall. A portion, whose amount depends on the powder optical characteristic, of the photons reflected by the particles may be lost through the front or back walls. It is well known that in most of the gas-solid fluidised beds formation of gas cavities (bubbles) occurs. These bubbles are quite transparent and form the most significant contribution to light transmitted through the bed. The bubble sizes depend on gas flow rate and particle characteristics, which are therefore important variables that affect the performance of a fluidised photoreactor. Another parameter to consider is the gap between the flat walls, i.e. the internal bed thickness. A small thickness will enhance the exposure of the particles to irradiation, but will also increase the amount of light transmitted through the bed as well as decrease the effective reactor volume. On the other hand, high bed thicknesses result in a poor utilisation of the catalyst mass. The particles reflectance influences the absorption and in turn the transmission of light. These variables have therefore to be studied.

To the aim of collecting data for future photoreactor modelling, Rizzuti and Yue [23, 24] used this type of photoreactor for measuring light transmission through the bed during fluidisation. Due to the complexity of this experimental arrangement, these measurements have been performed without the complication of chemical kinetics: the flowing gas was pure nitrogen. The amount of light transmitted through the irradiated fluidised bed was measured at different reactor levels by a photocell. The experimental variables studied comprised gas flow rate (varying from 2 to more than 10 times that for minimum fluidisation), mean particle size (ranging from 0.138 to 0.275 mm), and internal bed thickness (from 2.1 to 4.9 mm). Three types of particles were used, having different reflectance spectra, with the purpose of examining the effect of powder reflectance on transmission of light. They were γ-alumina, silica gel and cobalt-molybdenum doped γ-alumina. The relation found between transmittance and the studied variables is

$$E_t = a + b e_b d_p^{0.5} d_r^{-1}$$

in which e_b is the bed voidage, d_p the mean particle diameter and d_r the reactor thickness. The two constants a and b have equal values for the two catalysts having similar reflectance spectra, alumina and silica gel, and a different one for the Co-Mo doped γ-alumina. This suggests that the reflectance of solids could be a factor affecting light transmission.

Rizzuti et al. [25] studied this influence in an experimental arrangement similar to that of the previous works [23, 24], but performing integral light transmittance measurements, by a photocell detecting all the light transmitted through the bed. The catalyst used was again γ-alumina., whose reflectance spectrum was varied by impregnating two samples with inorganic salts ($Fe(NO_3)_3$ and $CuSO_4$) solutions and successively drying. In this way, by utilising four wavelengths for irradiating the white and "coloured" powders, the experiments have been performed at 12 values of reflectance, ranging from 15% to 82%. Only a qualitative analysis has been performed, from which it arises that at minimum fluidisation conditions light transmission does not occur for values of reflectance below about 65%. Moreover only the influence of reflectance on transmission is observed, while the wavelength does not seem to have any influence on the relation between transmittance and flow rate above minimum fluidisation conditions. The influence of reflectance on transmission can be explained with the presence of a "cloud" of particles inside the bubbles, with the expansion of the "dense phase" above minimum fluidisation conditions (i.e. the simple two-phase model is not valid, as found by Yue et al. [26]). The independence on wavelength suggests that a simple "single scattering" approach to the optical properties of both phases is sufficiently accurate, thus simplifying both the interpretation of data on this subject and the description of a photoreactor model.

The structure of flat fluidised beds has also been investigated by Brucato et al. [27] and Yue et al. [26] by means of light transmittance measurements. These authors pointed out that the main contribution to light transmittance in such beds arises from bubbles which are large enough to bridge the gap between reactor walls. So they suggested that the total bubble voidage of the fluidised bed should be treated as having two portions: "visible" and "invisible" bubble voidages. They also showed that a Lambert-Beer type of equation could be applied to account for the dependence of bed light transmittance on bed thickness.

Yue and Rizzuti [28] pointed out the importance of wall effects in the design of this type of flat photoreactors especially when the ratio of bubble depth along the transmission direction to reactor thickness has a high value.

The bed transmittance and reflectance during fluidisation have been measured by Iatridis et al. [29] in an experimental arrangement which allowed the simultaneous measurement of these two quantities. The flat photoreactor was positioned between two integrating spheres, collimated with an irradiating lamp and a monochromator. The two spheres had ports for collecting the transmittance and reflectance signals and apertures for allowing the passage of photons. Four reactors, with internal thickness ranging from 2.45 to 5.3 mm, and two photocatalysts have been used. The first was five cuts of zeolite 13X particles with mean diameters comprised between 0.116 and 0.231 mm. The second was three cuts of Co-Mo doped alumina with the same range of mean diameters. The static bed height was kept at 90 mm, and nitrogen was used to fluidise the particles. The measurements have been performed at various reactor heights as in [23] and [24], and at a wavelength of 560 nm.

The transmittance data have been correlated with the following equation:

$$E_t = a + b\left(\frac{H - H_{mf}}{H_{mf}}\right) d_p^{0.5} d_r^{-1}$$

in which H is the bed height during fluidisation and H_{mf} is the bed height at minimum fluidisation conditions. This equation is similar to that previously found [24], as bed expansion may be considered to be proportional to bed voidage.

The reflectance data have been correlated with the following equation:

$$E_r = c + d\left(\frac{H}{H_{mf}}\right)^{0.5} d_p^{0.125}$$

with the constants c and d having different values for the two photocatalysts.

Once the combined fraction of incident energy absorbed and reflected by the empty reactor (E_w) and the values of the four constants are known, the bed absorbance, i.e. the total amount of radiation which is retained in the reactor corresponding to the maximum energy available for reaction, may be calculated by an energy balance. This is an important result particularly for the design of this type of photoreactors.

3.7 SOME OTHER STUDIES

Many other interesting investigations have been performed. Some of them are reported and shortly described in the following.

Scattering phenomena in an annular photoreactor have been modelled by utilizing a Monte Carlo approach for solving the radiant energy balance equation [30]. The distribution of the local rate of energy absorption has been obtained as a function of some parameters. The results show that scattering lowers the doable conversion in all cases except when completely segregated flow, together with high optical thickness, occurs.

With regard to the direct utilization of solar irradiation, the main problems encountered in the design of solar chemical reactors have been reviewed by Villermaux [31]. A gas-solid endothermic chemical process has been performed by Badie et al. [32] in two solar irradiated reactors: a rotary kiln, and a fluidised bed irradiated from the top with concentrated sun rays. In this process the solar radiation is utilized as a heat source for increasing the temperature of the reacting system, not as a "reagent" as in the case of photocatalysis. The treatment temperature varied in both reactors from 600°C to 1300°C. In the fluidised bed temperature homogeneity could be reached in a rather large volume, while a strong thermal gradient was present in the rotary kiln. A model of radiant heat transfer in the first type of reactor (fluidised bed) together with a comparison with experimental data has been proposed by Flamant [33]. He determined the temperature profile, total emissivity, flux density distribution, and effective mean penetration distance in a high temperature solar fluidised bed. The model proposed involves conductive, convective, and radiative heat transfer phenomena in a porous medium for describing the high-temperature heat exchange in this type of photoreactor. The model describes properly the behaviour of beds of silicon carbide and chamotte at incipient fluidisation conditions, while it was not precise in the case of beds of zirconia and silica sand. The goal of the investigation of Hunt et al. [34] is the understanding of optical, thermodynamic, and chemical processes in solar heated particle suspensions. As in the two previous papers [32] and [33] the aim of sunlight utilization is that of providing energy for endothermic chemical reactions. Spectral extinction efficiencies have been calculated from Mie theory for various particle sizes. The authors developed also an equation for calculating the heat transfer from a particle to the surrounding gas, by means of a simple model for conductive heat transfer. The experiments have been performed simulating concentrated sunlight by means of a high intensity radiant source. The laboratory scale photoreactor utilized particles, entrained by the gas flow, irradiated in a small reaction zone

A three phase (gas-liquid-solid) photoreactor has been used by Khan et al. [35, 36] for carrying out two photochemical processes: water cleavage and synthesis of ammonia, both over titanium exchanged zeolites. Although the study presents interesting results, particularly because visible light radiation has been used, no emphasis has been given to the role played by the type of photoreactor used. The photosynthesis of ammonia over doped titanium dioxide catalysts has also been studied by Augugliaro et al. [37, 38] in a gas-solid fluidised bed photoreactor of cylindrical shape. The reactor was mounted in the focal axis of an elliptical reflecting chamber, with the lamp positioned in the other axis. Also in this case there was not any particular study on the photoreactor and only the advantages of fluidisation on radiation penetration inside the bed have been pointed out.

The interesting case of gas absorption with photochemical reaction has been treated by Mahajani and Sharma [39] with the classical diffusion-reaction equations. The photoreactor was of cylindrical geometry (wetted wall column) and the two cases of activation of a liquid reacting species and of activation of the dissolved gas have been treated. The assumptions made were: (a) application of film theory of mass transfer; (b) occurrence of the reaction entirely in the gas-liquid film; (c) rectilinear coordinate system for mass and photon transfer; (d) monochromatic radiation; (e) validity of Lambert-Beer law for photon transfer; and (f) activation step of first order with respect to light absorbed. The equations giving the specific rate of absorption as function of the process parameters have been developed, and the authors conclude that there is need for experimental work in this area. As a matter of fact this type of photoreactor is very interesting and promising also because photocatalyst particles may be entrained by the flowing liquid, thus realizing a three-phase photoreactor.

The paper by Otake et al. [40] concerns an investigation on a method for determining the light intensity profile in a heterogeneous system, which uses the equation for the light intensity profile for a homogeneous system. A rectangular bubble column reactor has been used for studying the dependence of the effective absorption coefficient and the operating conditions. The light intensity profile for a heterogeneous system containing air was expressed in terms of the diffuse-line-source model. The effective absorption coefficient was expressed as the sum of the effect of absorption by the liquid phase and the effect of reflection, refraction and transmission by the gas bubbles. These effects are related to the bubble diameter, the gas hold-up, the specific surface area, and the optical properties of the gas and the liquid. The simple correlating equation obtained is:

$$\mu_e = \mu_l \left(1 - \varepsilon_g\right) + k\,a$$

where μ_e is the effective absorption coefficient, μ_l the absorption coefficient for liquid phase, ε_g the gas hold-up, $a = 6\ \varepsilon_g/d_b$ the specific surface area, d_b the volume-surface mean bubble diameter, and k an empirical coefficient, function of the relative refractive indices of the gas and liquid and of the reflectivity and absorption coefficient of the dispersed phase particle. The photoreduction rate of potassium ferrioxilate with bubbling of nitrogen agreed closely with the theoretical calculations, thus confirming the validity of the proposed method.

4. Conclusions

Although numerous studies have been carried out on heterogeneous photochemical reactor modelling, a comprehensive predictive model has not been developed so far. The reason of this can be found in the complexity of the system, that requires a large number of factors to be taken into account and a great deal of parameters to be experimentally determined. Nevertheless, some cases of simplified approaches can be successfully applied to real situations, in particular for design purposes.

It is clear that there is a need of further experimental investigations, with the aim of both obtaining correct values of all the parameters needed for the formulation of models, as well as for validating the models themselves.

5. References

[1] Alfano, O. M., Romero, R. L. and Cassano, A. E. (1986) 'Radiation field modelling in photoreactors - I. Homogeneous media', Chem. Engng Sci. 41, 421-444.

[2] Alfano, O. M., Romero, R. L. and Cassano, A. E. (1986) 'Radiation field modelling in photoreactors - II. Heterogeneous media', Chem. Engng Sci. 41, 1137-1153.

[3] Rizzuti, L. and Brucato, A. (1988) 'Photochemical reactors engineering fundamentals', in M. Schiavello (ed.), Photocatalysis and Environment, Kluwer Academic Publishers, Dordrecht, pp. 623-636.

[4] Cassano, A. E. and Alfano, O. M. 'Photochemical engineering design methods. A rigourous approach for homogeneous systems', This book.

[5] Daroux, M., Parent, Y. and Klvana, D. (1980) 'A new reactor for the study of photocatalytic reactions', Chem. Eng. Commun. 4, 501-506.

[6] Parent, Y. (1975) 'Conception d'un réacteur continu pour l'étude de la photocatalyse hétérogène', M.Sc.A. Thesis, Ecole Polytechnique de Montréal.

[7] Daroux, M., Klvana, D., Duran, M. and Bideau, M. (1985) 'Photocatalytic oxidation of ethane over TiO_2', Canad. Jour. Chem. Eng. 63, 668-673.

[8] Matthews, R. W. (1987) 'Photooxidation of organic impurities in water using thin films of titanium dioxide', J. Phys. Chem. 91, 3328-3333.

[9] Matthews, R. W. (1987) 'Solar-electric water purification using photocatalytic oxidation with TiO_2 as a stationary phase', Sol. Energy 38, 405-412.

[10] Borgarello, E., Serpone, N., Liska, P., Erbs, W., Grætzel, M. and Pelizzetti, E. (1985) 'Photocleavage of hydrogen sulphide in alkaline aqueous media with a RuO_2-loaded CdS catalyst supported on a policarbonate matrix', Gazz. Chim. Ital. 115, 599-602.

[11] Serpone, N., Borgarello, E., Harris, R., Cahill, P., Borgarello, M. and Pelizzetti, E. (1986) 'Photocatalysis over TiO_2 supported on a glass substrate', Sol. Energy Mater. 14, 121-127.

[12] Al-Ekabi, H. and Serpone, N., (1988) 'Kinetic studies in heterogeneous photocatalysis. 1. Photocatalytic degradation of chlorinated phenols in aerated aqueous solutions over TiO_2 supported on a glass matrix', J. Phys. Chem. 92, 5726-5731.

[13] Simarro, R., Cervera-March, S., Giménez, J. and Escudero, J. C. (1987) 'A novel three-phase (S-L-G) photoreactor to perform kinetic studies', EPA Newsl. 29, 35-37.

[14] Simarro, R., Cervera-March, S. and Esplugas, S. (1985) 'Hydrogen photoproduction in a continuous flow system with UV-light and aqueous suspensions of $RuO_x/Pt/TiO_2$', Int. J. Hydrogen Energy 10, 221-226.

[15] Esplugas, S., Ibarz, A. and Vicente, M. (1983) 'Influence of lamp position on the performance of the annular photoreactor', Chem. Eng. Jour. 27, 107-111.

[16] Esplugas, S., Cervera, S. and Simarro, R. (1987) 'A reactor model for water photolysis experimental studies in the liquid phase with suspensions of catalytic particles', Chem. Eng. Comm. 51, 221-232.

[17] Escudero, J. C., Simarro, R., Cervera-March, S. and Giménez, J. (1989) 'Rate-controlling steps in a three-phase (solid-liquid-gas) photoreactor: a phenomenological approach applied to hydrogen photoproduction using Pt-TiO_2 aqueous suspensions', Chem. Engng Sci. 44, 583-593.

[18] Suidan, M. T. and Severin, B. F. (1986) 'Light intensity models for annular UV disinfection reactors', AIChE Journal 32, 1902-1909.

[19] Yokota, T., Takahata, Y., Nanjo, H. and Takahashi, K. (1989) 'Estimation of light intensity in a solid-liquid photoreaction system', J. Chem. Eng. Japan 22, 537-542.

[20] Yokota, T., Takahata, Y., Otho, Y. and Takahashi, K. (1989) 'Estimation of parameters used in a light scattering model for a solid-liquid photoreaction system', J. Chem. Eng. Japan 22, 548-551.

[21] Yokota, T., Takahata, Y., Hosoya, J, Suzuki, K. and Takahashi, K. (1989) 'Heterogeneous photosensitized oxygenation of 2,3-dimethyl-2-butene', J. Chem. Eng. Japan 22, 543-548.

[22] Yue, P. L., Khan, F. and Rizzuti, L. (1983) 'Photocatalytic ammonia synthesis in a fluidised bed reactor', Chem. Engng Sci., 38, 1893-1900.

[23] Rizzuti, L., and Yue, P. L. (1982) 'The measurement of light transmission through a fluidised photocatalytic reactor', Proc. 3rd Austrian-Italian-Yugoslavian Chem. Eng. Conf., Graz 14-16/9/1982 Vol.1, 550-557.

[24] Rizzuti, L. and Yue, P. L. 'The measurement of light transmission through an irradiated fluidised bed', Chem. Engng Sci. 38, 1241-1249.

[25] Rizzuti, L., Brucato, A. and Brucato, V. (1984) 'The influence of the catalyst reflectance on the light transmission through an irradiated fluidised bed', Proc. 9th Iberoamerican Symposium on Catalysis, Lisbona 16-21/7/1984, vol.II, 1549-1550.

[26] Yue, P. L., Rizzuti, L. and Augugliaro, V. (1986) 'Bubble phase voidage and dense phase voidage in two-dimensional fluidised beds', Chem. Engng Sci. 41, 171-177.

[27] Brucato, A., Brucato, V., Rizzuti, L. and Yue, P. L. (1984) 'Bubble voidage and bed expansion in flat fluidised beds', Proc. 4th Italian-Yugoslavian-Austrian Chem. Eng. Conf., Grado 24-26/9/1984, vol.1, 199-207.

[28] Yue, P. L. and Rizzuti, L. (1983) 'Light transmittance through irradiated fluidised photoreactors', Proc. 5th Yugoslavian-Austrian-Italian Chem. Eng. Conf., Portoroz 16-18/1986), 465-472.

[29] Iatridis, D., Yue, P. L., Rizzuti, L. and Brucato, A. (1990) 'The absorption of light energy in fluidised photoreactors', Chem. Engng J., in press.

[30] Spadoni, G., Bandini, E. and Santarelli, F. (1978) 'Scattering effects in photosensitized reactions', Chem. Engng Sci. 33, 517-524.
[31] Villermaux, J. (1979) 'Les réacteurs chimiques solaires', Entropie n. 85, 25-31.
[32] Badie, J. M., Bonet, C., Faure, M., Flamant, G., Foro, R. and Hernandez, D. (1980) 'Decarbonation of calcite and phosphate rock in solar chemical reactors', Chem. Engng Sci. 35, 413-420.
[33] Flamant, G. (1982) 'Theoretical and experimental study of radiant heat transfer in a solar fluidized-bed receiver', AIChE Journal 28, 529-535.
[34] Hunt, A. J., Ayer, J., Hull, P., Miller, F., Russo, R. and Yuen, W. (1985) 'Solar radiant processing of gas-particle systems for producing useful fuels and chemicals', Presented at the 23rd ASME/AIChE National Heat Transfer Conference (Denver, August 4-8, 1985), AIChE Symposium Series.
[35] Khan, F., Yue, P. L., Rizzuti, L., Augugliaro, V. and Schiavello, M. (1981) 'Photochemical synthesis of ammonia over zeolites', J.C.S. Chem. Comm. 20, 1049-1050.
[36] Khan, F., Yue, P. L., Rizzuti, L., Augugliaro, V. and Brucato, A. (1983) 'Photoassisted water cleavage and nitrogen fixation over titanium exchanged zeolites', I. E. C. Prod. Res. Dev. 22, 238-241.
[37] Augugliaro, V., Lauricella, A., Rizzuti, L., Schiavello, M. and Sclafani, A. (1982) 'Conversion of solar energy to chemical energy by photoassisted processes. I-Preliminary results on ammonia production over doped titanium dioxide catalysts in a fluidised bed reactor', Int. J. of Hydrogen Energy 7, 845-849.
[38] Augugliaro, V., D'Alba, F., Rizzuti, L., Schiavello, M. and Sclafani, A. (1982) 'Conversion of solar energy to chemical energy by photoassisted processes. II-Influence of the iron content on the activity of doped titanium dioxide catalyst for ammonia photoproduction', Int. J. of Hydrogen Energy 7, 851-855.
[39] Mahajani, V. V. and Sharma, M. M. (1981) 'Gas absorption with photochemical reaction', Chem. Engng Sci. 36, 595-597.
[40] Otake, T., Tone, S., Higuchi, K. and Nakao, K. (1983) 'Light-intensity profile in gas-liquid dispersion. Applicability of the effective absorption coefficient', Intern. Chem. Eng. 23, 288-297.

THE CURRENT STATUS OF SOLID STATE PHOTOVOLTAIC RESEARCH AT SERI

S. K. DEB
Solar Energy Research Institute
1617 Cole Boulevard
Golden, CO 80401 USA

ABSTRACT. As one of the nation's lead laboratories for renewable energy research, SERI conducts and manages advanced research and development on solid state photovoltaics for the U. S. Department of Energy. The main purpose of SERI's photovoltaic program is to conduct high risk, potentially high payoff research and development which will result in a scientific and technological knowledge base from which private enterprises can further product development. Consistent with this strategy, a multiple research approach has been developed that ensures greater probability of success. The activities are divided into the following areas: amorphous silicon, compound semiconductor thin films, III-V high efficiency cells, crystalline silicon, solid state theory, advanced measurement and characterization, and solar radiation research. In this paper, a brief overview of the current status of research and development in each of these technologies is given.

1. Introduction

The conversion of solar energy directly into electrical power by solid state photovoltaic (PV) devices provides one of the most promising energy supply options. In order for photovoltaics to make a significant contribution to global energy supply needs, it is imperative that this technology be able to provide electricity at a cost that is competitive with conventional methods of generating electricity. The U. S. Department of Energy's (DOE) Photovoltaic Program objective is to develop this technology to a point where it can become a viable energy supply option. Some of the key elements of the program strategy are to conduct fundamental research to increase our understanding of photovoltaic materials and devices, build the technology base, bring the cost of PV-generated electricity to a competitive level, and transfer the technology to industry and end users.

As a DOE laboratory, SERI's primary responsibility is to conduct fundamental and applied research on a variety of semiconductor materials and devices for PV applications. The main purpose of our PV program is to carry out potentially high-payoff research and development, which will result in a scientific and technological knowledge base that private enterprises can use for further product development. Consistent with this strategy, a multiple research approach that ensures greater probability of success has been developed. The activities are divided into the following areas amorphous silicon, polycrystalline thin films, III-V high efficiency cells, single crystal silicon, solid state theory, and advanced measurement and device technologies. The PV program at SERI has two components: the in-house research concentrates on the fundamental understanding of the material and device properties, and the subcontracted research focuses

E. Pelizzetti and M. Schiavello (eds.), Photochemical Conversion and Storage of Solar Energy, 575–591.

primarily on technology development. In this paper, a brief overview of the current status of our research accomplishments and major advances in technology development will be given.

2. Amorphous Silicon Thin Films

Hydrogenated amorphous silicon (a-Si:H) and its alloys constitute one of the leading thin film technologies for PV applications. Ever since the University of Dundee group[1] discovered that n- and p-type materials can be fabricated by glow-discharge decomposition of silane, and the RCA group[2] first fabricated a PV cell with this material, this field has experienced a phenomenal growth. Intense studies of the optimization and characterization of materials and device parameters have led to the development of a small-area, single-junction p-i-n type solar cell with an efficiency of 12.0%.[3] Figure 1 shows the efficiency as a function of time for a-Si:H solar cells and submodules.

In spite of this remarkable achievement, the ultimate success of this technology depends upon the solution of some key research issues. Based upon several government and industry studies, it is generally recognized that module efficiencies must be on the order of 15% before U. S. central power applications can be realized by the thin film technologies. In the case of

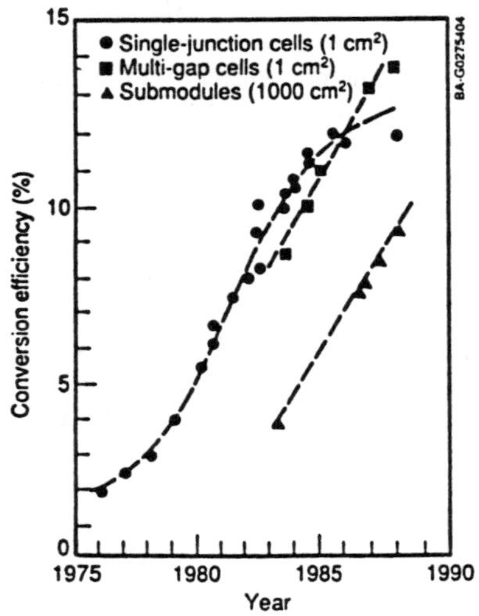

Figure 1. Efficiency versus time for amorphous silicon cells and modules.

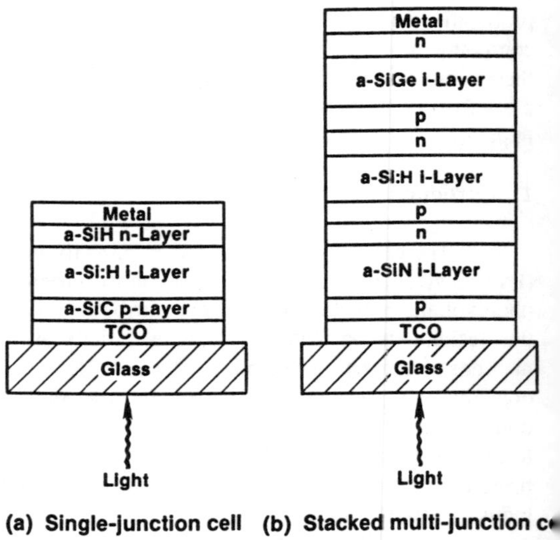

Figure 2. Device structure for amorphous silicon solar cells.

amorphous silicon, it appears that a single-junction a-Si:C:H/a-Si:H(i)/a-Si:H(n) p-i-n cell, shown in Figure 2(a), has been optimized to the extent that 10-12% efficiencies have been achieved in a number of laboratories. However, further dramatic improvements in efficiency are difficult to achieve without a multijunction approach. In this type of approach, low-energy photons are directed to cells made from smaller band-gap materials and high energy photons are directed to cells made from wide band-gap materials, and this results in greater utilization of the solar spectrum and hence increased conversion efficiency. The cells whose active semiconductors possess different band gaps are connected in series and stacked on top of each other with the largest band-gap materials facing the incident radiation. A typical multijunction amorphous silicon solar cell is schematically shown in Figure 2(b). In such a structure, the thicknesses and band gaps of the component layers are adjusted in such a way that the currents generated from each constituent cell are the same. Calculations made by Kuwano et al.[4] show that it is possible to achieve an efficiency around 24% with a combination of a-Si amorphous materials with 1.4-eV, 1.7-eV, and 2.0-eV band gaps. Clearly, there is a need to develop new types of silicon-based alloy materials with band gaps in the range 1.4-eV to 2.0-eV.

In a typical three-cell tandem device, a-Si:C:H and a-SiGe:H alloys have recently emerged as the most likely candidates for incorporation into such a device as high band-gap and low band-gap materials respectively. However, research into these alloys has until now proceeded along slightly different lines. The interest in a-SiC:H alloys has been primarily limited to that of a p-type window layer in a typical p-i-n type solar cell. However, the optimized deposition conditions for intrinsic a-SiC:H have not yet emerged and, consequently, it is not clear what the best properties of such a material could be. On the other hand, with a-Si:Ge:H the deposition parameters for the production of good intrinsic material, down to a band gap of 1.5 eV, is fairly well studied; promising results have been obtained using this material both in a single-junction and a multijunction approach. Another approach to low band-gap material is to alloy a-Si with Sn. It has been shown that by the addition of a small amount of Sn (less than 10%) the band gap can be lowered to 1.0 eV.[5] Unfortunately, the electronic properties degrade very quickly with the incorporation of Sn. It appears that in all cases new defect states are introduced, which increases the density of states within the gap. The nature of these defect states is poorly understood. In spite of all these limitations, major advances have been made recently on the fabrication of efficient solar cells using an a-SiGe:H alloy.

The amorphous silicon based technology has reached a degree of maturity such that inexpensive solar cells can be made. However, a key factor in determining the viability of this device for power production is the long-term stability. Staebler and Wronski[6] first observed that the dark and photoconductivity of a-Si alloys change after exposure to light. The effect was found to be reversible by annealing at temperatures above 150°C. Later work showed light-induced changes in many other properties, including the solar cell efficiency, which experienced degradation with prolonged exposure to light. Consequently, extensive studies have been made in recent years on the light-induced effects in a-Si and its alloy materials and devices prepared under a variety of deposition conditions. Four conclusions were reached: (i) The defects are generated primarily by the recombination of an electron-hole pair, and to a lesser extent by hole trapping. (ii) The defects create a wide range of gap states distributed in energy over the mobility gap. (iii) Different kinds of defects are produced. (iv) The structure of the material, impurities, and residual stress play an important role in creating and annealing defects. Two models are now being considered to explain the kinetics of light-induced effects: a bond-breaking model, which results in the creation of a dangling bond, and a negative correlation energy model. It has also been proposed that hydrogen plays an important role in the process. As yet, no definitive model identifying the exact nature of the defects has emerged. In spite of the fact that we do not clearly understand the nature of the defect states, considerable progress has been made in finding a practical solution to the degradation problem by a combination of engineering design and the use of improved materials. The use of a tandem cell configuration in which the top cell is very thin permits us to maintain a high field, which reduces recombination and minimizes degradation.

2.1. KEY RESEARCH ISSUES IN AMORPHOUS SILICON

In spite of extensive research done in laboratories throughout the world, there still remains several fundamental questions regarding the basic nature of amorphous silicon. The development of alloy materials with properties suitable for the fabrication of stable high efficiency devices will require a much better understanding of the role of the film microstructure, including microvoids, growth morphology, and hydrogen distribution on internal stresses to determine the minority carrier transport properties. Particular attention has to be paid to the band-tail density of states in the alloys, which must be reduced to a level close to that of a-Si:H to improve the device efficiency. The kinetics of film growth are very poorly understood for a-Si:H and even less understood for the alloys. It is important to identify the dominant species contributing to film growth during plasma deposition. A variety of deposition methods exist, such as plasma-assisted chemical vapor deposition (CVD) using both diode and triode configuration, photo-CVD, and magnetron sputtering. Carefully designed experiments need to be done to improve the understanding of: (i) the species in the plasma that are critical to film growth, (ii) surface and interface chemistry critical to the formation of the best quality film and devices, and (iii) the role of fluorine in the properties of alloy material, particularly Si:Ge:H. Fundamental work needs to be undertaken on the photo-degradation effect and its correlation with the microstructure of a-Si:H and other alloy films. Special attention should be given to various effects associated with impurities, hydrogen content, internal strain, and weak bonds.

2.2. SERI IN-HOUSE RESEARCH ON a-Si:H ALLOYS

Considerable work has been done in recent years, both at SERI and elsewhere, on the fabrication of high quality alloy materials. Unfortunately, it has been found in all cases that the electronic properties and the photoresponse characteristics severely degrade upon alloying. In general, the materials contain a significantly higher densities of states at the Fermi level, which makes it difficult to fabricate high efficiency solar cells. Therefore, one of the primary objectives of our research at SERI on these alloy materials has been to understand the reasons for this decrease in photoresponse on alloying so that improved alloy materials can be fabricated. We focused on the film microstructure as determined from infrared measurements of the local H-bonding configuration, because previous studies have linked microstructural effects to the reduced photoresponse of both a-Si:C:H and a-Si:Ge:H alloys.

We have recently developed Small Angle X-ray Scattering to study the size and density of microvoids in these materials. This has dramatically increased our knowledge of the microstructure of amorphous silicon. We have shown that even the best device-quality films contain 2-5 x 10^{19} microvoids, each consisting of 16-20 missing atoms. By combining these results within infrared (IR) absorption, we conclude that the interim surfaces of these voids are largely unhydrogenated. We have also shown that the presence of microvoids adversely affects the transport properties and the stability behavior of amorphous silicon alloys. Therefore, several innovative approaches to reducing microstructure by alternative deposition methods are being developed, such as photo-CVD, periodic deposition and etching-CVD, and ion beam rehydrogenation of hydrogen-depleted materials. The role of fluorine in a-Si:H is also being studied, with particular emphasis on photo-CVD materials and fluorine's role in controlling the microstructure. Special emphasis is given to the connection between optical and electrical properties and light-induced instability (the Staebler-Wronski effect) commonly observed in a-Si:H materials.

A promising new deposition technique, hot-wire catalytic decomposition of silane and germane, is being developed for fabricating superior amorphous silicon alloy materials. This technique has already led to the fabrication of intrinsic a-Si:H material with very low hydrogen content and superior transport properties. In addition, we are doing extensive theoretical modeling to establish the nature of defect states, to study the kinetics of annealing and producing metastable defects, and to establish the role of hydrogen in the metastability of a-Si:H. These modeling studies are leading to new cell designs and processing steps.

2.3 SUBCONTRACTED RESEARCH ON AMORPHOUS SILICON

The objectives of the subcontracted research in amorphous silicon are to improve and understand the optoelectronic properties of amorphous-silicon-based alloy materials and to improve the conversion efficiency and stability of single-junction and multijunction solar cells and submodules. The research is directed toward achieving the FY 1990 goals, which are 10% efficiency for single-junction and 13% for multijunction submodules of 900-cm^2 area.

The technical plan of the Amorphous Silicon Research Project (ASRP) is divided into two principal activities: (1) multidisciplinary research and (2) fundamental research. The multidisciplinary research activities comprise the government/industry cost-shared programs that have broad-based research teams located at the facilities of the individual companies. These teams perform directed research that covers aspects from starting materials to demonstration of proof-of-concept cells and submodules. Fundamental research, on the other hand, involves basic, higher-risk, and supporting research at universities and research laboratories that aids industry by advancing the technology base. Cost-shared multidisciplinary programs address issues concerning single-junction and multijunction devices. Research is also performed to advance the conversion efficiency and the stability of both small-area cells and large (900-cm^2 and up) submodules fabricated by plasma-enhanced CVD. The stability encompasses both intrinsic aspects, such as the light-induced effect, and extrinsic aspects, such as diffusion or corrosion as a result of the environment. The efficiency improvement involves work on improving light trapping, achieving higher conductivity of interconnectors, and minimizing area losses due to interconnections.

After the successful completion of the first government/industry program conducted from FY 1984 through FY 1987, the best single-junction a-Si cells had efficiencies as high as 11.7% and single-junction a-Si submodules had efficiencies up to 7.9%

A three-year government/industry program was started in FY 1987 with participation by ARCO Solar, Inc.; Chronar Corporation; Energy Conversion Devices, Inc.; and Solarex Corporation. Emphasis of this second program is placed on amorphous silicon multijunction technology, with the major goals being the achievement by 1990 of efficiencies of: (1) 18% for multijunction cells (1 cm^2), (2) 13% for multijunction submodules (900 cm^2), and (3) 10% for single-junction submodules (900 cm^2). Stability criteria, namely less than 5% degradation after the equivalent exposure to 2 months of insolation, are associated with the efficiency goals. The status of this program, now in its third year, is as follows.

ARCO Solar has achieved SERI-measured, aperture-area efficiencies of 9.3% for a single-junction, amorphous silicon, 844 cm^2 submodule and 11.3% for an amorphous silicon/copper-indium-diselenide stacked, four-terminal, 900-cm^2, 55-cell submodules. The best a-Si/$CuInSe_2$ 4-cm^2 tandem cell measured by ARCO Solar is 15.6%, whereas the best value measured by SERI on a different tandem cell is 14.5% (13.1%) measured at SERI and for a 4-ft^2 module 8.5% (aperture-area) efficiency. The highest $CuInSe_2$ cell measured by ARCO is 14.1% in efficiency.[7] For a 40-cm^2 submodule, Solarex achieved 9.5% (aperture-area) efficiency for single-junction amorphous silicon and 8.5% (aperture-area) for triple-junction a-Si/a-Si/a-SiGe.

A new three-year government/industry/partnership program will start in the February-March 1990 timeframe with funding from the FY 1990 budget. The principal objectives of this research are: (1) to conduct research on semiconductor materials and non-semiconductor materials to enhance two-terminal, multijunction, thin-film, large-area, all-amorphous-silicon-alloy device performance; (2) to develop high-efficiency, stable, reproducible, and low-cost multijunction PV modules based on all-amorphous materials; (3) to demonstrate in FY 1993 stable 12% (AM 1.5) aperture-area solar conversion efficiency for different-band-gap modules; and (4) to demonstrate FY 1993 stable 10% (AM 1.5) aperture-area solar conversion efficiency for same-band-gap modules. The modules will be at least 900 cm^2 in area and consist of at least two integrally stacked devices using all-amorphous-silicon-alloy materials.

In addition to the government/industry programs, fundamental research involves work on the light-induced effect, material deposition rates, alternate deposition methods, amorphous-silicon-based alloy materials, material and plasma characterization, and modeling. Summaries of these studies follow.

3. Polycrystalline Thin Film Solar Cells

3.1 $CuInSe_2$/CdS CELL

Starting from the mid-fifties, for a period of almost two decades, the polycrystalline thin film solar cell field was dominated by the Cu_2S/CdS heterojunction device. After years of extensive research and development, it was possible to achieve a 10.2% conversion efficiency in this type of device. Considerable process development work had been done, both on the laboratory scale and in the commercial production scale and this led to various low-cost process developments. In spite of the basic simplicity of cell fabrication and the promise of inexpensive, large-scale production, the stability of the device remained the most serious problem. Most of the stability problems were associated with Cu_2S, which under normal operating conditions underwent rapid degradation. The emergence of $CuInSe_2S$/CdS as a stable, high efficiency solar cell in recent years led to the demise of Cu_2S/CdS solar cells. The basic device structure is shown in Figure 3. It led to a higher reported efficiency of 11.9% in a 1-cm^2 CdZnS/$CuInSe_2$ cell developed by the Boeing Corporation.

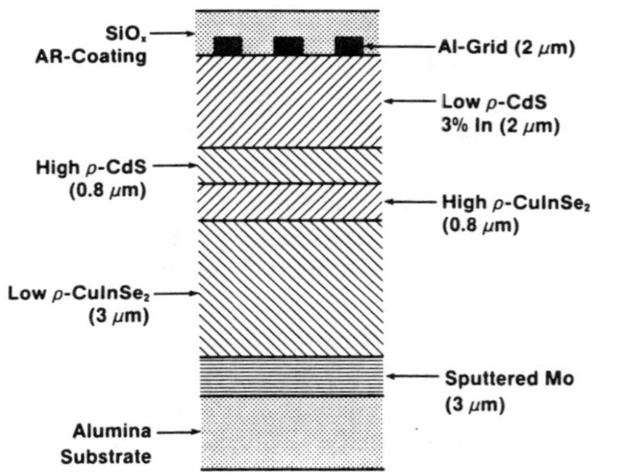

Figure 3. Device structure for CdS/$CuInSe_2$ solar cells.

The most successful method of fabricating thin films of $CuInSe_2$, the most crucial component in this device, is based on vacuum deposition from three separate sources of Cu, In, and Se and was developed by Mickelsen et al. at Boeing.[8] The fabrication process consists of the following sequential steps: (i) sputter deposition of approximately 3 micrometers (μ) of molybdenum metal on an alumina or glass substrate; (ii) three source evaporation of a low resistivity $CuInSe_2$ film (2-3 μ thick) at 450°C; (iii) three source evaporation of a high resistivity $CuInSe_2$ film (approximately 0.8 μ thick) at 450°C; (iv) evaporation of a CdS film (approximately 0.8 μ thick) at 150°C; (v) evaporation of a CdS film doped with In (approximately 2 μ thick) at 150°C; (vi) electron beam evaporation of an aluminum grid structure; and (vii) evaporation of SiO or MgF_2 antireflection coating. For an efficient device, it is extremely important to control the carrier type (p-type) and concentration in the optimum range (10^{16} - $10^{17}/cm^3$) by carefully controlling the stoichiometry of the film during deposition.

The most dramatic improvement of this technology occurred recently when ARCO Solar (now Siemens Solar Industries) reported achieving 14.1% active-area efficiency in a $CuInSe_2$ cell. This was accomplished by replacing the conventional, thicker CdS film (3 μ) with a combination of a thinner CdS layer (.05 μ) and a conductive ZnO layer: this is a proprietary low-cost $CuInSe_2$ fabrication process that may involve selenization of sputter deposited copper and indium films containing small amounts of gallium. The effectiveness of this technology was further demonstrated when they fabricated a square-foot submodule and achieved an 11.1% efficiency. More recently, they scaled up their module production size to a four-square-foot area and reported a record efficiency of 9.7% for this size of thin film module.[9] This landmark achievement has put $CuInSe_2$ devices in the forefront of all thin film technologies.

The performance characteristics of a typical $CuInSe_2$ thin film solar cell fabricated at SERI is shown in Figure 4. The most striking feature of this i-v curve is the attainment of a very high short-circuit current (approximately 40 mA/cm^2), which is very unusual for a polycrystalline material having a large density of grain boundaries and other imperfections. It appears that the crystalline imperfections are somehow passivated so that it is possible to collect almost 95% of the carriers generated by light. The areas that need further development are the improvement of open-circuit voltage (V_{oc}) and fill factor (FF). Since the band gap of $CuInSe_2$ is relatively low (1.04 eV), the highest V_{oc} one hopes to achieve is around 0.55 V at room temperature; this has already been achieved in a single-crystal $CuInSe_2$ cell. The highest FF one can reasonably achieve in this device should be around 75%, which means a modest increase over the current state-of-the-art FF of 70%. Based on this modest expectation, it is reasonable to expect that the cell efficiency should reach well over 15%.

As we have already mentioned, one of the most attractive features of a $CuInSe_2$/CdS cell is its promise for long-term stability. Initial testing done at Boeing of unencapsulated cells under simulated illumination at 25°C showed degradation over a period of 10,000 hours of exposure. Similarly, outdoor testing of an ARCO Solar $CuInSe_2$ module with efficiency over 8% has shown no degradation during 500 days of exposure.

3.2. SERI IN-HOUSE RESEARCH ON $CuInSe_2$ MATERIALS AND DEVICES

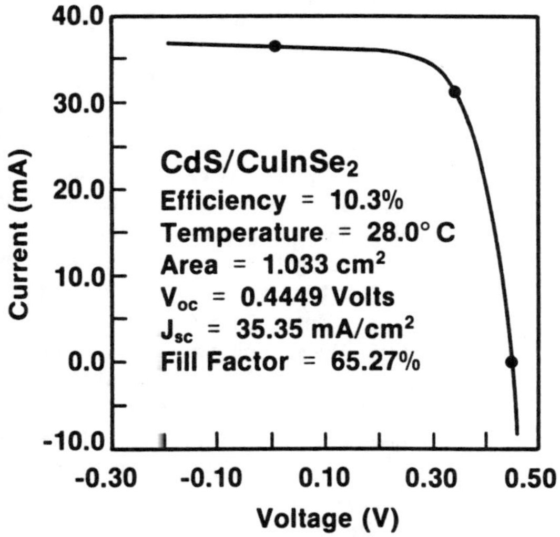

Figure 4. Performance Characteristics of a CdS/$CuInSe_2$ Cell.

The primary thrust of our in-house research in this area involves (i) fundamental studies of material properties, (ii) a basic understanding of the device physics, and (iii) fabrication of thin films of $CuInSe_2$ by a one-step electrodeposition process. The fundamental material properties are studied both on polycrystalline thin film and bulk single crystal, whereas the device physics are studied on polycrystalline thin films fabricated largely by co-evaporation.

3.2.1. *Material Properties*

In a PV device of this type, the most crucial component is the p-type $CuInSe_2$, which acts as the absorber. For an efficient device, it is extremely important to make it p-type with an optimum carrier concentration on the order of 10^{16}-10^{17}. The electrical properties of $CuInSe_2$ are dominated by native defects such as vacancies, interstitials, and antisite defects, whose concentrations are determined by deviations from the stoichiometric composition. The p-type conductivity is achieved either by cation vacancies or anion interstitials, although it is generally believed that copper vacancies are mainly responsible for the p-type conductivity in $CuInSe_2$ thin films. One of the most successful methods of fabricating thin films of $CuInSe_2$ is based on vacuum evaporation from separate elementary sources of copper, indium, and selenium. The carrier type and the concentrations are controlled by varying the Cu:In ratio while maintaining a constant selenium concentration. Figure 5 shows the experimentally determined carrier concentration and type as a function of the Cu:In ratio.[10] There is a narrow, V-shaped region in which p-type conductivity is observed, and within that region the carrier concentration changes by several orders of magnitude. This implies that in order to achieve high conversion efficiency it is important to maintain a rigid control of the stoichiometry.

The other material parameters important for PV applications are the optical absorption coefficient and the energy gap as a function of composition. As a part of the standard device fabrication procedures, a copper-poor layer of $CuInSe_2$ is deposited on a near-stoichiometric layer at deposition temperatures of 350°C and 450°C, respectively. It is important, therefore, to know the composition of the layer and its effect on the optical properties. Secondary mass spectroscopy (SIMS) profiling of the $CuInSe_2$ bilayer indicates that the bulk of the material assumes an average composition of the two layers, with the exception of a surface layer (about 0.6 μ thick) which is Cu-poor.[2] As a function of composition, the absorption coefficient increases until the Cu:In ratio approaches unity, and then decreases as the ratio increases past unity. The energy gap for these thin films, shown in Figure 6, ranges from 0.94 eV for nearly stoichiometric films deposited at 350°C, to 1.02 eV for copper-poor films deposited at 450°C.

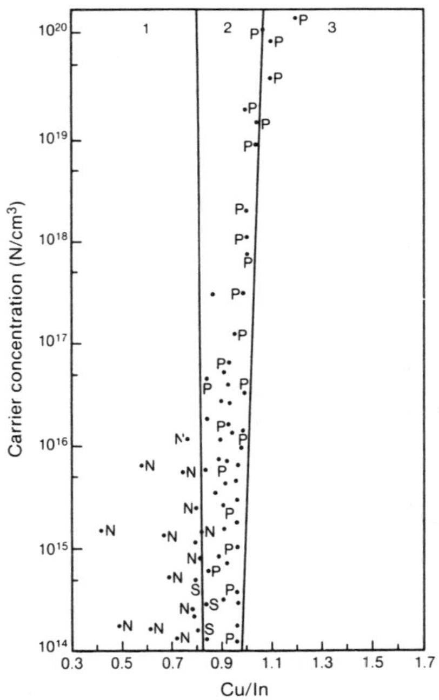

Figure 5. Carrier concentration versus composition of $CuInSe_2$ films.

Despite its strong absorption properties, the band-gap energy (0.94-1.02 eV) of $CuInSe_2$ is somewhat smaller than that required for optimum PV conversion efficiencies. Instead of $CuInSe_2$, its other ternary analogs, such as $CuGaSe_2$ (1.7 eV) and $AgInSe_2$ (1.2 eV), can be used to increase the band-gap and hence increase V_{oc}. The alternative approach is to form a quaternary $Cu(Ga,In)Se_2$ alloy of varying composition to modulate the band-gap to an optimum value. In fact, Boeing has already taken the latter approach and had achieved a 12.5% efficiency with a $ZnO/CdZnS/Cu(Ga,In)Se_2$ cell structure. We are doing a systematic study of the optical, electrical, and structural properties of thin films of $CuGaSe_2$ and $Cu(Ga,In)Se_2$ as a function of composition and deposition parameters to establish the optimum PV performance parameters.

Besides these compositional, optical, and electrical measurements, we are using a variety of other measurement techniques such as photoluminescence, steady-state capacitance, and conductance and thermally stimulated capacitance to derive information on the intrinsic defect levels in $CuInSe_2$ thin films.

3.2.2. *Device Optimization and Characterization*

In this area, our work primarily involves the fabrication of ternary thin film solar cell materials and devices to (i) understand the defect chemistry and charge transport that control the photovoltaic properties of these materials, (ii) understand the role of the oxidation process in thin films of $CuInSe_2$ and $CdS/CuInSe_2$, and (iii) modify these materials to optimize device performance. The basic device structure, shown in Figure 3 is fabricated by sequential deposition of the component layers in vacuum on an Al_2O_3/Mo substrate. One of the key steps in fabricating $CdS/CuInSe_2$ thin film solar cells is post-deposition annealing in oxygen at about 200°C. A systematic study has been carried out to determine the effect of heat treatment in oxygen. We have shown that $CuInSe_2$ thin films contain around 0.1 at.% oxygen, and that it affects the electrical properties of the film. Oxygen can be removed from the film by electron beam irradiation or treatment in a reducing atmosphere, and it can be incorporated into the film by annealing in air at an elevated temperature (about 200°C) or by treatment with a chemical oxidizing agent. The process is reversible. Based on a detailed study of redox processes, and from a variety of physical measurements involving infrared, X-ray photoemission, scanning tunneling microscopy, Hall-effect, and other methods, we have formulated a consistent defect chemical picture of the effect of the annealing process. The model centers on oxygen-induced neutralization of surface (or near surface) donor states associated with selenium vacancies. Extensive electron-beam induced current (EBIC) measurements[5] were made on $CdS/CuInSe_2$ devices as a function of post-deposition heat treatment to determine whether this type of device is a buried homojunction or a heterojunction.

3.2.3. *Electrodeposition of $CuInSe_2$*

We are developing a simple, one-step electrodeposition process for fabricating $CuInSe_2$ films that does not involve the selenization process. Near-stoichiometric and In-rich monolayers and bi-layer thin films of $CuInSe_2$ with chalcopyrite structure have been given by this method. X-ray diffraction studies show that the electrodeposition films are structurally identical to vacuum evaporated films. Careful control of the solution and the deposition conditions have been shown to result in device-quality $CuInSe_2$ thin films. However, the efficiency of a device fabricated from electrodeposited $CuInSe_2$ is still considerably lower than that of the evaporated films.

3.3. SUBCONTRACTED RESEARCH ON POLYCRYSTALLINE THIN FILM

The objective of the Polycrystalline Thin Film Program is to develop thin-film, flat-plate modules that meet DOE's long-term goals of reasonable efficiencies (15%-20%), very low cost (near $50/m^2), and long-term reliability (30 years). The approach relies on developing PV devices based on highly light-absorbing compound semiconductors such as $CuInSe_2$, CdTe, and their alloys. These semiconductors are fabricated as thin films (1-3 μm thick) with minimal material and processing costs.

Very high efficiencies have been achieved by these promising materials.[11] $CuInSe_2$ cells made by ARCO Solar and by Boeing were measured at SERI at 13.1% and 12.9% efficiency (active area), respectively. ARCO Solar reports achievement of a 14.1% efficiency which has not been confirmed. Other surpassing 10% efficiency in $CuInSe_2$ are SERI, International Solar Electric Technology, and Institute of Energy Conversion. Larger area $CuInSe_2$ devices have also been fabricated with very high efficiencies. ARCO Solar has made a 938-cm^2 (aperture-area)

$CuInSe_2$ module with 11.1% efficiency (10.4 W), and an 8.5% (4-ft^2) module, both measured at SERI.

$CuInSe_2$ shows good proven stability under controlled conditions (9000 of illumination), and initial outdoor tests (ARCO) are also promising. We have conducted a full year outdoor testing on two ARCO Solar $CuInSe_2$ panels with very little change in their efficiencies (under 4%). These are the first such tests on $CuInSe_2$ by an independent agency and show the great potential stability of $CuInSe_2$ panels.

Three laboratories (University of South Florida, Ametek, and Photon Energy) report CdTe cell efficiencies between 10.5% and 12.5%. Photon Energy has fabricated near-square-foot CdTe submodules measured outdoors at SERI at 7.3% efficiency (aperture-area). They also have the world record in CdTe cell efficiency (12.3%), measured at SERI. Innovative designs are now addressing past difficulties in contacting CdTe. We have recently begun testing encapsulated CdTe submodules provided by Photon Energy. After four months, they show no degradation. As with the $CuInSe_2$, these are the first independent tests of encapsulated devices made from this material. Since stability has been an identified issue with CdTe, these initial results are considered favorable.

The improved efficiencies and larger areas of $CuInSe_2$ and CdTe devices, and their apparent stability, are the major recent advances in these technologies, but polycrystalline cells require continued development to achieve 15%-20% conversion efficiencies. Two strategies are being used: development of single-junction cells and an innovative effort on two-junction cascade cells. Improvement of the single-junction technologies has been steady and reliable. This strategy remains the major focus of the task. Potentially achievable cell efficiencies approach 20%, and projections indicate the likelihood of fabricating modules of more than 15% efficiency. Developing two-junction, $CuInSe_2$-based cascade cells permits even more ambitious long-term efficiency goals. Combining two well-matched, single-junction modules into a cascade cell design could lead to module efficiencies of better than 20% at costs under \$60/m^2. The materials being investigated for the top cells include CdTe and ZnTe alloyed with Mn, Mg, Zn, and Hg.

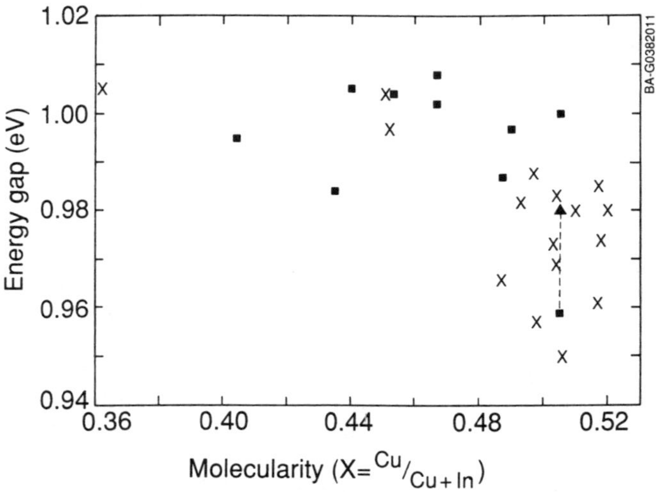

Figure 6. Bandgap of $CuInSe_2$ as a function of composition.

Developing scalable, low-cost fabrication methods is important in providing industry with a foundation for future large-area, high-throughput commercial processes. Research methods for fabricating polycrystalline cells include, for $CuInSe_2$, an e-beam evaporation/selenization method (International Solar Electric Technology), a reactive-sputtering and hybrid sputtering/evaporation method (University of Illinois), and evaporation (Boeing); for CdTe, close-spaced sublimation

(USF), evaporation (Institute of Energy Conversion), electrodeposition (Ametek), metal-organic chemical vapor deposition (MOCVD) (Georgia Tech and University of South Florida), and spraying (Photon Energy).

An initiative in the Polycrystalline Thin Film Program was begun with the release of an RFP in FY 1989. The objective of the solicitation was to stimulate progress in $CuInSe_2$ and CdTe submodule development and to deepen the U.S. participation in these promising technologies. The research community responded favorably, as reflected in the many excellent technical proposals received in response to the solicitation. The contracts resulting from that RFP will begin in spring 1990.

4. High Efficiency Single-Junction and Multijunction Cells

4.1 DOE/SERI PROGRAM

The objective of the High Efficiency Solar Cell Program is to evaluate and develop advanced PV technologies capable of energy conversion efficiencies in excess of 20% for flat-plate configurations and 30% for concentrator systems. These goals are discussed in the DOE Five-Year Research Plan as technology targets for the late 1990s. Even on this longer term horizon, it is difficult to envision a technology capable of achieving such high efficiencies without incorporating the demonstrated performance of crystalline III-V semiconductors. Thus, the High Efficiency Concepts task has become synonymous with III-V compound semiconductor research.

SERI's program of research in High Efficiency Concepts has approached the terrestrial PV goals from the direction of first demonstrating the feasibility of exceeding the efficiency targets to assure that production engineering trade-offs between performance and cost can be accommodated. Recent advancements by the community researching high efficiency technologies provide a high level of confidence that the efficiency goals can be readily met.

The achievement of 25% efficiency in commercial concentrator modules will likely require production cells with over 25% efficiencies. The system cost target for that efficiency corresponds to cell costs below $\$10/cm^2$. Several organizations have demonstrated single-junction GaAs cells at approximately this efficiency level. Boeing Corporation recently reported dramatic results, an efficiency over 35% (at 100 suns, and AM 1.5) in a GaAs/GaSb stacked concentrator cell assemblies designed for terrestrial applications.[12] However, a fivefold reduction in processing costs would be needed to meet the cell cost target for concentrator cells. Much of this reduction can be achieved by using larger wafers and higher through put deposition systems. Research supported by this program benefits future development by strengthening the understanding of basic mechanisms that affect the uniformity of doping, composition, and thickness over large-area wafers, from wafer to wafer, and from run to run. Efficient utilization of source materials and evaluation to potentially superior sources (cost, purity, control, safety, and other factors) are also important topics for research. Continued improvement in cell efficiency is also a critical factor in reaching cost effectiveness for the technology.

Flat-plate technologies have several advantages relative to concentrator technologies because the ability to utilize both the direct and diffuse components of the solar energy resource increases the geographical range of operation, simplifies system design and operation, and opens a variety of market opportunities for small installations. High efficiency modules can be achieved either through developing multi-junction devices or developing processes for low-cost deposition of single-crystal thin films. Two technologies have already reached performance levels consistent with the efficiency goals. One approach produces thin crystalline film separated from a reusable substrate in a process called cleavage of lateral epitaxial film for transfer (CLEFT). Kopin Corporation[13] developed the process and has reached efficiencies of 22.4%. Thin films of GaAs grown on silicon substrates developed by Spire Corporation,[14] are rapidly closing in on the 20% efficiency target, having improved from 11% to 19.9% efficiency in the last two years.

4.2. SERI IN-HOUSE RESEARCH ON HIGH EFFICIENCY SOLAR CELLS

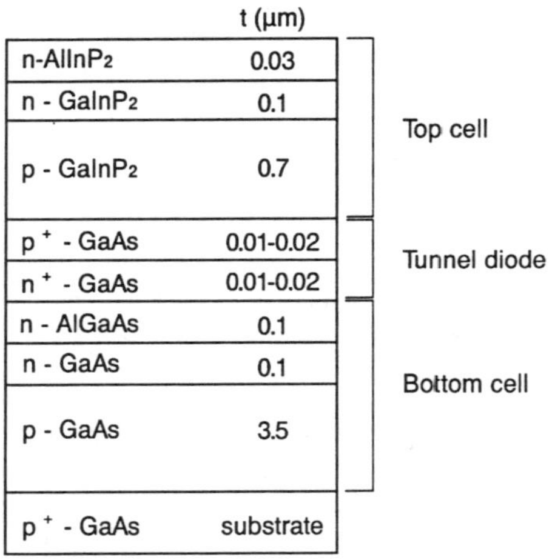

Figure 7. Device Structure for GaAs/$GaInP_2$ solar cell.

The objective of our in-house research program is to develop an understanding of, and solutions to, the problems inherent in the fabrication of high efficiency multijunction cascade solar cells. The cells under investigation are fabricated from III-V semiconductors (GaAs, GaP, GaAsP, InP, GaInP, and GaInAs). Research covers three main areas: (1) defect analysis and interaction, (2) crystal growth, and (3) multijunction device fabrication and characterization. The metal-organic chemical vapor deposition (MOCVD) technique is primarily used for the epitaxial growth of materials. Both lattice-matched and lattice-mismatched multijunction structures are studied for high efficiency solar cells. A novel concept of using superlattice structures in the multijunction devices is being developed. Measurement techniques including high resolution electron microscopy and photoelectroreflectance are routinely used for materials and device characterization.

For a monolithic, two-junction tandem device epitaxially deposited on a substrate that also serves as the bottom cell, the quality of the top cell generally will be adversely affected by any lattice mismatch between the bottom and top cell materials. Heretofore, current-matched semiconductors with band-gap energies of 1.1 eV and 1.7 eV have received the most attention. This combination of band gaps yields the maximum AM1.5 conversion efficiency, 36% [1]. GaAs has a band gap of 1.42 eV, and there are at least two III-V alloys with band gaps of 1.9 eV that are also lattice-matched to GaAs: $Al_{0.4}Ga_{0.6}As$ and $Ga_{0.5}In_{0.5}P$ (hereafter, $GaInP_2$). The AlGaAs/GaAs system has been studied extensively, and most recently Varian Associates has reported a record efficiency of 27.6% in a two-terminal monolithic multijunction cell.[15] We have chosen to investigate the $GaInP_2$/GaAs multijunction device in preference to GaAlAs particularly because $GaInP_2$ is relatively insensitive to the normal O_2 and H_2O background contaminations typical of MOCVD growth systems. A schematic of the $GaInP_2$/GaAs monolithic cascade cell is shown in Figure 7. The $GaInP_2$ and GaAs were grown in a vertical, air-cooled reactor at one atmosphere using MOCVD, the detailed aspects of which are described elsewhere.[16] The best efficiency measured to date for this device is 27.3% at 1 sun and AM 1.5.[17] This is the highest efficiency reported for a two-terminal, tunnel-junction interconnected tandem photovoltaic device. There are several parameters that affect the efficiency of this device structure. They include the electronic quality of the top and bottom cell materials, the band-gap and thickness of the top cell, the design of the antireflection coating, the tunnel junction interconnect, and the thickness and passivating properties of the window layers. Optimization of all of these parameters should lead to further improvements in cell efficiency.

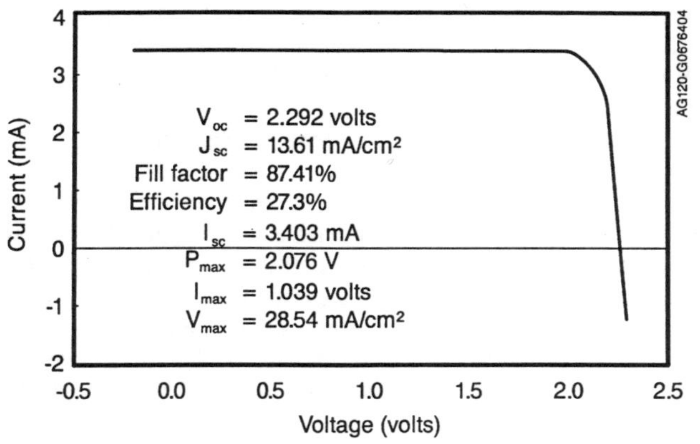

Figure 8. Performance Characteristics of a GaAs/GaInP$_2$ Solar Cell.

We have also shown that the surface recombination velocity of the GaInP$_2$/GaAs interface is less than 1.5 cm/s. As a result, bulk minority carrier lifetimes in the range of 10 to 14 μs have been observed in an undoped GaAs double heterostructure grown by MOCVD. These numbers are significantly better than those reported for AlGaAs/GaAs double heterostructures. The enhancement is attributed to cleaner GaInP$_2$/GaAs interface. We have been able to take advantage of these superior interface properties and fabricate record high efficiency (25.7%) GaAs single-junction solar cells using GaInP$_2$ as the window and back-surface hetero-interface layers.[18]

SERI is also working on other ternary and quaternary III-V alloy materials, such as GaInAs and GaInAsP for the development of high efficiency tandem solar cells. A monolithic, three-terminal tandem device incorporating an InP top cell and a low band-gap GaInAs bottom cell has been fabricated recently using an atmospheric MOCVD technique. This cell structure has yielded a record efficiency of 22.2% (AM0, 25°C). This is a promising PV device for achieving ultra-high efficiencies and high radiation resistance for space power applications.

Future research and development in this area will be primarily focussed on the reduction of cost. Technology for deposition of III-V compound semiconductors and their alloys by potentially low-cost processes is clearly the key factor for achieving the near term photovoltaic program goals. Research on characterizing materials and cells, analyzing loss mechanisms, demonstrating improved cell designs, and controlling and addressing safety issues related to the processes will contribute greatly to the desired technology base.

5. Crystalline Silicon Solar Cells

The last area we address here is the use of single-crystal and polycrystalline silicon for high efficiency solar cell applications. Among all the PV materials, the properties of silicon are the most widely studied and are conceptually, the simplest PV devices. It was the first terrestrial PV technology, is presently the most mature, and still holds the lion's share of the global PV market. The electricity generated by crystalline silicon is roughly five times more expensive than that generated by conventional means, which means a 20-fold decrease since 1975. This was largely possible by major technological innovations in materials and device processing as well as by increased automation and upgrading of manufacturing processes. Even though it is generally considered as a matured technology, major breakthroughs on various aspects of this technology continue to occur and should have a significant impact on cost reduction. About seven years ago, silicon solar cell efficiencies were in the range of 14-17% and, at that time, even the expert opinion was that the single crystal cell efficiencies had reached their practical limit. This has proven to be wrong, and cell efficiencies as high as 23.2% at one sun have been achieved by using advanced design concepts and high quality materials. This remarkable achievement was possible by two very innovative design concepts: one cell structure, called the PERC (passivated emiller and rear) cell, was developed by Green et al.[19] at the University of New South Wales (UNSW); the other structure is known as the point contact cell and was developed by Swanson et al.[20] at Stanford University. It is now believed that further improvement in cell design will lead to a single-crystal cell efficiency of 25% in the near future.

While single-crystal silicon continues to establish new records, substantial progress has also been made in other low-cost approaches such as continuous ribbon growth and various casting techniques for polycrystalline silicon. It seems possible that up to 17% efficiencies can be achieved in the ribbon materials, and that polycrystalline solar modules can reach efficiencies of 14%. An important development along this line is the recent achievement of 18% efficiency in laser textured, laser grooved polycrystalline silicon solar cells.[21] Another move toward significant cost reduction is Astropower Corporation's fabrication of 15.7% thin polycrystalline silicon films on a low-cost ceramic substrate.[22]

5.1. SERI IN-HOUSE RESEARCH ON CRYSTALLINE SILICON

Consistent with DOE program objectives, our in-house research addresses some of these issues particularly with respect to improving the material technology. Our efforts are directed primarily along the following lines: (i) We investigate crystal growth parameters, impurities, and crystallographic defects that influence minority carrier lifetime and solar cell performance in single-crystal silicon grown by the float-zone technique. We have grown various doped crystals with majority carrier concentrations in the range 10^{15}-10^{18}/cm^3 and achieved minority carrier lifetimes of 20 ms as shown in Figure 9.[23] These materials have been used to fabricate very high efficiency silicon solar cells. (ii) We are investigating methods for growing thin, narrow (~12mm), but very high quality (dislocation-free and high lifetime) silicon ribbons for potential use in long

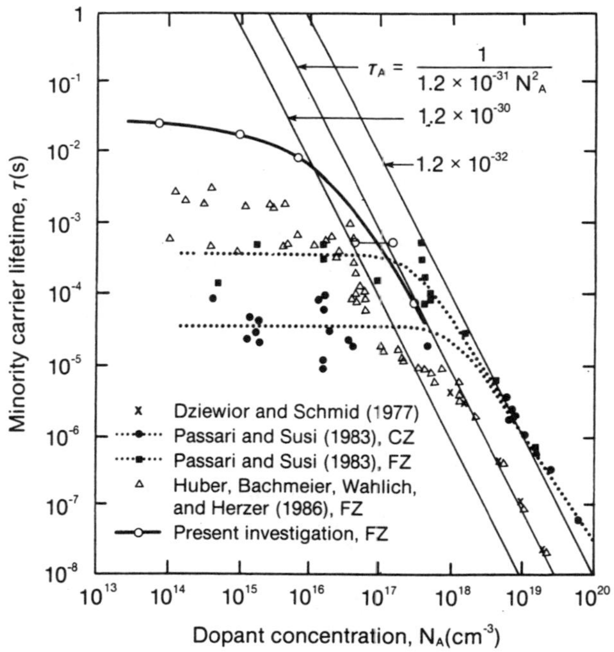

Figure 9 Minority Carrier Lifetime as a function of dopant concentration of FZ-grown silicon.

concentrator modules. (iii) We are exploring innovative techniques for growing <1mm grain-size silicon for use in multi-light-pass solar cells with a view to generating data for determining material property requirements. (iv) We are evaluating the potential of scaled-up cold crucible (electromagnetic confinement) technology for crystal growth and/or casting of silicon for high efficiency concentrator cells or high purity polycrystalline cell technologies.

5.2. THE DOE PROGRAM ON CRYSTALLINE SILICON

Although crystalline silicon technologies for material growth and device processing are considered to be mature for a variety of applications, there remain a host of issues about which the current understanding is only marginal. These issues pertain to the role of defects and impurities in altering the properties of silicon and silicon solar cells. It is generally recognized that the presence of impurities/defects in the substrate can degrade the cell performance; however, recent data show that in the presence of crystal defects, the influence of impurities on the cell performance is somewhat mitigated. It is also know that some defects can be passivated by presence of hydrogen or oxygen, although a detailed knowledge of these effects is certainly lacking. Clearly, these issues are critical to PV technology based on use of low-cost substrates which contain high concentrations of impurities and/or defects. The current research program was developed in collaboration with industry representatives to ensure that the industry research needs, pertaining to basic material issues, are adequately addressed. In this coordinated research effort between industry, universities, and SERI, the SERI role is primarily to support the area of test device fabrication/analysis in order to relate material characteristics directly to solar cell performance.

This research program is aimed at developing understanding of basic mechanisms related to influence of impurities and defects on the important PV parameters in silicon. In processes that can be applied to low-cost substrates, preferably as a part of solar cell fabrication process, in order to improve subcontracts that are supported by SERI in-house research participation. Major areas of research are: (1) mechanisms of hydrogen, oxygen interaction in low-cost silicon; (2) mechanisms of hydrogen passivation and kinetics of hydrogen diffusion; (3) effects of hydrogen on solar cells containing different types of crystal defects; (4) development of techniques for impurity characterization in silicon containing crystal defects; and (5) nondestructive testing of the photovoltaic parameters of commercial material. In the coming year an effort will be made to develop some low-cost cell processing techniques that incorporate the knowledge obtained from this research, to fabricate high-efficiency cells on commercial solar cell silicon.

6. Fundamental and Support Research

In addition to the specific technologies mentioned above, SERI conducts basic and applied research in theoretical solid state physics pertaining to the fundamental understanding of the different elements that constitute a PV solar cell. This includes elementary units such as characterization and prediction of electronic properties of relevant (pure) bulk semiconductors, surface properties, semiconductor alloys, and defect and impurity states in semiconductors. Furthermore, when needed, we provide new theoretical techniques that can address complex problems relevant to the field. Specifically, the group performs first-principles calculations (using only the atomic numbers of atoms present and their arrangement in space) on all aspects of actual or potential PV materials, addressing such issues as (1) structural, cohesive, and electronic properties of new crystalline ternary materials including superlattices; (2) analogous properties of ordered and disordered alloys, including thermodynamic properties; (3) detailed predictions about properties of common impurities in semiconductors of current technological interest; and (4) the development of theoretical techniques needed to deal with these problems in a numerically efficient way.

SERI also provides measurement, testing, characterization, device fabrication, and modeling support to research and development groups involved in solar cell technologies. It maintains a whole range of state-of-the-art facilities for material characterization, surface and interface analysis, device development, electro-optical characterization, cell performance measurements, and advanced module and submodule testing under simulated and actual sunlight conditions. We also conduct research on solar insolation, assessment, characterization, and instrumentation in support of all solar technologies.

7. Conclusions

The 1980s had been the most exciting period for solid state photovoltaic research and development. Remarkable progress has occurred in all technologies. The multiple path options pursued by the DOE program have paid their dividend handsomely. We enter the 1990s with a high degree of optimism that some of the remaining hurdles will be overcome and that, by the turn of the century, photovoltaics will make a strong impact as a global alternative energy resource.

This work was supported by the U. S. Department of Energy under Contract No. DE-AC02-83CH10093.

8. References

1. Spear, W. E., and LeComber, P. G. (1972) J. Non-Cryst. Solids, 8-10, 727.
2. Carlson, D. R., and Wronski, C. R. (1976) 12th IEEE PV Spec. Conf., Baton Rouge, B-893.
3. Catalano, A., Arya, R. R., Bennett, M., Yang, L., Morris, J., Goldstein, B., Fieselman B., Newton, J., and Wiedeman, S., (1989) Solar Cells, 27, 25-37.
4. Kuwano, Y., Ohnishi, M., Nishiwaki, H., Tsuda, S., Fukatsu, T., Enomoto, K., Nakashima, Y., and Tarui, H. (1982) Proc. 16th IEEE PV Specialists Conference, 1338-1343.
5. Williamson, D. L., Kerns, R. C., and Deb, S. K., (1984) J. Appl. Phys. 55, 2816.
6. Staebler, D. L., and Wronski, C. R. (1980) J. Appl. Phys., 51, 3262.
7. Mitchell, K. W., and Liu, H. I. (1988) Proc. 20th IEEE PV Specialists Conf., 26-30.
8. Mickelsen, R. S., and Chen, W. S. (1981) Proc. 15th IEEE PV Specialists Conf., 800.
9. Mitchell, K. W., Willett, D. R., Eberspacher, C., Ermer, J. H., Pier, D. N., and Pauls, K. (1990) 21st IEEE PV Specialists Conf., Orlando, Florida. Proc. to be published.
10. Noufi, R., Axton, R., and Deb, S. K. (1984) Appl. Phys. Letts, 45, 668.
11. Zweibel, K., Ullal, H. S., and Mitchell, R. L. (1990) 21st IEEE PV Specialists Conf., Orlando, Florida. Proc. to be published.
12. Fraas, L. M., Avery, J. E., Sundaram, V. S., Dinh, V. T., Davenport, T. M., and Yerkes, J. W. (1990) 21st IEEE PV Specialists Conf., Orlando, Florida. Proc. to be published.
13. McClelland, R. W., Bozler, C. O., and Fan, J. C. C. (1990) Appl. Phys. Lett., 37, 560.
14. Vernon, S. M., Tobin, S. P., Wojtczuk, S. J., Keavney, C. J., Bajgar, C., SanFacon, M. M., Daly, J. T., and Dixon, T. M. (1989) Solar Cells, 27, 107-120.
15. Chung, B.-C., and Virshup, G. F., Hikido, S., and Kaminar, N. R. (1989) Appl. Phys. Lett, 55, 1741.
16. Olson, J. M., and Kibbler, A. (1986) J. Cryst. Growth, 77, 182.
17. Olson, J. M., Kurtz, S. R., Kibbler, A., and Faine, P. (1990) Appl. Phys. Lett., 56, 623.
18. Olson, J. M., Kurtz, S. R., Kibbler, A. E. (1990) 21st IEEE PV Specialists Conf. Proc. to

be published.
19. Green, M. A., Wenham, S. R., Zhao, J., Zolper, J., and Blakers, A. W. (1990) 21st IEEE PV Specialists Conf., Orlando, Florida. Proc. to be published.
20. Sinton, R. A., Kwark, Y., Gan, J. Y., and Swanson, R. M. (1986) IEEE Electron Dev. Lett., EDL-7 No. 10, 567-569.
21. Narayanan S., Zolper, J., Wenham, S. R., Sproul, A. B., Yun, F., Chong, C. M., and Green, M. A. (1990) 21st IEEE PV Specialists Conf., Orlando, Florida. Proc. to be published.
22. Rock, M. L., Cunningham, D. W., Kendall, C. L., Hall, R. B., and Barnett, A. M. (1990) 21st IEEE PV Specialists Conf., Orlando, Florida. Proc. to be published.
23. Ciszek, T. F., Wang, T., Schuyler, T., and Rohatgi, A. (1989) J. Electchem. Soc., 136, 230.

(SDG/Italy2.wp1)

SOLAR-ASSISTED PHOTOCATALYSIS FOR WATER PURIFICATION: ISSUES, DATA, QUESTIONS

DAVID F. OLLIS
Department of Chemical Engineering
North Carolina State University
Box 7905
Raleigh, NC 27695-7905 USA

ABSTRACT. The current status of potential applicability of solar-assisted photocatalysis to water purification is considered by summarizing data germane to nine issues in reactor engineering design: these are (1) photocatalyst action spectrum, (2) rate dependence on intensity, (3) normal vs. concentrated solar photons, (4) reactant concentration influence, (5) catalyst disguises, (6) simultaneous homogeneous and heterogeneous photochemistry, (7) oxidant concentration, (8) mass transfer influences, and (9) simultaneous mass transfer and photon influences.

Introduction

Over the last decade, laboratory studies around the world have added heterogeneous photocatalysis to the list of advanced oxidation processes emerging for possible application in removal of organic and related contaminants from water. In the last several years, programs at the Solar Energy Research Institute (SERI) and Sandia National Laboratories in the USA have begun examining the potential application of solar energy for water purification using titanium dioxide photocatalysts, as this semiconductor was early shown capable of mineralizing trichloroethylene and trichloromethane (Ahmed and Ollis (1983)) via solar photons. The present paper reviews the progress in this area by considering nine issues from an engineering viewpoint.

E. Pelizzetti and M. Schiavello (eds.), Photochemical Conversion and Storage of Solar Energy, 593–622.

Issue #1: The photocatalyst action spectrum

As the vast majority of extant data concerns use of the anatase form of titanium dioxide (TiO_2), we focus on its properties as both pertinent and illustrative. The absorption spectrum of TiO_2 is characterized by a strong band gap edge at about 360 nm, with an exponential tail dropping to zero at about 400nm. In consequence, no more than 1.5-2.0% of the sea level solar spectrum is available for photocatalysis as summarized in Table 1. Patently, solar is a dilute resource: one m^2 of collected solar near-UV flux receives $6 \cdot 10^{-3}$ Einsteins/s (Table 1); thus with perfect absorption and a quantum efficiency of unity, it could process about 22 gram-mols of reactant per hour. A more typical photocatalytic quantum efficiency of 1% acting on a very easily oxidized solutes such as TCE or PCE would yield at most 0.22 gm-mol processed per m^2 per hr. A starting concentration of 1 ppm TCE in water could be processed at a rate of

$$\frac{0.22\text{g-mol}}{\text{m}^2\text{- hr}} \times \frac{130\text{gTCE}}{\text{g-molTCE}} \times \frac{10^6\text{gH}_2\text{O}}{1\text{gTCE}} \times \frac{1\text{L}}{10^3\text{gH}_2\text{O}} = \frac{28{,}600\text{L}}{\text{m}^2\text{- hr}}$$

This is a maximum estimate, as the actual quantum efficiency may decrease with decreasing TCE concentration.

A number of lamps, especially xenon and mercury-xenon, have been used in laboratory studies to simulate the solar near UV spectrum. As Figure 1 indicates, these lamps may have a substantial 290-310 contribution vs. sea level solar, and the use of such lamps to "simulate" solar must be considered carefully, as not only photocatalysis but also homogeneous photochemical activation of pollutant (e.g. phenol) or added oxidant (e.g., H_2O_2) may occur at these shortest wavelengths.

Issue #2: Rate dependence on intensity

Some of the earliest examinations of the light intensity influence on photocatalyzed reaction rate were those carried out by industrial laboratories associated with pigment and paint manufacture. An excellent example is the report of Egerton and King (1979), which examined use of the photocatalyzed oxidation of isopropanol (propan-2-ol) to acetone by pure rutile and two uncoated pigments. The reaction was carried out in pure

TABLE 1*

Light Source	near UV (300-400nm) Power (W/m^2)	Suns	Quanta ($Einsteins/m^2$-s)
Sun, D.N. AmI.5	22	1	$6 \cdot 10^{-3}$
1.4kW Xe lamp, old	2,240	86.6	0.5
1.4kW Xe lamp, new	5,980	230	1.4
1.6 kW Xe-Hg lamp, new	49,300	1870	11.2

*Blake et al, 1990.

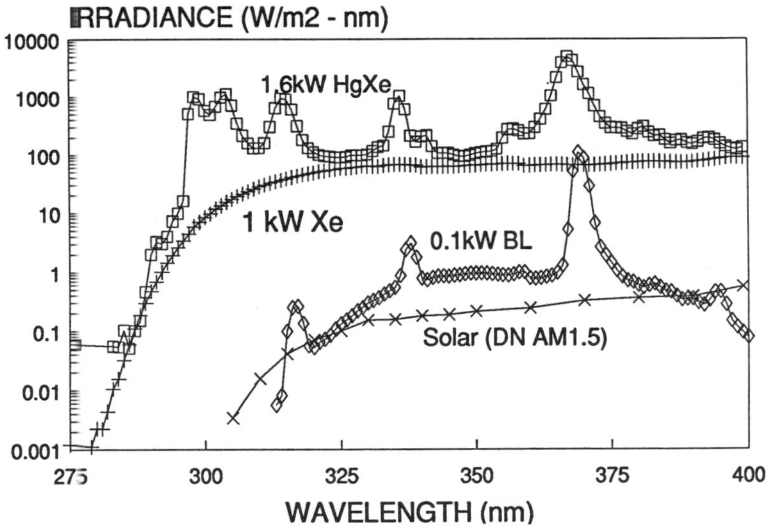

1. Spectral irradiance (W/m^2-nm) of solar source (x) and of several lamps used as solar simulators; 1.6 kW HgXe ( ), 1 kW Xe(+), and 0.1 kW black light (BL) (Turchi (1990)).

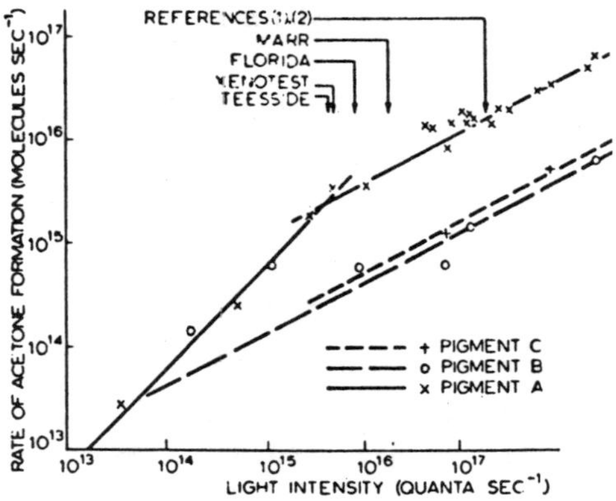
RATE OF ACETONE FORMATION (MOLECULES SEC^{-1})
LIGHT INTENSITY (QUANTA SEC^{-1})
MARR
FLORIDA
TEESSIDE
PIGMENT C
PIGMENT B
PIGMENT A

(2a)

(2b)

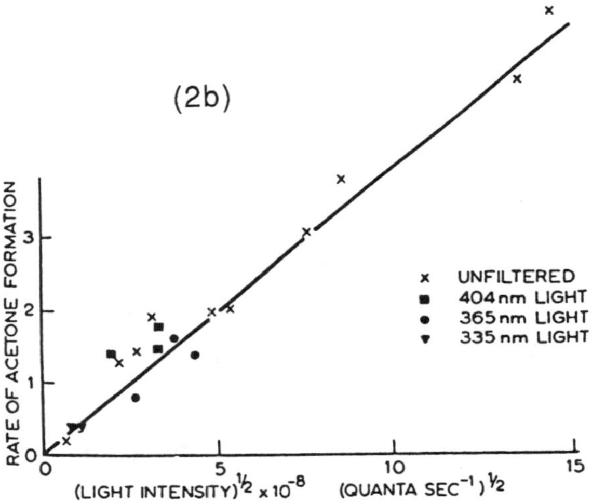
RATE OF ACETONE FORMATION
(LIGHT INTENSITY)$^{1/2}$ x 10^{-8} (QUANTA SEC^{-1})$^{1/2}$
UNFILTERED
404 nm LIGHT
365 nm LIGHT
335 nm LIGHT

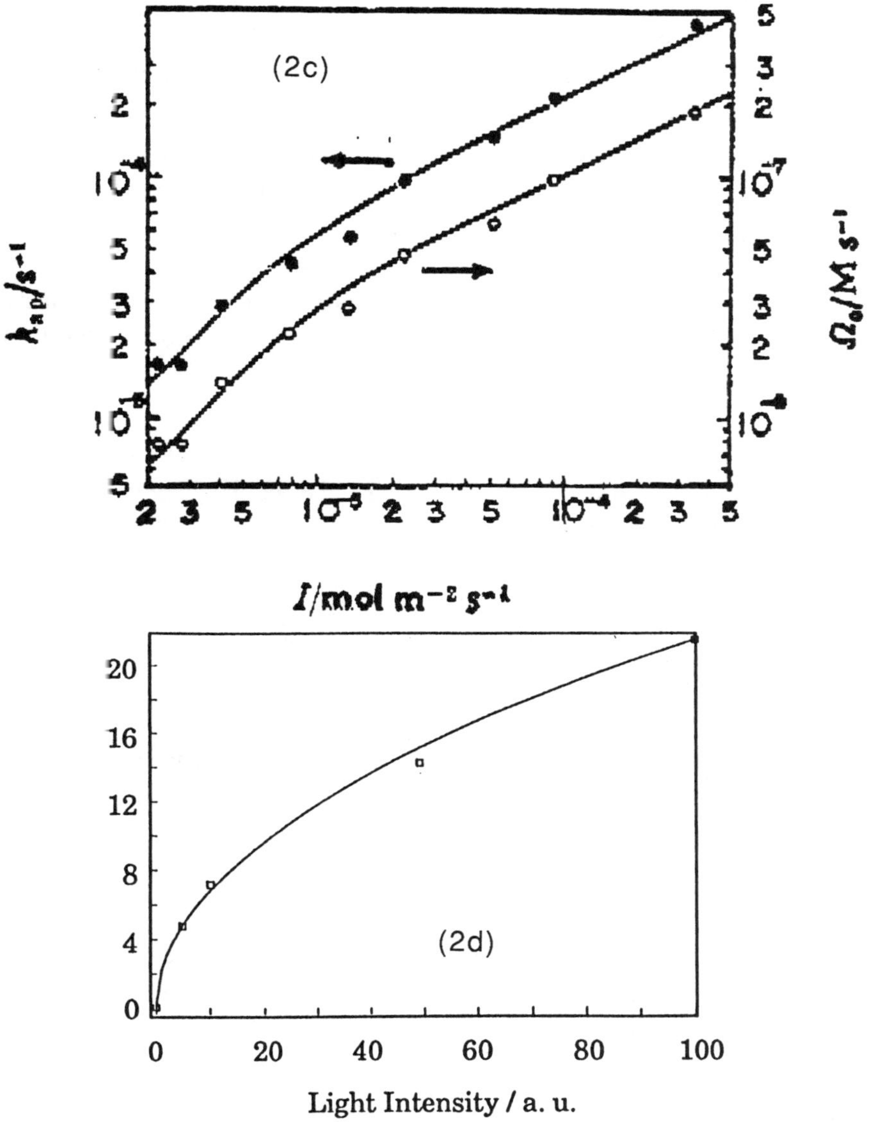

2. Photocatalyzed rate vs. intensity for (a) and (b) isopropanol (neat) oxidation by rutile (Egerton and King (1978)), (c) phenol oxidation in aqueous anatase slurries (Okamoto et al (1985)), and (d) trichloromethane (chloroform) disappearance in aqueous anatase slurry (Kormann et al (1990)). At or above 1 sun equivalent, all four figures show a strict half-order dependence of rate on intensity.

liquid isopropanol. Their rate data with pure rutile for acetone production vs. intensity showed a clear square root dependence for intensities at or above 10^{16} quanta/sec (Fig. 2a). Assuming a $10cm^2$ reactor receiving area, this incident rate is 10^{15} quanta/s-cm^2 or about 60% of the solar near UV flux of Table 1. Their full data plotted over 6 orders of magnitude of intensity found first order behavior below 5.10^{15} quanta /sec (Fig. 2b). Reaction intensity vs. rate data for impure TiO_2 pigments were consistent with half order dependence but less definitive (Fig. 2b).

Earlier work by Cundall, Rudham and coworkers(1978) indicated a first order intensity dependence, but Egerton and King(1979) claimed that Rudham and Ward later confirmed a square root dependence for 365nm photons.

At and above "one sun" near-UV fluxes, the half order reaction rate dependence on intensity was ascribed by Egerton and King(1979) to a dominance of electron-hole recombination over some other more productive step such as $h^+ + \cdot OH \rightarrow \bullet OH$, or $e^- + O_2 \rightarrow O_2^-$.

Similar results (Figure 2c) were obtained by Okamoto et al (1985) in a quantitative study of phenol photocatalyzed oxidation by anatase TiO_2 in an aqueous slurry reactor. At 0.5mM initial phenol (about 50ppm), the measured apparent first order rate constant k and the corresponding rate Ω_o showed first order intensity dependence below 10^{-5} einsteins/m^2-s (6.10^{14} photons/cm^2-s) and half order dependence above this value. (Fig. 2c). This transition intensity value for anatase is of the same magnitude as the range (6.10^{13} - 6.10^{15}) reported above by Egerton and King for rutile.

Very recent work by Kormann et al(1990) on anatase photocatalyzed degradation of trichloromethane (chloroform) aqueous solutions also shows a perfect half-order intensity dependence (solid curve: rate = $k\, I^{0.5}$). (Figure 2d).

It is instructive to compare the corresponding first order rate constants in Figures 2b and 2c. At the transition between 1st and 2nd order of Fig. 2a (100% isopropanol liquid), the rate of 2.10^{15} molecules/s occurs at an intensity of about 3.10^{15} quanta/s, providing a very high quantum efficiency of disappearance of ~ 2/3. The data of Figure 2c (aqueous dilute solution) provides a rate of (3.10^{-8} mol/L-S) ($6 \cdot 10^{23}$ molecules/mol) ~ $18 \cdot 10^{15} = 1.8 \cdot 10^{16}$ molecules/L·s at an intensity of 10^{-5} einsteins/m^2-s. Assuming a reactor cross section of $20cm^2$ (= 2.10^{-3} m^2 and a volume of 100 mL, the photon rate into the reactor was 10^{-8} Einstein/s vs. a measured reaction rate of 3.10^{-9} mol/s, resulting in an apparent quantum yield of about 30% for initial phenol disappearance.

Recent work at SERI is insufficient to provide a plausible discrimination between linear vs. square root dependencies of photocatalyzed TCE degradation rate vs. intensity. For example, TCE degradation studied with a 1kW Xenon arc lamp and filters, to yield only 300-450nm illumination in a recirculating batch reactor provided only three data points, which may be fit equally well with linear or square root dependencies (Figure 3). Similar conclusions apply with data for photocatalyzed degradation of Direct Red #79 dye. (Magrini and Webb (1990).

Similarly, Sandia data for TCE conversion in a long tubular photoreactor situated in a concentrating reflector trough provides insufficient data to resolve cleanly a half-order vs. first order reaction. Consider a simple plug flow reactor such that

$$v\frac{dC}{dZ} = -kI^{\alpha}C$$

or, integrating, $\frac{C}{C_o} = \exp(-kI^{\alpha}L/v)$

The residence time to reduce a 100ppb TCE feed to 5 ppb is given by $t^* = L/v$, where

$$\frac{5}{100} = .05 = e^{-kI^{\alpha}t^*}$$

Rearrangement gives $I^a t^* = \frac{-\ln .05}{k} = \text{constant}$

Figure 4 shows that either a = 1.0 or a = 1/2 may be used to represent plausibly the data.

Earlier literature work on rutile and anatase TiO_2 photocatalyzed rates demonstrate clear first order to half order transitions at intensities at or below the near-UV equivalent of one sun. The SERI and Sandia data are sparce enough to be unable to provide a clear discrimination. We conclude that all data to date are consistent with half-order intensity

(3a)

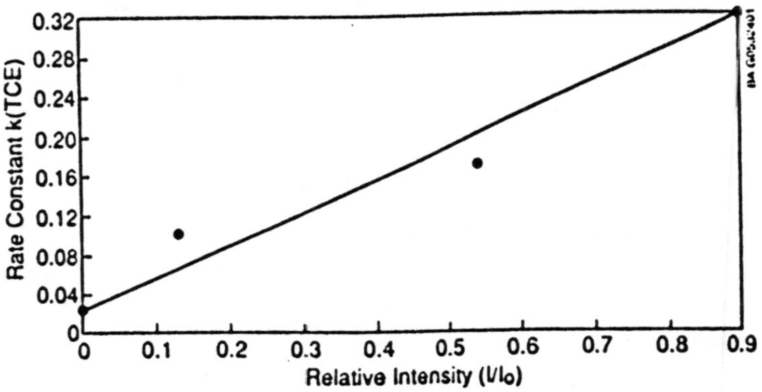

(3b)

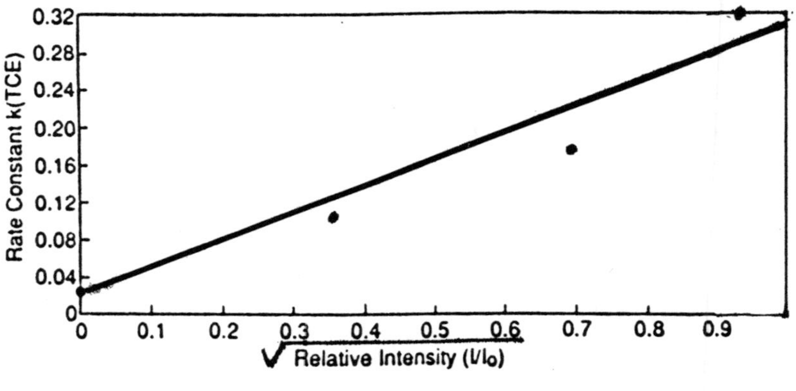

3. SERI data for trichloroethylene (TCE) disappearance: (a) rate vs. intensity (Magrini and Webb (1990)), (b) rate vs. $(\text{intensity})^{1/2}$.

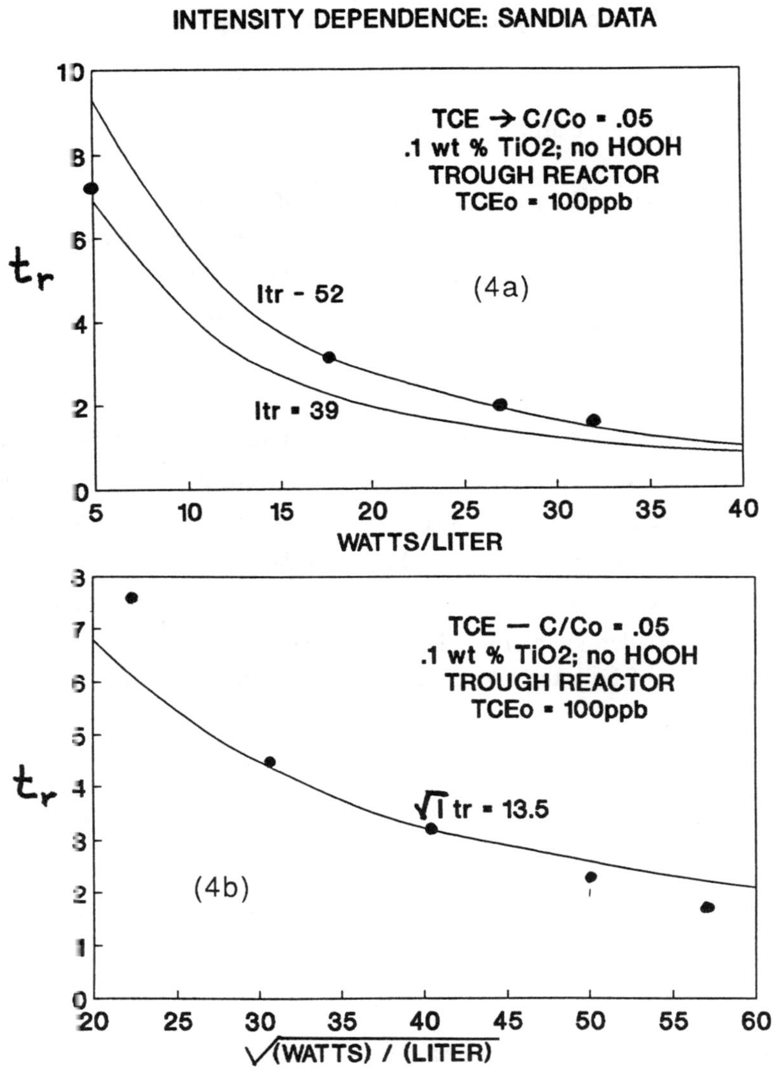

4. Sandia large trough data for TCE disappearance: (a) data vs. model for $(\text{intensity})^{1.0}(\text{time}) = \text{constant}$. (b) data vs. model for $(\text{intensity})^{0.5}(\text{time}) = \text{constant}$.

dependence at and above one sun near-UV equivalents.

Issue #3: Normal vs. Concentrated Solar Photons

The near UV spectrum is divided among direct and diffuse sources. The latter cannot be concentrated to a single point, the former can. Consider a flat plate collector cross section (Fig. 5a). Each point at the surface receives a direct contribution of I_d (photons/cm^2-s) and a diffuse contribution of I_s (same units). The total intensity of photons collected by the plate is $I_p = I_s + I_d$.

With a tracking concentrator, the perfect concentrator collects σI_d of the direct near UV (σ = reflectance) and (d/D) I_s of the diffuse. The total intensity I_t collected is now σI_d + (d/D)I_s. For an aluminum reflector with σ=0.8 and a reactor to reflector radial ratio of d/D=0.1 , the collected photon rate in the Figure 5b case is $0.8I_d + 0.1\ I_s$. If I_s = $.25I_d$ is used as the correct order of magnitude for near-UV intensities, then the convenience of a smaller, simpler tubular reactor using concentrated solar photons is bought at the price of both the additional reflector equipment and the reduction of available photons to

$$\frac{I_t(\text{trough})}{I_t(1\ \text{sun})}\ \frac{0.8 + 0.1(.25)}{1.+.25} = 66\%$$

of the original solar source availability. (Additional small corrections can be made for tubular reactor reflection and for diurnal variations of intensity ratios).

Issue #4: Influence of Reactant Concentration

The influence of reactant concentration on reaction rate is complex in several ways:

(i) a simple small reactant R which passes from R to CO_2 nearly instantaneously can be represented by a simple Langmuir rate (Turchi and Ollis (1989))

$$\frac{dC}{dt} = \frac{kKC}{(1 + KC)}$$

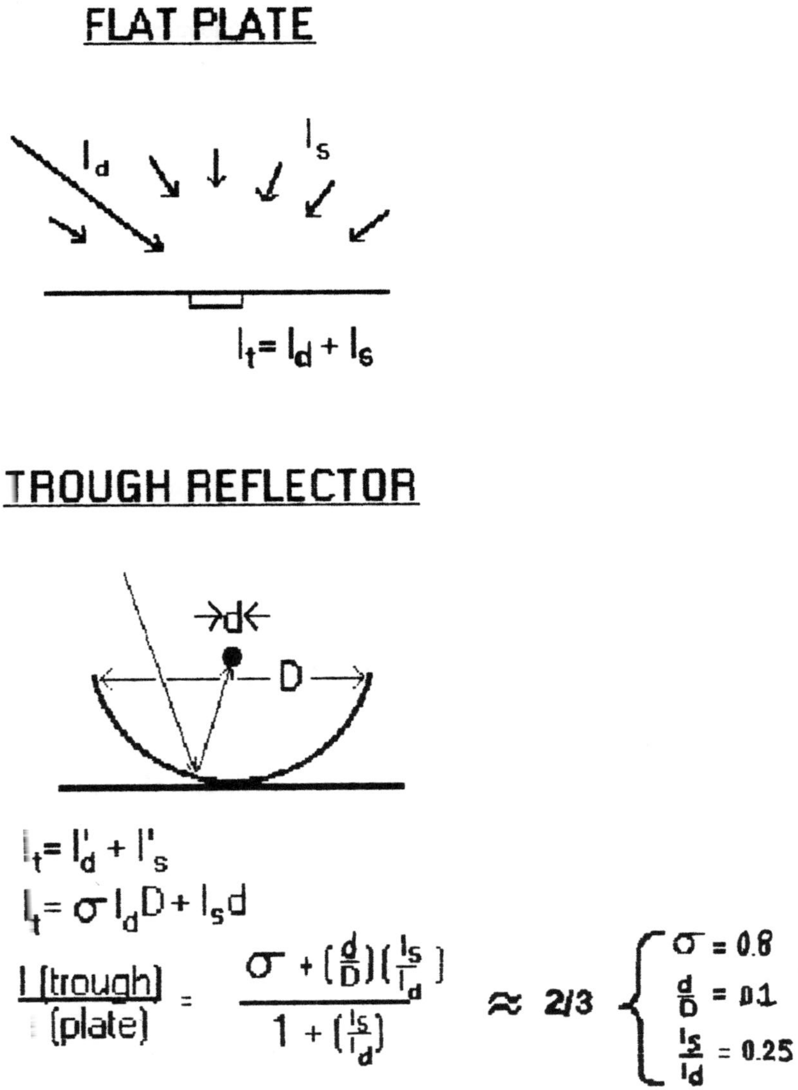

5. Schematic of reactor collection of direct and scattered (diffuse) near UV for (a) flat plate and (b) parabolic trough collector geometries. Plate collects $I_p = (I_d + I_s)$ whereas trough reactor collects $I_t \cong (\sigma I_d + (d/D) I_s)$.

(ii) as the reactant is made more complex, so that passage through one or more kinetically important intermediates occurs, the rate must include competitive adsorption by solvent and by intermediates, which slows the disappearance of reactant according to Serpone et al (1988).

$$\frac{dC}{dt} = \frac{-kKC}{1 + KC + \sum K_i C_i}$$

(all intermediates)

as well as that of each intermediate, C_i:

$$\frac{dC_i}{dt} = \frac{-k_i K_i C_i + k_{i-1} K_{i-1} C_{i-1}}{1 + KC + \sum K_i C_i}$$

When reactant mineralization is the treatment goal (no residual dissolved organic material), a useful measure of reaction lifetime might be achievement of a satisfactory mineralization, e.g., 95% generation of maximum possible CO_2. For simple reactants such as PCE, this corresponds essentially to the same reaction time as 95% PCE disappearance (Figure 6a). For more complex reactants, such as benzene and all other aromatics in particular (phenols, halophenols, PCBs, etc), the time required for 95% appearance of final CO_2 may be several multiples of the reactant 95% disappearance time (Figure 6b) (Turchi and Ollis (1989)). This same point was emphasized recently by Matthews(1990).

A kinetic scheme (Table 2) which accounts for half-order dependence at high intensity, first order at low intensity, and which is consistent with photoelectrochemical cathodic and anodic current transients has been proposed recently by SERI researchers (Peterson, Turner, and Nozik(1990)). This scheme also allows for H_2O_2 rate enhancement at low levels by electron trapping and rate diminution at higher H_2O_2 levels where hydroxyl radical consumption may become important (reactions 5, 7, Table 2).

These same authors report evidence for • OH in solution, whereas Bolton(1990) has suggested via ESR data that • OH exists primarily on the catalyst surface. At this

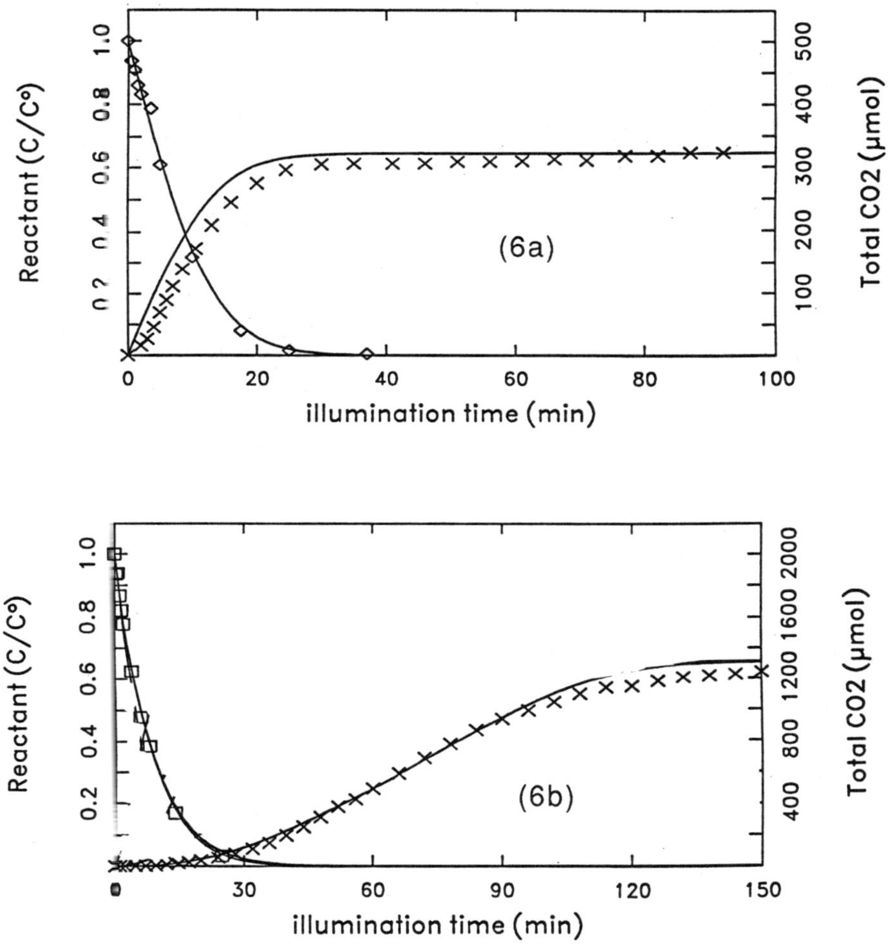

6. Reactant disappearance and mineralization product CO_2 appearance give similar or quite different 95% reaction times for simple (a, PCE) or complex (b. benzene) reactants, respectively. (Turchi and Ollis (1990)).

TABLE 2

(Peterson et al (1990))

Proposed mechanism for transient simulation of OH•

$TiO_2 + OH^- \rightarrow TiO_2^- + OH\bullet$	.01 sec^{-1}	
$TiO_2^- + OH\bullet \rightarrow TiO_2 + OH^-$	1×10^{10}	$M^{-1} sec^{-1}$
$2\ OH\bullet \rightarrow H_2O_2$	1×10^{10}	
$H_2O_2 + TiO_2^- \rightarrow TiO_2 + OH^- + OH\bullet$	2×10^7	
$OH\bullet + H_2O_2 \rightarrow H_2O + HO_2$	2×10^7	
$HO_2 + TiO_2^- \rightarrow TiO_2 + HO_2^-$	1×10^{10}	
$HO_2 + OH\bullet \rightarrow H_2O + O_2$	1×10^{10}	
$2\ HO_2 \rightarrow H_2O_2 + O_2$	3×10^7	
$HO_2 + H_2O_2 \rightarrow O_2 + H_2O + OH\bullet$	2×10^7	
$OH\bullet + Pol \rightarrow OH^- + Pol^+$	5×10^9	

conference, groups in France (Pouyet et al (1990)) and Ireland (Cunningham and El-Sayyed (1990)) have both proposed that a considerable fraction of pollutant degradation may occur by • OH radicals freed (desorbed) from the photocatalyst surface. Turchi and Ollis(1990a) have shown that regardless of whether both, one only, or neither of oxidant (• OH) and reductant (pollutant) are adsorbed on the catalyst at the moment of radical attack, the overall rate equation will still follow an apparent Langmuir-Hinshelwood form if there is only one dominant reaction mechanism.

Issue #5 : Catalyst disguises

Batch experiments with TiO_2 powders have often shown an initial lag period before the photocatalyzed rate achieves its full initial value. When fresh catalyst is used as a suspended slurry in tubular flow reactor operation , this lag effect manifests itself again an entry length or entry time as as seen in recent data from Sandia on a large scale tubular reactor operating in a continuous flow parabolic trough concentrator system (Figure 7a). For this figure of trichloroethylene (TCE) concentrations vs. residence time, several points deserve mention. First, in a once through operating mode, the reactor achieves TCE reduction to below 5ppb at all starting concentrations examined (118ppb to 5ppm). Second, the data may all be described by an equation of the form

$$\frac{1}{C_o}\frac{dC}{dt} = \frac{kK(C/C_o)}{1+K(C/C_o)}$$

When the original data are reported in a normalized fashion (inset to Figure 7a), a common curve appears to describe all results satisfactorily. Thus, independent of initial concentration, the apparent first order rate constant at t=0 is

$$k^{app}_{t=0} = \frac{kK}{1+K} = \frac{(0.7)(3.7)}{4.7} =.55 \text{ min}^{-1}$$

whereas later run data give apparent first order rate constant value of

$$k^{app} \text{ (large t)} = kK = 2.6 \text{ min}^{-1}$$

The continual increase in k^{app} over the first minute of light exposure is consistent with a catalyst activation phase. Earlier data of Pruden and Ollis(1983) indicate that TCE

TCE Destruction at Various Initial Concentrations

(7a)

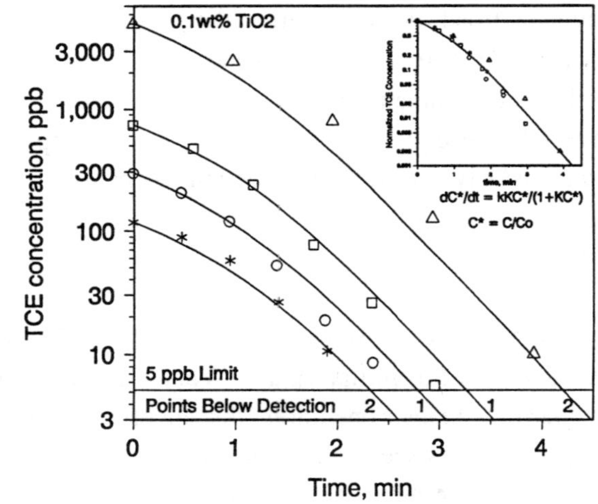

TCE Destruction With Hydrogen Peroxide

(7b)

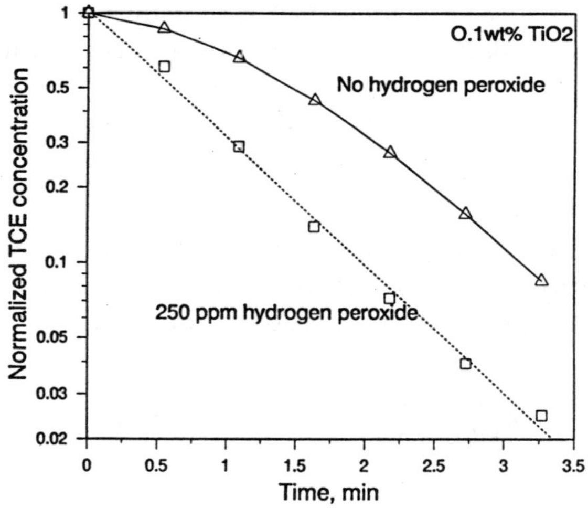

7. Sandia large parabolic trough flow photoreactor TCE concentration vs. time (a) various initial TCE feed concentrations (data) compared with model equation (main and inset) (solid curve, see text). (b) hydrogen peroxide addition in feed provides apparent first order TCE reaction at all times. (LDegussa P-25, 0.1 wt% slurry) (Pacheco et al (1990)). Reactor: 218 m 38mm ID borosilicate glass pipe set in 465 m^2 aperture parabolic reflector trough (Fig. 5b geometry).

conversion over an activated catalyst should be uniformly first order with invariant rate constant at these low TCE concentrations. A more complex kinetic picture for trichloromethane has recently been suggested by Kormann et al (1990).

The "activation" period is shortened or eliminated in the presence of added hydrogen peroxide, suggesting again a direct chemical modification of the surface or near-surface region of the catalyst-fluid interface (Figure 7b).

Issue #6: Simultaneous homogeneous and heterogeneous photochemistry

When the reactant also absorbs photons and may itself undergo homogeneous photolytic reactions, both the homogeneous and heterogeneous reaction must be included. These rates are coupled; since the catalyst and reactant each absorb photons, they determine in part the photon flux available to the other. This situation has been analyzed by Turchi and Ollis(1990b). Their results show that for sufficiently strong homogeneous absorption, adding photocatalyst will diminish the rate, while for weak homogeneous absorption, catalyst addition is favorable.

As an example, consider a very simple linear case of simultaneous catalyst and reactant absorption, where the variation of intensity I with distance Z into the liquid is

$$\frac{dI}{dZ} = -\underbrace{\alpha_C C_C}_{\text{catalyst}} - \underbrace{\alpha_R C_R}_{\text{reactant}}$$

The individual homogeneous and photocatalyzed rates are r_h and r_c, respectively, where

$$r_h = -k_h(\alpha_R C_R)\ (Z)\ \phi_h$$

and $$r_c = -k_c(\alpha_c C_c)\, I^{\beta}(Z)\ \phi_c\left[\frac{K_R C_R}{1 + K_R C_R}\right]$$

The total rate r_T is

$$r_T = r_h + r_c = \left[k_h (\alpha_R C_R) I \phi_n + \frac{k_c (\alpha_R C_R) I^\beta \phi_c K_R}{1 + K_R C_R} \right]$$

If β=1/2, as seems likely for heterogeneous photocatalysis (recall Issue #2 above), it is possible within a narrow concentration range for r_T to be proportional to $I^{1/2}$ at low intensities and to I at higher values.

Issue #7: Oxidant concentration

Photocatalyzed mineralization with TiO_2 is known to require at least molecular oxygen, and to give rate forms which vary with dissolved oxygen level (CO_2) according to a simple Langmuir-Hinshelwood form (Okamoto et al(1985), Szechowski (1986)):

$$\text{rate} = k(C) \cdot \frac{K_0 \cdot C_{O_2}}{1 + K_0 \cdot C_{O_2}}$$

A number of recent literature reports indicate an enhancement of photocatalytic rate upon addition of hydrogen peroxide. The role of hydrogen peroxide may be multiple; acting as both an accelerator and inhibitor (see reactions in Table 2 and discussion in Peterson et al(1990)). Further, hydrogen peroxide may be involved in catalyst activation, as suggested from Figure 7b.

H_20_2 could also be photolyzed directly (Pelizzetti et al (1990)) as occurs in UV/H_2O_2 advanced oxidation processes. This last case would also increase the rate over that measured in the absence of H_2O_2. Hydrogen peroxide photolysis is very sensitive to the shortest wavelength portion of the near UV. Indeed, the rate of H_2O_2 photolysis vs. wavelength indicates a critical need to know this 290-320nm solar portion extremely accurately in order to evaluate the reported H_2O_2 rate enhancements and their cause when solar illumination or solar simulators are used (recall Figure 1). H_20_2 induced homogeneous solar photochemical conversion of TCE has been reported in parabolic trough concentrator reactors (Pacheco (1990)).

Issue #8: The mass transfer problem

Most early photocatalyst work involved use of 0.1 wt% TiO_2 aqueous slurries. For the particle sizes of 0.2μm and 0.03μm as characterized earlier Fisher Chemical and Degussa individual TiO_2 particles, respectively, the average particle-particle spacings in a completely dispersed slurry were about 6μm and 1μm, respectively. The corresponding maximum diffusion path for a reactant to reach a photocatalyst particle was 3μm and 0.5μm or about 10^{-4} cm. The corresponding maximum possible diffusive time constant would thus be of the order of

$$t = \frac{X^2}{D} = \frac{\left(10^{-4}\text{cm}\right)^2}{10^{-5}\text{cm}^2/\text{S}} = 10^{-3}\text{S}$$

As the fastest reported photocatalyzed half life is 10-20 minutes at 10-100ppm levels, it is apparent that no reactant mass transfer limitation exists in slurry photoreactors using ≤ 1 sun equivalent lamps.

With immobilized catalysts, however, the small TiO_2 particles are permanently deposited on the surface of a convenient large carrier such as a glass bead (Serpone et at (1986)), glass mesh (Mathews (1987), or the inner wall of a coiled tube (Mathews (1987). For the bead and the coiled tube configuration, the maximum distance from bulk fluid to catalyst surface was of order 1mm, a value about 10^3 times larger than mentioned above. It is apparent that a mass transfer influence could easily arise when photocatalysts are immobilized.

Matthews(1987) reported apparent first order rate constants for salicylic acid disappearance at various flow rates in empty coiled tubes containing about 0.2mm thick TiO_2 coating on the inside wall. He found a marked dependence of reaction rate constants on fluid flow rate through the photocatalytic tube. The data were later plotted as reaction rate vs. flow rate and compared with exact analyses for mass transfer limited rates (Turchi and Ollis(1988)). The observed rates (Figure 8) were found to be from 60% to 90% of the maximum possible calculated mass transfer rate. Since the surface rate represents a "resistance" in series with diffusion of reactant from bulk to surface, the overall rate can be written in a form of Ohm's law,

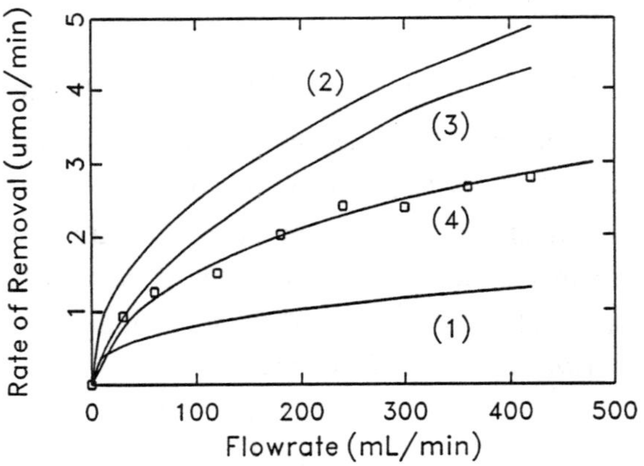

8. Mass transfer influence in coiled tube photoreactor. Data (□) (Matthews (1987)) compared with mass transfer limited rate for straight (1) and coiled tube with entrance condition (2) or log mean (3) driving force. Combined chemical wall reaction and mass transfer serial resistances (4). (Turchi and Ollis (1988)).

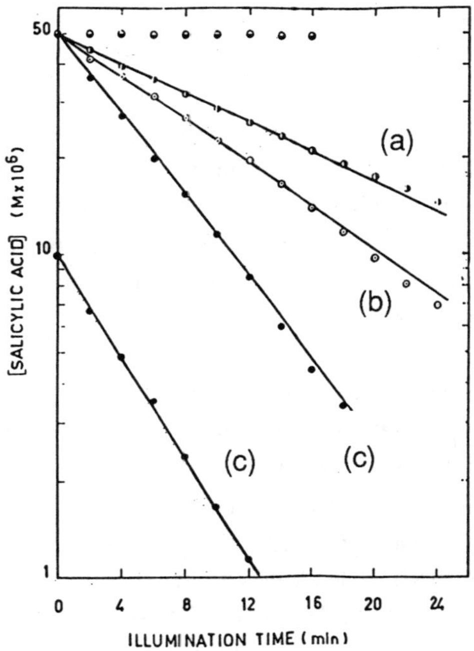

9. Salicylic acid concentration vs time in a coiled tube reactor. Data sets (a) 20W black light illumination (same conditions as Fig. 8, flow rate = 130 cm^3/min), (b and (c) same coil set in 0.35 m^2 parabolic trough reflector. (Matthews (1989)).

$$\text{Current} = \frac{\text{driving force}}{\Sigma \text{ of resistances}}$$

$$\text{or reaction rate} = \frac{1}{\dfrac{1}{\text{mass transfer conductance}} + \dfrac{1}{\text{surface rate conductance}}}$$

The calculated mass transfer influenced curve (4) in Figure 8 established very clearly the achievement of mass transfer dominated rate conditions even under mild illumination levels.

Inclusion of this same coilphotoreactor along the focal axis of a solar trough concentrator allowed higher intensities to be used. However, Fig. 8 indicates that at a flow rate of 130 mL/min, the rate can only approximately double to reach a completely mass transfer limited case , in nice agreement with the reactor-concentrator of Figure 9 data (Matthews 1989). [Compare black light coil data (Fig. 9) (same data as analyzed in Figure 8) with solar trough coil data (Fig. 9).]

Issue #9: Simultaneous photon and mass transfer influences

A second kind of mass transfer problem arises in excessively opaque suspensions. For example, a 0.1 wt% TiO_2 slurry, a common preparation, just provides opacity when used in a 1cm diameter tube or reactor annulus . Hence the effective light penetration depth is 1cm (Childs (1980)) Calculations based on the known absorbance and scattering power of TiO_2 particles support this result (Ayoub (1986) and Ayoub and Dranoff (1987), Figure 10).

If such a photocatalyst loading is now used in a larger reactor such as a 6 cm ID glass tube, only the outer 1cm is therefore appreciably illuminated. The behavior of a flow reactor in this circumstance will depend critically on the relative rates of photocatalyzed reaction vs the rates of fluid and mass exchange between the "illuminated" annular volume and the "dark" cylindrical core. This general circumstance was analyzed exhaustively years ago by Spadoni et at (1978).

A specific example of this situation appears in a simulation example carried out at SERI (M. Mehos and C. Turchi (1990)) for a flowing tube photoreactor using a parabolic trough reflector. The configuration is a large scale parabolic trough carrying a photocatalyst slurry inside a tubular reactor situated at the reflector focus. The

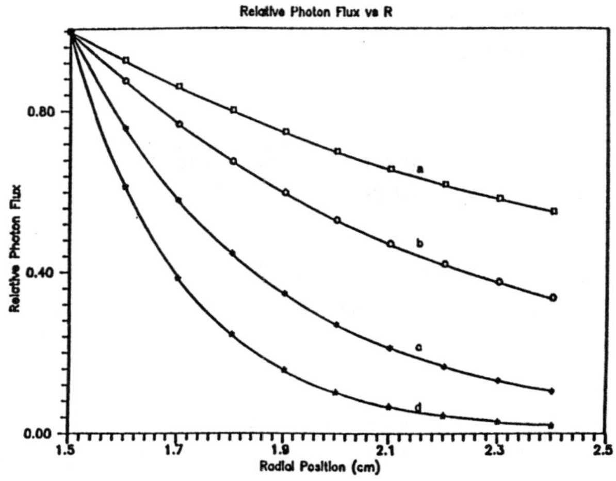

10. Calculated intensity profiles vs. radial position in an annular photoreactor. (Ayoub(1986)).

appreciable concentration power of the reflector results in asymmetric illumination delivery to the cylindrical photoreactor (Figure 11a) as well as inside the catalyst-free reactor (Figure 11b). The resultant reaction rate must also be strongly non-uniform inside the photoreactor. For example, using a pollutant-free kinetic scheme involving only eqns(1-3) of Table 2, a calculated distribution of hydroxyl radicals within a cross section of the flow reactor is given in Figure 11c. Radicals are seen to be present largely in the outermost 1 cm of the illuminated side of the reactor, in agreement with illumination profiles calculated previously by Ayoub (1986) for various TiO_2 reactor loadings (Figure10).

The quality of radial fluid mixing in such a reactor will be a key to predicting the overall system performance. For example, a processing rate of 25 gallons per minute in the Sandia tubular flow reactor corresponds to a Reynolds number of 25,000-50,000. For water with a Schmidt number $\simeq$ 1000, this operating condition in an empty straight tube produces [$D_r \cdot$ density/viscosity] = 1-10 or a radial dispersion coefficient D_r of 1-10 x (10^{-2} cm^2/s) = 10^{-2} to 10^{-1} cm^2/s. (Figure 12, Bischoff and Levenspiel (1962)). The relaxation time required to achieve appreciable radial uniformity is of the order of

$$t_r \simeq \left(\frac{d_t}{2}\right)^2 / D_r = (2\text{cm})^2 / \left(10^{-2} - 10^{-1}\right)\left(\text{cm}^2/\text{s}\right) \simeq \quad 40 \text{ to } 400 \text{ s.} \quad (.7 \text{ to } 6.7$$

minutes).

Since this mixing time is comparable to the fluid residence time of about 3 minutes, it is apparent that complete radial mass transfer is not an instantaneous process, and a mass transfer limitation could again easily arise. Under radial mass transfer limitation, the conversion observed for a non-uniformly illuminated catalyst slurry (Fig. 11b, c) and a wall immobilized catalyst mesh would be similar as conformed by the data of Fig. 13.

As indicated by both Spadoni et al (1978) and by opacity in Figure 11b, more uniform illumination could be achieved with no loss in light utilization if the slurry catalyst loading were reduced so that the product $C_{cat} \cdot R$ just produced opacity. For example, Ahmed and Ollis (1983) reported the achievement of maximum solar photocatalyzed rate in a tubular flow reactor at 0.3wt% TiO_2 and 6mm ID tubing to give opacity at (0.3)(0.6)=0.18 wt%-cm. Thus a larger 6cm ID glass tube would only require about 0.03wt% TiO_2 to achieve slurry opacity, and at the same time to provide a more uniform catalyst illumination.

The potential for wall-related mass transfer limitations which has been established for coiled, internally coated tubes (Figures 8 and 9) and is suggested by the radial

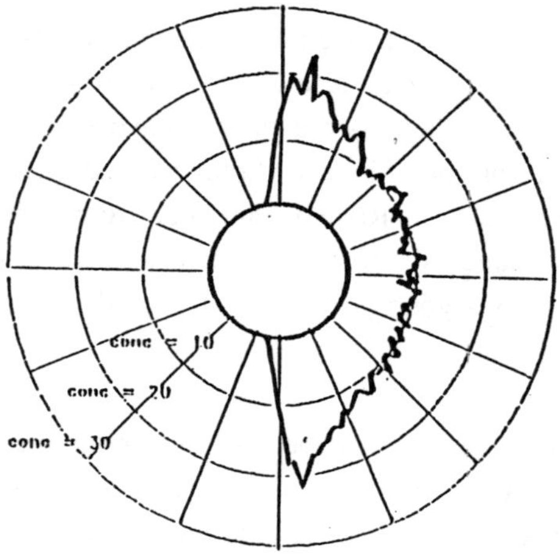

Solar Flux Distribution on SERI Receiver/Reactor (OPTTRF)

(11a)

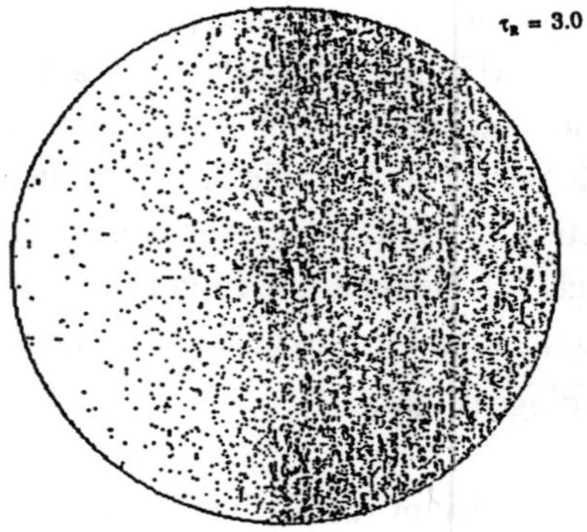

Distribution of Absorbed Radiation inside Receiver/Reactor Tube (MONTE3)

(11b)

DISTRIBUTION OF ABSORBED RADIATION INSIDE RECEIVER/ REACTOR TUBE FOR 0.1 WT% TiO2

(11c)

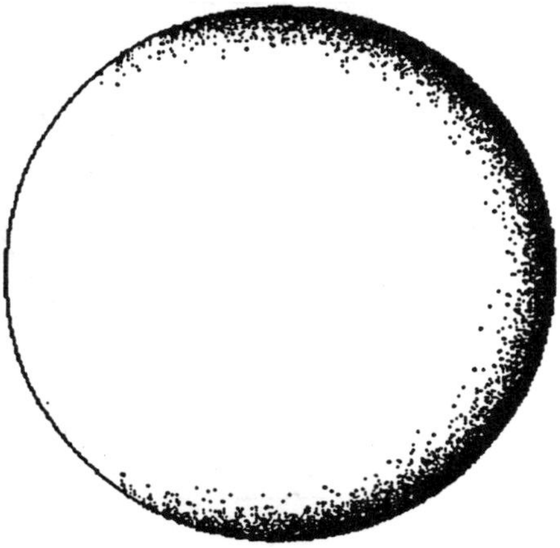

THREE-DIMENSIONAL PLOT OF OH RADICAL CONCENTRATION WITHIN A PARABOLIC TROUGH REACTOR

(11d)

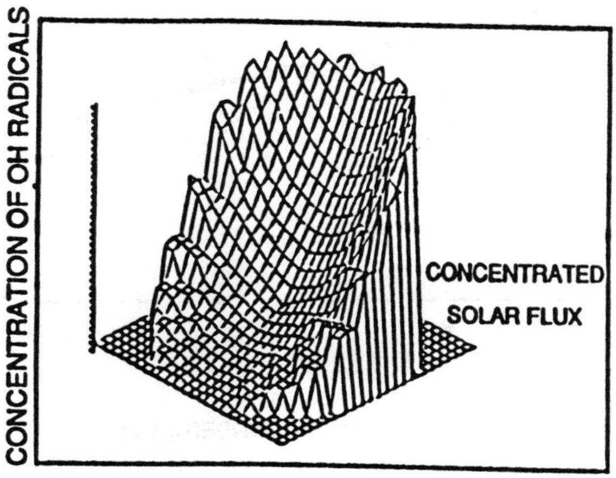

11. Calculated profiles for (a) intensity received vs. angle (b) intensity in empty tube reactor (c) intensity in 0.1 wt% TiO_2 slurry reactor and (d) hydroxyl radical concentration vs. angle and radial location for a reactant-free tubular photoreactor using a parabolic trough concentrator. (Mehos (1990)).

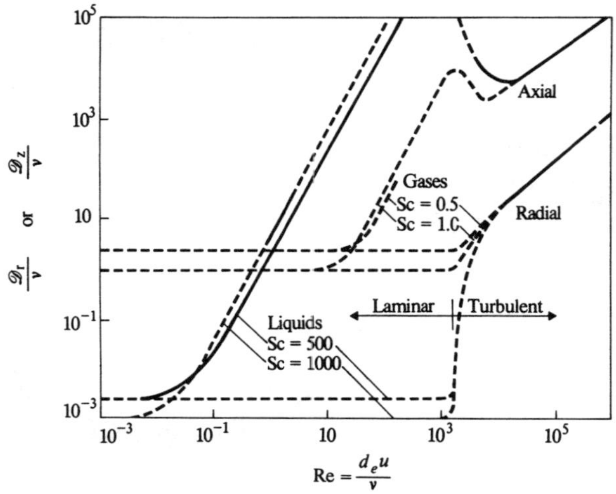

12. Axial and radial dispersion coefficients for liquid and gas flow in a straight, smooth tube vs. tube Reynolds number (Re $\equiv \rho v D/\mu$) (Bischoff and Levenspiel (1962)). For Re ~ 23,000 - 50,000, the corresponding axial dispersion coefficient is about $D_r \simeq 10^{-2}$ to 10^{-1} cm^2/s. (See text).

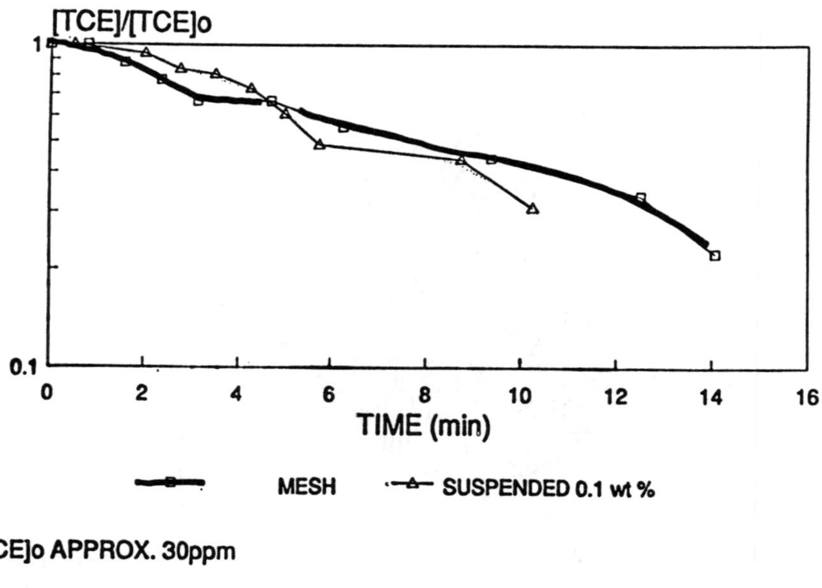

13. TCE conversion in recirculating flow parabolic trough reactor. (SERI data (1990) for slurry and wall mesh catalysts)).

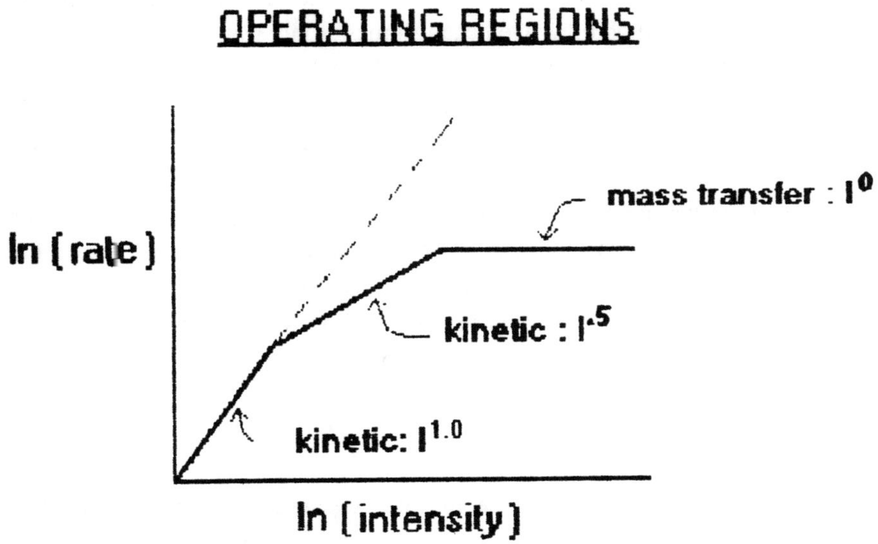

14. Rate vs. intensity indicating various operating regimen dependences on intensity order.

dispersion estimate in this section and the similar behavior of dense slurry (wall region illumination only) and immobilized photocatalyst mesh (only wall region active) (Figure 13) brings us to a final design consideration: as illumination increases from much less than one sun to 1-10 suns to quite concentrated, the rate influence by intensity will be expected to shift from first order to half order (recall Figures 2b and 2c) to zero order (the mass transfer limited rate of Figs. 8 and 9 is not a function of intensity). The qualitative behavior is sketched in Figure 14. The transition regions between first and half-order and half- and zero-order depend on intrinsic reaction kinetics of a given photocatalyst, and on the relative mass transfer and photocatalyzed reaction conductances, respectively. Operation in each regime would also logically lead to different dependencies of reaction quantum yield on intensity.

Overview

Nine engineering issues of importance in solar assisted photocatalysis for water purification have been summarized and discussed briefly. As much of the text was suggestive rather than definitive, much remains to be done in order to establish clearly the relative advantages and disadvantages of this process concept. It is hoped that this review will assist research and development groups everywhere in achieving more quantitative approaches to their own essays, and in identifying areas needing further study.

Acknowledgements

The author is pleased to thank several groups for permission to cite their work in advance of publication. The author also thanks SERI and SANDIA personnel for useful discussions.

References

1. Ahmed, S. and D.F. Ollis, Solar Energy, **32**, 597 (1984).
2. Aycub, P., PhD thesis, Northwestern University, 1986.
3. Aycub, P. and J. Dranoff, AIChE mtg, 1987, New York.
4. Bischoff, K. B. and O. Levenspiel, Chem. Eng. Sci., **17**, 257 (1962).
5. Blake, D. et al., Adv. Oxid'n. Symp., Toronto, Canada, June 4-5, 1990.
6. Bolton, J. R., Adv. Oxid'n. Symp., Toronto, Canada, June 4-5, 1990.
7. Childs, L.P., PhD thesis, Princeton University, 1980.
8. Cundall, R.B., B. Hulme, R. Rudham, and M.S. Salim, J. Oil Coll. Chem. Assoc., **61**, 351, 1978.
9. Cunningham, J. and G. Al-Sayyed, International Photochemical Society-8, Palermo, Italy, 1990.
10. Egerton, T. A. and C. J. King, J. Oil Coll., Chem. Assoc., **62**, 386(1979).
11. Kormann, et al., Environ. Science Technol., in press (1990).
12. Magrini, K. A. and J. D. Webb, ASME Int'l. Solar Energy Conf., 1990.
13. Mathew, R., J. Phys. Chem., **91**, 3328 (1987).
14. Matthews, R., Water Research, **24**, 653 (1990).

15a. Menos, M. and C. Turchi, SERI (private communication).

15b. Okamoto, K. Y. Yamamoto, H. Taneka, A. Itaya, Bull.Chem. Soc. Japan, **58**, 2023(1985).

16. Pacheco, J., Sandia Nat'l. Laboratory (private communication).

16a. Pachco, J., M. Prairie, L. Evans and L. Yellowhorse, 25th Intersociety Energy Conversion Eng'g. Conf., Aug 12-17, 1990, Reno, Nevada, USA.

17. Pelizzetti, E., E. Pramauro, C. Minero and N. Serpone, Waste Management, **10**, 65 (1990).
18. Percheroncier, J. P., G. Perichet, P. Meallier and B. Pouyet, International Photochemical Society-8, Palermo, Italy, 1990.
19. Peterson, M. W., J. A. Turner, and A. J. Nozik, J. Phys. Chem., (in press), 1990.
20. Pruden, L.P. and D.F. Ollis, J. Catalysis, **82**, 404 1983.
21. Sclafani, A., L. Palmisano, and E. Davis, New J. Chem, **14**, 266 (1990).
22. Sepone, N. and H. Al-Ekabi, J. Phys. Chem., **92**, 5726 (1988).
23. Spadoni, G., E. Bandini and F. Santarelli, Chem. Eng. Science, **33**, 517 (1978).
24. Szechowski, J., MS thesis, NCSU, 1986.
25. Turchi, C. and D.F. Ollis, J. Phys. Chem., **92**, 6852 (1988).
26. Turchi, C. and D.F. Ollis, J. Catalysis, **119**, 483 (1989).
27. Turchi, C. and D. F. Ollis, J. Catalysis, **122**, 178 (1990a).

28. Turchi, C., Solar Energy Research Institute, Golden Colorado.
29. Turchi, C. and D.F. Ollis, Chem. Eng. Science, (in press, 1990b).
30. Tyner, C., Sandia National Laboratories, 1990. (private communication).

PHOTOCHEMICAL ENGINEERING DESIGN METHODS. A RIGOROUS APPROACH FOR HOMOGENEOUS SYSTEMS.

Alberto E. Cassano and Orlando M. Alfano
INTEC. (CONICET and UNL)
Guemes 3450
3000 - Santa Fe
Argentina

ABSTRACT. The most salient aspects of a method for an "a priori" design of homogeneous photochemical reactors are described. It is based on the use of a conceptual approach extracted from the fundamentals of radiative transport. Results for six different applications show good agreement between model predictions -with no adjustable parameters- and bench scale experiments.

1. INTRODUCTION

The significance and eventual advantages of the photochemical route for the chemical process industries has been repeatedly reported during the past years (Doede and Walker, 1955; Cassano et al., 1967; Yue, 1985 a,b,c; Andre and Villermaux, 1986; etc.). More recently, the problems related to the analysis and design of photoreactors have been dealt with some detail. Considerable improvements have been reported in the area of homogeneous photoreactors as it can be deduced from several revisions published by Cassano and coworkers (Alfano et al., 1986 a; Cassano et al., 1986; De Bernardez et al., 1987; Cassano and Alfano, 1989). However, no parallel accomplishments have been obtained with heterogeneous photoreactors (Yue, 1985 c; Alfano et al., 1986 b).

In this paper we intend to describe in a concise way one available methodology for homogeneous (and pseudo-homogeneous) photoreactor design. Our scope it to show that the state-of-the-art in this area, as far as 'a priori" reactor design methods is concerned, has reached an almost equivalent level of development to that existing in conventional (thermal) reactors and that the main effort should be switched over to the proposal of novel designs and the analysis of heterogeneous photoreactors.

Reactor design is a complete procedure by which it can be defined:

1. Production specifications (flow rate or reaction time; inlet and outlet or initial and final concentrations).
2. Operating conditions (single or multiphase system; type of hydrodynamic operation (continuous flow, batch, semibatch, etc.), temperature/s, pressure/s and catalyst load if applicable).
3. The reactor geometry (shape and dimensions).

E. Pelizzetti and M. Schiavello (eds.), Photochemical Conversion and Storage of Solar Energy, 623–652.

4. The mixing requirements if applicable.
5. The heating and/or cooling requirements.
6. Construction materials, control and safety requirements.

In photochemical reactors we must add:

1. Lamp specifications (output power and its spectral distribution, shape and dimensions, operating requirements).
2. Geometrical arrangement of the radiation sources relative to the reactor location.
3. Radiation entry system into the reactor (manner, construction materials, cleaning procedures).
4. Reflector/s if applicable (arrangement, shape, dimensions).

Since lamp sizes and specifications cannot be changed in a continuous fashion, the design always implies an iterative procedure.

2. THE DISTINCT CHARACTERISTIC OF A PHOTOREACTOR

The obvious distinct characteristic of a photochemical reactor is the way by which a reaction is initiated. This peculiarity introduces significant changes and important complexities in the design. Let us consider as an illustrative example the single phase, monochlorination of ethane. The reaction mechanism is:

INITIATION	$Cl_2 + h\nu$	$\longrightarrow$	$2\ Cl\cdot$	Φ_P
PROPAGATION	$Cl\cdot + EtH$	$\rightleftharpoons$	$HCl + Et\cdot$	$k_2,\ k_2'$
	$Et\cdot + Cl_2$	$\rightleftharpoons$	$Cl\cdot + EtCl$	$k_3,\ k_3'$
TERMINATION	$M + Cl\cdot + Cl\cdot$	$\longrightarrow$	$Cl_2 + M$	k_4
	$Cl\cdot + Et\cdot$	$\longrightarrow$	$EtCl$	k_5
	$Et\cdot + Et\cdot$	$\longrightarrow$	$Et\text{-}Et$	k_6
	$Cl\cdot + wall$	$\longrightarrow$	end	k_7
	$Et\cdot + wall$	$\longrightarrow$	end	k_8

The reaction rates required for the mass conservation equations are written according to the mass action formulation. For example, using the following notation: $Cl_2(1)$, $Cl\cdot(2)$, EtH (3), EtCl (4), $Et\cdot(5)$, HCl(6), $C_4H_{10}(7)$ and $N_2(8)$, the rate of ethyl chloride appearance is:

$$\Omega_4 = k_3\ C_5\ C_1 - k_3'\ C_2\ C_4 + k_5\ C_2\ C_5 \qquad (1)$$

Similarly, one also needs, for example, an equation for the rate of reaction of the atomic and free radical species. For atomic chlorine:

$$\Omega_2 = 2\ \Omega_{init} - k_2\ C_2\ C_3 + k_2'\ C_6\ C_5 + k_3\ C_5\ C_1 - k_3'\ C_2\ C_4 - 2\ k_4\ C_2^2\ C_8 - k_5\ C_2\ C_5 \qquad (2)$$

Equation (2) shows the rate of initiation Ω_{init}. Every time that we need to use the rate of reaction of molecular or atomic chlorine, the initiation rate must be formulated. For single photon absorption reactions we express this rate in terms of the primary quantum yield and the local value of the volumetric rate of energy absorption (LVREA), i.e. $\Omega_{init} = \Phi p\ e^a$.

It is conceptually clear that the value of the LVREA surely depends on the characteristics of the radiation source. Hence an "a priori" photoreactor design must look for the way to incorporate the lamp specifications into the design equations; more particularly into the reaction rates that include the initiation rate. To do it we need: (i) a radiation balance, (ii) constitutive equations for radiation absorption and (iii) an emission model for the lamp.

3. THE RADIATION BALANCE

The employed definitions were described by Irazoqui et al. (1976) and a precise derivation can be found in Ozisik (1973) or Santarelli (1985). The equation for radiative transfer for coherent scattering has the following form:

$$\frac{dI_\nu(s,\underset{\sim}{\Omega},t)}{ds} + [\mu_\nu(s) + \sigma_\nu(s)]\ I_\nu(s,\underset{\sim}{\Omega},t) =$$

$$j^e_\nu(s,t) + \frac{1}{4\pi}\ \sigma_\nu(s) \int_{\Omega'=4\pi} p_\nu(\underset{\sim}{\Omega}' \to \underset{\sim}{\Omega})\ I_\nu(s,\underset{\sim}{\Omega}',t)\ d\Omega' \qquad (3)$$

In the L.H.S. the first term expresses the rate of change of the spectral specific intensity along the radiation path s; the second indicates the ways by which radiation is lost by two different mechanisms: absorption (coefficient μ_ν) and scattering-out (coefficient σ_ν). In the R.H.S. we have two ways of gaining radiation: the first term is the emission and the second corresponds to the scattering-in from all directions in space. The function p_ν is the phase function which describes the directional distribution of the scattered radiation. In homogeneous systems we are concerned mostly with absorption and, eventually, with emission.

With the following definitions:

$$\int_0^{4\pi} \mu_\nu(s)\ I_\nu(s,\underset{\sim}{\Omega},t)\ d\Omega = e^a \quad (4) \quad \text{and} \quad \int_0^{4\pi} I(s,\underset{\sim}{\Omega},t)\ \underset{\sim}{\Omega}\ d\Omega = \underset{\sim}{q}_\nu \quad (5)$$

it can be shown easily that in an isotropic medium the conservation of radiant energy is given by:

$$\underset{\sim}{\nabla} \cdot \underset{\sim}{q}_\nu = \int_0^\infty \int_0^{4\pi} j^e_\nu\ (s,t)\ d\Omega\ d\nu - \int_0^\infty e^a_\nu\ d\nu \qquad (6)$$

and the LVREA is simply:

$$e^a = \int_0^\infty \mu_\nu \left| \underset{\sim}{q_\nu} \right| d\nu = \int_0^\infty \mu_\nu \, q_\nu \, d\nu \qquad (7)$$

If radiation scattering must be accounted for, equation (6) does not provide enough information and one needs to resort to the solution of equation (3) which is written in terms of specific intensities.

4. THE CONSTITUTIVE EQUATIONS

The constitutive equation for absorption which has been incorporated already in equation (3), in a homogeneous medium is obtained by making use of the Lambert's and the Beer's equations:

$$W_\nu^a = - \mu_\nu \, I_\nu(s) \qquad (8) \qquad\qquad \mu_\nu = \sum_{i=1}^{N} \alpha_{\nu,i} \, C_i \qquad (9)$$

For gas-liquid systems the use of an effective absorption coefficient has been suggested by the Otaque et al.'s (1981) and the Yokota et al.'s (1981) correlations. Using the latter for example:

$$\mu_{eff,\nu} = \mu_\nu \, (1 + h \, \varepsilon_g) \quad (10) \qquad \text{with:} \quad h = \left(\frac{3.6}{d_b}\right)^{0.66} \qquad (11)$$

However they should be used with caution because the semi-empirical coefficients may have an unaccounted for dependence on the geometry of the system. The rigorous approach implies the use of the complete equation for radiative transfer and a theory for scattering, which includes the knowledge of the extinction coefficient (the sum of the absorption and scattering coefficients), the absorption coefficient and the phase function. Actually this is nothing else than the appropriate treatment for any heterogeneous system.

5. THE SOURCE MODEL

Several models have been proposed for photoreactor design. They have been described in the previously quoted revisions. A source model should relate the radiation output power and the physical dimensions of the lamp to the LVREA inside the reactor. Following Figure 1, the requirement is translated to the knowledge of the value of e^a at any point I inside the reactor. Rays come out of the lamp in every direction in space. By the same token, many rays arrive at any point inside the reaction space coming from many directions. In the figure it can be seen that, for example, one ray comes out from P_T ($q = q_e$), transverses the reactor wall at P_W ($\rho = \rho_o$ and $q = q_o$) and finally reaches the point of incidence P_I ($\rho = \rho_I$ and $q = q$). For all practical purposes from P_T to P_W there is no absorption. On the contrary, from P_W to P_I there is absorption and reaction. The conservation of radiation along the direction ρ in the absorbing medium gives:

$$\frac{1}{\rho^2}\frac{d}{d\rho}(\rho^2\, dq_\nu) = -\mu_\nu\, dq_\nu \qquad (12)$$

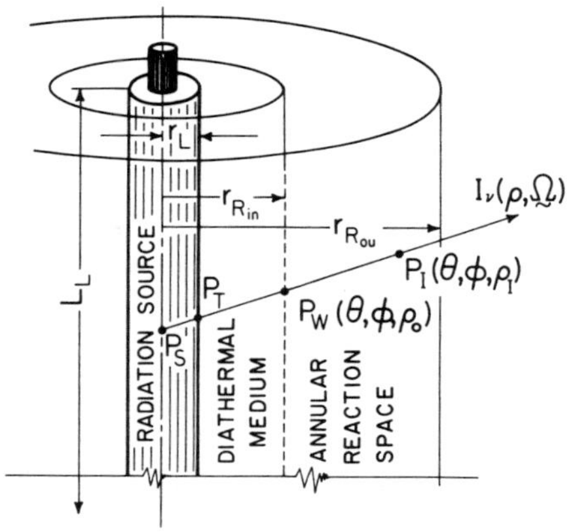

Figure 1

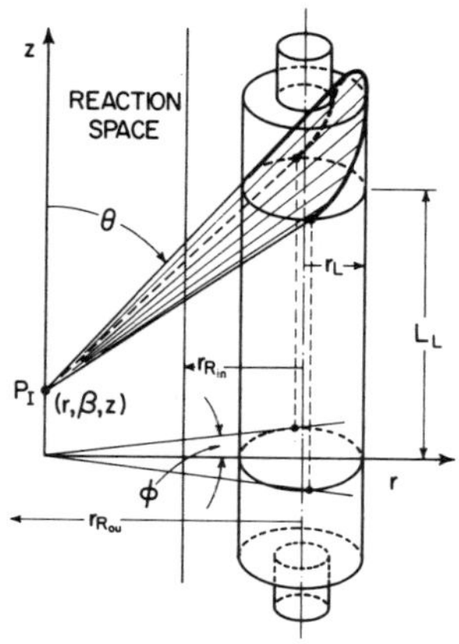

Figure 2

integrating from point P_W to point P_I:

$$dq_\nu = \frac{\rho_o^2\, dq_{\nu,o}}{\rho^2} \exp\left(-\int_{\rho_o}^{\rho_I} \mu_\nu\, d\rho\right) \qquad (13)$$

From P_T to P_W there is neither absorption nor emission, then:

$$\frac{1}{\rho^2}\frac{d}{d\rho}(\rho^2\, dq_\nu) = 0 \quad (14) \quad \text{and} \quad \rho_o^2\, dq_{\nu,o} = \rho_e^2\, dq_{\nu,e} \quad (15)$$

The value of q_e, can be obtained from the definition of the emission power density and application of the radiation conservation equation inside the lamp under the assumption of a purely emitting medium.

Strictly speaking the emission and absorption of radiation by a body constitute a bulk process; a surface never emits or absorbs radiation. When, using an approximation, one speaks of superficial processes, in order to maintain consistency with the definitions, the units of W^a and j^e must be changed accordingly.

$$j_\nu^e = \frac{dE_\nu^e}{d\Gamma_e\, d\Omega} \quad (16) \qquad dq_{\nu,e}\, \rho_e^2 = j_\nu^e\, d\Gamma_e \quad (17)$$

The value of dE_ν^e is expressed in terms of the lamp characteristics according to:

$$dE_{\nu,L} = j_\nu^e \, d\Gamma_e \, d\Omega \qquad (18) \qquad\qquad E_{\nu,L} = \int_0^{4\pi} j_\nu^e \, d\Omega \int_{\Gamma_e} d\Gamma_e \qquad (19)$$

Hence, for a cylindrical, tubular lamp, if one assumes isotropic and voluminal emission we finally get:

$$dq_\nu = \frac{E_{\nu,L}}{4\pi^2 \, r_L^2 \, L_L} \frac{1}{\rho^2} \exp\left(- \int_{\rho_o}^{\rho_I} \mu_\nu \, d\rho\right) d\Gamma_e \qquad (20)$$

Considering the whole extension of the lamp (Γ_e) and polychromatic radiation, the final relationship we are seeking for takes the form:

$$e^a = \int_0^\infty \frac{\mu_\nu \, E_{\nu,L}}{4\pi^2 \, r_L^2 \, L_L} \int_{\Gamma_e} \frac{1}{\rho^2} \exp\left(- \int_{\rho_o}^{\rho_I} \mu_\nu \, d\rho\right) d\Gamma_e \, d\nu \qquad (21)$$

Equation (21) can be applied equally to reactors directly irradiated (as for example, in annular type reactors) or to those that include reflected radiation (as for example in the classical elliptical reflector). It has to be changed slightly when fluorescent lamps with superficial emission are used. Also, it can be simplified (particularly due to the complications of the integration limits for Γ_e) when only direct radiation is present. Then, the line model of Dranoff and co-workers (Jacob and Dranoff, 1966) can be used with an acceptable error.

6. EVALUATION OF THE LVREA

Equation (20) looks rather simple. Actually, the application of the integration limits, particularly if they have to be treated in a very exact way is always complex. The problem has been dealt with in several papers, as for example Irazoqui et al. (1973), Cerdá et al. (1973, 1977), Romero et al. (1983), De Bernardez and Cassano (1985), Alfano et al. (1985), etc. We will describe here the case of a reactor with direct irradiation to illustrate the procedure. Each point inside the reactor receives a fraction of the lamp emission. Performing the integration in spherical coordinates:

$$q_\nu = \kappa_\nu \int_\phi \int_\theta \int_\rho d\rho \, d\theta \, d\phi \, \sin\theta \exp\left(- \int_{\rho_o(\theta,\phi)}^{\rho_I(\theta,\phi)} \mu_\nu \, d\rho\right) \qquad (22)$$

Starting from the point of reception (I) one must follow every ray arriving at I with a given direction in space (θ,ϕ) and find out if they are originated from the interior of the lamp boundaries. The limits are defined precisely by those rays that intersect or are tangent to the lamp boundaries. For the ρ coordinate:

$$\rho_{1,2}(\theta,\phi) = \frac{r\cos\phi \pm (r^2\cos^2\phi - r^2 - r_L^2)^{1/2}}{\sin\theta} \tag{23}$$

In the case of the limiting angles for the variable θ, one has to look for the common solution of the equation describing the ray and the equation of those parts of the circumferences that define the upper and lower boundaries of the lamp which are closer to the point under consideration (Figure 2). The results are:

$$\theta_1(\phi) = \tan^{-1}\frac{r\cos\phi - [r^2(\cos^2\phi - 1) + r_L^2]^{1/2}}{L_L - z} \tag{24}$$

$$\theta_2(\phi) = \tan^{-1}\frac{r\cos\phi - [r^2(\cos^2\phi - 1) + r_L^2]^{1/2}}{-z} \tag{25}$$

Finally the limits for the angle ϕ are obtained after imposing the condition that both values of the limits for ρ be equal ($\rho_1 = \rho_2$). Then:

$$-\phi_1 = \phi_2 = \cos^{-1}\frac{(r^2 - r_L^2)^{1/2}}{r} \tag{26}$$

The corrections required to account for the changes produced by the presence of both lamp ends and the existence of reaction wedges at the incoming and outgoing sections of the reactor tube that modify these approximate results, have been described in the references quoted above.

Even though the necessary trigonometric and algebraic manipulations are much more involved, the methodology is equally applicable for reflected radiation. The reasoning has been used by Cassano and co-workers with elliptical, parabolic and circular cross section reflectors and it has also been extended to multilamp reactors (Alfano et al., 1990) and multitube reactors (Tymoschuk, 1990).

7. DESIGN EQUATIONS

The design of a reactor requires the solution of the momentum, energy and mass conservation equations. In the case of a photoreactor, the photon balance must be added. The radiation balance can be treated separately from the thermal energy balance because the thermal effects of the photochemically useful energy are generally negligible and consequently one is concerned only with its kinetics effects. This asseveration does not mean that thermal effects in photochemical reactions do not need to be taken into account, and special attention must be exercised in highly exo or endothermic reactions. However, since most photochemical reactors never reach very high temperatures, generally, radiation emission inside the reactor can be neglected. Strictly speaking the mass balance must be written for monochromatic radiation, and one would have a set of different mass balances, one for each wavelength. In practice one assumes that the kinetics effects of the different wavelengths are additive and writes just only one mass balance for each species. The rationality and the details of an elegant way of treating polychromaticity have been described by Claria et al. (1988).

As usual the mass and energy balances are coupled through the reaction rate terms, and in a photochemical reactor this coupling extends to the radiation balance as it was indicated in section 2. There is only one practical exception to the latter simultaneous solution requirement: a system where the chemical species that absorbs radiation and participates in the reaction keeps its concentration constant, as it is the case of a true photosensitized reaction. When this is not the case, and the reactor is not perfectly mixed (for example in a continuous tubular flow reactor), the link between the attenuation of radiation and the extent of the reaction greatly increases the complexity of the design. The problem has been described thoroughly by Cassano and coworkers in the review papers listed in section 1.

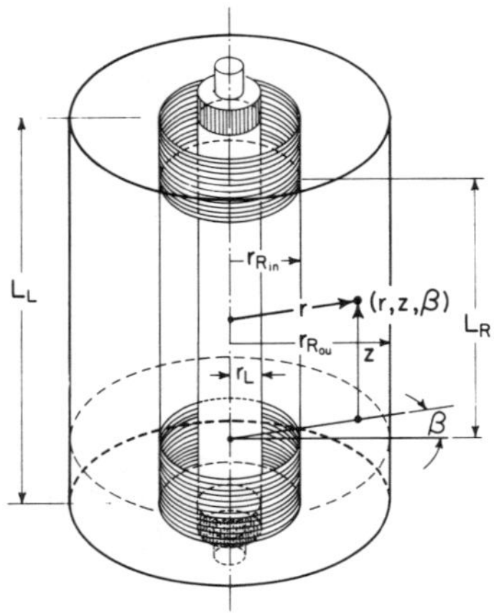

Figure 3

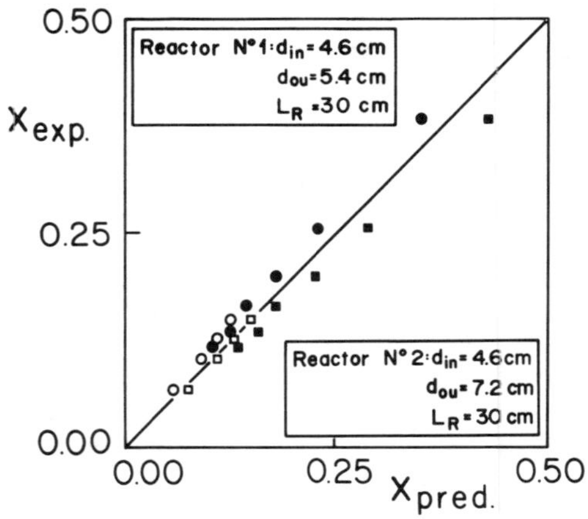

Figure 4

In all but a perfectly mixed reactor the solution of the momentum balance, generally uncoupled from the mass and energy equations, will be required. This situation poses on photochemical reactor design the same restrictions with regard to an "a priori" design than in any other reacting system. Turbulence and multiphase flows are still areas where the lack of proved predictions of the flow behavior in practical problems is most critical. As a direct consequence the modeling of mass and energy fluxes under these conditions have the same flaws in photochemical systems than in the whole area of chemical reaction engineering.

A complete and precise formulation of the conservation equations for multicomponent systems can be found in the book written by Slattery (1978). In order to keep the extension of this work within the required limitations, we will illustrate them for the case of a steady state, continuous flow, annular photoreactor where the monophotochlorination of ethane is conducted (Figure 3). The system is irradiated (only with direct radiation) by means of a tubular lamp of the arc type with polychromatic light. To avoid the inclusion of uncertainties in the expressions for the mass and energy fluxes, the flow will be assumed laminar and completely developed.

Under the following additional assumptions: (i) constant physical properties, (ii) azimuthal symmetry and (iii) negligible pressure drop, all of them relaxable at the expense of computing time, the solution of the Cauchy's equation produces the velocity profile. In dimensionless form:

$$U(\gamma) = 2\,\frac{1 - [\gamma(1-\chi)]^2 + \dfrac{1-\chi^2}{\ell n\,(1/\chi)}\,\ell n\,[\gamma(1-\chi)]}{1 + \chi^2 - [(1-\chi^2)/\ell n\,(1/\chi)]} \qquad (27)$$

where:

$$\chi = r_{R_{in}}/r_{R_{ou}}$$

The mass and energy balances may be written under two more assumptions: (i) neglect axial molecular fluxes when compared with the convective flow and (ii) diffusivities calculated according to the Stefan-Maxwell relationship. Then, for each species, the dimensionless expression is:

$$U(\gamma)\,\frac{\partial \Psi_i}{\partial \zeta} - \frac{1}{Ge}\,\frac{1}{\gamma}\,\frac{\partial}{\partial \gamma}\left(\frac{1}{Pe_i}\,\gamma\,\frac{\partial \Psi_i}{\partial \gamma}\right) = \Omega_i^* \qquad i = 1 \text{ to } 7 \qquad (28)$$

with the following initial condition:

$$\Psi_i(0,\gamma) = \Psi_i^\circ \qquad i = 1 \text{ to } 7 \qquad (28.a)$$

For the stable species, the boundary conditions are:

$$\frac{\partial \Psi_i}{\partial \gamma}\,(\zeta,\ \chi/1-\chi) = 0 \qquad j = 1,3,4,6,7 \qquad (28.b)$$

$$\frac{\partial \Psi_i}{\partial \gamma}\,(\zeta,\ 1/1-\chi) = 0 \qquad j = 1,3,4,6,7 \qquad (28.c)$$

For unstable, intermediate species that may undergo heterogeneous termination reactions, the boundary conditions are:

$$\frac{\partial \Psi_k}{\partial \gamma}\,(\zeta,\ \chi/1-\chi) = -\,Pe_k\,\Omega_{W,k}^* \qquad k = 2,5 \qquad (28.b')$$

$$\frac{\partial \Psi_k}{\partial \gamma}\,(\zeta,\ 1/1-\chi) = -\,Pe_k\,\Omega_{W,k}^* \qquad k = 2,5 \qquad (28.c')$$

The reaction rates are:

$$\Omega_1^* = -J - K_3\Psi_1\Psi_5 + K_3'\Psi_4\Psi_2 + K_4\Psi_2^2\Psi_8 \tag{29}$$

$$\Omega_2^* = 2J - K_2\Psi_2\Psi_3 + K_2'\Psi_5\Psi_6 + K_3\Psi_1\Psi_5 - K_3'\Psi_4\Psi_2 - 2K_4\Psi_2^2\Psi_8 - K_5\Psi_2\Psi_5 \tag{30}$$

$$\Omega_3^* = -K_2\Psi_2\Psi_3 + K_2'\Psi_5\Psi_6 \tag{31}$$

$$\Omega_4^* = K_3\Psi_1\Psi_5 - K_3'\Psi_4\Psi_2 + K_5\Psi_2\Psi_5 \tag{32}$$

$$\Omega_5^* = K_2\Psi_2\Psi_3 - K_2'\Psi_5\Psi_6 - K_3\Psi_1\Psi_5 + K_3'\Psi_4\Psi_2 - K_5\Psi_2\Psi_5 - 2K_6\Psi_5^2 \tag{33}$$

$$\Omega_6^* = K_2\Psi_2\Psi_3 - K_2'\Psi_5\Psi_6 \tag{34}$$

$$\Omega_7^* = K_6\Psi_5^2 \quad (35) \qquad \Omega_{W,2}^* = K_7\Psi_2 \quad (36) \qquad \Omega_{W,5}^* = K_8\Psi_5 \quad (37)$$

and the dimensionless rate of initiation is:

$$J = \int_0^\infty \frac{\Phi_{P,\nu}\, E_{L,\nu}\, \varepsilon_\nu\, T_{R,\nu}\, \Psi_1}{2\pi^2\, \gamma_L^2\, \zeta_L\, L_R \langle v_z\rangle (r_{R_{ou}} - r_{R_{in}})\, C_1^\circ} \int_{\phi_1}^{\phi_2} (\gamma_L^2 - \gamma^2 \sin^2\phi)^{1/2}\, d\phi \times \int_{\theta_1(\phi)}^{\theta_2(\phi)} \exp\left[- \int_{(\chi/1-\chi)(1/\sin\theta)}^{\gamma/\sin\theta} \varepsilon_\nu\, \Psi_1\, d\rho\right] d\theta \tag{38}$$

with:

$$\varepsilon_\nu = \alpha_\nu\, C_1^\circ\, (r_{R_{ou}} - r_{R_{in}}) \tag{39}$$

The thermal energy equation in dimensionless form is:

$$U(\gamma)\frac{\partial\tau}{\partial\zeta} - \frac{1}{Pe_T\, Ge}\frac{1}{\gamma}\frac{\partial}{\partial\gamma}\left(\gamma\frac{\partial\tau}{\partial\gamma}\right) = Q \tag{40}$$

with the following initial and boundary conditions:

$$\tau\,(0,\gamma) = 0 \tag{40.a}$$

$$\frac{\partial\tau\,(\zeta,\, \chi/1-\chi)}{\partial\gamma} = -Nu\,(\tau - 1) \tag{40.b}$$

$$\frac{\partial\tau\,(\zeta,\, 1/1-\chi)}{\partial\gamma} = -Nu\,(\tau - 1) \tag{40.c}$$

and where:

$$Q = \frac{L_R\ (-\Delta H_G)\ \Omega_G}{\rho\ Cp\ \langle v_z \rangle\ (T_c - T^\circ)} \qquad (41) \qquad\qquad \tau = \frac{T - T^\circ}{T_c - T^\circ} \qquad (42)$$

De Bernardez and Cassano (1986), Clariá et al. (1988) and Alfano and Cassano (1988 a,b) have shown that in most photochlorination reactions, the local or micro steady state approximation can be used for atomic chlorine and free radicals. Similarly they found that at atmospheric or higher pressures and in practical reactors, the heterogeneous reactions at the wall have negligible influence on the final exit conversions. Thus, this two simplifications can be used normally in photoreactor design.

The equations listed above show same peculiarities that should be pointed out:

1) The presence of the lamp output characteristics and dimensions, as well as the reactor wall transmittance as photoreactor unique parameters. If reflectors are present the appropriate equations will show also the reflection coefficient as an additional parameter.
2) The existence of a characteristic length ($r_{R_{ou}} - r_{R_{in}}$) that not only influences the hydrodynamics of the reactor but also has a strong effect on the radiation optical path expressed by the number ε_ν. Again this is particular of photoreactors.
3) The important fact that the absorption number, ε_ν (which includes the absorption coefficient, α_ν), and the concentration of the absorbing species Ψ_1, come into play twice: one time they affect the rate of initiation in a linear manner and the second time they participate in the exponent of a decreasing exponential function inside the attenuation integral. This is significant because it indicates that the concentration of the absorbing species has two different effects that act in opposite directions, given rise to the existence of an optimum value which is not independent of the characteristic length.
4) The concentration of the absorbing species that goes into the linear part of the initiation rate may not be the same than the one that must be included in the exponential part. In the linear part intervenes only that species that participates in the first step of the reaction (activation), whereas in the attenuation exponential, the concentration of all the species that absorb radiation must be taken into account. In this way we can include in the evaluation of the radiation field any form of inner filtering effect present in the system.

The set of equations (27) to (40) must be solved numerically and the final result provides the concentration and temperature fields (also the radiation field if desired) as a function of position. Obviously from this results one can obtain the output conversion of the reactor. It is also clear that these equations can be manipulated to render the optimal annular space and length for a given production (including selectivity if desired). For example one can resort to an optimization program as it was done for the methane photochlorination by De Bernardez and Cassano (1986).

8. APPLICATIONS

8.1. Annular photoreactor with simple kinetics

This first application presents the "a priori" modeling of a well known reaction carried out in a continuous annular photoreactor (Figure 3). The radiation field was described by resorting to the Extense Source model with isotropic Voluminal Emission (ESVE) proposed by Irazoqui et al. (1973). Upon the basis of the kinetic data published by Leighton and Forbes (1930), Forbes and Heidt (1934) and Heidt et al. (1970), we employed the photodecomposition of the complex formed by the aqueous solution of oxalic acid and uranyl sulfate.

The overall chemical reaction, that has a product distribution which is strongly dependent on the pH, may be written as follows:

$$UO_2^{2+} + H_2C_2O_4 + h\nu \rightarrow UO_2^{2+} + CO + CO_2 + H_2O$$

$$UO_2^{2+} + H_2C_2O_4 + h\nu \rightarrow UO_2^{2+} + HCOOH + CO_2$$

The reactor was modeled with the momentum, mass and radiation conservation equations described in section 7. The kinetics was formulated with a phenomenological approach following the oxalic acid concentration using the overall quantum yield according to:

$$\Omega_{ox} = - \Phi_{ox}\, e^{a} \tag{43}$$

The reaction was represented on the basis of the published quantum yields valid for oxalic acid conversions below 80%. However, since we are not dealing with a perfectly mixed system but a laminar flow reactor, the hypothesis of using a constant absorption coefficient loses validity. This is so because it is inevitable to reach local values of conversions that are well above 20%. At wavelenghts below the near UV range (precisely where the absorption is stronger), when the oxalic acid-uranyl sulfate molar ratio falls below 4, the absorption coefficient becomes a non-linear function of the oxalic acid concentration. In practice, when the conversion is higher than 20% the kinetic behavior shows a dependence upon the concentration of the oxalic acid, because the nature of the uranyl oxalate complex changes and so does the absorption coefficient.

We performed the experimental verifications with two reactors, the characteristics of which are indicated in Figure 4. We employed a 15 W input power germicidal lamp and an aqueous solution of 0.005 M in oxalic acid and 0.001 M in uranyl sulfate. Neither free, nor experimentally adjustable parameters were used. Details on the modeling, the apparatus, the operation and the obtained results may be found in De Bernardez and Cassano (1985).

Figure 4 illustrates, for both reactors, the quality of the predictions compared with the experimental results. In the figure, the black symbols correspond to Reactor No. 1 and the white ones to Reactor No. 2. Squares correspond to predictions computed with the lamp maximum output power supplied by the manufacturer and circles are the values using the average life-span output power value, also provided by the lamp manufacturer. It must be stressed that the observed good agreement can be obtained only when a variable absorption coefficient (a function of the oxalic acid concentration) was used. Thus, when very high conversions are obtained (and in the region close to the inside wall of the annular reactor they certainly are), even for this photosensitized reaction, the assumption of constant attenuation cannot be used.

8.2. Cylindrical reactor inside an elliptical reflector. Monochlorination of ethane

This section presents the results of an "a priori" modeling of a complex chain reaction, carried out in a system where direct and reflected radiation are present (Figure 5). We performed the monochlorination of ethane in the gas phase, using an excess of nitrogen as an inert. The kinetic mechanism and the values of the specific rate constant were taken from the literature. They comprise values for homogeneous and heterogeneous (at the wall) reactions (Noyes and Fowler, 1951; Pritchard et al., 1955; Gosselain et al., 1956; Chiltz et al., 1963). The full reaction sequence was shown in section 2.

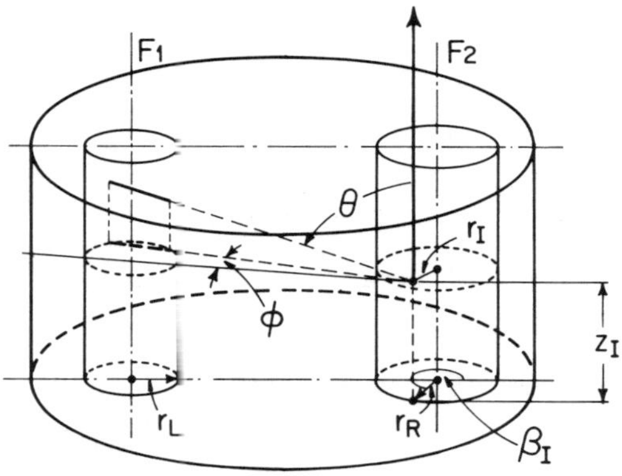

Figure 5

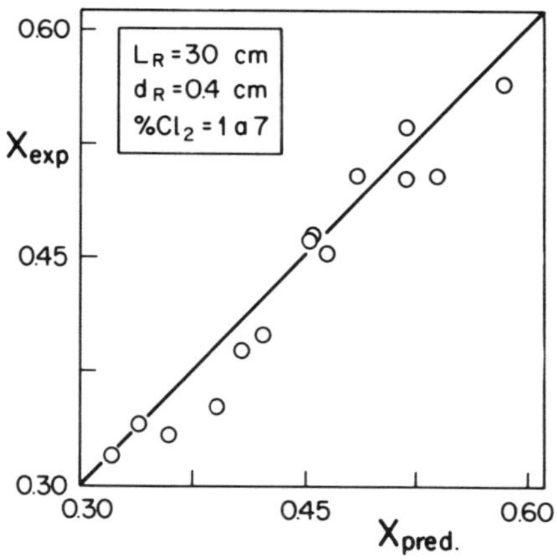

Figure 6

The radiation field was predicted with the ESVE model. The evaluation of the direct radiation is a straightforward application of the results described in section 6. On the other hand, the reflected radiation demands a much more elaborated treatment of the integration limits (Cerdá et al., 1973, 1977; Clariá et al., 1986).

Strictly speaking the reactor inside the elliptical reflector does not have angular symmetry unless $r_R/r_L < 0.5$ (De Bernardez and Cassano,

1982). This condition can be fulfilled only in laboratory or bench scale reactors. Otherwise, three dimensional mass and energy balances will have to be used.

For this system the design equations described in section 7, must be adapted for a cylindrical tube. The main changes are produced in:

1) The velocity profile, which is now the classical parabolic one.
2) The boundary condition at the inside wall of the annular reactor must be changed by a condition of symmetry at the cylindrical reactor center line, both for the mass and the energy balances.

Since the operation was almost isothermal (a large excess of nitrogen was used and conversions never exceeded 60%), the rigorous solution of the problem involves the numerical integration of a set of seven integro-partial differential equations.

The design parameters were taken from the literature (reflection coefficient for aluminum and absorptivities for chlorine, for example), or from manufacture's specifications (lamp characteristics, reactor and reflector dimensions, for example).

Firstly we performed computational experiments which indicated the possibility of disregarding the heterogeneous termination reactions and provided the range of applicability of the local steady state approximation for the short-lived intermediate species. Afterwards, the model was verified in bench-scale experiments with an emitting system made up of a lamp with 30 W of nominal input power (germicidal type) and a reflector built with an aluminum reflecting sheet with Alzac treatment. The reactants were carefully purified and the reactor operation was minutely controlled. Each experimental datum demanded a minimum of 8 hours of operation. The details may be found in Clariá et al. (1988).

Figure 6 portrays the "a priori" predictions (with neither free nor experimentally adjustable parameters), in comparison with the experimental results. The agreement is very good.

8.3. Cylindrical reactor inside an elliptical reflector. Non-isothermal and polychromatic polychlorination of methane

In this example we will show the modeling of the conjunction of a difficult irradiating system like the elliptical reflector and a complex kinetics represented by the successive photochlorination of methane, where attractive selectivity studies can be performed. The work was carried out with three different lamps. Nitrogen was used as an inert.

There is ample information on the kinetics of the system and one can use the results reported in the review by Chiltz et al. (1963) complemented with the data of Clyne and Stedman (1968) for the atomic chlorine recombination. Other authors reported data which showed some discrepancies with those of Chiltz et al. Consequently, the data were allowed to change within a maximum spread of 5% about the original values, on account of surely existing experimental errors.

The reaction is completely represented by a mechanistic sequence of 37 steps. Simulation studies performed by De Bernardez and Cassano (1986) have shown that some of them can be neglected without affecting the prediction of the concentration field. The simplified chain mechanism was then reduced to the following 22 steps:

INITIATION	$Cl_2 + h\nu$	$\longrightarrow$	$Cl\cdot$	Φ_P
PROPAGATION	$CH_4 + Cl\cdot$	$\rightleftarrows$	$CH_3\cdot + HCl$	k_2, k_2'
	$CH_3\cdot + Cl_2$	$\rightleftarrows$	$CH_3Cl + Cl\cdot$	k_3, k_3'
	$CH_3Cl + Cl\cdot$	$\rightleftarrows$	$CH_2Cl\cdot + HCl$	k_4, k_4'
	$CH_2Cl\cdot + Cl_2$	$\rightleftarrows$	$CH_2Cl_2 + Cl\cdot$	k_5, k_5'
	$CH_2Cl_2 + Cl\cdot$	$\rightleftarrows$	$CHCl_2\cdot + HCl$	k_6, k_6'
	$CHCl_2\cdot + Cl_2$	$\rightleftarrows$	$CHCl_3 + Cl\cdot$	k_7, k_7'
	$CHCl_3 + Cl\cdot$	$\rightleftarrows$	$CCl_3\cdot + HCl$	k_8, k_8'
	$CCl_3\cdot + Cl_2$	$\rightleftarrows$	$CCl_4 + Cl\cdot$	k_9, k_9'
HOMOGENEOUS TERMINATION	$Cl\cdot + CH_3\cdot$	$\longrightarrow$	CH_3Cl	k_{10}
	$Cl\cdot + CH_2Cl\cdot$	$\longrightarrow$	CH_2Cl_2	k_{11}
	$Cl\cdot + CHCl_2\cdot$	$\longrightarrow$	$CHCl_3$	k_{12}
	$Cl\cdot + CCl_3\cdot$	$\longrightarrow$	CCl_4	k_{13}
	$Cl\cdot + Cl\cdot + M$	$\longrightarrow$	$Cl_2 + M$	k_{14}

The design equations are the same than those used for the monochlorination of ethane with the following exceptions:

1) The thermal energy balance was required because in spite of using a relatively large excess of nitrogen, the operating conditions were such that the isothermal behavior could not be assumed.
2) The number of simultaneous integro-partial differential equations to be solved has been increased to 13.
3) Three different lamps have been used, all of them with an emission of a very wide range of polychromaticity, and very different output powers among them; additionally, one of the lamps is of the fluorescent type and it required the use of the Extense Source with Diffuse, Superficial Emission model (Yokota et al., 1976).

The lamps used in the experiments had the following input powers: 40, 360 and 1200 W. The runs were performed in a bench scale equipment at three different nominal temperatures (297.9 K, 312.8 K and 322.2 K). The feed molar ratio of chlorine to methane was changed from 0.5 to 10. Thus, the experimental verifications were performed under a broad range of operating conditions.

Not only total reactant conversions but individual stable product concentrations (selectivities) were subjected to test.

Figure 7 shows a compendium of all the experimental data, compared with the model predictions. The model also provides the complete temperature field. Figure 8 shows, for example, the mixing-cup (radial average) axial temperature profiles for runs performed with the three lamps when the feed molar ratio was 9.5. Further details on the subject can be found in Cabrera et al. (1990 a,b).

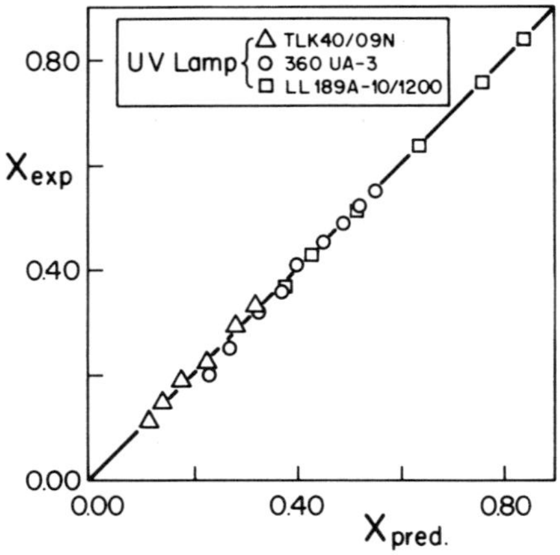

Figure 7

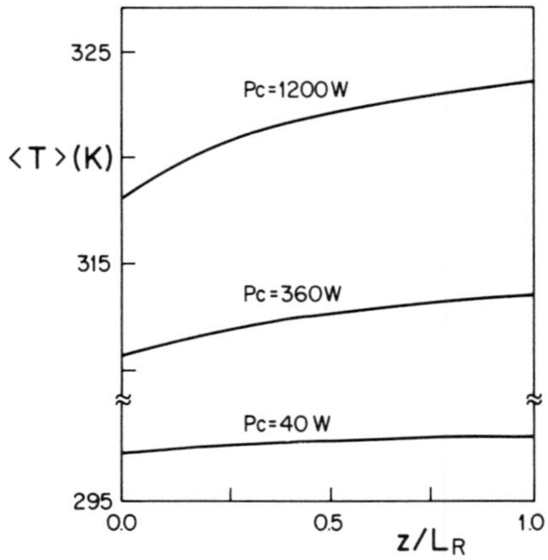

Figure 8

8.4. Gas-liquid tank photoreactor irradiated from the bottom. Chlorination of trichloroethylene and pentachloroethane

This section presents a heterogeneous gas-liquid photochemical reaction carried out inside a perfectly stirred tank, semibatch reactor. The reactor is irradiated from the bottom by means of a tubular source located at the focal axis of a cylindrical reflector of parabolic cross section (Figure 9). The trichloroethylene and pentachloroethane photochlorination reactions were studied.

Information about the reaction kinetics may be obtained from the data provided by the specific literature on gas and liquid phase chlorinations (Leermakers and Dickinson, 1932; Daiton et al., 1957; Walling, 1957; Poutsma and Hinman, 1964). The overall reaction may be represented by the following reactions:

$$C_2HCl_3 \xrightarrow{+Cl_2} C_2HCl_5 \xrightarrow{+Cl_2} C_2Cl_6$$

where the first step is an addition reaction and the second a substitution reaction.

The reaction mechanism including all the possible steps, is given by the following sequence:

INITIATION $\quad Cl_2 \longrightarrow 2\ Cl\bullet \quad \Phi_P$

PROPAGATION	$C_2HCl_3 + Cl\cdot$	$\rightleftarrows$	$C_2HCl_4\cdot$	k_2, k_2
	$C_2HCl_4\cdot + Cl_2$	$\rightleftarrows$	$C_2HCl_5 + Cl\cdot$	k_3, k_3'
	$C_2HCl_5 + Cl\cdot$	$\rightleftarrows$	$C_2Cl_5\cdot + HCl$	k_4, k_4'
	$C_2Cl_5\cdot + Cl_2$	$\rightleftarrows$	$C_2Cl_6 + Cl\cdot$	k_5, k_5'
TERMINATION	$2\ Cl\cdot$	$\longrightarrow$	Cl_2	k_6
	$2\ C_2HCl_4\cdot$	$\longrightarrow$	Products	k_7
	$2\ C_2Cl_5\cdot$	$\longrightarrow$	Products	k_8
	$Cl\cdot + C_2HCl_4\cdot$	$\longrightarrow$	Products	k_9
	$Cl\cdot + C_2Cl_5\cdot$	$\longrightarrow$	Products	k_{10}
	$C_2Cl_5\cdot + C_2HCl_4\cdot$	$\longrightarrow$	Products	k_{11}

The analysis considers that the existence of a nonuniform radiation field produces a nonuniform reaction rate. Hence, even if one assumes perfect mixing for stable reactants and products, concentration gradients of unstable species may be present. To consider these mixing states, a comparative study between the mean lifetime of all reacting species and a characteristic mixing time was performed. Two mixing states can be considered (Felder and Hill, 1969; 1970): i) Perfect mixing of all the species (Model 1), and ii) perfect mixing of the stable species and no mixing for atomic chlorine and free radicals (Model 2).

Using the macroscopic (Model 1) or microscopic (Model 2) steady-state approximation and introducing the long chain approximation, we obtained the following dimensionless mass balances in the liquid bulk (Alfano and Cassano, 1988 a):

Model 1

$$\frac{d\underset{\sim}{\Psi}}{d\tau} = \underset{\sim}{N}(\tau) - \underset{\approx}{B}(\tau) \cdot \underset{\sim}{\hat{\Psi}}(\tau) - \langle \underset{\sim}{g}(\underset{\sim}{x},\tau)\rangle + \underset{\sim}{t}(\tau) \tag{44}$$

$$\underset{\sim}{\Psi}(0) = \underset{\sim}{\Psi}^{\circ} \tag{45}$$

Model 2

$$\frac{d\underset{\sim}{\Psi}}{d\tau} = \underset{\sim}{N}(\tau) - \underset{\approx}{B}(\tau) \cdot \langle \underset{\sim}{\hat{\Psi}}(\underset{\sim}{x},\tau)\rangle - \langle \underset{\sim}{g}(\underset{\sim}{x},\tau)\rangle + \langle \underset{\sim}{t}(\underset{\sim}{x},\tau)\rangle \tag{46}$$

$$\underset{\sim}{\Psi}(0) = \underset{\sim}{\Psi}^{\circ} \tag{47}$$

To simplify the nomenclature of the arrays, the chemical species of the reacting system may be identified as follows: $HCl(\Psi_1)$, $Cl_2(\Psi_2)$, $C_2HCl_3(\Psi_3)$, $C_2HCl_5(\Psi_5)$, $C_2Cl_6(\Psi_6)$, $Cl\cdot(\hat{\Psi}_1)$, $C_2HCl_4\cdot(\hat{\Psi}_4)$ and $C_2Cl_5\cdot(\hat{\Psi}_5)$.

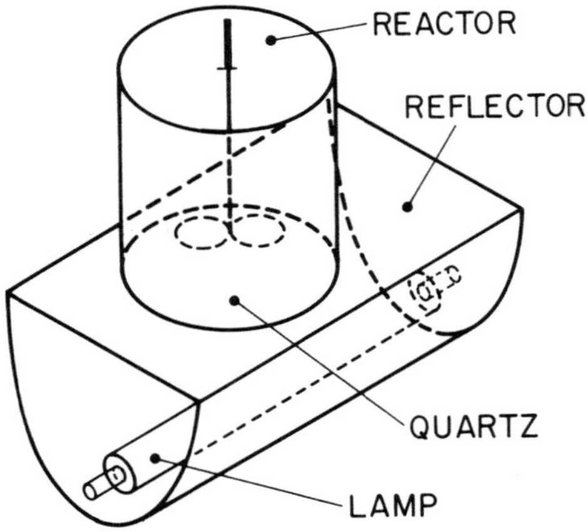

Figure 9

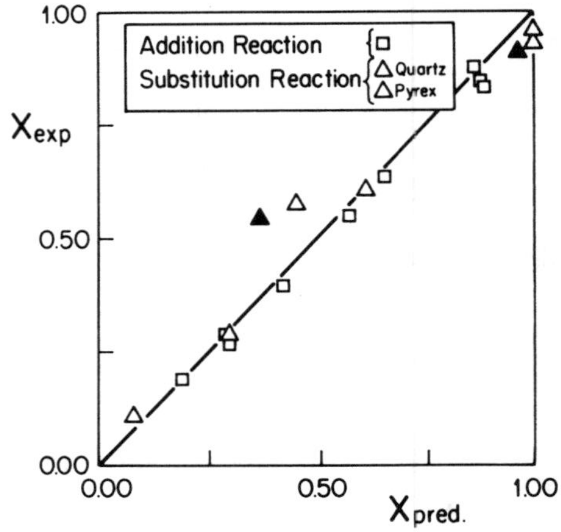

Figure 10

The gas-liquid absorption rate into the liquid bulk, is given by:

$$N_2(\tau) = K_L\ [\Psi_2^{int}(\tau) - \Psi_2(\tau)] \quad ; \quad N_i(\tau) = 0 \quad (i\neq 2) \tag{48}$$

where Ψ_2^{int} is the dimensionless chlorine concentration at the interface. For this case, it can be shown that the reaction in the liquid phase is slow enough to be negligible during the lifetime of the liquid surface elements at the gas-liquid interface ($t_D << t_r$). This condition was defined by Astarita (1967) and Astarita et al. (1983) as a slow reaction regime; if this is the case, the chemical absorption coefficient is equal to the physical absorption coefficient.

The arrays used in the models for the initiation, propagation and termination reaction rates, are defined as follows:

Initiation Reaction Array $\underset{\sim}{g}(\underset{\sim}{x},\tau)$

$$[\sum_\lambda J_\lambda\ \Psi_2(\tau)\ \Theta_\lambda(\underset{\sim}{x},\tau),\ 0,\ \ldots,\ 0]^T \tag{49}$$

Propagation Reaction Array $\underset{\approx}{B}(\tau)$

$$\begin{bmatrix} -K_4\Psi_5 & 0 & K_4'\Psi_1 \\ -K_3'\Psi_5 + K_5'\Psi_6 & K_3\Psi_2 & K_5\Psi_2 \\ K_2\Psi_3 & -K_2' & 0 \\ K_3'\Psi_5 + k_4\Psi_5 & -K_3\Psi_2 & -K_4'\Psi_5 \\ K_5'\Psi_6 & 0 & -K_5\Psi_2 \end{bmatrix} \tag{50}$$

Free Radical Concentration Arrays $\hat{\underset{\sim}{\Psi}}(\tau)$

$$\begin{bmatrix} \gamma^{1/2} \langle \sum_\lambda J_\lambda \ \Psi_2 \ \Theta_\lambda(\underset{\sim}{x},\tau)\rangle^{1/2} \\ \gamma^{1/2} \langle \sum_\lambda J_\lambda \ \Psi_2 \ \Theta_\lambda(\underset{\sim}{x},\tau)\rangle^{1/2}/\alpha \\ \gamma^{1/2} \langle \sum_\lambda J_\lambda \ \Psi_2 \ \Theta_\lambda(\underset{\sim}{x},\tau)\rangle^{1/2}/\beta \end{bmatrix} \quad (51)$$

$\hat{\underset{\sim}{\Psi}}(\underset{\sim}{x},\tau)$

$$\begin{bmatrix} \gamma^{1/2} [\sum_\lambda J_\lambda \ \Psi_2 \ \Theta_\lambda(\underset{\sim}{x},\tau)]^{1/2} \\ \gamma^{1/2} [\sum_\lambda J_\lambda \ \Psi_2 \ \Theta_\lambda(\underset{\sim}{x},\tau)]^{1/2}/\alpha \\ \gamma^{1/2} [\sum_\lambda J_\lambda \ \Psi_2 \ \Theta_\lambda(\underset{\sim}{x},\tau)]^{1/2}/\beta \end{bmatrix} \quad (52)$$

Termination Reaction Arrays $\underset{\sim}{t}(\tau)$ or $\underset{\sim}{t}(\underset{\sim}{x},\tau)$

$$[K_6(\hat{\Psi}_1)^2, \quad 0, \ \ldots, \quad 0] \quad (53)$$

where the following definitions were employed:

$$\gamma = K_6 + \frac{K_7}{\alpha^2} + \frac{K_8}{\beta^2} + \frac{K_9}{\alpha} + \frac{K_{10}}{\beta} + \frac{K_{11}}{\alpha\beta} \quad (54)$$

$$\alpha = \frac{K_3\Psi_2 + K_2'}{K_2\Psi_3 + K_3'\Psi_5} \quad (55) \qquad \beta = \frac{K_5\Psi_2 + K_4'\Psi_1}{K_4\Psi_5 + K_5'\Psi_6} \quad (56)$$

To complete the theoretical description of the model, it is necessary to introduce the radiation field expression in the initiation reaction array and in the free radical concentration array. The radiant field was evaluated by means of a proposed model which had been tested experimentally in previous instances (Alfano et al., 1985; 1986 c,d). The reader may refer to these papers for further details. The radiant field expressions for direct and indirect radiation at the point $\underset{\sim}{x}$ and at the instant τ are:

$$\Theta_{D,\lambda}(\underset{\sim}{x},\tau) = \frac{\langle T_R \rangle}{4\pi \ r_L} \int_{\phi_D} \Delta\rho'(\underset{\sim}{x},\phi) \int_{\theta_D} \exp\left[-\frac{\mu_\lambda^*(\tau) \ z_I}{\sin\theta \ \sin\phi}\right] d\theta \ d\phi \quad (57)$$

$$\Theta_{In,\lambda}(\underset{\sim}{x},\tau) = \frac{\langle \Gamma_{Rf} \rangle \langle T_R \rangle}{4\pi \ r_L} \int_{\phi_{In}} \Delta\rho_E'(\underset{\sim}{x},\phi) \int_{\theta_{In}} \exp\left[-\frac{\mu_\lambda^*(\tau) \ z_I}{\sin\theta \ \sin\phi}\right] d\theta \ d\phi \quad (58)$$

In equations (57) and (58) an effective attenuation coefficient (μ_λ^*) was employed. It takes into account the radiation field distortions due to the existence of gas bubbles. This coefficient has been correlated as a function of the optical properties of the liquid phase, the gas holdup, and the specific interfacial area (see equations 10 and 11).

The experimental verification of the model was carried out in a 2 liters stirred tank photoreactor built of Pyrex glass with a quartz bottom to provide a wall transparent to the entrance of the radiant energy. A 360 W arc lamp, producing polychromatic light was employed. Each experimental run demanded about six hours, of which the first four were used for the stabilization of the source, flushing of the whole device with nitrogen to eliminate impurities (mainly oxygen and moisture) and adjustment of the streams of chlorine and nitrogen. Further details regarding the device and experimental procedures may be found in Alfano and Cassano (1988 b).

Firstly, the reaction subregime was experimentally studied, thus determining a diffusive control for the trichloroethylene chlorination and a predominantly kinetic control for the pentachloroethane chlorination. Afterwards, we compared the experimental results with the theoretical predictions of both models and found an acceptable coincidence when considering the free radicals in a non-mixing state (Model 2). Figure 10 summarizes the main results obtained by applying Model 2.

8.5. Cylindrical tubular multilamp photoreactor with cylindrical reflectors

This example shows the modeling of a commercially available type of reactor used for water and wastewater treatment. The system has a cylindrical quartz tube irradiated from outside by cylindrical tubular lamps symmetrically located about the reactor. One cylindrical reflector of circular shape surrounds each lamp. Typically as in this example, four germicidal lamps are employed (Figure 11). Thus, each point inside the reactor receives radiation from four lamps and four reflectors. Usually the reactor is considerably longer than the lamp and the reflector, giving rise to the existence of important regions in the entrance and exit parts of the tube that are partially illuminated by the spherical modalities of the lamp emission (reaction wedges).

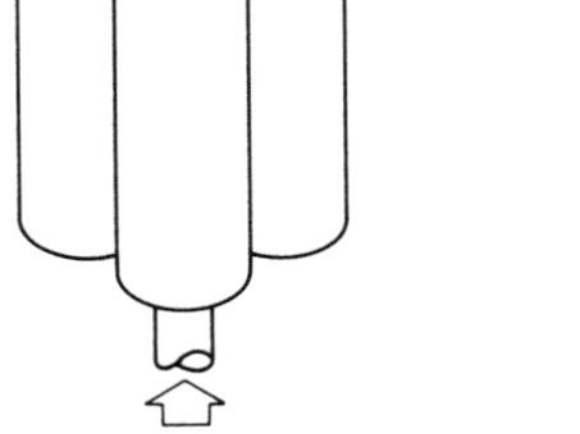

Figure 11

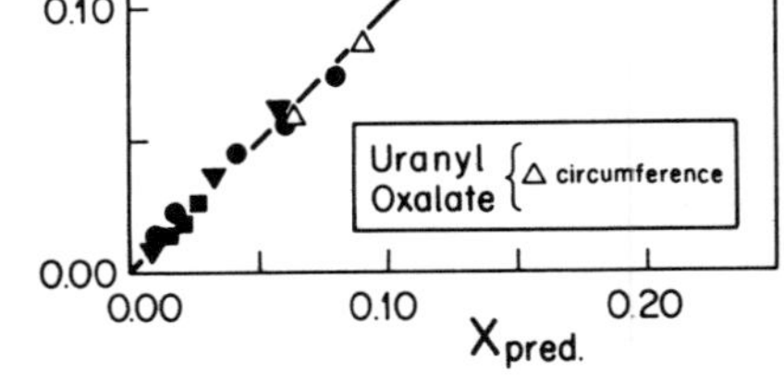

Figure 12

According to the physics of the problem, direct and indirect contributions from the four radiation sources must be added. However, at each plane (r,β) of the reactor, it is not required to compute the whole radiation field. Clearly, symmetry considerations allow us to reduce the analysis to a region limited by $0<r<r_R$, $0\leq\beta<\pi/4$, and $0\leq z\leq 2z_o+L_L$. Additionally, an approach was devised which allows the solution of the problem composed by one reactor, four lamps and four reflectors, by analyzing an equivalent system formed by one reactor, one lamp and one reflector only. With this procedure the value of the modulus of the radiant energy flux density vector coming from a single lamp can be obtained by using the ESVE model and the usual methods that were described in sections 5 and 6. For indirect radiation, only one reflection was considered; i.e., successive reflections after the first one were neglected. In addition, reflections that could be produced in reflectors others than the one specifically surrounding each lamp were also neglected. The radiation field was rigorously described taking into account also the following aspects: (1) The influence of the entrance and exit reaction wedges of the reactor on the overall production rate, which is incorporated into the design equations by the proper definitions of the limiting angles for θ, and (2) the restrictions imposed by the shape of the reflectors on the limiting angles for the variable ϕ, in the case of reflected radiation (Alfano et. al, 1990).

An experimental verification of the model (Vicente et al., 1990) was performed by using the well-known photodecomposition of the oxalic acid and uranyl sulfate reacting mixture (section 8.1). The expression for the overall reaction rate was represented by equation (43).

With the usual assumptions, the oxalic acid dimensionless mass balance can be written as follows:

$$U(\gamma)\frac{\partial\Psi_i}{\partial\zeta} = \frac{1}{\mathrm{Pe\ Ge}}\left[\frac{1}{\gamma}\frac{\partial}{\partial\gamma}\left(\gamma\frac{\partial\Psi_i}{\partial\gamma}\right) + \frac{1}{\gamma^2}\frac{\partial^2\Psi_i}{\partial\beta^2}\right] + \Omega_i^* \qquad (59)$$

The initial and boundary conditions are:

$$\Psi_i(0,\gamma,\beta) = 1 \qquad (59.a)$$

$$\frac{\partial\Psi_i}{\partial\gamma}(\zeta,0,\beta) = 0 \qquad (59.b)$$

$$\frac{\partial\Psi_i}{\partial\gamma}(\zeta,1,\beta) = 0 \qquad (59.c)$$

$$\frac{\partial\Psi_i}{\partial\beta}(\zeta,\gamma,0) = 0 \qquad (59.d)$$

$$\frac{\partial\Psi_i}{\partial\beta}(\zeta,\gamma,\pi/4) = 0 \qquad (59.e)$$

The reactor model was solved for a reacting solution with an initial concentration of 0.001 M in uranyl sulfate and 0.005 M in oxalic acid. Output conversion as a function of the volumetric flow rate using a two-dimensional mass balance model for the extreme values of $\beta = 0$ and $\beta = \pi/4$ were computed. Since differences were never larger than 2%, it was concluded that the additional numerical complexity required by the

3-D mass balance model is not justified.

The reactor used four 15 W germicidal lamps. The reflector and lamp lengths were 34 cm and the reactor dimensions were 100 cm long and 1.85 cm in diameter. Figure 12 shows a compendium of all the theoretical predictions and experimental results. It must be stressed that the observed good agreement was obtained by using the model with the 2-D mass balance and a variable attenuation coefficient (a function of the oxalic acid concentration). The radiation balance was always three-dimensional.

Due to the important contributions of the indirect radiation found in this system, Esplugas et al. (1990) studied the effects of the shape and dimensions of the reflecting mirrors on the reactor performance. For a given geometrical arrangement, it can be shown that: there exists only one ellipse with one focus located at the center of the reactor and the other at the center of the lamp; there exists only one parabola with its focus located at the center of the lamp, and there exist infinite circumferences each one with a different radius.

The flow of photons for each one of the geometries at issue was evaluated in order to characterize the Incidence Efficiency (Cerdá et al., 1977). From the computed results, it was found that for a better use of the radiation emitted by the lamps, it is convenient to employ elliptical reflectors. Besides, it was shown that reflectors of circular cross section may also reach a high efficiency provided that their location and dimensions are properly designed.

In order to compare the incidence efficiency theoretical predictions with the experimental results, two multilamp reactors with elliptical (maximum incidence efficiency) and parabolic (minimum incidence efficiency) reflectors were built. As a comparison basis between both geometries, the reference radius (radius of intersection of the different reflectors) and the reactor and lamp dimensions were kept constant. Several actinometric verifications in a flow reactor were performed by using the potassium ferrioxalate reaction (Parker, 1953; Hatchard and Parker, 1956). The overall chemical reaction may be written as follows:

$$2\ Fe^{3+} + C_2O_4{}^{2-} + h\nu \rightarrow 2\ Fe^{2+} + 2\ CO_2$$

Once more, the kinetics of this reaction can be described with a phenomenological approach as the one represented by equation (43).

The experimental system had four 30 W germicidal lamps. The reactor and reflector lengths were 59 cm and the lamp length was 81 cm. The radius of the reactor was 1 cm. The experiments were carried out using four elliptical or parabolic reflectors to measure the total radiation in each case, and with no reflecting mirrors to measure only direct radiation. Figure 12 also shows the comparison between model predictions and experimental results for elliptical and parabolic reflectors. It can be observed that in the two cases under study the agreement is very good.

8.6. Cylindrical multitube photoreactor, in an annular arrangement, with a circular reflector. Reaction of simple kinetics

The last system to be described in this review is an arrangement that can be used to perform photochemical reactions under moderate pressures. Thus, n small diameter, tubular reactors are located around a tubular lamp and enclosed by a circular reflector (Figure 13). The system could be used, for example, for the photochlorination of methyl chloride in the liquid phase and at pressures between 600 and 1000 kPa.

This work was developed using the ESVE model for the lamp. One must take into account the existence of direct and reflected radiation. The direct radiation can be determined with the usual procedures that were shown in section 6. The reflected radiation poses a problem of a similar degree of complexity than in the case of the elliptical reflector. Considering only a single reflection process, any reactor located in the annular arrangement can receive indirect radiation from that part of the reflector that is closer to it and from the opposite part of the circular mirror. Additionally, when analyzing the effect of the reflection produced by the reflector in one reactor tube, the blocking process produced by all the other reactor tubes of the system in the radiation that could be otherwise reflected must be incorporated, giving rise to the appearance of umbra and penumbra zones on the reflector surface. The complete model was used only in the analysis of the Incidence Efficiency but in any event, it is clear that with an appropriate design the reflected radiation should be minimized. (Placing as many reactors as possible and assigning to each one of them the required optical path for complete absorption.) Fortunately, neglecting second order effects (such as the mutual interaction by reflection among reactors, for example), the problem can be reduced to the modeling of a single tube and the results, with some additional effort, can be extended to the multitubular system.

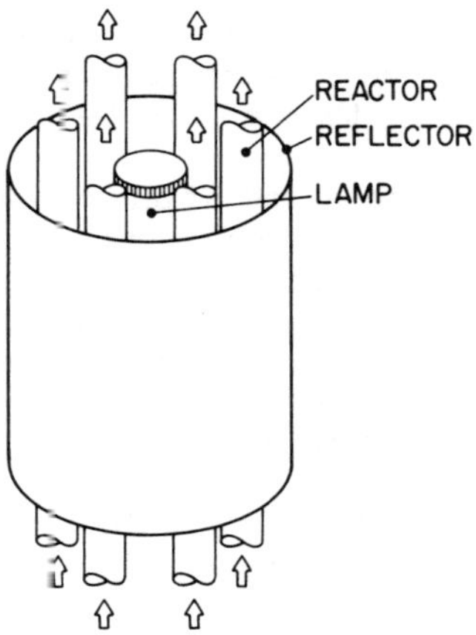

Figure 13

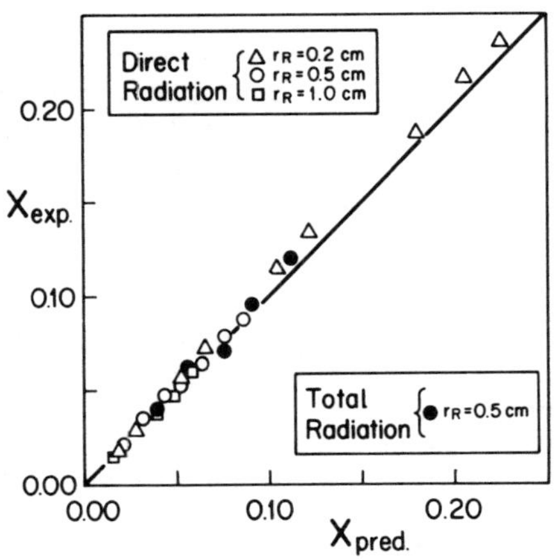

Figure 14

Simulated results assuming a reactor with constant attenuation showed the existence of strong angular asymmetries due to the decisive participation of the direct radiation in the total value of the radiation field. This situation translates to the mass and energy balances the unavoidable consideration of the angular coordinate to account for an angularly asymmetric reaction field.

The reaction used to test the model was the potassium ferrioxalate photodecomposition previously mentioned. The radiation field is a function of the actinometer concentration. Thus, as in the fifth example, there exists the coupling between the radiation attenuation and the reaction extent that the model must take into consideration.

The design equations are similar to the ones employed in reactors of circular cross section under laminar flow regime. The main difference appears in the mass balance because it must incorporate the angular component of the mass fluxes (see equation 59). Under normal operating conditions the system can be assumed to behave isothermically.

The experimental test in bench scale reactors used a 0.006 M solution of the ferric salt in three different reactors (0.2, 0.5 and 1.0 cm in diameter and 59.5 cm long). The radiation source was a germicidal UV arc lamp with 30 W of nominal input power (Tymoschuk, 1990).

Figure 14 depicts the comparison between theoretical predictions and experimental results. As in previous examples, the agreement is very satisfactory.

9. CONCLUSIONS

Through six different examples it has been shown that photoreactor design methods for homogeneous systems allow a correct "a priori" prediction of bench scale experimental results.

The method does not include adjustable parameters. Hence, it can be used equally for any reactor size. If the mass and thermal fluxes are predictable, and the intrinsic reaction kinetics is known, the proposed methodology produces trustfully results.

It seems clear that a similar effort should be placed to develop an equivalent state-of-the-art in heterogeneous photoreactors. As far as reactor design is concerned, the research lines that should be explored deeply, resorting to fundamental principles are:

a) Solution of the design equations coupled with a radiation balance in a reacting media when absorption and scattering are present.
b) Development of experimental techniques to measure the scattering coefficient and the phase function, particularly for systems with catalytic particles of practical sizes, and radiation in the UV range.
c) Formulation of workable constitutive equations for the photon-semiconductor particle interaction and the associated transport phenomena inside the solid.
d) Much better understanding of the photocatalytic intrinsic kinetics, aimed a providing mechanistically based rate expressions.

ACKNOWLEDGMENTS. The authors are grateful to Consejo Nacional de Investigaciones Científicas y Técnicas (CONICET) and to Universidad Nacional del Litoral (UNL) for their support to produce this work.

NOTATION

$\underset{\sim}{B}$ = propagation reaction array defined in eq. (50), dimensionless.
C = concentration, mole m^{-3}.
C_p = heat capacity, J $mole^{-1}$ K^{-1}.
d_b = bubble diameter, m.
e^a = local volumetric rate of energy absorption, einstein m^{-3} s^{-1}.
E = energy flow rate, einstein s^{-1}.
$\underset{\sim}{g}$ = initiation reaction array defined in eq. (49), dimensionless.
$\tilde{G}e$ = geometric number, dimensionless.
I = intensity of radiation, einstein s^{-1} m^{-2} sr^{-1}.
j^e = emission of radiative energy, einstein s^{-1} m^{-3} sr^{-1}.
J = rate of initiation step, dimensionless.
k_i = reaction rate constant for the i-step, s^{-1} (first order) or m^3 $mole^{-1}$ s^{-1} (second order).
K_i = reaction rate constant for the i-step, dimensionless.
K_L = ratio of reaction time to absorption time, dimensionless.
L = length, m.
$\underset{\sim}{N}$ = mass absorption rate array defined in eq. (48), dimensionless.
$\tilde{N}u$ = Nusselt number, dimensionless.
p = phase function, dimensionless.
Pe = Peclet number, dimensionless.
$\underset{\sim}{q}$ = radiation flux density vector, einstein m^{-2} s^{-1}.
$\tilde{Q}$ = heat of reaction defined in eq. (41), dimensionless.
r = radial coordinate, m, also radius, m.
s = linear coordinate along the direction of a ray, m.
t = time, s.
$\underset{\sim}{t}$ = termination reaction array defined in eq. (53), dimensionless.
$\tilde{T}$ = temperature, K.
U = velocity, dimensionless.
v = velocity, m s^{-1}.
W = gain or loss of radiant energy from the beam, einstein s^{-1} m^{-3} sr^{-1}.
X = fractional conversion, dimensionless.
z = axial coordinate, m.

<u>Greek Letters</u>

α = molar absorptivity, m^2 $mole^{-1}$.
β = cylindrical coordinate, rad.
γ = radial coordinate, dimensionless.
Γ = reflection coefficient, dimensionless; also emitting extension, dimension are source-model dependent.
ΔH = heat of reaction, J $mole^{-1}$.
ε = absorption number defined in eq. (39), dimensionless; also volume fraction, dimensionless.
ζ = axial coordinate, dimensionless.
θ = spherical coordinate, rad.
Θ = function defined by eqs. (57) and (58), dimensionless.
κ = characteristic property of the lamp emission, units depend on the characteristics of the lamp emission.

μ = absorption coefficient, m^{-1}.
ν = frequency, s^{-1}.
ρ = spherical coordinate, m; also density of the reacting mixture, kg m^{-3}.
σ = scattering coefficient, m^{-1}.
τ = time, dimensionless; also temperature, dimensionless.
T = transmittance, dimensionless.
ϕ = spherical coordinate, rad.
Φ = quantum yield, mole einstein^{-1}.
χ = ratio of the inner to the outer radius, dimensionless.
Ψ = concentration, dimensionless.
Ω = reaction rate, mole m^{-3} s^{-1}; also solid angle, sr.
Ω^* = reaction rate, dimensionless.

Subscripts

c = refers to the cooling fluid.
D = direct radiation; also characteristic diffusion time.
e = emission value.
eff = effective property.
exp = experimental value.
g = relative to the gas phase.
G = relative to a global property.
i = i-species.
in = relative to the inner wall of the reactor.
init = initiation value.
I = incident point.
In = indirect radiation.
L = lamp property.
o = at the surface of entrance of radiation.
ou = relative to the outer wall of the reactor.
ox = relative to oxalic acid.
pred = predicted value.
P = relative to a primary quantum yield.
r = characteristic reaction time.
R = reactor property.
Rf = reflector property.
W = wall property.
z = relative to the z-axis.
1 = lower limit of integration.
2 = upper limit of integration.
ν = frequency dependence.

Superscripts

a = absorbed value.
e = emission value.
int = interface value.
° = inlet value.
' = denotes reverse reaction.

Special symbols

$|\ |$ = indicates modulus of a vector.
$<\ >$ = indicates average value.
Δ = indicates an interval.
$\hat{}$ = denotes an atomic or a free radical species.

REFERENCES

Alfano, O.M., Romero, R.L. and Cassano, A.E. (1985) 'A cylindrical pho-toreactor irradiated from the bottom. I. Radiation flux density generated by a tubular source with a parabolic reflector', Chem. Eng. Sci. 40, 2119-2127.

Alfano, O.M., Romero, R.L. and Cassano, A.E. (1986a) 'Radiation field modelling in photoreactors - I. Homogeneous media', Chem. Eng. Sci. 41, 421-444.

Alfano, O.M., Romero, R.L. and Cassano, A.E. (1986b) 'Radiation field modelling in photoreactors - II. Heterogeneous media', Chem. Eng. Sci. 41, 1137-1153.

Alfano, O.M., Romero, R.L. and Cassano, A.E. (1986c) 'A cylindrical pho-toreactor irradiated from the bottom. II. Model for the local volu-metric rate of energy absorption with polychromatic radiation and their evaluation', Chem. Eng. Sci. 41, 1155-1161.

Alfano, O.M., Romero, R.L., Negro, A.C. and Cassano, A.E. (1986d) 'A cylindrical photoreactor irradiated from the bottom. III. Measurement of absolute values of the local volumetric rate of energy absorption. Experiments with polychromatic radiation', Chem. Eng. Sci. 41, 1163-1169.

Alfano, O.M. and Cassano, A.E. (1988a) 'Modeling of a gas-liquid tank photoreactor irradiated from the bottom. I. Theory', Ind. Eng. Chem. Res. 27, 1087-1095.

Alfano, O.M. and Cassano, A.E. (1988b) 'Modeling of a gas-liquid tank photoreactor irradiated from the bottom. II. Experiments', Ind. Eng. Chem. Res. 27, 1095-1103.

Alfano, O.M., Vicente, M., Esplugas, S. and Cassano, A.E. (1990) 'Radiation field inside a tubular multilamp reactor for water and wastewater treatment', Ind. Eng. Chem. Res., in press.

Andre, J.C. and Villermaux, J. (1986) 'Photochimie industrielle et genie de la reaction photochimique', Entropie No. 127, 4-7.

Astarita, G. (1967) Mass Transfer with Chemical Reaction, Elsevier, Amsterdam.

Astarita, G., Savage, D.W. and Bisio, A. (1983) Gas Treating with Chemical Solvent, J. Wiley and Sons, New York.

Cabrera, M.I., Alfano, O.M. and Cassano, A.E. (1990a) 'Selectivity stu-dies in the photochlorination of methane I. Reactor model and kinetic studies in a non-isothermal, polychromatic environment', to be published.

Cabrera, M.I., Alfano, O.M. and Cassano, A.E. (1990b) 'Selectivity stu-dies in the photochlorination of methane II. Effect of the radiation source output power and output spectral distribution', to be published.

Cassano, A.E., Silveston, P.L. and Smith, J.M. (1967) 'Photochemical reaction engineering', Ind. Eng. Chem. 59, 18-38.

Cassano, A.E., Alfano, O.M. and Romero R.L. (1986) 'Photoreactor engineering: analysis and design' in S. Whitaker and A.E. Cassano (eds.), Concepts and Design of Chemical Reactors, Gordon and Breach Science Pub., New York, pp. 339-512.

Cassano, A.E. and Alfano, O.M. (1989) 'Photoreactor design', in N.P. Cheremisinoff (ed.), Handbook of Heat and Mass Transfer, Vol. 3: 'Catalysis, Kinetics, and Reactor Engineering', Gulf, Houston, pp. 583-670.

Cerdá, J., Irazoqui, H.A. and Cassano, A.E. (1973) 'Radiation fields inside an elliptical photoreflector with a source of finite spatial dimensions', AIChE J. 19, 963-968.

Cerdá, J., Marchetti, J.L. and Cassano, A.E. (1977) 'Radiation efficiencies in elliptical photoreactors', Lat. Am. J. Heat and Mass Transfer 1, 33-63.

Clariá, M.A., Irazoqui, H.A. and Cassano, A.E. (1986) 'Modelling and experimental validation of the radiation field inside an elliptical photoreactor', Chem. Eng. J. 33, 119-131.

Clariá, M.A., Irazoqui, H.A. and Cassano, A.E. (1988) 'A priori design of a photoreactor for the chlorination of ethane', AIChE J. 34, 366-382.

Clyne, M.A.A. and Stedman, D.H. (1968) 'Recombination of ground state halogen atoms', Trans. Faraday Soc. 64, 2698-2707.

Chiltz, G., Goldfinger, P., Huybrechts, G., Martens, G. and Verbeke, G. (1963) 'Atomic chlorination of simple hydrocarbon derivatives in the gas phase', Chem. Rev. 63, 355-372.

Daiton, F.S., Lomax, D.A. and Weston, M. (1957) 'The kinetics of the gas-phase photochlorination of trichloroethylene', Trans. Faraday Soc. 53, 460-467.

De Bernardez, E.R. and Cassano, A.E. (1982) 'Azimuthal asymmetries in tubular photoreactors', Lat. Am. J. Heat and Mass Transfer 6, 333-351.

De Bernardez, E.R. and Cassano, A.E. (1985) 'A priori design of a continuous annular photochemical reactor. Experimental validation for simple reactions', J. Photochem. 30, 285-301.

De Bernardez, E.R. and Cassano, A.E. (1986) 'Methodology for an optimal design of a photoreactor. Application to methane chloroderivatives production', Ind. Eng. Chem. Process Des. Dev. 25, 601-612.

De Bernardez, E.R., Clariá, M.A. and Cassano, A.E. (1987) 'Analysis and design of photoreactors' in J.J. Carberry and A. Varma (eds.), Chemical Reaction and Reactor Engineering, Marcel Dekker Inc., New York, pp. 839-921.

Doede, C.M. and Walker, C.A. (1955) 'Photochemical engineering', Chem. Eng. 62, 159-178.

Esplugas, S., Vicente, M., Alfano, O.M. and Cassano, A.E. (1990) 'Effect of the reflector shape on the performance of multilamp photoreactors applied to pollution abatement', Ind. Eng. Chem. Res., in press.

Felder, R.M. and Hill, F.B. (1969) 'Mixing effects in chemical reactors. I. Reactant mixing in batch and flow systems', Chem. Eng. Sci. 24, 385-397.

Felder, R.M. and Hill, F.B. (1970) 'Mixing effects in chemical reactors. Chain reactions in batch and flow systems', Ind. Eng. Chem. Fundam. 9, 360-367.

Forbes, G.S. and Heidt, L.J. (1934) 'Optimum composition of uranyl oxalate solutions for actinometry', J. Am. Chem. Soc. 56, 2363-2365.

Gosselain. P.A., Adam, J. and Goldfinger, P. (1956) 'La specificité des halogénations atomiques', Bull. Soc. Chim. Belg. 65, 533-548.

Hatchard, C.G. and Parker, C.A. (1956) 'A new sensitive chemical actinometer - II. Potassium ferrioxalate as a standard chemical actinometer', Proc. R. Soc. A235, 518-536.

Heidt, L.J., Tregay, G.W. and Middleton, F.A. Jr. (1970) 'Influence of pH upon the photolysis of the uranyl oxalate actinometer system', J. Phys. Chem. 74, 1876-1882..

Irazoqui, H.A., Cerdá, J. and Cassano, A.E. (1973) 'Radiation profiles in an empty annular photoreactor with a source of finite spatial dimensions', AIChE J. 19, 460-467.

Irazoqui, H.A., Cerdá, J. and Cassano, A.E. (1976) 'The radiation field for the point and line source approximations and the three-dimensional source models: Applications to photoreactors', The Chem. Eng. Journal 11, 27-37.

Jacob, S.M. and Dranoff, J.S. (1966) 'Radial scale-up of perfectly mixed photochemical reactors', Chem. Eng. Prog. Symp. Ser. 64, 54-63.

Leermakers, J.A. and Dickinson, R.G. (1932) 'The photochlorination of tetrachloroethylene in carbon tetrachloride solution', J. Am. Chem. Soc. 54, 4648-4657.

Leighton, W.G. and Forbes, G.S. (1930) 'Precision actinometry with uranyl oxalate', J. Am. Chem. Soc. 52, 3139-3152.

Noyes, R.M. and Fowler, L. (1951) 'Mechanisms of chain termination in chlorine atom reactions', J. Am. Chem. Soc. 73, 3043-3045.

Otake, T., Tone, S., Higuchi, K. and Nakao, K. (1981) 'Light intensity profile in gas-liquid dispersion. Applicability of effective absorption coefficient', Kagaku Kogaku Ronbunshu 7, 57-63.

Ozisik, M.N. (1973) 'Radiative Transfer and Interactions with Conduction and Convection', J. Wiley & Sons, New York.

Parker, C.A. (1953) 'A new sensitive chemical actinometer - I. Some trials with potassium ferrioxalate', Proc. R. Soc. A220, 104-116.

Poutsma, M.L. and Hinman, R.L. (1964) 'Chlorination studies of unsaturated materials in nonpolar media. I. Solvent effects on radical addition of chlorine to chloroethylenes', J. Am. Chem. Soc. 86, 3807-3814.

Pritchard, H.O., Pyke, J.B. and Trotman-Dickenson, A.F. (1955) 'The study of chlorine atom reactions in the gas phase', J. Am. Chem. Soc. 77, 2629-2633.

Romero, R.L., Alfano, O.M., Marchetti, J.L. and Cassano, A.E. (1983) 'Modelling and parametric sensitivity of an annular photoreactor with complex kinetics', Chem. Eng. Sci. 38, 1593-1605.

Santarelli, F. (1985) 'Radiative transfer in photochemical processes', in M. Schiavello (ed.), Photoelectrochemistry, Photocatalysis and Photoreactors, D. Reidel Pub. Co., Dordrecht, Holland, pp. 549-559.

Slattery, J.C. (1978) 'Momentum, Energy, and Mass Transfer in Continua', R.E. Krieger Pub. Co., Huntington, New York.

Tymoschuk, A.R. (1990) 'Modelado de un Foto-reactor Multitubo Anular con Pantalla Reflectora para una Reaccion Homogenea Simple', Ph.D. Dissertation, Universidad Nacional del Litoral, Santa Fe, Argentina.

Vicente, M., Alfano, O.M., Esplugas, S. and Cassano, A.E. (1990) 'Design and experimental verification of a tubular multilamp reactor for water and wastewater treatment', Ind. Eng. Chem. Res., in press.

Walling, C. (1957) 'Free Radicals in Solution', J. Willey & Sons, New York.

Yokota, T., Iwano, T. and Tadaki, T. (1976) 'Light intensity in an annular photochemical reactor', Kagaku Kogaku Ronbunshu 2, 298-303.

Yokota, T., Iwano, T., Deguchi, H. and Tadaki, T. (1981) 'Ligth absorption rate in a bubble column photochemical reactor', Kagaku Kogaku Ronbunshu 7, 157-163.

Yue, P.L. (1985a) 'Introduction to the modelling and design of photoreactors', in M. Schiavello (ed.), Photoelectrochemistry, Photocatalysis and Photoreactors, D. Reidel Pub. Co., Dordrecht, Holland, pp. 527-547.

Yue, P.L. (1985b) 'Studies of photoreactions in homogeneous photoreactors', in M. Schiavello (ed.), Photoelectrochemistry, Photocatalysis and Photoreactors, D. Reidel Pub. Co., Dordrecht, Holland, pp. 561-573.

Yue, P.L. (1985c) 'Studies of photoreactions in heterogeneous photoreactors', in M. Schiavello (ed.), Photoelectrochemistry, Photocatalysis and Photoreactors, D. Reidel Pub. Co., Dordrecht, Holland, pp. 575-585.

AUTHOR INDEX

SUBJECT INDEX